PHYSICAL METHODS OF CHEMISTRY

Second Edition

Volume IIIB

DETERMINATION OF CHEMICAL COMPOSITION AND
MOLECULAR STRUCTURE—PART B

PHYSICAL METHODS OF CHEMISTRY

Second Edition

Editors: **Bryant W. Rossiter**
John F. Hamilton

PHYSICAL METHODS OF CHEMISTRY

Second Edition

Edited by

BRYANT W. ROSSITER

ICN Pharmaceuticals, Inc.
Costa Mesa, California

and

JOHN F. HAMILTON

Research Laboratories
Eastman Kodak Company
Rochester, New York

Volume IIIB
DETERMINATION OF CHEMICAL COMPOSITION AND MOLECULAR STRUCTURE—PART B

WILEY

A WILEY-INTERSCIENCE PUBLICATION

JOHN WILEY & SONS

New York • Chichester • Brisbane • Toronto • Singapore

Library of Congress Cataloging-in-Publication Data:
(Revised vol. 3, pt. B)

Physical methods of chemistry.

  "A Wiley-Interscience publication."
  Includes bibliographies and indexes.
  Contents: v. 1. Components of scientific instruments
and applications of computers to chemical research—
v. 2. Electrochemical methods—v. 3. Determination
of chemical composition and molecular structure.
  1. Chemistry—Manipulation. I. Rossiter, Bryant W.,
1931–   . II. Hamilton, John F.
QD61.P47 1986   542   85-6386
ISBN 0-471-85051-9 (v. 3B)

Printed in the United States of America

10 9 8 7 6 5 4 3 2 1

CONTRIBUTORS

WOLFRAM BAUMANN, Institut für physikalische Chemie, der Universität Mainz, Mainz, Federal Republic of Germany

KARIN D. CALDWELL, Department of Bioengineering, University of Utah, Salt Lake City, Utah

ALEX F. DRAKE, Department of Chemistry, Birkbeck College London, London, England

J. CALVIN GIDDINGS, Department of Chemistry, University of Utah, Salt Lake City, Utah

IRA B. GOLDBERG, Rockwell International Science Center, Thousand Oaks, California

GEORGE A. GRAY, Varian Associates, Palo Alto, California

ROY A. KELLER, Department of Chemistry, State University of New York College at Fredonia, Fredonia, New York

TED M. MCKINNEY, Rockwell International Science Center, Thousand Oaks, California

LAXMAN N. MULAY, Materials Science and Engineering Department, The Pennsylvania State University, University Park, Pennsylvania

INDUMATI L. MULAY, Materials Science and Engineering Department, The Pennsylvania State University, University Park, Pennsylvania

ELIZABETH V. PATTON, Research Laboratories, Eastman Kodak Company, Rochester, New York

JAMES N. SHOOLERY, Varian Associates, Palo Alto, California

PREFACE TO PHYSICAL METHODS OF CHEMISTRY

This is a continuation of a series of books started by Dr. Arnold Weissberger in 1945 entitled *Physical Methods of Organic Chemistry*. These books were part of a broader series, *Techniques of Organic Chemistry*, and were designated Volume I of that series. In 1970, *Techniques of Chemistry* became the successor to and the continuation of the *Techniques of Organic Chemistry* series and its companion, *Techniques of Inorganic Chemistry*, reflecting the fact that many of the methods are employed in all branches of chemical sciences and the division into organic and inorganic chemistry had become increasingly artificial. Accordingly, the fourth edition of the series, entitled *Physical Methods of Organic Chemistry*, became *Physical Methods of Chemistry*, Volume I in the new *Techniques* series. The last edition of *Physical Methods of Chemistry* has had wide acceptance, and it is found in most major technical libraries throughout the world. This new edition of *Physical Methods of Chemistry* will consist of eight or more volumes and is being published as a self-standing series to reflect its growing importance to chemists worldwide. This series will be designated as the second edition (the first edition, Weissberger and Rossiter, 1970) and will no longer be subsumed within *Techniques of Chemistry*.

This edition heralds profound changes in both the perception and practice of chemistry. The discernible distinctions between chemistry and other related disciplines have continued to shift and blur. Thus, for example, we see changes in response to the needs for chemical understanding in the life sciences. On the other hand, there are areas in which a decade or so ago only a handful of physicists struggled to gain a modicum of understanding but which now are standard tools of chemical research. The advice of many respected colleagues has been invaluable in adjusting the contents of the series to accommodate such changes.

Another significant change is attributable to the explosive rise of computers, integrated electronics, and other "smart" instrumentation. The result is the widespread commercial automation of many chemical methods previously learned with care and practiced laboriously. Faced with this situation, the task of a scientist writing about an experimental method is not straightforward.

Those contributing to *Physical Methods of Chemistry* were urged to adopt as their principal audience intelligent scientists, technically trained but perhaps inexperienced in the topic to be discussed. Such readers would like an introduction to the field together with sufficient information to give a clear understanding of the basic theory and apparatus involved and the appreciation for the value, potential, and limitations of the respective technique.

Frequently, this information is best conveyed by examples of application, and many appear in the series. Except for illustration, however, no attempt is made to offer comprehensive results. Authors have been encouraged to provide ample bibliographies for those who need a more extensive catalog of *applications*, as well as for those whose goal is to become more expert in a *method*. This philosophy has also governed the balance of subjects treated with emphasis on the *method*, not on the results.

Given the space limitations of a series such as this, these guidelines have inevitably resulted in some variance of the detail with which the individual techniques are treated. Indeed, it should be so, depending on the maturity of a technique, its possible variants, the degree to which it has been automated, the complexity of the interpretation, and other such considerations. The contributors, themselves expert in their fields, have exercised their judgment in this regard.

Certain basic principles and techniques have obvious commonality to many specialties. To avoid undue repetition, these have been collected in Volume I. They are useful on their own and serve as reference material for other chapters.

We are deeply sorrowed by the death of our friend and associate, Dr. Arnold Weissberger, whose enduring support and rich inspiration motivated this worthy endeavor through four decades and several editions of publication.

BRYANT W. ROSSITER
JOHN F. HAMILTON

Research Laboratories
Eastman Kodak Company
Rochester, New York
March 1986

PREFACE

This is the second of two volumes entitled *Determination of Chemical Composition and Molecular Structure*. Admittedly, this grouping of techniques is somewhat arbitrary and overlaps with a later volume entitled *Determination of Structural Features of Crystalline and Amorphous Solids*. Nonetheless, the ordering offers more consistent volumes for those interested in certain types of chemical determinations as opposed to the grouping of similar techniques used for a variety of measurements. This should make it easier for the reader to obtain from the library, or purchase at minimum cost, those parts of the treatise of greatest personal value.

Chapters have been written by world-class authors who are widely recognized in their fields. Authors have directed their writing to the competent, professional scientist who is interested in obtaining information provided by the technique, but who is perhaps not an expert in the use of the method. In each case, authors of chapters have supplied, either in the text or through liberal reference to monographs and other scientific literature, sufficient information for the investigator to apply the techniques successfully in the laboratory.

We acknowledge our deep gratitude to the contributors who have spent long hours over manuscripts. We greet a previous contributor, Professor Laxman N. Mulay, and welcome several new contributors to Volume IIIB: Professor Dr. Wolfram Baumann, Dr. Karin D. Caldwell, Dr. Alex F. Drake, Professor J. Calvin Giddings, Dr. Ira B. Goldberg, Dr. George A. Gray, Professor Roy A. Keller, Dr. Ted M. McKinney, Dr. (Mrs.) Indumati L. Mulay, Dr. Elizabeth V. Patton, and Dr. James N. Shoolery.

We are also extremely grateful to the many colleagues from whom we have sought counsel on the choice of subject matter and contributors. We express our gratitude to Mrs. Ann Nasella for her enthusiastic and skillful editorial assistance. In addition, we heartily thank the specialists whose critical readings of the manuscripts have frequently resulted in the improvements accrued from collective wisdom. For Volume IIIB they are Dr. G. L. Beyer, Dr. M. S. Burberry, Dr. H. Coll, Dr. G. J.-S. Gau, Dr. J. S. Lewis, Dr. L. E. Oppenheimer, Dr. J. R. Overton, Dr. E. V. Patton, Dr. O. E. Schupp, III, Dr. D. J. Williams, Dr. R. H. Young, and Dr. N. Zumbulyadis.

Bryant W. Rossiter
John F. Hamilton

Costa Mesa, California
Rochester, New York
August 1989

CONTENTS

PHYSICAL METHODS OF CHEMISTRY
Second Edition

Volume IIIB

DETERMINATION OF CHEMICAL COMPOSITION AND
MOLECULAR STRUCTURE—PART B

Chapter **1**

CHIROPTICAL SPECTROSCOPY

Alex F. Drake

1 INTRODUCTION

A major difference between the animate and the inanimate is the ability to self-replicate. This involves the unwinding of coiled genes. The situation would certainly be complicated if the helicity of these coils were not guaranteed. It is the integrity of chirality (handedness) of macromolecules' building blocks—the amino acids and sugars—that ensures the consistency of handedness on the macro scale. This selectivity for one enantiomer over another is a particular aspect of *chiral discrimination* [1, 2].

The physical and chemical properties of a racemic mixture and its two enantiomers are identical under conditions where discrete molecular units exist and the interacting medium is isotropic. Thus, all three would have the same molar extinction coefficient in a symmetric fluid phase as long as the Beer–Lambert law was obeyed. The only means of distinguishing them is through their differential interaction with a chiral agent. Spectroscopically, this chiral entity is circularly polarized light. It is the differential behavior of a particular enantiomer toward left and right circularly polarized light that is known as optical activity and its study, *chiroptical spectroscopy.*

Linearly (plane-) polarized light is composed of two equal intensity components, one left circular and the other right circular. In a chiral medium the refractive indices for left and right circularly polarized light (their velocities) are generally different and one component will be retarded (travel more slowly) so that the phase relationship between these two components will change. The plane of the linearly polarized light will continue to rotate on passing through the chiral medium until it finally emerges with a resulting rotation of its plane with respect to that of the incident radiation given by the expression

$$[\alpha] = (4\pi d/\lambda)[n_L - n_R] \qquad (1)$$

where $[\alpha]$ is the specific rotation; d is the path length; λ is the wavelength of radiation; and n_L and n_R are the refractive indices of left and right circularly polarized light, respectively. The wavelength dependence of optical rotation is known as *optical rotatory dispersion* (ORD).

In regions of absorption, these two circular components will be absorbed (or emitted) differentially by the opposite antipodes of a chiral pair. This is *circular dichroism* (CD). In line with conventional isotropic absorption spectroscopy the differential optical density ΔA is given by

$$\Delta A = (A_L - A_R) = \log(I_0/I_L) - \log(I_0/I_R) \qquad (2)$$

where I_0 is the intensity of incident radiation and $I_{L,R}$ and $A_{L,R}$ are, respectively, the transmitted intensities and optical densities for left and right circularly polarized light. On a molar basis, the differential molar decadic extinction coefficient is given through Beer's law as

$$\Delta\varepsilon = (\varepsilon_L - \varepsilon_R) = \Delta Acl \qquad (3)$$

where c is concentration in moles per liter and l is the path length in centimeters.

The differential absorption of left and right circularly polarized light by chiral molecules leads to changes in the ellipticity of the transmitted radiation, which justifies the alternative unit for CD, molar ellipticity $[\theta]$, with the relationship

$$[\theta] = 3300\Delta\varepsilon \tag{4}$$

This unit found favor because CD data were originally derived from ellipticity measurements, and it has similar dimensions to molecular rotation so that CD and ORD spectra can be displayed using similar graphic axes. Although CD can be determined from ellipticity data, the direct measurement of differential light intensities is in principle at least three times more sensitive than either ellipticity or polarimetric measurements [3, 4]. The differential optical density unit is, therefore, a closer description of the parameter that is normally measured by a modern CD spectrometer.

Another means of judging the magnitude of a CD peak is given by the *Kuhn dissymmetry factor* (*g*-factor), which can be defined [5] as the ratio of the differential and isotropic extinction coefficients at a particular wavelength

$$g_\lambda = (\Delta\varepsilon_\lambda/\varepsilon_\lambda) = (\Delta A_\lambda/A_\lambda) \tag{5}$$

Independent of concentration and path length, the g-factor is a measure of a chiral molecule's ability to discriminate between left and right circularly polarized light and as such is characteristic of both the molecule and the particular transition responsible for the absorption. The ability of a spectrometer to detect a certain g-factor is a good measure of performance insomuch as it includes an element of reliability as well as noise level. Thus in electronic chiroptical spectroscopy the ability to detect g-factors between 10^{-4} and 10^{-5} will cover almost all cases; in the vibration IR (infrared) technique the strongest simplest spectra have g-factors approximately 5×10^{-5}, and generally an order of magnitude increase in sensitivity is required. Theoretically, CD instruments must be capable of detecting differences in the intensities of two circularly polarized light beams, of parts per million (ppm) or less. This obviously puts a great constraint on instrument design, particularly on the polarization optics.

In the ORD mode the sign of the rotation at typically 589 nm is used to characterize the handedness of an enantiomer, while for CD the sign associated with a particular transition (or Cotton effect) is quoted. The relative merits of the two techniques are open to debate. Certainly it is easier to measure CD artifact free at high sensitivities, and being an absorption rather than a dispersion phenomenon, it yields results that are easier to analyze spectroscopically. However, optical rotation measurements, which are relatively simpler at low sensitivity, may be preferred for monitoring processes or labeling enantiomers.

Cotton [6] reported the first CD spectra in 1895. However, because of experimental difficulties, ORD remained the preferred technique for gathering optical activity data until 1960. A notable exception to this is found in the work of Kuhn [7], who reported CD spectra employing photographic detection and made important contributions to both the theoretical and practical aspects of chiroptical spectroscopy. In 1960 Djerassi's book [8], *Optical Rotatory Dispersion*, was published and Grosjean and Legrand [9] first described their CD spectrometer.

In 1965 Velluz, Legrand, and Grosjean [10] published their fundamental book *Optical Circular Dichroism, Principles, Measurements and Applications*, in which they detailed the design of the two CD spectrometers, *Dichrographes*, which they had constructed in the early 1960s. Both instruments employed polarization modulation. Indeed, until recently, the French held patent rights for their concepts with royalties being payable by American (Cary-Varian) and Japanese (Jasco) manufacturers. The Dichrographe owed its origin to the work of Billings [11], who had spent much time developing the Pockell's cell. The Pockell's cell was fabricated from a single crystal of ammonium dihydrogen phosphate. This permitted the automatic, routine measurement of CD at high sensitivity in the wavelength range of 615–185 nm.

At this time a NATO Summer School report, *Optical Rotatory Dispersion and Circular Dichroism in Organic Chemistry* [12], reflected the trend away from optical rotation measurements in favor of CD measurements. During the mid-1960s the recording spectropolarimeters produced by Cary-Varian and Jasco were the most sophisticated of their type, but the Dichrographe had arrived. Subsequently, CD has played an increasingly important role in the study of chiral molecules and their spectroscopy to the extent that ORD curves are now rarely measured. The proceedings [13] of the 1971 NATO Summer School in Optical Activity are a testimony to this fact.

From the mid-1960s, following the lead of Billardon and Badoz [14], reports began to appear [15–18] describing polarization modulators based on the photoelastic properties of transparent materials. These spawned a new generation of CD spectrometers, many of which were described [1] at the 1977 NATO Summer School in Optical Activity. Commercial spectrometers are now available that cover the conventional photomultiplier range (185–1000 nm), operating in both CD and linear dichroism modes. Efforts have also been successful in extending this range from vacuum ultraviolet circular dichroism (vacuum-UV CD) to infrared or vibrational circular dichroism (IR CD or vibrational CD), while other instruments have been described that measure circularly polarized differentials in the emission experiment [circularly polarized luminescence (CPL) and fluorescence-detected circular dichroism (FD CD)] and the scattering mode [Raman circular intensity differentials (Raman CID)].

2   GENERAL METHODS

The methods of producing circularly polarized light and the means of measuring circularly polarized differentials have been subjects in several reviews [10, 19–22]. As quoted by Mitchell [23], Cotton studied CD in two ways, "...(1) by direct experiments with circularly polarized light and (2) by ellipticity measurements...." The latter was the preferred scheme until the reports of Mitchell [23] and Velluz, Legrand, and Grosjean [10]. However, the general methods of measuring CD are best classified as *static* and *dynamic* *measurements*.

2.1   Static Measurements

Until 1950, following Bruhat's methods [24], typical CD measurements involved the optical train illustrated in Figure 1.1. Plane-polarized light passing through a chiral sample experiences a change in ellipticity that becomes detectable after passage through the analyzer, A. A defined manual rotation about the optic axis of one of these optical elements can then be related to the CD of the sample. Originally, these changes were monitored by the eye [19], although during the period 1930–1950 improvements in photographic detection methods allowed Kuhn [7] to increase the measurement sensitivity. A further improvement was realized [25, 26] by the introduction, into the optical train, of Faraday modulators coupled with photomultiplier detection. This was the basis of the Spectropol polarization spectrometer (manufactured by FICA, Yvelines, France) and the Bellingham and Stanley, Bendix Spectropolarimeter/CD spectrometers that were available in the 1960s.

Circular dichroism spectra were first expressed in terms of extinction coefficients by Bruhat [27], yet it was Mitchell [23] in 1950 who was the first to measure explicitly A_L and A_R spectrophotometrically. In Mitchell's apparatus (Figure 1.2) a Glazebrook prism–quarter-wave plate combination was oriented so that circularly polarized light of one hand was generated, with the other hand being created by a 90° rotation of the quarter-wave plate. The measurement of the two corresponding transmission spectra gave the CD spectrum. Replacement of the quarter-wave plate by a Babinet compensator provided [20] the basis of a CD attachment to the Shimadzu Seisakusho QV-50 spectrophotometer.

Historically, the sensitivity of static measurements was severely limited to g-factors of approximately 0.01 (ca. 0.001 in some cases) because of both inherent limitations of direct current measurements and artifactual signals from imperfect optics. Nevertheless, when Mitchell's scheme was followed, but the quarter-wave plate was replaced by a Fresnel rhomb, the first CD measurements in Mason's laboratories were made of magnetic dipole allowed transitions such as

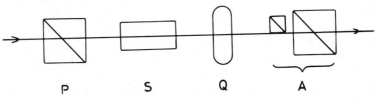

P   S   Q   A

Figure 1.1   Bruhat's apparatus for measuring CD: P, linear polarizer; S, sample; Q, quarter-wave plate; A, analyzer.

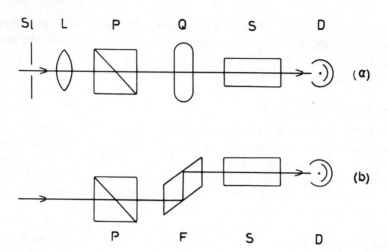

Figure 1.2 Mitchell's apparatus for measuring CD: (a) S1, slit; L, lens; P, linear polarizer; Q, quarter-wave plate; S, sample; D, photoelectric detector; (b) P, linear polarizer; F, Fresnel rhomb; S, sample; D, photoelectric detector.

the $n-\pi^*$ transition of ketones [28] and the $d-d$ transition of metal complexes [29]. The CD spectrum of calycanthine, an alkaloid with two aromatic groups stereochemically fixed such that they present a chiral array of transition electric dipoles, was also measured [30] and afforded the first correct nonempirical determination of absolute configuration by CD. This compound has since given strong signals [31] with other chiroptical techniques CPL, FD CD, and vibrational CD.

Progress into the near- and middle-IR was achieved [32, 33] by the modification of existing double-beam spectrophotometers, with circularly polarized light passing through both the sample and reference compartments.

The first measurements of CD in the vacuum-UV were reported in 1968 by Feinleib and Bovey [34], who described an instrument that apparently used the beam-splitting properties of a polarizing prism (Rochon or Wollaston). The two orthogonally plane-polarized light beams passed through a static quarter-wave plate and then onto two balanced photomultipliers, one detecting the transmitted left and the other the right circularly polarized light beam.

In summary, static measurements of CD are absolute; however, despite the ability to supply calibration standards with easily constructed instruments, their sensitivity is extremely limiting. Another drawback of static phase retarders is that they often have wavelength-dependent properties. Rosenheck and Doty [35] and Holtzwarth and colleagues [3, 36] have in fact taken advantage of this by using a fixed quartz Rochon prism–quartz wave-retardation plate assembly to make single point measurements of A_L and A_R at wavelengths corresponding to the quarter-wave condition (peaks and troughs on an apparently *sinusoidal* absorption spectrum). Extrapolation between these points gave the ΔA at various wavelengths. This device, manufactured by Rehovoth Instruments Ltd., became available [20] as an attachment to the Cary range of UV–Vis-absorption spectrometers.

2.2   Dynamic Measurements

Modulation of the polarization of radiation increases the scope of chiroptical spectroscopy by virtue of the intrinsic improvements in sensitivity generally associated with alternating current techniques. The means of achieving this can be classified in two ways as mechanical or electronic and square wave or sine wave.

Mechanical systems, involving typically the motorized rotation of one of the polarizing elements in a Mitchell-type arrangement, allowed the detection [37–39] of g-factors as low as 10^{-3}–10^{-4}. Higher sensitivities were rarely obtained because of the difficulty in achieving a fast, smooth, synchronous mechanical motion that maintained a particular reproducible time-dependent and wavelength-independent degree of polarization. At this point instruments employing Faraday modulation offered many advantages. However, the modern means of modulating light polarization involves an electronically driven device that replaces the Fresnel rhomb or wave plate of static and mechanical devices. To date two classes of device have proven successful: the electrooptic and the piezooptic (photoelastic) modulators. These involve applying onto an optical element an electric field or stress of sufficient power to induce a birefringence perpendicular to the optic axis that transforms it into a programmable multiwave plate.

3   POLARIZATION MODULATION

A linearly polarized light beam can be said to be the resultant of two orthogonal in-phase linear components. Consider this light beam, with its resultant axis oriented at 45° to the pressure axes, passing through a block of isotropic fused silica rendered birefringent by pressure exerted along the x or y axis, so that $n_x \neq n_y$ (Figure 1.3). One of its components will travel through the silica faster than the other. If $n_x > n_y$, then the x component of the light beam will travel the more slowly; that is, it will be retarded. As drawn in Figure 1.3a, an x component retardation of exactly quarter-wave ($\lambda/4$) means that the emergent light beam will be right circularly polarized ($-\lambda/4$, $3\lambda/4$, $7\lambda/4$, etc., retardation will lead to left circular polarization). A half-wave retardation, plus or minus, will cause a 90° rotation of the plane of linear polarization. Zero ($n_x = n_y$), or full-wave retardation, leaves the state of polarization of the light beam apparently unchanged. The general phase retardation, δ, leads to elliptically polarized light. This is the general principle of *photoelasticity* as first described by Brewster [40] in 1816. It was Fresnel [41] in 1820 who ascribed the phenomenon to differential velocities and Biot [42] in the same year who demonstrated that a glass plate positioned between crossed polarizers was capable of restoring light trans-

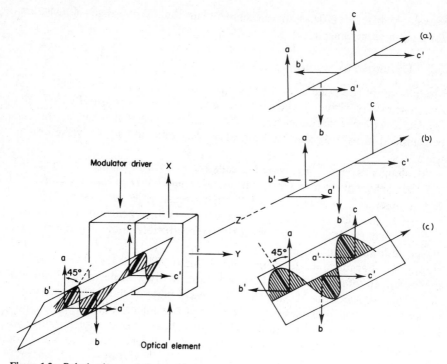

Figure 1.3 Polarization modulation: (a) $n_x > n_y$, $\delta = \lambda/4$; (b) $n_x < n_y$, $\delta = \lambda/4$; (c) $n_x \neq n_y$, $\delta = \lambda/2$.

mission when made to vibrate. The electrooptic modulators function in a similar fashion.

3.1 Electrooptic Modulators

The Pockell's cell used by Grosjean and Legrand [9] was composed of a Z-cut section of a single crystal of ammonium dihydrogen phosphate (ADP). When transparent electrodes are put on these faces and linearly polarized light, oriented at $\pm 45°$ to the crystallographic axes, is propagated down this unique axis, the device behaves as a retardation plate with the retardation at a particular wavelength being a function of the applied voltage. The voltage required to maintain the quarter-wave condition for CD measurements at a particular wavelength, λ nm, is given by

$$V = 1350 + 10.6(\lambda - 220) \qquad (6)$$

Obviously, these voltages are very high, particularly as the crystal section is only about 2 mm thick. Therefore, a long wavelength limit of approximately 600 nm is set by a driving voltage of approximately 5380 V with the short wavelength limit (ca. 185 nm) set by the transmission of ADP. Linear dichroism measurements that require twice these voltages to maintain a half-wave condition are, accordingly, severely wavelength limited.

The ADP crystal possesses D_{2d} symmetry, and light propagated along any direction other than the isotropic unique axis meets an inherently birefringent path. Therefore, the acceptance angle is critically small (1 or 2°), and the incident radiation must be ideally collimated.

Another disadvantage of the original ADP device was that it could have a restricted lifetime if it was used continuously in the visible region with high voltages or in a humid atmosphere. Attempts have been made to produce other electrooptic modulators; deuteration extends the long wavelength limit to approximately 1000 nm, while the use of transparent gold electrodes offers advantages over the traditional glycerol. The ideal device would be fabricated from a cubic crystal that has the appropriate electrical and transmission characteristics.

An advantageous property of the electrooptic modulator is that the driving voltage can be sine or square wave at a frequency dictated by the operator. The original commercial Jouan CD spectrometers employed sine-wave (50–60-Hz) modulation, while square-wave modulation has become an important feature [43] in Raman CID. The ability to function at low frequencies could be of interest in the IR where many detectors have a low-frequency response.

3.2   Piezooptic Modulators

For many years mechanical engineers [44] have used polarized light to detect stress patterns in materials, and many research workers [14–18] have recently put the effect of photoelasticity to use in designing polarization modulators.

These systems have been described under several names, such as photoacoustic, photoelastic, or stress modulators. However, they are perhaps best classified as piezooptic modulators, of which there are two main types: composite resonators and matched element resonators (see Figure 1.4).

The original piezooptic devices, as described by Billardon and Badoz [14], were composite resonators composed of a central block of optical material (e.g., silica or germanium) with ceramic piezoelectric plates at each end. Typically, blocks of brass, which act to anchor the motion, are joined to the plates. One of the piezoelectric plates drives the system, the other senses the modulation and stabilizes it through a feedback loop. Although these devices are relatively simple to fabricate because their dimensions are not critical, they suffer from the inherent fault of having maximum oscillation at the junction of the optical element and the piezoplate. This causes adhesion problems that can lead to static strain in the optical element and joint rupture during modulation.

The matched element resonators (Figure 1.4) rely on the sympathetic oscillation in an optical element that is in contact with a piezoelectric driver excited into its natural oscillation frequency. A typical matched two-element device is fabricated from a gold-plated, X-cut, quartz block that is adhered to a block of optical material such as silica or calcium fluoride (CaF_2), with the

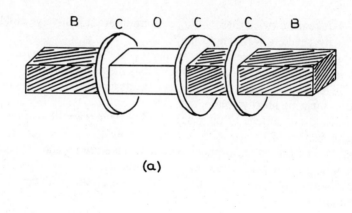

(a)

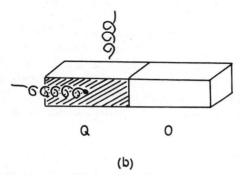

Q   O

(b)

Figure 1.4 Polarization modulators. (*a*) Composite resonator: B, brass blocks; C, ceramic piezoelectric plates; O, optical block. (*b*) Matched element resonator: Q, quartz driver; O, optical block.

dimensions of the latter being critically governed by the former. During modulation a standing wave is induced in the device with a nondestructive node at the junction of the two elements.

The optical element of both types can be chosen to be isotropic; hence, there will be a large acceptance angle. The choice of materials such as CaF_2 and the lower power requirements mean that the wavelength range for electronic polarization modulation is extended and that linear dichroism measurements are made generally possible. Sine wave in nature, these devices operate in the 10–50-kHz region. Obviously, there is an upper limit to the power at which these devices can operate. At very high powers the optical element tends to shatter, although it has been found that the provision of two drivers results [18] in reduced power requirements.

4 THE SIGNAL

The typical optical layout of a CD spectrometer using polarization modulation and a photomultiplier detector is given in Figure 1.5. The original linearly polarized light beam (intensity, I_0) is now considered to be composed of two

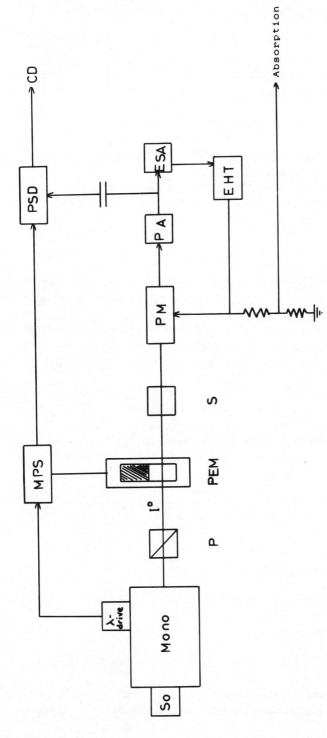

Figure 1.5 The CD spectrometer: So, source; Mono, monochromator; P, linear polarizer (Rochon prism); I^0, initial light intensity; PEM, photoelastic modulator; MPS, modulator power supply; S, sample; PM, photomultiplier; PA, preamplifier; ESA, error signal/servoamplifier; EHT, photomultiplier power supply; PSD, phase-sensitive detector.

11

circular components, one left and the other right:

$$I_0 = (I_L^0 + I_R^0)/2 \qquad (7)$$

The amount of these circular components in the elliptical light after their transmission through a general static phase retarder (wave plate) is given by

$$I_{L,R}^0 = (I_0/2)(1 \pm \sin \delta) \qquad (8)$$

When $\delta = \pi/2$, the quarter-wave condition is obtained with $I_L^0 = I_0$, $I_R^0 = 0$, and the emergent beam is pure left circular.

With polarization modulation, for example, a sinusoidal change in the pressure applied to the silica block described above, δ can be made to change sinusoidally with time

$$\delta = \delta_0(\sin \omega t) \qquad (9)$$

where δ_0 is peak phase retardation, ω is modulation frequency, and t is time. In this way the function of the piezooptic (or electrooptic) modulator can be described.

In a CD spectrometer, a photon flux I_0 photons per second, whose polarization is modulated, passes through an absorbing medium with optical density A. The transmitted photon flux reaching the detector, I_0 can be expressed as

$$I_{det} = I_0(10^{-A_L} + 10^{-A_R}) \qquad (10)$$

The optical density of a sample illuminated by I_0 photons per second and transmitting I_{det} photons per second is given by

$$A = \log(I_0/I_{det})$$
$$I_{det} = I_0/\log^{-1}A$$
$$= I_0 10^{-A}$$

With polarization modulation, (10) becomes

$$I_{det} = I_0\{[1 + \sin(\delta_0 \sin \omega t)]10^{-A_L} + [1 - \sin(\delta_0 \sin \omega t)]10^{-A_R}\} \qquad (11)$$

When (11) is rearranged,

$$I_{det} = I_0[(10^{-A_L} + 10^{-A_R}) + (10^{-A_L} - 10^{-A_R})\sin(\delta_0 \sin \omega t)] \qquad (12)$$

This time-dependent flux falls on a photosensitive element, which, in the case of a photomultiplier, produces a current whose magnitude is also dependent on the quantum efficiency, Q, of the photocathode

$$i_{det} = eQI_{det} \qquad (13)$$

This current is then amplified, factor G, by the dynode chain of the photo-multiplier such that the signal, $i_s = eQGI_{det}$, is given by

$$i_s = eQGI_0[(10^{-A_L} + 10^{-A_R}) + (10^{-A_L} - 10^{-A_R})\sin(\delta_0 \sin \omega t)] \qquad (14)$$

The signal is composed of both a time-dependent and a time-independent part; at a fixed wavelength within an absorption band the signal will have the general form given in Figure 1.6. Assuming G is frequency independent, these components can now be expressed as

$$i_{dc} = eQGI_0\{(10^{-A_L} + 10^{-A_R})\} \qquad (15a)$$

$$i_{ac} = eQGI_0[(10^{-A_L} - 10^{-A_R})\sin(\delta_0 \sin \omega t)] \qquad (15b)$$

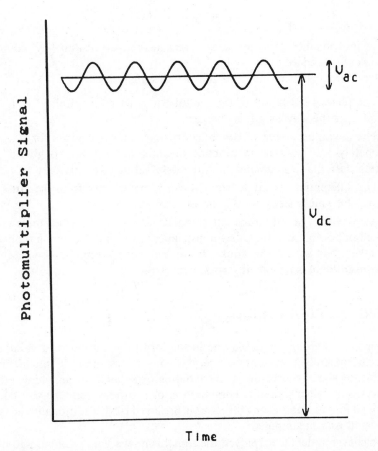

Figure 1.6 The photomultiplier signal, which has two parts: ac (V_{ac}) and dc (V_{dc}).

With the knowledge that $10^{-A} = e^{-2.303A}$, and $(e^{-A_L} - e^{-A_R})/(e^{-A_L} + e^{-A_R})$ $= \tanh[(A_L - A_R)/2] = (A_L - A_R)$ as $(A_L - A_R) \to 0$, the ratio of the two signal components can be shown to be

$$i_{ac}/i_{dc} = \tanh[2.303/2(A_L - A_R)]\sin(\delta_0 \sin \omega t) \tag{16}$$

$$= 1.1515\Delta A \, \sin(\delta_0 \sin \omega t) \tag{17}$$

If the power on the polarization modulator (the driving voltage) is set such that $\sin(\delta_0 \sin \omega t)$ is always maximal (0.58186)

$$i_{ac}/i_{dc} = 0.67\Delta A \tag{18}$$

Actually, the absolute determination of i_{ac} and i_{dc} either separately or as the ratio is not practicable. The result of a typical signal process is that a signal V_{ac}/V_{dc} proportional to i_{ac}/i_{dc} is generated and is registered on a chart recorder as V_{rec}

$$V_{rec} = V_{ac}/V_{dc} = k\Delta A \tag{19}$$

The proportionality constant, k, is determined experimentally by calibration with known standards.

Several important points emerge from this analysis:

1. The phase retardation of the modulator must be maintained at its peak value as the wavelength is changed.
2. The maximum value of the function $\sin(\delta_0 \sin \omega t)$ is given for δ approximately $110°$ so that in practice the modulator is set to give a phase retardation a little greater than the nominal quarter wave.
3. The differential optical density ΔA can be determined from the ratio $i_{ac}:i_{dc}$ only for values less than 0.1 for an error less than 1%.
4. A knowledge of only transmitted radiation is required, which means that it will suffice to use a single-beam instrument as a double beam in time device rather than to use the double-beam technique normally associated with conventional isotropic absorption spectroscopy.

5   SIGNAL PROCESSING

The signal a detector emits may be in the form of a current or a voltage; the photomultiplier is a current source. To convert this current into a voltage suitable for signal processing while avoiding impedance mismatching, either a load resistor–buffer amplifier combination or a current amplifier can be used. Although for various reasons the latter is preferred [45], traditionally the former has served as a preamplifier.

Whichever system is employed the signal is now a voltage composed of two parts, one related to the overall light level throughput and the other to the differential absorption (polarization modulation). Normally, with photomultiplier detection only the polarization of the radiation is modulated. Therefore, the signal components can be represented as V_{dc} and V_{ac}, respectively, and the CD is proportional to the ratio $V_{ac}:V_{dc}$. In the infrared, because of high background-derived signal contributions to V_{dc} (and, e.g., with photomultipliers at very low signal levels or phosphoresence measurements), it is also necessary to modulate (chop) the overall light level; this procedure presents two ac voltages to the signal processor.

The $V_{ac}:V_{dc}$ ratio can be determined in many ways. An obvious method is to use a ratiometer. Conceptually easy to appreciate and fast to respond, a ratiometer can present problems caused by a restricted dynamic range and offset errors. Ratioing before synchronous rectification of V_{ac} can be advantageous [46], although its frequency limitations (ca. 10 kHz) may render this method practical for use with only the slower polarization modulators. Dynamic range problems can be overcome by varying the signal levels before ratiometry. The ratiometer is an example of a feed-forward technique.

The preferred means of determining the $V_{ac}:V_{dc}$ ratio is through a feedback system, as illustrated in Figure 1.5. In this mode, both V_{ac} and V_{dc} are amplified to the same extent such that the V_{ac} presented to the synchronous rectifier (lock-in amplifier) is always with respect to the same overall light-level signal component V_{dc}. This can be achieved optically, mechanically, or electronically using a servosystem. The relative merits of these systems and ratiometry have been discussed at length elsewhere [47]. Only the method generally employed with a photomultiplier is described here.

The V_{dc} signal component is compared to some preset fixed voltage. The difference between these two voltages is referred to as an *error signal*. The error signal is used to control the high voltage applied to the dynode chain of the photomultiplier. This varies, independent of frequency, the amplitudes of V_{ac} and V_{dc} simultaneously until the resulting error signal is zero. In this way V_{ac} is always presented with respect to some constant-valued V_{dc} and is directly proportional to $(A_L - A_R)$. Several methods of implementing such a servosystem have been described [45, 48–50].

The V_{ac} signal component is fed to a lock-in amplifier, where it is synchronously rectified with reference to the frequency of polarization modulation. The input gain of the lock-in amplifier is set by calibration so that its output, when registered on a chart recorder, gives a deflection that can be related directly to the $(A_L - A_R)$ of the sample under investigation. A microprocessor can conveniently replace the chart recorder.

6 NOISE LEVEL CONSIDERATIONS

The limiting sensitivity of a CD spectrometer is set by fundamental factors (Figure 1.7) such as the performance of the detector and the statistical nature of any photon source. For the ideal spectrometer noise is derived from two major

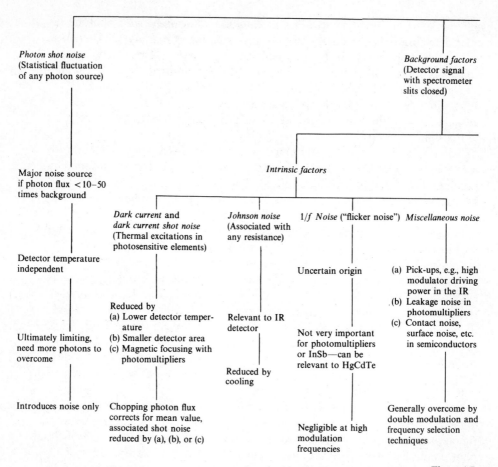

Figure 1.7

sources: photon shot noise and background signals with their associated shot noise. Background noise is all those signals that are registered on the detector when the exit slit of the monochromator is closed. Sharpe [51] has pointed out that if the required signal level from a detector is at least 10 times greater than the background level, the noise contribution of the latter can be neglected; however, it will be shown that this factor should be closer to 50 for the precise measurement of differential absorption.

6.1   Photon-Shot-Noise-Only Limitation

The photon shot noise ip_{shot} associated with the statistical fluctuation in the photon flux I_{det} derived from the original intensity I_0 with a measurement bandwidth Δf is [10]

$$ip_{shot} = (2eQGI_0\Delta f 10^{-4})^{1/2} \tag{20}$$

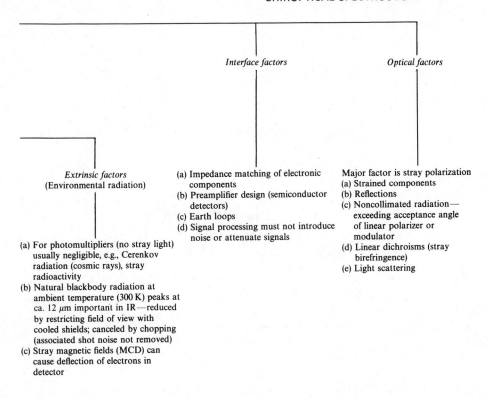

Interface factors   *Optical factors*

Extrinsic factors
(Environmental radiation)

(a) For photomultipliers (no stray light) usually negligible, e.g., Cerenkov radiation (cosmic rays), stray radioactivity
(b) Natural blackbody radiation at ambient temperature (300 K) peaks at ca. 12 μm important in IR—reduced by restricting field of view with cooled shields; canceled by chopping (associated shot noise not removed)
(c) Stray magnetic fields (MCD) can cause deflection of electrons in detector

(a) Impedance matching of electronic components
(b) Preamplifier design (semiconductor detectors)
(c) Earth loops
(d) Signal processing must not introduce noise or attenuate signals

Major factor is stray polarization
(a) Strained components
(b) Reflections
(c) Noncollimated radiation— exceeding acceptance angle of linear polarizer or modulator
(d) Linear dichroisms (stray birefringence)
(e) Light scattering

Some limitation factors in CD spectroscopy.

For CD measurements employing conventional lock-in amplifier synchronous detection with an output RC filter such that $\Delta f = 1/4\tau$, where τ is the measurement time constant in seconds, the signal-to-noise ratio, S/N, is given by

$$i_{ac}/ip_{shot} = S/N = 0.5678(QI_0\tau10^{-4})^{1/2}\Delta A \qquad (21)$$

Therefore, in this limit the sensitivity of a CD spectrometer is governed by the overall light-level throughput, the measurement time constant, and the detector quantum efficiency. Accordingly, a photomultiplier having an S-20 photocathode, with a peak quantum efficiency of approximately 20% at approximately 400 nm, offers approximately a sevenfold improvement in the S/N ratio for a given photon flux and time constant, compared to one with an S-1 photocathode at its peak wavelength approximately 800 nm (Q ca. 0.004). When a Xenon lamp is employed, the radiation powers at 400 and 800 nm will be roughly the same; that is, for the same spectral bandwidth (wavenumber, cm^{-1}) the photon

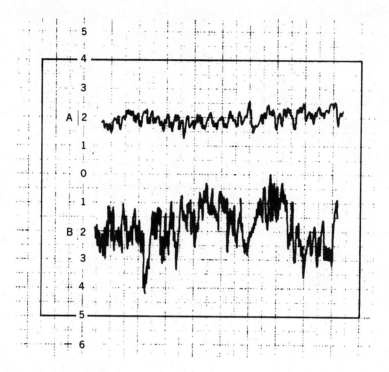

Figure 1.8   Jasco J40-CS spectropolarimeter noise level, where sensitivity $= 2.10^{-5}$ cm^{-1} and time constant $= 1$ s; (A) 400-nm, 1-nm spectral bandwidth; (B) 800-nm, 4-nm spectral bandwidth.

rate will be twice as great at the longer wavelength [I (photons per second) $\propto$ 1/frequency]. These figures predict that the noise level at 400 nm (spectral bandwidth 1 nm) should be five times better than at 800 nm (spectral bandwidth 4 nm), with a different detector but the same sensitivity and time constant. Such a position pertains to the Jasco J40-CS CD spectrometer (Figure 1.8).

The optimum value of S/N ratio is obtained for an optical density of 0.864, although Figure 1.9 indicates that there is little change between 0.4 and 1.6.

The S/N ratio is proportional to the square root of the time constant, or perhaps more practically the noise level (noise-to-signal ratio) is inversely proportional to $\tau^{1/2}$. Normalizing on $\tau = 0.25$ s, the relative reduction in noise with increasing time constant is

τ (s)	0.25	1	4	16	64
N/S	1.0	0.5	0.25	0.125	0.0625

Longer than 16 s, a realistic reduction in the noise level is given only for τ greater than 64 s, which is unreasonably long for a normal single-scan measurement. A fourfold increase in time constant with the commensurate reduction in scan speed is required to halve the noise level.

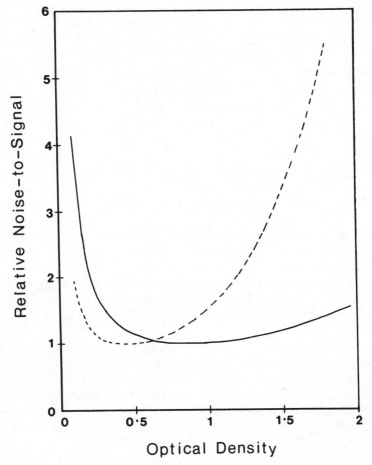

Figure 1.9 Plot of the effect of optical density on the noise-to-signal ratio: ———photon shot-noise-only limit, – – – background-only limit.

In a conventional spectrometer the entrance and exit slits are equal, thus halving the slit width leads to a fourfold reduction in light level and a doubling of the noise-to-signal ratio.

$$N/S \propto (I_{det})^{-1/2} \propto (1/\text{slit width}) \tag{22}$$

From (21) the minimum photon rate required to detect a particular ΔA (see table) can be calculated. These figures are based on unit quantum efficiency and should be increased by a factor of 10 ($I_{det} = I_0 10^{-4}$) if a sample having $A = 1.0$ is present. The transmitted photon rate can be estimated by the instrument's detector. The current emitted from a photomultiplier, i_{anode} ($= VR_{load}$ if a load resistor is used), which is kept constant in the servoloop, is given by

Sensitivity	Photon Rate (photons/s) I_{det}		
$(S/N = 1)$	$\tau = 1$ s	$\tau = 4$ s	$\tau = 16$ s
10^{-6}	3×10^{12}	7.75×10^{11}	2×10^{11}
10^{-5}	3×10^{10}	7.75×10^{9}	2×10^{9}
10^{-4}	3×10^{8}	7.75×10^{7}	2×10^{7}

$i_{anode} = eQI_{det}G$. If a substitution is made for G, the detector gain

$$I_{det} = i_{anode}[6.255(S/QM)10^{12}] \tag{23}$$

The cathode sensitivity $[S(\mu A/lm)]$, the overall sensitivity $[M(A/lm)]$, and Q can be found from the detector manufacturer's literature. Thus, for example, a typical EMI 6256 photomultiplier (S ca. $50\,\mu A/lm$; Q peak ca. 0.2) with a V_{dc} set level of $1\,\mu A$ (1 V with a 1-MΩ load resistor) achieves a limiting sensitivity of approximately 1×10^{-5} only if a time constant of 4 s is used and if an EHT of approximately 700 V (corresponds to I_{det} ca. 4×10^{10} photons per second) maintains the set level.

In summary, it is clear that if the required signal (V_{dc}) is 50 times greater than the background and the signal process is not introducing noise, then it is the photon shot noise that dictates the measurement noise level. An effective increase in light intensity, time scale of measurement, or detector Q is required to improve the instrument's sensitivity.

6.2   Background-Limited Detection

There are many sources of background signals, and they vary from intrinsic factors such as the dark current associated with detectors to extrinsic background radiation that is always present in the IR.

Nevertheless, the background signal can be represented, generally, as a photon rate I_B, although for an intrinsic source this is an "effective" parameter. The signal emitted from the detector can now be represented as

$$i_s = eQ(I_{det} + I_B) \tag{24}$$

and the associated shot noise is given by

$$ip_{shot} = [2e^2Q(I_0 10^{-4} + I_B)\Delta f]^{1/2} \tag{25}$$

Again, the S/N ratio for a CD measurement with $I_{det} = I_0 10^{-4}$ is

$$S/N = 0.5678(QI_0\tau 10^{-4})^{1/2}(1 + I_B/I_{det})^{-1/2}\Delta A \tag{26}$$

For given values of I_0, Q, A, τ, and ΔA

$$S/N \propto (1 + I_B/I_{det})^{-1/2} \tag{27}$$

From this expression it is clear that if the background radiation (I_B) or its effective value is a tenth of the required signal (I_{det}) the noise level will only be about 5% worse than in the photon-shot-noise limit, which is in agreement with Sharpe's conclusion [51]. The more critical result of significant background signal is an error in the measurement of the CD signal (V_{ac}); that is, if the dc signal contains a contribution from the background, the servosystem will maintain a level that includes a component with no related V_{ac}; consequently, the measured CD will be proportionally lower than its true value. For example, an I_B twice the I_{det} will lead to an acceptable doubling in the noise level, but an intolerable two-thirds drop in the measured CD. Chopping the overall light level will allow CD to be determined precisely in this case, but will not improve the S/N ratio.

Provided the true overall light level (I_{det}) is 50 times greater than the background level, then, at worst, the measured CD will only be approximately 2% low with no significant contribution (ca. 1%) from the background to the photon shot noise. If chopping is employed, then I_B can be as large as I_{det} without too great a deterioration in performance.

Intrinsic noise limitation is particularly relevant to the dark current in a photomultiplier

$$S/N = 0.5678(QI_0\tau 10^{-4})^{1/2}(1 + i_d/i_s)^{-1/2}\Delta A \tag{28}$$

The true V_{dc} signal should be approximately 50 times greater than the photomultiplier dark current to ensure good measurements. However, it is better to reduce intrinsic noise sources (and their associated shot noise) than to correct for them by chopping. This would involve cooling the detector, reducing its effective area, employing magnetic focusing, or counting photons.

The extrinsic noise limitation requires the photon rate I_B to be a real quantity; and in the IR, derived from natural blackbody radiation, it may be much greater than the required signal level. If $I_B \gg I_{det}$

$$S/N = 0.5678I_0 10^{-4}\Delta A(Q\tau/I_B)^{1/2} \tag{29}$$

In this case, although the dependence on Q and τ are the same as for the photon-shot-noise-only limit, the S/N ratio is now directly proportional to the photon flux, which makes it even more important to have a high light-level throughput even at the expense of resolution

$$S/N \propto (\text{slit width})^2 \tag{30}$$

The noise level is now minimal for A approximately 0.4, half that for the photon-shot-noise limit, and from the graph (Figure 1.9) it is clear that high optical densities will lead to a great increase in noise.

A further indication of the importance of photon flux in the IR is given by the realization that if $I_B = 10(I_{det})$, then three times more photons per second are required to produce the same S/N ratio as in the photon-shot-noise-limited case, all other things being equal.

6.3   Noise Level in the Circularly Polarized Luminescence Experiment

In many respects the circularly polarized luminescence (CPL) measurement is the *reverse* of the CD one. Preferentially, a chiral sample emits, on excitation with unpolarized light, photons of one hand. On passage through a polarization modulator set at the quarter-wave condition both photon types will experience a phase retardation. As in Figure 1.10, at the quarter-wave peaks in one half-cycle the left circular component will emerge from the polarization modulator plane-polarized parallel to the following linear polarizer and transmitted, while the right circular component will emerge perpendicularly polarized and absorbed. In the subsequent half-cycle the reverse will occur. Independent of the ellipticity of the emitted radiation, the polarization of the radiation passing through the analyzing monochromator will be linear, with an orientation fixed by the linear polarizer. Ideally, therefore, the polarization modulator–linear polarizer combination should be rotated about the optic axis until a position is found where polarization effects derived from the analyzing monochromator are negligible.

The signal at the detector will be of the form shown in Figure 1.10 (assuming positive dichroism), where $I_{\parallel}(I_L)$ and $I_{\parallel}(I_R)$ are the intensities of the light level derived from a particular circular component. If the emitted radiation were not polarized (left and right circular components having equal intensity), then $I_{\parallel}(I_L)$ and $I_{\parallel}(I_R)$ would be equal and there would be no ensuing V_{ac}. This corresponds to the baseline condition.

The intensity at the detector, in the absence of effects such as quenching, is given by

$$I_{det} = I_0 A \phi_f K \qquad (31)$$

where ϕ_f is the fluorescence quantum efficiency and K is an instrumental factor depending on the collection optics and the efficiency of the analyzing monochromator. As these latter two parameters can rarely be evaluated, it is difficult to equate instrument performance to the initial light intensity.

The total emission will have a left and a right component

$$I^{em} = (I_L^{em} + I_R^{em})/2 \qquad (32)$$

These two circular components enter the analyzing monochromator, after passage through the polarization modulator–linear polarizer combination as

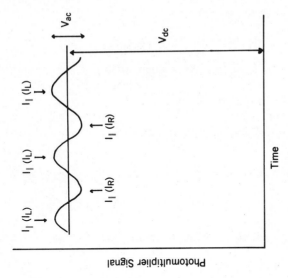

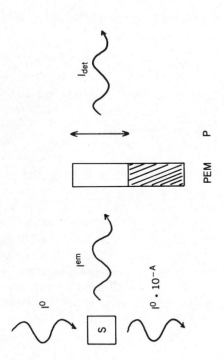

Figure 1.10 Analysis of the polarization of luminescence. A sample (S) with an absorbance A is excited by a photon flux I^0 photons per second and emits an intensity I^{em}. The intensity being detected after passage through a polarization modulator (PEM) and a linear polarizer (P) is I_{det}. At this point the photon flux has a single linear polarization ($I_{\parallel}$) parallel to the linear polarizer's ordinary axis. This intensity varies with polarization modulation from $I_{\parallel}(I_L)$ during the *left circular* modulator/linear polarizer condition to $I_{\parallel}(I_R)$ during the *right circular* condition. The eventual photomultiplier signal has an ac component (V_{ac}) and a dc component (V_{dc}).

23

$I_\parallel(I_L)$ and $I_\parallel(I_R)$, respectively, with intensities

$$I_\parallel(I_L) = (I^{em}/2)[1 + \sin(\delta_0 \sin \omega t)] \tag{33a}$$

$$I_\parallel(I_R) = (I^{em}/2)[1 - \sin(\delta_0 \sin \omega t)] \tag{33b}$$

The total emitted light throughput is

$$I_\parallel(I_L) + I_\parallel(I_R) = (I^{em}/2)[1 + \sin(\delta_0 \sin \omega t)] + (I^{em}/2)[1 - \sin(\delta_0 \sin \omega t)] \tag{34}$$

The photon flux at the detector, I_{det}, is composite with

$$I_{det} = (I^{em}/2) + [\Delta I^{em} \sin(\delta_0 \sin \omega t)] \tag{35}$$

The photomultiplier signal components i_{dc} and i_{ac} are given as

$$i_{dc} = eQGK(I^{em}/2) \tag{36a}$$

$$i_{ac} = eQGK[\Delta I^{em} \sin(\delta_0 \sin \omega t)] \tag{36b}$$

With the optimum modulation condition, $\delta_0 = 110°$, the parameter of interest, the emission dissymmetry factor, g_{em}, is given as

$$g_{em} = i_{ac}/i_{dc} = 1.164 \Delta I^{em}/I^{em} \tag{37}$$

In our experience, as indicated by Cavenett and Sowerby [52], the direct determination of this ratio using techniques as just described for CD spectroscopy is difficult at the start and end of a scan where I^{em} tends to zero (division by zero). Typically, therefore, i_{ac} and i_{dc} are measured individually and a calibrated ratio is then determined.

However, the S/N ratio analysis will be the same as for absorption CD and S/N will be directly proportional to ΔI and proportional to the square root of detector quantum efficiency, total emission throughput, and time constant. The conclusions concerning the value of these parameters will be the same as for CD. Thus one comes to the dilemma discussed in Section 6.4. If the sample has strong fluorescence, differential luminescence can be detected even if it has a low value; on the other hand, for weak fluorescers only those with an associated strong g-factor can be monitored. To measure a weak CD (small signal) a minimum photon rate for given Q and τ is essential. Hence, unlike conventional isotropic fluorescence, it is generally much more difficult to detect CPL than CD.

6.4 Conclusion

Since the production of polarized light has been optimized and modern electronic components have such low noise characteristics, the question arises as to what is the lowest g-factor that can be detected. When asked how a CD spectrometer can be made more sensitive, it is important to establish that: (1) the spectrometer has been optimally calibrated (see Section 7) and (2) the signal-processing circuitry is optimal. There should be no signal deterioration caused by pick-ups, earth loops, or poor component matching, for example, impedance mismatch.

As in the foregoing discussion detection is limited by two major factors: photon shot noise and background signal with its associated shot noise. Therefore, it is not surprising that in our experience the question regarding ultimate sensitivity requires qualification. There are two limits: (1) For a given high light level, what is the smallest g-factor that can be detected? and (2) For a given g-factor, what is the lowest light level permissible?

For a photomultiplier, detecting CD in the mid-visible, photon shot noise is the limiting factor and limit 1 is relevant. In a luminescence experiment or in the IR, background signals mean that limit 2 is relevant. These two factors have subtle consequences regarding the tactics to employ in improving a CD spectrometer. If the aim is to measure a lower g-factor, then it is necessary to increase the photon flux or time scale of measurement. If the light level is low, for example, high-resolution work or low-quantum efficiency in fluorescence, then cooling the detector to remove background signals will be beneficial. In the Raman CID experiment the g-factors associated with vibrations are approximately 10^{-3} and the photon shot noise is less important than measuring g-factors of 10^{-5} in a CPL measurement at the same light level.

The best compliment for an instrument is that its sensitivity is photon-shot-noise-limited; that is, the signal-processing circuitry does not introduce noise. On such an instrument sensitivity will be increased by increasing the overall light throughput, the time scale of measurement, and the quantum efficiency of the detector. This would be the classical approach to the problem. Obviously, these parameters can be effectively increased by techniques such as Fourier transform spectroscopy [46, 53–56], photon counting [57], diode-array detection [58, 59], piezooptic dispersion [60], and photoacoustic spectroscopy [61]. Photon shot noise is a statistical fluctuation requiring statistical methods to reduce the uncertainty it generates; suitable computer-assisted techniques employing such things as signal averaging [62], mathematical smoothing [63, 64], "random" light chopping [65], and Kohlman filters are now beginning to be applied. These techniques have advantages and disadvantages.

7   CALIBRATION OF A CIRCULAR DICHROISM SPECTROMETER

Before calibration the CD spectrometer must be "tuned" correctly. The detection electronics should be phase locked to a correctly functioning polarization modulator. The plane polarizer–polarization modulator orientation must be correct, and the latter must be programmed such that the required phase retardation of the polarized components of the radiation is maintained at every wavelength. The instrumentalist must control four factors: (1) polarizer–modulator orientation, (2) lock-in amplifier phase, (3) any active tuned filters, and (4) polarization modulator power. Experience will permit the operator to set approximate values for these. Then, with any strong dichroic sample in the light beam, fine tuning can be achieved so that a maximal CD is registered on a chart recorder.

The power on the modulator must be optimized and programmed to give a maximal CD at every wavelength. Ideally, this involves at least three samples {[Co(en)$_3$]Cl$_3$ or pantoyl lactone at approximately 200 nm, epiandrosterone at approximately 300 nm, [Co(en)$_3$]Cl$_3$ at approximately 500 nm, and nickel tartrate at approximately 800 nm}. The resulting plot of polarization modulator drive voltage against wavelength (nanometers) should be a straight line.

The V_{ac} gain in the system can now be adjusted to give a required deflection on the chart recorder corresponding to a known CD. Standards for chiroptical spectroscopy can be classified as primary, secondary, or special optical.

7.1   Primary Standards

The direct measurement of the ε_L and ε_R coefficients of a chiral compound will give an absolute primary standard. The CD measurement of [Co(en)$_3$]Cl$_3$ by McCaffery and Mason [66] is a good example of this. Despite this, CD spectrometers have generally been calibrated by pseudoprimary standards whose CD has been determined by a Kronig–Kramers transform of the ORD Cotton effects of isolated electronic transitions. Epiandrosterone and 10-camphorsulfonic acid fall into this class—the former in a solvent that is known to deteriorate (dioxan) and the latter depending on the presence of water of crystallization. Meguro and colleagues [67, 68] have surveyed these standards and have made useful suggestions regarding their values. Schippers and Dekkers [69] have described an interesting method of generating primary standards. In their experiment the photomultiplier output from a CD spectrometer employing 50-kHz piezooptic polarization modulation is fed to a phase-locked photon counter. The ratio of the difference and the sum of the left and right circularly polarized counts is equivalent to the ΔA of the test solution.

7.2   Secondary Standards

After a CD spectrometer has been calibrated, secondary standards can be created. This is particularly important for "overlap" calibration of spectrometers that extend the wavelength range. For example, the value of ΔA at 230 nm of a [Co(en)$_3$]Cl$_3$ solution can be carried from a commercial to a vacuum-UV spectrometer. Similarly, in our laboratories cross calibration of an IR CD spectrometer with a Jasco J40-CS spectropolarimeter with [Cu(spartein)Cl$_2$] at 940 nm has led to values for the vibrational CD spectrum of camphor that are in good agreement with those of Cheng and colleagues [70, 71] based on a special optical device. This result emphasizes the value of ensuring that a CD spectrometer is correctly tuned before calibration.

7.3 Special Optical Devices

Obviously, the use of an optical device that creates a known dichroism is important; it avoids uncertainties related to concentration and optical and chemical purity always associated with chemical standards. Three devices have been described that involve a judicial combination of two wave plates. Jasperson and Schnatterly [16] calibrated their ellipsometer with a plane-polarizer statically stressed silica block combination; Steinberg and Gafni [72] calibrated their CPL spectrometer using a "variable" plane-polarizer and a quarter-wave plate; Cheng and colleagues [70, 71] calibrated their IR CD spectrometer with a multi-half-wave plate and a plane-polarizer combination. (Marzott [73] has described an improved version of this technique.)

The device of Gafni and Steinberg, although constructed for a 90° fluorimeter, can be replaced after calibration by a light source (strongly fluorescing achiral sample) and a suitable dichroic sample, which in turn becomes a secondary standard.

8 CHIROPTICAL SPECTROMETERS

Fast-polarization modulation has become standard in chiroptical spectrometers since the seminal papers of Legrand and colleagues [10]. Within its working range the performance of an electrooptic modulator is not inherently inferior to one that is piezooptic; in fact, the ability of the former to square-wave modulate can be a positive advantage. It is the durability and adaptability (greater acceptance angle, relatively low modulation power, and spectral transparency) of the latter that has helped to widen the scope of chiroptical spectroscopy. Therefore, the low noise (high sensitivity) now achievable in the visible- and near-UV is attributable to both developments in signal processing and the use of good optics (generally based on double monochromators) with more intense light sources to ensure a greater light throughput. High-intensity, stable xenon lamps (450 W) driven by power supplies having internal servosystems to stabilize the discharge current are used with back reflectors to enhance light collection. A xenon lamp that is fabricated† within an ellipsoidal mirror is now available.

An interesting feature introduced in the Jouan Mark II Dichrograph is the design of the double monochromator that is constructed around two dispersing cultured quartz prisms with orientations such that they become equivalent to split components of a Rochon prism.

†Xenon illuminator, originally available from EMI-Varian, now from ILC Technology, Sunnyvale, CA 94086.

A modern chiroptical spectrometer can operate in several modes:

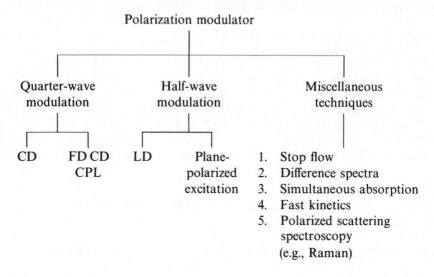

8.1   Vacuum Ultraviolet Circular Dichroism

Results in this field have been well reviewed [50, 74–76]. In line with (21) the biggest challenge associated with the measurement of CD in the vacuum-UV is the achievement of a sufficiently intense, stable photon throughput. Therefore, much consideration has gone into the design of the lamp. Much success has been achieved [77] with a Hinteregger source whose performance has been improved by constructing the cathode of stainless steel with a stainless-steel needle to stabilize the discharge. The H_2 pressure is allowed to be as high as 10–12 torr, which permits an electrical power of approximately 1.5 kW (2.5 kV, 600 mA). Resolution has been sacrificed in favor of photon flux, so spectral bandwidths of approximately 1–2 nm are usual, which means there are large slit widths. Accordingly, a MgF_2 window (CaF_2 solarizes more readily) is used to isolate the lamp from the monochromator and a second pumping system. This window also serves to prevent sputtering products from entering the monochromator and act as a barrier in case the water-cooled capillary of the lamp fractures. Calcium fluoride (or LiF_2) is used as an exit window (MgF_2 is birefringent). Brahms and colleagues [78] employed an Original Hanau D_2 lamp (D200F) with light collection improved by the use of CaF_2 lenses.

To avoid light losses because of focusing elements most monochromators used to date have employed a single grating as the only dispersive and focusing element within the monochromator. This has been achieved with either a 1-m ($f/10$) normal reflection configuration (e.g., McPherson 225) or an $f/10$ grazing incidence configuration (Jobin-Yvon LHT 30). Jobin-Yvon claims that the LHT 30 (with torroidal grating) has lower reflection losses that are caused by grating coatings, negligible depolarization effects, and a configuration that is particularly adaptable for use with a synchrotron beam.

Four types of linear polarizer have been used [50, 79–82]: a double Rochon (difficult to fabricate), a conventional Rochon, a Wollaston, and a Biotite single-surface reflection polarizer. The three prism polarizers are ideally fabricated from MgF_2 (a silica Rochon polarizer can transmit to ca. 165 nm). In practice the Wollaston and Biotite devices, although they lead to movement of the optic axis, have the better transmission properties with the latter giving the highest light throughput below approximately 170 nm. Monochromators with $f/10$ optics give a solid angle of radiation that is compatible with the acceptance angle of the polarizers (a lower f-number to increase the light-gathering capacity necessitates the use of lenses), and polarizers can be located before the exit slit oriented so that the extraordinary beam is cut off by the natural slit height. Alignments in a convergent rather than in a divergent beam are always more reliable. In this way the optical path can be made as short as possible, and the need for any external focusing optics is avoided. Nitrogen is effectively transparent in the vacuum-UV, and it is most convenient to maintain the monochromator under vacuum with a nitrogen atmosphere in the sample compartment. To date the best choice of polarization modulator has been the CaF_2 piezooptic device (LiF_2 may give shorter wavelength penetration).

Traditionally, detectivity in the vacuum-UV has been achieved by coating the window of a conventional UV/Vis photomultiplier with a scintillator (e.g., sodium salicylate) although solar-blind devices (Cd/Te photocathodes) with polycrystalline MgF_2 windows offer advantages.

A recommended signal-processing scheme, pertinent to photomultipliers in general, is given in Figure 1.5. A current preamplifier, with a minimum 100-kHz frequency response (for linear dichroism measurements) feeds both a servo- and a lock-in amplifier. The latter is phase locked to the polarization modulator so that its output is directly proportional to the differential optical density. The servoamplifier, of a type described in Section 5, is coupled to an EHT supply that varies the gain of the photomultiplier. The best dc level, current amplifier gain, and servoresponse will depend to a large extent on circuit design; and a condition is required such that servoaction is as fast as possible without signal deterioration from signal attenuation or induced "noise." In this way lamp fluctuations can be compensated for to a certain extent. A tuned, active filter associated with the lock-in input augments the frequency rejection character-istics of the lock-in and allows the ac signal component to be registered more readily on an oscilloscope. With a top quality lock-in amplifier this active filter will not necessarily lead to noise reduction in the final differential optical density measurement.

In general vacuum-UV absorption spectra are dominated by electric dipole allowed, magnetic dipole forbidden electronic transitions, and accordingly high g-factors are not anticipated. Certainly, g-factors of about 10^{-4} should be readily detectable down to 160 nm; below this wavelength a steady decrease in overall photon flux becomes more limiting. The shortest wavelength at which CD spectra have been reported is approximately 130 nm. However, the two major factors limiting CD detection in the vacuum-UV remain the overall light level throughput and lamp stability.

An alternative means of maintaining a high stable photon flux is offered by the synchrotron. Encouraging experiments have been initiated by Snyder [83, 84] at the synchrotron ring in Wisconsin. The high-intensity synchrotron beam is ready-plane polarized to a very high degree, certainly for radiation near the optic axis of the radiation; therefore, with due care this polarization can be maintained through a monochromator and presented to a conventional photoelastic modulator, and experiments will proceed as described previously. Relatively high-resolution work is possible and will yield more detailed spectra, particularly of Rydberg transitions. In this context magnetic CD measurements of fundamental molecules such as methane, ethylene, and acetylene should prove rewarding.

8.2 Infrared Circular Dichroism Spectroscopy

The measurement of CD from the visible to 1150 nm ($8500 \, cm^{-1}$) is readily achieved, as described previously, using a photomultiplier with an S-1 photocathode. Beyond this wavelength, the need to redesign the servosystem for semiconductor detectors and to compensate for high extrinsic background noise signals has introduced [46, 54–56, 85, 86] many new ideas into the field of chiroptical spectroscopy instrumentation.

Although the first successful measurements of pure vibrational CD by Holtzwarth [87–90] were achieved by using a Globar source, it soon became apparent that the traditional IR sources (Nernst or Globar) do not offer sufficient intensity for truly reliable low-noise spectra. It is also clear from Section 6 that high extrinsic background signals in the IR mean that a much higher photon flux is required than for an equivalent performance in the photon-shot-noise-limited case. Therefore, in the near-IR to approximately 2000-cm^{-1}, high-wattage (650-W) tungsten lamps have been used, while the newly introduced sapphire-windowed xenon illuminators are even more successful with output down to the sapphire cut off (ca. $1600 \, cm^{-1}$). Recently, vibrational CD has been observed [91] to approximately $500 \, cm^{-1}$ using a carbon furnace source [86, 92].

The layout of the IR CD spectrometer constructed at King's College, London, is illustrated in Figure 1.11. Typically, radiation is dispersed by an $f/6$ monochromator fitted with order-sorting filters on the exit slit (reduced pyrolytic effects from intense light sources) and a mechanical light chopper at the entrance slit. Spectral bandwidths are typically about $10 \, cm^{-1}$ at $3000 \, cm^{-1}$. As will be discussed, all optics are best maintained on the optic axis with one BaF_2 lens to focus onto the monochromator entrance slit, another at the exit slit to render the radiation collimated for the polarization optics and the sample, and a third to focus down onto the detector element.

Several plane polarizers have been used as progress into the IR was made. Obviously, a quartz Rochon or Calcite Glan prism is good to approximately $1800 \, cm^{-1}$. Beyond this a MgF_2 or $LiIO_4$ Rochon or wire-grid polarizer is necessary. To achieve good light throughput IR optics traditionally have

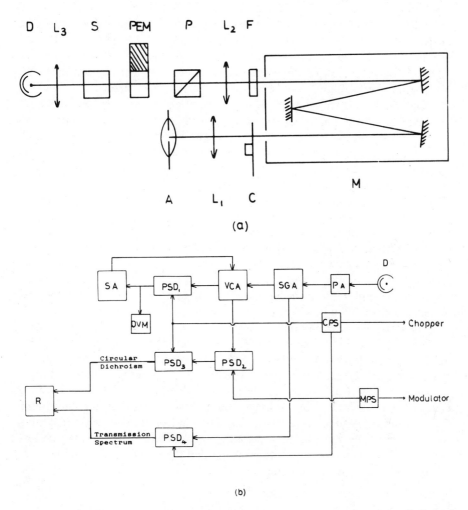

Figure 1.11   The IR CD spectrometer. (*a*) Optical layout: A, radiation source (arc lamp); L_1, L_2, and L_3, lenses; C, mechanical chopper; M, monochromator; F, filter; P, linear polarizer; PEM, photoelastic modulator; S, sample; D, detector. (*b*) Signal process: D, detector; PA, preamplifier; SGA, switch-gain amplifier; VCA, voltage-controlled amplifier; PSD_1, PSD_2, PSD_3, and PSD_4, phase-sensitive detectors (lock-in amplifiers); SA, servoamplifier; DVM, digital voltmeter; R, dual-channel chart recorder; CPS, chopper power supply; MPS, modulator power supply.

relatively large apertures and low *f*-numbers, hence, the importance of the second lens. The wire-grid polarizer appears to be the most successful (large aperture) although it may need to be tilted slightly to prevent artifactual polarizations that are derived from multireflections from reaching the detector. The relatively high driving powers needed to maintain quarter-wave (or half-wave) retardation in the IR are too high for the CaF_2 modulators used in the visible and ultraviolet; therefore, CaF_2 and ZnSe modulators specifically for use in the IR have been developed. Holtzwarth and colleagues have been successful in reaching approximately $1000 \, \text{cm}^{-1}$, in both the CD [93, 94] and LD [95] modes, with a germanium modulator of the Badoz type.

In retrospect, somewhat unnecessarily, the detector construction rather than its inherent performance has been the main source of artifacts during the development of this technique. The frequency of polarization modulation (20–100 kHz) limits the choice of detector, typically, photovoltaic or photoconductive semiconductors such as InAs, InSb, and HgCdTe have been employed. To reduce intrinsic noise, IR detectors generally have small areas ($1.0 \times 0.5 \, \text{mm}$). Therefore, the final focusing element is critical. For optimum detectivity these devices are maintained at liquid nitrogen temperature in a Dewar flask, and a very convenient way of achieving this is to use a glass Dewar flask with a bottom-mounted sapphire window and an ellipsoidal mirror (see Figure 1.12). Sapphire is naturally birefringent; and, coupled to the polarizing effects of reflection, it is not surprising that problems occurred. Replacement of the ellipsoidal mirror with a lens-plane 45° mirror assembly was not a solution. Although measurement artifacts can be reduced by the introduction, after the sample, of a sapphire Wollaston prism or a second polarization modulator, in a manner akin to Billings' [11] depolarization experiments, their elimination is best achieved by relearning the lessons from traditional UV/Vis CD spectroscopy. There should be as few optical components between the plane polarizer, polarization modulator, and detector as possible; for a photomultiplier, there are generally none. Therefore, the best configuration is achieved with a strainfree CaF_2 lens that is focusing directly onto a photosensitive element, which is housed in a side-windowed Dewar flask that has a strainfree CaF_2 window. Artifacts are also easier to control if relatively large area detectors, typically $4.0 \times 4.0 \, \text{mm}$, are employed at the expense of noise performance. This is probably related to ensuring that the entire transmitted beam is always falling on the active element.

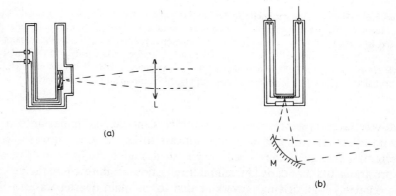

(a)

(b)

Figure 1.12 Encapsulated IR detectors: (a) side mounted; L, lens; (b) bottom mounted; M, ellipsoidal mirror.

The final feature of artifact removal lies in the design of the signal processing circuitry. The rejection of frequencies not phase-locked by a lock-in amplifier (e.g., 100 kHz for a 50-kHz polarization modulation) is often not sufficient, and it may be prudent to include tuned, active filters into the signal processing before synchronous rectification. Adjustment of the position of optics can often reduce these unwanted signal components by effectively reducing stray ellipticities such as those generated by multireflections. The absolute signal levels are extremely small, less than picoamps, so "pick ups" can be very critical. Earth loops must be avoided. With 50-kHz polarization modulation, particularly at high powers in the IR, it is easy to have a 50-kHz signal, not related to light level or differential absorption, that sits on the desired signal. As the overall light level throughput drops, so generally does its associated V_{ac}, while the "parasitic" V_{ac} remains effectively unchanged in absolute magnitude, but relatively larger with respect to the true V_{ac}. As the result of servoaction both 50-kHz signal components are amplified but there comes a time when the parasite signal becomes manifest. A following effect ensues. To an extent, the adoption of a double demodulation technique reduces some of these effects, but it does not eradicate them. The best solution is to ensure that the 0-V line follows a tree pattern as closely as possible.

Unlike photomultipliers, semiconductor detectors have inherent resistance and capacitance characteristics that are critical with respect to preamplifier design. The preamplifier should neither attenuate the 50–100-kHz signal nor should it add noise. Inherently better than photoconductive devices, photovoltaic devices are to be preferred although they do have a tendency to "blow" irreversibly into the photoconductive mode. To prevent this problem and to ensure optimum signal capture a preamplifier based on an audio transformer has been successfully employed [96] in our laboratories for a particular InSb detector. Unfortunately, there is no general preamplifier design for this class of radiation detector.

Many methods have been described in the literature for determining, in the infrared, the ratio of the differential absorption and overall light throughput signal components; the one preferred in these laboratories involves the use of a *voltage-controlled amplifier*. In this scheme the preamplified signal is sent to a switched-gain amplifier that amplifies both signal components, in the same sense as changing the load resistor or current preamplifier gain on a photomultiplier, to bring the signal levels into the working range (dynamic range) of the servosystem. The polarization-modulated signal component is then amplified–attenuated by an amplifier whose gain is controlled by the error signal generated by referring the synchronously rectified overall light level signal component to a preset level. The ensuing high-frequency ac signal (50–100 kHz) has an amplitude that is directly proportional to the differential absorption. For differential absorptions of about 10^{-5} or less this signal can be of the same order as the "pick-up" signal on the 0-V line, and care is obviously required in its rectification. The addition of tuned, active filters in the signal-processing line improves the rejection of unwanted frequencies, but the important factor is the use of a double-demodulation technique.

The ac signal emitted by the servosystem is first passed through a lock-in amplifier with a sub-millisecond time constant, demodulated with respect to the polarization modulation and then to a second lock-in amplifier with the effective instrument time constant (1–16 s) that synchronously rectifies the signal with respect to the light-level chopper (200–800 Hz). This dc signal is proportional to the differential absorption and is sent directly to either a chart recorder or the A/D convertor of a microprocessor.

A fourth lock-in amplifier samples the signal between the switched-gain amplifier and the servosystem, which permits the transmission spectrum to be monitored simultaneously.

8.3   Polarized Emission Spectroscopy

In conventional nonchiral emission spectroscopy there are two general modes of operation. The fluorescence at a fixed wavelength can be monitored as a function of the excitation wavelength; this is termed the *excitation spectrum*. Alternatively, the wavelength dependence of the emission (fluorescence or phosphorescence) can be measured with respect to a particular excitation wavelength; this is the *fluorescence* (or *phosphorescence*) spectrum. When achiral molecules are in a state of restricted motion, photoselection can occur, and the degree of linear polarization in the emission can be measured. If this quantity is measured and the excitation wavelength is changed, the technique is referred to as *plane-polarized excitation spectroscopy*. The wavelength dependence of this quantity can also be monitored in the fluorescence spectrum.

It is necessary to include the prefix "plane-" to the names of the latter two techniques because it is now possible to make similar measurements with chiral molecules exciting with or analyzing for circularly polarized light. The chiroptical analog of excitation spectroscopy is *fluorescence-detected circular dichroism* (FD CD) and of luminescence spectroscopy it is *circularly polarized luminescence* (CPL).

Light focused on a sample generates luminescence that is emitted over a sphere centered on the sample. Collection of this radiation for analysis can then be made at any angle with respect to the excitation beam. The angle chosen in practice, that is, the configuration of the spectrometer, is the result of many considerations. In the nonchiral experiment, with the achiral molecule dissolved in a fluid medium giving no photoselection, the emission will be homogeneously distributed around this sphere and generally a right-angle spectrometer configuration is chosen to reduce the amount of excitation radiation (stray light) entering the analyzing monochromator. A laser source, being both highly monochromatic and well collimated would be preferred for a straight-through configuration. Once photoselection occurs, however, the choice of spectrometer configuration will be dictated by other factors [97–100].

It is worth noting that in conventional emission spectroscopy optics can be chosen to maximize the collection of fluorescence, but in plane- or circular-polarized fluorescence the polarization of the fluorescence is the factor of interest; therefore, only a small fraction of the emission can reasonably be collected for analysis. This combined with the inherent low-quantum efficiency of many compounds places a severe shot noise limitation on the measurements. Obviously, emission can be increased by increasing the excitation intensity, but the real limits of lamp power, excitation bandwidth and sample photolysis, coupled with restricted collection area means that in general the sensitivity of a polarized fluorescence experiment will be lower than in an absorption experiment.

8.3.1   Polarized Excitation Spectroscopy

These experiments involve [99, 100] the modulation of the polarization of the excitation radiation, and a conventional absorption CD spectrometer can be used. In excitation spectroscopy, the sample effectively acts as its own "wavelength shifter" in a way analogous to the use of a scintillator on a photomultiplier window to detect vacuum-UV radiation. An alternative method of measuring absorption data, the major potential of the FD CD experiment, lies in the ability to measure the absorption CD of only fluorescent species in a complex mixture such as tryptophan in a protein.

The typical apparatus for either plane- or circular-polarized excitation spectroscopy using a photoelastic polarization modulator is given in Figure 1.13. Once the instrument is set up and calibrated in the conventional absorption mode, the detector can be moved to a 90° configuration without any loss in principle. The instrument is designed to operate in the "servomode"; that is, with calibration, it directly registers $g_{em} = (I_L - I_R)/(I_L + I_R)$ in the FD CD experiment and the mean polarization $P = (I_\parallel - I_\perp)/(I_\parallel + I_\perp)$ in the plane-polarized experiment.

For the former the photoelastic modulator is set to be driven so that at its natural frequency, left–right alternate circular polarization is obtained at the peaks and troughs of the motion. In the latter the modulator passes light polarized perpendicular to the original plane at the peaks and troughs but

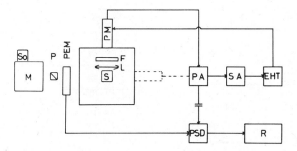

Figure 1.13 The FD CD spectrometer: So, radiation source; M, monochromator; P, linear polarizer; PEM, photoelastic modulator; S, sample; L, lens; F, filter; PM, photomultiplier; PA, preamplifier; SA, servoamplifier; EHT, photomultiplier power supply; PSD, phase-sensitive detector; R, chart recorder.

parallel when the modulator motion is zero; this is an effective doubling in modulation frequency. As described previously, the electronics should be tuned accordingly.

The main sensitivity limiting factor is the light throughput, which in the circular-polarized excitation experiment can be offset by increasing the collection area with mirrors and/or using a signal carrier [101] in the form of a strong, noninteracting independent fluorescer in the sample system.

8.3.2   Polarized Fluorescence Spectroscopy

An instrument that measures polarized fluorescence spectra [22, 52, 57, 102–105] effectively acts to analyze the polarization characteristics of a light source (the emitting sample) in a manner akin to an ellipsometer. The typical instrumental block diagram is given in Figure 1.14. The source would be a high-wattage xenon lamp (e.g., xenon illuminator) or a laser, which combined with low f-number, large collection monochromators ensures that sufficient dc intensity reaches the detector for differential fluorescence measurements, strong emitters such as lanthanides permit relatively high resolution; but, in general, CPL is a low-resolution technique.

The signal processing is essentially the same as has already been described, although in our experience the measurement is just as well undertaken without a dc-set level maintaining servosystem such as that described by Cavenett and Sowerby [52]. The location of the baselines is more reliable when I^{em} and $(I_L - I_R)$ are monitored on separate pens of a dual-channel chart recorder and the g- or p-factor determined subsequently.

As with polarized excitation spectroscopy the polarization modulator can be set for circularly or linearly polarized light with a suitable instrument configuration. Despite its restrictive acceptance angle a Glan prism is preferable to a nonfluorescent UV polaroid on the basis of better transmission properties.

The sensitivity of the CPL experiment is very much limited by the generally low light level throughputs caused by both low-quantum efficiencies and a

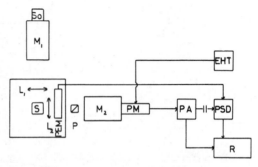

Figure 1.14   The CPL spectrometer: So, radiation source; M_1, excitation monochromator; L_1 and L_2, lenses; S, sample; PEM, photoelastic modulator; P, linear polarizer; M_2, emission monochromator; PM, photomultiplier; PA, preamplifier; PSD, phase-sensitive detector; EHT, photomultiplier power supply; R, dual-channel chart recorder.

restricted collection area. Attempts to overcome this by increasing either the excitation flux or the time scale of the measurement are generally self-limiting because of the photolysis that often occurs. Possibly the means of improving the sensitivity of this form of chiroptical spectroscopy will be through the techniques employed in Raman spectroscopy, namely, photon counting [43, 57–59] or array detection.

The excitation of a racemic mixture with circularly polarized light generally leads to a preferential excitation of one optical isomer, and the CPL of a racemic mixture can be measured [106].

8.4   Raman Circular Intensity Differentials

In many respects the instrumentation in Raman spectroscopy [43, 58, 59, 107, 108] is similar to that employed in excitation spectroscopy, although the analyzed light is scattered rather than emitted.

Because Raman scattering, even with laser irradiation, is usually so low, techniques involving an effective increase in the time scale of the measurement to overcome the general shot-noise limitation must be employed. Accordingly, the first published Raman CID spectra were recorded [109–111] using photon-counting techniques. Even so it is fortunate that the conventional Raman experiment involves visible radiation. At these wavelengths, the anticipated g-factors for vibrations can be 100 times greater than in the complementary IR techniques; consequently, the limiting overall light level can be many orders of magnitude less.

The block diagram of the instrument described by Barron and Buckingham [43] is illustrated in Figure 1.15. The laser source, an argon ion of 1-W power at

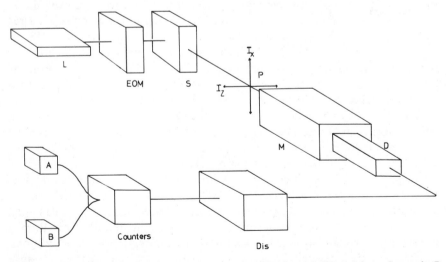

Figure 1.15   The Raman CID spectrometer: L, laser; EOM, electrooptic modulator; S, sample; P, linear polarizer; M, monochromator; D, detector; Dis, discriminator; A and B, counters. The linear polarizer **P** can be oriented to transmit along an axis parallel to the laser beam giving an intensity I_Z, or perpendicular to this giving intensity I_X.

488 nm, is ready linearly polarized and it is orientated so that the polarization axis is at 45° to the birefringence axis of the polarization modulator. To avoid coherent plane-polarized components in the polarization modulation cycle, an electrooptic modulator (KDP type) is employed and is driven in the square-wave mode at typically 900 Hz.

The z component of the light scattered at 90° is generally selected by a static linear polarizer as this is the component most likely to be free of artifacts derived from the photoselection phenomenon associated with conventional Raman spectroscopy. The scattered photons are now detected and counted after their passage through a monochromator. Photons arriving during the polarization modulator's right-circular period are counted in channel A, while channel B stores the counts during the left-circular period. Once statistically significant counts have been accumulated, the difference between the two channels $(A - B = I_R - I_L)$ and the sum $(A + B = I_R + I_L)$ are read. The CID spectra can be recorded as either $(I_z^R - I_z^L)$ or $\Delta_z = (I_z^R - I_z^L)/(I_z^R + I_z^L)$. The CIDs are reported in terms of $(I_R - I_L)$ rather than $(I_R + I_L)$ to ensure sign compatibility with the corresponding absorption technique IR CD (cf. FD CD).

The z-component-selected Raman CID is known as the depolarized CID spectrum, while the x component is the polarized CID. The latter is very difficult to measure free of artifacts although the expectation that it will be weaker and more difficult to interpret than the former means that this experiment may not be very important. It would also appear that the Stokes and the anti-Stokes CID spectra are similar, although the latter obviously requires a longer counting time.

In this way, using a single detector, a Raman CID measurement in the wavelength range from 1500 to 100 cm^{-1} with an instrumental spectral bandwidth of 10 cm^{-1} can take 24 h. A significant improvement over single-detector photon counting can be obtained by the use of a vidicon or an array detector. Such a system has been described by Brocki and colleagues [58, 59, 112]. In the latter instrument, as in the original design, the sample is illuminated by square-wave polarization-modulated (1.27-Hz) radiation derived from an argon ion laser, and the I_z-selected scattered radiation is presented to a monochromator for analysis. The "whole" spectrum is detected at the monochromator exit by an intensified vidicon tube [EG & G Instruments, Princeton Applied Research Optical Multichannel Analyzer (PAR OMA)]. The resulting digital signal stored in 500 channels is swept 12 times (33 ms/channel) during each left- and right-polarization period. To further ensure the absence of artifacts the three sweeps that occurred during the polarization switching were rejected. Typically, approximately 1.4×10^4 sweeps were accumulated, and the information was built up in two memory banks related to $I_R - I_L$ and $I_R + I_L$, respectively. Thus 350-cm^{-1} portions (with 1000 points/point) of the Raman CID spectrum are measurable in a reasonable time. Despite these modifications, the Raman CID measurement can still take 10 h for the 1500–100-cm^{-1} range albeit at high resolution. Clearly, compared to the single-detector technique, this multiband system offers many advantages: (1) higher resolution for the same time scale, (2) higher sensitivity with the same spectral resolution and measurement time scale, and (3) much faster detection for the same spectral resolution and sensitivity.

9   MISCELLANEOUS TECHNIQUES

There are obviously many specialized applications of chiroptical spectroscopy that cannot be fully discussed in the space of this chapter. Some of them are listed with relevant references.

9.1   Magnetic Circular Dichroism

In principle, all substances become optically active in a magnetic field. This phenomenon, referred to as the *Faraday effect*, is detectable by essentially the same instrumentation as used to construct a conventional chiroptical spectrometer [113–116]. Magnetic optical activity has been detected in electronic transitions in the absorption [117] and the emission [118] modes, in infrared vibrations [119], and in Raman bands [120].

9.2   Linear Dichroism

Given sufficient power a polarization modulator can be induced to produce alternate cycles of parallel and perpendicular linear polarization. Although this will be at twice the frequency of the quarter-wave (circular polarization) mode, a CD spectrometer can be *tuned* to detect linear dichroism. This facility is now available on the Jasco range of spectropolarimeters. References [22, 93, 94, 121] should be consulted for a review of linear dichroism.

9.3   Simultaneous Absorption Measurements

For various reasons it is often advantageous to monitor ordinary light absorption during a CD measurement; this is done routinely on the IR CD spectrometer at Birkbeck College. Three papers, specifically devoted to this subject, have been published [122–124].

9.4   Fast Circular Dichroism Measurements

References [60, 125–128] offer a thorough discussion of fast CD measurements.

9.5   Differential Absorption-Flattening and Differential Light-Scattering Effects

Optically active large solute molecules, suspensions, and solid-state samples can present CD spectra that are distorted by the effects of absorption-flattening [129, 130] and light scattering [131, 132]. Maestre and colleagues [131] have studied the angular dependence of differential light scattering and have drawn conclusions about macromolecular structure.

References

1. S. F. Mason, Ed., *Optical Activity and Chiral Discrimination*, Reidel, Dordrecht, Holland, 1979.

2. S. F. Mason, *Molecular Optical Activity and the Chiral Discriminations*, Cambridge University Press, Cambridge, 1982.

3. G. Holtzwarth, *Rev. Sci. Instrum.*, **36**, 59 (1965).

4. G. Holtzwarth, W. B. Gratzer, and P. Doty, *Biopolym. Symp.*, **1**, 389 (1964).

5. W. Kuhn, *Trans. Faraday Soc.*, **26**, 293 (1930).

6. A. Cotton, *C. R. Acad. Sci.*, **120**, 989 and 1044 (1895).

7. W. Kuhn, *Annu. Rev. Phys. Chem.*, **9**, 417 (1958).

8. C. Djerassi, *Optical Rotatory Dispersion*, McGraw-Hill, New York, 1960.

9. M. Grosjean and M. Legrand, *C. R. Acad. Sci.*, **251**, 2150 (1960).

10. L. Velluz, M. Legrand, and M. Grosjean, *Optical Circular Dichroism, Principles, Measurements, and Applications*, Academic, New York, 1965.

11. B. H. Billings, *J. Opt. Soc. Am.*, **37**, 738 (1947); **39**, 797 (1949); **42**, 12 (1952).

12. G. Snatzke, Ed., *Optical Rotatory Dispersion and Circular Dichroism in Organic Chemistry*, Heyden, London, 1967.

13. F. Ciardelli and P. Salvadori, Eds., *Fundamental Aspects and Recent Developments in Optical Rotatory Dispersion and Circular Dichroism*, Heyden, London, 1973.

14. M. Billardon and J. Badoz, *C. R. Acad. Sci.*, **263**, 139 (1966); **262**, 1672, (1966).

15. J. C. Kemp, *J. Opt. Soc. Am.*, **59**, 950 (1969).

16. S. N. Jasperson and S. E. Schatternley, *Rev. Sci. Instrum.*, **40**, 761 (1969).

17. L. F. Mollenauer, D. Downie, H. Engstrom, and W. B. Grant, *Appl. Opt.*, **8**, 661 (1969).

18. J. C. Cheng, L. A. Nafie, S. D. Allen, and A. I. Braunstein, *Appl. Opt.*, **15**, 1960 (1976).

19. T. M. Lowry, *Optical Rotatory Power*, Dover, New York, 1964.

20. F. Woldbye, in G. Snatzke, Ed., *Optical Rotatory Dispersion and Circular Dichroism in Organic Chemistry*, Heyden, London, 1967, p. 85.

21. P. Crabbe and A. C. Parker, "Optical Rotatory Dispersion and Circular Dichroism," in A. W. Weissberger and B. W. Rossiter, Eds., *Physical Methods of Chemistry, Techniques of Chemistry*, Part IIIC, Vol. I, Wiley-Interscience, New York, 1972.

22. K. W. Hipps and G. A. Crosby, *J. Phys. Chem.*, **83**, 555 (1979).

23. S. Mitchell, *Nature (London)*, **166**, 434 (1950); *J. Sci. Instrum.*, **34**, 89 (1957).

24. G. Bruhat, *Bull. Chim. Soc. Fr.*, **47**, 251 (1930).

25. M. Billardon and J. Badoz, *C. R. Acad. Sci.*, **248**, 2466 (1959).

26. M. Grosjean, A. Lacam, and M. Legrand, *Bull. Soc. Chim. Fr.*, 1495 (1959).

27. G. Bruhat, *Ann. Phys. Paris*, **3**, 232 (1915).

28. S. F. Mason, *Mol. Phys.*, **5**, 343 (1962).

29. A. J. McCaffery and S. F. Mason, *Proc. Chem. Soc. London*, 388 (1962); *Mol. Phys.*, **6**, 359 (1963).

30. S. F. Mason, *Proc. Chem. Soc. London*, 362 (1962).

31. C. J. Barnett, A. F. Drake, and S. F. Mason, *Bull. Chim. Soc. Belg.*, **88**, 853 (1979).

32. R. J. Dudley, S. F. Mason, and R. D. Peacock, *J. Chem. Soc. Chem. Commun.*, 1084 (1972); *Trans. Faraday Soc.*, **71**, 997 (1975).

33. W. A. Eaton and W. Lovenburg, *J. Am. Chem. Soc.*, **92**, 7195 (1970).

34. S. Feinleib and F. A. Bovey, *J. Chem. Soc. Chem. Commun.*, 978, (1968).

35. K. Rosenheck and P. Doty, *Proc. Natl. Acad. Sci. USA*, **47**, 1775 (1961).

36. G. Holtzwarth, W. B. Gratzer, and P. Doty, *J. Am. Chem. Soc.*, **84**, 3194 (1962).

37. R. Grinter, M. J. Harding, and S. F. Mason, *J. Chem. Soc. A*, 667 (1970).

38. S. F. Mason and R. D. Peacock, *J. Chem. Soc. Dalton Trans.*, 226 (1973).

39. S. Allen, O. Schnepp, and E. F. Pearson, *Rev. Sci. Instrum.*, **41**, 1136 (1970).

40. D. Brewster, *Philos. Trans. R. Soc. London*, 156 (1816).

41. A. Fresnel, *Ann. Chim. Phys.*, **15**, 379 (1820).

42. M. Biot, *Ann. Chim. Phys.*, **13**, 151 (1820).

43. L. D. Barron and A. D. Buckingham, *Ann. Rev. Phys. Chem.*, **26**, 381 (1975).

44. E. G. Coker and L. N. G. Filon, *Photoelasticity*, Cambridge University Press, Cambridge, 1931.

45. F. A. Modine, *Rev. Sci. Instrum.*, **50**, 386 (1979).

46. M. F. Russel, M. Billardon, and J. P. Badoz, *Appl. Opt.*, **11**, 2375 (1972).

47. A. F. Drake, *J. Phys. E*, **19**, 170 (1986).

48. E. Krausz and G. Cohen, *Rev. Sci. Instrum.*, **48**, 1506 (1977).

49. J. H. Obbink and A. M. F. Hezemans, *J. Phys. E*, **10**, 769 (1977).

50. E. S. Pysh, *Ann. Rev. Biophys. Bioeng.*, **5**, 63 (1976).

51. J. Sharpe, EMI Symposium 1964 revised 1970, EMI document reference R/P021Y70.

52. B. C. Cavenett and G. Sowerby, *J. Phys. E*, **8**, 365 (1975).

53. D. Fournier, A. C. Boccara, and J. P. Badoz, *Appl. Opt.*, **21**, 74 (1982).

54. L. A. Nafie and M. Diem, *Appl. Spectrosc.*, **33**, 130 (1979).

55. L. A. Nafie, M. Diem, and D. W. Vidrine, *J. Am. Chem. Soc.*, **101**, 496 (1979).

56. L. A. Nafie and D. W. Vidrine, *Fourier Transform Infrared Spectrosc.*, **3**, 83 (1982).

57. P. H. Schippers, A. Van den Beukel, and H. P. L. M. Dekkers, *J. Phys. E*, **15**, 945 (1982).

58. W. Hug and H. Surbeck, *Chem. Phys. Lett.*, **60**, 186 (1979).

59. W. Hug, A. Kamatari, K. Srinivasan, H. J. Hanson, and H. R. Sliwka, *Chem. Phys. Lett.*, **76**, 469 (1980).

60. M. Hatano, T. Nozawa, T. Murakami, T. Yamamoto, M. Shigehisa, S. Kimura, T. Takakuwa, N. Sakayanagi, T. Yano, and A. Watanabe, *Rev. Sci. Instrum.*, **52**, 1311 (1981).

61. D. Fournier, A. C. Boccara, and J. P. Badoz, *Appl. Phys. Lett.*, **32**, 640 (1978).

62. Y. P. Myer and L. H. MacDonald, *J. Am. Chem. Soc.*, **89**, 7142 (1967).

63. A. Savitzky and M. J. E. Golay, *Anal. Chem.*, **36**, 1627 (1964).

64. G. W. Trott and J. H. Beynon, *Int. J. Mass. Spectrom. Ion Phys.*, **31**, 37 (1979).

65. G. F. Kirkbright and R. M. Miller, *Anal. Chem.*, **55**, 502 (1983).

66. A. J. McCaffery and S. F. Mason, *Mol. Phys.*, **6**, 359 (1963).

67. T. Konno, H. Meguro, and T. Tuzimura, *Anal. Biochem.*, **67**, 226 (1975).

68. K. Tuzimura, T. Konno, H. Meguro, M. Hatano, T. Murakami, K. Kashiwabara, K. Saito, Y. Kondo, and T. M. Suzuki, *Anal. Biochem.*, **81**, 167 (1977).

69. P. H. Schippers and H. P. L. M. Dekkers, *Anal. Chem.*, **53**, 778 (1981).

70. G. A. Osborne, J. C. Cheng, P. J. Stephens, *Rev. Sci. Instrum.*, **44**, 10 (1973).

71. L. A. Nafie, T. A. Keiderling, and P. J. Stephens, *J. Am. Chem. Soc.*, **98**, 2715 (1976).

72. I. Z. Steinberg and A. Gafni, *Rev. Sci. Instrum.*, **43**, 409 (1972).

73. C. A. Marzott, doctoral dissertation, University of Minneapolis, MN, 1979.

74. O. Schnepp, "Natural and Magnetic Circular Dichroism Spectroscopy in the Vaccuum Ultraviolet," in S. F. Mason, Ed., *Optical Activity and Chiral Discrimination*, Reidel, Dordrecht, Holland, 1979, p. 87.

75. A. F. Drake and S. F. Mason, *J. Phys. Paris Colloq.*, **39**, 212 (1978).

76. W. C. Johnson, Jr., *Ann. Rev. Phys. Chem.*, **29**, 93 (1978).

77. W. C. Johnson, Jr., *Rev. Sci. Instrum.*, **42**, 1283 (1971).

78. S. Brahms, J. Brahms, G. Spach, and A. Brack, *Proc. Natl. Acad. Sci. USA*, **74**, 3208 (1977).

79. W. C. Johnson, Jr., *Rev. Sci. Instrum.*, **35**, 1375 (1964).

80. D. L. Steinmetz, W. G. Phillips, M. Wirick, and F. F. Forbes, *Appl. Opt.*, **6**, 1001 (1967).

81. A. Gedanken and M. Levy, *Rev. Sci. Instrum.*, **48**, 1661 (1977).

82. M. B. Robin, N. A. Keubler, and Y. H. Pao, *Rev. Sci. Instrum.*, **37**, 922, (1966).

83. P. A. Snyder and E. M. Rowe, *Nucl. Instrum. Methods*, **172**, 345 (1980).

84. P. A. Snyder, P. A. Lund, P. N. Schatz, and E. M. Rowe, *Chem. Phys. Lett.*, **82**, 546 (1981).

85. P. J. Stephens and R. Clark, "Vibrational Circular Dichroism: The Experimental Viewpoint," in S. F. Mason, Ed., *Optical Activity and Chiral Discrimination*, Reidel, Dordrecht, Holland, 1979, p. 263.

86. T. A. Keiderling, *Appl. Spectrosc. Rev.*, **17**, 189 (1981).

87. E. C. Hsu and G. Holtzwarth, *J. Chem. Phys.*, **59**, 4678 (1973).

88. G. Holtzwarth, E. C. Hsu, H. S. Mosher, T. R. Faulkner, and A. Moscowitz, *J. Am. Chem. Soc.*, **96**, 251 (1974).

89. T. R. Faulkner, A. Moscowitz, B. Holtzwarth, E. C. Hsu, and H. S. Mosher, *J. Am. Chem. Soc.*, **96**, 252, (1974).

90. I. Chabay and G. Holtzwarth, *Appl. Opt.*, **14**, 454 (1975).

91. T. A. Keiderling, personal communication.

92. C. N. Su, V. J. Heintz, and T. A. Keiderling, *Chem. Phys. Lett.*, **73**, 157 (1980).

93. G. Holtzwarth, I. Chabay, and N. A. W. Holtzwarth, *J. Chem. Phys.*, **58**, 4816 (1973).

94. T. Kursar and G. Holtzwarth, *Biochemistry*, **15**, 3352 (1976).

95. E. C. Hsu and G. Holtzwarth, *J. Am. Chem. Soc.*, **95**, 6902 (1973).

96. C. J. Barnett and A. F. Drake, unpublished work.

97. I. Z. Steinberg and B. Ehrenberg, *J. Chem. Phys.*, **61**, 3382 (1974).

98. I. Tinoco, B. Ehrenberg, and I. Z. Steinberg, *J. Chem. Phys.*, **66**, 916 (1977).

99. D. H. Turner, "Fluorescence-Detected Circular Dichroism," in C. H. W. Hirs and S. N. Timasheff, Eds., *Methods in Enzymology*, Vol. XLIX, Part G, Academic, New York, 1978, Section II.7 p. 199.

100. I. Tinoco, "Circular Dichroism and Fluorescence Detected Circular Dichroism of Macromolecules," in S. F. Mason, Ed., *Optical Activity and Chiral Discrimination*, Reidel, Dordrecht, Holland, 1979, p. 57.

101. B. P. Dorman, J. E. Hearst, and M. F. Maestre, *Methods Enzymol.*, **27**, 767 (1973).

102. I. Z. Steinberg and A. Gafni, *Rev. Sci. Instrum.*, **43**, 409 (1976).

103. I. Z. Steinberg, "Circularly Polarized Luminescence," in C. H. W. Hirs and S. N. Timasheff, Eds., *Methods in Enzymology*, Vol. XLIX, Part G, Academic, New York, 1978, Section II.7, p. 179.

104. F. S. Richardson and J. P. Riehl, *Chem. Rev.*, **77**, 773 (1977).

105. F. S. Richardson, "Circular Polarization Differentials in the Luminescence of Chiral Systems," in S. F. Mason, Ed., *Optical Activity and Chiral Discrimination*, Reidel, Dordrecht, Holland, 1979, p. 189.

106. H. P. L. M. Dekkers, C. A. Emeis, and L. J. Oosterhoff, *J. Am. Chem. Soc.*, **91**, 4589 (1969).

107. L. D. Barron, "Raman Optical Activity," in S. F. Mason, Ed., *Optical Activity and Chiral Discrimination*, Reidel, Dordrecht, Holland, 1979, p. 219.

108. L. D. Barron, *Molecular Light Scattering and Optical Activity*, Cambridge University Press, Cambridge, 1982.

109. L. D. Barron, M. P. Bogaard, and A. D. Buckingham, *J. Am. Chem. Soc.*, **95**, 603 (1973).

110. L. D. Barron, M. P. Bogaard, and A. D. Buckingham, *Nature (London)*, **241**, 113 (1973).

111. L. D. Barron and A. D. Buckingham, *J. Chem. Soc. Chem. Commun.*, 152 (1974).

112. T. Brocki, M. Moskovits, and B. Bosnich, *J. Am. Chem. Soc.*, **102**, 495 (1980).

113. J. Badoz, "Optical Activity Induced by a Magnetic Field," in G. Snatzke, Ed., *Optical Rotatory Dispersion and Circular Dichroism in Organic Chemistry*, Heyden, London, 1967, p. 389.

114. B. Briat, "Faraday-Effect Spectroscopy," in C. Ciardelli and P. Salvadori, Eds., *Fundamental Aspects and Recent Developments in Optical Rotatory Dispersion and Circular Dichroism*, Heyden, London, 1973, p. 375.

115. A. C. Gilby and P. E. R. Tatham, *J. Phys. E*, **2**, 1004 (1969).

116. B. Holmquist and B. L. Valle, "Magnetic Circular Dichroism," in C. H. W. Hirs and S. N. Timasheff, Eds., *Methods in Enzymology*, Vol. XLIX, Part G, Academic, New York, 1978, Section II.7, p. 149.

117. P. N. Shatz and A. J. McCaffery, *Q. Rev. Chem. Soc.*, **23**, 552 (1969).

118. R. A. Shatwell and A. J. McCaffery, *J. Phys. E*, **7**, 297 (1974).

119. L. D. Barron, *Chem. Phys. Lett.*, **46**, 579 (1977).

120. T. A. Keiderling, *J. Chem. Phys.*, **75**, 3639 (1981).

121. R. Gale, A. J. McCaffery, and R. Shatwell, *Chem. Phys. Lett.*, **17**, 416 (1972).

122. J. C. Collingwood, P. Day, R. G. Denning, P. N. Quested, and T. R. Snellgrove, *J. Phys. E*, **7**, 991 (1974).

123. R. E. Koning, R. M. E. Vliek, and P. J. Zandstra, *J. Phys. E*, **8**, 710 (1975).

124. C. J. Barnett, A. F. Drake, and S. F. Mason, *J. Chromatogr.*, **202**, 239 (1980).

125. M. Anson and P. M. Bayley, *J. Phys. E*, **7**, 481 (1974).

126. P. M. Bayley, S. R. Martin, and M. Anson, *Rev. Sci. Instrum.*, **48**, 953 (1977).

127. B. Greunewald and W. Knoche, *Rev. Sci. Instrum.*, **49**, 797 (1978).

128. H. P. Bachinger, H. P. Eggenberger, and G. Hanisch, *Rev. Sci. Instrum.*, **50**, 1367 (1979).

129. L. M. N. Dysens, *Biochim. Biophys, Acta*, **19**, 1 (1956).

130. G. Holtzwarth, D. G. Gordon, J. E. McGinnes, B. P. Dorman, and M. F. Maestre, *Biochemistry*, **13**, 126 (1974).

131. I. Tinoco, C. Bustamante, and M. F. Maestre, *Ann. Rev. Biophys. Bioeng.*, **9**, 109 (1980); *Trends Biochem. Sci. Pers. Ed.*, **8**, 41 (1983); *Proc. Natl. Acad. Sci. USA*, **80**, 3568 (1983).

132. Y. Taniguchi and Y. Shimura, *Bull. Chem. Soc. Jpn.*, **55**, 754 and 2847 (1982).

Chapter **2**

DETERMINATION OF DIPOLE MOMENTS IN GROUND AND EXCITED STATES

Wolfram Baumann

1   INTRODUCTION

In his tables of experimental dipole moments [1, 2] McClellan indicates that about 0.1% of all papers cited in *Chemical Abstracts* deal with dipole moments in some way. His statement includes the early 1930s, when chemists first became interested in dipole moments, up to about 1970; however, the current situation is similar.

This is because the dipole moment of a molecule in a considered state (the ground state or any excited state) is an observable that is a measure of the molecule's charge distribution in this state. Since the charge distribution determines the interaction of molecules over the whole scale of phenomena observed in the wide field of chemistry, chemists adopted all improvements in electronic instrumentation and in the development of physical methods to determine this quantity precisely.

Quite naturally, people began determining ground-state dipole moments. With the increasing number of determined ground-state dipole moments it was rationalized that the dipole moments of larger molecules can be reproduced by a vectorial addition of so-called group (and bond) moments, thus reflecting structural subunits. For example, the dipole moment of benzene is zero, from symmetry. The dipole moment of chlorobenzene is $(5.64 \pm 0.1) \times 10^{-30}$ C·m [3]. Thus very roughly, the group moment of —Cl is also 5.64×10^{-30} C·m. By vectorial addition, *o*-dichlorobenzene and *m*-dichlorobenzene should have a dipole moment of 9.77 and 5.64×10^{-30} C·m, respectively, which is in agreement with literature values [3] to within 1.5 and 2%. For a more detailed discussion, see [4].

While the group moment additivity law is approximately applicable where the subunits are electronically weakly coupled, it fails with molecules that show strong intramolecular charge transfer, as with the merocyanine dyes $(CH_3)_2$—N—$(CH$=$CH)_x$—CH=$C(CN)_2$, where x is from 0 to 4 [5]. Here the ground-state dipole moment is much larger than the sum of the dipole moments of the terminating groups and increases considerably with the chain length. For more details, see Section 4.7.2. Similar results have been reported for other chainlike molecules, for example, in [4, 6, 7].

For comparison, a dipole consisting of two elementary point charges of opposite sign separated by 1.4×10^{-10} m (which is about 1 aromatic C—C bond length) has a dipole moment of 22.4×10^{-30} C·m.

As another example, the value found for the dipole moment of so-called electron-donor–acceptor complexes greatly supported the idea of a charge transfer from the donor to the acceptor molecules [8, 9]. Tetracyanoethylene (TCNE) and hexamethylbeznene (HMB) are molecules with zero-ground-state dipole moments. In solution they form a colored 1:1 complex, the dipole moment of which is about 7×10^{-30} C·m [10], in the ground state, and increases to about 22×10^{-30} C·m in the first excited charge-transfer state [10]. In this same paper a $(HMB)_2$—TCNE 2:1 complex was shown to have a zero-ground-state dipole moment, which supports a staggered structure HMB—TCNE—HMB.

In principle, the accuracy of values resulting from liquid-phase measurements is much less than those from gas-phase measurements because it is necessary for the molecular model to take into account solute–solvent interactions in liquid-phase measurements. Hence highly accurate gas-phase methods have been developed.

If Stark splitting is used with microwave spectroscopy and molecular beam resonance methods [11], precise dipole moments of small molecules in their electronic ground state in well-defined vibrational states can be determined, and small isotope effects on the electric dipole moment can be shown as well as the dipole moment of van der Waals complexes. The examples given in Table 2.1 are compiled from [12–16]. Regrettably, such highly precise measurements can only be obtained for fairly small molecules that have sufficiently high vapor pressure and may be vaporized without decomposition.

Since about 1960 investigators have focused on excited-state phenomena. Here, for example, the structure of exciplexes [17], which are stable only in their electronically excited state, on a nanosecond time scale in solutions, was shown to be determined by a charge transfer between the donor and acceptor moieties [18].

Their excited-state dipole moments have been determined first from solvent-shift measurements [19] and later from the field effect on their fluorescence [20, 21] to be about 40 to 50×10^{-30} C·m, which is a value that represents nearly one electron charge transfer over the intermolecular distance.

As another example, from the electric field effect on the $S^1 \leftarrow S_0$ absorption of azulene, it has been shown [22] that the dipole moment of the first excited state

Table 2.1 Highly Accurate Dipole Moment Values of Some Small Molecules[a]

Molecule	$\mu/(10^{-30}\,C\cdot m)$	Remarks	Reference
HC≡C—CN	12.44767(3)		[12]
	12.4296(3)	$v_6 = 1$	[12]
	12.4169(7)	$v_7 = 1$	[12]
$^1H—C≡C—^2H$	0.07869(36)	$v_4 = 1$	[13]
	0.18760(10)	$v_5 = 1$	[13]
(H—C≡C—H)—NH$_3$	6.6283(53)		[14]
^{1}HCN—CO$_2$	10.6964(107)		[15]
^{2}HCN—CO$_2$	10.7670(97)		[15]
$^{16}O^{12}C^{32}S$	2.38538(10)		[16]
$\mu(^{16}O^{13}C^{32}S)/\mu(^{16}O^{12}C^{32}S)$	1.00017(2)		[16]
$\mu(^{16}O^{12}C^{34}S)/\mu(^{16}O^{12}C^{32}S)$	1.00031(2)		[16]

[a]Unless otherwise noted, molecules are in their ground states.

is opposite to the ground-state dipole moment, fully in accordance with earlier theoretical predictions [23, 24].

Dipole moments of molecules in excited states have been determined in the gas phase, too, from various observed effects of an electric field on absorption and fluorescence spectra in the visible and ultraviolet spectral region. The selected results given in Table 2.2 are taken from [25–29].

Starting from Debye's [30] basic work on dielectrics, the determination of molecular ground-state dipole moments has experienced continuous development accompanied by improvements of models that relate the primarily measured bulk properties of the system under consideration to the permanent dipole moment of an isolated molecule. The choice of an appropriate model and a related and adequate experimental method depends on the kind of question we present for the considered system. Related remarks will be given in the succeeding, respective sections.

Table 2.2 Some Gas-Phase Dipole Moments of Molecules in Their First Excited Singlet State

Molecule	$\mu/(10^{-30}\,C\cdot m)$	Reference
Formaldehyde	5.20 ± 0.23	[25]
Formaldehyde	4.67 ± 0.33	[26]
Aniline	8.17 ± 0.33	[27]
p-Nitroanisole	50.4 ± 3.3	[28]
9-Cyanoanthracene	17.4 ± 1	[29]
2,6,N,N-Tetramethyl-4-cyanoaniline	38.9 ± 2.2	[29]

The conversion factors for dipole moment values given in Debye and for polarizabilities given in 10^{-24} cm^3 to SI units are 10^{-30} C·m $= 3.33564$ D, and 10^{-40} C V^{-1} m$^2 = 1.11265 \times 10^{-24}$ cm^3. These values are used in this chapter whenever dipole moments and polarizabilities from literature were converted to SI units.

2   BULK AND MOLECULAR PROPERTIES

Very recently Liptay and colleagues [31] discussed the basic problem of all experimental investigation considered in this chapter, that is, the relationships between molecular quantities and some bulk properties. This was done in a very rigorous way, starting from some very general requirements to be fulfilled by macroscopic systems and quantities. The following discussion follows closely that in [31]. Basic to all determinations of dipole moments in condensed phases is that some bulk quantities are to be determined.

The bulk quantities related to the field of interest here may be divided into extensive and intensive quantities. For definition of these terms consider the value γ of a macroscopic quantity of a system. Lump together l such identical systems. If the value γ_s of that macroscopic quantity of the whole system is given as

$$\gamma_s = l^g \gamma \tag{1}$$

the macroscopic quantity is defined as intensive for $g = 0$ and as extensive for $g = 1$. For example, temperature T and pressure p are intensive quantities; the amount of substance n_J or the mass m_J of the species J in a phase are extensive.

If an extensive bulk property of a single phase with $J = 1, \ldots, E$ components is given as a differentiable function Φ of n_J and T, p, the partial derivative

$$\Phi_J = (\partial\Phi/\partial n_J)_{n'_J, T, p} \tag{2}$$

is called a *partial molar quantity* (PMQ) of species J. Subscript n'_J means that the differentiation must be completed with respect to n_J, while the amount of substance of all other species $I \neq J$ in the phase is kept constant. Liptay points out the importance of the PMQs for the analysis of experimental data of multicomponent single-phase systems, for example, for dilute solutions of one solute species in a solvent or solvent mixture. Generally all mathematical analyses of experimental data with respect to one component result in PMQs, but do not give molecular quantities.

Values for PMQs are determined according to their definitions (2) if the extensive quantity ϕ is accessible experimentally. Then, in practice instead of the amount of substance n_J, a concentration variable is used, for example, the mass fraction of species J, and is defined as

$$w_J = \frac{n_J M_J}{m_0} = \frac{m_J}{m_0} \tag{3}$$

where M_J is the molar mass of species J in the phase and m_0 is the total mass of the phase. With this concentration variable the PMQ (2) is rewritten to give

$$\phi_J = \left[\frac{(1 - w_J)M_J}{m_0} \right] \left(\frac{\partial \phi}{\partial w_J} \right)_{n'_J, T, p} \tag{4}$$

Thus, to determine the PMQ ϕ_J a set of experimental data $\{\phi, w_J\}$ at constant pressure and temperature and at constant values of all $n_{I \neq J}$ is necessary.

To derive the values of such molecular quantities as the dipole moment or the polarizability of a molecule from macroscopic data, that is, from PMQs, a physical model must necessarily be used. Generally it is assumed that in a single-phase multicomponent system ϕ can be represented as

$$\phi = \sum_{J=1}^{K} \overline{\pi_J} N_J \tag{5}$$

where N_J is the number of molecules or ions of species J and $\overline{\pi_J}$ is the average (over species) of the contribution π_J of one molecule of species J to the bulk quantity ϕ. The value of $\overline{\pi_J}$ is calculated from molecular data by using a molecular model. Summation must be done over all species $J = 1, \ldots, K$, for example, in a simple solution over the solute and the solvent. With

$$\varphi_J = N_A \overline{\pi_J} \tag{6}$$

where N_A is Avogadro's number, a new molar quantity φ_J is introduced, called *model molar quantity* (MMQ) of species J.

The introduction of φ_J and the amount of substance n_J of species J into (5) yields

$$\phi = \sum_{J=1}^{K} \varphi_J n_J \tag{7}$$

If it is assumed that the n_J are mutually independent (i.e., no chemical reaction occurs), the PMQ ϕ_J is derived from (7)

$$\phi_J = \varphi_J + \sum_{I=1}^{K} n_I \left(\frac{\partial \varphi_I}{\partial n_J} \right)_{n'_J, T, p} \tag{8}$$

Equation (8) is most important as it shows that the PMQs ϕ_J and the MMQs φ_J generally are not identical.

To proceed from the purely macroscopic PMQ ϕ_J to the MMQ φ_J, information about the second term on the right side of (8) is necessary. Generally this term can be derived from the foregoing molecular model. In special cases of some proper models this term, or at least its limit with $n_J \rightarrow 0$, is zero. In all cases φ_J or $\lim_{n_J \rightarrow 0} \varphi_J$ can be derived from the experimentally determined PMQ ϕ_J or from $\lim_{n_J \rightarrow 0} \phi_J$, respectively. Then, on the basis of the particular model chosen, the value of those molecular quantities that contribute to the bulk quantity ϕ can be calculated from φ_J.

3 DETERMINATION OF THE PERMANENT DIPOLE MOMENT OF MOLECULES IN THEIR GROUND STATES FROM DIELECTRIC INVESTIGATIONS

3.1 General Remarks

The determination of the ground-state dipole moment of a molecule from the dielectric properties of either the pure phase or the dilute solutions is the standard laboratory method used to determine electric dipole moments since Debye's [30] basic work.

Various methods have been proposed that differ in the molecular model used or in the approximations introduced. As a consequence the values of the permanent dipole moment of molecules given in the literature differ markedly. To compare such different values of dipole moments or to analyze the results with respect to the power of the molecular model used, it is necessary to reproduce the primary experimental results or at least the respective PMQs in every paper dealing with the determination of dipole moments. For a general survey on dipole moments and dielectrics the reader is referred to a selection of articles and textbooks [4, 32–44].

3.2 Theoretical Foundation

For this discussion an isotropic phase will have volume V in a uniform static electric field $\mathbf{E}_a$ at constant T and p. No chemical reactions are taken into account. For more details see [31, 45]. A static electric field $\mathbf{E}_a$ induces a total electric moment $\mathbf{P}V$ in the phase.

$$\mathbf{P}V = \varepsilon_0(\varepsilon - 1)\mathbf{E}_a V \tag{9}$$

where $\mathbf{P}$ is the dielectric polarization, ε_0 is the permittivity of the vacuum, and ε is the relative permittivity (dielectric constant) of the phase.

An electric field $\mathbf{E}_a'$ at optical frequencies, by analogy, induces a total electric moment $\mathbf{P}'V$ in the phase

$$\mathbf{P}'V = \varepsilon_0(n^2 - 1)\mathbf{E}_a' V \tag{10}$$

where n is the refractive index of the phase. On a microscopic scale the respective quantity is the total electric moment of a molecule.

Following the scheme just sketched, two experimentally accessible extensive scalar quantities $\mathscr{L}$ and $\mathscr{L}'$ of the phase can be defined, starting from the potential energy PVE_a of the phase in the external electric field and using (9) and (10)

$$\mathscr{L} = \frac{1}{\varepsilon_0} \lim_{E_a^2 \to 0} \left[\frac{\partial}{\partial E_a^2} (E_a PV) \right] = (\varepsilon - 1)V \tag{11}$$

$$\mathscr{L}' = \frac{1}{\varepsilon_0} \lim_{E_a'^2 \to 0} \left[\frac{\partial}{\partial E_a'^2} (E_a' P'V) \right] = (n^2 - 1)V \tag{12}$$

and from these the PMQs $\mathscr{L}_J$ and $\mathscr{L}'_J$ follow

$$\mathscr{L}_J = \left[\frac{\partial}{\partial n_J} (\varepsilon - 1)V \right]_{n'_J, T, p} \tag{13}$$

$$\mathscr{L}'_J = \left[\frac{\partial}{\partial n_J} (n^2 - 1)V \right]_{n'_J, T, p} \tag{14}$$

Here and hereafter these derivatives with respect to n_J are to be evaluated for constant $n_{J' \neq J}$, T, and p. For the definition of corresponding MMQs it is assumed that the electric moment of the entire phase PV, or $P'V$, can be reproduced by

$$PV = \sum_{J=1}^{K} \overline{\mathbf{p}_J} N_J \tag{15}$$

$$P'V = \sum_{J=1}^{K} \overline{\mathbf{p}'_J} N_J \tag{16}$$

where $\overline{\mathbf{p}_J}$ is the average electric dipole moment of a molecule in a static electric field, and $\mathbf{p}'_J$ is the corresponding quantity averaged at optical frequencies. From (15) and (16) it follows that $\mathscr{L}$ and $\mathscr{L}'$ may be represented by

$$\mathscr{L} = \frac{1}{\varepsilon_0} \lim_{E_a^2 \to 0} \left(\frac{\partial}{\partial E_a^2} \sum_{J=1}^{K} E_a \overline{\mathbf{p}_J} N_J \right) \tag{17}$$

$$\mathscr{L}' = \frac{1}{\varepsilon_0} \lim_{E_a'^2 \to 0} \left(\frac{\partial}{\partial E_a'^2} \sum_{J=1}^{K} E_a' \overline{\mathbf{p}'_J} N_J \right) \tag{18}$$

and the MMQs ζ_J and ζ'_J corresponding to $\mathscr{L}_J$ and $\mathscr{L}'_J$ are then given as

$$\zeta_J = \frac{N_A}{\varepsilon_0} \lim_{E_a^2 \to 0} \left[\frac{\partial}{\partial E_a^2} (E_a \overline{\mathbf{p}_J}) \right] \tag{19}$$

$$\zeta'_J = \frac{N_A}{\varepsilon_0} \lim_{E_a'^2 \to 0} \left[\frac{\partial}{\partial E_a'^2} (E_a' \overline{\mathbf{p}'_J}) \right] \tag{20}$$

The relationships between $\mathscr{L}_J$ and ζ_J and $\mathscr{L}'_J$ and ζ'_J are given by (8), with $\phi_J = \mathscr{L}_J$ or $\mathscr{L}'_J$ and $\varphi_J = \zeta_J$ or ζ'_J.

To derive dipole moments from experimentally determined values of $\mathscr{L}$ and $\mathscr{L}'$ or of $\mathscr{L}_J$ and $\mathscr{L}'_J$ a particular model must be introduced. With all molecular models it is assumed that the orientation distribution of molecules in the ground state in an externally applied electric field $\mathbf{E}_a$ or $\mathbf{E}'_a$ can be described by the (normalized) Maxwell–Boltzmann distribution functions v_g^E and $v_g^{E'}$ in the low- and high-frequency limits.

$$v_g^E = \exp\left(\frac{-W_J^E}{kT}\right)\bigg/ \int \exp\left(\frac{-W_J^E}{kT}\right) d\tau \tag{21}$$

$$v_g^{E'} = \exp\left(\frac{-W_J^{E'}}{kT}\right)\bigg/ \int \exp\left(\frac{-W_J^{E'}}{kT}\right) d\tau \tag{22}$$

where W_J^E and $W_J^{E'}$ are those parts of the potential energy of a molecule of species J that depend on the external electric field $\mathbf{E}_a$ or $\mathbf{E}'_a$, respectively. Integration must be performed over all orientations τ. With these orientation distribution functions the average electric dipole moments $\overline{p_J}$ and $\overline{p'_J}$ and the MMQs ζ_J and ζ'_J can be calculated

$$\overline{\mathbf{p}_J} = \mathbf{E}_a \int (\mathbf{E}_a \mathbf{p}_J) \exp\left(\frac{-W_J^E}{kT}\right) d\tau \bigg/ E_a^2 \int \exp\left(\frac{-W_J^E}{kT}\right) d\tau \tag{23}$$

$$\overline{\mathbf{p}'_J} = \mathbf{E}'_a \int (\mathbf{E}'_a \mathbf{p}'_J) \exp\left(\frac{-W_J^{E'}}{kT}\right) d\tau \bigg/ E_a'^2 \int \exp\left(\frac{-W_J^{E'}}{kT}\right) d\tau \tag{24}$$

$$\zeta_J = \frac{N_A}{\varepsilon_0} \lim_{E_a^2 \to 0} \left[\frac{\partial}{\partial E_a^2} \int \mathbf{E}_a \mathbf{p}_J \exp\left(\frac{-W_J^E}{kT}\right) d\tau \bigg/ \int \exp\left(\frac{-W_J^E}{kT}\right) d\tau \right] \tag{25}$$

$$\zeta'_J = \frac{N_A}{\varepsilon_0} \lim_{E_a'^2 \to 0} \left[\frac{\partial}{\partial E_a'^2} \int \mathbf{E}'_a \mathbf{p}'_J \exp\left(\frac{-W_J^{E'}}{kT}\right) d\tau \bigg/ \int \exp\left(\frac{-W_J^{E'}}{kT}\right) d\tau \right] \tag{26}$$

In static electric fields, the total electric dipole moment $\mathbf{p}_J$ generally is represented as

$$\mathbf{p}_J = \boldsymbol{\mu}_{gJ}^0 + \boldsymbol{\alpha}_{gJ}^0 \mathbf{E}_J + \text{higher order terms} \tag{27}$$

where

$$\boldsymbol{\alpha}_{gJ}^0 = \mathbf{A}_{gJ}^0 + \mathbf{B}_{gJ}^0 \tag{28}$$

Higher order terms in the development with respect to $\mathbf{E}_a$ and $\mathbf{E}'_a$ generally are neglected. The permanent electric dipole moment of the free species J is $\boldsymbol{\mu}_{gJ}^0$. The static electric polarizability tensor α_{gJ}^0 is considered as a sum of a purely electronic polarizability $\mathbf{A}_{gJ}^0$ and a nuclear polarizability $\mathbf{B}_{gJ}^0$. For electric fields at optical frequencies far below the first absorption band, the total electric dipole moment $\mathbf{p}_J'$ is represented as

$$\mathbf{p}_J' = \mathbf{A}_{gJ}^0 \mathbf{E}_J' \tag{29}$$

since molecular motions and intramolecular nuclear motions cannot follow the external field at these frequencies.

The various molecular models used differ mainly in the description of the internal electric field $\mathbf{E}_J$; also, the simplest model neglects the induced part of the total dipole moment, which is the second right-side term of (27).

3.3   A Gas Phase Consisting of One Species G

With a sufficiently dilute gas the internal electric field $\mathbf{E}_G$ is equal to the externally applied electric field $\mathbf{E}_a$.

For a gas phase consisting of only one species G, according to (11) and (12), it is on a macroscopic scale

$$\mathscr{L} = (\varepsilon - 1)V \tag{30}$$

and

$$\mathscr{L}' = (n^2 - 1)V \tag{31}$$

and on a microscopic scale with (17) and (18)

$$\mathscr{L} = \frac{1}{\varepsilon_0} \lim_{E_a^2 \to 0} \frac{\partial}{\partial E_a^2} (\overline{\mathbf{E}_a \mathbf{p}_G} \, N_G) \tag{32}$$

and

$$\mathscr{L}' = \frac{1}{\varepsilon_0} \lim_{E_a'^2 \to 0} \frac{\partial}{\partial E_a'^2} (\overline{\mathbf{E}_a' \mathbf{p}_G'} \, N_G) \tag{33}$$

To perform the integrations in (23) to (26), W_J^E and $W_J^{E'}$ must be represented in more detail. In the simplest case of a dilute gas discussed here, W_J^E is given

$$W_J^E = -\mathbf{\mu}_{gG}^0 \mathbf{E}_a - \tfrac{1}{2} \mathbf{E}_a \alpha_{gG}^0 \mathbf{E}_a \tag{34}$$

and

$$W_J^{E'} = -\tfrac{1}{2} \mathbf{E}_a' \mathbf{A}_{gG}^0 \mathbf{E}_a' \tag{35}$$

From (32) and (33), with (34) and (35), (23) and (24), the quantities $\mathscr{L}$ and $\mathscr{L}'$ follow as defined from a microscopic molecular model basis, where $\operatorname{tr} \boldsymbol{\alpha}_{gG}^0$ is the trace of the polarizability tensor $\boldsymbol{\alpha}_g^0$

$$\mathscr{L} = \frac{N_G}{3\varepsilon_0}\left[\frac{(\mu_{gG}^0)^2}{kT} + \operatorname{tr}\boldsymbol{\alpha}_{gG}^0\right] \tag{36}$$

$$\mathscr{L}' = \frac{N_G}{3\varepsilon_0}\operatorname{tr}\mathbf{A}_{gG}^0 \tag{37}$$

where W_J^E/kT and $W_J^{E'}/kT$ have been assumed to be small compared to one, in (34) and (35). Then from (30) and (36) it follows that

$$\frac{(\mu_{gG}^0)^2}{kT} + \operatorname{tr}\boldsymbol{\alpha}_{gG}^0 = \frac{3\varepsilon_0 M_G}{\rho N_A}(\varepsilon - 1) \tag{38}$$

and from (31) and (37)

$$\operatorname{tr}\mathbf{A}_{gG}^0 = \frac{3\varepsilon_0 M_G}{\rho N_A}(n^2 - 1) \tag{39}$$

where ρ is the mass density of the phase and M_G the molar mass of species G.

According to (38), from a set of experimental data $(\rho, \varepsilon)_{T,p}$, $(\mu_{gG}^0)^2/kT + \operatorname{tr}\boldsymbol{\alpha}_{gG}^0$ can be determined.

Generally from the temperature dependence of $(\varepsilon - 1)/\rho$, μ_{gG}^0 and $\operatorname{tr}\boldsymbol{\alpha}_{gG}^0$ can be separated. Values of μ_{gG}^0 of small molecules have been reported with less than 1% error. Table 2.3 reproduces a small selection of dipole moments taken from [46]. Whenever the range of temperature variation is restricted by experimental limitations, an attempt to eliminate the separation problem by combining (38) and (39) with some assumptions on polarizabilities is made.

It has been discussed in some detail elsewhere ([32] and references cited therein) that

$$\mathbf{B}_{gG}^0 = c_{AB}\mathbf{A}_{gG}^0 \tag{40}$$

where usually the value of c_{AB} is assumed to be from 0.05 to 0.15, however, exceptionally high values up to 0.50 are discussed in [47] in reference to some acetylacetonato complexes. See Table 4.1 of [32]. Thus, from a set of experimental data $(\rho, n^2, \varepsilon)_{T,p}$ with such an assumption for c_{AB}, the permanent dipole moment is determined from gas-phase measurements as

$$(\mu_{gG}^0)^2 = \frac{3kT\varepsilon_0 M_G}{\rho N_A}[(\varepsilon - n^2) - c_{AB}(n^2 - 1)] \tag{41}$$

Table 2.3 Ground-State Dipole Moments of Some
Molecules Determined From the Temperature
Dependence of the Dielectric Constant[a]

Molecule	$\mu_{gG}^0/(10^{-30}\,\mathrm{C \cdot m})$
Trifluoromethane	5.30
Formaldehyde	7.57
Chloromethane	6.20
Methanol	5.631
Pentachloroethane	0.31
1,1-Dichloroethane	6.84
Acetaldehyde	8.94
Dimethylsulfone	14.7
Acetone	9.37
1-Nitropropane	12.41
1-Chlorobutane	6.80
o-Dichlorobenzene	7.20
Di-n-butyl ether	3.87
Isopropylbenzene	2.17

[a]Adapted from [46].

As the molecular model used in this paragraph is well established and well
proven and has been introduced without critical assumptions on the internal
electric field $\mathbf{E}_G$, values of experimental dipole moments determined according
to this method are very reliable, in principle. But the uncertainty introduced by
an arbitrary choice of c_{AB} obviously may be quite important with small dipole
moments. Then the temperature-variation method should be used.

The most severe limitation for this method is that most of the molecules of
interest cannot be vaporized without decomposition. Also, a chemist is frequently more interested in the behavior of molecules in solution.

3.4 A Solution of a Solute G in a Pure Solvent S—Liptay's Method

Section 3.2 presented the basic framework of dielectric investigations, which
closely followed Liptay's work [31, 45]. In this section a solution of one solute
species $J = G$ in a pure solvent $J = S$ is considered without discussing further
the penetrating details given by Liptay. He showed that starting from (8) the
following relation holds:

$$\zeta_G^* = \mathscr{L}_G^* - \frac{\rho^*}{M_S}[\mathscr{L}_G^* - (\varepsilon^* - 1)V_G^*]\left(\frac{\partial}{\partial\varepsilon}\zeta_S\right)_{T,p}^*$$

$$- \frac{M_G}{M_S}\sum_{\alpha_\kappa}\left(\frac{\partial}{\partial\alpha_\kappa}\zeta_S\right)_{\alpha'_\kappa,T,p}^*\left(\frac{\partial\alpha_\kappa}{\partial w_G}\right)_{n_S,T,p}^* \qquad (42)$$

Starred quantities denote the limit for the pure solvent ($w_G \rightarrow 0$); V_G^* is the molar volume $(\partial V/\partial n_G)_{n_s, T, p}$; α_κ values are some other bulk properties. Depending on the molecular model the second right-side term of (42) is small or even zero. With the models used in this chapter this term is calculated as zero or may be neglected, thus

$$\zeta_G^* = \mathscr{L}_G^* - [\mathscr{L}_G^* - (\varepsilon^* - 1)V_G^*]\frac{\rho^*}{M_s}\left(\frac{\partial}{\partial \varepsilon}\zeta_s\right)_{T,p}^* \tag{43}$$

The term $(\partial \zeta_s/\partial \varepsilon)_{T,p}^*$ depends only on the solvent; therefore, it may be considered as a property of the solvent that can be calculated on the basis of the same molecular model used to evaluate dipole moments from ζ_G^*.

Quite an analogous equation relates ζ_G' to PMQs

$$\zeta_G'^* = \mathscr{L}_G'^* - [\mathscr{L}_G'^* - (n^{*2} - 1)V_G^*]\frac{\rho^*}{M_s}\left(\frac{\partial}{\partial n^2}\zeta_s'\right)_{T,p}^* \tag{44}$$

Dilute solutions or preferably the infinite dilution limit is used since all solute–solute interactions (dimer formation, e.g.) then may be neglected.

The extensive quantities $\mathscr{L}$, $\mathscr{L}'$, and V cannot be measured directly or with sufficient accuracy from an experiment. However, some related generalized densities can be measured; namely,

$$\frac{\mathscr{L}}{m_0} = \frac{\varepsilon - 1}{\rho} \tag{45}$$

$$\frac{\mathscr{L}'}{m_0} = \frac{n^2 - 1}{\rho} \tag{46}$$

$$\frac{V}{m_0} = \frac{1}{\rho} \tag{47}$$

where m_0, $\mathscr{L}$, $\mathscr{L}'$, and V are extensive quantities. Then, (48) is valid in the binary system considered here

$$\frac{\varepsilon - 1}{\rho} = \frac{\mathscr{L}_s}{M_s} + \left(\frac{\mathscr{L}_G}{M_G} - \frac{\mathscr{L}_s}{M_s}\right)w_G \tag{48}$$

as well as similar equations for $(n^2 - 1)/\rho$ and $1/\rho$.

Taking into account that the PMQs also depend on the mass fraction w_G, Liptay developed the generalized densities into power series with respect to w_G

$$\frac{\varepsilon - 1}{\rho} = \frac{\varepsilon^* - 1}{\rho^*} + \left(\frac{\mathscr{L}_G^*}{M_G} - \frac{\varepsilon^* - 1}{\rho^*}\right)w_G + \text{higher order terms} \tag{49}$$

$$\frac{n^2 - 1}{\rho} = \frac{n^{*2} - 1}{\rho^*} + \left(\frac{\mathscr{L}_G^{\prime *}}{M_G} - \frac{n^{*2} - 1}{\rho^*}\right) w_G + \text{higher order terms} \qquad (50)$$

$$\frac{1}{\rho} = \frac{1}{\rho^*} + \left(\frac{V_G^*}{M_G} - \frac{1}{\rho^*}\right) w_G + \text{higher order terms} \qquad (51)$$

Thus from a set of experimental data $\{\varepsilon, n^2, \rho, w_G\}$ the PMQs $\mathscr{L}_G^*$, $\mathscr{L}_G^{\prime *}$, and V_G^* can be derived from a multiple regression analysis (least-squares fit) according to (49)–(51). Significance tests reveal that higher order terms may be neglected for sufficiently dilute solutions in the range of measuring errors. Therefore, linear regressions are usually adequate.

The values of the PMQs estimated in such a manner are then used to calculate the MMQs ζ_G^* and $\zeta_G^{\prime *}$ according to (43) and (44).

To determine the values of the permanent electric dipole moment and of some polarizabilities of the solute, ζ_G^* and $\zeta_G^{\prime *}$ should be considered the limits of ζ_G and ζ_G' given in (25) and (26). This can be done on the basis of the particular molecular model used.

3.4.1   The Lorentz Model

The Lorentz model is widely used although its limited applicability has been discussed [28, 36, 45, 48].

In the Lorentz model [36] the electric fields $\mathbf{E}_G$ and $\mathbf{E}_G'$ are given

$$\mathbf{E}_G = \frac{\varepsilon^* + 2}{3} \mathbf{E}_a \qquad (52)$$

$$\mathbf{E}_G' = \frac{n^{*2} + 2}{3} \mathbf{E}_a \qquad (53)$$

Then according to (25) and (26) ζ_G and ζ_G' can be calculated where W_G^E and $W_G^{E'}$ are given as

$$W_G^E = -\frac{\varepsilon^* + 2}{3} \boldsymbol{\mu}_{gG}^0 \mathbf{E}_a - \frac{1}{2}\left(\frac{\varepsilon^* + 2}{3}\right)^2 \mathbf{E}_a \boldsymbol{\alpha}_{gG}^0 \mathbf{E}_a \qquad (54)$$

and

$$W_G^{E'} = -\frac{1}{2}\left(\frac{n^{*2} + 2}{3}\right)^2 \mathbf{E}_a' \mathbf{A}_{gG}^0 \mathbf{E}_a' \qquad (55)$$

Then, ζ_G^* and $\zeta_G^{\prime *}$ become

$$\zeta_G^* = \frac{\varepsilon^* + 2}{9\varepsilon_0} N_A \left[\frac{(\mu_{gG}^0)^2}{kT} + \text{tr } \boldsymbol{\alpha}_{gG}^0\right] \qquad (56)$$

$$\zeta_G^{\prime *} = \frac{n^{*2} + 2}{9\varepsilon_0} N_A \operatorname{tr} \mathbf{A}_{gG}^0 \tag{57}$$

To connect the MMQs with the measured PMQs according to (43) and (44) the terms

$$\frac{\rho^*}{M_S} \left(\frac{\partial}{\partial \varepsilon} \zeta_S \right)_{T,p}^* \quad \text{and} \quad \frac{\rho^*}{M_S} \left(\frac{\partial}{\partial n^2} \zeta_S' \right)_{T,p}^*$$

must also be determined on the basis of the Lorentz model; deriving ζ_S^* and $\zeta_S^{\prime *}$ from equations analogous to (56) and (57) and using (13) and (14) leads to [19]

$$\frac{\rho^*}{M_S} \left(\frac{\partial}{\partial \varepsilon} \zeta_S \right)_{T,p}^* = \frac{\varepsilon^* - 1}{\varepsilon^* + 2} \tag{58}$$

and

$$\frac{\rho^*}{M_S} \left(\frac{\partial}{\partial n^2} \zeta_S' \right)_{T,p}^* = \frac{n^{*2} - 1}{n^{*2} + 2} \tag{59}$$

With these equations (43) and (44) simplify to

$$\zeta_G^* = \frac{3}{\varepsilon^* + 2} \mathscr{L}_G^* + \frac{(\varepsilon^* - 1)^2}{\varepsilon^* + 2} V_G^* \tag{60}$$

$$\zeta_G^{\prime *} = \frac{3}{n^{*2} + 2} \mathscr{L}_G^{\prime *} + \frac{(n^{*2} - 1)^2}{n^{*2} + 2} V_G^* \tag{61}$$

As in the general, preceding discussion, from experimental values of the PMQs $\mathscr{L}_G^*$, $\mathscr{L}_G^{\prime *}$, and V_G^* the MMQs ζ_G^* and $\zeta_G^{\prime *}$ can be calculated according to (60) and (61).

Therefore from $\zeta_G^{\prime *}$ the optical polarizability $\operatorname{tr} \mathbf{A}_{gG}^0$ can be derived; by combination of ζ_G^* and $\zeta_G^{\prime *}$ (56) and (57) and with additional information or some assumption on $\mathbf{B}_{gG}^0$, the permanent dipole moment μ_{gG}^0 can be calculated.

The Lorentz model fails with polar and even slightly polar solvents. Hence, the foregoing simple Lorentz-model-based formula should be used only if the measurement is done in nonpolar solvents or if no discussion on absolute values is based on the determined values for μ_{gG}^0. Otherwise, the Onsager-model-based evaluation scheme revealed in Section 3.4.2 is preferred.

3.4.2   The Onsager Model

Consider a solute molecule of species G in a homogeneous and isotropic solvent S with relative permittivity ε and refractive index n. In the well-defined Onsager model [49] a solute molecule is approximately treated as a polarizable point dipole in the center of a cavity (or equivalently as a homogeneous dipole density in the whole cavity), and the cavity is about the size of the molecule. This cavity is approximated by an ellipsoid [50, 51], or more restricted by a spheroid, or, as used most often, by a sphere. The total electric field affecting the solute molecule in the center of the ellipsoid is represented as a sum

$$\mathbf{E}_G = \mathbf{E}_{GH} + \mathbf{E}_{GR} \tag{62}$$

where $\mathbf{E}_{GH}$ is the cavity field, which is effective in the empty cavity; $\mathbf{E}_{GR}$ is the reaction field [49] and is caused by the polarization of the dielectric surrounding the cavity, which is induced by the total dipole moment $\mathbf{p}_G$ of the solute. As determined by Onsager [49] and Scholte [50] with static external electric fields

$$\mathbf{E}_{GH} = \mathbf{f}_e \mathbf{E}_a \tag{63}$$

$$\mathbf{E}_{GR} = \mathbf{f} \mathbf{p}_G \tag{64}$$

where the total dipole moment $\mathbf{p}_G$ is a sum of the permanent dipole moment $\mathbf{\mu}_{gG}^0$ and a reaction field induced moment, as already defined in (27)

$$\mathbf{p}_G = \mathbf{\mu}_{gG}^0 + \mathbf{\alpha}_{gG}^0 \mathbf{E}_G \tag{65}$$

Index g is used to denote the ground state.

Tensors $\mathbf{f}_e$ and $\mathbf{f}$ are given in the following equations by their principal axes elements $i = x, y, z$

$$f_{ei} = \frac{\varepsilon}{\varepsilon - \kappa_i(\varepsilon - 1)} \tag{66}$$

$$f_i = \frac{1}{4\pi\varepsilon_0} \frac{3}{a_x a_y a_z} \frac{(\varepsilon - 1)(1 - \kappa_i)\kappa_i}{(1 - \kappa_i)\varepsilon + \kappa_i} = \frac{1}{4\pi\varepsilon_0 a_x a_y a_z} g_i \tag{67}$$

where a_x, a_y, a_z are the main half-axes of the ellipsoidal cavity, and

$$\kappa_i = \frac{a_x a_y a_z}{2} \int_0^\infty (s + a_i^2)^{-1} [(s + a_x^2)(s + a_y^2)(s + a_z^2)]^{-1/2} \, ds \tag{68}$$

Obviously $\mathbf{f}$ depends on the form and size of the cavity (or the solute), but $\mathbf{f}_e$ depends only on its form. Values for κ_i for selected half-axes ratios are given as an appendix to [52].

If (63) to (65) are substituted into (62), then

$$\mathbf{E}_G = (\mathbf{I} - \mathbf{f}\mathbf{\alpha}_{gG}^0)^{-1}(\mathbf{f}\mathbf{\mu}_{gG}^0 + \mathbf{f}_e \mathbf{E}_a) \tag{69}$$

where $\mathbf{I}$ is a unit tensor.

For optical frequencies a similar equation holds

$$\mathbf{E}'_G = (\mathbf{I} - \mathbf{f}'\mathbf{A}^0_{gG})^{-1}\mathbf{f}'_e\mathbf{E}_a \tag{70}$$

where the principal axes elements of the tensors $\mathbf{f}'_e$ and $\mathbf{f}'$ are given as

$$f'_{ei} = \frac{n^2}{n^2 - \kappa_i(n^2 - 1)} \tag{71}$$

$$f'_i = \frac{1}{4\pi\varepsilon_0} \frac{3}{a_x a_y a_z} \frac{(n^2 - 1)(1 - \kappa_i)\kappa_i}{(1 - \kappa_i)n^2 + \kappa_i} = \frac{1}{4\pi\varepsilon_0 a_x a_y a_z} g'_i \tag{72}$$

The tensors $\mathbf{f}_e$, $\mathbf{f}'_e$, $\mathbf{f}$, and $\mathbf{f}'$ are simplified when a spherical cavity ($\kappa_1 = \kappa_2 = \kappa_3 = \frac{1}{3}$) with radius a_w is adopted, which often may be considered as a sufficient approximation.

$$\mathbf{f}_e = f_e \cdot \mathbf{I} = \frac{3\varepsilon}{2\varepsilon + 1} \mathbf{I} \tag{73}$$

$$\mathbf{f}'_e = f'_e \cdot \mathbf{I} = \frac{3n^2}{2n^2 + 1} \cdot \mathbf{I} \tag{74}$$

$$f = \frac{1}{4\pi\varepsilon_0} \frac{2}{a_w^3} \frac{\varepsilon - 1}{2\varepsilon + 1} = \frac{1}{4\pi\varepsilon_0} \frac{2}{a_w^3} g \tag{75}$$

$$f' = \frac{1}{4\pi\varepsilon_0} \frac{2}{a_w^3} \frac{n^2 - 1}{2n^2 + 1} = \frac{1}{4\pi\varepsilon_0} \frac{2}{a_w^3} g' \tag{76}$$

Liptay and associates [6, 53] have further extended this model by taking into account the fluctuation of the electric field, caused by time- and space-dependent solvent dipole orientation within the solvent shell around solutes. It will not be considered here.

With E_G and E'_G the part of the potential energy W_G^E dependent on the external electric field can be calculated [25]

$$W_G^E = -\boldsymbol{\mu}_g \mathbf{f}_e \mathbf{E}_a - \tfrac{1}{2}\mathbf{E}_a \mathbf{f}_e \boldsymbol{\alpha}_g \mathbf{f}_e \mathbf{E}_a \tag{77}$$

$$W_G^{E'} = -\tfrac{1}{2}\mathbf{E}'_a \mathbf{f}'_e \mathbf{A}_g \mathbf{f}'_e \mathbf{E}'_a \tag{78}$$

where

$$\boldsymbol{\mu}_g = (\mathbf{I} - \mathbf{f}\boldsymbol{\alpha}^0_{gG})^{-1}\boldsymbol{\mu}^0_{gG} \tag{79}$$

and

$$\mathbf{A}_g = (\mathbf{I} - \mathbf{f}'\mathbf{A}_{gG}^0)^{-1}\mathbf{A}_{gG}^0 \tag{80}$$

and

$$\boldsymbol{\alpha}_g = (\mathbf{I} - \mathbf{f}\boldsymbol{\alpha}_{gG}^0)^{-1}\boldsymbol{\alpha}_{gG}^0 \tag{81}$$

With (65), (77), (25), and (26), (82) results

$$\zeta_G^* = \frac{N_A}{3\varepsilon_0}\left[\frac{1}{kT}\boldsymbol{\mu}_g\mathbf{f}_e\boldsymbol{\mu}_g + \mathrm{tr}\,\mathbf{f}_e\boldsymbol{\alpha}_g\right] \tag{82}$$

which can be given directly in the pure solvent limit; analogously,

$$\zeta_G'^* = \frac{N_A}{3\varepsilon_0}\,\mathrm{tr}\,\mathbf{f}_e'\mathbf{A}_g \tag{83}$$

To determine ζ_G^* and $\zeta_G'^*$ from the experimentally determined PMQs of (43) and (44), the Onsager cavity parameters must be assumed as concentration independent, and the terms $(\partial\zeta_S/\partial\varepsilon)_{T,p}^*$ and $(\partial\zeta_S/\partial n^2)_{T,p}^*$ must also be calculated on the basis of the Onsager model. Spherical approximation may be sufficient, since the second right-side terms of (43) and (44) are normally much smaller than the first terms. Then

$$\frac{\rho^*}{M_S}\left(\frac{\partial\zeta_S}{\partial\varepsilon}\right)_{T,p}^* = \frac{\varepsilon - 1}{\varepsilon(2\varepsilon + 1)} + \frac{3N_A\rho\varepsilon\,\mathrm{tr}[(\mathbf{I} - \mathbf{f}\boldsymbol{\alpha}_g^0)^{-2}(\boldsymbol{\alpha}_g^0)^2]}{2\pi\varepsilon_0 M_S a_{wS}^3(2\varepsilon + 1)^3} \tag{84}$$

$$\frac{\rho^*}{M_S}\left(\frac{\partial\zeta_S'^*}{\partial n^2}\right)_{T,p}^* = \frac{n^2 - 1}{n^2(2n^2 + 1)} + \frac{3N_A\rho n^2\,\mathrm{tr}[(\mathbf{I} - \mathbf{f}'\mathbf{A}_{gS}^0)^{-2}(\mathbf{A}_{gS}^0)^2]}{2\pi\varepsilon_0 M_S a_{wS}^3(2n^2 + 1)^3} \tag{85}$$

where the asterisks have been omitted for convenience.

Values of $(\rho^*/M_S)(\partial\zeta_S/\partial\varepsilon)_{T,p}^*$ and $(\rho^*/M_S)(\partial\zeta_S/\partial n^2)_{T,p}^*$ must be determined only once for each solvent and then are used with all determinations in said solvent. Table 2.4 gives values for these quantities as reported in [45]. Furthermore, the values are not very different in nonpolar solvents so that a dipole moment in another nonpolar solvent may be determined, for example, by using the average value of these quantities from Table 2.4 as a first starting point, if the specific values of $(\rho^*/M_S)(\partial\zeta_S/\partial\varepsilon)_{T,p}^*$ and $(\rho^*/M_S)(\partial\zeta_S'/\partial n^2)_{T,p}^*$ are known for the considered solvent.

Thus, with these values characteristic for the particular solvent and the PMQs $\mathcal{L}_G^*$, $\mathcal{L}_G'^*$, and V_G^* determined in the same solvent, the MMQs ζ_G^* and $\zeta_G'^*$ can be determined according to (49)–(51). From their values according to (82)

Table 2.4 Values for $(\rho^*/M_S)(\partial\zeta_S/\partial\varepsilon)^*_{T,p}$ and $(\rho^*/M_S)(\partial\zeta_S'/\partial n^2)^*_{T,p}$ for Some Solvents[a]

Solvent	$\dfrac{\rho^*}{M_S}\left(\dfrac{\partial}{\partial\varepsilon}\,\zeta_S\right)^*_{T,p}$	$\dfrac{\rho^*}{M_S}\left(\dfrac{\partial}{\partial n^2}\,\zeta_S'\right)^*_{T,p}$
Cyclohexane	0.174	0.176
Benzene	0.191	0.189
Carbon tetrachloride	0.183	0.176
Carbon disulfide	0.188	0.188

[a]Adapted from [45].

and (83), the permanent dipole moment and the optical polarizability tr $\mathbf{A}^0_{gG}$ can be determined after making some assumptions about the size of the cavity and the nuclear polarizability, as discussed in Section 3.3.

The Onsager model obviously needs two parameters ε and a_w in spherical approximation and four, ε, a_x, a_y, and a_z, in ellipsoidal approximation. By selecting the parameters a_w or a_x, a_y, and a_z the strength of the interaction between solute and solvent molecules is obviously determined. Since Onsager a great deal of brainstorming has been applied [28, 37, 53–62] either to finding a general theoretical procedure to estimate values of these parameters for a given molecule in a given solvent, or to modifying this model.

To determine a_w experimentally, a comparison of values for dipole moments and polarizabilities determined from gas-phase measurements and from solutions can be made with small molecules and in rare cases with somewhat larger ones; however, only a few methods exist for larger dipolar molecules such as typical dyes [28, 29]. As a quintessence, today the main axes $2a_\lambda(\lambda = x, y, z)$ are estimated to be between

$$2a_{G\lambda} \leqslant 2a_\lambda \leqslant 2a_{G\lambda} + 2a_{S\lambda} \tag{86}$$

This relation expresses the idea that the size of the cavity is better described by an "interaction sphere" that comprises at least part of the next solvent shell than by a sphere (or an ellipsoidal) of the solute's size. Hence for polar solvents the value of $2a_\lambda$ should be at the upper limit and for nonpolar solvents, at the lower limit of the relation (86). See [45] for further remarks.

Here $2a_{G\lambda}$ are the main axes estimated from the geometry of a molecule G also taking into account van der Waals radii of the atoms.

A few years ago, Rettig [55] presented his microstructural model, where $1/a_w^3$ is replaced by a factor A, which is calculated from the real charge distribution of the solute interacting with the homogeneous surrounding solvent using charge distributions from quantum chemical calculations. Although this extension is an advancing idea, it is not known currently whether there is a real improvement in the estimation of the Onsager parameter $1/a_w^3$ (or A) because of the accuracy obtained from using quantum chemical methods with large molecules.

3.4.3   Comparison of the Onsager and Lorentz Models

Often the results from dielectric measurements in nonpolar solvents are in fairly good agreement, if μ_{gG}^0 values are compared that have been derived on the basis of the poorer Lorentz model and on the better defined spherical Onsager model.

Assume that

1. $\varepsilon = n^2 \approx 2$ (nonpolar solvent)

2. $\dfrac{1}{4\pi\varepsilon_0}\dfrac{\alpha_{gS}^0}{a_{wS}^3} = \dfrac{1}{4\pi\varepsilon_0}\dfrac{\alpha_{gG}^0}{a_{wG}^3} = \dfrac{\varepsilon - 1}{\varepsilon + 2} = \dfrac{1}{4}$ (very arbitrary)

3. $3(\mathrm{tr}\,\alpha_{gS}^2) = (\mathrm{tr}\,\boldsymbol{\alpha}_{gS})^2$ (isotropic polarizability)

4. $V_S^* = N_A \dfrac{4\pi}{3} a_{wS}^3$ (a_{wS} from the solvent's molar volume)

5. $V_G^* \ll \mathscr{L}_G^*$ (most often valid, if $\mu_{gG}^0 > 3 \times 10^{-30}\,\mathrm{C \cdot m}$)

Then, in the spherical Onsager model

$$\frac{\rho^*}{M_S}\left(\frac{\partial}{\partial\varepsilon}\,\zeta_S\right)_{T,p}^* = \frac{\rho^*}{M_S}\left(\frac{\partial}{\partial n^2}\,\zeta_S'\right)_{T,p}^* \approx \frac{\varepsilon - 1}{2\varepsilon + 1} = \frac{n^2 - 1}{2n^2 + 1} = 0.2 \tag{87}$$

Hence, using (60) and (65) for the evaluation of μ_{gG}^0 based on the Lorentz model and (43), (79), and (82) for the evaluation of μ_{gG}^0 based on the Onsager model, it can easily be seen that at $\varepsilon = 2$, values of μ_{gG}^0 should differ by only some percentage.

With the ground-state polarizability known from gas-phase measurements or Kerr effect measurements (and eventually some assumptions on $\mathbf{B}_{gG}^0$), $\alpha_{gG}^0/4\pi\varepsilon_0 a_{wG}^3$ can be calculated and is often around 0.25, a value equal to that derived from the foregoing assumption 2 with $\varepsilon = 2$. Kriebel and Labhart [28] report an experimental value $\alpha_{gG}^0/4\pi\varepsilon_0 a_{wG}^3$ of p-nitroanisole to be 0.18.

Hence, application of the Lorentz formula is often justified when evaluating data from dielectric measurements in nonpolar solvents. Indeed, its application gives evaluation procedures that are very simple to handle. On the other hand, be aware that it is a very approximative model in solutions and breaks down, especially in polar solvents, where the Lorentz-field factor $(\varepsilon + 2)/3$ diverges with ε, in contrast to the cavity-field factor $(\varepsilon - 1)/(2\varepsilon + 1)$.

3.5   Other Methods

Other methods usually are based implicitly on the Lorentz model. This is true, too, for the well-known method introduced by Halverstadt and Kumler in 1942 [63]. According to their procedure, $d\varepsilon/dw_G$ and $d(1/\rho)/dw_G$ are determined experimentally and extrapolated to infinite dilution. Then the molar polarization P_G is calculated according to their theory

$$P_G = \frac{M_G}{\rho_S}\frac{3}{(\varepsilon_S + 2)^2}\frac{d\varepsilon}{dw_G} + M_G\left[\frac{1}{\rho_S} + \frac{d}{dw_G}\left(\frac{1}{\rho}\right)\right]\frac{\varepsilon_S - 1}{\varepsilon_S + 2}. \tag{88}$$

Similarly, the molar refraction R_{DG} at the sodium D line can be determined from the dependence of n^2 of a solution on w_G

$$R_{DG} = \frac{M_G}{\rho_S} \frac{3}{(n_{DS}^2 + 2)^2} \frac{dn^2}{dw_G} + M_G \left[\frac{1}{\rho_S} + \frac{d}{dw_G}\left(\frac{1}{\rho}\right)\right] \frac{n_{DS}^2 - 1}{n_{DS}^2 + 2} \tag{89}$$

From P_G and R_{DG} the dipole moment μ_{gG}^0 is calculated

$$\mu_{gG}^0 = \left[(P_G - R_{DG})\frac{9\varepsilon_0 kT}{N_A}\right]^{1/2} \tag{90}$$

n_{DS}^2 is equal to ε_S in good approximation with nonpolar solvents. Then, $d/dw_G(1/\rho)$ need not be determined, and μ_{gG}^0 is calculated as

$$\mu_{gG}^0 = \left[\frac{M_G}{\rho_S} \frac{3}{(\varepsilon_S + 2)^2}\left(\frac{d\varepsilon}{dw_G} - \frac{dn_D^2}{dw_G}\right)\frac{9\varepsilon_0 kT}{N_A}\right]^{1/2} \tag{91}$$

Various other methods have been used, and Higasi's [64] formula is mentioned because its simplicity induces one to use it

$$\mu_{gG}^0 = \gamma(d\varepsilon/dw_G)^{1/2} \tag{92}$$

where γ is a solvent-dependent constant that also implies the Lorentz model. The assumptions made for the verification of (92) are discussed in [65], where it is easily seen that without further knowledge the application of such simplified formulas may lead to considerably erroneous results, especially with small dipole moments. Hence, applications of such formulas are not recommended.

A few recent papers are mentioned that discuss this topic more rigorously. As recently as 1968 Liptay and associates [66] used the Onsager–Scholte model for the theoretical and experimental determination of dipole moments. Then, in 1980 Exner [67] compared some older methods and Myers and Birge [52, 68] and Bossis [69] pointed out that results derived from dielectric investigations using ellipsoidal cavities for the solute in the Onsager model give better results, apparently without having been aware of [66].

The reader is also referred to a review article [47] on dipole moments of many metal complexes that were determined by dielectric loss measurements [70].

3.6   Experimental

From the discussion in Section 3.4 it follows that $(\varepsilon - 1)/\rho$ and $(n^2 - 1)/\rho$ of the solution considered must be measured at different mass fractions w_G of the solute G, at a well-defined temperature. This means that the values of ε, n, and ρ should be measured simultaneously with each given mass fraction. Following is a discussion of the principles of the measuring devices, and remarks, details, and practical hints about the apparatus are given.

But first, an estimation is given using (91) to demonstrate the necessary measuring accuracy.

For estimation assume the following values: $M_G = 0.100 \, \text{kg}$; $\rho = 10^3 \, \text{kg/m}^3$; $\varepsilon_S = 2$; $T = 300 \, \text{K}$; $\mu_{gG}^0 = 10^{-29} \, \text{C} \cdot \text{m}$. With these values, $(d\varepsilon/dw_G) - (dn_D^2/dw_G) = 29.2$; and since dn_D^2/dw_G typically is about 0.8, $d\varepsilon/dw_G = 30$.

To avoid nonlinearity w is usually chosen smaller than 10^{-3}. Then with a desired accuracy of 1% ε must be measured with a relative accuracy of 3×10^{-4}. Since $d\varepsilon/dT$ is about $2 \times 10^{-3} \, \text{K}^{-1}$ with standard nonpolar solvents, temperature must be stabilized to better than 0.1 K.

Even better stability is necessary if smaller dipole moments or less soluble compounds are to be determined. Also, the temperature coefficient of the measuring apparatus (the cell), usually cannot be neglected if $\Delta\varepsilon$ of 10^{-4} or smaller has to be determined.

Hence, two water bath thermostats in series and a temperature-controlled room are recommended for high-precision measurements.

3.6.1   Method to Measure the Relative Electric Permittivity of Solutions

The scope of this chapter does not encompass the variety of devices presented in the literature that are proposed and used to measure relative electric permittivity. Reference [40] gives a valuable overview of the methods used. This chapter is restricted to the measurement of the relative electric permittivity ε of solutions of polar solutes in nonpolar solvents. To be sure that at any time the orientation distribution of the solute molecules may be described by a Boltzmann distribution law, as assumed in Section 3.2, ε must be measured under quasi-static conditions, that is, with frequencies between zero and some frequency v_{max} that depends on the situation of the solution. Typically with polar dye molecules in nonpolar, nonviscous solvents at room temperature v_{max} may be up to 100 MHz. It is a good compromise to use $v = 1\text{--}10 \, \text{MHz}$.

With all methods that are relevant here, the solution to be investigated is placed into a *capacitor*, hitherto called a *cell*. From electrostatics, the capacitance C of such a cell is given as

$$C = C_0 + \varepsilon C_1 \tag{93}$$

where C_0 and C_1 are constants of the cell.

Hence, after having once calibrated the cell with some appropriate solvents, which gives the values of C_0 and C_1, ε may be determined from the measurement of the capacitance C of the cell filled with the solution to be investigated.

The standard method for determining C incorporates the cell into an oscillator circuit, the resonance frequency of which is given as

$$v_1 = \frac{1}{2\pi[L(C + C_x)]^{1/2}} \tag{94}$$

where L is the inductance of the oscillator circuit, and C_x is a small capacitance given by an adjustable, highly stable capacitor, which carries a calibrated scale.

The frequency v_1 is compared to a standard frequency v_2, which is delivered from an oscillator that comprises electronic parts in such a way that both oscillators show the same temperature coefficient. Substituting (93) into (94) gives

$$v_1 = \frac{1}{2\pi[L(C_0 + \varepsilon C_1 + C_x)]^{1/2}} \tag{95}$$

Thus, to achieve resonance of v_1 and v_2 at various ε, C_x must be adjusted. Under these circumstances, C_x depends linearly on ε, which can be used for simple two- (or a few-) point calibrations with, for example, dry nitrogen and cyclohexane.

The condition $v_1 = v_2$ can be indicated in various ways. But usually the frequencies v_1 and v_2 are mixed, and the difference frequency $v_1 - v_2$ is observed directly or is compared to a low-frequency standard v_3, where, for example, a stable certain Lissajous curve may be used on an oscilloscope to define a reference condition for all measurements. When $v_1 - v_2 = v_3$, the curve is generally an ellipse. Obviously, $v_2 - v_1 = v_3$ gives the same reference conditions. Hence, we should not confuse the two situations.

We have discussed an absolute procedure. However, if nonpolar solutions are handled, very low concentrations of water, which generally might be accepted from the air or from the cell walls, would seriously disturb the results of a series of absolute measurements because of the very high relative permittivity of water. Also, high long-term stability is necessary. Thus it is recommended that measurements be made of the difference between the relative permittivity of the solution and the pure solvent in a completely closed system. This can be done by filling the cell with a weighted quantity of the solvent. Also a weighted amount of the compound to be investigated is introduced into the cell chamber above the surface of the solvent, situated in a small weighing vessel on a small bench mounted on a sealed feed-through turnaround axes.

The cell is tightly closed, and the solvent is allowed to reach the temperature of the cell as given by the thermostat. All water is solved from the cell walls, for example, so that after a relatively short time the reading of the condensor C_x is constant; that is, the relative permittivity of the solvent has adopted a constant value. The process may be accelerated with a magnetic stirrer. The stable reading value naturally should be very close to the exact value of the solvent. Then the vessel containing the compound to be measured is tipped into the solution by turning the small bench around its feed-through axis. The subsequent solution process also is accelerated by a magnetic stirrer. Usually after some minutes a stable reading of the capacitor C_x indicates the end of the solvation process. The difference between that and the former reading is used together with the mass fraction for further evaluation, where ε^* of the solvent is taken from the literature to determine by the preceding procedure values of ε of the solution from the difference $\varepsilon - \varepsilon^*$.

Adequately accurate absolute values of the refractive index and the density of the solution prepared in this way in the cell can be determined from independent experiments in separate apparatus since the refractive index n and the mass density ρ of solutions are not as sensitive as relative permittivity to small amounts of water because of the high ε value of water.

Figure 2.1 shows an experimental setup that is based on a commercial unit[†] and used with some modifications. The glass vessel used to dissolve the compound is mounted from below to the measuring cell. The whole system is

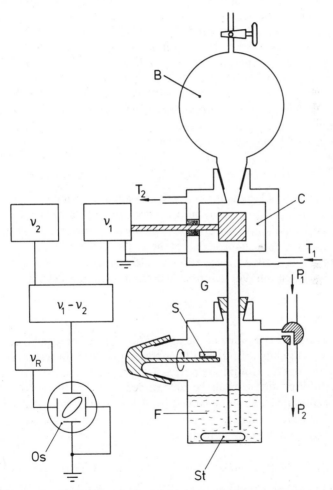

Figure 2.1   B, glass bulb, about 1000 cm³; C, measuring cell (liquid capacitor); T_1, T_2, connections to water bath thermostat; G, glass tube; S, sample in a weighing vessel; F, flask to prepare the solution; St, magnetic stirrer; P_1, P_2, pressure inlet and outlet; v_1, v_2, high-frequency oscillators; $v_1 - v_2$, mixer; v_R, reference oscillator at 50 Hz; Os, oscilloscope with Lissajous curve.

†Model DMO1 of WTW, 812 Weilheim, Federal Republic of Germany.

closed by a glass bulb of about $1000 \, cm^3$ volume on the upper side of the cell. The solvent or the solution is pumped up into the cell by dry nitrogen pressure. The pure solvent and the solution are pumped up and allowed to flow down several times to flush the system thoroughly until the reading of C_x is constant. Obviously the measurement is taken at a pressure about 0.1 bar higher than laboratory conditions, which because $d\varepsilon/dp \approx 10^{-4} \, bar^{-1}$ and because of the reproducibility of that overpressure, does not cause a problem. Dissolved N_2 in the solution also does not produce errors since the effect of N_2 saturation is below the sensitivity threshold in nonpolar solvents or at least is canceled in difference measurements, as described.

The commercial unit used carefully avoids many problems that would arise with home-built units, such as with cell stability, inductance of the connecting lines between the oscillator and the cell, locking effects between the two oscillators because of insufficient shielding if the resonance method is used, and so on. The sensitivity that can be achieved is about 10^{-5} with ε, which is sufficient for chemical applications. Refer to [40] and the literature cited therein, to improve sensitivity, but remember problems may arise when trying to establish sufficient temperature control!

3.6.2   Method to Measure the Mass Density of Solutions

The mass density of liquids is generally measured by using pycnometers. Even with highly sophisticated large-bulb pycnometers the accuracy achieved with this method is relatively poor. In 1969, Kratky and co-workers [71] described a method that utilizes the dependence of the mechanical resonance frequency v_0 of a U-shaped, thin-walled quartz tube on the density ρ of a solution filling the quartz tube. Equation (96) describes the relationship between ρ and v_0

$$\rho = A(v_0^{-2} - B) \tag{96}$$

where A and B are constants of the apparatus. The accuracy of a commercially built unit† is very high. If sufficiently good temperature control is provided and only a small interval around the pure solvent value is used, to assure adequate linearity, solution density can be determined to better than $10^{-5} \, g/cm^3$. An absolute calibration can be achieved by using liquids with well-known mass densities that cover a small density range around the density value of the solvent used to determine the dipole moments.

3.6.3   Method to Measure the Refractive Index of Solutions

Standard Abbé-type refractometers are used often, but seldom with good results, to measure the refractive index of solutions since the thin layer of the solution on its prism rapidly changes its concentration because of solvent vaporization.

---

†Anton Paar KG, A-8054 Graz, Austria.

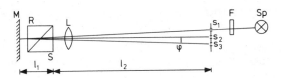

Figure 2.2 Principle of a differential refractometer: Sp, light source; F, optical filter; L, lens; M, mirror; S, sample chamber; R, reference chamber.

Hence, a closed cell must be involved. Modified Pulfrich refractometers can be used that usually also give higher sensitivity than Abbé-type instruments.

However, differential refractometers seem to be better suited to these measurements. The well-known principle is shown in Figure 2.2. The light coming from a filtered tungsten filament lamp or from a spectral line source passes through an entrance slit s_1, a lense L, and the cell. The cell is divided at 45° into a sample chamber S and a reference chamber R. Input and output windows of the cell are positioned to be as parallel as possible.

The light is then reflected by a mirror M and again passes through both chambers of the cell and the lens L, which will give a slit s_1 image at position s_2 if both cell chambers are filled with the solvent. Now let the refractive index n_R of the solvent in the reference chamber R be different from the solution with refractive index n_S in the sample chamber S where

$$n_S - n_R \ll n_R \tag{97}$$

The image of s_1 will then be found at position s_3, and it can be shown that the tangent of the deviation angle φ is

$$\tan \varphi = 2 \frac{n_S - n_R}{n_R} \tag{98}$$

which is valid if $l_1 \ll l_2$.

The cell may be of all-glass or all-quartz composition, and it will be closed by a Teflon stopper. If the difference between the positions s_3 and s_2 is measured optoelectronically and with $l_2 \approx 1$ m, a sensitivity with the measurement of $(n_S - n_R)/n_R$ of better than 10^{-5} can be achieved, which is fully sufficient for the purpose discussed here. With a typical value of the temperature coefficient of the refractive index $dn/dT = 4 \times 10^{-4}$ K^{-1}, a temperature control to about 0.01 K must be provided.

There are commercially available refractive index detectors for liquid chromatography that show even much higher sensitivity [72, 73]. They, too, may be adapted for the purpose considered here, but the calibration generally depends on the solvent used, in a nonlinear way.

3.7   Evaluation Schemes

It is convenient to have simple evaluation schemes for the determination of dipole moments from dielectric measurements. Hence, such schemes are given for the most common procedures.

DIPOLE MOMENT OF A PURE GASEOUS SPECIES FROM DIELECTRIC MEASUREMENTS

EVALUATION SCHEME I

1.   Determine $(\varepsilon - 1)/\rho$ as a function of T.
2.   Perform a regression analysis of $(\varepsilon - 1)/\rho$ on $1/T$ according to (38).
3.   Calculate μ_{gG}^0 and $\text{tr}\,\alpha_{gG}^0$ from the regression (least-squares) coefficients.

EVALUATION SCHEME II

1.   Determine experimentally the values of ε, n^2, and ρ of the gas at given T.
2.   Make some assumption about $c_{AB} = \mathbf{B}_{gG}^0/\mathbf{A}_{gG}^0$.
3.   Calculate the value of μ_{gG}^0, according to (41).

DIPOLE MOMENT OF A SOLUTE FROM DIELECTRIC MEASUREMENTS IN NONPOLAR SOLVENTS USING THE LORENTZ MODEL

EVALUATION SCHEME

1.   Determine $\mathscr{L}_G^*$, $\mathscr{L}_G'^*$, and V_G^* experimentally at constant T and p according to (49)–(51) from the dependence of $(\varepsilon - 1)/\rho$, $(n^2 - 1)/\rho$, and $1/\rho$ on the mass fraction of a solute in a nonpolar solvent, where $\varepsilon^* \approx n^{*2}$.
2.   Calculate ζ_G^* and $\zeta_G'^*$ with values from step 1, according to (60) and (61).
3.   Calculate value of $\text{tr}\,\mathbf{A}_{gG}^0$ according to (57) from step 2.
4.   Make assumption on $\mathbf{B}_{gG}^0$ (e.g., $\mathbf{B}_{gG}^0 \ll \mathbf{A}_{gG}^0$).
5.   Calculate value of μ_{gG}^0 according to (56) using results from steps 3 and 4.

DIPOLE MOMENT OF A SOLUTE FROM DIELECTRIC MEASUREMENTS IN NONPOLAR SOLVENTS USING THE ONSAGER MODEL

EVALUATION SCHEME

1.   Determine $\mathscr{L}_G^*$, $\mathscr{L}_G'^*$, and V_G^* experimentally at constant T and p according to (49)–(51) from the dependence of $(\varepsilon - 1)/\rho$, $(n^2 - 1)/\rho$, and $1/\rho$ on the mass fraction of the solute.
2.   Take the appropriate values of $(\rho^*/M_S)(\partial\zeta_S/\partial\varepsilon)_{T,p}^*$ and $(\rho^*/M_S)(\partial\zeta_S'/\partial n^2)_{T,p}^*$ from Table 2.4; if these are unknown, use an estimate according to Table 2.4.
3.   Calculate ζ_G^* and $\zeta_G'^*$ with values from steps 1 and 2 according to (43) and (44).
4.   Determine a_w or a_x, a_y, a_z from the size of the solute (Section 3.4.2).
5.   Calculate the value of $\text{tr}\,\mathbf{A}_{gG}^0$ from $\zeta_G'^*$ (step 3) according to (80) and (83).
6.   Make an assumption about $\mathbf{B}_{gG}^0$ (e.g., $\mathbf{B}_{gG}^0 \ll \mathbf{A}_{gG}^0$).
7.   Calculate values μ_{gG}^0 according to (79) and (82) from step 3 using results from steps 5 and 6.

4   DIPOLE MOMENTS OF MOLECULES IN THEIR GROUND AND EXCITED STATES FROM ELECTROOPTICAL ABSORPTION MEASUREMENTS

4.1   General Remarks

With dielectric measurements in solutions the permanent ground-state dipole moment of a solute can be determined, mainly from the change of the macroscopic polarization of a solution caused by the concentration of the solute under investigation. This method requires rather high concentrations of the solute whenever small dipole moments are to be determined. Because of the poor solubility of larger molecules, insufficiently accurate results are often obtained when this method is used on their ground-state moments.

If in such cases a solution concentration can be reached with an order of one optical density in the considered spectral range, electrooptical absorption measurements (EOAM) may be used. Naturally, there are other drawbacks with this method that will be discussed in the following sections.

Other electric moments, such as the dipole moment of Franck–Condon excited states, and in favorable cases ground state and Franck–Condon excited state polarizabilities, too, can also be determined by electrooptical measurements.

4.2   Brief Survey on the History of Electrooptical Absorption Measurements

In 1940, Kuhn and co-workers [74] realized that a solution of dipolar solutes should show a dichroism in the absorption in an external electric field, because of the orientation anisotropy induced by the electric field. This effect has been verified experimentally; however, further study was dormant for about 20 years. Independently Labhart [75] and Czekalla [76] studied this effect in the early 1960s and found that to describe fully the effect observed, an absorption band wavenumber shift caused by the difference between ground and excited Franck–Condon (FC) state dipole moments must be taken into account [77]. With further development of the experimental setup based on Labhart's method [78], additional refinement of the molecular model was necessary. This was accomplished by Liptay and associates [53, 54, 79, 80] within the first 10 years after the idea was revived. Also, other groups [7, 81–92] started some related work in solutions as well as in polymer matrices, sometimes using a less elaborated theoretical framework. Others [93] used the basic idea with large molecules (polymers, too) and dipole moments. For a survey of related literature and effects the reader is referred to the book of Fredericq and Houssier [94] and to some other review articles [48, 95–98]. The direct relationship between EOAM and the Kerr effect has been pointed out by Kuball and Singer [99].

4.3 Theory of Electrooptical Absorption Measurements

4.3.1 Model and Partial Molar Quantities With Electrooptical Absorption Measurements

Consider a solution of a solute species G in a pure solvent S. With Lambert's law being valid

$$\log(q_{00}/q) = \sigma l \tag{99}$$

where q_{00} is the photon current entering the phase and q is that leaving it after having traversed a path length l, the absorption coefficient σ is defined. An external electric field $\mathbf{E}_a$ is applied to this solution with volume V in the electric field. Then a change of the optical density σl of the phase is observed, which for symmetry reasons depends on the square of the electric field in an originally isotropic solution. On the macroscopic scale an extensive quantity ^aY is defined

$$^aY = {}^aMV \tag{100}$$

where

$$^aM = \lim_{E_a^2 \to 0} \left(\frac{\partial \sigma^E}{\partial E_a^2} \right)_{n_J, T, p} \tag{101}$$

from which the corresponding PMQs ^aY_J can be defined

$$^aY_J = \left[\frac{\partial}{\partial n_J} ({}^aMV) \right]_{n'_J, T, p} \qquad J = G, S \tag{102}$$

where V is known and ^aM can be determined by a suitable experimental setup to be described in Section 4.8. Hence, the PMQs ^aY_J can be determined according to (4). In particular, $^aY_G^*$ can be determined.

$$^aY_G^* = \left(\frac{^aM}{c_G} \right)^* \tag{103}$$

To proceed further, corresponding MMQs ^ay_J must be defined using a molecular model. The same basic molecular model just introduced with dielectric investigations has been used and extended by Liptay and co-workers [53, 54, 79, 80, 96] to represent the MMQs by intrinsic molecular quantities such as dipole moments.

With this model ^aY is represented according to (7) by

$$^aY = \sum {}^ay_J n_J \qquad J = G, S \tag{104}$$

where the ^ay_J are the MMQs corresponding to ^aY_J.

To relate $^aY_G^*$ to $^ay_G^*$, (8) is used. If the investigations are done in a spectral region where the solvent S does not absorb, it follows from the arguments given in [100] that

$$^aY_G^* = {}^ay_G^* \tag{105}$$

In the dilute system considered, the MMQs ^ay_G are given as [100]

$$^ay_G^* = \kappa_G^* L_G^* + \kappa_G^* s_E^* \tag{106}$$

where

$$L_G^* = \lim_{E_a^2 \to 0} \frac{1}{\kappa_G^{*E}} \left(\frac{\partial \kappa_G^{*E}}{\partial E_a^2} \right)_{n_J, T, p} \tag{107}$$

describes the relative change of the molar decadic extinction coefficient κ_G^{*E} with an external electric field, and

$$s_E^* = \lim_{E_a^2 \to 0} \frac{1}{\rho^E} \left(\frac{\partial \rho^E}{\partial E_a^2} \right)_{n_J, T, p} \tag{108}$$

describes the electrostriction. Labhart [48] has shown that usually the effect from electrostriction can be disregarded.

Then, combining (105), (107), and (108) yields

$$\left(\frac{{}^aM}{c_G} \right)^* = \kappa_G^* L_G^* \tag{109}$$

and with the validity of Beer's law

$$\sigma^* = c_G \kappa_G^* \tag{110}$$

it follows:

$$L_G^* = \left(\frac{{}^aM}{\sigma} \right)^* \tag{111}$$

Hence, from the measured quantities ^aM and σ the quantity L_G^* can be determined.

4.3.2   The Specific Molecular Model Used With Electrooptical Absorption Measurements

Whenever the effect of an external electric field on a solute molecule is studied, one basic problem is to relate the local electric field effective on a solute to the externally applied electric field $\mathbf{E}_a$. Necessarily, a model must be used. As noted in Section 3.4.2, the Onsager model extended by Scholte and Liptay is well defined and currently has proven to be the most reliable; it is also used with EOAM.

The more specific ideas that define a model for the effect of that local field on a solute molecule have had to be developed.

In the development of today's theoretical EOAM framework three separate effects have been used to define the model [77, 79, 80]:

1. The orientation distribution of solute molecules in an electric field is described by the Boltzmann distribution law, as discussed in Section 3.2.
2. The ground and FC excited states related to a considered absorption band are perturbed by the perturbation Hamiltonian $-\mu E$, which generally causes the absorption band to shift. A careful detailed discussion is necessary [80] since the local electric field perturbing the ground state is different from that perturbing the FC excited state.
3. The same perturbation Hamiltonian must also be included in the time-dependent perturbation theory and yields a field-dependent effective transition moment, that is, a field-dependent absorption intensity, which is described by transition polarizabilities.

With this model, Liptay worked out the theoretical framework of EOAM, the result of which, with respect to the MMQ $^a y_G^*$ or to L_G^*, is given in Section 4.3.3.

4.3.3 The Quantity L_G^*

In a spectral region where an isolated vibronic transition can be investigated, L_G^* is given according to Liptay's theory [100]

$$L_G^*(\chi, \tilde{\nu}) = {}^a Dr(\chi) + ({}^a E/6)s(\chi)$$
$$+ [{}^a Fr(\chi) + {}^a Gs(\chi)]^a t(\tilde{\nu})$$
$$+ [{}^a Hr(\chi) + {}^a Is(\chi)]^a u(\tilde{\nu}) \tag{112}$$

where

$$r(\chi) = (2 - \cos^2\chi)/5 \tag{113}$$

$$s(\chi) = (3\cos^2\chi - 1)/5 \tag{114}$$

$$^a t(\tilde{\nu}) = \frac{1}{hc_0(\kappa/\tilde{\nu})}\left(\frac{\partial}{\partial\tilde{\nu}'}\frac{\kappa}{\tilde{\nu}'}\right)_{\tilde{\nu}'=\tilde{\nu}} \tag{115}$$

$$^a u(\tilde{\nu}) = \frac{1}{2h^2 c_0^2(\kappa/\tilde{\nu})}\left(\frac{\partial^2}{\partial\tilde{\nu}'^2}\frac{\kappa}{\tilde{\nu}'}\right)_{\tilde{\nu}'=\tilde{\nu}} \tag{116}$$

As the angle between the external electric field and the electric field vector of a linearly polarized light beam χ is used for measuring absorption and is also an experimental variable, as is the wavenumber $\tilde{v}$ of the incident light wave; h is Planck's constant; c_0 is the velocity of light. The quantities $^at(\tilde{v})$ and $^au(\tilde{v})$ are accessible through the first and second derivative of an ordinary absorption spectrum taken from the solution under investigation.

The terms ^aD to ^aI are given as

$$^aD = (2/kT)\mathbf{m}_a\mathbf{f}_e^2\mathbf{m}_a'\boldsymbol{\mu}_g + \text{tr}(\mathbf{f}_e\mathbf{m}_a')^2 \tag{117}$$

$$^aE = (1/kT)^2[3(\mathbf{m}_a\mathbf{f}_e\boldsymbol{\mu}_g)^2 - \boldsymbol{\mu}_g\mathbf{f}_e^2\boldsymbol{\mu}_g] + (1/kT)(3\,\mathbf{m}_a\mathbf{f}_e^2\boldsymbol{\alpha}_g\mathbf{m}_a - \text{tr}\,\mathbf{f}_e^2\boldsymbol{\alpha}_g)$$
$$+ (6/kT)(\mathbf{m}_a\mathbf{f}_e\boldsymbol{\mu}_g\cdot\text{tr}\,\mathbf{f}_e\mathbf{m}_a' + \boldsymbol{\mu}_g\mathbf{f}_e^2\mathbf{m}_a'\mathbf{m}_a) + 3(\text{tr}\,\mathbf{f}_e\mathbf{m}_a')^2$$
$$+ 3\,\text{tr}(\mathbf{f}_e\mathbf{m}_a')^2 + Q^{(1)}(E_f) \tag{118}$$

$$^aF = (1/kT)\boldsymbol{\mu}_g\mathbf{f}_e^2\Delta^a\boldsymbol{\mu} + \tfrac{1}{2}\,\text{tr}\,\mathbf{f}_e^2\Delta^a\boldsymbol{\alpha} + 2\mathbf{m}_a\mathbf{f}_e^2\mathbf{m}_a'\Delta^a\boldsymbol{\mu} + Q^{(2)}(E_f) \tag{119}$$

$$^aG = (1/kT)(\mathbf{m}_a\mathbf{f}_e\boldsymbol{\mu}_g)(\mathbf{m}_a\mathbf{f}_e\Delta^a\boldsymbol{\mu}) + \tfrac{1}{2}\mathbf{m}_a\mathbf{f}_e^2\Delta^2\boldsymbol{\alpha}\mathbf{m}_a + \mathbf{m}_a\mathbf{f}_e\Delta^a\boldsymbol{\mu}\cdot\text{tr}\,\mathbf{f}_e\mathbf{m}_a'$$
$$+ \Delta^a\boldsymbol{\mu}\mathbf{f}_e^2\mathbf{m}_a'\mathbf{m}_a + Q^{(3)}(E_f) \tag{120}$$

$$^aH = \Delta^a\boldsymbol{\mu}\mathbf{f}_e^2\Delta^a\boldsymbol{\mu} + Q^{(4)}(E_f) \tag{121}$$

$$^aI = (\mathbf{m}_a\mathbf{f}_e\Delta^a\boldsymbol{\mu})^2 + Q^{(5)}(E_f) \tag{122}$$

where the following definitions have been used:

$$\boldsymbol{\mu}_g = (\mathbf{I} - \mathbf{f}\boldsymbol{\alpha}_g^0)^{-1}\boldsymbol{\mu}_g^0 \tag{123}$$

$$\boldsymbol{\alpha}_g = (\mathbf{I} - \mathbf{f}\boldsymbol{\alpha}_g^0)^{-1}\boldsymbol{\alpha}_g^0 \tag{124}$$

$$\Delta^a\boldsymbol{\mu} = (\mathbf{I} - \mathbf{f}'\boldsymbol{\alpha}_a^{OFC})^{-1}(\mathbf{I} - \mathbf{f}\boldsymbol{\alpha}_a^{OFC})(\mathbf{I} - \mathbf{f}\boldsymbol{\alpha}_g^0)^{-1}(\mathbf{I} - \mathbf{f}'\boldsymbol{\alpha}_g^0)(\boldsymbol{\mu}_a^{FC} - \boldsymbol{\mu}_g) \tag{125}$$

$$\Delta^a\boldsymbol{\alpha} = (\mathbf{I} - \mathbf{f}'\boldsymbol{\alpha}_a^{OFC})^{-1}(\mathbf{I} - \mathbf{f}\boldsymbol{\alpha}_g^0)^{-2}(\mathbf{I} - \mathbf{f}'\boldsymbol{\alpha}_g^0)(\boldsymbol{\alpha}_a^{OFC} - \boldsymbol{\alpha}_g^0) \tag{126}$$

$$\boldsymbol{\mu}_a^{FC} = (\mathbf{I} - \mathbf{f}\boldsymbol{\alpha}_a^{OFC})^{-1}\boldsymbol{\mu}_a^{OFC} \tag{127}$$

The terms $\boldsymbol{\mu}_g^0$ and $\boldsymbol{\alpha}_g^0$ are the permanent dipole moment and the polarizability of the isolated species G under consideration in its ground state, respectively; $\boldsymbol{\mu}_a^{OFC}$ and $\boldsymbol{\alpha}_a^{OFC}$ are the respective molecular quantities in its FC excited state; $\mathbf{m}_a$ is a unit vector in the direction of the transition moment $\boldsymbol{\mu}_{ga}$ involved in the considered absorption process,

$$\mathbf{m}_a' = \frac{\boldsymbol{\alpha}_{ga}}{|\boldsymbol{\mu}_{ga}|} \tag{128}$$

which is a standardized tensor determined mainly by the transition polarizability tensor $\boldsymbol{\alpha}_{ga}$ [101].

The tensors $\mathbf{f}_e$, $\mathbf{f}$, and $\mathbf{f}'$ have already been defined by (66), (67), and (72) when an ellipsoidal cavity is used with the Onsager model, and by (73), (75), and (76) when a spherical cavity is used.

The terms $Q^{(i)}(E_f)$ are functions of the fluctuation of the electric field. They are discussed in [53] in detail. Equations (117)–(122) differ slightly from those in [100], but only in notation.

4.4   Determination of the Terms ^aD to ^aI

According to (111), measuring $(^aM/\sigma)^*$ in dilute solutions as a function of χ and $\tilde{\nu}$ gives a set of values $L^*_G(\chi, \tilde{\nu})$.

A multilinear regressional analysis (least-squares fit) according to (112), where $^at(\tilde{\nu})$ and $^au(\tilde{\nu})$ have been determined from an absorption spectrum, yields the terms ^aD to ^aI as regression coefficients.

Up to that point, EOAM is a time-consuming but straightforward experimental method.

From the six independent quantities ^aD to ^aI, which are related to many more molecular quantities, it is immediately evident that it is impossible to proceed further with the evaluation of electric moments without additional information or assumptions.

Thus, obviously, there is no general way to further evaluation, and most often some general approximations must be used.

4.5   Determination of Dipole Moments

4.5.1   Often-Applied General Approximations

When sufficiently large dipole moments are to be determined in nonpolar solvents, it is often assumed that all terms caused by the polarizability and the transition polarizability of the solute and the effects of the fluctuation of the local field can be neglected, since the neglected terms amount to only some percentages of the terms ^aE to ^aI. This may be seen

1. For the ground state polarizability: through estimation of the polarizability tensor from group polarizabilities [46].
2. For the transition polarizability: a negligible term ^aD, which is most often found, leads to the assumption that the corresponding terms in ^aE may be neglected also.
3. For the fluctuation-dependent terms: at least in hydrocarbon solvents the fluctuation of the electric field is negligible.

For further details and exceptions see [53, 54]. With these simplifications, the following approximative terms $^aE_0$ to $^aI_0$ result from EOAM

$$^aE_0 = (1/kT)^2[3(\mathbf{m}_a \mathbf{f}_e \boldsymbol{\mu}_g)^2 - \boldsymbol{\mu}_g \mathbf{f}_e^2 \boldsymbol{\mu}_g] \tag{129}$$

$$^aF_0 = (1/kT)\boldsymbol{\mu}_g \mathbf{f}_e^2 \Delta^a\boldsymbol{\mu} \tag{130}$$

$$^aG_0 = (1/kT)(\mathbf{m}_a \mathbf{f}_e \boldsymbol{\mu}_g)(\mathbf{m}_a \mathbf{f}_e \Delta^a \boldsymbol{\mu}) \tag{131}$$

$$^aH_0 = \Delta^a \boldsymbol{\mu} \, \mathbf{f}_e^2 \Delta^a \boldsymbol{\mu} \tag{132}$$

$$^aI_0 = (\mathbf{m}_a \mathbf{f}_e \Delta^a \boldsymbol{\mu})^2 \tag{133}$$

From this set of five terms generally it is possible to calculate all molecular quantities that have not been neglected up to this point, namely, μ_g, $\Delta^a \mu$, $\not\prec (\mathbf{m}_a, \boldsymbol{\mu}_g)$, $\not\prec (\boldsymbol{\mu}_g, \Delta^a \boldsymbol{\mu})$, and $\not\prec (\mathbf{m}_a, \Delta^a \boldsymbol{\mu})$. It is noteworthy that from $\not\prec (\boldsymbol{\mu}_g, \Delta^a \boldsymbol{\mu})$ it may be shown whether the dipole moment increases or decreases with excitation.

4.5.2 Often-Necessary Additional Data

Regrettably, ^aH and ^aI often cannot be determined with sufficient accuracy. Then further information, usually supplied by symmetry, is needed. For example, with a molecular species belonging at least approximately to a point group with one axis of rotation, transitions may be studied where $\mathbf{m}_a \parallel \boldsymbol{\mu}_g \parallel \Delta^a \boldsymbol{\mu}$ is fulfilled in good approximation. This often happens with molecules showing large internal charge transfer. Under such conditions, from a set of data $\{^aE, {}^aF, {}^aG\}$, μ_g and $\Delta^a \mu$ can be calculated without knowledge of ^aH and ^aI.

Conversely, if the effective ground-state dipole moment μ_g is known from dielectric measurements—even when the investigated molecule does not show an axis of rotation but instead it may be assumed from symmetry considerations that $\mathbf{m}_a$, $\boldsymbol{\mu}_g$, and $\Delta^a \boldsymbol{\mu}$ span a plane—then $\Delta^a \mu$ and the angles $\not\prec (\mathbf{m}_a, \boldsymbol{\mu}_g)$, $\not\prec (\boldsymbol{\mu}_g, \Delta^{ap})$, and $\not\prec (\mathbf{m}_a, \Delta^a \boldsymbol{\mu})$ can be calculated from a set of data $\{^aE, {}^aF, {}^aG\}$ by using (129)–(131). Note that $\mathbf{f}_e$ (as well as $\mathbf{f}$ and $\mathbf{f}'$) is often used in a spherical approximation. However, the ellipsoidal approximation is better with long chain molecules, for example, although, strictly speaking, it is applicable only if the principal axes of all tensors involved are parallel.

4.6 Determination of Polarizabilities

It is beyond the scope of this chapter to discuss the special cases where polarizabilities can be determined. Obviously, such determinations are especially favored with molecules that do not show permanent dipole moments. Then from experimental ^aA, ^aF, and ^aG the terms $3\mathbf{m}_a \mathbf{f}_e^2 \alpha_g \mathbf{m}_a - \mathrm{tr}\, \mathbf{f}_e^2 \alpha_e^2 \Delta^a \alpha$, and $\mathbf{m}_a \mathbf{f}_e^2 \Delta^a \alpha \mathbf{m}_a$ can be calculated. For further examples the reader is referred to [6, 96].

Polarizabilities of molecules that possess a large permanent dipole moment can also be determined from EOAM in glassy frozen solutions [74, 102, 103] or from measurements in polymer films [7, 82, 83]. Generally, however, one must be very careful with the preparation of such frozen solutions in order to avoid microcrystallization of the solute.

A method by which polarizabilities and permanent dipole moments of a solute are determined simultaneously is given in Section 4.7.

4.7   Evaluation of the Permanent Dipole Moment of the Free Molecule From μ_g and $\Delta^a\mu$

4.7.1   Evaluation Using Estimates for the Cavity Parameters

Let μ_g and $\Delta^a\mu$ be derived from an EOAM experiment; μ_g and $\Delta^a\mu$ have been defined by (123) and (125)

$$\mu_g = (1 - f\alpha_g^0)^{-1}\mu_g^0 \tag{134}$$

$$\Delta^a\mu = (1 - f'\alpha_a^{OFC})^{-1}(1 - f\alpha_a^{OFC})(1 - f\alpha_g^0)^{-1}(1 - f'\alpha_g^0)(\mu_a^{FC} - \mu_g) \tag{135}$$

Therefore, the values of μ_g^0 and μ_a^{OFC} can be calculated from the experimental values of μ_g and $\Delta^a\mu$ only when $f\alpha_g^0$, $f\alpha_a^{OFC}$, $f'\alpha_g^0$, and $f'\alpha_a^{OFC}$ are known or can be estimated with sufficient reliability. Using the spherical Onsager model with radius a_w, this means that α_g^0/a_w^3 and α_a^{OFC}/a_w^3 must be known or a reliable estimator must exist. An estimate of a_w^3 can be obtained from molecular dimensions (see Section 3.4.2) and α_g^0, or as an estimate the mean polarizability $\frac{1}{3}\mathrm{tr}\,\alpha_g^0$, is known from dielectric investigations or from the Kerr effect or may be estimated from group polarizabilities [101]. Hence, μ_g^0 can be calculated from μ_g.

With μ_a^{OFC} the situation is more critical, since in addition a value for α_a^{OFC}, or as an estimate at least $\frac{1}{3}\mathrm{tr}\,\alpha_a^{OFC}$, is necessary. Estimation of α_a^{OFC} from an assumed additivity of excited-state group polarizabilities is impossible. Instead, excited-state polarizabilities can be determined (as mentioned in Section 4.6) from EOAM in glassy frozen solutions [6, 103] or in polymer matrices [82, 23]. As an alternative one may assume that $\alpha_a^{OFC} \approx \alpha_g^0$, which seems to be an acceptable approximation with many compounds, considering Table 3 in [96]. But exceptionally large excited-state polarizabilities are listed there, too. For example, the long axis component of the polarizability of all-trans retinal has been reported to increase with excitation from 84 to 280×10^{-40} CV^{-1} m^2 [7] and that of tetraphenyltetradecaheptaen from 246 to 1218×10^{-40} CV^{-1} m^2 [6]. Instead of using assumptions for α_a^{OFC}, an upper limit may be given for μ_a^{OFC}. In nonpolar solvents f equals f' in good approximation. Therefore, from (135) it follows that

$$\Delta^a\mu = \mu_a^{FC} - \mu_g \tag{136}$$

Thus at least μ_a^{FC} can be calculated and may be considered as an upper limit for μ_a^{OFC}, according to (127), unless α_a^{OFC} is negative, which is rarely, if ever, the case with molecules.

4.7.2   Evaluation Through the Solvent Dependence of μ_g and $\Delta^a\mu$

As the field developed it became desirable to have methods to determine α_a^{OFC} or at least α_a^{OFC}/a_w^3. In a recent paper [5] it has been shown that from values of μ_g and $\Delta^a\mu$ determined experimentally in various solvents, μ_g, α_g^0/a_w^3, μ_a^{OFC}, and

α_a^{OFC}/a_w^3 can be calculated. This can be seen if (123) and (125) are rewritten in the following way, where the ellipsoidal Onsager–Scholte model has been used:

$$\frac{1}{\mu_g} = \frac{1}{\mu_g^0} - \frac{1}{4\pi\varepsilon_0} \frac{1}{\mu_g^0} \frac{\alpha_{gzz}^0}{a_x a_y a_z} g_z \tag{137}$$

and

$$^aS = \mu_a^{OFC} + \frac{1}{4\pi\varepsilon_0} \frac{\alpha_{azz}^{OFC}}{a_x a_y a_z} {}^aT \tag{138}$$

where

$$^aS = \Delta^a\mu \frac{1 - f_z \alpha_{gzz}^0}{1 - f_z' \alpha_{gzz}^0} + \mu_g \tag{139}$$

$$^aT = \Delta^a\mu \frac{1 - f_z \alpha_{gzz}^0}{1 - f_z' \alpha_{gzz}^0} g_z' + \mu_g \cdot g_z \tag{140}$$

Here it is assumed that μ_g, $\Delta^a\mu$, and a principal axis of all the tensors lie along the z direction; g_z and g_z' depend on ε and n according to (67) and (72).

Hence, $\alpha_{gzz}^0/(a_x a_y a_z)$ and μ_g^0 result from a linear regression of $1/\mu_g$ on g_z and $\alpha_{azz}^{OFC}/(a_x a_y a_z)$, and μ_a^{OFC} result from another regression of ^aS on ^aT, where $(1 - f_z \alpha_{gzz}^0)(1 - f_z' \alpha_{gzz}^0)^{-1}$ is determined from the results of the first regression based on (137).

Tables 2.5 and 2.6 show an example from [5], where merocyanine dyes $(CH_3)_2N$—$(CH{=}CH)_x$—$CH{=}C(CN)_2$ defined as CNx (where x is a number from 0 to 4) have been investigated. A comparison of μ_g and μ_a^{FC} with μ_g and $\Delta^a\mu$ immediately shows the role played by polarizability with such measurements of solutions. Table 2.6 shows that if $\alpha_{gzz}^0/(a_x a_y a_z)$ is compared with $\alpha_{azz}^{OFC}/(a_x a_y a_z)$, α_{azz}^{OFC} is about 2.5 times greater than α_{gzz}^0.

This example is certainly somewhat unusual with respect to the discussion of these large polarizabilities, since the static polarizability measured with EOAM

Table 2.5 Experimental Values[a] of μ_g and $\Delta^a\mu$ for the Merocyanine Dye CN4

Solvent	μ_g	$\Delta^a\mu$
Cyclohexane	49.1 ± 1.2	58.2 ± 9.3
Decalin	52.0 ± 0.9	61.3 ± 6
Diisopropyl ether	59.2 ± 1.1	85.2 ± 5.3
Dioxane	61.2 ± 1.3	115 ± 7

[a]All values given in $(10^{-30}\,C \cdot m)$

Table 2.6 Dipole Moments and Polarizabilities of
the Merocyanine Dye CN4

$\mu_g^0/(10^{-30}\,\mathrm{C\cdot m})$	39.2 ± 1.4
$\mu_a^{0FC}/(10^{-30}\,\mathrm{C\cdot m})$	50.2 ± 2.8
$(\alpha_g^0/\alpha_x a_y a_z)(10^{-10}\,\mathrm{C\,V^{-1}\,m^{-1}})$	1.33 ± 0.25
$(\alpha_a^{0FC}/a_x a_y a_z)(10^{-10}\,\mathrm{C\,V^{-1}\,m^{-1}})$	3.22 ± 0.22

might show a considerable amount of nuclear polarizability B_{gG}^0, because of a solvent-dependent (i.e., field-dependent!) geometrical structure. Nevertheless, it shows how carefully one must use assumptions like $\alpha_a^{0FC} \approx \alpha_g^0$.

The absolute value found for $\alpha_{gzz}^0/(a_x a_y a_z)$ for CN4 can be compared with a value given by Kriebel and Labhart [28] for the ground state of *p*-nitroanisole, which was determined from a similar electrooptical absorption measurement to be $\alpha_g^0/a_w^3 = 0.16 \times 10^{-10}\,\mathrm{CV^{-1}\,m^{-1}}$, in spherical approximation.

4.8   Experimental Method and Setup

4.8.1   Method to Determine Values for $L_G^*(\chi, \tilde{\nu})$

In the early 1960s it was determined that the experimental method used by Labhart [75, 78] is superior to that used by Czekalla [76] and thus today it is used exclusively. Basic to this method is that the external electric field is modulated at a frequency ω from which a small modulation of the absorption coefficient follows that can be analyzed with respect to ^aM or Y_G^*.

Consider a measuring cell of length l containing a solution of the solute G to which an external electric field $\mathbf{E}_a$ is applied by means of a voltage across two parallel plate electrodes; q_{00} photons per second, polarized at angle χ with respect to the electric field, enter the cell; q photons per second leave it.

If it is assumed that Lambert–Beer's law (99) is valid, then

$$q^E = q_{00} \exp(-2.3\sigma^E l) \tag{141}$$

From this by differentiation it follows that

$$\frac{\partial}{\partial E^2} q^E = -q_{00} \exp(-2.3\sigma^E l) \cdot 2.3l \frac{\partial}{\partial E_a^2} \sigma^E \tag{142}$$

Hence

$$^aM = \lim_{E_a^2 \to 0} \left(\frac{\partial \sigma^E}{\partial E_a^2}\right)_{T,p} = \lim_{E_a^2 \to 0} \left(-\frac{1}{2.3l}\frac{1}{q^E}\frac{\partial}{\partial E_a^2} q^E\right) \tag{143}$$

from which follows the experimentally accessible PMQ using (103)

$$Y_G^* = \left(\frac{{}^aM}{c_G}\right)^* = \lim_{E_a^2 \to 0} \left(-\frac{1}{2.3 c_G l} \frac{1}{q^E} \frac{\partial}{\partial E_a^2} q^E\right) \tag{144}$$

As the effects to be measured are very small, the effect is modulated by modulation of the external electric field

$$E_a = E_{a0} + E_{a\omega} \cos \omega t \tag{145}$$

Hence

$$\begin{aligned} E_a^2 &= E_{a0}^2 + E_{a\omega}^2 \cos^2 \omega t + 2 E_{a0} E_{a\omega} \cos \omega t \\ &= E_{a0}^2 + \tfrac{1}{2} E_{a\omega}^2 + \tfrac{1}{2} E_{a\omega}^2 \cos 2\omega t + 2 E_{a0} E_{a\omega} \cos \omega t \end{aligned} \tag{146}$$

Then, develop q^E to powers of E_a^2

$$q^E = (q^E)_{E_a^2=0} + \left(\frac{\partial}{\partial E_a^2} q^E\right)_{E_a^2=0} E_a^2 + \frac{1}{2} \left[\frac{\partial^2}{\partial (E_a^2)^2} q^E\right]_{E_a^2=0} E_a^4 + \cdots \tag{147}$$

With sufficiently small external electric fields the expansion can always be interrupted after the second right-side term in (147). This holds true for all electric fields that solutions can withstand, with molecules having dipole moments of up to 100×10^{-30} C·m. For larger dipole moments, as with polymers, the external field strength should be reduced. Then, introducing E_a^2 from (146) into (147) reveals that q^E shows a component q_0^E that is unmodulated, a component with amplitude q_ω^E that is modulated with frequency ω, and a component with amplitude $q_{2\omega}^E$ modulated with frequency 2ω. It is

$$q_0^E = q^{E=0} \tag{148}$$

$$q_\omega^E = 2 E_{a0} E_{a\omega} \frac{\partial}{\partial E_a^2} q^E \bigg|_{E_a^2=0} \tag{149}$$

$$q_{2\omega}^E = \frac{1}{2} E_{a\omega}^2 \frac{\partial}{\partial E_a^2} q^E \bigg|_{E_a^2=0} \tag{150}$$

Combining (148) and (149) yields

$$\frac{1}{q^{E=0}} \left(\frac{\partial}{\partial E_a^2} q^E\right)_{E_a^2=0} = \frac{q_\omega^E}{2 E_{a0} E_{a\omega} q_0^E} \tag{151}$$

and with (144)

$$Y_G^* = -\frac{1}{2.3c_G l} \cdot \frac{q_\omega^E}{q_0^E} \cdot \frac{1}{2E_{a0}E_{a\omega}} \tag{152}$$

If q_ω^E, q_0^E are measured at the given E_{a0} and $E_{a\omega}$, and (103) and (109) are then used, the quantity L_G^* can be derived from the experiment according to the relation

$$L_G^* = -\frac{1}{2.3\kappa c_G l} \frac{q_\omega^E}{q_0^E} \frac{1}{2E_{a0}E_{a\omega}} \tag{153}$$

or, if the absorbance A is used,

$$A = \kappa c_G l \tag{154}$$

$$L_G^* = -\frac{1}{2.3A} \cdot \frac{q_\omega^E}{q_0^E} \cdot \frac{1}{2E_{a0}E_{a\omega}} \tag{155}$$

As q_ω^E and q_0^E are components of one light beam that is detected by one photomultiplier, the quotient of the respective components i_ω^E and i_0^E of the photomultiplier's anode current is independent of the sensitivity of the photomultiplier and of variations of the light intensity, and it is

$$\frac{q_\omega^E}{q_0^E} = \frac{i_\omega^E}{i_0^E} \tag{156}$$

and thus

$$L_G^* = -\frac{1}{2.3A} \frac{i_\omega^E}{i_0^E} \frac{1}{2E_{a0}E_{a\omega}} \tag{157}$$

L_G^* can be measured without knowing the weighed concentration, since only the absorption A has to be measured.

Instead of using the ω-modulated component i_ω^E, the component modulated with 2ω might be used, which is most often done with measurements in frozen solutions [102, 103] or in polymer films [7, 83]. Then L_G^* would be determined using the following equation [104]:

$$L_G^* = -\frac{1}{2.3A} \frac{i_{2\omega}^E}{i_0^E} \frac{2}{E_{a\omega}^2} \tag{158}$$

In this way dipole alignment by an average dc field is avoided.

4.8.2   Experimental Setup to Determine i_ω^E/i_0^E

From the definition of L_G^* in Section 4.3.3 it follows that an experimental determination of the terms ^aD to ^aI needs a set of values for $L_G^*(\chi, \tilde{v})$ determined as a function of χ and $\tilde{v}$. Hence, the experimental setup must be constructed so that the wavenumber of the linearly polarized light can be set to all values between 10,000 and 50,000 cm^{-1} (200–1000 nm), with an optical bandwidth that can be adjusted to the molecular system under investigation. This means a continuum light source is necessary. Also, the angle χ between the externally applied electric field and the electric field vector of the light beam must be adjustable, which can be achieved by a linear polarizer. As determined by reviewing (133) and (112), the experimental setup must be able to measure values of i_ω^E/i_0^E from the photomultiplier current as a function of χ and $\tilde{v}$.

From an independent experiment with a precise conventional spectral photometer the absorption spectrum $A(\tilde{v})$ and the derivatives $^at(\tilde{v})$ and $^au(\tilde{v})$, defined according to (114) and (116), must be determined.

Figure 2.3 shows the experimental setup used today in Mainz, which is a more modern version of that described in [104] or [105]. A similar experimental setup is used by Causley, Scott, and Russell [106]. Although Figure 2.3 is self-explanatory, some additional remarks should be given.

1.   The introduction of a modulated high-voltage power supply to electrooptical emission measurements [107] and today to EOAM, also, made it possible to construct brass-bodied measuring cells with less isolation problems, thus considerably simplifying shielding and making handling of the cell safer, since the cell body can be grounded. Also, the temperature control capability

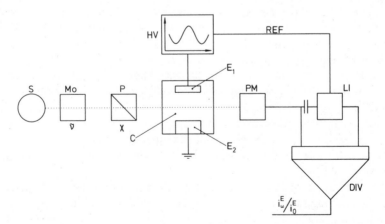

Figure 2.3 Diagram of the apparatus used for EOAM: HV, high-voltage power supply; S, light source (high-pressure xenon); Mo, monochromator; P, polarizer to adjust the angle χ; C, measuring cell with two parallel plate electrodes E_1 and E_2; PM, photomultiplier; LI, lock-in amplifier; REF, reference signal to the lock-in system; DIV, dividing module.

with all-metal cells is very important and is necessary if equilibrium systems like electron donor–acceptor complexes are investigated [10] or if high- or low-temperature experiments are performed.

2.   Photon statistics place a limit on the further sensitivity of the measuring system. With a given photon current of q photons per second and a measuring time interval τ, the noise of the measurement of $q\tau$ photons is defined as the square root of the variance σ_v^2 of the Poisson distribution that describes the photon statistics of incoherent light sources, where

$$\sigma_v = (q\tau)^{1/2} \tag{159}$$

In the considered experiment the signal is $q_\omega^E\tau$, and the noise stems from the total photon current $q_0^E\tau$. Hence, the signal-to-noise ratio (S/N) with the determination of q_ω^E is given as

$$S/N = q_\omega^E\tau(q_0^E\tau)^{-1/2} \tag{160}$$

Substituting q_ω^E from (155) yields

$$S/N = 2.3A(q_0^E\tau)^{1/2}\,2E_{a0}E_{a\omega}L_G^* \tag{161}$$

and with Lambert–Beer's law

$$S/N = 2.3A\exp(-2.3A/2)\,2E_{a0}E_{a\omega}L_G^*(q_{00}\tau)^{1/2} \tag{162}$$

Note that with all practical instruments, τ^{-1} is much lower than the bandwidth of the lock-in system. From (112) to (122) one estimates that L_G^* should be evaluated to a precision of about $10^{-20}\,\text{m}^2/\text{V}^2$ at $S/N = 1$. When a field strength of $E_{a0} = E_{a\omega} = 5 \times 10^6$ V/m and solutions with $A = 1$ are used and it is realized that the function $A[\exp(-2.3A/2)]$ has a broad maximum at $A = 2/2.3$, an estimate for the necessary number of photons to be observed $q_{00}\tau \approx 5 \times 10^{13}$ can be obtained.

A high-pressure xenon arc lamp delivers about 10^{15} photons/s·cm·sr at about $40,000\,\text{cm}^{-1}$. With 20% quantum efficiency of the photo-multiplier and 20% overall transmission of the optical path, about 4×10^{13} photons/s·cm·sr are effective for statistics.

This value determines the choice of the monochromator with respect to its light-guiding power and also sets a limit to the cell geometry. For further discussion see [104].

The calibration of the cell (field strength–electrode space, path length) is performed through use of a special standard cell described in [104].

4.8.3   Remarks on the Determination of $^at(\tilde{v})$ and $^au(\tilde{v})$

Basic to precise values for $^at(\tilde{v})$ and $^au(\tilde{v})$, defined mainly as the first and second derivatives of the absorption spectrum by (115) and (116), is a high-quality absorption spectrum. Modern high-level spectral photometers with double monochromators are adequate for this job because they have high intrinsic accuracy with the absolute calibration, reproducibility, and linearity of the wavelength-adjusting electronics and mechanics and with the reproducibility and linearity of the light-measuring logarithmic path. However, to get the best results the noise must also be kept down to some 10^{-4} absorption units, which often can be achieved only by averaging over some runs of the same spectrum. Then, in a stepwise procedure a polynomial $G(\tilde{v}_i)$ of at least third degree is fitted to an odd number m of consecutive measuring points $\{\kappa(\tilde{v}_i)/\tilde{v}_i;\ \tilde{v}_i\}$ of the absorption spectrum, and the first and second derivatives of $G(\tilde{v}_i)$ are then calculated from this polynomial at the center point of the considered m points. The entire procedure is then stepped through the whole spectrum. The number of points m used for least-squares fitting the absorption curve can be chosen arbitrarily, for example, from 5 to 51, and should be adapted to the spectral resolution that is necessary on one side and that will give the smoothest $^a t(\tilde{v})$ and $^a u(\tilde{v})$ on the other side. In this way, the choice of m will depend heavily on the system under investigation and on the spectral photometer's minimum step width with the wavenumber drive.

The accuracy of the values of $^a H$ and $^a I$ resulting from a regression analysis according to (112) depends greatly on the quality of the second derivatives, which are more difficult to determine accurately; hence, great care should be taken with the determination of the absorption spectrum. The first and second derivatives delivered automatically by high-level commercial spectral photometers are not sufficiently accurate unless a smoothing procedure, as previously described, is applied.

4.8.4   Remarks on the Solvents Used

To obtain values from EOAM with high S/N ratios, the solvent must be prepared to withstand high electric fields without breaking down. This is especially critical with polar solvents. Standard procedures [108] are used to prepare clean solvents. Whenever this procedure is applicable, solvents are also dried under reflux conditions over Na–K alloy. When this is impossible, the solvent is effectively dried by flowing through a small column filled in series with basic and neutral alumina. This procedure is followed immediately before preparing the solution. For example, the solvents given in Table 2.7, together with their relative permittivities, have been used with maximum field strengths, which are also indicated in Table 2.7.

With solvents prepared very carefully and solutions kept under dry nitrogen even higher breakdown voltages have been achieved.

4.9   Some Ground and Electronically Excited-State Dipole Moments Determined by Electrooptical Absorption Measurements

Since about 1960 EOAM has been used to determine dipole moments (and polarizabilities, too), and the values of the dipole moments reported have been evaluated from the measurements using various simplifying assumptions

Table 2.7 Solvents Already Used With EOAM

Solvent	$(E_{a0} + E_{a\omega})/(10^7 \text{ V/m})$	$\varepsilon(25°C)$
n-Hexane	1.5	1.888
n-Heptane	1.5	1.92
Cyclohexane	1.5	2.015
Methylcyclohexane	1.5	2.02
Decalin	1.5	2.18
Benzene	1.2	2.274
Dioxane	1.5	2.209
Carbon tetrachloride	1.2	2.23
Fluorobenzene	1.0	5.42
o-Dichlorobenzene	0.5	9.93
o-Dichlorotoluene	0.5	4.72
Tetrahydrofuran	0.5	7.39
Benzotrifluoride	0.7	9.035
Diisopropyl ether	0.7	3.88
n-Propyl ether	0.7	3.39
n-Butyl ether	0.6	3.06
n-Amyl ether	0.6	2.77
2-Chlorobutane	0.7	6.99
1,2-Dichloroethane	0.5	10.36

(e.g., about polarizabilities) and differently elaborate molecular models. Hence, although the values in Table 2.8 are not exactly comparable, they do show how dipole moments, especially those of electronically excited molecules, are related to their structures. In the ground state the mentioned uncertainties are small, usually less than 15% in nonpolar solvents, while in electronically excited states the amount of reaction field induced moments (caused by polarizabilities) may be much larger than 15%, thus introducing a larger uncertainty to the permanent excited-state dipole μ_a^{0FC} when it is to be calculated from μ_a^{FC} or from μ_g and $\Delta^a\mu$, even with measurements in nonpolar solvents. Refer to Section 4.7 for additional details.

5   DIPOLE MOMENTS OF MOLECULES IN THEIR GROUND AND FLUORESCENT EXCITED STATES FROM ELECTROOPTICAL EMISSION MEASUREMENTS

5.1   General Remarks

The determination of ground-state and FC excited-state dipole moments was discussed in the previous sections. It is well known that FC excited states do not necessarily show the same properties as respective equilibrium states. For example, CO_2 is a linear molecule in its ground state and so does not have a dipole moment; it is also a linear molecule in its first FC excited singlet state, but

Table 2.8 Some Dipole Moments Determined From EOAM[a]

Compound	$\mu_g^0/(10^{-30}\,C \cdot m)$	$\mu_a^{0FC}/(10^{-30}\,C \cdot m)$	Reference
4-Dimethylamino-4'-nitrostilbene	23.8	88.4	[109]
4-Amino-4'-nitrostilbene	21.7	73.4	[109]
4-Dimethylamino-4'-cyanostilbene	23.2	63.4	[109]
4-Dimethylamino-4'-nitrodiphenyl	22.0	80.0	[109]
p-Nitrodimethylaniline	22.8	50.0	[109]
p-Nitroaniline	21.0	46.7	[109]
p-Nitrosodimethylaniline	21.5	38.4	[109]
1-(4-Dimethylaminophenyl)- 2-nitroethylene	25.5	60.0	[109]
1-Nitro-3,5-diaminobenzene	16.7	39.4	[105]
3,5-Dinitroaniline	19.3	39.4	[105]
Carbazol	6.3	10.3	[105]
3,6-Dinitrocarbazol	22.0	63.4	[105]

| | 40.4 | 67.0 | [66] |

| | 37.4 | 70.7 | [66] |

Uracil	13.0	19.0	[110]
Thymine	13.0	17.5	[110]
Pyridin-N-oxide	14.0	10.0	[111]
4-Cyano-N,N-dimethylaniline	22.3	39.2 (S_2)	[112]
4-Dimethylamino-4'-nitrostilbene	27.4	64.0	[102]
$(CH_3)_2$—N—$(CH{=}CH)_x$—CH$=$C$(CN)_2$			
$\quad x = 0$	24.5		[5]
$\quad x = 1$	31.3	39.6	[5]
$\quad x = 2$	38.0	55.0	[5]
$\quad x = 3$	41.8	56.8	[5]
$\quad x = 4$	39.2	50.2	[5]
Crocetin dimethyl ester	8.0	11.3	[6]
Retinal	17.7	23.3	[7]

Table 2.8 (*continued*)

Compound	$\mu_g^0/(10^{-30}\,\mathrm{C\cdot m})$	$\mu_a^{0FC}/(10^{-30}\,\mathrm{C\cdot m})$	Reference
Complex: hexamethylbenzene–			
tetracyanoethylene (1 : 1)	7.0	19.8	[10]
(2 : 1)	0	19.8	[10]
Complex: stilbene–tetracyanoethylene	4.7	38.7	[113]
Azulene	0.8	−0.23	[114]
Azulene	0.8	−1.88	[22]

[a]Unless otherwise noted, the Onsager model has been used, and the first excited singlet state was investigated. Some ground-state dipole moments have been determined from dielectric measurements.

it is a bent molecule in the respective relaxed (equilibrium) excited state that shows a dipole moment, which is perpendicular to the equilibrium ground-state molecular axis. Changes of the dipole moment after excitation because of rotational isomerization are also discussed for double fluorescent molecules like N,N-dimethyl-4-cyanoaniline [115].

Hence, comparing dipole moments of molecules in their FC excited and in their excited equilibrium states presumably should reveal changes of molecular structure after excitation. Also, the dipole moments of molecules that are stable only in the excited state, such as exciplexes, cannot be studied by dielectric measurements or by EOAM.

In these cases electrooptical emission measurements (EOEM) can be performed when the fluorescence process is studied in an external electric field. With the fluorescence process starting from the excited equilibrium state and terminating with the FC ground state, EOEM gives a set of dipole moments in these states.

5.2 Brief Survey on the History of Electrooptical Emission Measurements

When Czekalla (University of Würzburg) started to develop EOAM, he also thought about the effect of an external electric field on the fluorescence.

It was well known that the fluorescence of crystals is generally anisotropic because of the anisotropic orientation distribution of most molecules in crystals. Thus a solution should emit anisotropic fluorescence if an anisotropic orientation distribution is imposed on that solution by an external electric field. Czekalla decided to investigate the influence of an external electric field on the degree of anisotropy of the fluorescence [76, 109, 116–118], and in 1963 Liptay [119] gave an improved theoretical foundation to this method. Presumably because of experimental difficulties, this method has not been followed further in Würzburg. The method was revived in 1966 by Weber [120] and in 1971 by Weill and Hornick [121, 122].

As experience with EOAM increased, Baumann and Deckers [123] began in 1977 developing a related method called electrooptical emission measurements (EOEM), where the influence of an external electric field on the fluorescence intensity is investigated. Independently, Groenen and van Velzen [21] began the same type of experiment. In Section 5.3 is a short discussion on the theory of EOEM, which closely follows that presented in [123]. The reader is referred to some related papers that study the fluorescence and phosphorescence of crystals and also solutions, usually with special points of view or with simplified theoretical foundations [124–129].

5.3   Theory of Electrooptical Emission Measurements

5.3.1   Partial Molar Quantities With Electrooptical Emission Measurements

Consider a solution of a species G in a pure solvent S. In a fluorescence experiment q fluorescence photons per second are observed when a dilute solution with $\sigma l \ll 1$ is studied

$$q = \eta 2.3\sigma l\phi_f \tag{163}$$

where η takes into account the fluorescence quantum efficiency and the spectral sensitivity of the photodetector as well as the overall light-guiding power of the fluorometer and the excitation intensity; l is the path length of the excitation beam in the solution; and ϕ_f is a normalized fluorescence band shape function, where

$$\int_{band} \phi_f(\tilde{v})\, d\tilde{v} = 1 \tag{164}$$

In an external electric field, q depends on the electric field caused by a field dependence of σ, ϕ_f, and η.

The quantity ^eP that may be determined primarily with EOEM is then defined as

$$^eP = \lim_{E_a^2 \to 0} \left(\frac{\partial}{\partial E_a^2} \eta^E \sigma^E \phi_f^E \right)_{n_J, T, p} \tag{165}$$

Superscript e denotes the emission process. The extensive quantity ^eY is deduced from ^eP, with the system volume designated V

$$^eY = {}^ePV \tag{166}$$

The PMQs follow

$$^eY_J = \left(\frac{\partial}{\partial n_J} {}^ePV \right)_{n_J', T, p} \tag{167}$$

In particular when only species G is excited and fluoresces,

$$^eY_G^* = \left(\frac{^eP}{c_G}\right)^*$$ (168)

where c_G is the concentration of species G.

With (163) and (165) it follows from (168) that

$$^eY_G^* = (2.3c_Gl)^{-1} \lim_{E_a^2 \to 0} \left(\frac{\partial}{\partial E_a^2} q^E\right)_{T,p}$$ (169)

Depending on the experiment, $^eY_G^*$ can be determined with polarized or unpolarized light at a given wavenumber $\tilde{v}$ with more or less optical bandwidth, or it may even be determined with an optical bandwidth large enough to cover the entire fluorescence band. The specific kind of experiment conducted depends on the experimenter and will influence the molecular model to be used, that is, the specific MMQ.

5.3.2   The Definition of the Molecular Model and the Model Molar Quantity $^ey_G^*$

The MMQ $^ey_G^*$ must be defined on the basis of a suitable molecular model. The molecular model used is the same as that introduced into EOAM by Liptay (see Section 1.3.2), because it has been proven useful for describing the observed effects and also for allowing the results to be completely comparable on a molecular model basis.

Then, with a dilute solution that has a light beam of sufficiently low intensity for fluorescence excitation (which is sufficiently low with all conventional light sources!) and also sufficiently low solution optical density at the excitation wavenumber,

$$^ey_G^* = \eta_G^*\kappa_G^*\phi_f^*(L_G^* + M_G^*)$$ (170)

M_G^* (do not confuse with relative molecular mass M_G!) is a quantity dependent on molecular parameters and is discussed in detail below, and

$$L_G^* = L_G^*(\chi_0, \tilde{v}_0)$$ (171)

is a constant for EOEM that is given by (112) at the angle χ_0 and the excitation wavenumber $\tilde{v}_0$, as both determined by the excitation light beam.

For the given system, it follows from (8) that

$$^eY_G^* = {}^ey_G^*$$ (172)

With the definition

$$^eX_G^* = L_G^* + M_G^*$$ (173)

and by combining (169) and (173),

$$\lim_{E_a^2 \to 0} \frac{1}{q^E} \left(\frac{\partial}{\partial E_a^2} q^E \right)_{T,p} = {}^eX_G^* \tag{174}$$

Baumann and Deckers [123] discussed two limiting cases for the theoretical calculation of M_G^*:

1. The excited-state lifetime of the molecule investigated is very short compared with its reorientation relaxation time and with the solvent reorganization time. The orientation distribution of the fluorescent molecules is given by their original ground-state distributions, and the solvent cage is also the same as that in the ground state. This limiting case is unrealistic for most solutions.

2. The excited-state lifetime is long enough that the solute excited species and the solvent cage can reorganize before a considerable amount of fluorescence occurs. Then because of generally different electric moments in the ground and excited states, the orientation distribution of the excited species differs from the original one in the ground state, and for the same reasons the solvent cage is also different. This limiting case is quite realistic for many solutions. Obviously, as a limiting case it will never give an exact description of the real situation, which has also been pointed out in [120]. Comparison of fluorescence decay times with dielectric relaxation times [30] and also the measurement of the degree of fluorescence anisotropy [116, 130] helps to determine whether this limiting case can be used.

5.3.3   The Quantity M_G^*

In all subsequent discussions the limiting case (case 2, Section 5.3.2) is stated to be present. Then from theory [123] it follows that M_G^* depends on φ and $\tilde{\nu}$

$$M_G^* = M_G^*(\varphi, \tilde{\nu}) \tag{175}$$

where φ is the angle between the external electric field and the electric field vector of the fluorescence light beam as selected by an analyzing polarizer, and $\tilde{\nu}$ is the wavenumber observed in the fluorescence light beam as selected by a monochromator.

From [123] it follows that

$$\begin{aligned} M_G^*(\varphi, \tilde{\nu}) = {}^eD\, r(\varphi) + ({}^eE/6)s(\varphi) \\ + [{}^eF\, r(\varphi) + {}^eG\, s(\varphi)]^e t(\tilde{\nu}) \\ + [{}^eH\, r(\varphi) + {}^eI\, s(\varphi)]^e u(\tilde{\nu}) \end{aligned} \tag{176}$$

where

$$r(\varphi) = (2 - \cos^2 \varphi)/5 \tag{177}$$

$$s(\varphi) = (3\cos^2 \varphi - 1)/5 \tag{178}$$

$$^e t(\tilde{v}) = \left(\frac{hc_0 q}{\tilde{v}^3}\right)^{-1} \left[\frac{\partial}{\partial \tilde{v}'} \frac{q}{\tilde{v}'^3}\right]_{\tilde{v}'=\tilde{v}} \tag{179}$$

$$^e u(\tilde{v}) = \left(\frac{2h^2 c^2 q}{\tilde{v}^3}\right)^{-1} \left[\frac{\partial}{\partial \tilde{v}'} \frac{q}{\tilde{v}'^3}\right]_{\tilde{v}'=\tilde{v}} \tag{180}$$

where $^e t(\tilde{v})$ and $^e u(\tilde{v})$ are accessible through the first and second derivative of a corrected fluorescence spectrum $q(\tilde{v})$ of the solution under investigation.

The terms $^e D$ to $^e I$ are given as

$$^e D = (2/kT)\mathbf{m}_e \mathbf{f}_e^2 \mathbf{m}'_e \boldsymbol{\mu}_a + \text{tr}(\mathbf{f}_e \mathbf{m}'_e)^2 \tag{181}$$

$$^e E = (1/kT)^2 [3(\mathbf{m}_e \mathbf{f}_e \boldsymbol{\mu}_a)^2 - \boldsymbol{\mu}_a \mathbf{f}_e^2 \boldsymbol{\mu}_a] + (1/kT)(3\mathbf{m}_e \mathbf{f}_e \boldsymbol{\alpha}_a \mathbf{m}_e - \text{tr} \, \mathbf{f}_e^2 \boldsymbol{\alpha}_a)$$
$$+ (6/kT)(\mathbf{m}_e \mathbf{f}_e \boldsymbol{\mu}_a \cdot \text{tr} \, \mathbf{f}_e \mathbf{m}'_e + \boldsymbol{\mu}_a \mathbf{f}_e^2 \mathbf{m}'_e \mathbf{m}_e) + 3(\text{tr} \, \mathbf{f}_e \mathbf{m}'_e)^2$$
$$+ 3\text{tr}(\mathbf{f}_e \mathbf{m}'_e)^2 \tag{182}$$

$$^e F = (1/kT)\boldsymbol{\mu}_a \mathbf{f}_e^2 \Delta^e \boldsymbol{\mu} + \tfrac{1}{2} \text{tr} \, \mathbf{f}_e^2 \Delta^e \boldsymbol{\alpha} + 2\mathbf{m}_e \mathbf{f}_e^2 \mathbf{m}'_e \Delta^e \boldsymbol{\mu} \tag{183}$$

$$^e G = (1/kT)(\mathbf{m}_e \mathbf{f}_e \boldsymbol{\mu}_a)(\mathbf{m}_e \mathbf{f}_e \Delta^e \boldsymbol{\mu}) + \tfrac{1}{2}\mathbf{m}_e \mathbf{f}_e \Delta^e \boldsymbol{\alpha} \mathbf{f}_e \mathbf{m}_e$$
$$+ \mathbf{m}_e \mathbf{f}_e \Delta^e \boldsymbol{\mu} \, \text{tr} \, \mathbf{f}_e \mathbf{m}'_e + \Delta^e \boldsymbol{\mu} \mathbf{f}_e^2 \mathbf{m}'_e \mathbf{m}_e \tag{184}$$

$$^e H = \Delta^e \boldsymbol{\mu} \mathbf{f}_e^2 \Delta^e \boldsymbol{\mu} \tag{185}$$

$$^e I = (\mathbf{m} \mathbf{f}_e \Delta^e \boldsymbol{\mu})^2 \tag{186}$$

where fluctuations of the local electric field have been completely ignored.

The following definitions have been used in (181) to (186)

$$\boldsymbol{\mu}_a = (\mathbf{I} - \mathbf{f}\boldsymbol{\alpha}_a^0)^{-1}\boldsymbol{\mu}_a^0 \tag{187}$$

$$\boldsymbol{\alpha}_a = (\mathbf{I} - \mathbf{f}\boldsymbol{\alpha}_a^0)^{-1}\boldsymbol{\alpha}_a^0 \tag{188}$$

$$\Delta^e \boldsymbol{\mu} = (\mathbf{I} - \mathbf{f}'\boldsymbol{\alpha}_g^{\text{OFC}})^{-1}(\mathbf{I} - \mathbf{f}\boldsymbol{\alpha}_g^{\text{OFC}})(\mathbf{I} - \mathbf{f}\boldsymbol{\alpha}_a^0)^{-1}(\mathbf{I} - \mathbf{f}'\boldsymbol{\alpha}_a^0)(\boldsymbol{\mu}_a - \boldsymbol{\mu}_g^{\text{FC}}) \tag{189}$$

$$\Delta^e \boldsymbol{\alpha} = (\mathbf{I} - \mathbf{f}'\boldsymbol{\alpha}_g^{\text{OFC}})^{-1}(\mathbf{I} - \mathbf{f}\boldsymbol{\alpha}_a^0)^{-2}(\mathbf{I} - \mathbf{f}'\boldsymbol{\alpha}_a^0)(\boldsymbol{\alpha}_a^0 - \boldsymbol{\alpha}_g^{\text{OFC}}) \tag{190}$$

$$\boldsymbol{\mu}_g^{\text{FC}} = (\mathbf{I} - \mathbf{f}\boldsymbol{\alpha}_g^{\text{OFC}})^{-1}\boldsymbol{\mu}_g^{\text{OFC}} \tag{191}$$

where $\boldsymbol{\mu}_a^0$ and $\boldsymbol{\alpha}_a^0$ are the permanent dipole moment and the polarizability of the isolated species under consideration in its fluorescent excited state; $\boldsymbol{\mu}_g^{\text{OFC}}$ and $\boldsymbol{\alpha}_g^{\text{OFC}}$ are the respective molecular quantities in the FC ground state reached by the fluorescence process; $\mathbf{m}_e$ is a unit vector in the direction of the transition dipole moment $\boldsymbol{\mu}_{ga}$; and $\mathbf{m}'_e$ is a standardized transition polarizability, defined analogously to $\mathbf{m}'_a$ (128).

The tensors $\mathbf{f}_e$, $\mathbf{f}$, and $\mathbf{f}'$ have already been defined. In principle, the molecular size and form are different in the equilibrated excited state from that in the ground state. However, the effect is far below the error introduced into $\mathbf{f}$ and $\mathbf{f}'$ through the estimation of a_w.

5.4 Determination of ^eD to ^eI

Combining (173) and (176) gives

$$
\begin{aligned}
^eX^*_G(\varphi, \tilde{v}) = {} & L(\chi_0, \tilde{v}_0) + {}^eD/3 + \tfrac{1}{6}({}^eE - 2{}^eD)s(\varphi) \\
& + [{}^eF\, r(\varphi) + {}^eG\, s(\varphi)]^e t(\tilde{v}) \\
& + [{}^eH\, r(\varphi) + {}^eI\, s(\varphi)]^e u(\tilde{v})
\end{aligned}
\tag{192}
$$

Thus the quantity $^eX^*_G(\varphi, \tilde{v})$ must be measured as a function of φ and $\tilde{v}$. This is done with at least two angles $\varphi = 0$ and $90°$ and with wavenumber steps of about $100\,\text{cm}^{-1}$ depending on the system being used.

For example, the spectrum of the function $^eX^*_G(\varphi, \tilde{v})$ for the molecule 4-dimethylamino-4'-nitrostilbene measured in cyclohexane is given as a function of $^e t(\tilde{v})$ with two angles φ in Figure 2.4.

The overall slope in Figure 2.4 reflects the shift of the whole fluorescence spectrum and is formally the result of the terms that depend on $^e t(\tilde{v})$, while the loops and the curvature reflect the vibrational structure and are the result of the terms depending on $^e u(\tilde{v})$. Hence, a plot of $^eX^*_G(\varphi, \tilde{v}) - [{}^eH\, r(\varphi) + {}^eI\, s(\varphi)]^e u(\tilde{v})$ versus $^e t(\tilde{v})$ should give a straight line; see the dashed lines in Figure 2.4. The intersects on the ordinate are then related to the field-induced anisotropy of the

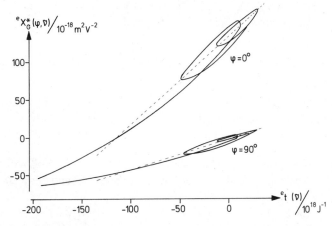

Figure 2.4 Plot of $^eX^*_G(\varphi, \tilde{v})$ of 4-dimethylamino-4'-nitrostilbene over $^e t(\tilde{v})$ for two angles.

fluorescence, described by the terms $\frac{1}{3}{}^eD + L$ and ${}^eE$, and from the slopes ${}^eF$ and ${}^eG$ they can be determined.

A multilinear regression analysis according to (192), where ${}^et(\tilde{\nu})$ and ${}^eu(\tilde{\nu})$ have been determined from the fluorescence spectrum, yields the terms $\frac{1}{3}{}^eD + L$ to ${}^eI$ as regression coefficients.

Further analysis of the molecular quantities $\frac{1}{3}{}^eD + L$ to ${}^eI$ suffers from the same problems mentioned in Section 4.5.3.

5.5   Determination of Dipole Moments

Previously only molecules with very large dipole moments in their fluorescent states have been investigated by EOEM. This is mainly because of the good S/N ratio that normally is achieved with such molecules. Then the polarizability and the transition polarizability terms can be considered negligibly small as compared to the dipole moment terms in (181) to (185). With these assumptions the following approximative terms ${}^eE_0$ to ${}^eI_0$ result from EOEM

$$
{}^eE_0 = (1/kT)^2[3(\mathbf{m}_e\mathbf{f}_e\boldsymbol{\mu}_a)^2 - \boldsymbol{\mu}_a\mathbf{f}_e^2\boldsymbol{\mu}_a] \tag{193}
$$

$$
{}^eF_0 = (1/kT)\boldsymbol{\mu}_a\mathbf{f}_e^2\Delta^e\boldsymbol{\mu} \tag{194}
$$

$$
{}^eG_0 = (1/kT)(\mathbf{m}_e\mathbf{f}_e\boldsymbol{\mu}_a)(\mathbf{m}_e\mathbf{f}_e\Delta^e\boldsymbol{\mu}) \tag{195}
$$

$$
{}^eH_0 = (\mathbf{f}_e\Delta^e\boldsymbol{\mu})^2 \tag{196}
$$

$$
{}^eI_0 = (\mathbf{m}_e\mathbf{f}_e\Delta^e\boldsymbol{\mu})^2 \tag{197}
$$

From this set of quantities μ_a; $\Delta^e\boldsymbol{\mu}$; $\not\prec (\boldsymbol{\mu}_a, \Delta^e\boldsymbol{\mu})$; $\not\prec (\mathbf{m}_e, \Delta^e\boldsymbol{\mu})$; and $\not\prec (\mathbf{m}_e, \boldsymbol{\mu}_a)$ can be determined. As with EOAM, ${}^eH$ and ${}^eI$ often cannot be determined with sufficient accuracy. Further information may be deduced from symmetry, as described in Section 4.5.2. Then with $\mathbf{m}_e \| \boldsymbol{\mu}_a \| \Delta^e\boldsymbol{\mu}$, for example, μ_a and $\Delta^e\mu$ can be calculated from the three quantities ${}^eE$, ${}^eF$, and ${}^eG$ only.

5.6   Evaluation of the Permanent Dipole Moment of the Free Molecule

Let $\boldsymbol{\mu}_a$ and $\Delta^e\boldsymbol{\mu}$, which are defined by (187) and (189), be given from an EOEM measurement. It is apparent from (187), (189), and (191) that μ_a^0 and μ_g^{0FC} can only be calculated if terms like $f\alpha_a^0$ are either known or can be estimated reliably.

Regrettably, α_a^0 is unknown, and α_g^{0FC} cannot be taken as α_g^0 without care. Hence, not even μ_a^0 can be determined without assumptions or further information. As an upper limit for μ_a^0, the value μ_a determined in nonpolar solvents can be used, where induced parts are relatively small. As a general method, that given in Section 4.7.2 is applicable with EOEM, too. For this purpose (187) and (189) are rewritten to give

$$\frac{1}{\mu_a} = \frac{1}{\mu_a^0} - \frac{1}{4\pi\varepsilon_0} \frac{1}{\mu_a^0} \frac{\alpha_{azz}^0}{a_x a_y a_z} g_z \tag{198}$$

$$^eS = \mu_g^{OFC} + \frac{1}{4\pi\varepsilon_0} \frac{\alpha_{gzz}^{OFC}}{a_x a_y a_z} {}^eT \tag{199}$$

where

$$^eS = \Delta^e\mu \frac{1 - f_z\alpha_{azz}^0}{1 - f_z'\alpha_{azz}^0} + \mu_a \tag{200}$$

$$^eT = \Delta^e\mu \frac{1 - f_z\alpha_{azz}^0}{1 - f_z'\alpha_{azz}^0} g_z' + \mu_a g_z \tag{201}$$

and where $\mu_a \parallel \Delta^e\mu$ has been used, per definition in the z direction, which is valid with most molecules showing large internal charge transfer. Therefore, from a linear regression of $1/\mu_a$ determined in different solvents on g_z, $\alpha_{azz}^0/(a_x a_y a_z)$ and μ_a^0 result; and from another regression of ^eS on ^eT the results are $\alpha_{gzz}^{OFC}/(a_x a_y a_z)$ and μ_g^{OFC}, where $(1 - f_z\alpha_{azz}^0)(1 - f_z'\alpha_{azz}^0)^{-1}$ is determined using the results from the first regression according to (198). Changes of g_z caused by changes of the molecular geometry on excitation are too small to influence results noticeably.

This method has not been verified with standard EOEM pending the development of integral electrooptical emission measurements (see Section 6) because: (1) standard EOEMs in different polar solvents require much time, and (2) some experimental feeling for the maximum applicable field strengths and the drying procedures had to be developed first.

5.7 Experimental Method and Setup

5.7.1 Method to Determine Values for $^eX_G^*(\varphi, \tilde{\nu})$

The experimental method for the determination of $^eX_G^*(\varphi, \tilde{\nu})$ has been described in [131]. In a solution of the solute G that is irradiated by an excitation source, q fluorescence quanta per second are observed. If an external electric field is applied to this solution, the number of fluorescence quanta q^E is dependent on the electric field strength. Then develop q^E to powers of E_a^2

$$q^E = q^E\bigg|_{E_a^2=0} + \frac{\partial}{\partial E_a^2} q^E\bigg|_{E_a^2=0} E_a^2 + \frac{1}{2} \frac{\partial^2}{(\partial E_a^2)^2} q^E\bigg|_{EE_a^2=0} E_a^4 + \cdots \tag{202}$$

The external electric field is modulated according to (146). When E_a^2 from (146) is introduced into (202), it is shown that q^E has a component q_0^E that is unmodulated, a component with amplitude q_ω^E that is modulated with frequency ω, and higher harmonics. It is

$$q_0^E = \lim_{E_a^2 \to 0} q^E \tag{203}$$

$$q_\omega^E = 2E_{a0}E_{a\omega} \lim_{E_a^2 \to 0} \frac{\partial}{\partial E_a^2} q^E \tag{204}$$

From the quotient of (203) and (204) follows:

$$\lim_{E_a^2 \to 0} \left(\frac{1}{q^E} \frac{\partial}{\partial E_a^2} q^E \right) = \frac{1}{2E_{a0}E_{a\omega}} \frac{q_\omega^E}{q_0^E} \tag{205}$$

and with (174)

$$^e X_G^* = L_G^* + M_G^* = \frac{1}{2E_{a0}E_{a\omega}} \frac{q_\omega^E}{q_0^E} \tag{206}$$

According to this equation M_G^* can be determined if L_G^* is known from a separate EOAM experiment. When using a photomultiplier, anode currents i_ω^E and i_0^E correspond to the photon currents q^E. Then

$$^e X_G^* = L_G^* + M_G^* = \frac{1}{2E_{a0}E_{a\omega}} \frac{i_\omega^E}{i_0^E} \tag{207}$$

can be determined irrespective of variations in the excitation intensity and independent of photomultiplier sensitivity.

5.7.2 Experimental Setup to Determine i_ω^E / i_0^E

Equation (207) precisely defines the demands made on the experimental setup. From (192) it follows that a set of values $^e X_G^*(\varphi, \tilde{\nu})$ must be determined as a function of φ and $\tilde{\nu}$.

A standard fluorescence spectral photometer may be used, with high intensity excitation sources like high-pressure mercury lamps at the 500-W level and with provision to vary the angle φ using an analyzing polarizer. Obviously, the fluorescence cell must be constructed so that the electric field is perpendicular to the fluorescence light beam; otherwise, φ could not be varied (with respect to the applied field!).

A practical cell design has been given in [131]. It is a brass-body cell with the advantages discussed in Section 4.8.2. Figure 2.5 shows the cell used currently with EOEM in Mainz.

From an independent experiment with a precision quanta-corrected fluorescence spectral fluorometer, the terms $^e t(\tilde{\nu})$ and $^e u(\tilde{\nu})$ are determined as the

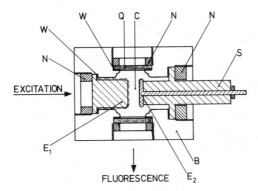

Figure 2.5 An all-brass measuring cell used with EOEM: B, brass body; Q, quartz glass window; N, ring nut; W, Teflon washer; S, quartz or ceramic stud; E_2, hot electrode; E_1, grounded electrode, quartz covered with SnO_2; C, sample compartment.

derivatives of a relative, but quanta-corrected, fluorescence spectrum. This experiment is crucial because of the transmission characteristic of the fluorescence channel. Standard procedures [132, 133] may be used, where for a standard emission spectrum an incandescent lamp with an iodine cycle may be used, such as an EGG-type 597-1 calibrated to an NBS standard.

Figure 2.6 shows the experimental setup for the EOEM experiment, which is an updated version of that shown in [131]. This diagram is self-explanatory; as a transmittant electrode material, SnO_2 is used on a quartz rod.

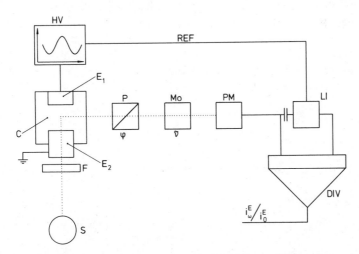

Figure 2.6 Diagram of the apparatus used for EOEM: S, source (high-pressure mercury); F, excitation filter; C, cell; E_1, hot electrode; E_2, grounded electrode, light transmitting; P, polarizer; Mo, monochromator; PM, photomultiplier; LI, lock-in amplifier; DIV, dividing module; REF, reference signal to the lock-in system; HV, high-voltage power supply.

5.7.3   Additional Remarks

It is unnecessary to repeat discussion of signal-to-noise ratio optimization. The ideas presented in Section 4.8.2 may be used with some additional modification for the noise analysis of EOEM, or [5] may be referenced.

The determination of $^e t(\tilde{v})$ and $^e u(\tilde{v})$ is accomplished by simply transforming the procedure discussed in Section 4.8.3 to fluorescence spectra.

Obviously the solvents used must show the same degree of purity, especially with respect to water, to avoid electric breakdown, see Section 4.8.4. In addition, to maintain the validity of (172), the solvents are not allowed to fluoresce or to contain impurities that are fluorescent in the spectral region under investigation.

5.8   Comparison of Results From Electrooptical Absorption and Emission Measurements

In nonpolar solvents for which $f \approx f'$, from (125) and (189), it follows that

$$\mu_a^{FC} = \mu_g + \Delta^a \mu \tag{208}$$

$$\mu_g^{FC} = \mu_a - \Delta^e \mu \tag{209}$$

With rigid molecules, μ_a determined from EOEM should be equal to μ_a^{FC} determined from EOAM, as should μ_g and μ_g^{FC}.

This has been proven with 4-dimethylamino-4'-nitrostilbene [102], and the results are shown in Table 2.9.

The results from EOAM, EOEM, and dielectric investigations agree very well. They also fit the result obtained from a time-resolved microwave conductivity technique reported by de Haas and Warman [134]. This means that on the basis of the primary molecular model the description of the effect is well elaborated. Also, it is a slight indication that the actual lifetime of this molecule in cyclohexane is much larger than all reorientation relaxation times.

Table 2.9 Comparison of Results From EOAM and EOEM on 4-Dimethylamino-4'-nitrostilbene in Cyclohexane[a]

$\mu_a/(10^{-30}$ C·m)	88.2 ± 0.7
$\mu_a^{FC}/(10^{-30}$ C·m)	89.5 ± 2
$\mu_g/(10^{-30}$ C·m)	$26.7 \pm 0.8 \ (26.3 \pm 0.4)^b$
$\mu_g^{FC}/(10^{-30}$ C·m)	27.8 ± 5.7

[a]Add about 3% error because of calibration.
[b]Value from dielectric measurements.

6 DIPOLE MOMENTS OF MOLECULES IN THEIR FLUORESCENT EXCITED STATES FROM INTEGRAL ELECTROOPTICAL EMISSION MEASUREMENTS

6.1 General Remarks

Standard electrooptical emission measurements (EOEM) have been proven to give reliable results on solution systems. As mentioned previously there are two severe drawbacks. To achieve a good S/N ratio, especially with molecules showing low quantum efficiencies, long measuring periods are to be applied as well as high excitation photon currents, and photochemistry problems may arise from both. This limits considerably the practical applicability range of the method. For example, the investigation of complex systems like the study of the concentration dependence of the term ^eD to ^eI with exciplexes has been observed but could not be evaluated quantitatively because of photochemical degradation. Hence, only preliminary results have been presented [20]; however, they are in good agreement with those obtained by Groenen and van Velzen [21] from similar measurements.

As another example, investigations into the role played by polarizabilities in the term μ_a of those molecules that show twisted intramolecular charge-transfer (TICT) state formation require results from EOEM in various solvents to perform an evaluation procedure as in Section 5.6. For the measurement of molecules with solvent-dependent quantum efficiencies of the fluorescence of about 0.1 or less, long measuring periods are required; however, rapid photochemical reactions in some solvents may occur unless the solution is renewed frequently or a flow-through cell is used. Therefore the need arose for some modification of EOEM to meet the requirement of short measuring times and less irradiation intensity. Starting from this point, integral electrooptical emission measurements (IEOEM) have been developed [107].

6.2 Theory With Integral Electrooptical Emission Measurements

The aforementioned demands may only be fulfilled if the entire fluorescence band is studied integrally, whereby the integration must be done optically; that is, the total fluorescence intensity of a fluorescence band must be measured as dependent on an external electric field. This method [107] is called *integral electrooptical emission measurement* (IEOEM).

Let the system and the premises for the system under investigation be the same as discussed in Section 5.3.2, case 2. Then a similar PMQ $^iY_G^*$ is basic to the experiment, but other MMQs must first be defined to meet the intention of the experiment. Naturally, this will place demands on the experiment and thus on the PMQ $^iY_G^*$.

According to the model used with IEOEM, the MMQ $^iy_G^*$ describing the IEOEM experiment is the integral of the MMQ $^ey_G^*$ of the standard EOEM experiment over the entire fluorescence band. With (167) it follows that

$$^{i}y_{G}^{*} = \int_{band} {}^{e}y_{G}^{*}\,d\tilde{v} = \kappa_{G}^{*}\left[\int_{band} \phi_{f}^{*}\eta_{G}^{*}L_{G}^{*}\,d\tilde{v} + \int_{band} \phi_{f}^{*}\eta_{G}^{*}M_{G}^{*}\,d\tilde{v}\right] \qquad (210)$$

Integration has been done in [107] assuming η_{G}^{*} independent from $\tilde{v}$. In a recent paper [135] the integration has been done for wavenumber dependent η_{G}^{*}. Then

$$^{i}y_{G}^{*} = \kappa_{G}^{*}\int \eta_{G}^{*}\phi_{f}^{*}\,d\tilde{v}(L_{G}^{*} + {}^{i}M_{G}^{*}) \qquad (211)$$

where

$$^{i}M_{G}^{*} = \tfrac{1}{3}{}^{i}D + \tfrac{1}{6}{}^{i}E\,s(\varphi) \qquad (212)$$

and

$$^{i}D = {}^{e}D + 15\,{}^{e}F\,{}^{i}t \qquad (213)$$

$$^{i}E = {}^{e}E - 2\,{}^{e}D + 30(3\,{}^{e}G - {}^{e}F)^{i}t \qquad (214)$$

The terms ${}^{e}D$, ${}^{e}E$, ${}^{e}F$, ${}^{e}G$ have already been defined in (181)–(184), and $s(\varphi)$ is given in (178).

$$^{i}t = \int \phi_{f}^{*}\eta_{G}^{*}\,{}^{e}t(\tilde{v})\,d\tilde{v}\bigg/\int \phi_{f}^{*}\eta_{G}^{*}\,d\tilde{v}$$

$$= \int q(\tilde{v})^{e}t(\tilde{v})\,d\tilde{v}\bigg/\int q(\tilde{v})\,d\tilde{v} \qquad (215)$$

The respective PMQ ${}^{i}Y_{G}^{*}$ is given by integration of ${}^{e}Y_{G}^{*}$ over all wavelengths

$$^{i}Y_{G}^{*} = \int {}^{e}Y_{G}^{*}\,d\tilde{v} \qquad (216)$$

and with (169)

$$^{i}Y_{G}^{*} = \frac{1}{2.3c_{G}l E_{a}^{2}}\lim_{E_{a}^{2}\to 0}\frac{\partial}{\partial E_{a}^{2}}\int q^{E}\,d\tilde{v} \qquad (217)$$

With (211) and (172) and with the definition

$$^{i}X_{G}^{*} = L_{G}^{*} + {}^{i}M_{G}^{*} \qquad (218)$$

follows:

$$\lim_{E_a^2 \to 0} \frac{\partial}{\partial E_a^2} \int q^E \, d\tilde{\nu} = 2.3 c_G l \kappa_G^* \int \eta_G^* \phi_f^* \, d\tilde{\nu}^i X_G^*$$

$$= {}^i X_G^* \int q \, d\tilde{\nu} \tag{219}$$

Hence

$$\lim_{E_a^2 \to 0} \frac{1}{\int q^E \, d\tilde{\nu}} \frac{\partial}{\partial E_a^2} \int q^E \, d\tilde{\nu} = {}^i X_G^* \tag{220}$$

6.3   Evaluation of Experimental Data

According to (218) with (212) ${}^i X_G^*$ is a function of φ at constant excitation parameters χ_0 and $\tilde{\nu}_0$

$$^i X_G^* = {}^i X_G^*(\varphi) \tag{221}$$

Thus with the experiment

$$\lim_{E_a^2 \to 0} \frac{\partial}{\partial E_a^2} \int q^E(\tilde{\nu}) \, d\tilde{\nu} \bigg/ \int q \, d\tilde{\nu}$$

must be determined as a function of φ to give ${}^i X_G^*(\varphi)$.

$$^i X_G^*(\varphi) = \tfrac{1}{3}(3 L_G^* + {}^i D) + (\tfrac{1}{6}{}^i E) s(\varphi) \tag{222}$$

Consider the following linear combination $X3$ defined as

$$X3 = 6^i X_G^*(\varphi = 0°) - 18^i X_G^*(\varphi = 90°) \tag{223}$$

From (222), (213), and (214) it follows that:

$$X3 = {}^e E - 6^e D - 12 L_G^* + 30(3^e G - {}^e F)^i t - 60^e F \, {}^i t \tag{224}$$

Then the term dependent on ${}^i t$ can be neglected compared to ${}^e E - 6^e D - 12 L_G^*$ if

1. ${}^i t$ is small enough, which, according to (215), implies that the photo-multiplier response is flat and ϕ_f is field independent.
2. ${}^e G = {}^e F$, which is the case with polar fluorescent molecules that show

$m_e \| \mu_a \| \Delta^e\mu$. This is a restriction, but with many such compounds this symmetry condition is fulfilled at least in good approximation.

With at least one of the preceding assumptions valid, it follows with (182) and (214) that

$$(\zeta_e \mu_a)^2 = X\ 3\ (kT)^2/2 \tag{225}$$

if $\mu_a^2/(kT)^2$ is sufficiently larger than L_G^* and all polarizability terms in (182). Further evaluation of $f_e \mu_a$ with respect to μ_a^0 can be done according to (198) if μ_a has been determined in various solvents with different polarities. First results have been published in [107], where a sufficiently flat detector response has yet been assumed; and a comparison of results from IEOEM on the molecules 9-cyanoanthracene (9-CA) and 2,6,N,N-tetramethyl-4-cyanoaniline in various solvents with those from IEOEM performed in the gas phase is given in [29] and shows the power of the method and also the applicability of the Onsager model. The original results from [29] are reproduced in Table 2.10 for the molecule 9-CA. For demonstration, these results have been recalculated with the Lorentz field and are also represented in Table 2.10.

   The agreement between values for the excited-state dipole moment extrapolated from solution data using the spherical Onsager model and the gas-phase values is very good. Also the value found for α_a^0/a_w^3 appears reasonable and compares well to the respective ground-state quantity reported by Kriebel and Labhart [28]. Note that dioxane behaves like a solvent with $\varepsilon \approx 5$!

   Although no reasonable values for μ_a ($=\mu_a^0$) are found using the Lorentz

Table 2.10 Evaluation of Excited-State Dipole Moments of 9-CA From IEOEM Results in Various Solvents and in the Gas Phase[a]

	Onsager Model	Lorentz Model
	$\mu_a/(10^{-30}\ C\cdot m)$	$\mu_a/(10^{-30}\ C\cdot m)$
Gas phase	17.4 ± 1	
Heptane	19.4 ± 0.1	16.7
Decalin	19.9 ± 0.1	17.4
Di-n-amyl ether	20.8 ± 0.1	16.6
Di-n-butyl ether	21.0 ± 0.1	16.1
Diisopropyl ether	21.3 ± 0.1	14.4
Dioxane	22.4 ± 0.1	19.5
Fluorobenzene	22.6 ± 0.1	12.5
Trifluorotoluene	23.5 ± 0.1	9.1
[b]$\mu_a^0/(10^{-30}\ C\cdot m)$	17.3 ± 0.2	
$[\alpha_a^0/(4\pi\varepsilon_0 a_w^3)]/(10^{-10}\ C\ V^{-1}\ m^{-1})$	0.30 ± 0.06	

[a]Adapted from [29]. Errors are statistical. Add about 4% for calibration errors.
[b]Results after evaluation of tabulated information.

model for polar solvents, acceptable agreement with the gas-phase value in nonpolar solvents is observed.

In addition the theory of IEOEM has been extended recently to fourth-order effects in the electric field [135]. With the molecule 4-dimethylamino-4'-nitrostilbene the dipole moment μ_a has been determined from the (very small) fourth-order effect to be within 10% agreement with the second-order effect. This result shows that the used molecular model is very well suited for the description of the electric field effect on the fluorescence of solute molecules.

6.4   Experimental Setup

The standard experimental setup for IEOEM (Figure 2.7) is very similar to that described for EOEM (Figure 2.6) and differs primarily because of the lack of a monochromator and the addition of a cut-off filter to reduce stray light.

Equation (207) applies

$$^iX_G^*(\varphi) = \frac{1}{2E_{a0}E_{a\omega}} \frac{i_\omega^E}{i_0^E} \tag{226}$$

where the photocurrents i_ω^E and i_0^E result from the entire fluorescence spectrum.

Generally, it is desirable to have ^it as small as possible. This means η_G^* should be nearly independent of $\tilde{v}$ in the wavelength region of interest. Hence, a photodetector should be used with flat quantum response. Silicon photodiodes are the ideal choice, but as they do not show an inherent current amplification,

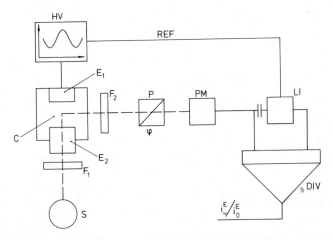

Figure 2.7   Diagram of the apparatus used for IEOEM: S, source (high-pressure mercury); F_1, excitation filter; C, measuring cell; E_1, hot electrode; E_2, grounded electrode, light transmitting; F_2, cutoff filter; P, polarizer; PM, photomultiplier; HV, high-voltage power supply; LI, lock-in amplifier; REF, reference signal to the lock-in system; DIV, dividing module.

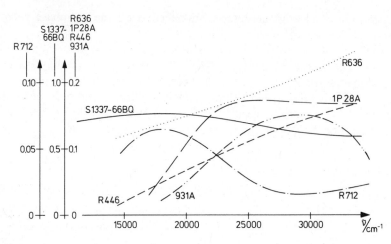

Figure 2.8 Quantum efficiency η_e of the photocathode of some photomultipliers and a silicon photodiode.

high-impedance, current-to-voltage converters must be used that are most sensitive to spurious signals from the high-voltage source.

GaAs-type photocathodes are fairly good choices, and depending on the spectral range used, other cathode materials can be used also. Figure 2.8 shows a plot of the quantum efficiency of various cathode materials over the wavenumber. A GaAs cathode, such as the R636, or above 23,000 cm^{-1}, the 1P28, is the best choice if a silicon photodiode cannot be used, although it shows a nearly linear dependence of the quantum efficiency on the wavenumber in the range of interest here. To assure that the quantum efficiency's deviations from flatness do not cause great error, the IEOEM experiment can be repeated with cut-off filters that cut part of the fluorescence spectrum. In all cases studied to date, the data obtained are relatively insensitive to such manipulations. Also, comparison of data obtained from EOEM and from IEOEM performed with a model R636 photomultiplier and with a model R446 photomultiplier (both Hamamatsu) show perfect agreement for model R636 with only slight deviations for model R446; and IEOEMs with R636, R446, 1P22, and 1P28 have been compared with respect to the terms connected with $^i t$ according to (213) and (214) where quantitative agreement between theory and experiment has been found [136].

By some modification of the experimental setup a direct reading dipolemeter for the dipole moments of excited states could be built up with some assumptions using IEOEM [107]. However, when methods involve complicated molecular models and also assumptions, even standard assumptions, the PMQs, MMQs, and assumptions should not be mixed; instead, determine PMQs or some simply related, well-defined quantities that may then be treated further, naturally also using assumptions.

6.5   Some Representative Results From Integral Electrooptical Emission Measurements

Integral electrooptical emission measurements almost exclusively depend on the orientating effect of the applied electric field, since only a small residual band shift remains, manifested in the terms coupled to $^i t$.

This makes the basic model most simple and hence reliable. Table 2.11 shows some results from IEOEM taken from [107] and [136]. Previous values have been recalculated from the newer and more elaborate theory presented in [136] and in this chapter. The EOEM results, when available, are given for comparison, as well as other literature values.

Comparison of the values in the first and second rows of Table 2.11 demonstrates the influence of the terms coupled to $^i t$ in (215) and gives some feeling for errors introduced by insufficient models. However, a 10% relative accuracy is often sufficient for many chemistry-related problems, for example, as when comparing the two fluorescent states in molecules like N,N-dimethyl-amino-4-cyanoaniline [137].

Table 2.11 Some Dipole Moments μ_a Determined by IEOEM and by Other Methods[a]

	Reference					
	[107][b]	[107][c]	[136]	[116]	[137][d]	[134]
4-Dimethylamino-4'-nitrostilbene	96.9	106.2	103.6	96.9		77.4
2-Amino-7-nitrofluorene	82.0	87.6	85.9	75.2		
4-Amino-4'-nitrobiphenyl	87.1	91.5		85.4		
4-Dimethylamino-4'-nitrobiphenyl	89.6	99.7		90.2		
4-Amino-4'-cyanobiphenyl	59.7	61.7		61.6		
1-(4-Dimethylaminophenyl)-2-nitroethylene	67.3	64.7		68.1		
4-(9-Anthryl)-3,5,N,N-tetramethylaniline	59.7	64.1			64.2	

[a]Given in $(10^{-30}\,\mathrm{C\cdot m})$. Statistical errors are about $2 \times 10^{-30}\,\mathrm{C\cdot m}$; add about 4% error because of calibration errors.
[b]From IEOEM, original values.
[c]From IEOEM, recalculated after Baumann and Bischof [136].
[d]From EOEM.

7 DIPOLE MOMENTS OF MOLECULES IN EXCITED STATES FROM SOLVENT-SHIFT MEASUREMENTS

7.1 History and General Remarks

The determination of excited-states dipole moments of large polar molecules from solvent-shift measurements was developed mainly between 1955 and 1965. Earlier, experimenters had realized that the wavenumber of the absorption maximum of a solute depends on the solvent. Starting from this observation, two lines of treatment of the experimental material became evident.

1. The influence of the solvent on the absorption spectra is described by empirical polarity parameters, for example, by Kosower's Z value [138–140] or by the $E_T(30)$ value of Dimroth and colleagues [141].
2. The influence of a solvent on the wavenumber of electronic transitions is described by a suitable model that relates molecular quantities to well-defined solvent parameters, like the relative permittivity and the refractive index.

The empirical way has merely a qualitative and macroscopically descriptive nature. Because this chapter is concerned with the determination of molecular quantities, only the line of history that relates the experimentally observed solvent dependence of the wavenumber of electronic transitions to molecular electric moments is discussed.

In 1948 Coggeshall and Lang [142] noticed that the change of the dipole moment with excitation plays an important role with the solvent-induced spectral shift. But it was only some years later that the first determinations of dipole moments of molecules in excited states were published by Lippert [143] and by Mataga and colleagues [144]. Quantitative theories were first developed in 1954 by Bayliss and McRae [145] and Ooshika [146]; other papers on this topic were then presented by McRae [147], Lippert [148], Bakshiev [149], Bilot and Kawski [150], and Liptay [151, 152]. The approaches used have been classical electrostatics or quantum mechanical perturbation treatments, or mixtures of both.

Liptay [151, 152] gave the most complete theoretical treatment of the topic by classical electrostatics as well as by quantum mechanical perturbation. He discusses very thoroughly the differences among the various theoretical formulas given in the literature up to that time, the differences resulting mainly from the neglects or the assumptions of some authors. Later, Liptay and Walz [53] also studied the fluctuation terms of the local field. Other papers are Basu [153]; Abe and colleagues [154–156]; Baur and Nicol [157]; Nicol and colleagues [158]; Suppan [159, 160]; Mazurenko and Bakshiev [161]; Amos and Burrows [162], who gave a broad review of earlier literature; and Fischer-Hjalmars and colleagues [163], who presented a different approach. There are many experimental papers based on these theoretical papers.

Almost all cited papers describe the solvent–solute interaction by the interaction of the total (permanent plus induced) dipole moment of the solute with the local electric field, which is described by Onsager's reaction field. Since this procedure uses the same basic well-defined model as the methods discussed in the foregoing sections of this chapter, it is also used here. Thus the results of the method can be compared and further treated and mixed, always on the same well-established molecular model basis. For additional details the reader is referred to Liptay [151, 152].

7.2 Theory of Solvent-Shift Measurements

7.2.1 General Remarks

In the preceding sections the clear separation of experimentally accessible quantities and model quantities, which are dependent on molecular quantities as well as on the physical model, has been stressed. So, too, will this be stressed when we consider the system to be studied and define the primarily determined experimental quantity.

Consider a solution of solute molecules G and solvent molecules S in which, for the sake of simplicity, only molecules G may absorb or fluoresce. The primarily observed quantity with solvent-shift measurements is defined as the wavenumber $\tilde{v}_{a,s}$ corresponding to the maximum of the absorption spectrum $\kappa_S(\tilde{v})/\tilde{v}$ of species G in a solvent S, where $\kappa_S(\tilde{v})$ is the absorption coefficient of molecules G in a solvent S at the wavenumber $\tilde{v}$. By analogy the respective quantity $\tilde{v}_{e,s}$ with solvent-shift measurements of the fluorescence is defined as that wavenumber corresponding to the maximum of the fluorescence quantum spectrum $\phi_S(\tilde{v})/\tilde{v}^3$, where $\phi_S(\tilde{v})$ is measured as quanta per second and wavenumber interval. Use $\kappa(\tilde{v})/\tilde{v}$ instead of $\kappa(\tilde{v})$ and $\phi_S(\tilde{v})/\tilde{v}^3$ instead of the normalized fluorescence spectrum $\phi_S(\tilde{v})$, since these quantities are proportional to the Franck–Condon envelope on the basis of the square of the respective transition moment.

In Section 7.2.2, two models are briefly described that relate molecular quantities to the solvent dependence of the wavenumber of maximum absorption or fluorescence.

7.2.2 Approach by Classical Electrostatics

Consider the energy W_{gS} of a molecule in its ground state g, solved in a solvent S. The correspondent energy of the free molecule is W_{g0}.

Assume that only classical electrostatic interactions and dispersion interactions must be taken into account. Particularly, specific solvent–solute interactions, such as hydrogen bonding or complex formation, or solute–solute interactions, such as dimer formation, are not considered.

Then W_{gS} is represented by a sum of energy terms that resembles the steps of the physical model for introducing a solute into a solution

$$W_{gS} = W_{g0} + W_{gD} + W_{gL} + W_{gC} + W_{gP} + W_{gR} \qquad (227)$$

where W_{gD} is a dispersion interaction term; W_{gL} is representative of librational interactions; W_{gC} is the energy required for formation of a cavity in the solvent to accept the solute; W_{gP} is the energy required to polarize the solvent around the cavity correspondent to the total dipole moment $\mathbf{p}_{gG}$ of the solute in the center of the cavity. This polarization causes a reaction field $\mathbf{E}_{Rg}$ in the cavity; and W_{gR} is the interaction energy of the total dipole moment of the solute in the cavity with the reaction field $\mathbf{E}_{Rg}$.

Then, using pure electrostatics

$$W_{gS} = W_{g0} + W_{gD} + W_{gL} + W_{gC} + \tfrac{1}{2}\mathbf{p}_{gG}\mathbf{E}_{Rg} - \boldsymbol{\mu}_g^0\mathbf{E}_{Rg} - \tfrac{1}{2}\mathbf{E}_{Rg}\boldsymbol{\alpha}_g^0\mathbf{E}_{Rg} \quad (228)$$

With the Onsager model as discussed in Section 3.4.2,

$$W_{gS} = W_{g0} + W_{gD} + W_{gL} + W_{gC} - \tfrac{1}{2}\boldsymbol{\mu}_g^0\mathbf{f}(\mathbf{I} - \mathbf{f}\boldsymbol{\alpha}_g^0)^{-1}\boldsymbol{\mu}_g^0 \quad (229)$$

For consistency, $\mathbf{f}$ is written in its general form as a tensor, although there are only very few applications of solvent-shift measurements where the evaluation procedure employs an ellipsoidal approximation. The principal axes of all tensors involved are assumed to be parallel. The spherical approximation is generally used in literature, and it is adequate at the present level of accuracy. In principle, this procedure can be used to evaluate the free energy W_{aS} of the solute in a corresponding FC excited state

$$W_{aS} = W_{a0} + W_{aD} + W_{aC} + W_{aL} + W_{aP} + W_{aR} \quad (230)$$

The discussion of W_{aP} and W_{aR} is much more complicated for an FC excited state than for an (equilibrium) ground state because all effects on orientation polarization are caused by ground-state properties, whereas those dependent on the electronic polarization are different. The result then is [151, 152]:

$$\begin{aligned}
W_{aS} =\; & W_{a0} + W_{aD} + W_{aL} + W_{aC} - \tfrac{1}{2}\boldsymbol{\mu}_a^{OFC}\mathbf{f}'(\mathbf{I} - \mathbf{f}'\boldsymbol{\alpha}_a^{OFC})^{-1}\boldsymbol{\mu}_a^{OFC} \\
& - \boldsymbol{\mu}_a^{OFC}(\mathbf{f} - \mathbf{f}')(\mathbf{I} - \mathbf{f}'\boldsymbol{\alpha}_a^{OFC})^{-1}(\mathbf{I} - \mathbf{f}\boldsymbol{\alpha}_g^0)^{-1}\boldsymbol{\mu}_g^0 \\
& + \tfrac{1}{2}\boldsymbol{\mu}_g^0(\mathbf{f} - \mathbf{f}')(\mathbf{I} - \mathbf{f}'\boldsymbol{\alpha}_a^{OFC})^{-1}(\mathbf{I} - \mathbf{f}\boldsymbol{\alpha}_g^0)^{-2}(\mathbf{I} - \mathbf{f}\boldsymbol{\alpha}_a^{OFC})\boldsymbol{\mu}_g^0
\end{aligned} \quad (231)$$

Hence, the energy difference $\Delta W_{ag,S}$ between the considered FC excited state and the equilibrium ground state g in a solvent S is given as

$$\Delta W_{agS} = W_{aS} - W_{gS} \quad (232)$$

It is usually assumed that

$$W_{aC} = W_{gC} \quad (233)$$

$$W_{aL} = W_{gL} \quad (234)$$

and it is defined

$$W_{a0} - W_{g0} = \Delta W_{ag0} \tag{235}$$

and from a quantum mechanical treatment [151, 152] it follows in good approximation that

$$W_{aD} - W_{gD} = -hc_0 Df' \tag{236}$$

where $hc_0 D$ stands for a dispersion interaction term given by (93) in [151]. Then, with (230) and (229),

$$\Delta W_{agS} = \Delta W_{ag0} - hc_0 Df' + \tfrac{1}{2}(\mu_a^{OFC} - \mu_g^0)F'(I - f'\alpha_a^{OFC})^{-1}(\mu_a^{OFC} - \mu_g^0)$$
$$- (\mu_a^{OFC} - \mu_g^0)f(I - f\alpha_g^0)^{-1}\mu_g^0 - \mu_g^0 f(I - f'\alpha_a^{OFC})^{-1}(I - f\alpha_g^0)^{-2}$$
$$\times (\alpha_a^{OFC} - \alpha_g^0)[\tfrac{1}{2}f(I - f'\alpha_g^0)\mu_g^0 + f'(I - f\alpha_g^0)(\mu_a^{OFC} - \mu_g^0)] \tag{237}$$

By interchanging the ground- and FC-excited-state labeling, the correspondent formula for the solvent-dependent energy difference related to a fluorescence process from state e to state g can be written. However, this is possible only if the fluorescence originates from an excited state well equilibrated with respect to internal relaxation processes of the solute G and also to the surrounding solvent cage. This is most often the case, at least in sufficient approximation, for solvent-shift measurements. But be careful when studying a molecule with a short fluorescence decay time in a viscous solvent!

In the case of fluorescence from an excited equilibrium state, the following equation holds, where analogous assumptions and labeling are used as with (237)

$$\Delta W_{geS} = \Delta W_{ge0} - hc_0 Df' + \tfrac{1}{2}(\mu_a^0 - \mu_g^{OFC})f'(I - f'\alpha_g^{OFC})^{-1}(\mu_a^0 - \mu_g^{OFC})$$
$$- (\mu_a^0 - \mu_g^{OFC})f(I - f\alpha_a^0)^{-1}\mu_a^0 - \mu_a^0 f(I - f'\alpha_g^{OFC})^{-1}(I - f\alpha_a^0)^{-2}$$
$$\times (\alpha_a^0 - \alpha_g^{OFC})[\tfrac{1}{2}f(I - f'\alpha_a^0)\mu_a^0 - f'(I - f\alpha_a^0)(\mu_a^0 - \mu_g^{OFC})] \tag{238}$$

Here the same dispersion interaction term $hc_0 Df'$ has been used as with the absorption process (237), which is possible in good approximation in most cases [152].

The same basic electrostatic model has been used by Koutek [164] to test two other local field models in comparison to the Onsager model. He shows that the Block and Walker (BW) model [61] and the Wertheim (W) model [62] appear comparable to the Onsager (O) model, generally with $(\mu_a^0)^{BW} > (\mu_a^0)^O > (\mu_a^0)^W$.

7.2.3   Quantum Mechanical Perturbation Treatment

The system consisting of one solute molecule G surrounded by N solvent molecules S is described by the Hamiltonian

$$\mathscr{H} = \mathscr{H}_G + \sum_{i=1}^{N} \mathscr{H}_{Si} + \mathscr{H}' \tag{239}$$

where $\mathscr{H}_G$ is the Hamiltonian describing the free molecule G, $\mathscr{H}_{Si}$ that describing the free molecules S_i, and $\mathscr{H}'$ is the interaction Hamiltonian. It is assumed that the wave functions of neighboring molecules do not overlap. Then $\mathscr{H}'$ may be described by a dipole interaction operator.

Assuming that the perturbation by dipole interaction is sufficiently small, the energy of the whole system with the solute in its ground state or FC excited state can be calculated and from this, by subtraction of the energy of the pure solvent system, would follow the energy of the solute G in a solvent S, in the ground and FC excited states [151, 152].

Careful comparison of the resultant somewhat cumbersome formulas with the classical formulas shows a complete term-by-term agreement, if some terms in the quantum mechanical formula are defined as reaction field terms. Moreover, using minor approximations, an explicit expression for the solvent shift of electronic transitions from the dispersion interaction was obtained that has been used in (237).

7.2.4   Observation of ΔW_{agS} or ΔW_{geS}

The theory sketched in Section 7.2.2 relates some molecular electric moments to a solvent-dependent energy difference between two well-defined states, a and g or e and g. The corresponding macroscopically observable quantity is an absorption or fluorescence wavenumber. What is the appropriate wavenumber to be observed with solvent-shift measurements?

From (237) or (238) it follows that when observing a transition $a \leftarrow g$ or $g \leftarrow e$, for example, a 0,0 transition, the same transition must be followed throughout the entire scale of solvents to keep ΔW_{ag0} constant in (237) or ΔW_{ge0} in (238), which is necessary to make full quantitative use of the other terms in (237) or (238). If typical solvent shifts of about $1000-5000$ cm^{-1} are observed, a certain point of the FC envelope corresponding to the observed transition must be reproduced in all solvents to greater than about 100 cm^{-1}.

First, assume that the FC factor determining the shape of the absorption or fluorescence band is insensitive to solvent variations; the only variation that will occur is the solvent-dependent shift of the FC envelope. This means, considering transition probability, that an arbitrary point, say the wavenumber corresponding to the maximum of the absorption spectrum $\kappa_S(\tilde{\nu})/\tilde{\nu}$ or of the fluorescence spectrum $\phi_S(\tilde{\nu})/\tilde{\nu}^3$, with ϕ_S given as quanta per second and wavenumber interval, must be used as the experimental quantity $\tilde{\nu}_S$ with

$$hc_0\tilde{\nu}_{aS} = \Delta W_{agS} \tag{240}$$

or

$$hc_0\tilde{v}_{fs} = \Delta W_{ges} \tag{241}$$

On the other hand, the FC factor is usually solvent dependent. Hence, researchers sought the best way to define an experimental quantity that would be the least sensitive to such changes.

Procedures, such as defining arbitrarily some weighed center of the absorption band to be $\tilde{v}_{aS}$, cannot improve the results unless information about the solvent dependence of the FC envelope is available.

Therefore simply use the wavenumber corresponding to the maximum of the fluorescence spectrum $\phi_S(\tilde{v})$, or better, $\phi_S(\tilde{v})/\tilde{v}^3$, and of the absorption spectrum $\kappa_S(\tilde{v})$, or better, $\kappa_S(\tilde{v})/\tilde{v}$, knowing that a precise assumption has been used that might introduce a minor error.

7.3 Experimental Application of the Theoretical Formulas

7.3.1 General Remarks

With solvent-shift measurements $\tilde{v}_S$ must be determined as discussed in Section 7.2.4. For this purpose the absorption and, if necessary and possible, the quanta-corrected fluorescence spectrum of the compound to be investigated are measured in many solvents that cover the entire range of f values that are accessible experimentally. In most cases, W_{agS} and W_{geS} depend much less on f' than on f because f' varies only slightly with the whole scale of solvents.

Figure 2.9 shows the function $g(\varepsilon)$, which is proportional to $f(\varepsilon)$, according to (75). As can easily be seen, the range from about $\varepsilon = 1.8$ to about $\varepsilon = 5$ is most important. Regrettably, only a few solvents exist that may be used generally in the range between 2 and 3.

Specific interactions between the solute and the solvent must be very carefully avoided, especially with solvent-shift measurements on the fluorescence. Such

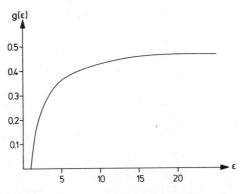

Figure 2.9 The function $g(\varepsilon) = (\varepsilon - 1)/(2\varepsilon + 1)$.

interactions include exciplex formation or emission from different excited states in different solvents or from impurities contained in the solvent.

To evaluate a set of values of $\tilde{v}_S$ that is dependent on the solvent with respect to μ_a^0 or other molecular quantities, values for f and f', or equivalently g and g' [which are mere solvent parameters as defined in (75) and (76)] are required. The spherical Onsager model is usually used to describe the reaction field. Hence, Table 2.12 may be useful. It comprises values of ε, n, g, g', $(g - g')$, and $(g - g'/2)$ for many solvents at 25°C, which is the temperature often used with solvent-shift measurements. However, an ellipsoidal approach may give better results, especially for long molecules.

Evaluation is simplified considerably or is even possible only with appropriate assumptions about polarizabilities, which will be discussed in the following sections.

Table 2.12 Properties of Some Solvents at 25°C, Useful for Solvent-Shift Measurements

Solvent	ε	n	g	g'	$g - g'$	$g - \frac{1}{2}g'$
n-Hexane	1.888	1.372	0.186	0.185	0.001	0.093
2-Methylhexane	1.914	1.382	0.189	0.189	0.001	0.095
Heptane	1.914	1.385	0.189	0.190	−0.001	0.095
3-Methylhexane	1.926	1.386	0.191	0.190	0.001	0.096
Octane	1.940	1.395	0.193	0.193	−0.001	0.096
Nonane	1.969	1.403	0.196	0.196	0.000	0.098
Decane	1.986	1.410	0.198	0.199	−0.000	0.099
Cyclohexane	2.015	1.424	0.202	0.203	−0.002	0.100
Methylcyclohexane	2.018	1.420	0.202	0.202	0.000	0.100
Isoprene	2.100	1.419	0.212	0.202	0.010	0.111
trans-Decalin	2.160	1.467	0.218	0.217	0.001	0.109
Dioxane	2.209	1.420	0.223	0.202	0.021	0.122
Dioxane[a]	6.000		0.385		0.183	0.375
Cyclohexene	2.220	1.445	0.224	0.210	0.014	0.119
Carbon tetrachloride	2.259	1.458	0.228	0.214	0.014	0.121
Benzene	2.274	1.498	0.230	0.227	0.003	0.116
Benzene[a]	5.000		0.363		0.137	0.251
Toluene	2.379	1.494	0.239	0.225	0.014	0.127
Styrene	2.430	1.544	0.244	0.240	0.004	0.124
Triethylamine	2.640	1.400	0.261	0.195	0.066	0.164
Amyl ether	2.770	1.410	0.271	0.199	0.072	0.171
Butyl ether	3.060	1.397	0.289	0.194	0.095	0.192
n-Propyl ether	3.390	1.378	0.307	0.187	0.120	0.214
Acetal	3.800	1.368	0.326	0.184	0.142	0.234
Isopropyl ether	3.880	1.366	0.329	0.183	0.146	0.237
Phentole	4.224	1.505	0.341	0.229	0.112	0.227
Anisole	4.330	1.514	0.345	0.231	0.113	0.229
Diethyl ether	4.330	1.352	0.345	0.178	0.167	0.256
m-Dichlorobenzene	5.040	1.543	0.365	0.240	0.125	0.245

Table 2.12   (*continued*)

Solvent	ε	n	g	g'	$g - g'$	$g - \frac{1}{2}g'$
Fluorobenzene	5.420	1.463	0.373	0.216	0.157	0.265
Ethyl benzoate	6.000	1.495	0.385	0.226	0.159	0.272
Ethyl acetate	6.020	1.370	0.385	0.184	0.201	0.293
Acetic acid	6.200	1.370	0.388	0.184	0.204	0.296
Aniline	6.770	1.583	0.397	0.250	0.146	0.272
2-Chlorobutane	7.070	1.395	0.401	0.193	0.208	0.304
Ethyl formate	7.160	1.359	0.402	0.180	0.222	0.312
Tetrahydrofuran	7.390	1.403	0.405	0.196	0.209	0.307
Dichloromethane	9.000	1.325	0.421	0.167	0.254	0.338
Benzotrifluoride	9.035	1.407	0.421	0.198	0.224	0.323
o-Dichlorobenzene	9.930	1.549	0.428	0.241	0.187	0.307
1,2-Dichloroethane	10.360	1.443	0.431	0.210	0.221	0.326
Pyridine	12.300	1.507	0.441	0.229	0.212	0.327
1-Butanol	17.100	1.397	0.457	0.194	0.263	0.360
Capronitrile	17.263	1.405	0.458	0.197	0.261	0.359
2-Propanol	18.300	1.375	0.460	0.186	0.274	0.367
Valeronitrile	19.710	1.395	0.463	0.193	0.270	0.366
1-Propanol	20.100	1.384	0.464	0.189	0.274	0.369
Acetone	20.700	1.356	0.465	0.179	0.285	0.375
Butyronitrile	23.280	1.382	0.468	0.189	0.280	0.374
Ethanol	24.300	1.359	0.470	0.180	0.289	0.380
Benzonitrile	25.200	1.527	0.471	0.235	0.236	0.353
Propionitrile	27.870	1.364	0.474	0.182	0.291	0.382
Methanol	32.630	1.327	0.477	0.168	0.309	0.393
Acetonitrile	36.050	1.342	0.479	0.174	0.305	0.392
Dimethylformamide	37.700	1.427	0.480	0.204	0.276	0.378
Acrylonitrile	38.000	1.384	0.481	0.189	0.291	0.386
Water	78.540	1.333	0.490	0.171	0.319	0.405
Formamide	109.500	1.447	0.493	0.211	0.282	0.388

[a]Denotes microscopic relative permittivity, as often used to better describe the reaction field of these quadrupolar- or *π-cloud*-interacting molecules.

7.3.2 Evaluation of Molecular Data Neglecting All Polarizabilities

The total neglect of polarizabilities gives very simple formulas that are derived from (237) and (238)

$$hc_0\tilde{v}_{aS} = \Delta W_{ag0} - hc_0 Df' - \tfrac{1}{2}(\mu_a^{OFC} - \mu_g^0)f'(\mu_a^{OFC} - \mu_g^0)$$
$$- (\mu_a^{OFC} - \mu_g^0)f\mu_g^0 \tag{242}$$

$$hc_0\tilde{v}_{fS} = \Delta W_{ge0} - hc_0 Df' + \tfrac{1}{2}(\mu_a^0 - \mu_g^{OFC})f'(\mu_a^0 - \mu_g^{OFC})$$
$$- (\mu_a^0 - \mu_g^{OFC})f\mu_a^0 \tag{243}$$

Since f' does not change very much over the range of solvents, a linear regression of $\tilde{v}_{aS}$ or $\tilde{v}_{fS}$ on g gives $\mu_g^0(\mu_a^{OFC} - \mu_g^0)/a_w^3$ or $\mu_a^0(\mu_a^0 - \mu_g^{OFC})/a_w^3$, where the spherical Onsager model has been used, as usual.

Respective plots of $\tilde{v}_{aS}$ or $\tilde{v}_{fS}$ against g often show considerable scatter, which is caused by specific interactions that have not been taken into account in the theoretical approach. As these interactions often are very similar in the ground and excited states, their effects are canceled when the difference $hc_0(\tilde{v}_{aS} - \tilde{v}_{fS})$ is considered. If corresponding absorption and fluorescence bands are considered with rigid molecules,

$$\mu_a^0 = \mu_a^{OFC} \tag{244}$$

$$\mu_g^0 = \mu_g^{OFC} \tag{245}$$

Then,

$$hc_0(\tilde{v}_{aS} - \tilde{v}_{fS}) = \Delta W_{ag0} - \Delta W_{ge0} - (f - f')(\mu_a^0 - \mu_g^0)^2 \tag{246}$$

From a linear regression of $\tilde{v}_{aS} - \tilde{v}_{fS}$ against $g - g'$ follows $(\mu_a^0 - \mu_g^0)^2/a_w^3$. A value for $|\mu_a^0 - \mu_g^0|$ can be determined if an estimate for a_w^3 exists. Obviously, the error introduced by estimation of a_w^3 is transformed directly to the error for $|\mu_a^0 - \mu_g^0|$; for example, a 20% uncertainty in estimation of a_w adds an uncertainty of about 30% to $|\mu_a^0 - \mu_g^0|$.

Section 3.4.2 and, by some remarks, [165] offer an estimation of a_w^3. Remember, that from $|\mu_a^0 - \mu_g^0|$ a value for μ_a^0 can be given only if the value of μ_g^0 and some information on the angle between μ_a^0 and μ_g^0 are given, for instance, from symmetry!

Another simplification is given with the solvent dependence of exciplexes, where $\mu_g^{OFC} = 0$. Then,

$$hc_0\tilde{v}_{fS} = \Delta W_{ge0} - hc_0 Df' - (f - \tfrac{1}{2}f')(\mu_a^0)^2 \tag{247}$$

Using this formula, many dipole moments of exciplexes have been estimated since the first work of Beens, Knibbe, and Weller [19].

7.3.3 Evaluation of Molecular Data Neglecting the Change of Polarizability Upon Excitation

When the last term in (237) or (238) is neglected, it is assumed that $f(\alpha_a^0 - \alpha_g^0)$ is small compared to 1, which is an assumption less severe than to assume $\alpha_a^0 = \alpha_g^0$. With this neglect the following equations hold

$$hc_0\tilde{v}_{aS} = \Delta W_{ag0} - hcDf' - \tfrac{1}{2}(\mu_a^{OFC} - \mu_g^0)f'(1 - f'\alpha_a^{OFC})^{-1}(\mu_a^{OFC} - \mu_g^0)$$
$$- (\mu_a^{OFC} - \mu_g^0)f(1 - f\alpha_g^0)^{-1}\mu_g^0 \tag{248}$$

$$hc_0\tilde{v}_{fs} = \Delta W_{ge0} - hcDf' + \tfrac{1}{2}(\boldsymbol{\mu}_a^0 - \boldsymbol{\mu}_g^{0FC})\mathbf{f}'(\mathbf{l} - \mathbf{f}'\boldsymbol{\alpha}_g^{0FC})^{-1}(\boldsymbol{\mu}_a^0 - \boldsymbol{\mu}_g^{0FC})$$
$$- (\boldsymbol{\mu}_a^0 - \boldsymbol{\mu}_g^{0FC})\mathbf{f}(\mathbf{l} - \mathbf{f}\boldsymbol{\alpha}_a^0)^{-1}\boldsymbol{\mu}_a^0 \qquad (249)$$

Since f' is only weakly solvent dependent, when the definitions:

$$f(1 - f\alpha_g^0)^{-1} = \frac{1}{2\pi\varepsilon_0 a_w^3} F_g(g) = \frac{1}{2\pi\varepsilon_0 a_w^3}\left(g^{-1} - \frac{\alpha_g^0}{2\pi\varepsilon_0 a_w^3}\right)^{-1} \qquad (250)$$

$$f(1 - f\alpha_a^0)^{-1} = \frac{1}{2\pi\varepsilon_0 a_w^3} F_a(g) = \frac{1}{2\pi\varepsilon_0 a_w^3}\left(g^{-1} - \frac{\alpha_a^0}{2\pi\varepsilon_0 a_w^3}\right)^{-1} \qquad (251)$$

are used, regressions of $\tilde{v}_{aS}$ or $\tilde{v}_{fS}$ on $F_g(g)$ or $F_a(g)$ give results on $(\boldsymbol{\mu}_g^{0FC} - \boldsymbol{\mu}_g^0)\boldsymbol{\mu}_g^0/a_w^3$ or $(\boldsymbol{\mu}_a^0 - \boldsymbol{\mu}_g^{0FC})\boldsymbol{\mu}_a^0/a_w^3$, where further evaluation again depends on an estimate for a_w^3 and on the knowledge of the angle between $\boldsymbol{\mu}_a^0$ and $\boldsymbol{\mu}_g^0$.

Before the regression is performed, values of α_g^0/a_w^3 and α_a^0/a_w^3 must be estimated. The former can be estimated using α_g^0 or $\mathrm{tr}\,\boldsymbol{\alpha}_g^0$ from either dielectric or Kerr effect measurements along with an estimate for a_w^3; the latter is more difficult to estimate since values for α_a^0 exist only for very few molecules [96]. Thus, it is often assumed that $\alpha_a^0 \approx \alpha_g^0$, which today generally seems erroneous. Hence, if polarizabilities cannot be neglected, the determination of $(\boldsymbol{\mu}_a^0 - \boldsymbol{\mu}_g^{0FC})\boldsymbol{\mu}_a^0/a_w^3$ is not possible with sufficient accuracy.

On the other hand, the nonlinear dependence of $F_g(g)$ or $F_a(g)$ on α_g^0/a_w^3 or α_a^0/a_w^3 suggests another least-squares procedure that gives terms like $(\boldsymbol{\mu}_g^{0FC} - \boldsymbol{\mu}_g^0)\boldsymbol{\mu}_g^0/a_w^3$ and $(\boldsymbol{\mu}_a^0 - \boldsymbol{\mu}_g^{0FC})\boldsymbol{\mu}_a^0/a_w^3$, for example, simultaneously. For this purpose the regression of $\tilde{v}_{fS}$ on, for example, $F_a(g)$ according to (249), is sought, depending on the α_a^0/a_w^3 that gives the best fit. Obviously, such procedures are suggested where plots of $\tilde{v}_{fS}$ or $\tilde{v}_{aS}$ against g show a nonlinear dependence. This was first recognized by Baumann and colleagues [137], who demonstrated the power of this procedure with the molecule 4-(9-anthryl)-N,N-dimethylaniline. However, currently there is some discussion ([166] and references therein) about whether the fluorescence of this compound is perhaps a superposition of two distinct bands. Then, of course, the procedure followed in [137] is based on the wrong model. Also, Wermuth and colleagues [167] found a linear dependence of $\tilde{v}_{fS}$ on $F_a(g)$ only, when an assumed value α_a^0/a_w^3 was taken into account.

Finally, by such a procedure it is possible to determine (with relatively large errors) the polarizability α_a^0 from the term $\alpha_a^0/2\pi\varepsilon_0 a_w^3$, which is derived from the optimization process, if a_w^3 is estimated. In addition, if from the system considered the validity of (244) and (245) may be drawn from values of $(\boldsymbol{\mu}_g^{0FC} - \boldsymbol{\mu}_g^0)\boldsymbol{\mu}_g^0/a_w^3$ and $(\boldsymbol{\mu}_a^0 - \boldsymbol{\mu}_g^{0FC})\boldsymbol{\mu}_a^0/a_w^3$, as determined from this procedure, the ratio of these two values gives

$$\frac{(\boldsymbol{\mu}_a^{0FC} - \boldsymbol{\mu}_g^0)\boldsymbol{\mu}_g^0/a_w^3}{(\boldsymbol{\mu}_a^0 - \boldsymbol{\mu}_g^{0FC})\boldsymbol{\mu}_a^0/a_w^3} = \frac{\mu_g^0\,\cos\,\measuredangle\,(\boldsymbol{\mu}_a^0 - \boldsymbol{\mu}_g^0, \boldsymbol{\mu}_g^0)}{\mu_a^0\,\cos\,\measuredangle\,(\boldsymbol{\mu}_a^0 - \boldsymbol{\mu}_g^0, \boldsymbol{\mu}_a^0)} \qquad (252)$$

So with some knowledge from symmetry, the ratio of μ_g^0 and μ_a^0 that is only weakly dependent on a_w^3 can be determined by using the functions $F_a(g)$ and $F_g(g)$.

A similar procedure that completely neglects polarizabilities has been discussed by Suppan [168].

Some experimental refinement has been given by Gryczyński and Kawski [169] who used one solvent at different temperatures, thus preparing dielectrics of different relative permittivity while presumably maintaining constant specific interactions. The resultant experimental wavenumbers have a smaller spread but nevertheless may be better analyzed in some cases with respect to small dipole moments.

7.4 An Example

As an example, the solvent effect on the maximum of the broad almost unstructured fluorescence band of 4-(9-anthryl)-2,6,N,N-tetramethylaniline (2,6-DM-ADMA) is studied [170]. Table 2.13 shows the wavenumber of the fluorescence maximum in various solvents.

Table 2.13 The Wavenumber of the Fluorescence Maximum of 2,6-DM-ADMA in Various Solvents

Solvent	$\tilde{\nu}_{max}/(10^5\,\mathrm{m}^{-1})$
n-Hexane	23.8
Cyclohexane	23.7
Cyclohexene	23.5
Decalin	23.5
Ethylbenzene	22.5
Di-n-butyl ether	23.0
Di-n-propyl ether	23.0
Diisopropyl ether	22.5
Diethyl ether	22.5
Dioxane	21.5
Fluorobenzene	21.1
Ethyl acetate	20.0
Ethyl benzoate	20.0
2-Chlorobutane	21.3
Benzotrifluoride	20.8
Dichloromethane	19.6
o-Dichlorobenzene	20.8
1,2-Dichloroethane	19.4
n-Butanol	19.4
Butyronitrile	18.2
Ethanol	18.4
Benzonitrile	18.4
Acetonitrile	18.2

Since no polarizability data are available for this new compound, all polarizability terms are neglected, and (243) can be used to evaluate the excited-state dipole moment. Furthermore, $f'/2$ does not vary much over the whole line of solvents (as in Section 7.3.2). Hence, the simplest evaluation may use (243) in a further simplified manner

$$hc_0\tilde{v}_{fs} = \text{constant} - f(\mu_a^0 - \mu_g^{OFC})\mu_a^0 \tag{253}$$

or with (75)

$$hc_0\tilde{v}_{fs} = \text{constant} - \frac{1}{2\pi\varepsilon_0}\frac{(\mu_a^0 - \mu_g^{OFC})\mu_a^0}{a_w^3}g \tag{254}$$

In this case a linear regression of the wavenumber of the fluorescence maximum with respect to g yields

$$(\mu_a^0 - \mu_g^{OFC})\mu_a^0/a_w^3 = (2.0 \pm 1.8) \times 10^{-30}\,\text{C}^2/\text{m}$$

The term μ_g^{OFC} may be assumed to be equal to μ_g^0 of 4-(9-anthryl)-N,N-dimethylaniline in good approximation. Hence μ_g^0 is taken from [137] to be $4 \times 10^{-30}\,\text{C}\cdot\text{m}$, parallel to μ_a^0.

The spherical Onsager parameter may be chosen as roughly $a_w = 6 \times 10^{-10}\,\text{m}$, which fits the molecular volume. Then μ_a^0 results, and $\mu_a^0 = (69.4 \pm 2.9) \times 10^{-30}\,\text{C}\cdot\text{m}$.

If a_w were taken as $7 \times 10^{-10}\,\text{m}$, which is also acceptable for this molecule, then $\mu_a^0 = (86.9 \pm 3.6) \times 10^{-10}\,\text{C}\cdot\text{m}$. This example shows the crucial dependence of dipole moment values determined from solvent-shift measurements on the cavity parameter a_w.

The uncertainty caused by statistical scatter is much less than that caused by the uncertainty over the selection of an appropriate value for a_w. Nevertheless, an important conclusion can be drawn: this large dipole moment may be understood as a strong charge transfer from the dimethylamino group to the anthryl moiety.

7.5 Concluding Remarks

Solvent-shift measurements of at least the absorption can be performed in any laboratory that is equipped with a spectral photometer. Further apparatus is unnecessary. This chapter was written to show how strongly the experimental simplicity of solvent-shift measurements rivals the complexity of necessary assumptions to simplify the molecular model quantity and thus to be able to deduce an excited-state dipole moment. Despite the drawbacks imposed by necessary assumptions, the method can be used especially to compare the excited-state dipole moments of related compounds.

8   DIPOLE MOMENTS OF MOLECULES IN THEIR GROUND AND EXCITED STATES DETERMINED BY OTHER METHODS

8.1   The Stark Effect

Consider a symmetric top molecule in a given state x with dipole moment μ_G^0, the state x being defined by its rotational quantum numbers J, K, and M, which are the quantum numbers of the total angular momentum and of the components of the angular momentum in the direction of the symmetric top axis and of the externally applied electric field $\mathbf{E}_a$, respectively.

The Stark interaction energy W_{St} can be represented by

$$W_{St} = -\mu_{xG}^0 K M \mathbf{E}_a / J(J + 1) - \tfrac{1}{2}\mu_{xG}^0 \mathbf{E}_a^2 F(J, K, M) \tag{255}$$

where $F(J, K, M)$ is a well-defined function of the quantum numbers J, K, M [171]. Hence, the Stark effect splits a given state into a certain number of rotational levels and consequently a considered transition into several lines, the number of which depends on the considered transition.

The Stark-split transitions may be observed:

1.   In the microwave region as pure rotational transitions, where both levels involved are in the same vibrational and electronic ground state. Therefore, a quantitative observation of the Stark effect on the rotational transition yields the ground-state dipole moment.

2.   In the rotational vibrational spectrum, where a transition between levels of different vibrational quantum numbers is observed. Hence, in this case the electronic ground-state dipole moment of the considered vibronic state can be deduced from the Stark effect.

3.   In the electronic spectrum, where the ground-state and excited-state dipole moments of the considered vibronic state can be deduced from the Stark effect.

Some basic problems arise:

1.   The assignment of the observed rotational lines must be known; this is often difficult with larger molecules in electronically excited states.

2.   The measurements must be done in the gas phase at low pressure to get the necessary resolution. In standard experiments SF_6 at $0.1-1$ bar is sometimes used to avoid electric breakdown at the necessary high fields. Since Stark splittings usually are less than 0.1 cm^{-1}, pressure broadening at 1 bar would easily obscure the Stark effect and also Doppler broadening might rival with Stark shift.

3.   The intensity of a transition is usually field dependent, too. This effect may disturb the quantitative evaluation of the Stark splitting of absorption or emission lines, if the observed Stark splitting cannot be resolved adequately.

4.   Nuclear quadrupole coupling can affect the Stark effect [172].

5. Polarizabilities may add considerable interaction energy to (255) thus at least in part determining the Stark splitting [67, 173].

These effects may introduce considerable error into the determination of molecular dipole moments from observed Stark splittings. Hence various methods were investigated to overcome at least part of these problems.

8.1.1 Molecular Beam Electric Resonance Technique

There are unique advantages to the spectroscopic investigation of molecules in a molecular beam (emitted from a suitable source into vacuum) rather than in a molecular gas.

Ideally there is no pressure broadening, since the single molecules proceed essentially without collisions. If the absorbed or emitted radiation is observed perpendicular to the beam, there is no velocity component of the molecules in the observation direction, and no Doppler effect is observed. Thus, dipole moments determined from the Stark effect in a molecular beam experiment are highly accurate. Some examples were given in Section 1, as determined with the molecular beam electric resonance technique applied to rotational transitions and described, for example, in [11, 174–177].

8.1.2 Laser-Induced Stark Quantum Beat Spectroscopy

Brieger and co-workers [178, 179] reported the observation of quantum beats in the exponential decay of the electronically excited $A^1\Sigma^+$ state of ^{7}LiH in various sublevels, using laser-induced Stark quantum beat spectroscopy [180] with a molecular beam.

Excitation is performed by a short dye laser pulse, and the decay of the fluorescence intensity q from the investigated state with lifetime τ is observed.

$$q = (q_0 + q_\omega \cos \omega t) \exp(-t/\tau) \qquad (256)$$

where q_0 and q_ω are constant and weak modulated intensity terms, respectively. The only beat frequency ω observed with an $R(0)$ line excitation is caused by the splitting of the sublevels $M' = 0$ and $M' = \pm 1$ and is connected to the dipole moment $\mu_{v'}^0$ in a vibrational state with quantum number v' according to [178]

$$\omega = \frac{3}{10} \frac{(\mu_{v'}^0)^2 E_a^2}{B_{v'} h^2 c_0} \qquad (257)$$

where $B_{v'}$ is the rotational constant.

Results from [178, 179] are given in Table 2.14. They compare well with results from a zero-field-molecular-level crossing experiment in an electric field [181], and the strong dependence of $\mu_{v'}^0(A^1\Sigma^+)$ on the vibrational quantum number also agrees with theoretical predictions [182].

Table 2.14   Dipole Moments of ^{7}LiH and ^{7}LiD in the Electronically Excited $A^1\Sigma^+$ State

Vibrational Quantum Number v'	$\mu_{v'}^0 (\times 10^{-30})(\text{C} \cdot \text{m})$	
	^{7}LiH	^{7}LiD
2	4.87 (23)	
4	1.87 (13)	
5	0.297 (46)	
6		1.43 (7)

8.1.3   Stark Effect on the Rotational Fine Structure of an Electronic Transition

In principle, observing the rotational fine structure of an electronic transition should yield the ground-state and the respective electronically excited-state dipole moments, since Stark splitting is determined by the difference of the respective Stark interaction energy terms according to (255). Hence, in a most straightforward experiment the rotational fine structure must be resolvable. This restricts the application to fairly simple small molecules that should be measured in the gas phase at low pressure. Freeman and Klemperer [25] reported the dipole moment of formaldehyde in its $^1A_2$ excited state to be $(5.20 \pm 0.23) \times 10^{-30}$ C·m.

With molecules in the gas phase, the Stark splitting is not much larger than the Doppler width of individual rotational lines at room temperature. This means, that the Stark components of a line are not completely separated but most often only blendings of these components are observed. Hence, computer programs have been used to simulate the expected line profile. This was accomplished with only one parameter, namely, the excited-state dipole moment. Its value was then determined as the parameter of best fit [25, 183–187]. For most of the molecules investigated, only the absolute value of the change of the dipole moment on excitation can be determined [188–190].

By a similar fitting procedure Buckingham and co-workers [191] determined the dipole moment of formaldehyde in its $^3A_2$ triplet state to be $(4.3 \pm 0.1) \times 10^{-30}$ C·m, and Clouthier and Kerr [192] have reported the dipole moment of thioformaldehyde in its $^3A_2$ state to be $1.90(10) \times 10^{-30}$ C·m.

A close-lying state of symmetry $^1B_2$ has been found in the vicinity of the 0–0 band of the $^1B_2 \leftarrow {}^1A_1$ ($\pi^* \leftarrow \pi$) transition in aniline [27, 193], although the orbital origin of this perturbing state cannot yet be elucidated. The dipole moment of both states and the transition dipole moment μ_{12}^0 between these states can be determined.

$$\mu^0[^1B_2(\pi^*\pi)] = (8.2 \pm 0.3) \times 10^{-30} \,\text{C} \cdot \text{m}$$

$$\mu^0(\text{perturbing state}) = (20.0 \pm 1.3) \times 10^{-30} \,\text{C} \cdot \text{m}$$

$$\mu_{12}^0 = (1.7 \pm 1.3) \times 10^{-30} \,\text{C} \cdot \text{m}$$

8.2   Dielectric Loss Measurements

Consider a complex dielectric constant ε^*

$$\varepsilon^* = \varepsilon - i\varepsilon'' \tag{258}$$

where ε is the relative permittivity (or dielectric constant) and ε'' the so-called dielectric loss factor. Then the dielectric loss tangent $\tan \Theta$ is defined as

$$\tan \Theta = \varepsilon''/\varepsilon \tag{259}$$

The dielectric loss factor ε'' is related to the conductivity σ_c

$$\sigma_c = \omega \varepsilon_0 \varepsilon'' \tag{260}$$

where ω is the measuring frequency. Hence, by measuring the capacity and the conductivity of a condenser filled with the solution under investigation, ε and σ_c, and therefore the loss tangent, can be determined.

If a solute G with concentration given by its mass fraction w_G is dissolved in a suitable nonpolar solvent, a linear dependence of the loss tangent $\tan \Theta$ on the solute's concentration is observed with dilute solutions.

Hence, the following quantity Δ may be used as the PMQ characteristic for this experiment

$$\Delta_G = (d/dw) \tan \Theta \tag{261}$$

The corresponding MMQ ϑ_G is given following Debye's theory and using the Lorentz field [5, 194, 195].

$$\vartheta_G = \frac{(\varepsilon + 2)^2}{\varepsilon \varepsilon_0} \frac{(\mu_{gG}^0)^2}{27kT} \frac{N_A \rho}{M_G} \frac{\omega \tau_r}{1 + (\omega \tau_r)^2} \tag{262}$$

The combination of (261) and (262) yields

$$\frac{d}{dw} \tan \Theta = \frac{(\varepsilon + 2)^2}{\varepsilon \varepsilon_0} \frac{(\mu_{gG}^0)^2}{kT} \frac{N_A \rho}{M_G} \frac{\omega \tau_r}{1 + (\omega \tau_r)^2} \tag{263}$$

Measurement of the frequency dependence of $(d/dw) \tan \Theta$ yields μ_{gG}^0 and the orientational relaxation time τ_r.

The measurement usually is done using an apparatus given in principle by Roberts and von Hippel [70] in 1946. They introduced the sample into a waveguide and calculated $\tan \Theta$ from the field strength pattern in the waveguide. Since this is a closed cavity, the calculation is precise; consequently, the determination of dipole moments by this method is reliable.

In Table 2.15 are shown results on some metal complexes, which were taken from [47], where a larger quantity of metal complex dipole moments have been

Table 2.15 Ground State Dipole Moments of Some Metal Complexes $Met(RCS\!\!=\!\!CHCOR')_x$ From Dielectric Loss Measurements[a]

	$\mu_{gG}^0/(10^{-30}\ C\cdot m)$		
Compound	R = Ph R' = Ph	R = Ph R' = CF$_3$	R = p-Methylphenyl R = CF$_3$
Square planar, x = 2			
Ni	7.60	15.51	17.85
Pd	7.20	16.28	18.48
Pt		16.34	18.75
Cu	8.81	13.11	15.44
Octahedral, x = 3			
Fe	11.01	18.38	19.15
Co	6.84	18.05	20.51
Tetrahedral, x = 2			
Zn	9.07	9.17	10.71

[a]Adapted from [47].

reported together with relaxation times and with the respective dipole moment values found from standard dielectric measurements.

The dipole moment values from dielectric loss measurements are about 1–$3 \times 10^{-30}\ C\cdot m$ less than those determined from (quasi-static) dielectric measurements, which is attributed to atomic polarization that does not contribute to the dielectric loss signal [47].

8.3   Time-Resolved Microwave Conductivity Measurements

During the early 1980s, de Haas, Warman and co-workers [134, 196] introduced a new method to determine dipole moment changes upon excitation. They called this method the *time-resolved microwave conductivity (TRMC) technique*. It is founded on the principles sketched in Section 8.2.

The additivity of the contributions to the loss tangent or, respectively, to the conductivity caused by several species J in a microwave waveguide cell is assumed. Then the macroscopically observable loss tangent tan Θ defined in (259) has its microscopic counterpart analogous to (262), if one species J each with number density (molecules per volume) N_J are dissolved

$$\vartheta = \frac{(\varepsilon + 2)^2}{\varepsilon\varepsilon_0} \frac{1}{27kT} \sum_{J=1}^{1} \left[(\mu_J^0)^2 \frac{\omega\tau_{rJ}}{1 + (\omega\tau_{rJ})^2} N_J \right] \qquad (264)$$

Obviously, ionic contributions were neglected, but can be combined into a complete theory according to [134].

Let the species G to be investigated be dissolved in a solvent S and filled into the cell. Then a certain loss tangent $\tan \Theta$ or conductivity σ_c is measured as in Section 8.2. The unique idea of TRMC is to pump the solute species G at least in part to an excited state, using a short laser pulse.

In the simplest case, this state may be the first singlet-excited state S_1. Immediately after the excitation pulse, a change $\Delta \tan \Theta$ in the loss tangent $\tan \Theta$ or a change $\Delta\sigma_c$ in the conductivity is observed because of the change of the dipole moment upon excitation and also the change of the number density of the solute in both the ground and excited states. Then, $\Delta \tan \Theta$ is given as

$$\Delta \tan \Theta = \frac{(\varepsilon + 2)^2}{\varepsilon\varepsilon_0 27kT} [(\mu_{aG}^0)^2 - (\mu_{gG}^0)^2]N_S \frac{\omega\tau_r}{1 + (\omega\tau_r)^2} \qquad (265)$$

where N_S is the number density of molecules in the excited state S_1. Also, it was assumed that the relaxation term $\omega\tau_r/[1 + (\omega\tau_r)^2]$ is the same for the solute in its ground and excited states.

Hence, from measurement of the change of the loss tangent or the conductivity with the excitation, the dipole moment μ_{aG}^0 can be determined if μ_{gG}^0, the rotational relaxation time τ_r, and the number density N_S of excited-state molecules are known. Excited-state dipole moments up to $200 \times 10^{-30}\,C \cdot m$ have been reported with rigid molecules [197].

If in addition the time behavior of $\Delta\sigma_c$ after the excitation pulse can be followed, even more complex systems can be investigated when, for example, relaxation from S_1 in part goes into a triplet state. Then the time behavior of the system can be measured and fitted to that expected from the respective system of rate equations.

The experimental setup for this technique is described in some detail in [134]. Using it, the dipole moment of 4-dimethylamino-4′-nitrostilbene (DMANS) has been determined in its S_1 state to be $79 \times 10^{-30}\,C \cdot m$, which agrees with results from other techniques [102, 109]; in addition, the dipole moment of the first triplet state T_1 was estimated to be $64 \times 10^{-30}\,C \cdot m$.

Similar results have been reported for molecules related to p-cyano-N,N-dimethylaniline [198] and 9,9′-bianthryl [199] that are currently in controversy about their fluorescence behavior [115], and for some bichromophoric rodlike molecules [200].

The great advantage of this method is that in principle dipole moments of large molecules in difficultly accessible states can be determined. The drawback with such more complex systems is that there are many unknown quantities some of which can be determined from the time behavior of the loss tangent or the conductivity and some of which must be taken from separate experiments. References [134, 196] give a thorough discussion of this technique as used with DMANS.

References

1. A. L. McClellan, *Tables of Experimental Dipole Moments*, Vol. 1, Freeman, San Francisco, CA, 1963.

2. A. L. McClellan, *Tables of Experimental Dipole Moments*, Vol. 2, Rahara Enterprises, El Cerrito, CA, 1974.

3. R. C. Weast, Ed., *Handbook of Chemistry and Physics*, 65th ed., CRC Press, Boca Raton, FL, 1985.

4. V. I. Minkin, O. A. Osipov, and Y. A. Zhdanov, *Dipole Moments in Organic Chemistry*, Plenum, New York, 1970.

5. W. Baumann, *Z. Naturforsch.*, **38**, 995 (1983).

6. W. Liptay, G. Walz, W. Baumann, H.-J. Schlosser, H. Deckers, and N. Detzer, *Z. Naturforsch.*, **26a**, 2020 (1971).

7. M. Ponder and R. Mathies, *J. Phys. Chem.*, **87**, 5090 (1983).

8. G. Briegleb, *Elektronen-Donator-Akzeptor-Komplexe*, Springer, Berlin, 1961.

9. R. Foster, Ed., *Molecular Complexes*, Paul Elek, London, 1973.

10. W. Liptay, T. Rehm, D. Wehning, L. Schanne, W. Baumann, and W. Lang, *Z. Naturforsch.*, **37a**, 1427 (1982).

11. C. H. Townes and A. L. Schawlow, *Microwave Spectroscopy*, McGraw-Hill, New York, 1955.

12. R. L. DeLeon and J. S. Muenter, *J. Chem. Phys.*, **82**, 1702 (1985).

13. M. D. Marschall and W. Klemperer, *J. Chem. Phys.*, **81**, 2928 (1984).

14. G. T. Fraser, K. R. Leopold, and W. Klemperer, *J. Chem. Phys.*, **80**, 1423 (1984).

15. K. R. Leopold, G. T. Fraser, and W. Klemperer, *J. Chem. Phys.*, **80**, 1039 (1984).

16. F. H. De Leeuw and A. Dymanus, *Chem. Phys. Lett.*, **7**, 288 (1970).

17. M. Gordon and W. R. Ware, Eds., *The Exciplex*, Academic, New York, 1975.

18. H. Leonhardt and A. Weller, *Ber. Bunsenges. Phys. Chem.*, **67**, 791 (1963).

19. H. Beens, H. Knibbe, and A. Weller, *J. Chem. Phys.*, **47**, 1183 (1967).

20. W. Baumann, H. Bischof, and J. C. Fröhling, *J. Lumin.*, **24–25**, 555 (1981).

21. E. J. J. Groenen and P. N. Th. van Velzen, *Mol. Phys.*, **35**, 19 (1978).

22. W. Baumann, *Chem. Phys.*, **20**, 17 (1977).

23. J. N. Murrell, *Theory of the Electronic Spectra of Organic Molecules*, Wiley, New York, 1963, pp. 247–254.

24. R. Pariser, *J. Chem. Phys.*, **25**, 1112 (1956).

25. D. E. Freeman and W. Klemperer, *J. Chem. Phys.*, **45**, 52 (1966).

26. N. J. Bridge, D. A. Haner, and D. A. Dows, *J. Chem. Phys.*, **48**, 4196 (1968).

27. J. R. Lombardi, *J. Chem. Phys.*, **56**, 2278 (1972).

28. A. Kriebel and H. Labhart, *Z. Phys. Chem.*, *NF*, **92**, 247 (1974).

29. H. Bischof, W. Baumann, N. Detzer, and K. Rotkiewicz, *Chem. Phys. Lett.*, **116**, 180 (1985).

30. P. Debye, *Polar Molecules*, Hirzel, Leipzig, Germany, 1929; Chemical Catalogue Co., New York, 1929.

31. W. Liptay, D. Wehning, J. Becker, and T. Rehm, *Z. Naturforsch.*, **37a**, 1369 (1982).

32. C. P. Smyth, *Dielectric Behaviour and Structure*, McGraw-Hill, New York, 1955.

33. C. P. Smyth, "Determination of Dipole Moments," in A. Weissberger, Ed., *Physical Methods of Organic Chemistry*, Part III, 3rd ed., Interscience, New York, 1960.

34. J. W. Smith, *Electric Dipole Moments*, Butterworths, London, 1955.

35. O. Exner, *Dipole Moments in Organic Chemistry*, Thieme, Stuttgart, 1975.

36. W. F. Brown, "Dielectrics," in S. Flügge, Ed., *Handbuch der Physik*, Vol. 17, Springer, Berlin, 1956.

37. C. J. F. Böttcher and P. Bordewijk, *Theory of Electric Polarization*, Vol. 1, Elsevier, Amsterdam, 1973.

38. C. J. F. Böttcher and P. Bordewijk, *Theory of Electric Polarization*, Vol. 2, Elsevier, Amsterdam, 1978.

39. R. J. W. Le Fèvre, *Dipole Moments; Their Measurement and Application*, Methuen, London, 1953; Wiley, New York, 1953.

40. A. Chelkowski, *Dielectric Physics*, Elsevier, Amsterdam, 1980.

41. T. T. Weatherly and R. Williams, "Methods of Experimental Physics," in D. Williams, Ed., *Molecular Physics*, Vol. 3, Academic, New York, 1962.

42. H. A. Lorentz, *Theory of Electrons*, Teubner, Leipzig 1909; *The Theory of Electrons*, Dover, New York, 1952.

43. L. V. Lorenz, *Ann. Physik*, **11**, 70 (1880).

44. H. A. Lorentz, *Ann. Physik*, **9**, 641 (1880).

45. W. Liptay, J. Becker, D. Wehning, W. Lang, and O. Burkhard, *Z. Naturforsch.*, **37a**, 1396 (1982).

46. D'Ans-Lax, *Taschenbuch für Chemiker und Physiker*, Springer, Berlin, 1970.

47. S. E. Livingstone, *J. Organomet. Chem.*, **239**, 143 (1982).

48. H. Labhart, "Electrochromism," in I. Prigogine, Ed., *Advances in Chemical Physics*, Vol. XIII, Interscience, New York, 1967.

49. L. Onsager, *J. Chem. Phys.*, **58**, 1486 (1936).

50. Th. G. Scholte, *Recl. Trav. Chim. Pays-Bas Belg.*, **70**, 50 (1951).

51. Th. G. Scholte, *Physica*, **XV**, 437 (1949).

52. A. B. Myers and R. R. Birge, *J. Chem. Phys.*, **74**, 3514 (1981).

53. W. Liptay and G. Walz, *Z. Naturforsch.*, **26a**, 2007 (1971).

54. W. Liptay, *Ber. Bunsenges. Phys. Chem.*, **80**, 207 (1976).

55. W. Rettig, *J. Mol. Struct.*, **84**, 303 (1982).

56. E. L. Pollock and B. J. Alder, *Phys. Rev. Lett.*, **39**, 299 (1977).

57. C. J. F. Böttcher, *Physica*, **IX**, 945 (1942).

58. Th. G. Scholte and F. C. DeVos, *Recl. Trav. Chim. Pays-Bas*, **72**, 625 (1953).

59. V. A. Gorodyskii, L. F. Kardashina, and N. G. Bakshiev, *Russ, J. Phys. Chem.*, **49**, 641 (1975).

60. R. Diguet, P. Bonnet, and J. Varriol, *J. Chim. Phys.*, **76**, 15 (1979).

61. M. Block and S. M. Walker, *Chem. Phys. Lett.*, **19**, 363 (1973).

62. M. S. Wertheim, *Mol. Phys.*, **25**, 211 (1973).

63. I. E. Halverstadt and W. D. Kumler, *J. Am. Chem. Soc.*, **64**, 2988 (1942).

64. K. Higasi, *Bull. Inst. Phys. Chem. Res. (Tokyo)*, **22**, 805 (1943).

65. B. Krishna and K. K. Srivastava, *J. Chem. Phys.*, **32**, 663 (1960).

66. W. Liptay, B. Dumbacher, and H. Weisenberger, *Z. Naturforsch.*, **23a**, 1601 (1968).

67. O. Exner, *Collect. Czech. Chem. Commun.*, **46**, 1002 (1981).

68. A. B. Myers and R. R. Birge, *J. Am. Chem. Soc.*, **103**, 1881 (1981).

69. G. Bossis, *Mol. Phys.*, **47**, 1317 (1982).

70. S. Roberts and A. von Hippel, *J. Appl. Phys.*, **17**, 610 (1946).

71. O. Kratky, H. Leopold, and H. Stabinger, *Z. Angew. Phys.*, **4**, 273 (1969).

72. W. Baumann, *Z. Anal. Chem.*, **284**, 31 (1977).

73. R. P. W. Scott, *Liquid Chromatography Detectors*, Elsevier, Amsterdam, 1986.

74. W. Kuhn, H. Dührkop, and H. Martin, *Z. Phys. Chem. Abt. B*, **45**, 121 (1940).

75. H. Labhart, *Chimia*, **15**, 20 (1961).

76. J. Czekalla, *Chimia*, **15**, 26 (1961).

77. W. Liptay and J. Czekalla, *Z. Naturforsch.*, **15a**, 1072 (1960).

78. H. Labhart, *Tetrahedron*, **19**, Suppl. 2, 223 (1963).

79. W. Liptay and J. Czekalla, *Ber. Bunsenges. Phys. Chem.*, **65**, 721 (1961).

80. W. Liptay, *Z. Naturforsch.*, **20a**, 272 (1965).

81. A. I. Kornilov, A. N. Shchapov, and F. P. Chernyakovskii, *Russ, J. Phys. Chem.*, **54**, 1055 (1980).

82. Å. Davidsson, *Chem. Phys.*, **35**, 413 (1978).

83. Å. Davidsson, L. B. Å. Johansson, *J. Phys. Chem.*, **88**, 1094 (1984).

84. S. H. Lin, *J. Chem. Phys.*, **62**, 4500 (1975).

85. A. R. Hill and M. M. Malley, *J. Mol. Spectrosc.*, **40**, 428 (1971).

86. M. A. Kurzmack and M. M. Malley, *Chem. Phys. Lett.*, **21**, 385 (1973).

87. E. E. Havinga and P. van Pelt, *Ber. Bunsenges. Phys. Chem.*, **83**, 816 (1979).

88. R. Mathies and A. C. Albrecht, *Chem. Phys. Lett.*, **16**, 231 (1972).

89. C. A. G. O. Varma and L. J. Oosterhoff, *Chem. Phys. Lett.*, **8**, 1 (1971).

90. C. A. G. O. Varma and L. J. Oosterhoff, *Chem. Phys. Lett.*, **9**, 406 (1971).

91. J. D. Scott and B. R. Russell, *J. Chem. Phys.*, **63**, 3243 (1975).

92. G. C. Causley, D. D. Altenloh, and B. R. Russell, *Chem. Phys. Lett.*, **89**, 213 (1982).

93. K. Yamaoka and E. Charney, *J. Am. Chem. Soc.*, **94**, 8963 (1972).

94. E. Fredericq and C. Houssier, *Electric Dichroism and Electric Birefringence*, Clarendon, Oxford, 1973.

95. Ch. T. O'Konski, Ed., *Molecular Electro-Optics*, Dekker, New York, 1976.

96. W. Liptay, "Dipole Moments and Polarizabilities of Molecules in Excited Electronic States," in E. C. Lim, Ed., *Excited States*, Vol. 1, Academic, New York, 1974.

97. W. Liptay, "Optical Absorption in an Electric Field," in Ch. T. O'Konski, Ed., *Molecular Electro-Optics*, Dekker, New York, 1976.

98. R. M. Hochstrasser, *Acc. Chem. Res.*, **6**, 263 (1973).

99. H. G. Kuball and D. Singer, *Ber. Bunsenges. Phys. Chem.*, **73**, 403 (1969).

100. W. Liptay and J. Becker, *Z. Naturforsch.*, **37a**, 1409 (1982).

101. W. Liptay, *Z. Naturforsch.*, **21a**, 1605 (1966).

102. W. Baumann, H. Deckers, K.-D. Loosen, and F. Petzke, *Ber. Bunsenges. Phys. Chem.*, **81**, 799 (1977).

103. K. Seibold, H. Navangul, and H. Labhart, *Chem. Phys. Lett.*, **3**, 275 (1969).

104. W. Baumann, *Ber. Bunsenges. Phys. Chem.*, **80**, 231 (1976).

105. W. Liptay, W. Eberlein, H. Weidenberg, and O. Elflein, *Ber. Bunsenges. Phys. Chem.*, **71**, 548 (1967).

106. G. C. Causley, J. D. Scott, and B. R. Russell, *Rev. Sci. Instrum.*, **48**, 264 (1977).

107. W. Baumann and H. Bischof, *J. Mol. Struct.*, **84**, 181 (1982).

108. A. Weissberger, Ed., *Techniques of Organic Chemistry, Organic Solvents*, Vol. VII, 2nd ed., Interscience, New York, 1955.

109. J. Czekalla and G. Wick, *Z. Elektrochem.*, **65**, 727 (1961).

110. K. Seibold and H. Labhart, *Biopolymers*, **10**, 2063 (1971).

111. K. Seibold, G. Wagnière, and H. Labhart, *Helv. Chim. Acta*, **52**, 789 (1969).

112. W. Baumann, *Z. Naturforsch.*, **36a**, 868 (1981).

113. C. J. Eckhardt, *J. Chem. Phys.*, **56**, 3947 (1972).

114. H. Sauter and A. C. Albrecht, *Chem. Phys. Lett.*, **2**, 8 (1968).

115. Z. R. Grabowski, K. Rotkiewicz, A. Siemiarczuk, D. J. Cowley, and W. Baumann, *Nouv. J. Chim.*, **3**, 443 (1979).

116. J. Czekalla, *Z. Elektrochem.*, **64**, 1221 (1960).

117. J. Czekalla and K.-O. Meyer, *Z. Phys. Chem. (Frankfurt am Main)*, **27**, 185 (1961).

118. J. Czekalla, W. Liptay, and K.-O. Meyer, *Ber. Bunsenges. Phys. Chem.*, **67**, 465 (1963).

119. W. Liptay, *Z. Naturforsch.*, **18a**, 705 (1963).

120. G. Weber, *J. Chem. Phys.*, **43**, 521 (1965).

121. G. Weill and C. Hornick, *Biopolymers*, **10**, 2029 (1971).

122. C. Hornick and G. Weill, *Biopolymers*, **10**, 2345 (1971).

123. W. Baumann and H. Deckers, *Ber. Bunsenges. Phys. Chem.*, **81**, 786 (1977).

124. R. M. Hochstrasser and L. J. Noe, *J. Mol. Spectrosc.*, **38**, 175 (1971).

125. R. M. Hochstrasser and L. J. Noe, *Chem. Phys. Lett.*, **5**, 489 (1970).

126. R. M. Hochstrasser and L. J. Noe, *J. Mol. Spectrosc.*, **42**, 197 (1972).

127. J. Bullot and A. C. Albrecht, *J. Chem. Phys.*, **51**, 2220 (1969).

128. J. R. Braun, T. S. Lin, F. P. Burke, and G. J. Small, *J. Chem. Phys.*, **59**, 3595 (1973).

129. S. J. Sheng and D. M. Hanson, *J. Chem. Phys.*, **60**, 368 (1974).

130. W. Liptay, "Transition Moments and the Anisotropy of Fluorescence and Phosphorescence," in O. Sinanoğlu, Ed., *Modern Quantum Chemistry*, Part 3, Academic, New York, 1965.

131. H. Deckers and W. Baumann, *Ber. Bunsenges. Phys. Chem.*, **81**, 795 (1977).

132. C. A. Parker, *Photoluminescence of Solutions*, Elsevier, Amsterdam, 1968.

133. R. Lippert, W. Nägele, I. Seibold-Blankenstein, W. Steiger, and W. Voss, *Z. Anal. Chem.*, **170**, 1 (1959).

134. M. P. de Haas and J. M. Warman, *Chem. Phys.*, **73**, 35 (1982).

135. H. Bischof and W. Baumann, *Z. Naturforsch.*, **40a**, 874 (1985).

136. W. Baumann and H. Bischof, *J. Mol. Struct.*, **129**, 125 (1985).

137. W. Baumann, F. Petzke, and K.-D. Loosen, *Z. Naturforsch.*, **34a**, 1070 (1979).

138. E. M. Kosower, *J. Am. Chem. Soc.*, **80**, 3253 (1958).

139. E. M. Kosower, *Molecular Biochemistry*, McGraw-Hill, New York, 1962.

140. E. M. Kosower, *An Introduction to Physical Organic Chemistry*, Wiley, New York, 1968.

141. K. Dimroth, Ch. Reichardt, Th. Siepmann, and F. Bohlmann, *Liebigs Ann. Chem.*, **661**, 1 (1963).

142. N. D. Coggeshall and E. M. Lang, *J. Am. Chem. Soc.*, **70**, 3283 (1948).

143. E. Lippert, *Z. Naturforsch.*, **10a**, 541 (1955).

144. N. Mataga, Y. Kaifu, and M. Koizumi, *Bull. Chem. Soc. Jpn.*, **29**, 465 (1956).

145. N. S. Bayliss and E. G. McRae, *J. Phys. Chem.*, **58**, 1002 (1954).

146. Y. Ooshika, *J. Phys. Soc. Jpn.*, **9**, 594 (1954).

147. E. G. McRae, *J. Phys. Chem.*, **61**, 562 (1957).

148. E. Lippert, *Ber. Bunsenges. Phys. Chem.*, **61**, 962 (1957).

149. N. G. Bakshiev, *Opt. Spectrosc. (USSR) (English trans.)*, **10**, 379 (1961).

150. L. Bilot and A. Kawski, *Z. Naturforsch.*, **17a**, 621 (1962).

151. W. Liptay, *Z. Naturforsch.*, **20a**, 1441 (1965).

152. W. Liptay, "The Solvent Dependence of the Wavenumber of Optical Absorption and Emission," in O. Sinanoğlu, Ed., *Modern Quantum Chemistry*, Part 2, Academic, New York, 1965.

153. S. Basu, *Adv. Quantum Chem.*, **1**, 145 (1964).

154. T. Abe, *Bull. Chem. Soc. Jpn.*, **38**, 1314 (1965).

155. T. Abe, Y. Amako, T. Nishioka, and H. Azumi, *Bull. Chem. Soc. Jpn.*, **39**, 845 (1966).

156. T. Abe, *Bull. Chem. Soc. Jpn.*, **41**, 1260 (1968).

157. M. E. Baur and M. Nicol, *J. Chem. Phys.*, **44**, 3337 (1966).

158. M. Nicol, J. Swain, Y.-Y. Shum, R. Merin, and R. H. H. Chen, *J. Chem. Phys.*, **48**, 3587 (1968).

159. P. Suppan, *J. Chem. Soc. A*, 3125 (1968).

160. P. Suppan, *Spectrochim. Acta*, **A24**, 1161 (1968).

161. T. Mazurenko and N. G. Bakshiev, *Opt. Spectrosc. (USSR)*, **28**, 490 (1970).

162. A. T. Amos and B. L. Burrows, *Adv. Quantum Chem.*, **7**, 289 (1973).

163. J. Fischer-Hjalmars, A. Henriksson-Enflo, and Ch. Hermann, *Chem. Phys.*, **24**, 167 (1977).

164. B. Koutek, *Collect. Czech. Chem. Commun.*, **49**, 1680 (1984).

165. L. S. Prabhumirashi, *Spectrochim. Acta*, **39A**, 91 (1983).

166. A. Siemiarczuk, *Chem. Phys. Lett*, **110**, 437 (1984).

167. G. Wermuth, W. Rettig, and E. Lippert, *Berg. Bunsenges. Phys. Chem.*, **85**, 64 (1981).

168. P. Suppan, *Chem. Phys. Lett.*, **94**, 272 (1983).

169. I. Gryczyński and A. Kawski, *Z. Naturforsch.*, **30a**, 287 (1975).

170. N. Detzer, W. Baumann, B. Schwager, J.-C. Fröhling, and C. Brittinger, *Z. Naturforsch.*, **42a**, 395 (1987).

171. J. H. Shirley, *J. Chem. Phys.*, **38**, 2896 (1963).

172. A. D. Buckingham and P. J. Stephens, *Mol. Phys.*, **7**, 481 (1964).

173. M. Brieger, *Chem. Phys.*, **89**, 275 (1984).

174. N. F. Ramsay, *Molecular Beams*, Clarendon, Oxford, 1956 and 1963.

175. J. S. Muenter, *J. Chem. Phys.*, **48**, 4544 (1968).

176. R. S. Freund and W. Klemperer, *J. Chem. Phys.*, **43**, 2422 (1965).

177. T. R. Dyke, G. R. Tomasevich, and W. Klemperer, *J. Chem. Phys.*, **57**, 2277 (1972).

178. M. Brieger, A. Hese, A. Renn, and A. Sodeik, *Chem. Phys. Lett.*, **76**, 465 (1980).

179. M. Brieger, A. Renn, A. Sodeik, and A. Hese, *Chem. Phys.*, **75**, 1 (1983).

180. A. Hese, A. Renn, and H. S. Schwede, *Opt. Commun.*, **20**, 385 (1977).

181. P. J. Dagdiagian, *J. Chem. Phys.*, **73**, 2049 (1980).

182. K. K. Docken and J. Hinze, *J. Chem. Phys.*, **57**, 4928 and 4936 (1972).

183. D. E. Freeman and W. Klemperer, *J. Chem. Phys.*, **45**, 52 (1966).

184. J. R. Lombardi, D. Campbell, and W. Klemperer, *J. Chem. Phys.*, **46**, 3482 (1967).

185. J. R. Lombardi, *J. Chem. Phys.*, **50**, 3780 (1969).

186. D. E. Freeman, J. R. Lombardi, and W. Klemperer, *J. Chem. Phys.*, **45**, 58 (1966).

187. K. T. Huang and J. R. Lombardi, *J. Chem. Phys.*, **52**, 5613 (1970).

188. Ch. Y. Wu and J. R. Lombardi, *J. Chem. Phys.*, **55**, 1997 (1971).

189. Ch. Y. Wu and J. R. Lombardi, *J. Chem. Phys.*, **54**, 3659 (1971).

190. K. T. Huang and J. R. Lombardi, *J. Chem. Phys.*, **51**, 1228 (1969).

191. A. D. Buckingham, D. A. Ramsay, and J. Tyrrell, *Can. J. Phys.*, **48**, 1242 (1970).

192. D. J. Clouthier and C. M. L. Kerr, *Chem. Phys.*, **80**, 299 (1983).

193. J. R. Lombardi, *Chem. Phys.*, **28**, 41 (1978).

194. P. Debye, *Physik. Z.*, **35**, 101 (1934).

195. R. J. W. Le Fèvre and E. P. A. Sullivan, *J. Chem. Soc.*, 2873 (1954).

196. J. M. Warman, M. P. de Haas, A. Hummel, C. A. G. O. Varma, and P. H. M. van Zeyl, *Chem. Phys. Lett.*, **87**, 83 (1982).

197. J. M. Warman, M. P. de Haas, M. N. Paddon-Row, E. Cotsaris, N. S. Hush, H. Oevering, and J. W. Verhoeven, *Nature* (*London*), **320**, 615 (1986).

198. R. J. Visser, P. C. M. Weisenborn, C. A. G. O. Varma, M. P. de Haas, and J. M. Warman, *Chem. Phys. Lett.*, **104**, 38 (1984).

199. R.-J. Visser, C. M. Weisenborn, P. J. M. van Kan, B. H. Huizer, C. A. G. O. Varma, J. M. Warman, and M. P. de Haas, *J. Chem. Soc. Faraday Trans. 2*, **81**, 689 (1985).

200. G. F. Mes, B. de Jong, H. J. van Ramesdonk, J. W. Verhoeven, J. M. Warman, M. P. de Haas, and L. E. W. Horsman-van den Dool, *J. Am. Chem. Soc.*, **106**, 6524 (1984).

Chapter **3**

STATIC MAGNETIC TECHNIQUES AND APPLICATIONS

Laxman N. Mulay and Indumati L. Mulay[†]

[†]This author contributed the section on "Applications to Biosystems" and to the organization of the text, including a survey of the literature.

1   INTRODUCTION

The principles of static magnetic techniques and their multifarious applications are reviewed in this chapter. The static techniques imply the measurement of magnetic susceptibility and magnetization. Although the two are closely related, they give different types of information depending on the properties of the systems under study. Furthermore, these parameters are considered to be "static" because they generally do not involve any "relaxation effects" in the sense that a magnetic system comes into equilibrium with an applied field and the thermal energy in a few seconds. Indeed, the well-known diamagnetic and paramagnetic systems and the magnetically ordered ferromagnetic, ferrimagnetic, and antiferromagnetic systems discussed in the text can reach such equilibrium in a few seconds; as such the theories of magnetic susceptibility and of magnetization do not involve relaxation time or frequency parameters.[†] On the other hand, NMR (nuclear magnetic resonance spectroscopy), EPR or ESR (electron paramagnetic or spin resonance spectroscopy), and Mössbauer spectroscopy of certain nuclei involve transitions between a ground state and their excited states. Thus, these spectroscopies do involve certain relaxation processes. In the first two categories (NMR, EPR or ESR) the energy levels of certain nuclei and unpaired electrons are produced by an external field; such levels are attributed to Zeeman splitting, which was first observed in optical spectra in an applied field. The transitions between levels are brought about by electromagnetic radiation of appropriate frequencies. In Mössbauer spectroscopy the energy levels of certain nuclei (e.g., the well-known ^{57}Fe and ^{119}Sn) are intrinsic to particular systems. Gamma rays of high frequency from radioactive sources, on which a Doppler velocity is superimposed, bring about the transitions between the two states. The s-electron density or wave function of electrons at the site of the nuclei produce an *internal field*. Even here, an external field can superimpose a Zeeman splitting on the intrinsic separation between the ground and excited states. Thus, all these spectroscopies, which are dependent on an *external* or *an internal plus an external splitting* caused by the corresponding magnetic fields, do involve a relaxation process, ranging from the very low to the very high frequencies. For this reason, these spectroscopies are *dynamic* techniques. Hence, while succinctly reviewing the realm of static methods in this chapter, when necessary, allied fields in magnetics are referenced because all branches of magnetism, including the above spectroscopies, are intricately related.

A systematic evolution of magnetics may be traced to the investigations of Michael Faraday, which started about 1800. He based his investigations on the earlier research of Ampère, Oersted, Arago, Biot, and Savart. Faraday showed that all matter is somehow magnetic; that is, matter is either attracted or repelled by a magnetic field. We now know that the former category embraces paramagnetism, ferromagnetism, and ferrimagnetism; and the latter corresponds to diamagnetism.

In the realm of magnetics, Faraday's genius not only established the laws of induction governing the relationship between electricity and magnetism, but it also foresaw the close relationship between magnetism and the effect of a magnetic field on plane-polarized light. His efforts were unsuccessful because of the experimental limitations of his time; nevertheless, they planted the seed for the subsequent discovery of the Zeeman effect, which in turn directed the research on electron paramagnetic and nuclear magnetic resonance (EPR or ESR and NMR) spectroscopy. Advances in EPR spectroscopy prompted work on *optical pumping*, which has produced the epoch-making science and tech-

---

[†]Measurements using alternating current (ac) magnetic fields do involve time-dependent parameters (see Section 6.5.2).

nology of the laser (light amplification by stimulated emission of radiation). This was preceded by the microwave-amplification device, the *maser*.

Faraday not only discovered the laws of electrolysis, well known to all chemists, but he also showed in 1845 that dried blood is diamagnetic and noted that he "must try recent fluid blood." His observations thus opened entirely new fields of magnetochemical research on the components of blood and on various systems of biomedical significance.

The history of magnetics is indeed fascinating and was reviewed by Stoner [1] and by Bhatnagar and Mathur [2], who in 1935 introduced the term *magnetochemistry* in the very first English book on the subject. The rigor of the mathematical analysis of magnetics was developed in the last century by Poisson (1820), Weber (1854), and Ewing (1890). The theories of Maxwell, Langevin, Honda, Oxley, and Stoner (1850–1930) established the present-day basis for a quantitative interpretation of magnetic properties of atoms in terms of electronic structure. In 1916 G. N. Lewis showed the relationship between magnetism, electrons, valence, and the chemical bond. This constitutes the basis for chemical interpretations of magnetic susceptibility. Magnetization measurements on magnetically ordered systems and subdomain clusters also give vital information on such systems.

Mulay [3] gave a brief sketch of the very early history of magnetism, starting with references to the lodestone (leading stone), which are found in the *Vedas*, the most ancient religious scriptures of the Hindus, dating back to some thousand years BC. However, a thorough historical account of magnetism, ranging all the way from ancient history to the modern theoretical development is found in *The Theory of Magnetism* by Mattis [4].

Several workers made outstanding theoretical contributions to many areas of magnetism since about 1930. These include Van Vleck, Néel, Anderson, Slater, Goodenough, Griffith, and Figgis. In the magnetochemical area the work of Pauling, Selwood, Klemm, Nyholm, J. Lewis, and several of their colleagues are well recognized by magneticists. Thus far, no reference was made to the accomplishments of the French school, notably those of P. Curie, Pascal, and Pacault, because such works occupy a unique position in the history of magnetochemistry. As a matter of fact, the application of Pascal's constants to diamagnetism and of the Curie–Weiss law to a study of paramagnetism were primarily responsible for the early development of magnetochemistry.

In recent years, Mulay and Boudreaux [5,[†] 6[‡]] have extensively covered topics dealing with the theory and applications of diamagnetism and paramagnetism. From 1962 to 1984 Mulay and Mulay wrote biannual reviews on topics such as the instrumentation and applications of magnetic susceptibility. Their reviews also list about 200 references to new books, chapters, and topical

---

[†]With reference to Table 5.2, page 292 of this book, one should read "$\chi_m = \chi/\rho$," where χ_m and χ are the mass and volume magnetic susceptibilities and ρ is the density. On page 301 the correct expression for μ_{eff} is $\mu_{eff} = \sqrt{n(n + 2)}$.

[‡]On page 488 the correct expression for μ_{eff} is $\mu_{eff} = \sqrt{n(n + 2)}$.

surveys on magnetism that appeared during this time. Until 1980, Mulay and Mulay wrote reviews under the title *Magnetic Susceptibility: Instrumentation and Applications*...; however, the title of their 1982 and 1984 reviews was changed to *Magnetometry: Instrumentation and Applications*... because the term *magnetometry* signifies not only the measurement of magnetic susceptibility (χ), but also that of magnetization (M). The two parameters are closely related, as discussed in Section 3. The χ and M parameters yield entirely different information for a wide variety of materials. Occasionally, the term *magnetometry* is used to signify apparatus for measuring the static field in permanent and electromagnetics. Most workers prefer the obvious terminology, the *gaussmeter*, for such devices.

It is shown that the static magnetic techniques, which involve the measurement of magnetic susceptibility or magnetization as a function of both the field and temperature, are *quick, nondestructive,* and may be performed on a *microscale*. In many instances the experiments may be conducted with inexpensive home-built apparatus that lends itself to automatic recording if desired. These advantages make the technique attractive for routine analysis and for basic and applied research.

2 PHYSICAL BASIS

The magnetic behavior of bulk matter often is explained in terms of the magnetic properties of the constituent molecules and atoms. In the ultimate analysis, one must consider the magnetic properties of the fundamental particles of matter. A discussion along these lines needs a thorough understanding of the electronic structure of atoms, in terms of the language of physics in general, and of spectroscopic nomenclature in particular. The reader should clarify the concepts of similar-sounding but often confusing terms, such as momentum and moment.

An attempt is made here to present the physical basis of magnetic parameters by considering simply the magnetic effects arising from the two distinct types of motion of the electron, namely, its orbital rotation and spinning around itself. These effects often are expressed in terms of the turning moments or magnetic moments, although one does not know, in the classical sense, the precise magnitude of the strength of the poles of an electron and the distance separating them.[†]

In the following discussion the orbital and spin quantum numbers are expressed in units of $h/2\pi$, where h is the Planck constant.

If l is the orbital quantum number, that is, the angular momentum of the

---

[†]Often the magnetic susceptibility arising from electrons is termed the *electronic susceptibility* to distinguish it from the susceptibility of the nucleus; the latter is of a very small magnitude (ca. 10^{-10}). The susceptibility of the nucleus is detectable at very low temperatures and may be ignored relative to the electronic susceptibility.

electron, the magnetic moment μ_l for the orbital motion is given by

$$\mu_l = \sqrt{l(l + 1)} \cdot \frac{eh}{4\pi mc}$$

$$= \sqrt{l(l + 1)} \text{ Bohr magnetons (BM)}$$

$$= \sqrt{l(l + 1)} \times 0.927 \times 10^{-20} \text{ erg Oe}^{-1}$$

This orbital magnetic moment can be expressed as a vector opposite to $l*$ since the electronic charge is negative.

Now, an electron spinning around an axis may be said to behave like a tiny magnet and to give rise to a magnetic moment. A theoretical value for the magnetic moment that is caused by the spin of the electron cannot be derived, as nothing is known about the shape of an electron or its charge distribution. However, to obtain agreement with experimental results, using wave mechanics, the following spin magnetic moment μ_s has been assigned to the electron.

$$\mu_s = 2\sqrt{s(s + 1)} \cdot \frac{eh}{4\pi mc}$$

$$= 2\sqrt{s(s + 1)} \text{ BM}$$

$$= 1.62 \times 10^{-20} \text{ erg Oe}^{-1}$$

Here $s = \frac{1}{2}$ represents the spin angular momentum of the electron. In vectorial presentation, μ_s and the mechanical moment(s) are in *opposite* directions.

The observed magnetic moment results from a combination of the orbital and spin moments. The resultant moment may be calculated theoretically by a vectorial addition of the two, taking into account various coupling mechanisms that arise in a system containing many electrons.

The effect of a magnetic field on different systems containing electrons is now considered. Whether this effect is one of attraction or of repulsion between the system and the applied field depends on the presence or absence of an unpaired electron(s) in the system. For instance, the ferric ion (Fe^{3+}) has five unpaired electrons in its d shell, and a free radical such as DPPH (2,2-diphenyl-l-picrylhydrazyl) has one unpaired electron somewhere inside the molecule. Hence, such systems have a permanent magnetic moment to start with and will be attracted appreciably toward an applied field. The magnetic moment of most paramagnetic systems may be expressed by the *spin only* formula mentioned earlier and correlated with the magnetic susceptibility, which is defined under Section 3.8. There exist systems, comprising practically all inorganic and organic compounds, that do not contain any unpaired electrons. In these, the magnetism caused by the spin of one electron is canceled in some fashion by that of another spinning in an opposite direction; as such these systems do not have a permanent magnetic moment. Therefore, ordinarily no effect from an applied field is expected. However, a very feeble yet significant repulsion is observed. This diamagnetic behavior is attributed entirely to the *induced* effect of the magnetic field on the orbital motion of the (paired) electrons, and the susceptibility in this case may be correlated with the radii of these orbits. A single unpaired electron also contributes to diamagnetism; however, in this case its strong paramagnetism masks its feeble diamagnetism.

According to the classical theory, an electron carrying a negative charge and moving in a circular orbit is equivalent to a circular current. If a magnetic field is applied perpendicularly to the plane of the orbit, the revolving electron experiences a force along the radius, the direction of which depends on that of the magnetic field and of the moving electron. Application of the well-known Lenz's law, which predicts the direction of motion of a current-carrying conductor placed in a magnetic field, to this situation shows that the system as a whole is repelled away from the applied field. An elaborate mathematical picture is presented by the Larmor theorem, which describes the behavior of a system of particles, all having the same ratio of charge to mass (e/m), in the presence of a constant uniform field. According to this theorem: "The superimposed field leaves the form of the orbits and their inclination to the magnetic line of force, as also the motion in the orbit, unaltered and merely leads to the addition of a uniform 'precession' of the orbit about the direction of lines of force." Let us clarify the meaning of precession as applied to the orbital and spinning motions of an electron. Of these, the latter is rather easy to visualize by its comparison with the behavior of a spinning top or a gyroscope. As shown in Figure 3.1,[†]

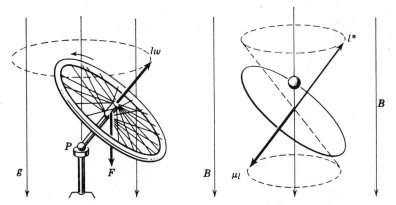

Figure 3.1   A mechanical top precessing in a uniform gravitational field g is analogous to an electron orbit precessing in a uniform magnetic field. Figure from *Atomic Age Physics* by Henry Semat and Harvey E. White, copyright © 1959 by Holt, Rinehart and Winston, Inc., renewed 1987 by Barbara Semat, Edith Semat Kemp, and Harvey E. White, reprinted by permission of the publisher.

[†]Figures 3.1–3.5 and appropriate descriptions are based on an excellent textbook by Semat and White [6]. We recommend this textbook for a clear understanding of the vector model of the atom.

when a top originally spinning erect around an axis is subjected to an external force *F*, it does not topple over completely but continues to spin around its own axis *P*, while this axis continues to rotate in an orbit around the axis *P*. This characteristic motion of the axis is termed the *precession*. This spin precession indeed arises in the case of a free (unpaired) electron subjected to a magnetic field and forms the basis of the EPR or ESR spectroscopy techniques. Similarly, the precession of spins of certain nuclei, such as the protons, forms the basis, for example, of NMR spectroscopy. It should now be possible to visualize the precession of the entire orbit of an electron around the direction of the applied field, as shown in Figure 3.2. It is this type of orbital precession of an electron that gives rise to diamagnetism and forms the core of the mathematical theory of diamagnetism, which is discussed in detail by Mulay and Boudreaux [6].

We now turn to a brief discussion of how the spin *s** and orbital *l** (mechanical) moments interact and how they are space quantized, that is, how they assume fixed positions with each other. Vector diagrams (which must not be taken too literally) are often useful in representing such space quantization, as they employ the right-hand rule of mechanics (Figure 3.2). The term *j** may be a combination of *s** and *l** and is given by

$$j^* = \sqrt{j(j + 1)}\,\frac{h}{2\pi}$$

with $j = \frac{1}{2}, \frac{3}{2}, \frac{5}{2}, \frac{7}{2}$.

The number *j*, known as the *inner quantum number*, has two values from the vector diagram for all values of *l*, except *l* = 0. These correspond to the doubling of all energy levels except the *S* levels in atoms, with but one valence electron. In all atoms, the total angular momentum *J* of each *completed* shell and subshell is

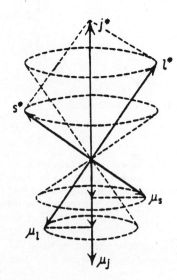

Figure 3.2 The electron spin and orbital moments precess around their resultant angular momentum *j**. Figure from *Atomic Age Physics* by Henry Semat and Harvey E. White, copyright © 1959 by Holt, Rinehart and Winston, Inc., renewed 1987 by Barbara Semat, Edith Semat Kemp, and Harvey E. White, reprinted by permission of the publisher.

nonexistent; thus, it is zero. The reader should become familiar with *Hund's rules*, which describe the ground state of individual atoms in terms of well-known spectroscopic term symbols.

Although it is easy to see *how* the corresponding magnetic moments μ_l and μ_s can be vectorially combined to give μ_j, one must consider the *mechanism* for their combination; in other words, what type of magnetic interactions give rise to j and in turn lead to its space quantization in a magnetic field. This is illustrated in Figure 3.3.

It will be observed in Figure 3.3 that the magnetic field at the electron orbit is zero because of its orbital motion. The field lines become more circular as they near the orbit and ultimately vanish; however, there is an effective magnetic field at the electron. This field stems from the presence of a positively charged nucleus. With reference to Figure 3.3, if an observer imagines that he is riding with the electron around the orbit, the observer (while imagining himself to be stationery) will see the positively charged nucleus moving around him in an orbit of the same size as that of the electron. Thus, the magnetic field H at the spinning nucleus will be the same as that of the positive charge moving around the electron orbit. Now one may visualize that the spinning electron with its moment μ_s will precess around the spin field. Finally, all moments will precess around j^*, as shown in Figure 3.3.

Except for s-electron orbits in which $l = 0$, there are two possible directions for spin s^* and orbit l^* to form j^*. These two orientations produce two closely spaced energy levels. From the energy-level diagrams of atoms containing one valence electron, we find the level with the lower of the two j values lying deeper than the other. Thus, for a p electron with $l = 1$, the two energy states allowable, $j = \frac{1}{2}$ and $j = \frac{3}{2}$, will have slightly different energies, the level $j = \frac{1}{2}$ lying deeper than the level $j = \frac{3}{2}$. In the same manner, the level $j = \frac{3}{2}$ will lie below the level $j = \frac{5}{2}$ for a d electron in which $l = 2$.

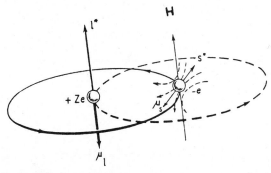

Figure 3.3 Diagram showing the orbital and spin angular moments above and the magnetic moments below. Figure from *Atomic Age Physics* by Henry Semat and Harvey E. White, copyright © 1959 by Holt, Rinehart and Winston, Inc., renewed 1987 by Barbara Semat, Edith Semat Kemp, and Harvey E. White, reprinted by permission of the publisher.

We now understand the reason for this. Figure 3.4 shows that the position of the electron spin axis is most stable when it is almost nearly parallel to the orbital field B. To roll the electron axis over from allowable position $j = l - \frac{1}{2}$ to allowable position $j = l + \frac{3}{2}$ requires exerting a torque through an angle θ and hence work. For this reason, the doublet state $j = l - \frac{1}{2}$ should lie lowest in the diagram.

Figure 3.4 depicts doublet levels in their relative positions labeled with their proper designations. Each capital letter signifies the orbital l value, and the subscripts are the j values.

The arrows in Figure 3.4 signify the transitions between energy levels in an atom. These transitions produce emissions of light and correspond to spectrum lines observed. The dotted arrows indicate *forbidden lines*. They show where transitions *cannot* take place. Thus, we may say that selection rules operate within the atom.

The first selection rule covers the orbital quantum number and stipulates that in any transition l may change by one—and *only* one—unit (Figure 3.5). Transitions may occur from $l = 3$ to $l = 2, l = 2$ to $l = 1, l = 1$ to $l = 0; l = 0$ to $l = -1, l = -1$ to $l = -2$, and so on; but they are forbidden from $l = 3$ to $l = 1$, $l = 2$ to $l = 0$, and so on. This may be expressed by the formula

$$\Delta l = \pm 1 \quad \text{or} \quad -1$$

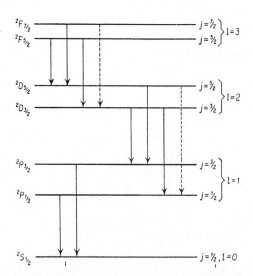

Figure 3.4 Diagram of typical energy levels showing the level designations and quantum numbers as well as allowed transitions. Figure from *Atomic Age Physics* by Henry Semat and Harvey E. White, copyright © 1959 by Holt, Rinehart and Winston, Inc., renewed 1987 by Barbara Semat, Edith Semat Kemp, and Harvey E. White, reprinted by permission of the publisher.

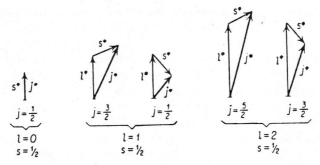

Figure 3.5   Electron spin-orbit quantization according to quantum mechanics. Figure from *Atomic Age Physics* by Henry Semat and Harvey E. White, copyright © 1959 by Holt, Rinehart and Winston, Inc., renewed 1987 by Barbara Semat, Edith Semat Kemp, and Harvey E. White, reprinted by permission of the publisher.

The second rule deals with the inner quantum number and may be written

$$\Delta j = +1, 0, \quad \text{or} \quad -1$$

From the diagram, we can see that the transitions $j = \frac{7}{2}$ to $j = \frac{3}{2}$, and so on, are forbidden.

In later discussions on paramagnetism we use a dimensionless parameter known as the *Landé spectroscopic splitting factor g*, which is often called the *g value* and is given by

$$g = 1 + \frac{S(S + 1) + J(J + 1) - L(L + 1)}{2J(J + 1)}$$

Although it is difficult to ascribe any physical significance to g, it may be looked on as a factor that uniquely describes a particular combination of S, L, and J values for a particular ground (or an excited) state of an ion or atom. This factor has helped to reconcile the observed and theoretical magnetic moments, which are discussed under Section 3 on definitions. It is easy to see that for a *free* spin with $L = 0$; that is, $J = S$, the g value is equal to 2. (Electron paramagnetic resonance spectroscopy establishes this value to be 2.00037.) Similarly, when the net spin S is equal to *zero* ($S = 0$, as with a paired set of electrons ↑↓), $J = L$ and $g = 1$.

3   DEFINITIONS, UNITS, AND BASIC LAWS

The questions: What is a magnetic field? and Why does a magnet attract iron? cannot be answered precisely and simply. Hence, using the approach of several texts on magnetism and considering the limitations of this discussion, it suffices to accept the phenomena and the physical basis of magnetism as commonly observed; therefore, certain concepts are defined within its own domain and in relation to the properties attributed to the fundamental particles of matter. Although it is relatively easy to describe basic concepts qualitatively, difficulties arise in defining them quantitatively, because the question of units must then be addressed. At first glance, magnetic units may appear to be the incomprehensible creations of "electromagneticists"; however, with sufficient patience and practice, various systems of units can be reconciled rather easily.

There has been indeed a great divergence among the systems of units employed not only by workers in various disciplines such as chemistry, physics, and electrical engineering, but also among groups from various countries such as England, France, Germany, Japan, India, the United States, and Russia.

In this chapter somewhat rigorous definitions of magnetic terms, including those presented earlier, are stated to help the reader to understand their significance and the new system of units relevant to magnetochemical investigations.

There are several centimeter-gram-second (cgs) systems of units. The most common system is the Gaussian system based on electrostatic units (esu) and electromagnetic units (emu). The introduction of the meter-kilogram-second (mks) system and other attempts to "rationalize" the approach have led to considerable confusion. For a clear understanding of this vexing situation reference should be made to the following publications that have appeared since 1965: Quickenden and Marshall [7], McGlashan [8], and reports by IUPAC (the International Union of Pure and Applied Chemistry) and IOS (International Organization for Standardization). The relatively new SI (Système International) units correspond essentially to the rationalized mks (or the Georgi) system, in which the units for length and mass are the meter and the kilogram, respectively [8]. A good discussion of the SI units for electricity and magnetism is given by Davies [9] and by others [5, 6, 10–13].

In the following sections, the intricate relationships among *all* systems of units are not discussed; instead, the SI units and their symbolisms are stated at appropriate places and are compared to the cgs system, which corresponds to the "unrationalized" (Gaussian) cgs–emu system. The cgs system appears to have been followed by most magnetochemists. Table 3.1 correlates magnetochemical formulas in the SI and other unit systems.

3.1 Magnetic Dipole

A *magnetic dipole* is a macroscopic or microscopic magnetic system in which the north and south poles of a magnet, equal and opposite in character, are separated by a short but definite distance. A magnetic dipole tends to orient itself parallel to an applied magnetic field in the same way that an electric dipole behaves in an electric field.

Table 3.1 Frequently Used Symbols, Nomenclature, and Factors for Conversion from cgs–emu (Gaussian) to mks–SI (Systéme International) Units in Magnetism[a–c]

Symbol	Definition	Symbol	Definition
A	Ampere (or ampere turns)	k	Boltzmann constant
B	Magnetic flux or induction	κ	Susceptibility per unit volume = $\chi \cdot \rho$ (χ is the mass susceptibility)
β or μ_B	Bohr magneton ($eh/4\pi mc$) (9.2731×10^{-11} erg G^{-1})		
c	Velocity of light	kg	Kilogram
cm	Centimeter	m	Meter ($= 100$ cm) (also used for pole strength)
χ	General symbol for magnetic susceptibility		
		m	Mass in general or mass of the electron
e	Charge on the electron	M or I	Magnetization
e	Electron	mol (or M)	Molar (or atomic)
g	Gram	N	Avogadro number
g	Landé splitting factor	N	Newton
g'	Gravitational constant	Oe	Oersted
G	Gauss	ϕ	Flux quantum ($hc/2e$)
h	Planck's constant	ρ	Density (g cm^{-3} or kg m^{-3})
h	Unit for inductance	s	Second
H	Magnetic field (i.e., intensity or strength of the field)	σ	Magnetization g^{-1} = emu g^{-1}
		T	Tesla
		v	Volume
H	H is sometimes used for henry	Wb	Weber (unit for magnetic pole) = 10^8 Maxwells (unit for B)
j or J	Exchange interaction or magnetic polarization (M); also for joules		

147

Table 3.1 (continued)

	Multiply Number for		Conversion from Gaussian to SI Units	To Obtain Number for
Gaussian Quantity	Unit	by	SI Quantity	Unit
Flux density, B	G	10^{-4}	Flux density, B	$T = Wb\ m^{-2}$ ($Vs\ m^2$)
Magnetic field strength, H	Oe	$10^3/4\pi$	Magnetic field strength, H	$A\ m^{-1}$
Volume susceptibility, κ (dimensionless)	emu cm^{-3}	4π	Rationalized volume susceptibility, κ	Dimensionless
Mass susceptibility, χ_ρ	emu g^{-1} ($\equiv cm^3\ g^{-1}$)	$4\pi \times 10^{-3}$	Rationalized mass susceptibility, κ_ρ	$m^3\ kg^{-1}$
Molar susceptibility, $\chi_{mole}{}^d$	emu mol^{-1} ($\equiv cm^3\ mol^{-1}$)	$4\pi \times 10^{-6}$	Rationalized molar susceptibility, κ_{mole}	$m^3\ mol^{-1}$
Magnetization, M or I (per cm^3)	G or Oe	10^3	Magnetization, M	$A\ m^{-1}$

Quantity	emu unit	Conversion	SI quantity	SI unit
Magnetization, M	μ_B/atom or μ_B/form. unit, etc.[b]	1	Magnetization, M	μ_B/atom or μ_B/form. unit, etc.[e]
Magnetic moment of a dipole, μ	erg G^{-1} or Oe·cm^3	10^{-3}	Magnetic moment of a dipole, m_i	A·m^2
Demagnetizing factor, N	Dimensionless	$1/4\pi$	Rationalized demagnetizing factor, N	Dimensionless
Bohr magneton, β or μ_B	erg G^{-1} (or erg Oe^{-1})	10^{-3}	Bohr magneton, β or μ_B	A·m^2

[a]In emu, the permeability μ of free space is equal to 1 and is dimensionless; whereas in SI, $\mu = 4\pi \times 10^{-7}\,\mathrm{kg \cdot ms^{-2} \cdot A^{-2}}$. However, the relative permeability $\mu_r = (\mu_{obs}/\mu_0)$ equals 1 and is dimensionless in both systems.

[b]For complete details, see chapters on "Units in Magnetism," by Mulay [5, 6].

[c]$\pi = 3.14$; $4\pi = 12.56$; $1/4\pi = 0.0796$; $10^3/4\pi = 79.61$.

[d]Also called atomic susceptibility. Molar susceptibility is preferred since atomic susceptibility has also been used to refer to the susceptibility per atom.

[e]Natural units, independent of unit system. However, the numerical value of the Bohr magneton does depend on the unit system.

3.2   Unit Pole

Dirac (cf. [5, 6]) proposed the concept of a unit (north or south) pole having a unit magnetic integer charge v of 3×10^{-18} emu. This concept corresponded to the existence of isolated electrical $(+)$ or $(-)$ charges. However, for several years the concept of a unit magnetic monopole was regarded as "ficticious" because it could not be isolated. A brief description of the "hunt for the monopole" in the lunar surface and other attempts are given in [5, 6] and in recent articles [14–17]. In view of the growing evidence the concept of the monopole can no longer be regarded as ficticious. The concept of the monopole (which was described as ficticious in many standard textbooks over the last century) *did help* to develop other useful quantitative aspects of magnetism. A unit pole is defined as one that repels an equal and similar pole, placed 1 cm away in vacuum, with a force of 1 dyn. The force between two poles is governed by Coulomb's law, which is discussed in Section 3.5. In the cgs–emu system no *specific* name was used for the unit monopole (north or south).

3.3   Pole Strength (m)

The strength (the attractive or repulsive power) of a magnet is measured by the number of unit poles to which each pole of the magnet is equivalent.

3.4   Intensity or Strength of a Magnetic Field (H)

If a unit pole is placed at a fixed point in vacuum in a magnetic field, it is acted upon by a force that is taken as a measure of the intensity or strength of the magnetic field. It follows from the preceding definitions that unit magnetic intensity exists at a point, where the force on a unit pole is 1 dyn. This unit magnetic intensity was formerly called the *gauss* and is used even today by many manufacturers and users of magnets. According to the recommendations of the International Conference on Physics, London, 1934, the term *oersted* is used instead. Some writers use the abbreviation Oe. A smaller unit, the gamma (γ), is equivalent to 10^{-5} Oe. In SI units the magnetic field strength is expressed as ampere (turns) per meter (A/m). The value in oersteds is multiplied by $10^3/4\pi$ to convert it to SI units. The ampere (turns) per meter concept is obviously based on electromagnetic induction and helps to unify both the electrical and magnetic parameters ($4\pi = 12.56$). Thus 1 Oe = 79.6 A/m, and 1 A/m = 0.0126 Oe.

Magnetic field intensity or the magnetizing force is also measured by the space rate of variation of magnetic potential. The unit was designated as the gilbert per centimeter and is the same as the oersted.

3.5   Magnetic Permeability (μ_0, μ, μ_r)

As explained previously, the magnetic permeability μ of a substance is best understood in terms of Coulomb's law, defining the force f between poles m_1 and m_2 separated by a distance r (in centimeters)

$$f = \frac{m_1 m_2}{\mu_0 r^2}$$

where μ_0 (a proportionality constant) has the same significance as the specific inductive capacity, or the dielectric constant k in the electrostatic phenomena. (In SI units the denominator is multiplied by 4π.)

In systems other than SI units the permeability μ_0 of vacuum is taken as 1 and is a dimensionless quantity. In SI units it has the dimensions kg ms^{-2} A^{-2} and is measured in terms of henrys per meter. (The relative permeability of a substance, defined by $\mu_r = \mu/\mu_0$, naturally becomes a dimensionless quantity in all systems of units.)

A rigorous definition of μ_0 from the force equation is obtained from the Biot–Savart law, which establishes the magnitude and direction of induction $d\bar{B}$ at any arbitrary point in a field H set up by a conductor dl carrying a current i. (In general a solenoid of n turns with unit radius carrying a current i produces a field $H = ni$.)

The dimensions kg ms^{-2} A^{-2} for μ_0 in SI are obtained by considering that the

$$\text{force (newtons)} = \text{Wb}^2/\mu_0 r^2$$

giving

$$\mu_0 = \text{Wb}^2/\text{N} \cdot \text{m}^2$$

The force on a unit pole (webers) in a field H is given by $f = \text{Wb} \cdot H$. Thus 1 Wb = N/A m^{-1}. Hence, $\mu_0 = (\text{N/A m}^{-1})^2/(\text{N m}^2) = \text{N/A}^2$.

Since the dimensions for the newton (in terms of mass $\times$ acceleration) are kg ms^{-2} in SI units, $\mu_0 = $ kg ms^{-2} A^{-2} in SI units, and $\mu_0 = 4\pi \times 10^{-7}$ H (henrys)$\cdot$m^{-1} as shown in standard texts (Bleany and Bleany [12]).

3.6   Magnetic Flux Density (Magnetic Induction B) and Flux

As shown in Figure 3.5, when a substance is placed in a field of H oersteds, then B, the magnetic induction (expressed in gauss), is given by the sum of the applied field H plus a contribution $4\pi M$, where M (also expressed in gauss) is designated as the intensity of magnetization caused by the substance. Quantitively, the induction B is visualized as the density of lines of force per unit area A in the substance

$$B = H + 4\pi M$$

The factor 4π arises from Gauss' law, which considers the number of lines of force emanating from a unit magnetic pole enclosed at the center of a sphere of 1-cm radius. Since the surface area of this sphere is 4π cm^2, it is assumed that 4π

lines of force emanate from a unit pole. Thus for a pole of strength m, $4\pi m$ Mx emanate from the surface.

The magnetic induction is also defined by

$$B = \mu_0 H$$

where μ_0 is the magnetic permeability and is dimensionless in the cgs–emu system on the assumption that B and H are measured in the same units (gauss). Its dimensions in the SI system were discussed in Section 3.5.

Combining these equations, one obtains

$$\mu_0 = B/H = 1 + 4\pi\kappa$$

where κ is termed the volume magnetic susceptibility, which is described in Section 3.8.

The magnetic flux density (flux per unit area) or magnetic induction in SI units is expressed as the tesla (T) or weber per square meter (Wb m^{-2}). The value in gauss is multiplied by 10^{-4} to express it as T or Wb m^{-2}. (Thus, the tesla is a much larger unit, $1\,T = 10^4$ G.)

In electrical engineering literature, the magnetic flux was expressed in units of maxwell (or the line); another unit, Wb $= 10^8$ Mx was also used.

3.7   Magnetization or the Intensity of Magnetization (I, M)

The amount of pole induced over unit area represents the magnetization. We discuss the concept of spontaneous magnetization I_{sp} under ferromagnetism (Section 4.8.4). Sometimes the symbol M is used in place of I; thus,

$$M \text{ (or } I) = m/A$$

where m is the induced pole strength over a total area of A cm^2 (Figure 3.6). Magnetization represents the magnetic polarization induced in a material by an applied field H. An alternative definition for magnetization is obtained by

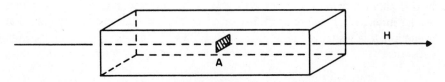

Figure 3.6 Magnetization induced in unit area inside a sample in a magnetic field. Reprinted with permission from L. N. Mulay, *Magnetic Susceptibility*, Interscience, New York, 1966, a reprint monograph based on L. N. Mulay, in I. M. Kolthoff and P. J. Elving, Eds., *Treatise on Analytical Chemistry*, Part 1, Vol. 4, Interscience, New York, 1963, Chap. 38. Copyright © 1966 by John Wiley & Sons, Inc.

multiplying both the numerator and the denominator by the distance l; this gives

$$M \text{ (or } I) = \frac{m}{A} = \frac{ml}{Al} = \frac{\mu}{\text{volume}}$$

or the magnetic moment per unit volume. The magnetic moment is defined in Section 3.11. In electrostatics the electric polarization P (corresponding to the magnetic polarization J) is defined as the electric dipole moment induced per unit volume. Correspondingly, most physicists and electrical engineers think of magnetization as the induced moment on a per unit basis; for instance, as Bohr magnetons per atom or ion. Their approaches stem from the basic concept that at the atomistic level the magnetization arises from the number of particles (atoms or ions) per unit volume and that each species (atoms or ions) possesses a moment μ_A that can be expressed in units of ergs per oersted or in terms of the Bohr magneton (see Section 3.11).

In the cgs–emu system the magnetization is commonly expressed by: gauss = $\mu \, \text{cm}^{-3}$; therefore, *gauss* has the dimensions cm^{-3} inherently incorporated in its significance. When calculating the magnetization per gram (denoted by σ), one divides gauss by the density $(\text{g} \cdot \text{cm}^{-3})$. Thus, $\sigma = \text{G/g} \cdot \text{cm}^{-3} = \text{G} \cdot \text{cm}^3/\text{g}$, which yields a peculiar term $\text{G} \cdot \text{cm}^3$ in the numerator. In the unrationalized system, the use of this term is circumvented by simply replacing it with emu. The σ is expressed as emu/g in most magnetics literature. Although emu stands for the *electromagnetic* (*system of*) *units*, its use has been firmly rooted in the literature, as though emu itself is a unit. Bleany and Bleany [12] and others have listed emu as a unit not only for magnetization, but also for the Bohr magneton! The unit for the Bohr magneton (explained under Section 3.11) is ergs per oersted or ergs per gauss. Hence, some authors use the unit $\text{ergs} \cdot \text{oersted} \cdot \text{cm}^{-3}$ for magnetization.

In SI units, magnetization is expressed as ampere (turns) per meter (A m^{-1}); for conversion to this system, the value is multiplied by 10^3.

Several authors, including Crooks [18] proposed that magnetization should be designated appropriately by the symbol B, which has been used generally for total induction. They chose B because it is the field *induced* by a current flowing in a conductor; thus, it should unify the concepts of electricity and magnetism.

Occasionally magnetization M or I is thought of in terms of magnetic *polarization* for which the symbol J was used in earlier literature. If the unrationalized cgs–emu value of M is expressed in terms of gauss or oersted, then its value in SI units is obtained by multiplying by the factor $4\pi \times 10^{-4}$ and the unit T (tesla) is used. Hence, if the magnetization is expressed initially as $4\pi M$, then one simply multiplies this value by 10^{-4} to obtain the corresponding J magnetization value in tesla units. The symbol J is used in the magnetics literature for the total angular momentum $(S + L)$ and also for the magnetic exchange interaction. Traditionally, J stands for the joule, a unit of energy equal to 10^7 ergs.

In Section 4.8.4 on ferromagnetism, ferrimagnetism, and paramagnetism, we illustrate the interconversion of the magnetization (σ, emu per gram of the material) and the moment (μ, expressed as Bohr magnetons per ion or atom). This illustration is applicable to paramagnetism, also, when measurements of σ are available. The χ_g is then calculated from σ/H.

The magnetization is thus considered in terms of the (change in) energy of a system induced by a magnetic field. Wertz and Bolton [19] and Flygare and colleagues,[†] for example [20], defined the magnetization I per unit volume as ergs per gauss cubic centimeter without making a rigorous distinction between the gauss and the oersted. They point out [20] that this definition in turn is equal to the *units* (or the *dimensions*) of the gauss, which can be seen by substituting $g^{1/2}$ cm$^{-1/2}$ s$^{-1/2}$ for the gauss.

$$\frac{\text{erg}}{\text{G} \cdot \text{cm}^3} = \frac{\text{g cm}^2 \text{ s}^{-2}}{g^{1/2} \text{ cm}^{-1/2} \text{ s}^{-1} \text{ cm}^3} = g^{1/2} \text{ cm}^{-1/2} \text{ s}^{-1}$$

The symbolism used by various groups of workers for the magnetization is also confusing and can be inferred only by a careful analysis of the context. Quite often σ refers to the magnetization (or the *moment*) per gram and M to the molar magnetization (i.e., $\sigma \times$ molecular weight). These symbols are used generally in ferromagnetism and in related areas.

3.8   Volume Magnetic Susceptibility (κ or χ_v)

The intensity of magnetization I induced at any point in a body is proportional to the strength of the applied field H

$$I = \kappa \cdot H \qquad \text{or} \qquad \kappa = I/H$$

where κ is a constant of proportionality depending on the material of the body. It is called the magnetic susceptibility per unit volume or the *volume susceptibility*, and may be defined qualitatively as the extent to which a material is susceptible to induced magnetization. For an isotropic body the susceptibility is the same in all directions. However, for anisotropic crystals, the susceptibilities along the three principle magnetic axes (κ_1, κ_2, κ_3) are different, and measurements on their powder samples give the average κ of the three values $\kappa = (\kappa_1 + \kappa_2 + \kappa_3)/3$.

The volume susceptibility is often designated by the symbols χ or χ_v and refers to the susceptibility per cubic centimeter. Since it represents a ratio of the induced magnetization to the applied field (measured in the same units), it is considered to be a dimensionless quantity. This value multiplied by 4π gives the

[†]Several publications from Flygare's group use ergs per square gauss mole as units for molar susceptibility.

corresponding volume susceptibility in SI units and is also considered to be dimensionless. Thus in SI units κ is 12.56 times larger than in cgs.

Despite the general dimensionless character of the volume susceptibility κ, one finds in the magnetics literature that κ is often expressed as emu per cubic centimeter (in the cgs units) and as mks per cubic meter (in SI units).

The volume susceptibility is related to the permeability in the following manner:

$$\kappa = \frac{\mu_r - 1}{4\pi} \quad \text{(cgs units)}$$

$$= \mu_r - 1 \quad \text{(SI units)}$$

Considering that the magnetization I per unit volume may be assumed to have the dimensions of ergs per gauss cubic centimeters and the field H is expressed in gauss, the volume susceptibility $\kappa(=I/H)$ has been expressed by a few workers [19, 20] in the so-called *units* ergs per square gauss cubic centimeters. Again, remembering the dimensions for ergs and gauss as explained in Section 3.7, we find that

$$\kappa = (\text{g cm}^2 \text{ s}^{-2})(\text{g}^{1/2} \text{ cm}^{-1/2} \text{ s}^{-1})^{-2} \text{ cm}^{-3}$$

$$= (\text{g cm}^2 \text{ s}^{-2})(\text{g}^{-1} \text{ cm}^1 \text{ s}^2) \text{ cm}^{-3}$$

$$= (\text{cm}^3) \text{ cm}^{-3}$$

This result reiterates the dimensionless character of κ.

3.9 Per Gram Magnetic Susceptibility (χ, χ_m, χ_g)

This is also known as the *specific* or *mass susceptibility* and sometimes is represented by symbols such as χ_m. The symbol χ is usually used for the mass susceptibility and is related to the volume susceptibility κ in the following manner:

$$\chi = \frac{\kappa}{(\rho)_{\text{SI}}} \quad \text{In SI units, the density } \rho \text{ is expressed in units of kilograms per cubic meter.}$$

or

$$\chi = \frac{\kappa}{(\rho)_{\text{cgs}}} \quad \text{In the cgs system, the density } \rho \text{ is expressed in grams per cubic centimeter.}$$

The mass susceptibility in the above cgs system when multiplied by $4\pi \times 10^{-3}$ gives the corresponding value in SI units ($4\pi = 12.56$). Since it essentially represents the reciprocal of density, the SI unit for this quantity is cubic meters per gram. Similarly in the cgs system the unit is cubic centimeters per gram. (In exceptional cases [19] ergs per square gauss gram is also used for the mass susceptibility, which in reality has the same dimensions as cubic centimeters per gram.) The ergs $G^{-2} g^{-1}$ unit stems from considerations of changes in energy in the material when a field is applied.

Thus, another equivalent definition of χ_m is obtained by considering how the energy of the material is altered by the applied magnetic field, rather than by the field induced in it. If W is the energy per mole of the substance then,

$$\chi_m = -\frac{1}{H} \cdot \frac{\partial W}{\partial H}$$

Thus χ_m can be expressed in cgs as ergs per square oersted per mole or as ergs per square gauss per mole.

Section 3.14 gives a brief discussion of magnetic susceptibility anisotropy.

3.10 Molar Magnetic Susceptibility (χ_M)

This term is often called the *susceptibility per mole*. It is usually designated by the symbol χ_M and is given by

$$\chi_M = \chi \times M = \kappa \times \frac{M}{\rho}$$

Here κ is the volume susceptibility and M is the molar mass (molecular weight). In SI units M is expressed in kilograms per mole and ρ, the density in kilograms per cubic meter. Thus, χ_M is expressed in cubic meters per mole (in SI units). Similarly in the cgs system, M is expressed in grams per mole, and ρ, in grams per cubic centimeter so that χ_M is given in cubic centimeters per mole. Hence, multiplying (χ_M) cgs by $4\pi \times 10^{-6}$ gives the corresponding (χ_M) in SI units (cubic meters per mole).

These conversions are illustrated by the molar magnetic susceptibility of the trivalent ytterbium ion [9]. In the foregoing cgs system Yb^{3+} has $\chi_M = 8550 \times 10^{-6} \, cm^3 \, mol^{-1}$. In the SI units it has the value $0.1091 \times 10^{-6} \, m^3 \, mol^{-1}$, which shows the relatively small magnitudes of χ_M encountered in this new system.

For the benefit of *nonchemists*, since the atomic weight (and the *ionic weight* in the case of ions) can be defined in exactly the same manner as the molar mass M, there is generally no need to define separately the so-called atomic (or ionic) susceptibility. Thus, the phrase *molar susceptibility of atom* A (*or ion* A) corresponds to the so-called atomic susceptibility (or to the susceptibility per gram-atom) stated in most magnetochemical literature

$$\chi_A = \chi \times \text{atomic weight}$$

3.11   Magnetic Moment and the Bohr Magneton (μ_B or β)

Magnetic moment is a term that is probably most widely known to chemists but one whose physical significance is least understood. Like the *moment of force*, the magnetic moment refers to the turning effect produced under certain conditions. When a magnetic dipole is placed in a magnetic field, it experiences a turning effect, which is proportional to a specific character termed the *magnetic moment*. If a field of strength H acts on a dipole N–S of length l and strength m, the N and S poles of the dipole will experience forces equal to $+mH$ and $-mH$, respectively (Figure 3.7). These two equal and opposite forces constitute a "couple" (of forces) whose turning moment M' is given by

$$M' = \text{force} \times \text{distance}$$

$$= mH \times l \sin \theta$$

$$= \mu H \sin \theta$$

where θ is the angle between the magnetic dipole and the direction of the applied field.

Thus, the quantity $\mu\,(=ml)$ defines the magnetic moment (Figure 3.7) and serves as a measure of the turning effect. It is expressed in dyne centimeter per oersted or ergs per oersted. Although no practical unit for the magnetic moment is formulated, experiments with the basic electrical and magnetic properties of fundamental particles revealed the existence of a fundamental unit or magnetic moment, the Bohr magneton. This is just as real a quantity as the charge of an electron and may be placed among the *universal constants*. Often abbreviated

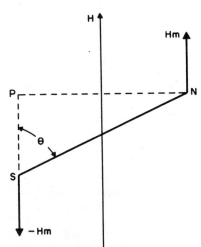

Figure 3.7   Forces $+Hm$ and $-Hm$ on a magnetic dipole (N–S) showing the *moment* or turning effect. Reprinted with permission from L. N. Mulay, *Magnetic Susceptibility*, Interscience, New York, 1966, a reprint monograph based on L. N. Mulay, in I. M. Kolthoff and P. J. Elving, Eds., *Treatise on Analytical Chemistry*, Part 1, Vol. 4, Interscience, New York, 1963, Chap. 38. Copyright © 1966 by John Wiley & Sons, Inc.

BM or given the symbol β or μ_B, this unit is equal to $eh/4\pi mc$ in the cgs system. Here e and m are, respectively, the charge and the mass of the electron, h is the Planck constant (6.62×10^{-27} erg·s), and c is the velocity of light (2.997×10^{10} cm s^{-1}). When the values of these quantities are introduced, $\mu_B = 9.274 \times 10^{-21}$ ergs Oe^{-1} in cgs units. Some authors associate a negative sign with the Bohr magneton since it arises from a negative electronic charge. [This is said to facilitate its distinction from the corresponding nuclear magneton (μ_n), which has the value $+5.05 \times 10^{-24}$ erg Oe^{-1}.] In early work the Weiss magneton was used for the moment of the electron; 1 Bohr magneton is equivalent to 4.97 Weiss magnetons.

In SI units the Bohr magneton μ_B is given by

$$\mu_B = eh/4\pi m = 9.274 \times 10^{-24} \text{ A·m}^2$$

3.12 Effective Magnetic Moment (μ_{eff} or p_{eff})

As mentioned previously and with reference to Figure 3.8, one may picture a slow precision of the total angular momentum J about the field direction H, and, in turn, visualize the precession of the resultant moment μ_R (OA') about J. Thus, the projection OC of μ_R on J represents the effective behavior of μ_R and as such is called the *effective magnetic moment* μ_{eff}.

The effective moment μ_{eff} is often defined in terms of J, which is the vector sum of the orbital quantum number L and the spin quantum number S

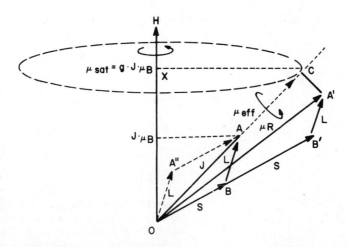

Figure 3.8 Vector diagram showing the significance of the effective magnetic moment μ_{eff}: The resultant moment μ_R arises from $2S + L$. The projection of μ_R on the J axis gives its effective value μ_{eff}. Reprinted with permission from L. N. Mulay and E. A. Boudreaux, Eds., *Theory and Applications of Molecular Diamagnetism*, Wiley-Interscience, New York, 1976. Copyright © 1976 by John Wiley & Sons, Inc.

$$\mu_{\text{eff}} = g\sqrt{J(J+1)}\,\mu_{\text{B}}$$

where g is the Landé splitting factor and μ_{B} is the Bohr magneton.

Since S, L, and J are the same in cgs and SI, it follows that the μ_{eff} obtained in this manner or from the well-known spin-only formula

$$\mu_{\text{eff}} = \sqrt{n(n+2)}$$

should be the same when μ_{B}, the Bohr magneton, is adopted as a unit in the two systems.

Consider the Van Vleck [21] relation between the temperature-dependent susceptibility χ_A of a magnetically dilute paramagnet and the effective moment of the associated electron

$$\chi_A = N\mu_0\mu_{\text{eff}}^2/3kT$$

$$\mu_{\text{eff}} = \sqrt{3k/N\mu_0\mu_{\text{B}}^2} \times \sqrt{\chi_A T}$$

where μ_0 represents permeability, and k and N are the Boltzmann constant and the Avogadro number, respectively. If the numerical values are inserted in appropriate cgs units the result is in Bohr magnetons

$$\mu_{\text{eff}} = 2.828\sqrt{\chi_A(\text{cgs})T}\,\mu_{\text{B}}$$

Correspondingly, the insertion of the values in the SI units yields μ_{eff} in Bohr magnetons by the relation

$$\mu_{\text{eff}} = 797.5\sqrt{\chi_A(\text{SI})T}\,\mu_{\text{B}}$$

Although the Bohr magneton μ_{B} is expressed in units of ergs per oersted in the cgs system, and as $A \cdot m^2$ (ampere meter squared) in SI units, fortunately it turns out that the effective magnetic moment μ_{eff} gives the same numerical value in both systems when expressed in units of the Bohr magneton μ_{B} by considering μ_{B} to be a *fundamental unit* or a *universal constant*.

Thus, in the case of the Yb^{3+} ion, the μ_{eff} is obtained in the following manner in the two systems, from its susceptibility, χ_A at 300 K

$$\mu_{\text{eff}}(\text{cgs}) = 2.828\sqrt{\chi_A(\text{cgs})T}\,\mu_{\text{B}}$$

$$= 2.828\sqrt{8550 \times 10^{-6} \times 300}\,\mu_{\text{B}}$$

$$= 4.53\,\mu_{\text{B}}$$

Similarly

$$\mu_{eff}(SI) = 797.5\sqrt{\chi_A(SI)T}\,\mu_B$$
$$= 797.5\sqrt{0.1091 \times 10^{-6} \times 300}\,\mu_B$$
$$= 4.53\,\mu_B$$

Therefore, it is advantageous to report μ_{eff} in units of the Bohr magneton μ_B; that is, $\mu_B = 9.274 \times 10^{-24}\,A \cdot m^2$. However, it is easy to use the following relation if there is a need to express μ_{eff} in units of $A \cdot m^2$

$$\mu_{eff}(SI) = 7.394 \times 10^{-21}\sqrt{\chi_A(SI) \cdot T}\,A \cdot m^2$$

Since the moment μ is defined by $\mu = m \times l$, where m is the pole strength of the magnetic dipole and l is the distance between the poles, a unit called weber-meter is used for the magnetic moment. One weber-meter $= (1/4\pi) \times 10^{10}\,G \cdot cm^3$. Furthermore, since μ is defined by $M = \mu/vol$ or $\mu = M \times vol$, it follows that μ may be expressed in units such as oersted cubic centimeters. Selwood [22] and others have adopted the oersted cubic centimeters or gauss cubic centimeters unit for the Bohr magneton (β). Thus, $\beta = 9.27 \times 10^{-21}$ ergs Oe^{-1} or $Oe\,cm^{-3}$. The equivalence of the two units is obtained by carrying out an analysis of the fundamental dimensions of the erg and of the oersted

$$\text{ergs } Oe^{-1} = (g\,cm^2\,s^{-2})(g^{1/2}\,cm^{-1/2}\,s^{-1})^{-1}$$
$$= g^{1/2}\,cm^{5/2}\,s^{-1}$$

and

$$Oe\,cm^3 = (g^{1/2}\,cm^{1/2}\,s^{-1})(cm^3)$$
$$= g^{1/2}\,cm^{5/2}\,s^{-1}$$

Furthermore, the unit oersted cubic centimeters for μ can be justified by considering the fundamental definition for magnetization; namely, $M = \mu/vol$. Since M is expressed in oersteds and vol in cubic centimeters, it follows that $\mu = M \cdot vol = Oe \cdot cm^3$.

3.13 The Saturation Moment (μ_{sat}, μ_s)

The saturation moment is based on the concept of observing the maximum value of the moment by aligning all the dipoles with the field, by maximizing it (to say 40,000 Oe), and by minimizing the randomization effects of the thermal energy by going down to perhaps 4 K. Thus, one can obtain the very high value of 10,000 for the ratio $H:T$, for saturation purposes, which are absent under ordinary conditions of magnetochemical research. Here, the ratio $H:T$ has a relatively small value of 70, which corresponds to maximum fields of 21,000 Oe and a temperature around 300 K.

In view of this concept of the saturation moment μ_{sat}, one expects the μ_{sat} (corresponding to a complete alignment of dipoles) to be greater than the μ_{eff}, which represents only a small fractional alignment of the spins. Quite surprisingly, in the literature

$$\mu_{sat} < \mu_{eff}$$

For example, μ_{sat} for the Fe^{3+} ion with $n = 5$ unpaired spins is just n; that is, $\mu_{sat} = 5\mu_B$, whereas, $\mu_{eff} = \sqrt{n(n+2)}\mu_B = 5.96\mu_B$.

The reason for this apparent paradox lies in the basic definition of μ_{sat}. Again, with reference to Figure 3.8, μ_{sat} is defined as the maximum component $gJ \cdot \mu_B$ on the direction of the field produced by gJ, whereas μ_{eff} is defined as the projection of the resultant moment μ_R on J (i.e., OC). It is now easy to see from this vectorial presentation why μ_{sat} is less than μ_{eff}. The need for such descriptions stems from considerations of complete space quantization of angular moments in a field.

3.14   Magnetic Anisotropy

The previous definitions pertain to isotropic media, in which B, H, and I have the same direction and thus the volume susceptibility κ is a scalar quantity that is dimensionless. However, in many crystalline materials B and H are not parallel and thus their relation is more complex. If we choose a set of Cartesian axes, B and H have components along each, which are related as follows:

$$B_x = \mu_{11}H_x + \mu_{12}H_y + \mu_{13}H_z$$
$$B_y = \mu_{21}H_x + \mu_{22}H_y + \mu_{23}H_z$$
$$B_z = \mu_{31}H_x + \mu_{32}H_y + \mu_{33}H_z$$

where the quantities μ_y are components of the permeability tensor. Correspondingly, it is possible to define components χ_{ij} of the susceptibility tensor for anisotropic (single-crystal) substances. Hence, the anisotropic components of the per gram susceptibility (χ_1, χ_2, χ_3) and the anisotropic components of the moment (μ_1, μ_2, μ_3) and the associated g values, such as $g_\parallel$ and $g_\perp$, which refer to the observed value when the applied field H is parallel ($\parallel$) to or perpendicular ($\perp$) to a given axis of the crystal, correspond to the components of volume susceptibility ($\kappa_1, \kappa_2, \kappa_3$).

3.15   Basic Laws

The fundamental inverse square law of Coulomb and Gauss' law, to which reference was made earlier, are the fundamental laws for electromagnetism. In later sections we consider the Curie–Weiss law encountered in paramagnetism and ferromagnetism. In this section we discuss laws of a more direct practical application.

3.15.1   Wiedemann's Law of Additivity

The mass susceptibility χ of a mixture of components with susceptibilities $\chi_1, \chi_2, \ldots, \chi_n$ and weight fractions $p_1, p_2, \ldots, p_n$ may be expressed by

$$\chi = \chi_1 p_1 + \chi_2 p_2 + \cdots + \chi_n p_n$$

The law is obeyed quite closely by mechanical mixtures and solutions of diamagnetic substances in which little or no interaction occurs either between molecules or ions of the components or between these and the solvent. This stipulation demands caution when deducing the susceptibility of a solute from that of the solution. The application of the law to solid or liquid solutions containing paramagnetic ions becomes even more difficult as the interactions among or between ions and the solvent become quite pronounced. Therefore, in such cases it is necessary to ascertain that the system is *magnetically dilute*.

Table 3.2   Atomic Susceptibility Constants[a]

Atom	$\chi\,(\times 10^{-6})$	Atom		$\chi\,(\times 10^{-6})$
Ag	-31.0	N	Open chain	-5.57
Al	-13.0		Closed chain (ring)	-4.61
As(III)	-20.9		Monamides	-1.54
As(V)	-43.0		Diamides and imides	-2.11
B	-7.0	Na		-9.2
Bi	-192.0	O	Alcohol or ether	-4.61
Br	-30.6		Aldehyde or ketone	$+1.73$
C	-6.0		Carboxylic $=$ O in esters and acids	-3.36
Ca	-15.9		3O atoms in acid anhydrides	-11.23
Cl	-20.1	P		-26.30
F	-6.3	Pb(II)		-46.0
H	-2.93	S		-15.0
Hg(II)	-33.0	Sb(III)		-74.0
I	-44.6	Se		-23.0
K	-18.5	Si		-20.0
Li	-4.2	Sn(IV)		-30.0
Mg	-10.0	Te		-37.3
		Tl(I)		-40.0
		Zn		-13.5

[a]Many of the values listed are taken from Figgis and Lewis [24]. Prasad and Mulay [23] provide values for As and Sb in different organometallic compounds; Mulay and Boudreaux [5, 6] provide additional information.

Table 3.3 Constitutive Correction Constants[a]

Group	$\lambda(\times 10^{-6})$
C=C, ethylenic linkage	+5.5
C≡C, acetylenic linkage	+0.8
C=C—C=C, diethylenic linkage	+10.6
Ar—C≡C—Ar	+3.85
CH_2=CH—CH_2—, allyl group	+4.5
Ar—C≡C—	+2.30
C in one aromatic ring (e.g., benzene)	−0.24
C in two aromatic rings (e.g., naphthalene)	−3.1
C in three aromatic rings (e.g., pyrene)	−4.0
$\diagup\!\!\!\diagdown$C—Br, monobromo derivative	−4.1
BrC—CBr, dibromo derivative	+6.24
$\diagup\!\!\!\diagdown$C—Cl, monochloro derivative	+3.1
ClC—CCl, dichloro derivative	+4.3
CCl_2	+1.44
—$CHCl_2$	+6.43
$\diagup\!\!\!\diagdown$C—I, monoiodo derivative	+4.1
C=NR	+8.2
RC≡N	+0.8
RN≡C	0.00
C=N—N=C, azines	+10.2
RC≡C—C(=O)R′ or RC≡C—C(=O)OR′	+0.8
C bound to other C atoms with 3 bonds and in α, γ, δ, or ε position with respect to —C=O	−1.3
C bound to other C atoms with 4 bounds and in α, γ, δ, or ε position with respect to —C=O	−1.54
C bound to other C atoms with 3 or 4 bonds in β position with respect to a —C=O group	−0.5
—N=N—, azo group	+1.8
—N=O	+1.7

[a]Groups containing only carbon atoms are listed first, then groups containing other atoms in the order of the symbols of these other atoms. For constitutive correction constants for cyclic systems, see Table 3.4.

3.15.2   Additivity of Atomic Constants

Pascal studied many gases and organic, metalloid, and complex compounds. The data collected by him and the constants derived for atomic susceptibilities on the assumption of an additivity law indeed stand out as a pioneering contribution to magnetochemistry. According to him, the molecular (molar) susceptibility χ_m of a compound can be expressed by

$$\chi_m = \sum n_A \cdot \chi_A + \lambda$$

where n_A is the number of atoms of susceptibility χ_A in the molecule and λ is a constitutive correction constant depending on the nature of chemical binding between the atoms. For ions, it is assumed that

$$\chi_m = \chi_{\text{cation}} + \chi_{\text{anion}}$$

which invalidates the need for a correction constant.

However, the derivation of Pascal's constants is purely empirical and is the result of a judicious mathematical juggling of numbers. A book on diamagnetism by Mulay and Boudreaux [5] gives a thorough discussion both for the theoretical significance of the constitutive correction constant λ and of alternate methods for calculating the molar susceptibility from atomic constants other than those suggested by Pascal.

Table 3.4   Constitutive Correction Constants for Cyclic Systems[a]

Structure	$\lambda(\times 10^{-6})$	Structure	$\lambda(\times 10^{-6})$
Benzene	−1.4	Piperidine	+3.0
Cyclobutane	+7.2	Pyramidon	0.0
Cyclohexadiene	+10.36	Pyrazine	+9.0
Cyclohexane	+3.0	Pyrazole	+8.0
Cyclohexene	+6.9	Pyridine	+0.5
Cyclopentane	0.0	Pyrimidine	+6.3
Cyclopropane	+7.2	α-Pyrone or γ-pyrone	−1.4
Dicyclohexyl (C_6H_{11})$_2$	+7.8	Pyrrole	−3.5
Dioxane	+5.5	Pyrrolidine	0.0
Furan	−2.5	Tetrahydrofuran	0.0
Imidazole	+8.0	Thiazole	−3.0
Isoxazole	+1.0	Thiophene	−7.0
Morpholine	+5.3	Triazine	−1.4
Piperazine	+7.0	Urazol	0.0

[a]Most of the values listed are taken from Mulay and Boudreaux [5]. For the constitutive correction constants corresponding to other structural features, see Table 3.5.

Table 3.5 Diamagnetic Susceptibilities per Gram-Ion[a] (All Values -1×10^{-6}); Underlying Dimagnetism of Paramagnetic Ions Indicated by *

Ion	Susceptibility	Ion	Susceptibility	Ion	Susceptibility	Ion	Susceptibility
Ag^+	24	$*Eu^{2+}$	22	Nb^{5+}	9	Se^{4+}	8
$*Ag^{2+}$	24?	$*Eu^{3+}$	20	$*Nd^{3+}$	20	Se^{6+}	5
Al^{3+}	2	F^-	11	$*Ni^{2+}$	12	SeO_3^{2-}	44
As^{3+}	9?	$*Fe^{2+}$	13	O^{2-}	12	SeO_4^{2-}	51
As^{5+}	6	$*Fe^{3+}$	10	OH^-	12	Si^{4+}	1
AsO_3^{3-}	51	Ga^{3+}	8	$*OS^{2+}$	44	SiO_3^{2-}	36
AsO_4^{3-}	60	Ge^{4+}	7	$*Os^{3+}$	36	$*Sm^{2+}$	23
Au^+	40?	Gd^{3+}	20	$*Os^{4+}$	29	$*Sm^{3+}$	20
Au^{3+}	32	H^+	0	$*Os^{6+}$	18	Sn^{2+}	20
B^{3+}	0.2	Hf^{4+}	16	Os^{8+}	11	Sn^{4+}	16
BF_4^-	39	Hg^{2+}	37	P^{3+}	4	Sr^{2+}	15
BO_3^{3-}	35	$*Ho^{3+}$	19	P^{5+}	1	Ta^{5+}	14
Ba^{2+}	32	I^-	52	PO_3^-	30	$*Tb^{3+}$	19
Be^{2+}	0.4	I^{5+}	12	PO_3^{3-}	42	$*Tb^{4+}$	17
Bi^{3+}	25?	I^{7+}	10	Pb^{2+}	28	Te^{2-}	70
Bi^{5+}	23	IO_3^-	50	Pb^{4+}	26	Te^{4+}	14
Br^-	36	IO_4^-	54	$*Pd^{2+}$	25	Te^{6+}	12
Br^{5+}	6	In^{3+}	19	$*Pd^{4+}$	18	TeO_3^{2-}	63
BrO_3^-	40	$*Ir^+$	50	Pm^{3+}	27[b]	TeO_4^{3-}	55
C^{4+}	0.1	$*Ir^{2+}$	42	$*Pr^{3+}$	20	Th^{4+}	23
CN^-	18	$*Ir^{3+}$	35	$*Pr^{4+}$	17	$*Ti^{3+}$	9
CNO^-	21	$*Ir^{4+}$	29	$*Pt^{2+}$	40	Ti^{4+}	5
CNS^-	35	$*Ir^{5+}$	20	$*Pt^{3+}$	33	Tl^+	34
CO_3^{2-}	34	K^+	13	$*Pt^{4+}$	28	Tl^{3+}	31
Ca^{2+}	8	La^{3+}	20	Rb^+	20	$*Tm^{3+}$	18
Cd^{2+}	22	Li^+	0.6	$*Re^{3+}$	36	$*U^{3+}$	46

165

Table 3.5 (continued)

Ion	Susceptibility	Ion	Susceptibility	Ion	Susceptibility	Ion	Susceptibility
*Ce^{3+}	20	Lu^{3+}	17	*Re^{4+}	28	*U^{4+}	35
Ce^{4+}	17	Mg^{2+}	3	*Re^{6+}	16	*U^{5+}	26
Cl^{-}	26	Mn^{2+}	14	Re^{7+}	12	U^{6+}	19
Cl^{5+}	2	Mn^{3+}	10	*Rh^{3+}	22	*V^{2+}	15
ClO_3^{-}	32	*Mn^{4+}	8	*Rh^{4+}	18	*V^{3+}	10
ClO_4^{-}	34	*Mn^{6+}	4	*Ru^{3+}	23	*V^{4+}	7
*Co^{2+}	12	*Mn^{7+}	3	*Ru^{4+}	18	V^{5+}	4
*Co^{3+}	10	*Mo^{2+}	31	S^{2-}	38?	*W^{2+}	41
*Cr^{2+}	15	*Mo^{3+}	23	S^{4+}	3	*W^{3+}	36
*Cr^{3+}	11	*Mo^{4+}	17	S^{6+}	1	*W^{4+}	23
*Cr^{4+}	8	*Mo^{5+}	12	SO_3^{2-}	38	*W^{5+}	19
*Cr^{5+}	5	MO^{6+}	7	SO_4^{2-}	40	W^{6+}	13
Cr^{6+}	3	N^{5+}	0.1	$S_2O_8^{3-}$	78	Y^{3+}	12
Cs^{+}	31	NH_4^{+}	11.5	Sb^{3+}	17?	Yb^{2+}	20
Cu^{+}	12	NO_2^{-}	10	Sb^{5+}	14	*Yb^{3+}	18
*Cu^{2+}	11	NO_3^{-}	20	Sc^{3+}	6	Zn^{3+}	10
*Dy^{3+}	19	Na^{+}	5	Se^{2-}	48?	Zn^{4+}	10
*Er^{3+}	18						

[a]Most values are taken from Selwood [25].
[b]The value for Pm^{3+} (referred to as II^3 in the original) is from A. V. Jagannadham, *Proc. Rajasthan Acad. Sci.*, **1**, 6 (1950).

Most researchers still prefer to use Pascal's constants because they provide an easy way to obtain diamagnetic susceptibility corrections, which cannot otherwise be estimated for paramagnetic systems such as the free radicals or ions in solution, and they allow for comparisons between the theoretical and experimental values of susceptibilities of atoms, ions, and so on. Prasad and Mulay [23] showed conclusively that the magnitude of the susceptibility of a cation depends on the nature of the anion and vice versa. The following generalizations apply to this dependence.

1.   The susceptibility of cations in the same group may be expressed as a simple function of the total number of their electrons; the susceptibility of a cation generally decreases with increasing valence of the anion.

2.   The susceptibility of an anion in combination with a group of chemically related cations is generally a fixed quantity, and it decreases with the increasing valence of the group of cations with which it combines.

Averaged values, listed in Tables 3.2 and 3.3, may be used for all practical considerations in which minor variations in the susceptibilities of ions, depending on the nature of oppositely charged ions, are not of prime importance [cf. 5, 6].

4   TYPES OF MAGNETISM OBSERVED

4.1   Summary of Magnetic Phenomena

Figure 3.9 and Table 3.6 summarize the common and special types of magnetic behavior described in later sections.

4.2   Atomic Diamagnetism

In diamagnetic substances the permeability $\mu_0 < 1$ and the intensity of induced magnetization I, hence, the susceptibilities (κ, χ, etc.), are negative. The substance is diamagnetic, and it causes a reduction in the lines of force as shown in Figure 3.10a, which is equivalent to the substance producing a magnetic flux in a direction opposite to the applied field. Thus, if the substance is placed in an inhomogeneous field, the substance will move to a region of the lowest field, and the net effect will manifest itself as one of *repulsion*. Therefore, the susceptibilities are shown with a negative sign and the per gram susceptibility is usually about -1×10^{-6} cgs units. According to the classical theory, an electron carrying a negative charge and moving in a circular orbit is equivalent to a circular current. If a magnetic field is applied perpendicularly to the plane of the orbit, the revolving electron will experience a force along the radius, the direction of which depends on that of both the magnetic field and the moving electron. Application of the well-known Lenz's law, which predicts the direction of motion of a current-carrying conductor placed in a magnetic field to this situation, shows

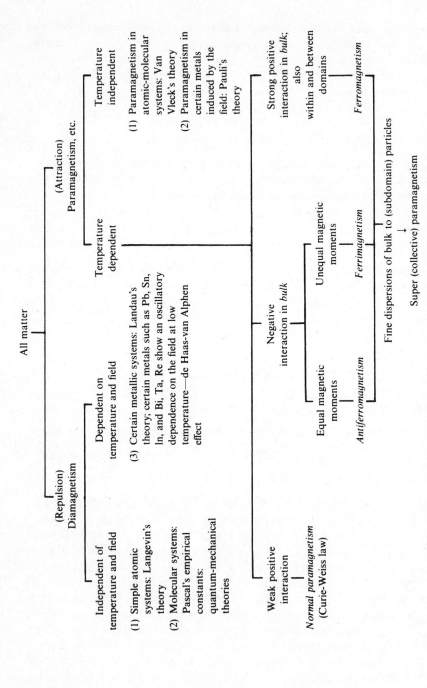

Figure 3.9 Types of magnetism. Reprinted with permission from L. N. Mulay and E. A. Boudreaux, Eds., *Theory and Applications of Molecular Diamagnetism*, Wiley-Interscience, New York, 1976. Copyright © 1976 by John Wiley & Sons, Inc.

Table 3.6 Common Types of Magnetic Behavior

Type	Effect of External Field on Substance	Examples	Comments on Origin	Magnitude of Specific Susceptibility χ, at 20°C	Dependence of Susceptibility	
					Temperature	Field
Diamagnetism (atomic, molecular)	Feeble repulsion $I < H$	Most inorganic compounds except those containing ions of transition elements; organic compounds except free radicals	Caused by orbital motion of electrons; hence, it is a universal property; most perceptible when all electrons are "paired," that is, when they have no permanent "spin" moment	Negative and very small (ca. 1×10^{-7}); attributable to change in state of aggregation of system with temperature	None theoretically; small dependence	None
Paramagnetism	Attraction $I > H$	Salts and certain complexes of transition elements; "odd" electron molecules like NO_2; O_2; free radicals such as triphenylmethyl	Caused by spin and orbital momentum of (unpaired) electrons; the system contains permanent magnetic dipoles (moments) with no interaction	Positive and small (ca. 1000×10^{-4}); sufficiently large to mask the underlying diamagnetism	$\chi = 1/T$, (Curie law); or $\chi = 1/(T + \theta)$ (Curie–Weiss law)	None; saturation is very difficult
Ferromagnetism	Intense attraction $I \gg H$	Metals like iron, cobalt, nickel, and their alloys; EuO	Caused by "domains" or lattice of particles containing electrons with parallel spins; positive interaction among dipoles	Positive and very large (ca. 1×10^2)	Complex; beyond a certain temperature (Curie point), magnetization I drops and shows paramagnetic behavior	Described by hysteresis curves; saturation is easy
Collective or super paramagnetism (behavior intermediate between paramagnetism and ferromagnetism)	Attraction $I > H$	Catalyst in which fine particles of nickel, etc., are dispersed on silica or alumina; also particles of antiferromagnetic or ferrimagnetic materials	Particles too small to constitute "domains," but exhibit "exchange" effects between dipoles within a "cluster"	Positive and very large (ca. 1×10^2)	$\chi = 1/(T + \theta)$ (Curie–Weiss law)	χ increases with fields; "saturation effects" intermediate between para and ferro; generally I is linear with H/T prior to saturation

Table 3.6 (*continued*)

Type	Effect of External Field on Substance	Examples	Comments on Origin	Magnitude of Specific Susceptibility χ, at 20°C	Dependence of Susceptibility — Temperature	Dependence of Susceptibility — Field
Landau diamagnetism (free electrons)	Feeble repulsion; oscillatory dependence of χ on H is sometimes observed (de Haas–van Alphen effect)	Cu, Cu–Ni alloy (e.g., 5-cent coin), Bi, graphite	Noncancellation of reflected angular momenta of electrons at the metallic boundary	-1×10^{-7}	Diamagnetic susceptibility $(-\chi)$ decreases with temperature	
Temperature-independent (Van Vleck) paramagnetism	Feeble attraction	$KMnO_4$, Co(III) amines	Atom with upper state separated from ground state by energy interval large compared to kT; system has no permanent magnetic moment	Positive and very small (ca. 1×10^{-6})	None	
Pauli (free-electron) paramagnetism	Feeble attraction	Metallic K and Na (vapors)	Paramagnetism of an "electron gas"	Positive and very small (ca. 1×10^{-6})	Very slight, generally for vapors $\chi = 1/T$ (Curie law)	
Antiferromagnetism	Feeble attraction	$KNiF_3$, MnSe, MnO	Two lattices having electron spins in one lattice antiparallel to those in another lattice; negative interaction among magnetic dipoles	Positive and very small ($\approx 1 \times 10^{-5}$)	Complex up to a critical temperature (antiferromagnetic) Curie point or Néel temperature; magnetization increases with temperature, then decreases	

170

Ferrimagnetism (shows hysteresis)	Attraction	$FeCr_2O_4$, $2 \cdot Fe_2O_3$, $BaFe_{12}O_{19}$	Interpenetrating lattices with unequal number of electrons with antiparallel spins; simultaneous unequal $(+)$ and $(-)$ interaction among dipoles	Positive and small (ca. 1×10^{-5})	Positive dependence
Metamagnetism (may be regarded as a special case of antiferromagnetism with low Néel temperature; it shows field-strength dependence)	Feeble attraction	$NiCl_2$ or $CoCl_2$ at liquid He temperature	Parallel or antiparallel alignment of moment in domains	Positive and small (ca. 1×10^{-3})	Positive dependence

171

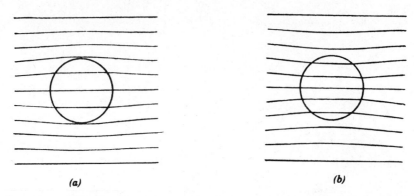

(a)   *(b)*

Figure 3.10   (a) Diamagnetic body in a magnetic field showing that its permeability μ' to the field is less than 1; (b) paramagnetic body in a field with μ' greater than 1. Reprinted with permission from L. N. Mulay and E. A. Boudreaux, Eds., *Theory and Applications of Molecular Diamagnetism*, Wiley-Interscience, New York, 1976. Copyright © 1976 by John Wiley & Sons, Inc.

that the system as a whole will be repelled by the applied field. An elaborate mathematical picture is presented by the Larmor theorem, which describes the behavior of a system of particles, all having the same ratio of charge to mass, in the presence of a constant uniform field. According to this theorem, "the superimposed field leaves the form of the orbits and their inclination to the magnetic line of force, as also the motion in the orbit, unaltered and merely leads to the addition of a uniform 'precession' of the orbit about the direction of lines of force." The precession is visualized easily in terms of the gyroscopic motion of a spinning top. Since an electron is said to experience a "spinning" motion around its own axis and an "orbital" motion around the positively charged nucleus, it is important to distinguish between the precession of the spin and the precession of the orbit. This spin precession indeed arises in the case of a free (unpaired) electron subjected to a magnetic field and forms the basis of the EPR techniques. Similarly, the precession of spins of certain nuclei, such as the protons, forms the basis of NMR methods. It should be noted here that it is the orbital precession of an electron that gives rise to diamagnetism and forms the core of the mathematical theory of diamagnetism.

According to the classical theory of Langevin, the susceptibility per gram atom is given by (cf. Van Vleck [21])

$$\chi_A = -\frac{Ne^2}{6mc^2}\sum \overline{r_i^2}$$

where e and m refer to the charge and mass of the electron, respectively; c is the velocity of light; N is Avogadro's number; and r_i^2 is the sum of the mean square radii of the orbit of the ith electron projected perpendicular to the direction of the applied field. Substitution of appropriate values gives

$$\chi_A = -0.792 \times 10^{-6}\sum \overline{r^2} \qquad \text{cgs units}$$

when $\Sigma \overline{r^2}$ is expressed in a_0^2, where a_0 is the Bohr (hydrogen atom) radius, and

$$\chi_A = -2.382 \times 10^{-10} \sum \overline{r^2} \qquad \text{cgs units}$$

when $\Sigma \overline{r^2}$ is expressed in square centimeters.

Thus, atomic diamagnetism depends only on the effective radii of the electronic orbits; the electrons in the outermost orbits contribute most to the atomic susceptibility. Langevin's equation also shows diamagnetism to be independent of temperature, but this is true in practice only to a first approximation. Theoretically, the negative sign associated with diamagnetic susceptibilities is a consequence of the Larmor precession, and as such, all atomic systems possess the universal property of diamagnetism.

Important here is that a single electron has a *permanent magnetic moment* arising from *its spin* and *also* from *its orbital motion*. In diamagnetic materials all electron spins are "paired up" so that there is no resultant magnetic moment; that is, $\mu = 0$. (The concept of the moment is explained in detail under paramagnetism.) Thus, the feeble diamagnetic susceptibility (-1×10^{-6} cgs units) must be regarded as a manifestation of an electromagnetic effect *induced* by the *external field on the orbital motions of* "spin-paired" *electrons*. Since even the simplest atom, namely, hydrogen, has one electron orbiting around the proton, the orbital motion will give rise to the (induced) diamagnetic susceptibility, which has been theoretically calculated, using Van Vleck's equation, to be $\chi_A = -2.37 \times 10^{-6}$ cgs units. However, the permanent moment arising from the same electron, which gives rise to a high paramagnetic susceptibility (described in the next section), masks this feeble diamagnetic susceptibility. Thus, although diamagnetism is a universal property, in the sense that all atoms experience *induced* effects from the orbital motion of their electrons, this property may be easily masked by the permanent moments arising from the unpaired spins of electrons.

The result of Van Vleck's quantum mechanical treatment for the atomic susceptibility χ_A may be mentioned here briefly before proceeding to a discussion of molecular susceptibility.

$$\chi_A = -\frac{Ne^2}{6mc^2}\left\{\frac{h^2}{4\pi^2ze^2m}\right\}^2\left[\frac{5}{2}n^4 - \frac{3}{2}n^2l(l+1) + \frac{1}{2}n^2\right]$$

where n and l are the principal and subsidiary quantum numbers, respectively; z is the atomic number; h is Planck's constant of action; and other terms have the same meanings as before.

4.3   Molecular Diamagnetism

Langevin's theory (cf. [5]) is strictly applicable to mononuclear systems; this implies that the electrostatic potential should be symmetrical about an axis parallel to the magnetic field. This stipulation is easily fulfilled by atoms;

however, in considering molecules, the classical treatment may be extended only to linear molecules in the Σ state, on the following lines. The magnetic field is presumed to act along the molecular axis in the z direction so that the cloud of electrons may rotate freely around this axis with an angular velocity of $eH/2mc$. This gives

$$\chi_M = -\frac{Ne^2}{6mc^2} \sum \overline{r^2}$$

Langevin's theory cannot be applied to other molecules, as the electric field acting on the electrons ceases to be spherically symmetrical. This is so because, in the classical treatment, if one refers the motion of electrons adequately to a rotating frame of coordinates, one cannot apply the equations of motion of a nonrotating system to a rotating one. Here the fixed nuclei, other than the one taken as the origin, produce an electric field that changes with rotation around the origin. Any electrical asymmetry about the direction of the magnetic field hinders free circulation of the electrons and thus reduces the total diamagnetism.

Van Vleck [21] treated this particular problem in terms of quantum mechanics, and the reduction in diamagnetism is ascribed to a mixing of the wave function for the ground state with that of some of the excited states brought about by the magnetic field. He considers the lowering of the energy $E_n - E_0$ to correspond to the reduction in the total diamagnetic susceptibility. This gives the following equation, which replaces the previous classical formula

$$\chi_M = -\frac{Ne^2}{4mc^2} \sum_i \overline{x_i^2 + y_i^2} + \frac{Ne^2}{2m^2c^2} \sum_{n \neq 0} \frac{|0|m_z|n)|^2}{E_n - E_0}$$

Here, in the second term, the summation is over the excited states n, and the numerators and denominators represent the squares of the matrix elements of the electronic angular momentum component m_z and the excitation energies, respectively.

In a generalized form for a polyatomic molecule with no resultant spin, Van Vleck derived the following equation for its molar susceptibility. In this, the frequency of various transitions and the moments arising therefrom are considered.

$$\chi_M = -\frac{Ne^2}{6mc^2} \sum \overline{r^2} + \frac{2}{3} N \sum \frac{|m_0(n; n')|^2}{h\nu(n'; n)}$$

where $m_0(n; n')$ is a nondiagonal element of the matrix for the angular momentum of the system, $h\nu(n'; n)$ is the frequency corresponding to the $(n'; n)$ transition, and the other terms have their usual meanings.

The second term is always positive and is often called the temperature-independent paramagnetic term, which is discussed in Section 4.5. This should not be confused with the temperature-dependent paramagnetism arising from a permanent dipole, which is discussed in Section 4.4. It should be apparent that the temperature-independent paramagnetism corresponds to the decrease in diamagnetism that was discussed previously.

In most cases, for instance in nitrous oxide, the second term vanishes. This reduces Van Vleck's equation to

$$\chi_M = -\frac{Ne^2}{6mc^2}\sum \overline{r^2}$$

which is identical with Langevin's equation.

Practically all organic and inorganic compounds, with the exception of free radicals and salts of the transition metals lanthanide and actinide elements, are diamagnetic. Diamagnetic susceptibility of simple atomic and molecular systems is independent both of temperature and of the applied field. Usually any significant changes in diamagnetic susceptibility with temperature can be attributed to a change in the physical or chemical structure of the material. For example, the small increase (ca. 0.6%) in the diamagnetic susceptibility of water with increasing temperature arises from the corresponding changes in its hydrogen bonding.

Although the per gram diamagnetic susceptibilities of most compounds are about -1×10^{-6} cgs units, in exceptional cases, such as graphite, the average isotropic per gram susceptibility may be as large as -7×10^{-6} cgs units. This behavior is attributed to the semimetallic properties of graphite and in particular to the behavior of π electrons spread over many aromatic benzenelike rings of carbon atoms. Since the *diamagnetism* in *semimetallic but otherwise* "*molecular*" systems is becoming increasingly important both for technological applications and theoretical studies, we discuss the *metallic* behavior in Section 4.6.

Various diamagnetic atomic susceptibility constants, diamagnetic corrections for certain ligands, the diamagnetism of ions (including the underlying diamagnetism of certain paramagnetic ions) and the constitutive correction constants λ are given in Tables 3.2–3.5 and 3.7. For certain cyclic systems $\lambda = 0$.

Table 3.7 Diamagnetic Corrections for Various Ligands

Ligand	$\chi(\times 10^{-6})$
Dipyridyl	-105
Phenanthroline	-128
o-Phenylenebisdimethylarsine	-194
Water	-13

4.4   Normal Temperature-Dependent Paramagnetism of Simple Systems

In paramagnetic compounds the permeability μ is greater than 1 and produces an increase in the density of lines of forces (Figure 3.10b). Thus, the intensity of magnetization I and the per gram and molar susceptibilities are positive. This implies that the substance is producing a flux in the same direction as the applied field, and when the substance is placed in an inhomogeneous field it will tend to move to regions of the highest field, which shows there is an attraction between the two. The susceptibilities are shown with a positive sign; the per gram susceptibility χ is numerically much greater than in the diamagnetic case and ranges between 100 to 1000 $\times$ 10^{-6} cgs units.

Paramagnetism occurs among the ions of the transition metals, lanthanides and actinides, as well as in free radicals. With paramagnetics, the intensity of magnetization I is directly proportional to the applied field. Thus, the susceptibility is independent of the field, H; however, it varies inversely with the temperature. This behavior is understood by assuming that each atom or molecule acts either like a magnetic dipole or as a microscopic bar magnet. When placed in a field, these intrinsic dipoles experience a turning or an aligning effect or may be said to be space quantized so that they assume one of a limited number of orientations relative to the field. This turning effect is visualized as the magnetic moment μ arising from the effect of the magnetizing force H on a magnetic dipole. The concept is further explained and defined in Section 5 with regard to its units. The magnetic moment is expressed in units of the Bohr magneton, the common symbols for which are μ_B, BM, or β. In the cgs system the Bohr magneton is 0.92731 $\times$ 10^{-20} erg Oe^{-1} and in SI it is 0.92731 $\times$ 10^{-24} A$\cdot$m^2. It is customary to calculate the moment from (1) $\mu = 2.8279\sqrt{\chi T}\,\mu_B$ in the cgs system and from (2) $\mu = 7.9774 \times 10^2\sqrt{\chi T}\,\mu_B$ in SI units. These expressions arise from a consideration of Langevin's theory of paramagnetism and the Curie law, which describe the temperature dependence of an ideal paramagnetic substance. The cgs and SI expressions (1) and (2) give the *same numerical value* for μ when the Bohr magneton μ_B is used as a unit.

The tendency of the dipoles to orient with the field is counteracted by the randomizing effects produced by the thermal energy, which in the room-temperature region (as measured by kT, where k is the Boltzmann constant) is several times greater than the magnetic energy of the dipoles. Hence, very weak magnetic polarization (and magnetic susceptibilities) are observed in many paramagnetics, in which each dipole is able to orient itself essentially "parallel" to the field independently of others. The paramagnetic systems with no interaction between the dipoles are called *magnetically dilute*. Experimentally, the randomization effect just described is reflected in a decrease in susceptibility with increasing temperature. This observation is known as the Curie law and was first stated in terms of the molar susceptibility χ_M:

$$\chi_M = \frac{C}{T}$$

where C is the Curie constant and is characteristic of every paramagnetic substance.

The observed molar susceptibility consists of contributions from the atomic paramagnetic χ_P and the (sum of the) atomic diamagnetic components χ_D. Hence, $\chi_M = \chi_P + \chi_D$, and $\chi_P = \chi_M - \chi_D$. Since the diamagnetic components are *negative* in sign, the diamagnetic correction amounts to *adding* its numerical value to χ_M. Typical values of diamagnetic constants are listed in Tables 3.2 to 3.5 and 3.7 for easy reference. The Curie law is better described in terms of the atomic susceptibility of the paramagnetic species:

$$\chi_A = \frac{C}{T}$$

for most ideally magnetically dilute substances. Subsequent studies showed that in other nonideal situations, a more exact relationship is followed:

$$\chi_A = \frac{C}{T - \theta}$$

where θ is designated as the Weiss constant, and this law is known as the Curie–Weiss law [26].

Several authors [24] used the opposite convention for θ. Thus, this expression is written as

$$\chi_A = \frac{C}{T + \theta}$$

A theoretical interpretation of the Curie law is found in Langevin's theory (cf. [6]), which shows that the molar susceptibility χ_M is given by

$$\chi_M = \frac{N\mu^2\beta^2}{3kT}$$

where μ is the permanent (effective) moment associated with the paramagnetic species (ion, atom, or the molecule), and T is the absolute temperature; the moment is expressed in units of the Bohr magneton β. The permanent moment μ arises from the spin and orbital contributions of the unpaired electron.

For normal "magnetically dilute" paramagnetics, the moment μ_{eff} becomes a parameter of fundamental significance, since it does not (significantly) vary with temperature, whereas the susceptibility varies inversely as the temperature. Therefore, it is helpful to have tables of such μ_{eff} (often designated as the p_{eff} or the Bohr magneton numbers) for various paramagnetic ions. Values for the 3d transition metal ions, the rare earth ions, and so on, are collected in Tables 3.8 to 3.10. In the section following these tables the quantum mechanical significance of the Langevin equation is outlined.

Table 3.8 Comparison of Theoretical and Experimental Values of μ_{eff} for the $3d$ Transition Ions; K^+ and Ca^{2+} Belonging to the Same Fourth Period Are Also Included[a]

Electrons in $3d$ Shell	Ion	Ground State	S	L	J	$\mu^2_{\text{eff}} = g^2J(J+1)$	μ^2_{eff} (Experimental)[b]	Spin Only[c] $\mu^2_{\text{eff}} = 4S(S+1)$	μ_{eff}	μ_{eff} (Experimental Range)[d]
0	K^+, Ca^{2+}, Sc^{3+} Ti^{4+}, V^{5+}	$^1S_0$	0	0	0	0	0	0	0	0
1	Ti^{3+}, V^{4+}	$^2D_{3/2}$	$\frac{1}{2}$	2	$\frac{3}{2}$	2.4	2.9	3	1.73	1.77–1.79
2	V^{3+}	$^3F_2$	1	3	2	2.67	6.8	8	2.83	2.76–2.85
3	V^{2+}, Cr^{3+}	$^4F_{3/2}$	$\frac{3}{2}$	3	$\frac{3}{2}$	0.6	14.8	15	3.87	3.68–4.00
4	Cr^{2+}, Mn^{3+}	$^5D_0$	2	2	0	0	23.3[e]	24	4.90	4.80–5.06
5	Mn^{2+}, Fe^{3+}	$^6S_{5/2}$	$\frac{5}{2}$	0	$\frac{5}{2}$	35	34.0	35	5.92	5.2–6.0
6	Fe^{2+}	$^5D_4$	2	2	4	45	28.7	24	4.90	5.0–5.5
7	Co^{2+}	$^4F_{9/2}$	$\frac{5}{2}$	3	$\frac{9}{2}$	44	24.0	15	3.87	4.4–5.2
8	Ni^{2+}	$^3F_4$	1	3	4	31.3	9.7	8	2.83	2.9–3.4
9	Cu^{2+}	$^2D_{5/2}$	$\frac{1}{2}$	2	$\frac{5}{2}$	12.6	3.35	3	1.73	1.8–2.2
10	Cu^+, Zn^{2+}	$^1S_0$	0	0	0	0	0	0	0	0

[a]Reprinted, with permission, from B. I. Bleany and B. Bleany, *Electricity and Magnetism*, Oxford University Press, London, 1965.

[b]The values of μ^2_{eff} (at 300 K) are for double sulfates of the type $M'''M'(SO_4)_2$, $6H_2O$, or $M''M'(SO_4)_2$, $12H_2O$, where M' is a monovalent diamagnetic ion, M'' is a divalent paramagnetic ion, and M''' is a trivalent paramagnetic ion. In these salts the distance between nearest paramagnetic ions is at least 6 Å, and interaction is negligible. Potassium chrome is thus $KCr(SO_4)_3 \colon 12H_2O$.

[c]Since μ^2_{eff} ("spin only") $= 4S(S+1)$ and $s = \frac{1}{2}$ quantum unit for one electron, μ_{eff} ("spin only") $= \sqrt{n(n+2)}$, where n is the number of unpaired electrons.

[d]Range of values observed for μ_{eff} [25].

[e]Value for $CrSO_4$, $6H_2O$; no double sulfate of Cr^{2+} was measured.

Table 3.9 Comparison of Theoretical and Measured Values of the Effective Magnetic Moments (μ_{eff}^2) for Trivalent Rare Earth Ions[a]

| Electrons in 4f Shell | Trivalent Ion | Ground Spectroscopic State | Theoretical Values | | | | | Average Experimental Value of μ_{eff}^2 | Theoretical Value of μ_{eff} [c] |
			S	L	J	g	$\mu_{eff}^2 = g^2 J(J+1)$ [b]		
0	La	$^1S_0$	0	0	0		0	0	0
1	Ce	$^3F_{5/2}$	$\frac{1}{2}$	3	$\frac{5}{2}$	$\frac{6}{7}$	6.43	6	2.56
2	Pr	$^3H_4$	1	5	4	$\frac{4}{5}$	12.8	12	3.62
3	Nd	$^4I_{9/2}$	$\frac{3}{2}$	6	$\frac{9}{2}$	$\frac{8}{11}$	13.1	12	3.68
4	Pm	$^5I_4$	2	6	4	$\frac{3}{5}$	7.2		2.83
5	Sm	$^6H_{5/2}$	$\frac{5}{2}$	5	$\frac{5}{2}$	$\frac{2}{7}$	0.71(2.5)	2.4	1.55–1.65
6	Eu	$^7F_0$	3	3	0		0(12)	12.6	3.40–3.51
7	Gd	$^8S_{7/2}$	$\frac{7}{2}$	0	$\frac{7}{2}$	2	63	63	7.94
8	Tb	$^7F_6$	3	3	6	$\frac{3}{2}$	94.5	92	9.7
9	Dy	$^6H_{15/2}$	$\frac{5}{2}$	5	$\frac{15}{2}$	$\frac{4}{3}$	113	110	10.6
10	Ho	$^5I_8$	2	6	8	$\frac{5}{4}$	112	110	10.6
11	Er	$^4I_{15/2}$	$\frac{3}{2}$	6	$\frac{15}{2}$	$\frac{6}{5}$	92	90	9.6
12	Tm	$^3H_6$	1	5	6	$\frac{7}{6}$	57	52	7.6
13	Yb	$^2F_{7/2}$	$\frac{1}{2}$	3	$\frac{7}{2}$	$\frac{8}{7}$	20.6	19	4.5
14	Lu	$^1S_0$	0	0	0		0	0	0

[a] Reprinted with permission, from B. I. Bleaney and B. Bleany, *Electricity and Magnetism*, Oxford University Press, London, 1965.
[b] The values given in parentheses for Sm and Eu are those calculated by Van Vleck allowing for population of excited states with higher values of J, at $T = 293$ K.
[c] For quick reference the values of μ_{eff} [5] are reported here along with the variation in the μ_{eff} for Sm and Eu in earlier literature.

Table 3.10 Bohr Magneton Numbers for the Palladium and Platinum Group Elements (Valence State, 3+)

Ion	μ
Ru	2.09
Rh	0.06
Pd	0.07–0.13
Os	0.27–0.50
Ir	0.11
Pt	0.0

4.5   Temperature-Independent Paramagnetism of Molecular Systems

As mentioned in Section 4.3, to understand molecular diamagnetism it is necessary to consider the effects of the temperature-independent paramagnetism. Many chemists use the abbreviation TIP to designate this magnetic behavior, which is also known popularly as the Van Vleck paramagnetism. In the Langevin–Debye formalism a term $N\alpha$ is added to the Langevin paramagnetism to account for magnetic effects other than those arising from the permanent moment of a molecule. Thus, the Langevin–Debye formula can be written in the following simple form:

$$\chi_M = \frac{N\mu^2\beta^2}{3kT} + N\alpha$$

Van Vleck deduced an expression similar to the Langevin–Debye formula on the basis of quantum mechanics, which describes all aspects of paramagnetism

$$\chi_M = \frac{N\mu^2\beta^2}{3kT} + \frac{2}{3}N\sum_{n'\neq n}\frac{|m^0(n;\,n')|^2}{h\nu(n';\,n)} - \frac{Ne^2}{6mc^2}\sum\overline{r^2}$$

where the first term is identical with that of Langevin's expression for paramagnetism and is dependent on temperature. The symbols in the second and the third terms have the same meanings as those given earlier. The second term represents the contribution of the high-frequency elements to paramagnetism and is independent of temperature, whereas the third term, also independent of temperature, accounts for the underlying diamagnetism of the paramagnetic molecule. This term may be neglected in most cases.

Let us now discuss the physical significance of the temperature-independent term in Van Vleck's equation. This is presented by Bates [27]; Figure 3.8

provides further clarification. Imagine a resultant magnetic moment μ_R precessing regularly around J (the sum of the spin S and orbital angular momentum vectors) in the absence of a field; however, when J is made to precess about an applied field H, the precession of μ_R ceases to be symmetrical and gives rise to a small increase in the magnetization parallel to H. An analysis of this situation shows this increase to be independent of temperature for a given L, S, J state.

Van Vleck's expression refers only to atoms with Russell–Saunders coupling, that is, only when the spin angular momenta s of the electrons can combine to form S and the orbital angular momenta l of electrons combine independently to give L. Application of Van Vleck's theory to various paramagnetic systems is rather complex, but it accounts partly for departures from the Curie–Weiss law in specific cases. This theory depends entirely on the magnitude of the spin multiplet intervals as compared to the Boltzmann distribution factor kT, and is described in a quote from Van Vleck's [21].

We shall classify a state as "normal" if its Boltzmann distribution factor $e^{-W/kT}$ is appreciably different from zero, i.e., if its excess of energy over the very lowest state is either smaller than or comparable with kT. An "excited" state is one which has such a small Boltzmann factor that its probability of being occupied is negligible, and whose energy thus exceeds the energies of the normal states by an amount large compared to kT. An energy-level diagram illustrating graphically the delineation into normal and excited states is given in [Figure 3.11a].

In order for the Langevin–Debye formula to be valid, it is vital for the electrical moment to involve no "medium frequency" elements, which involve energy changes of the same order of magnitude as kT. Thus here and throughout the remainder of the volume, the equipartition allowance kT of energy enters as the unit for determining whether an energy change is "large" or "small" for our purposes, or in other words, whether a frequency is "high" or "low." It is essential that the spacing between consecutive normal states or energy levels be small compared to kT. In [Figure 3.11] an interval such as $b–c$ must be much less than kT. It is not necessary to demand that the energy-difference between two widely separated normal states, such as $a–c$ in [Figure 3.11], be small compared to kT, as

$n' \neq n$

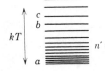

kT c
     b

$n' = n$

     a

Figure 3.11(a) Energy-level diagram. Reprinted with permission from L. N. Mulay and E. A. Boudreaux, Eds., *Theory and Applications of Molecular Diamagnetism*, Wiley-Interscience, New York, 1976. Copyright © 1976 by John Wiley & Sons, Inc.

ordinarily there will be selection principles which require that the matrix elements connecting two normal states be zero, or at least very small, unless the two states are adjacent, or nearly so (cf. the familiar selection rule, $\Delta j = 0, \pm 1$ for the inner quantum number, as a special example). It is clear that it is impossible to require that the energy-differences of two widely separated normal levels such as a–c be small compared to kT, as the equipartition theorem demands that at high temperatures the average excess of rotational energy over the very lowest state be kT itself. At very low temperatures the "unit" kT will become much smaller and the separation between adjacent normal states will become comparable with kT.

The final results of Van Vleck's calculations are

1.   *Multiplet intervals that are small compared to kT.*   Here, the high-frequency elements of the paramagnetic moment are absent. This gives

$$\chi_M = \frac{N\beta^2}{3kT}[4S(S + 1) + L(L + 1)]$$

This expression is used for calculating susceptibilities of ions of most transition-group elements (see Table 3.8).

2.   *Multiplet intervals that are large compared to kT.*

$$\chi_M = \frac{Ng^2\beta^2 J(J + 1)}{3kT} + \frac{N\beta^2}{6(2J + 1)}\left[\frac{F(J + 1)}{hv(J + 1; J)} - \frac{F(J)}{hv(J + 1; J)}\right]$$

where β is the Bohr magneton and

$$F(J) = \frac{1}{J}[(S + L + 1)^2 - J^2][J^2 - (S - L)^2]$$

This expression is used for most ions of the rare earth elements (see Table 3.9).

3.   *Multiplet intervals that are comparable to kT.*   In this special case, the effect of the quantum number J is comparable to kT; and it is necessary to consider that the system containing N atoms comprises groups of atoms N_{J_1}, $N_{J_2}, \ldots$, with different values for J. The distribution of atoms among various groups is governed by the Boltzmann temperature factor and is proportional to $(2J + 1)\exp(-W_j^0/kT)$, where W_j is the energy of precession.

$$\chi_M = N \sum_{J = |L - S|}^{L + S} \frac{\{[g_j^2\beta^2 J(J + 1)/3kT] + \alpha_j\}(2J + 1)\exp(-W_j^0/kT)}{\sum (2J + 1)\exp(-W_j^0/kT)}$$

This expression is used particularly for ions of samarium and europium.

Van Vleck's theory and recent modifications (cf. [21]) are quite satisfactory in that they explain a number of experimental facts. According to the theory, Curie's law should be obeyed in the limiting cases where the multiplet intervals are large or small compared to kT, which neglects the temperature-independent paramagnetic contribution to susceptibility that arises from high-frequency elements. When the multiplet intervals are comparable to kT, a Boltzmann distribution of the normal states occurs, which results in serious departures from the Curie–Weiss law.

4.6 The Landau Diamagnetism and Pauli Paramagnetism of Metallic and Semimetallic Systems

This section is written especially to provide reasonable information on the magnetic properties of certain aromatic systems, such as graphite, that display semimetallic behavior. It is well known that the valence electrons in a metal are not confined to a particular atom (that is, to a positive ion) but are free to wander throughout the metallic lattice. These electrons are called the *itinerant electrons*, to distinguish them from the electrons localized at particular sites (ions, ligands, etc.) and "between bonds." The localized electrons define various properties of the well-formulated (*stoichiometric*) compounds, with which the chemist is most familiar. The electrons in a metal behave like a free electron gas (FEG) in the same manner as molecules behave in a molecular gas, but with important differences in their statistical mechanics. In the simplest form of the FEG model, no interactions should occur among the electrons.

The magnetic susceptibility of an electron gas consists of two terms: the (*Pauli*) *paramagnetic* contribution from the spins of the electrons and the (*Landau*) *diamagnetic* contribution from their translational or orbital motion. Considering the limitations and scope of this chapter, it is difficult to compare here the Maxwell–Boltzmann statistics with the Fermi–Dirac statistics, which is crucial to a thorough understanding of the behavior of metallic and semimetallic systems. Therefore, the reader should refer to books by Mulay and Boudreaux [5, 6] and to references cited therein for this information. However, we briefly consider the behavior of an electron gas.

If the electron gas had obeyed the Boltzmann statistics as applied in the derivation of the Langevin equation for normal paramagnetics (in a magnetically dilute environment), one would expect a paramagnetic volume susceptibility (κ_p) of about 10^{-4} cgs/cm^3 from the following relation

$$\kappa_p = \frac{N'\mu_{\text{eff}}^2}{3kT}$$

where N' is the number of conduction electrons per unit volume, and $\mu_{\text{eff}}^2 = 2\sqrt{2(s+1)}\,\mu_B$. Here the spin quantum number for an electron $s = \frac{1}{2}$ quantum units, and μ_B is the Bohr magneton.

Quite surprisingly, certain metals like sodium and potassium have $\kappa_p = 10^{-6}$ cgs cm^{-3}; their susceptibilities are almost independent of temperature in contrast to the behavior expected from the above Langevin–Curie law. In 1927 Pauli first showed that this discrepancy between theory and experiment is removed by the application of the Fermi–Dirac statistics. Indeed, there exist striking differences between the simple behavior of molecules in a molecular gas, described by the Boltzmann statistics, and that of electrons in an electron gas, which necessitates the application of the Fermi–Dirac statistics. A clarification of these aspects follows.

In the Langevin paramagnetism, molecules with a permanent moment act as dipoles and when placed in a field H assume a range of orientations from 0 to π. If θ is the angle the dipole makes with the field, then each value of θ represents an energy state $-\mu H \cos \theta$ and the molecules are distributed among various energy states according to the Boltzmann distribution law; that is, for a uniform distribution of energy states, the number of particles in each state falls off exponentially with thermal energy, which tends to randomize the orientations of the dipoles. Thus, the relation between the energy states $n(E)$ and E is given by

$$n(E) = A_0 \exp(-E/kT)$$

where E is the (magnetic) energy state equal to $-\mu H \cos \theta$, A_0 is a constant, and other symbols have their usual meanings. This Boltzmann distribution behavior for molecular magnets or dipoles is shown in Figure 3.11a in terms of $f(E)$, which in turn depends on the energy states $f(E)$. The probability of speed of molecules and their energies are shown in Figure 3.11b.

In contrast to the Boltzmann statistics, in the Fermi–Dirac statistics the number of electrons in any one energy state is limited to two, that is, one electron with spin "up" and the other with a spin "down" according to the Pauli exclusion principle. The corresponding $f(E)$ follows the step function shown in Figure 3.11c with electrons filling each of the two states two at a time up to a

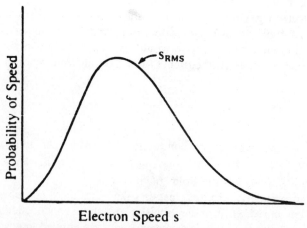

Figure 3.11(b) The Maxwell–Boltzmann distribution curve. The distribution of vector velocities in velocity space B. The equilibrium probability curve for scalar speed s, without regard for the direction of that speed. The most probable speed is $(2k_0 T/m)^{1/2}$, but the speed for the average energy is $s_{RMS} = (3k_0 T/m)^{1/2}$.

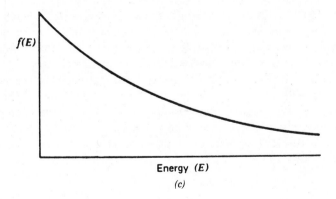

Energy (E)

(c)

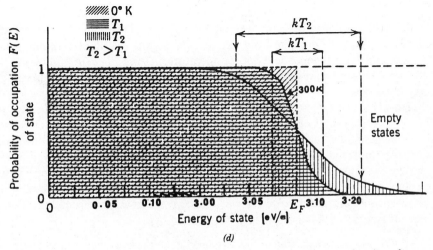

(d)

Figure 3.11c,d (c) Probability of occupation $f(E)$ for a molecular gas as a function of energy according to Boltzmann's law. (d) Probability of occupation $f(E)$ derived from the Fermi–Dirac statistics. (Note the break in the energy axis between 0.10 and 3.00 eV.) The Fermi level (or energy, E_F) at 300 K corresponding to 3.10 eV is appropriate for a dense electron *gas* such as metallic sodium. All the levels below E_{F_0} are completely filled with *paired-up* electrons, according to the Pauli exclusion principle; all levels above E_F are empty. When the temperature is increased from T_1 to T_2 more electrons spill over into the empty band. Quite often similar diagrams are shown, turned anticlockwise through 90°, to depict the *lower-filled* and *upper-empty* states. One can then visualize that only the electrons near the top of the filled band are affected by thermal or magnetic energy. For sodium at room temperature, occupancy is complete up to 0.95 E_F and zero beyond 1.05 E_F.

maximum energy known as the Fermi energy (E_F). At finite temperatures some of the electrons near the "top" of the energy band are excited by thermal agitation to slightly higher levels, and this behavior is shown by the dotted line in Figure 3.11c. This expression describes the Fermi–Dirac statistics

$$f(E) = \frac{1}{\exp[(E - E_F)/kT] + 1}$$

The Fermi energy E_F simply indicates the level *below* which all states are *filled* and *above* which the states are *empty*. The electrons in the lower energy states (or at the bottom of the energy band) are not virtually influenced by the field or temperature changes. Only those electrons in an energy interval (or band) of width kT near the Fermi level are affected by energy changes, because of either thermal or magnetic effects. Thus, only a small fraction kT/E_F or (T/T_F) contributes to the magnetic susceptibility; here T_F, the Fermi temperature, is defined simply by $kT_F = E_F$ (see Figure 3.11d).

A modification of this equation gives the final result for the Pauli paramagnetic susceptibility κ_p

$$\kappa_p = \frac{3N'^2\beta}{2kT_F}$$

Since E_F is a constant for a given metal (and very high), T_F is also a constant and has values in the range from 10^4 to 10^5 K. Thus, the Pauli paramagnetic susceptibility is independent of temperature. Using $T_F \simeq 10^4$ and $N' = 10^{22}$ cm^{-3} for the number of conduction electrons, we get $\kappa_p \simeq 10^{-6}$ emu cm^{-3}, which is in good agreement with the experiment.

Thus far, we neglected the interaction of the electrons with the ion cores. It can be taken into account by considering the effective mass m^* of an electron. This is a very useful concept, not only in Pauli paramagnetism of electron spins, but also in the Landau diamagnetism arising from the orbital motion of electrons in a metal. The inertia of such a moving electron in a metal appears to be different from the inertia of an electron moving in free space. This inertia, known as the *effective electron mass m^**, is a convenient way to account for these various forces and has been defined by Kittel [28]:

> m^* is the mass a free electron would need in order for the velocity increment under an applied impulse to be equal to the actual velocity increment of the conduction electron of the same impulse.

The value of m^* can then be used to obtain general information about the energy bands in metals and semiconductors.

The Pauli paramagnetic susceptibility per unit volume κ_p of the metal is then given by

$$\kappa_p = \frac{4\pi m^*}{h^2}\left(\frac{3N}{\pi}\right)^{1/3}\beta^2$$

Indeed, more sophisticated equations were derived by taking into consideration the exchange effects and the correlation effects between the electrons with parallel and antiparallel spins. By using the (estimated) values of m^* obtained from experiments such as the cyclotron resonance and the Hall effect, values of κ_p have been derived that are in good agreement with theory. Alternately, values of m^* deduced from the measured susceptibility κ_p yielded valuable information about the energy band profiles for various metals. It is essential to include corrections for the diamagnetism of ion cores and a contribution from the Landau diamagnetism of electrons. The Landau diamagnetic contribution discussed below *numerically* turns out to be just one-third the paramagnetic susceptibility. Thus, the total susceptibility of metals like sodium and potassium is described more precisely by

$$\kappa_{\text{total}} = \kappa_p + \kappa_{\text{dia}}^{\text{core}} + \kappa_{\text{dia(L)}}$$

where the three contributions represent the Pauli paramagnetism, the underlying diamagnetism of ion cores, and the Landau diamagnetism. All contributions are comparable in magnitude (ca. 10^{-6} emu cm^{-3}) and metals can be paramagnetic or diamagnetic, depending on which contribution is greater.

The Landau diamagnetism arises from the translational motion of the conduction or the itinerant electrons and should be distinguished from the classical Langevin diamagnetism of simple atomic (and molecular) systems presented in previous sections. The Landau susceptibility arises specifically because the angular moments of electrons, which are reflected at the boundary of the enclosure containing the free electrons (walls of the crystal), and the angular momenta of the electrons, which are able to follow closed paths inside such enclosure, do not cancel as they do in the classical case. The motion of electrons is said to follow helical paths, and the Landau diamagnetism can be explained only on the basis of quantum-mechanical concepts. The quantization of angular momenta of electrons in an applied field gives rise to highly degenerate discrete energy levels, that is, levels with the same energy.

The quantization is alternately described as a periodic variation in the Fermi surface, and this description is used to explain the de Hass–Van Alphen effect observed in certain diamagnetics, notably in graphite, bismuth, lead, indium, and so on. This effect is manifested as an oscillatory variation in the diamagnetic susceptibility at low temperatures when the applied field is gradually increased.

In the classical Langevin theory for the simple atomic systems, the diamagnetic susceptibility arises primary from the precessional effects of the entire electronic orbits in a field H and is proportional to $\Sigma \overline{r^2}$, which represents the average of the square of radii of various orbits projected in a plane perpendicular to H. Since no specific orbital radii can be conceived for the itinerant electrons in an electron gas, recourse is taken to considering the effective mass m^* of conduction electrons and by applying the Fermi–Dirac statistics. The Landau diamagnetic (volume) susceptibility $\kappa_{\text{dia(L)}}$ is then given by

$$\kappa_{\text{dia(L)}} = -\frac{4\pi m^*}{3h^2}\left(\frac{3N}{\pi}\right)^{1/3}\mu_B^2\left(\frac{m^2}{m^{*2}}\right)$$

A comparison of this equation with that for the Pauli paramagnetism shows the remarkable result that

$$\kappa_{\text{dia(L)}} = -\frac{1}{3}\left(\frac{m^2}{m^{*2}}\right)\cdot \kappa_p$$

One is likely to be perplexed by the factor μ_B (or β) in the preceding expression. It does not in any way indicate the existence of "permanent moments" (or dipoles) in the system. It should be further noted that $\mu_B = (eh/4\pi \cdot mc)$ in the paramagnetic case is replaced by $\mu_B = (eh/4\pi \cdot m^*c)$ in the diamagnetic situation. In the paramagnetic case the permanent moment of the spin is equal to μ_B, regardless of whether the electrons are free, and this aspect does not involve the effective mass m^*. In the theory of Landau diamagnetism, the factor μ_B enters simply as a substitution term replacing other constants.

4.7   Comparison of Different Types of Diamagnetism and Paramagnetism

The different magnetic phenomena we have discussed that do not involve magnetic ordering (i.e., coupling of unpaired spins) can be best understood by studying their comparisons in Figure 3.12.

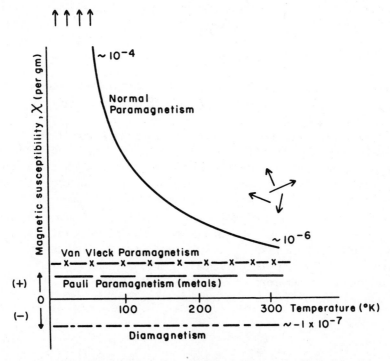

Figure 3.12   Dependence of mass magnetic susceptibility on temperature, for the diamagnetic, and various paramagnetic phenomena. The overall orientation of spins, as shown for normal paramagnetism, depicts the decrease in susceptibility by their thermal randomization. Reprinted with permission from L. N. Mulay and E. A. Boudreaux, Eds., *Theory and Applications of Molecular Diamagnetism*, Wiley-Interscience, New York, 1976. Copyright © 1976 by John Wiley & Sons, Inc.

4.8 Magnetically Ordered Systems: Antiferromagnetism, Ferrimagnetism, and Ferromagnetism

4.8.1 Origin of Magnetic Ordering

An earlier description of the normal paramagnetism of atoms (or ions) indicates that under "magnetically dilute conditions" an individual paramagnetic atom (or ion) behaves independently of its neighbors. Other relatively more complex phenomena arise when magnetic interactions occur between neighboring paramagnetic centers within solids that are electrical insulators (e.g., MnO and Fe_3O_4). This situation becomes even more complicated in metals that display electrical conductivity that is caused by the presence of itinerant electrons (e.g., Fe, Co, Ni). Quite often the interaction between the localized (paramagnetic) electrons and the itinerant electrons is responsible for unusual magnetic and electrical properties. In magnetically ordered systems "cooperative" effects exist. Thus, because of strong magnetic coupling between spins, any perturbation experienced by one spin (from an applied field H and/or temperature T) is virtually transmitted to every other spin in the system, whereas in paramagnetism one spin can behave independently of every other spin because the interactions, if any, are indeed very weak.

4.8.2 Antiferromagnetism in Ionic and Molecular Solids

In certain compounds (e.g., MnO, α-Fe_2O_3) the magnetic susceptibility increases with temperature up to a critical point called the *Néel temperature of the antiferromagnetic Curie temperature*, beyond which the susceptibility decreases in the normal paramagnetic fashion (Figure 3.13). This phenomenon is called *antiferromagnetism* and, along with others such as ferrimagnetism and ferromagnetism, is described as a cooperative phenomenon because the electron spins act collectively and cooperatively as a group because of magnetic interactions. These phenomena are contrasted from the "magnetically dilute" situation in paramagnetism where there is no or very little interaction between spins [5, 6].

The occurrence of the Néel point may be visualized as follows. At very low temperatures, an alternating "up-and-down" type spin arrangement is established in the lattice in the absence of any external field such that an array of parallel (plus) spins in one sublattice almost cancels the moments of antiparallel (minus) spins in another sublattice (Figure 3.14). A strong negative magnetic interaction is said to exist between the plus and minus spins, which decreases the tendency of all spins to be magnetized with decreasing temperature. As the temperature is increased, the uncoupling of magnetic spins facilitates greater and greater magnetization up to the Néel point; then, as with normal paramagnetism, with a further increase in temperature beyond this point, the susceptibility decreases because of the thermal randomization of spins (Figure 3.13).

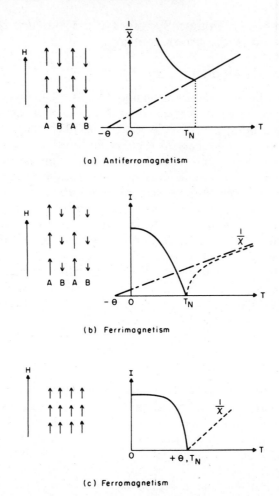

Figure 3.13   A schematic comparison of the three magnetic phenomena (*a*) In antiferromagnetism the spins at *A* sites (or sublattice) are antiparallel to those at *B*, below the Néel transition T_N. (*b*) In ferrimagnetism, the net spin at *A* sites (long arrows) is twice that at *B* (short arrows) and antiparallel to *B*. (*c*) The all-*parallel* ordering is within a ferromagnetic domain. The magnetic susceptibility χ or magnetization *M* is determined by the effect of the field *H* and the temperature on the ordering of spins; θ is the Weiss *molecular* (*internal field*) *constant*.

Since an "up-and-down spin order" exists below the Néel temperature, it is easy to visualize further that with measurements on single crystals the susceptibility must depend on the direction of the applied field, that is, on the angle between the easy axis of magnetization of the crystal and the field (Figure 3.15). Thus, if the field is applied parallel to the easy axis, the total magnetization *M* is simply equal to $M_A - M_B$, which is the difference between the magnetizations of the two sublattices A and B. *Parallel susceptibility* $\chi_{\parallel}$, among other things, depends on the molecular field constants N_{AA} for near-neighbor and on N_{AB}

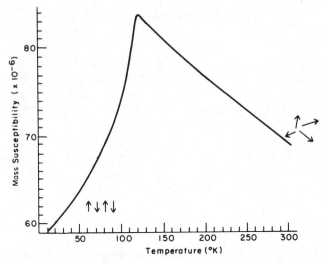

Figure 3.14 Mass magnetic susceptibility (10^{-6} cgs units) as a function of temperature for a typical antiferromagnet (polycrystalline MnO). The relative orientation of electric spins depicts the changes in the susceptibility as a function of temperature. Above the Néel temperature T_N the system behaves like a paramagnet.

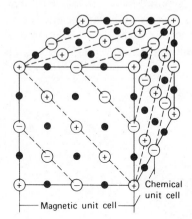

Figure 3.15 Arrangement of ions for an antiferromagnetic face-centered cubic lattice such as MnO. The circles with plus and minus signs represent the magnetic ions (Mn^{2+}) with moments parallel and antiparallel to the preferred direction, respectively. The dark spheres represent the oxygen ions. One magnetic unit cell is shown. Reprinted with permission from A. H. Morrish, *Physical Principles of Magnetism*, Wiley-Interscience, New York, 1965. Copyright © 1965 by John Wiley & Sons, Inc.

next-near-neighbor interactions. The theory predicts that at absolute zero $\chi_\parallel = 0$. The physical significance of this is that, in the approximation of the molecular field theory all atomic moments are either parallel or antiparallel to the applied field at $T = 0$ K. The field is then unable to exert any torque on the magnetic dipoles; this makes the induced magnetization zero.

When the applied field is perpendicular to the easy axis of magnetization, it is essentially able to produce a torque on both sublattice moments or magnetization, that is, $M_A + M_B$. Furthermore, the theory shows that the magnetization and hence the *perpendicular susceptibility* $\chi_\perp$ simply depends on N_{AB}, which represents the molecular field constant for near-neighbor interactions. Thus, $\chi_\perp$ remains independent of temperature up to the Néel temperature. The observed susceptibilities for a single crystal of MnF_2 are shown in Figure 3.15.

The $\chi_\perp$ component increases slightly with increasing field below the Néel temperature. This behavior is attributed to the *crystalline anisotropy energy*, which may be visualized as the "additional work" required to magnetize a material along a certain direction compared to along an easy direction.

By using measurements on powdered (i.e., polycrystalline) samples, which are conducted more frequently than measurements on single crystals, the mass susceptibility of a powder χ^P is easily correlated to the anisotropic mass susceptibilities by the following equation:

$$\chi^P = \tfrac{1}{3}\chi_\parallel + \tfrac{2}{3}\chi_\perp$$

Above the Néel temperature the distinction between $\chi_\perp$ and $\chi_\parallel$ vanishes and as pointed out before, a strict paramagnetic or Curie–Weiss behavior sets in. The plots of $1/\chi$ versus temperature along with a schematic of spin ordering for an antiferromagnet are given in Figure 3.16a. The extrapolation of the linear part of the curve gives a negative Weiss constant θ. This is generally greater than the values encountered in several of the paramagnetic systems.

So far, we considered ionic solids such as MnO and MnF_2, in which a long-range magnetic ordering exists. This suggests that every paramagnetic ion is magnetically coupled to every other ion by various exchange interactions, as follows; the most common interaction is through the intervening anion, such as O=

$$\cdots Mn\!-\!O\!-\!Mn\!-\!O\!-\!Mn\!-\!O \cdots$$

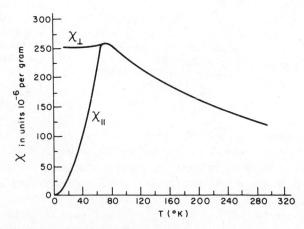

Figure 3.16 Mass magnetic susceptibility of a single crystal of manganese fluoride MnF_2 parallel and perpendicular to the tetragonal axis, which is the easy axis of magnetization.

In another class of compounds, especially in polynuclear (i.e., metal cluster) coordination compounds that crystallize as *molecular solids*, it is possible to confine paramagnetic ions within *molecules*, as shown in Figure 3.17.

Antiferromagnetism is then displayed by the molecules because the paramagnetic ions can be coupled antiferromagnetically. There is usually little or no interaction among neighboring molecules.

Chemists refer to this behavior as *intramolecular antiferromagnetism*, a term that has been used extensively by several workers (cf. Earnshaw [29]) and appears to have been rooted firmly in magnetochemical literature. Since the molecular solids do not involve any long-range magnetic ordering as in MnO, physicists do not like to use this term. However, a few physicists have paid sufficient attention to the unique magnetic properties of polynuclear complexes of copper [30] and appear to have overcome the nomenclature problem by referring to this phenomenon as the *magnetic properties of metal cluster compounds*. Mulay and Danley [31] coined the phrase *constrained antiferromagnetism* for short-range coupling of spins.

4.8.3   Ferrimagnetism in Ionic Solids

Although the phenomenon of ferrimagnetism was carefully investigated only a few decades ago, it can be described more appropriately with reference to antiferromagnetism outlined in the previous section. For organizational and pedagogical reasons the phenomenon of ferromagnetism, which has been known since ancient times, will also be considered in relation to antiferromagnetism and ferrimagnetism in Section 4.8.4.

Ferrimagnetism is the term proposed by Néel to describe the behavior of a class of compounds called ferrites. As in antiferromagnetism, magnetic ions occupy two types of sites, A (minus spin) and B (plus spin); however, the number of ions at the A site are different from those at the B site. For example, in a ferrite such as $MgFe_2O_4$, which has the *spinel* structure (named after the mineral $MgAl_2O_4$), there exist 64 tetrahedral and 32 octahedral sites; of these 8 tetrahedral (A) and 16 octahedral (B) sites are occupied by the paramagnetic Fe^{3+} ions (Figure 3.18). Since the number of magnetic ions (of the same type or of different types) and also the magnitude of spins of individual ions are different

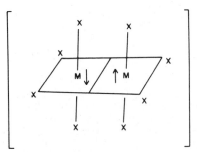

Figure 3.17 A coordination complex with ligands (x) at each corner of the two fused octahedra shows the possibility of *trapping* two transition metal ions (M). This gives *constrained* or *intramolecular* antiferromagnetic exchange interactions between their spins in certain cases.

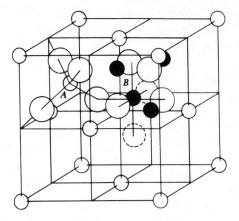

Figure 3.18 The unit cell of the spinel lattice. The large spheres represent oxygen ions, the small light spheres are ions in tetrahedral A sites, and the small dark spheres are ions in octahedral B sites. For only two octants are the positions of all the ions shown. The other octants have one or the other of these two structures and are arranged so that no two adjacent octants have the same configuration. Reprinted with permission from A. H. Morrish, *Physical Principles of Magnetism*, Wiley-Interscience, New York, 1965. Copyright © 1965 by John Wiley & Sons, Inc.

on the A and B sites, a resultant magnetization can be produced below a critical temperature with an external field. This magnetization is called *spontaneous magnetization* I_{sp}. Ferrites can be magnetized by applying an external field H; the variation in the magnetization with H is shown in Figure 3.19. Now, if we start with a ferrite, which is already magnetized, an increase in the temperature agitates the ordering of spins and results in a decrease of spontaneous magnetization. Above the critical temperature, called the *Curie point* (which is analogous to the Néel temperature for antiferromagnets), the spins are randomized completely and thus spontaneous magnetization disappears. The variation of spontaneous magnetization I_{sp} with temperature naturally depends on the structure and the specific magnetic ordering in ferrites. However, a typical plot for I_{sp} (measured at a constant saturating field H) versus temperature is shown in Figure 3.13*b*. The corresponding variation of the reciprocal susceptibility $1/\chi$ (χ is calculated from I/H) with temperature is shown by the dashed line, which makes a negative intercept on the T axis, thus giving rise to a negative Weiss constant. Above the Curie point the behavior is paramagnetic. Thus ferrimagnetism is similar to antiferromagnetism with regard to a negative Weiss constant, the existence of two different sites for spins, and negative exchange interactions between neighboring ions [5, 6].

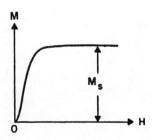

Figure 3.19 Magnetization M as a function of the applied field H for a ferrimagnet or a ferromagnet. The maximum value M_s is called the saturation (or spontaneous or intrinsic) magnetization.

Ferrimagnetic materials (or ferrites) can be magnetized permanently like the well-known ferromagnetic materials; however, antiferromagnets do not have this property. Ferrimagnets show hysteresis curves similar to those of the ferromagnets. Ferrites have been synthesized in numerous chemical compositions with equally interesting crystal chemistry. In addition to the well-known *spinels*, there are the *inverse spinel*, the *garnet*, and the *magnetoplumbite* structures, which define their magnetic properties. These properties also depend on the macroscopic structure, that is, on their particle profiles [32]. Ferrites that show *narrow* and *broad* hysteresis curves are classified as *soft* and *hard* magnetic materials, respectively. The hard materials can be magnetized permanently.

4.8.4   Ferromagnetism in Metals and Nonmetals

The phenomenon of ferromagnetism falls under the category of attraction between a material and an applied field; however, the forces of attraction are very great as compared with those of paramagnetism. Ferromagnetism is quite field dependent, and ferromagnetic substances show typical hysteresis curves, as illustrated in Figure 3.20. Hence, many ferromagnetic properties are measured at saturation, that is, by using high applied fields that cannot bring about a further increase in the intensity of induced magnetization at a given temperature. Ferromagnetic materials, when heated, lose their magnetism gradually. Beyond a critical temperature, called the *Curie temperature* or the *Curie point*, they behave as feeble paramagnets (Figure 3.20a).

In nature, ferromagnetism is restricted to only a few metals such as Fe, Co, and Ni. Several alloys (Alnico, $SmCo_5$) and exceptional oxides such as CrO_2 and EuO also exhibit ferromagnetism.

In ferromagnetism the spins are parallel; that is, all spins are, say, positive within a microscopic "volume" called the *domain*. A positive-exchange interaction exists among the spins. As in previous cases, such magnetic ordering is upset by thermal agitation, so that the spontaneous magnetization (expected from all parallel spins) decreases with increasing temperature. It becomes zero at the corresponding Curie point, as shown in Figure 3.13c. The corresponding susceptibility (ca. 10^{-6}) varies with temperature in the paramagnetic sense, above the Curie point; thus, as in ferrimagnetism, $1/\chi$ versus T gives a straight line except that it makes a positive intercept on the T axis with a positive Weiss constant (Figure 3.16c).

Although the existence of spontaneous magnetization without an external field is postulated on the basis of the presence of a *resultant spin*, a ferromagnetic or ferrimagnetic material in its *bulk state* does not display magnetization. The bulk material exists in an unmagnetized state because it consists of domains; the direction of spontaneous magnetization of each varies from domain to domain in the bulk state such that the resultant magnetization for all domains in the bulk state of the material is zero. Fortunately, this magnetization can be changed from zero to the expected maximum value of spontaneous magnetization I_{sp} by applying an external field, as shown in Figure 3.19. Furthermore, in

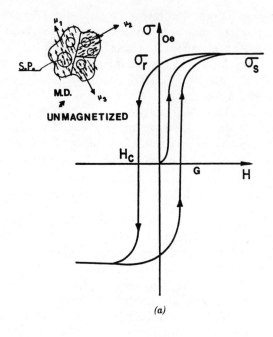

(a)

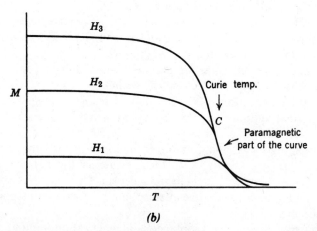

(b)

Figure 3.20 (*a*) Schematic drawing of per gram magnetization σ with the applied field H for a ferromagnetic material, showing a narrow hysteresis when impurities, mechanical strain, and so on are absent. The coercive force H_c required to bring the remanence σ_r to zero is very small (ca. 0.1 G) for such materials. The inset shows ideal superparamagnetic SP clusters, as subdomain components with different volumes of a multidomain (MD) large particle, with randomly oriented moments (μ_1, μ_2, μ_3) in the unmagnetized state. (*b*) Intensity of magnetization M versus temperature at three different fields for a ferromagnetic substance.

both ferrimagnetic and ferromagnetic materials we are blessed by the well-known phenomenon of hysteresis. Thus, by reducing the applied field H, we do not trace back the upgoing curve, but instead follow the hysteresis path shown in Figure 3.20a. The negative field $-H$ simply indicates a reversal in the direction of the applied field. Hence, the hysteresis phenomenon gives important magnetic parameters such as the coercive force (H_c) and remanence (I_r), which are very important in the design of permanent magnets. The maximum value of magnetization produced by an external field in the bulk material is known as the *saturation magnetization*, and this is said to correspond to the spontaneous magnetization postulated at the domain level. The remanence is expressed usually in a dimensionless form by considering the ratio $I_r : I_s$.

It was pointed out before that ferrimagnetism and ferromagnetism are quite field dependent, and such materials show typical hysteresis curves, as illustrated in Figure 3.20. Hence, many ferrimagnetic and ferromagnetic properties are measured at saturation, that is, by using high applied fields that cannot bring about any further increase in the intensity of induced magnetization at a given temperature. When heated, ferrimagnetic and ferromagnetic materials gradually lose their magnetism. Beyond a critical temperature, called the Curie temperature or the Curie point, they behave as regular paramagnetics.

In ferrimagnetism and ferromagnetism, it is necessary to distinguish between the effective moment, $\mu_{\text{eff}} = \sqrt{n(n+2)}\beta$ (Bohr magnetons), and the saturation moment μ_s, which is the maximum component of the magnetic moment in the direction of the applied field, given by $\mu_s = n$, where n is the number of the unpaired electrons. The saturation moment is obtained from specific magnetization σ that is studied as a function of both temperature and applied field strength. Extrapolation of the σ versus the T curve to $T = 0$ gives values for σ_0 that correspond to *different fields*. When these values are plotted against the reciprocal field $1/H$ and extrapolated to zero (equivalent to finding σ_0 at infinite or saturation field), the value for σ_0, ∞, that is, true saturation magnetization at $T = 0$ and $H = \infty$, is obtained.

In a ferrimagnetic material with the formula unit Fe_3O_4 (magnetite, with an inverse spinel structure) half the Fe^{3+} ions in the tetrahedral sites cancel the other half in the octahedral sites. Thus, $\mu_s = n = 5$ BM are completely canceled for this $3d^5$ ion and only the Fe^{4+} ion is left, with a $\mu_s = n = 4$ for the $3d^4$ ion. Hence, the theoretical saturation moment for this formula unit is 4. Since there are 8 such formula units per unit cell, the Bohr magneton number is 32, which is in good agreement with the experimental value. This example illustrates the usefulness of saturation versus the effective moments in magnetically ordered materials and the normal paramagnetics.

Spontaneous magnetization I_{sp} of iron at both room temperature and 0 K is 1707 and 1752 G, respectively. (Here, gauss refers to gauss per cubic centimeter; thus, I_{sp} must be regarded as the magnetization per unit volume.) Surprisingly, when the relative magnetization I_{sp}/I_0 (where I_0 is the true saturation magnetization at infinite field and absolute zero) is plotted against T/T_c (where T_c is the Curie temperature), the variation with temperature appears the same for many ferromagnets. The saturation magnetization I_s, which represents a complete alignment of all spins, can be produced *in principle* at any temperature other than 0 K by applying an infinite field.

As indicated before, in ferromagnetism and ferrimagnetism, the saturation moment $\mu_s = n$, is an important parameter and not the μ_{eff} that is used for paramagnets. We illustrate the calculation of the saturation moment μ_s, sometimes referred to as the *Bohr magneton number*, for iron (in the bulk state) from the saturation magnetization at 0 K.

The Fe^0 atom is represented by the $3d^5\, 4s^2$ configuration in which there are *four* unpaired spins. Therefore, a saturation Bohr magneton number, $\mu_s = n = 4$, for an isolated Fe^0 atom would be expected. However, a $\mu_s = 2.2$ is found from measurements on bulk iron because of the band structure of metallic iron, in which there is an overlap of the $3d$ and the $4s$ bands. Textbooks on solid-state physics should be read to understand the distinction between the *band structure* of metallic and semimetallic systems and the *bonding* in nonmetallic compounds.

As stated earlier, the saturation magnetization (I_0 or M_0) per unit volume arises from the number η' of constituent (identical) particles (atoms), each having a moment μ_A. Thus

$$I_0(\text{or } M_0) = \eta' \mu_A \beta$$

where β is the Bohr magneton, which will be expressed in appropriate units in a later equation. Thus, if $M_0 = 1752\,\text{G}$, to find μ_A we must first find the number of particles η' per unit volume. Now, if MW (g mol^{-1}) is the molar weight of the material, it will contain 6.022×10^{23} atoms per mole (which is the Avogadro number). If ρ is the density of the material, then ρ grams (g) will contain ($\rho \times 6.022 \times 10^{23}/\text{MW}$) atoms. However, these ρ grams correspond to a volume of 1 cm^3; hence

$$\eta' = \frac{\rho(\text{g cm}^{-3}) \times 6.022\,(\text{mol}^{-1})}{\text{MW (g mol}^{-1})}$$

Hence, μ_A is given by

$$\mu_A = M_0/\eta'\beta$$

Using the following values for iron, we calculate that $M_0 = 1752\,\text{G}$, MW $= 55.85$ g mol^{-1}, and $\rho = 7.895$ g cm^{-3}; and expressing $\beta = 9.27 \times 10^{-21}$ G$\cdot$cm^3, we obtain

$$\mu_{\text{Fe}} = \frac{1.752 \times 10^3\,\text{G} \times 5.585 \times 10\,\text{g mol}^{-1}}{7.895\,\text{g cm}^{-3} \times 6.022 \times 10^{23}\,\text{mol}^{-1} \times 9.27 \times 10^{-21}\,\text{G}\cdot\text{cm}^3}$$

$$= 2.22$$

Since the units such as gauss, cubic centimeter, gram, and reciprocal moles in the numerator cancel the corresponding units in the denominator, $\mu_{Fe} = 2.22$ is, as it should be, a *dimensionless quantity*. For this reason, Selwood [22] and others refer to μ_A as the *Bohr magneton number*, the key word *number* here immediately implies that the magnetic moment is a dimensionless quantity. Despite this aspect, most writers continue to say, for instance, that the moment per atom of iron is 2.22 BM. The rationale for using the somewhat uncommon unit gauss cubic centimeter for the Bohr magneton is presented in Section 3.13. This Bohr magneton number is for one atom of Fe^0 in the bulk state.

In general the magnetization observed per gram σ under a given set of conditions can be converted to the corresponding moment μ and expressed as the Bohr magneton number by following exactly the same procedures as above.

4.8.5   Magnetic Anisotropy in Ferromagnetism and Ferrimagnetism

There are several types of magnetic anisotropy (see Cullity [32]) encountered in ferromagnetism and ferrimagnetism. Of the following, the first two are easy to visualize. These are important in subsequent discussions on superparamagnetism (next section) and in thermomagnetic applications with regard to the characterization of certain metallic catalysts. The types of anisotropy are:

1.   *Crystal anisotropy*, also known as *magnetocrystalline anisotropy*. In many single crystals, imagine (a) the easy, (b) the medium, and (c) the hard axes of magnetization. In the case of iron, for instance, the directions $\langle 111 \rangle$, $\langle 110 \rangle$, and $\langle 100 \rangle$ correspond to the hard, medium, and easy directions, respectively.

2.   *Shape anisotropy.*   This depends on spherical or elongated (needlelike) small particles. Usually, in the latter case, the easy axis of magnetization may lie along the long axis (a) of the elongated particles, whereas the hard axis may be said to lie along the short axis (b) of the ellipsoid. The shape of the particles is largely responsible for the demagnetization factor.

3.   *Stress anisotropy.*   Mechanical stress can indeed cause changes in the magnetic properties of a system.

4.   *Exchange anisotropy* and anisotropy induced by (a) *magnetic annealing*, (b) *plastic deformation*, and (c) *irradiation* [32]. These types of anisotropy are not discussed here.

4.8.6   Superparamagnetism

The phenomenon of *superparamagnetism* displayed by very small particles (say, less than about 200 Å) is fascinating indeed and is important in the realm of heterogeneous catalysts. Therefore, if one were to "grind" materials that are ferromagnetism, ferrimagnetism, or antiferromagnetic in their bulk states to very fine single domain particles, such particles will display superparamagnetism. Thus, even at the microscopic level a type of magnetic ordering prevails. Hence, in a superparamagnetic particle, a parallel arrangement of spins ($\uparrow\downarrow$) or an antiparallel ($\uparrow\downarrow$) arrangement may continue to exist even at the subdomain level; these alignments are usually referred to as *ferromagnetic* and *antiferromagnetic ordering*, respectively; as such we included the superparamagnetic category of materials under the larger heading of magnetically ordered materials and showed it as a subdivision of such materials in Figure 3.9.

Superparamagnetic particles (or clusters) are very small and consist essentially of single domains with little or no magnetic interaction between neighbors. When a single-domain particle is small enough, the presence of thermal energy becomes an important consideration, since the thermal energy kT can act to randomize the magnetic moments of clusters with respect to an applied magnetic field H. Hence, superparamagnetic particles are thermally unstable in the magnetic sense. An assembly of such particles, therefore, can be treated as those of a paramagnetic system. Thus, the Brillouin function, applicable to normal paramagnetism, also applies. Here, since a large number of spins are imagined as "packed" in a small volume of the cluster, the total angular momentum J simply signifies an addition of all such spins. Since the orbital momenta (l) are quenched (especially in metallic systems), $s = j$ and $J = \Sigma\, s_i$. This situation should be contrasted with the significance of J approaching infinity in the classical interpretation of the Brillouin function applied to normal paramagnetism, where $J \to \infty$ implied its taking on an infinite number of orientations with respect to the applied field H (see [32]). Thus, although the significance of $J \to \infty$ is different for superparamagnetics, for large J, the Brillouin function passes over asymptotically to the Langevin function (with the saturation moment $\mu = gJ\beta$).

$$\frac{\sigma}{\sigma_s} = L(X) = \lim B_j(y) = \cot\frac{\mu_c}{kT} - \frac{kT}{\mu_c}$$

In this case, the magnetic moment μ_c is much larger than that of an ordinary paramagnet; indeed, it may describe a particle or cluster that can contain up to 10^5 magnetically coupled atoms. On the other hand (even small) hysteresis effects normally associated with (the magnetically *soft* and *hard*) ferrimagnetic and ferromagnetic materials will be absent; thus, there will be no remanence σ_r or coercive force H_c. Hence, as an *oversimplification*, it is possible to view a superparamagnetic system as the intermediate of a pure paramagnet and a pure ferrimagnet or ferromagnet (Figure 3.21). Generally, particle–particle interactions are neglected, although efforts are increasingly being made to consider this effect (see references in Collins, Dehn, and Mulay [33].

In addition to the (ideal) ferromagnetic "multidomain" and subdomain superparamagnetic structures (for which $\sigma_r = 0$ and $H_c = 0$) we consider the subdomain anisotropic particles, which can have contributions to their magnetic energy arising from their shape and crystal anisotropies. Subdomain anisotropic particles are thermally stable in the magnetic sense (Morrish [26]) and are characterized by significant hysteresis during magnetization, thus giving measurable values for H_c and σ_r. The contributions H_c and σ_r caused by the

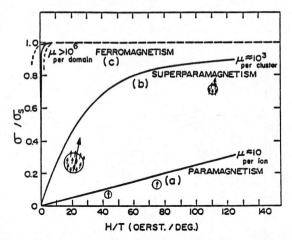

Figure 3.21 Schematic representation of (a) paramagnetism, (b) supermagnetism SP, and (c) ferromagnetism; variation of relative σ/σ_s magnetization as a function of the ratio of magnetic field to temperature. Large particles with a net large moment saturate relatively easily at lower H/T as compared to small particles, which saturate at a higher H/T ratio. The SP particles are "thermally" unstable; others are stable.

shape of elongated single-domain (SD) anisotropic particles far outweigh any contributions from magnetocrystalline anisotropy.

Kneller and Luborsky [34] carried out an elegant theoretical analysis of the contributions of each of these structures to H_c and σ_r as a function of the range of particle sizes. They further substantiated their theory with supporting data obtained for iron–cobalt alloy particles (20–300 Å) dispersed in mercury. Their approach and contributions by other workers have advanced our understanding of the particle-profile criteria that determine transitions from the (1) superparamagnetic to single-domain anisotropic and from (2) the single-domain anisotropic to multidomain particles. A schematic curve showing results of these transitions is given in Figure 3.22 and is based on Kneller and Luborsky [34].

The magnetic behavior of the particles depends not only on their critical size as just described, but also on the *blocking temperature* T_B. In the parlance of magnetism T_B is a critical blocking temperature above which the effects of thermal agitation (proportional to kT) dominate the orienting effects of the applied magnetic field H. Thus, superparamagnetic particles begin to display the σ versus H hysteresis loops with significant observable values for H_c and σ_r when cooled below T_B. The subdomain anisotropic particles are thermally stable in the magnetic sense below T_B (Morrish [26]).

For superparamagnetic particles, it is *assumed* that their spontaneous magnetization I_{sp} is the same as that for bulk material. For instance, for Fe, I_{sp} at room temperature is approximately 1707 G (per cubic centimeter), which increases slightly to 1752 G (per cubic centimeter) at 4 K ($\simeq 0$ K). Furthermore, the moment μ_c of a superparamagnetic cluster with average volume $\bar{V}$ is given by

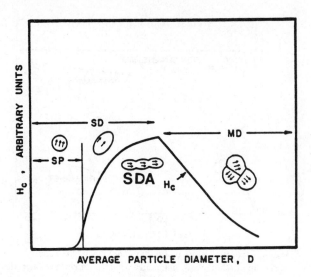

Figure 3.22 Variation of coercivity H_c with average particle diameter: SD, single domain; MD, multidomain; SP, superparamagnetic; and SDA, single domain anisotropic. Superparamagnetic region is thermally unstable; SDA and MD are thermally stable.

$\mu_c = I_{sp} \cdot \bar{V}$. This equation forms the *basis* for obtaining the $\bar{V}$ of superparamagnetic particles in the low-field (LF) and in the high-field (HF) region of the Langevin function stated earlier. Appropriate *deviations* are given by Selwood [22], with numerical and graphical examples for calculating the average volumes $\bar{V}_{LF}$ and $\bar{V}_{HF}$ for Ni particles on SiO_2 substrate by assuming the behavior of *large* and *small* particles in the low field (LF) and in the high field (HF). Generally, the particles are assumed to be spherical; that is, $\bar{V} = \frac{4}{3}\pi r^3$, with radius r.

The per gram magnetization σ of a sample is measured as a function of the field H at various constant measuring temperatures T_m. The corresponding per gram saturation magnetization σ_s is generally obtained by plotting values of σ at the lowest cryogenic temperature (ca. 4 K) as a function of $1/H$ and extrapolating $1/H$ to zero to obtain σ_s from the intercept on the σ axis. This procedure is tantamount to finding the true σ_s at $H \to \infty$ and $T_m \to 0$. Thus, plots of the relative magnetization σ/σ_s (which becomes a dimensionless quantity) are plotted as a function of H/T_m (in units of oersteds per degrees kelvin). A *good superposition* of the *data points* is a *working definition* for the superparamagnetic behavior of *clusters* of magnetic (ferromagnetic, ferrimagnetic, and antiferromagnetic) species.

Since large particles are saturated rather easily, their average volume $\bar{V}_{LF}$ is found from the initial slope of σ/σ_s versus the H/T_m curve. Thus for *low fields*, in general

$$\bar{V}_{LF} = \frac{4}{3}\pi r_{LF}^3 = \frac{(3k/I_{sp})(\sigma/\sigma_s)}{(H/T_m)}$$

Since small particles reach saturation at high fields, the HF approximation [22] gives

$$\bar{V}_{HF} = \frac{4}{3}\pi r_{HF}^3 = \frac{k}{I_{sp}}\left[\frac{1}{1-(\sigma/\sigma_s)}\right]\bigg/\frac{H}{T_m}$$

Thus, the extreme limits of $\bar{r}_{LF}$ (large particles) and $\bar{r}_{HF}$ (small particles) are obtained for a very narrow particle-size distribution of a superparamagnetic system. An example of such a system consisting of Ni particles on $(Al_2O_3 + graphite)$ substrate is given under Section 8.3.5.

In recent years, Desai (personal communication to L. N. Mulay by Desai) proposed equations for deriving particle-size distribution from magnetic measurements by assuming a *log normal* distribution function.

5   TYPES OF INFORMATION OBTAINABLE FROM MAGNETIC MEASUREMENTS

The scope of information that can be obtained from magnetic measurements is outlined next; the practical procedures to be followed for obtaining such information are given later.

5.1   Qualitative Aspects

By using any standard method, it is possible to determine whether the magnetic susceptibility of a material under test is dependent on (1) the temperature, (2) the field strength, or (3) both factors. This yields the qualitative information as to whether the material shows diamagnetically, paramagnetically, or magnetically ordered behavior. In geochemical studies a simple magnet test reveals the nature of a mineral, that is, *ferrous* (ferromagnetic) or nonferrous character. The application of magnetic separation of minerals is well known. High gradient magnetic separation (HGMS) rapidly evolved as a tool for the separation of pollutants, such as pyrite, from coal and impurities, such as the iron oxides, from kaolin.

5.2   Quantitative Aspects

Thermomagnetic analysis, which is discussed separately, yields valuable information on the particle-size distribution and activity of catalysts and on structural and electronic phase transitions. By measuring magnetization and/or permeability factors such as the composition of alloys, the degree of precipitation of a metal in its alloy, and the carbon content in steel can be determined. The susceptibility of a paramagnetic system that includes solid and gaseous mixtures, solutions, and colloids and dispersed media, such as glass, yields the concentration of a transition metal, rare earth, or a paramagnetic gas involved in any of these systems.

A study involving diamagnetic and paramagnetic measurements of organometals undergoing dissociation proves useful in detecting the presence and measuring the concentration of free radicals, the constants for the dissociation process, and so on. Similar studies in biological areas furnish the analysis of certain components of blood, whereas diamagnetic measurements alone are useful in characterizing and ascertaining the purity of organic and inorganic compounds and, to a moderate extent, in studying polymerization processes.

Diamagnetism of superconducting materials, when studied as a function of temperature, provides a useful probe for pinpointing the critical temperature, below which the material becomes superconducting. In recent years the advent of high-resolution NMR spectroscopy has resulted in a revival of investigations of diamagnetism of matter in its different physical states. Magnetic anisotropy studies have been helpful in elucidating the electronic structure of both molecular solids (such as the organometals) and magnetic materials.

6   TECHNIQUES FOR MAGNETIC MEASUREMENTS

6.1   General Comments

Here, we describe a wide variety of techniques for the measurement of magnetic susceptibility, which range from the well-known classical force-type methods (such as the Gouy and the Faraday) to the most modern and relatively less-known techniques based on unusual phenomena associated with superconductivity; for instance, the SQUID (superconducting quantum interference device) susceptometers. A section on the measurement of susceptibility under pressure is also included. Under each category is a succinct outline of the working principles and only the "working equations" that are used to derive the susceptibility or the magnetization from experimentally measurable parameters. Readers interested in understanding the derivations of these equations must read the original papers and/or the many theoretical works on magnetism. In describing the modifications of these techniques, truly novel concepts and the types of instrumentation that can be employed easily in laboratories operating under a limited budget are addressed.

Pedagogically, there are included a few references to magnetic apparatus, which may be useful for classroom demonstration or for laboratory experiments at the undergraduate level. Descriptions of auxiliary instrumentation for the control and measurement of temperature are also included. Readers should refer to [35–46] for information on other modifications that were not incorporated here because of space limitations.

6.2   Comparison of Various Techniques

A summary of various techniques and their advantages and limitations are given in Table 3.11. Descriptions of the individual techniques and their modifications are given in Sections 6.3 to 6.5.

6.3   Principle of Force Methods

Magnetic susceptibilities are generally measured by the so-called *uniform field* and *nonuniform field methods*. The principles common to the conventional force methods are described first, and the details of instrumentation are deferred to later sections.

A body experiences an orienting effect in a magnetic field depending on its magnetic anisotropy and, to a lesser degree, on its shape factor. This effect is directly proportional to a product of its volume susceptibility χ, its volume V, and the applied field H. This body will experience a linear displacing force if the field is made nonuniform with a gradient dH/dX in the X direction; the force is given by

$$f = \kappa V H \frac{dH}{dX}$$

6.3.1   Uniform Field (Gouy) Method

In the uniform field method, developed by the French physicist Gouy (cf. [3]), a long cylindrical sample is suspended such that one end lies in a region of strong uniform field and the other end lies in a region of negligible field. Thus even here a "nonuniform" field is produced over the entire sample as in the Faraday method. This is accomplished by a setup (Figure 3.23) that permits a direct measurement of force. In this case, integration of $VH \cdot dH/dX$ over all layers between the limits of maximum field H and negligible field ($H \simeq 0$) gives, for force, the relation

$$f = \tfrac{1}{2}\kappa H^2 A$$

where A is the cross-sectional area of the sample. This equation tacitly assumes that the atmosphere surrounding the sample has a negligible susceptibility and that the field at one end of the tube is negligible compared with that at the other. If these conditions are not fulfilled so that the atmosphere has a susceptibility of κ_0 and the field at the other end of H_0, then the equation for f becomes

$$f = \tfrac{1}{2}(\kappa - \kappa_0)(H^2 - H_0^2)A$$

The magnitude of the earth's magnetic field, which is about 0.4 Oe acting at one end of the sample, is considered negligible for all practical purposes, compared with fields of a few thousand oersted applied at the other end. Even fields of about 100 Oe may be neglected if H is about 10,000 Oe. A uniform field in the pole gap is necessary for a practical consideration. If the field is not uniform, the sample will experience a horizontal force, and it may tend to move toward one of the pole forces and thus vitiate an exact measurement of the vertical force.

Table 3.11 Summary of Important Aspects for Measuring Magnetic Parameters

Method	General Field Requirements	Applicable to	Physical Nature of Sample	Approximate Minimum and Convenient Size of Sample	χ_g Accuracy	Temperature Control
Gouy	Union field; recommended and easily available range with electromagnets is 3000–15,000 Oe; permanent magnets up to 5000 Oe	Diamagnetics and paramagnetics only	Powdered solids pure liquids, and solutions (adaptable for measuring χ of a gas surrounding a known sample)	0.5 g Solids, 5 mL liquids (macroscale); few milligrams or micrograms can be handled in special apparatus	Generally ±1% may be improved to ±0.1% (separate density measurement required; accuracy depends on packing)	Is possible over a wide range; from liquid helium or liquid hydrogen temperatures to several hundred degrees may be obtained
Quincke	Uniform field; recommended and easily available range with electromagnets is 3000–15,000 Oe; permanent magnets up to 5000 Oe	Diamagnetics and paramagnetics only	Pure liquids and solutions (adaptable for measuring χ of a gas above the meniscus of a known liquid)	ca. 5 mL	Generally ±0.1%	Limited range depending on fp or bp of system
Rankine	Low fields; 15–100 Oe	Diamagnetics and paramagnetics only	Pure liquids and solutions (adaptable to flow system and gases)	ca. 2 mL	Generally ±0.1%	Limited range depending on fp or bp of system
Faraday	Field strength range same as for Gouy balance, but giving nonuniform field with a constant $H \cdot dH/dX$	Diamagnetic, paramagnetic, and ferromagnetic materials	Generally useful for powdered solids (liquids may be handled in special containers)	Few mg (microtechniques are also available)	±0.1%	Temperature control is easily possible over a wide range (same as for Gouy technique)

Method	Field requirements	Applicability	Sample form	Sample size	Accuracy	Temperature control
Change in flux methods (ac including rf and dc)	External fields not required except in a study of ferromagnetics	Generally to diamagnetics and paramagnetics; ferromagnetics may be studied in a special apparatus	Solids and liquids	0.5 g Solids; 5 mL liquids	Accuracy generally better than ±0.1% but depends on electronic characteristics	Temperature control over a wide range is rather difficult with rf method, but is adaptable in other inductance methods
Vibrating sample magnetometer (VSM)	Uniform fields up to 20,000 Oe for studies on magnetic materials	Mostly ferromagnetics and ferrimagnetics and very strongly paramagnetics; not useful for normal diamagnetic and paramagnetic materials	Solids only; cannot be used for adsorption studies, and so on.	Few mg	Working accuracy ca. ±1% [VSM gives σ as a $f(H)$]	Temperature control is possible over a very wide range with a reliable cryostat
SQUID magnetometer	Fields ca. 10^5 Oe (with a superconducting magnet) and a Josephson junction	All types; excellent for diamagnetics and weakly paramagnetics	Solids and liquids	Few mg	+0.01 (10^{-13} emu g^{-1})	Temperature control from 2 to 300 K is possible

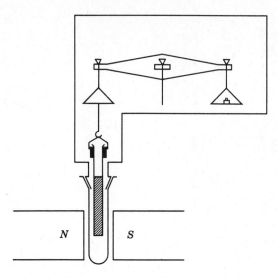

Figure 3.23   Principle of the Gouy balance; N and S stand for the north and south polarities.

The Gouy method is particularly suited for the measurement of diamagnetic and paramagnetic susceptibilities of samples that are powdered solids, liquids, and solutions. It cannot be used successfully for ferromagnetic, superparamagnetic, and related measurements because a magnetic saturation of the entire sample cannot be attained. Bates [27], however, describes its applicability for such measurements, although the method is rather cumbersome. Techniques such as those employed by the vibrating sample magnetometer are excellent for ferromagnetic and ferrimagnetic systems.

The theory requires a sample that is in the form of a long homogeneous cylinder. With solids this is accomplished only to an approximation by powdering the sample and packing it uniformly in a glass tube of uniform diameter. The reproducibility of packing is rather difficult, and although corrections (Section 7.9.4) for the air pockets can be introduced, the overall accuracy cannot exceed $\pm 1\%$ for measurements on solids. Another limitation is introduced because some of the loosely bound particles in the magnetic field may orient preferentially because of their magnetic anisotropy.

More accurate measurements to within $\pm 0.1\%$ can be obtained with liquids, which do not pose the problems of packing, anisotropic effects, and so on. Usually 0.5- to 1-g samples of solids, liquids, and solutions of moderate concentration are required to obtain results with an analytical balance of sufficient accuracy. The use of a microbalance furnishes more refined measurements and allows the use of very small samples.

The Gouy method always provides a measurement of *volume susceptibility*. Therefore, conversion to mass susceptibility requires an independent measurement of density; hence, the accuracy of this measurement also determines the accuracy of the mass susceptibility data. The Gouy method is especially suited for making absolute measurements since it is possible to measure the field H and the cross-sectional area A of the sample tube with great accuracy; however, this gives only the volume susceptibility χ or κ, as discussed earlier.

6.3.2 Nonuniform Field (Faraday) Method

The Faraday method and its principle are illustrated in Figure 3.24. If a sample of mass m with mass susceptibility χ_m is placed in a nonuniform field H with a gradient dH/dX, it will be subjected to a force along χ, given by

$$f = m\chi_m H \frac{dH}{dX}$$

The nonuniform field may be obtained by inclining the flat poles of a magnet or by using pole pieces of special design (Figures 3.25 and 3.26). Clarification of the terminology *constant gradient* and *constant force* poles is offered in Section 7.10.1.

The *Faraday* method is most *versatile* and is applicable to the measurement of susceptibilities encountered in all magnetic phenomena. With superparamagnetic and ferromagnetic measurements, a magnetic saturation of the sample can be achieved because the entire sample is enclosed in the field.

The method is particularly suitable for solids, which may be powdered and compressed into tablet form if necessary and placed at the region of maximum $H \cdot dH/dX$ in the magnetic field. The powdering and compressing technique

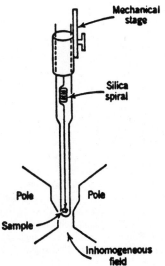

Figure 3.24 Faraday balance using a vertical suspension from a silica spiral spring. Reprinted with permission from P. W. Selwood, *Magnetochemistry*, Wiley-Interscience, New York, 1956. Copyright © 1956 by John Wiley & Sons, Inc.

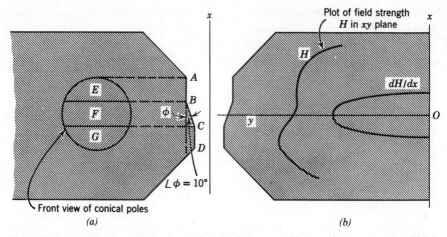

Figure 3.25   Details of the pole gap for the Faraday balance used in one of our apparatus, which is similar to Sucksmith's design (a) $AB = CD$ = projection of BC on axis parallel to AB and CD; that is, heights of sections E, F, and G are equal; (b) cross section of mild steel conical pole tips.

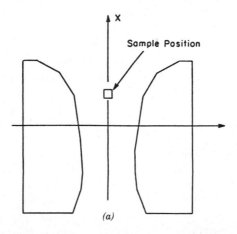

Figure 3.26   (a) Profile of 6.5-in.-diameter pole faces. The original poles give a region of uniform $H \cdot dH/dX$ (constant force) were supplied by Spectromagnetic Industries, Hayward, CA (now a division of the Varian Associates, Palo Alto, CA). (b) Details of the Heyding 4-in.-diameter poles used in another Faraday balance to give a region of uniform $dH \cdot H/dX$ (constant force). Reprinted with permission from L. N. Mulay, *Magnetic Susceptibility*, Interscience, New York, 1966, a reprint monograph based on L. N. Mulay, in I. M. Kolthoff and P. J. Elving, Eds., *Treatise on Analytical Chemistry*, Part 1, Vol. 4, Interscience, New York, 1963, Chap. 38. Copyright © 1966 by John Wiley & Sons, Inc. (c) The upper portion shows the Lewis gradient coil sets as they are mounted in the gap of the large magnet. The lower portion shows a view looking toward the left pole face. The arrows on the coils show suitable current directions for obtaining a downward directed field gradient. The gap in the large magnet, between the 30.5-cm-diameter pole faces, is 6.3 cm. The three-layer gradient coil sets shown leave clearance for a 2.6-cm-diameter tube down the center along the z axis. (Available from George Associates, P.O Box 960, Berkeley, CA 94701.)

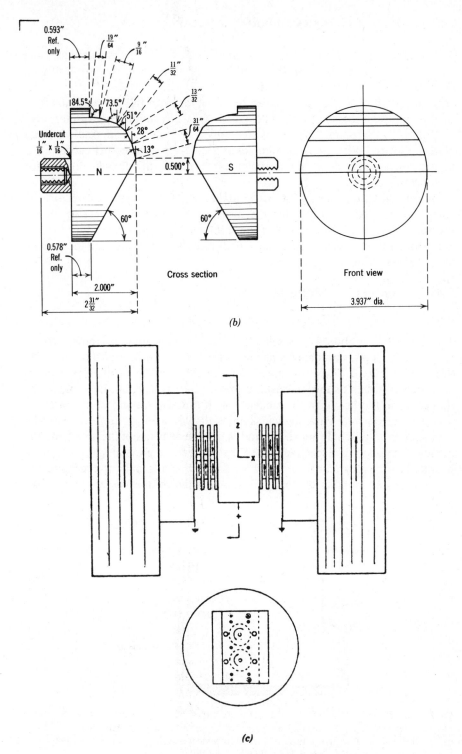

0.593"
Ref.
only

$\frac{19}{64}$"

$\frac{9}{16}$"

$\frac{11}{32}$"

$\frac{13}{32}$"

84.5°   73.5°

51°

Undercut
$\frac{1}{16}$" x $\frac{1}{16}$"

28°

$\frac{31}{64}$"

13°

0.500"

N

S

60°

60°

0.578"
Ref.
only

2.000"

$2\frac{31}{32}$"

Cross section

Front view

3.937" dia.

(b)

Z

x

+

(c)

Figure 3.26 (*continued*)

211

destroys the anisotropic effects of individual particles. With a few milligrams of sample, a very accurate (to within $\pm 0.1\%$) measurement of directly furnished mass susceptibility can be obtained.

Susceptibilities on liquids can also be measured with this method; however, they are quite cumbersome to handle. Efforts to enclose liquids in capsules small enough to be surrounded by the limited region of uniform field gradient cause difficulties with both sealing the samples and calibrating the empty capsule. It is indeed possible to fabricate miniature quartz "bottles" with airtight stoppers for such measurements.

In general, the accuracy of susceptibility measurements depends on the particular method employed for the measurement of force and also on the precision with which the setting of the sample container is reproduced at the exact point of reference between the poles during the calibration and the making of final measurement on a sample.

6.4   Derived Force Methods

6.4.1   Quincke Method

The Quincke method [47] is related to the Gouy technique and is strictly applicable to liquids and solutions and, with some modifications, to gases. As shown in Figure 3.27, the magnetic force acting on the capillary sample is measured in terms of the hydrostatic pressure. In the actual experiment, the change Δh in the height of the meniscus with the field off and on is measured with a cathetometer. Paramagnetic liquids display an increase in height; the diamagnetic liquids, a decrease. If ρ and ρ_0 are the densities of the liquid and the gas above the liquid, respectively, then the hydrostatic pressure $g'(\rho - \rho_0)\Delta h$ is balanced by the magnetic force $(1/2H^2)(\kappa - \kappa_0)$, where H is the applied field, κ is the volume susceptibility of the liquid, and κ_0 is the volume susceptibility of the

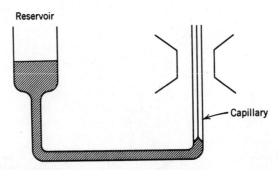

Reservoir

Capillary

Figure 3.27 Principle of the Quincke balance. Reprinted with permission from L. N. Mulay, *Magnetic Susceptibility*, Interscience, New York, 1966, a reprint monograph based on L. N. Mulay, in I. M. Kolthoff and P. J. Elving, Eds., *Treatise on Analytical Chemistry*, Part 1, Vol. 4, Interscience, New York, 1963, Chap. 38. Copyright © 1966 by John Wiley & Sons, Inc.

gas over the meniscus. With solutions exposed to air, the gas consists of a mixture of air and the vapor of the liquid. Hence

$$\frac{2g'(\rho - \rho_0)\Delta h}{H^2} = (\kappa - \kappa_0)$$

Here g' is the gravitational constant. From this χ_m, the susceptibility per gram is given by

$$\chi_m = \frac{\kappa}{\rho} = \frac{2g\Delta h}{H^2} + \chi_0 \frac{\rho_0}{\rho}$$

where χ_0 and ρ_0 are the mass susceptibility and the density of the gas over the meniscus, respectively.

For practical purposes, the second term can be ignored when the reservoir is of a large diameter compared with the capillary and the susceptibility of the gas χ_0 is small. A simpler relation is obtained by neglecting the second term

$$\chi_m = \frac{2g\Delta h}{H^2}$$

which is advantageous because an independent measurement of the liquid density is not required. It is not usually necessary to find the value of the applied field H, since the factor $2g/H^2$ can be eliminated by taking measurements on a sample (subscript s) and on a reference (subscript r) under identical conditions:

$$\frac{\chi_s}{\chi_r} = \frac{\Delta h_s}{\Delta h_r}$$

Fields of about 25,000 Oe are recommended. Accuracy of about 1% in the susceptibility measurement is easily obtained. This is comparable to that of the Gouy method, particularly at room temperature, which can be regulated to a high degree of accuracy.

It follows that the susceptibility χ_0 of a gas above the meniscus also may be determined by the Quincke method. In practice, a reference liquid such as water in an atmosphere of hydrogen gas may be used in the initial experiment. Hydrogen gas has a susceptibility of only 0.02% of water (χ_0). The experiment is repeated with a sample gas. If Δh_0 and Δh_g are the changes in the height of the meniscus of the liquid in the two experiments produced by the same magnetic field, the susceptibility χ of the gas is given by

$$\chi = \chi_0 \frac{\Delta h_0 - \Delta h_g}{\Delta h_0}$$

6.4.2   Modifications of the Quincke Method for Gases, Liquids, and Titrations

Several modifications of the Quincke method applicable to liquids and gases are described in the literature [2, 25].

A modification useful for magnetic titrations involving solutions of paramagnetic ions is described by Graybill and co-workers [48]. The authors carefully designed an apparatus in which the titrating assembly is made an integral part of the Quincke apparatus. Thus the susceptibility of the mixed solutions can be monitored continuously at successive stages of the titration.

The authors point out several advantages of their technique over the Gouy method, which has been adapted by many workers for titrations. The modified Quincke method does not provide the convenience of mixing solutions in the magnetic apparatus; however, all susceptibility measurements are performed at room temperature. The measurement of susceptibility as a function of temperature provides valuable information on the nature of the species formed in solution. With the Gouy method (Section 6.3.1) we can obtain the measurement of solutions over a limited temperature range more easily than with the Quincke method. To this extent the Gouy method has some advantages over the Quincke method.

Another point to be considered is that several sophisticated techniques (potentiometric, spectrophotometric, etc.) are now available for carrying out titrations in the classical sense and for the elucidation of structures of species in solution. Hence, it is necessary to show clearly that any magnetic method of titrations will provide information, at least in special cases, not obtainable from other instrumental methods.

6.4.3   Rankine Method

The measurement of susceptibility of a gas entails several difficulties. The volume susceptibilities of liquids and solids range from 10^{-6} to 10^{-3} cgs units per cubic centimeter, whereas the susceptibilities of gases and vapors are found to be much smaller (ca. 10^{-10} cgs units per cubic centimeter at STP). Measurements on a compressed gas, which is expected to have a larger volume susceptibility, are limited by the degree to which it may be compressed and the size of the vessels that may be used for the determinations. Furthermore, most gases are diamagnetic and even a trace (1 part in 1000) of oxygen (from air), which is markedly paramagnetic ($\kappa = 0.162 \times 10^{-6}$ cgs units), is enough to vitiate the measurements of diamagnetic susceptibilities by about 10%. Therefore, all gases and vapors must be purified and particularly freed from oxygen prior to measurement.

Many methods for measuring the susceptibility of gases are related to the Gouy and Faraday techniques. Sone [49], for instance, used a partitioned glass tube with the Gouy technique; one part contained air under pressure of known susceptibility and the other end was evacuated and sealed off. Air was then replaced by the gas, and an optical system with a sensitive balance was used to measure the forces on the sample. Stossel [50] also used the Gouy principle for measurement of gases; for the measurement of small forces, however, it appears that the sensitivity and stability of a balance must be stretched to extreme limits. On the other hand, modifications to the Faraday method have been accomplished with ease. The methods introduced by Glaser [51], Bitter [52], Havens [53], Lallemand [54], Néel [55], Reber and Boeker [56], Vaidyanathan [57], and Efimov [58] fall into this category, and each has its characteristic advantages. Reviews of these procedures and several other modifications are found in the literature [2, 25]. However, considering the general usefulness for routine susceptibility measurements and analytical applications, the Rankine method [59] and its modifications are considered here.

Instead of keeping the magnet fixed and observing the displacement of the sample, the sample in the Rankine method is fixed and the displacement of a small permanent magnet is observed (Figure 3.28). A bar magnet placed parallel to a sample surface induces a polarity on the surface that exerts a force on the magnet and displaces it. As in other methods, the force is an attraction for a paramagnetic sample and a repulsion for a diamagnetic one. The force is measured with a torsion arm suspended from a quartz fiber and is designed to minimize the effect of the earth's magnetic field and other stray fields. The method is quite sensitive and uses a magnet of only a few hundred oersteds. The method is used for measuring the susceptibilities of a sample relative to the known susceptibility of a reference.

An important modification of the Rankine balance is described in detail by Bockris and Parsons [60]. The major improvements are (1) the use of tungsten

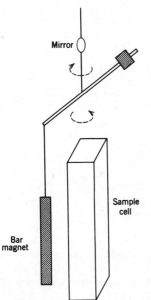

Mirror

Sample cell

Bar magnet

Figure 3.28 Principle of the Rankine balance. The quartz fiber with the mirror twists in either direction, depending on whether the sample in the cell is diamagnetic or paramagnetic. Reprinted with permission from L. N. Mulay, *Magnetic Susceptibility*, Interscience, New York, 1966, a reprint monograph based on L. N. Mulay, in I. M. Kolthoff and P. J. Elving, Eds., *Treatise on Analytical Chemistry*, Part 1, Vol. 4, Interscience, New York, 1963, Chap. 38. Copyright © 1966 by John Wiley & Sons, Inc.

fibers that facilitate construction of the equipment and the sensitivity, (2) the use of a photoelectric cell for measurement of small deflections, and (3) the use of a small powerful Alomax magnet. With improved permanent magnetic materials such as the samarium cobalt ($SmCo_5$) alloy now available, the sensitivity could be increased.

The balance was constructed particularly for investigating free radicals of half-life time of more than 1 min. Volume susceptibility changes of 0.0004×10^{-6} cgs units could be followed. Because of this sensitivity, the balance is quite useful for studying the small diamagnetic and paramagnetic susceptibilities of gases.

The apparatus and its details are shown in Figures 3.29 and 3.30, and the original paper [60] provides further details. The entire system is placed in a glass enclosure and is evacuated. Samples (liquids or gases) are changed easily through the inlet and outlet connected to the flat-faced cell, which is held rigidly

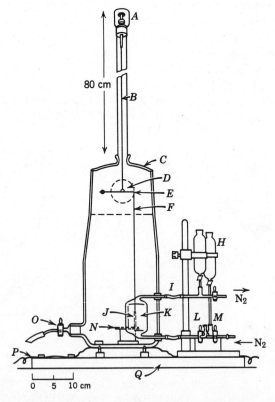

Figure 3.29 Bockris and Parson's modification of the Rankine balance: A, torsion head; B, 6.5-μm tungsten fiber; C, lid; D, lens; E, beam; F, 4.0-μm tungsten fiber; H, solution reservoirs; I, ground wire; J, magnet; K, cell; L, oxidizing agent; M, solution inlet; N, damping plate; O, to pump; P, balancing wire; and Q, slate slab.

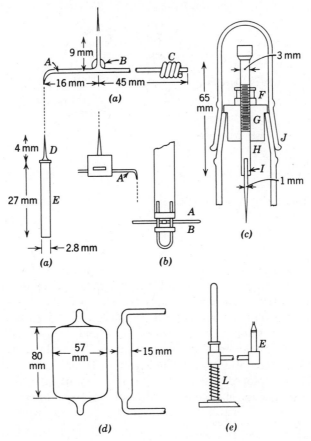

Figure 3.30 Details of Bockris and Parson's modification of the Rankine balance (a) beam and magnet, (b) beam support, (c) quartz suspension, (d) cell, and (e) magnet support. Key: A, 1-mm-quartz rod beam; B, aluminized plane mirror; C, silver wire counterpoise; D, silver pin; E, magnet; F, locking nuts; G, Tulnol bushing; H, brass rod; I, quartz rod; J, B29 joint; and L, spring.

inside the vessel. A light beam reflected from the mirror on the torsion head is focused onto a photocell, which is operated differentially. When the beam illuminates both parts of the cell equally, no deflection is observed on the galvanometer. Small displacements of the balance produce proportionate deflection on the galvanometer.

The apparatus is calibrated by filling the cell with, for example, oxygen or nitrogen. Benzene, water, and acetone also may be used.

6.4.4 Modified Rankine Balance for Solutions

The original study of the magnetic moment of iron in biological systems such as hemoglobin and related materials—by Haurowitz and Kittel [61], Coryell and co-workers [62], Pauling [63, 64], and Pauling and co-workers [65, 66]—led to

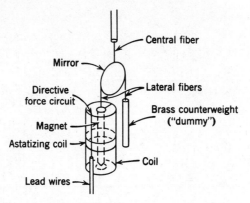

Figure 3.31   Brill and co-workers' [67] *susceptometer* for studying kinetics of solution. This is a modification of the Rankine balance.

many magnetochemical studies of biological systems. A fast and sensitive instrument for the measurement of the magnetic susceptibilities of rapid biochemical reactions has been developed recently by Brill and co-workers [67]. In our opinion it represents a major accomplishment in instrumentation. The instrument was developed for measuring the rapid changes in the magnetic moment of iron-containing proteins during their chemical reactions.

A part of the apparatus is shown in Figures 3.31 and 3.32. A differential cell (Figure 3.33) is employed. The interchange of solutions between the two half-cells produces twice the change in the force on the magnet that is produced by

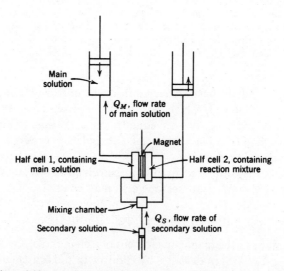

Figure 3.32   Schematic of a flow system for the Brill susceptometer.

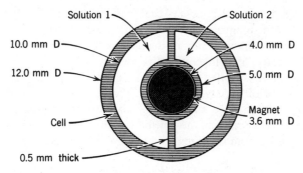

Figure 3.33 Details of a cell used in Brill susceptometer.

one-half cell, as in the original Rankine balance. A nonmagnetic counterweight of the same mass and moment of inertia as the magnet is suspended from a lateral fiber similar to the one shown in Figure 3.31. The small Alnico bar hangs inside the two-compartment cell to which it would be stiffly coupled in the absence of a magnetic field produced by the astatizing coil. The stiffness and damping effects maintain the symmetrical situation between the coil and the cell and prevent a coupling between the magnet and the earth. These effects are added by a servomechanism to decrease the response time of the suspension. Several precautions are taken to discriminate against vibrational disturbances and accelerations of the earth. Displacements of the magnet as small as 10^{-8} cm are detected by an optical system and recorded on a fast Esterline Angus recording milliammeter. The original article [67] must be consulted for a detailed analysis of the problems involved in the measurement of susceptibilities of a flow system and the steps taken to overcome major difficulties.

The instrument described can detect in one measurement a change in volume magnetic susceptibility of 5×10^{-12} emu cm^{-3}. The response time has been adjusted to a fraction of a second using an electronic servosystem. The time of resolution of the flow system was found to be limited to 5 s by the flushing time of the cell. This low resolution of the flow system has been attributed to the sharpness of the corners of the cell and to their remoteness from the entrance and exit tubes. A feasibility is indicated of accomplishing (during a 1-s measurement) a time resolution of hundredths of a second and of detecting a change of 1.5×10^{-12} emu cm^{-3} corresponding to the rms Brownian force that arises from air resistance to the motion of the magnet. The method is indeed very promising for studying chemical and biochemical reactions.

6.4.5 Gordon Force-Balance Method

The force-balance method, developed by Gordon [68], is quite ingenious. It not only provides the obvious advantage of microdetermination, but also yields directly the specific (or mass) susceptibility, which is also true of the Faraday technique. Quite strikingly, however, determination of the susceptibility does not involve an actual measurement of the "mass" of the sample, since the force of gravity balances the magnetic force and both depend linearly on mass.

We have seen a demonstration of the new method, and we are impressed by its potential. Because of its ingenuity, this and another method described in the next section will be designated the Gordon method. It is a null method in which basically the force of gravity balances the force on a sample because of an inhomogeneous field. For magnetically anisotropic crystals, this method measures the largest principal susceptibility. The vertical component Fy of a specially designed inhomogeneous field (Figure 3.34) is given by

$$Fy = -m\chi H^2 (2a)^{-1} \sin(2y/a)$$

where m is the mass of the sample, χ is the specific susceptibility, H is the maximum component of the field, and a is the scale factor (a constant of the instrument) to which the field is constructed. It is further proven that the vertical component of the force is not a function of the field components x and z, which is very convenient when designing the experiment.

The field at which the sample falls because of gravity is given by

$$mg' = \frac{\pm m\chi H^2}{2a}$$

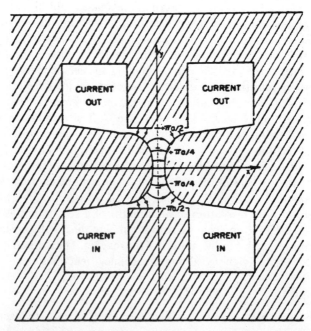

Figure 3.34 Schematic of the special electromagnet with doorknob-type poles for the Gordon method. The shaded portions are iron. The field gives a scalar potential $\mu = -aH \sin h$ $(n/a) \times \cos(y/a)$ (see text).

where g' is the gravitational constant and the $(+)$ and $(-)$ signs apply to paramagnetics and diamagnetics, respectively.

The apparatus consists of a specially designed electromagnet with two doorknoblike poles and two flat poles at right angles to them. In this quadrupolar gap a small melting point-type capillary tube, containing a small powdered sample (ca. 10^{-4} g) is placed. By varying the current and hence the field, a condition of "equilibrium" or balance is reached. Fields of about 16 kOe are generally employed. The critical value of the magnet current at which the sample falls is said to be reproducible to a few tenths of 1%, provided the magnet is cycled on and off a few times at the current appropriate to the measurement.

Gordon [68] discusses several aspects of handling and measuring the maximum anisotropic susceptibility of microcrystals. In my opinion, his techniques mark a significant and a revolutionary advance in the measurement of magnetic susceptibility; further modifications and detailed description of the techniques are necessary before they can be applied universally to diamagnetic and paramagnetic materials.

6.4.6   Gordon Density-Gradient Method

Gordon [68] also describes an extension of the previous method to the measurement of paramagnetic susceptibility of compounds in solutions. Solutions containing less than 10^{-8} mol of solute were investigated, and the triumph of this method lies in an increase in the sensitivity to more than 10^4 over the classical Gouy method. The magnetic forces arising from an inhomogeneous field acting on the sample are balanced against the density gradients in the solution.

For very dilute solutions, the density and magnetic susceptibility are assumed to be functions of molar concentration. If a specially produced inhomogeneous field with a constructional parameter a is used, the difference in vertical magnetic forces acting at the two layers of the solution where the concentrations of the layers are C_1 and C_2 is given by

$$Fy(C_2) - Fy(C_1) = (C_2 - C_1)\frac{H^2}{2a}\frac{\partial \chi}{\partial c} - \frac{\partial \rho}{\partial c}$$

The critical change occurs when

$$\frac{H^2}{2a}\frac{\partial \chi}{\partial c} = g'\frac{\partial \rho}{\partial c}$$

and it is independent of the concentrations.

The chief uncertainty in the technique is not in the critical field required for inversion of mixing but in estimating $\partial \rho/\partial c$, which represents the density-concentration gradient. For ionic salts $\partial \rho/\partial c$ is approximately M, the molecular weight, whereas for organic (nonpolar) compounds a better approximation is given by $\partial\rho/\partial c$ of approximately $(1 - \rho_0/\rho_1)M$, where ρ_0 and ρ_1 are the densities of the solvent and the solute, respectively.

In the experimental procedure, the layers of pure solvent and solution are formed by injecting either one in a capillary tube and allowing the other component to float. The magnet current and hence the field is controlled until a mixing of the two is observed. The use of a dye to facilitate observation is suggested, and no corrections are necessary for the susceptibility of the solvent.

6.4.7   Other Force Micromethods

An ingeneous method is available [69] for studying biological and chemical processes that occur within small particles 1–100 μ in diameter. The force on a single diamagnetic 10-μ particle is about 10^{-9} dyn in an inhomogeneous field. Forces of this nature are measured in terms of the velocity the particle assumes in hydrodynamic motion, depending on the viscosity of the medium. A microscope is used for measuring the velocity (of about $1\,\mu\mathrm{s}^{-1}$) in an inhomogeneous field of a particle suspended in a medium. Susceptibility differences as small as 0.04×10^{-6} were measured for blood cells in salt solutions and also for polystyrene latex. In our opinion this represents a significant advance in microtechniques. Another method [70] employs a torsion balance with very delicate quartz suspension fibers (0.5-μ diameter) for measuring susceptibilities of particles about 1 μ in diameter. A stereomicroscope is used to observe the displacement of the sample in a pulsed field of 2000 Oe in 1/60 s. Appreciable motion of the suspension system is avoided during this small duration.

Another unique method also uses high-pulsed fields [71]. The magnetic force [71] on the sample is converted to a stress wave through the apparatus and excites two piezoelectric crystals, which generate a voltage that is then measured accurately.

6.5   Change in Flux Methods

Changes in flux may be measured by direct current, alternating current, and radio frequency *induction* methods. In these cases, the principle is to measure the change in inductance of a solenoid in zero applied field with the introduction of a sample, which then changes the magnetic flux. In a fourth category, we classify methods in which the sample is vibrated in the field and the change in flux is measured. The first three methods have some advantages over the Gouy and Faraday methods in that no bulky and expensive magnets are needed and no cumbersome procedures for measuring force are involved. However, in the discussion to follow it becomes clear that each method has limitations, which restrict its use to highly specialized areas. McGuire and Flanders and Oguey [72] give good reviews of *change in flux* methods.

6.5.1 Direct Current Induction Methods

Among the many induction methods, a technique developed by Barnett [73] is simple and effective. A field of about 75–200 Oe is produced by the long magnetizing coil, which encloses identical secondary coils 1 and 2, connected in opposition. The sample is displaced quickly from coil 1 to coil 2, which changes the mutual inductance of the two coils. This is measured in terms of the deflection produced by the induced current in the shunted ballistic galvanometer. An identical effect may be obtained by reversing the current in the magnetizing coil. The deflection is proportional to the volume susceptibility of the material (Figure 3.35).

Unfortunately, the method requires samples as large as 15 cm³ and detects volume susceptibilities of about 10^{-5}. It is apparent that only the susceptibilities of those ferromagnetics that are easily saturable below 400 Oe can be determined by this method. Selwood developed a similar apparatus for thermomagnetic analysis. It is essentially based on the well-known *extraction method*, first developed by Weiss and Forrer (cf. [25]). Other modifications are described by Mulay and Mulay [36–46].

6.5.2 Alternating Current Induction Methods

BROERSMA'S APPARATUS

A technique was developed by Broersma [74] especially for the measurement of diamagnetic susceptibilities. In this, the effect of a sample on the flux produced by a primary coil carrying an alternating current is studied. The induced emf is given by

$$V = -i\omega M_0 I(1 + f_0 K) - \frac{i\omega\phi_0 f_0 \kappa}{2\sqrt{2}}$$

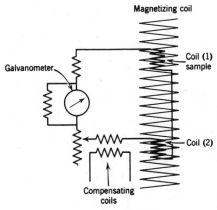

Magnetizing coil

Galvanometer

Coil (1) sample

Coil (2)

Compensating coils

Figure 3.35 A simple, effective induction method, as described by Barnett [75]. Reprinted with permission from P. W. Selwood, *Magnetochemistry*, Wiley-Interscience, New York, 1956. Copyright © 1956 by John Wiley & Sons, Inc.

where ϕ_0 is the flux produced by the current I, of frequency ω; f_0 refers to the maximum filling factor of the vessel, determined from its shape; and κ is the volume susceptibility. The coefficient of mutual inductance

$$M_0 = \mu_0 n N_s S_s$$

and is determined by the physical characteristics of the coils. Here, n is the number of primary turns per meter, N_s is the total number of turns in the secondary coil, S_s is the cross-sectional area, and μ_0 is the permeability under vacuum. The value of M_0 is adjusted to 0.25 mH, f_0 is made 0.45, and a current I of 50 A is used. A description of the apparatus and its operation are given in Broersma's paper [74] and in Figures 3.36 and 3.37.

Using the inductance apparatus, Broersma measured the susceptibility of many organic compounds. As indicated previously, the average time for a single measurement was about 8 min; the torsion (Faraday) or Gouy methods required about 25 min. The method is particularly suited for measurements on diamagnetic materials [74].

Broersma describes an inductance potentiometer similar to his earlier design for measurement of piezosusceptibility [75]. The new apparatus consists of a Delrin chamber connected to an oil-carrying steel tube. A hand pump produces pressures to 1000 bars on the sample (75 mm^3) that is placed in a plastic container. The sample chamber is placed in a horizontal solenoid that forms a part of the inductance potentiometer. An obvious limitation is that it cannot be used for measurements over a range of temperatures. Norder utilized features of Broersma's and Selwood's apparatus for magnetic susceptibility measurements during adsorption and desorption processes [76].

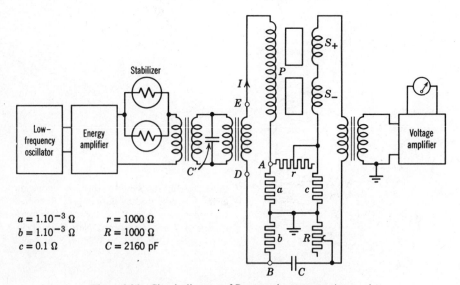

Figure 3.36 Circuit diagram of Broersma's apparatus (see text).

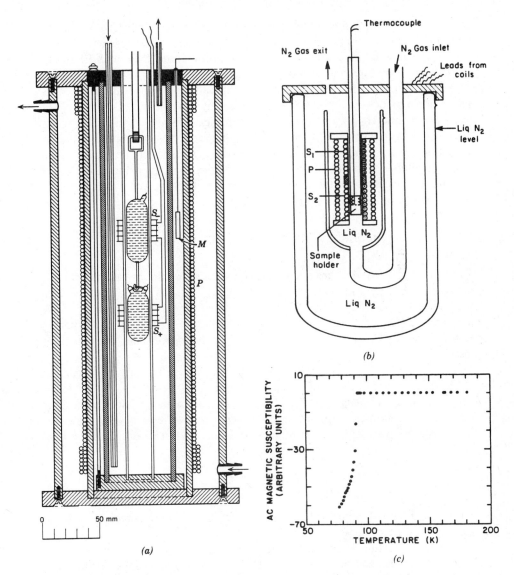

Figure 3.37 (*a*) Inductance apparatus of Broersma. The original drawing was supplied by Professor Broersma. (*b*) Inductance coil and a cryostat (77–300 K) by Mulay and co-workers. (*c*) Ac susceptibility of the oxide superconductor $Y_1Ba_2Cu_3O_{7-\delta}$ as a function of temperature. See Figure 3.36 for symbols.

MODIFICATION OF ALTERNATING CURRENT INDUCTION METHODS

Mulay, Cao, and Klemkowsky [76] recently described an inexpensive inductance coil and a cryostat for measuring the alternating current susceptibility of the new oxide superconductors (Figure 3.37*c*). A unique feature of the system is that it can be operated from 77 K upwards, whereas the conventional gas-flow cryostats reach a lowest temperature of about 90 K only. These authors [76] present typical diamagnetic susceptibility results for the $Y_1Ba_2Cu_3O_{7-\delta}$ compound (Figure 3.37c) and discuss several advantages of their system. Further comments on the new oxide materials are given under Section 8.3.1 in the subsection "New Superconducting Materials."

A mutual inductance bridge and a cryostat for low-temperature measurements of magnetic susceptibilities are described by Ericson and co-workers [77], and by Fritz and co-workers [78]. Other modifications are found in reviews by Mulay and Mulay [36–46]. A commercial apparatus that was based on the ac induction method and marketed by Cryotronic Industries (Clifton, NJ) is no longer available. However, an important modification that helped to improve the intrinsic electronic instability appeared in the literature and was reviewed in a 1982 article by Mulay and Mulay [45].

6.5.3   Radio Frequency Methods

OSCILLATOR METHOD

During the past two decades a few radio frequency techniques were developed, notably by Effemy and co-workers in England [79] and by Pacault and co-workers in France [80]. Joussot-Dubien modified the latter technique for low-temperature (70 K) measurement [81]. The technique of the English workers is reviewed here [79], since the French workers did not discuss the details of instrumentation and operation.

When placed in the tuning coil of a high-frequency critical oscillator ($2.5\,MHz\,s^{-1}$), a substance of volume susceptibility κ causes a change in its frequency, which is measured by the heterodyne beat method, that is, by beating the oscillator against another of fixed frequency. If the frequency change is Δf, the cross-sectional area of the tuning coil is A and the test substance is A'; then,

$$\kappa = \frac{1}{2\pi}\frac{\Delta f}{f}\frac{A}{A'}$$

The original paper [79] provides details of circuitry and of the precautions to be observed during the operation.

The ultimate limitation on sensitivity is imposed by the stability of the oscillator, which must be about 1 in 10^8 Hz. This necessitates a control of the temperatures of both the critical oscillator and the sample at 30°C within ±0.001°C. Thus it is almost impossible to measure susceptibilities at different temperatures.

The apparatus is quite adequate for measurement of diamagnetic and paramagnetic susceptibilities near room temperature, although it was designed especially to measure the changes in susceptibility of liquids of about 0.0004×10^{-6} that are caused by the generation in situ of short-lived (0.05 s) free radicals at very low concentrations (ca. $0.005\,M$). Since some of the conventional magnetic methods do not permit a study of short-lived free radicals, the radio frequency method is unique for such work. The authors [79] point out other possibilities of adapting the same technique for the precise measurements of dielectric constants, paramagnetic relaxation studies, and radio frequency titration.

NUCLEAR MAGNETIC RESONANCE TECHNIQUES

The reader is probably familiar with the principles of NMR absorption spectroscopy. A few applications are based on the broadline (or, so-called *low-resolution*) technique, which is used to study molecular motion in solids such as ferrocene (cf. Mulay and Attalla [82]).

Generally, judging from the number of papers published in this area, the use of broadline techniques [83, 84] for measuring susceptibility is not popular. However, high-resolution NMR spectroscopy, which is used extensively by organic chemists for structural elucidation, is very successful. Several such applications are described by Mulay and Mulay [36–46]. A typical example, based on the *shape-factor* NMR method for measuring magnetic susceptibility of pure liquids, solutions, and suspensions, is outlined by Mulay and Haverbusch [85]. Their paper cites references to earlier developments in the field.

As shown in Figure 3.38, a small glass sphere and cylinder assembly containing a standard sample of high-volume diamagnetic susceptibility (e.g.,

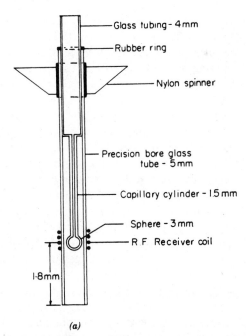

(a)

Figure 3.38 (*a*) Mulay's assembly [85] showing the concentric sphere and cylinder enclosed in the high-resolution NMR tube. (*b*) Calibration curve showing the separation η (chemical shift between the sample and a reference) as a function of the volume susceptibility of the sample.

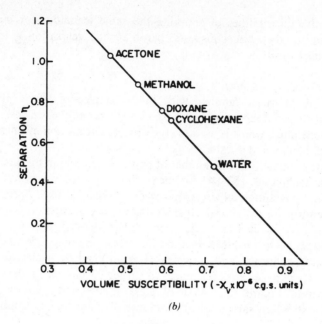

(b)

Figure 3.38 *(continued)*

bromoform, κ or $\chi_v = -0.948 \times 10^{-6}$ cgs unit) is placed inside a standard 5-mm NMR tube. A sphere (bulb) can be easily glassblown at the end of a long organic chemist's melting point glass capillary tube. The entire sphere and part of the cylinder are placed in such a way that they are enclosed by the radio frequency receiver coil in the probe of a high-resolution NMR apparatus such as the A-60 spectrometer of Varian Associates (Palo Alto, CA), which operates at 60 MHz. (With the advent of very high-resolution NMR apparatus operating at 100- and 300-MHz frequencies, the sensitivity of the susceptibility technique can be enhanced.) The NMR tube contains the sample (liquid or solution) to be analyzed. The standard material in the sphere and in the cylinder experiences different magnetic fields because of both the differences in the "shape (demagnetization) factors" of the vessels and the shielding effects from the magnetic susceptibility of the surrounding sample. This yields two proton resonance signals (peaks) arising from the standard material separated by a splitting η (measured in parts per million, hertz, or milligauss) and is given by

$$\eta = \delta_{\text{cyl,ref}} - \delta_{\text{sph,ref}} = (g_{\text{cyl}} - g_{\text{sph}})(\chi_{v_{\text{ref}}} - \chi_{v_{\text{sample}}})$$

where δ is the chemical shift (in parts per million, hertz, etc.), χ_v is the volume susceptibility in cgs units, g is the geometrical shape factor, and subscripts cyl and sph refer to the cylinder and sphere, respectively. For ideal geometry the shape factor difference $g_{\text{cyl}} - g_{\text{sph}}$ is expected to be $(2\pi/3) - 0$ or 2.094; in practice, distortions of shape and incomplete coverage of the sphere and cylinder in the radio frequency coil cause deviations. However, the distortions are included in the instrumental parameter and are compensated for by constructing a calibration curve (Figure 3.39) for the splitting η versus the magnetic susceptibility of a large number of known compounds. The susceptibility of a sample is found by interpolation from such a calibration curve. The method gives values of susceptibility to within $\pm 0.001 \times 10^{-6}$ cgs unit; however, the more sensitive Faraday method is the most versatile and suitable for refined measurements.

In a paper by Douglas and Fratiello [86] sources of error in the *concentric tube* type methods are discussed.

Deutsch and co-workers [87] gave a good review of high-resolution NMR methods for measuring the magnetic susceptibility of diamagnetic liquids and a careful analysis of the principles. They used H_3PO_4, CH_3COOH, CF_3COOH, and $(CH_3)_4Si$ as reference compounds and obtained the volume susceptibilities of several organic liquids. As with other methods, susceptibilities to three significant figures were obtained; for example, for bromoform $\chi_v = -0.948 \times 10^{-6}$ cgs unit was obtained [87]. Recent developments by Evans and others were reviewed by Mulay and Mulay [36–45].

Evans [88] has described a concentric tube method for determining the *paramagnetic susceptibilities of solutions*. Orrel and Sik [89] pointed out that *this method*, and its modifications, based on an equation derived in 1955 is *incorrect*. The authors give a new theoretical analysis for the concentric tube technique, important references to paper and to NMR tests, and a correct equation for obtaining paramagnetic susceptibilities of solutions. Their paper [89] *must be consulted* by all workers using this technique. It should be noted that the sphere in an external tube approach developed by Mulay and co-workers (cf. [85]) for diamagnetic systems differs from the concentric tube method.

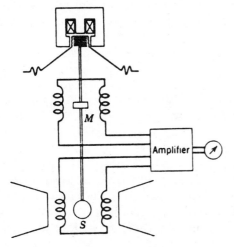

Figure 3.39 Principle of the vibrating sample magnetometer VSM. The sample S is vibrated vertically in an external field, with a transducer. The changing flux induces a voltage in the pickup coils surrounding the sample. This is measured relative to the feedback assembly M. Reprinted with permission from S. Chikacumi and S. Charap, *Physics of Magnetism*, Wiley, New York, 1964. Copyright © 1964 by John Wiley & Sons, Inc.

6.5.4   Vibrational Methods

MICROVIBRATIONAL METHODS

Yousef and co-workers [90] developed a successful microvibrational method for measuring susceptibilities. This appears to be the first reported instance of a change in flux that is produced by vibrating the sample in a steady field. The method further opened an entire area of instrumentation, which has produced sophisticated equipment now known as the *vibrating-sample magnetometers* (VSM).

In the original technique [90] a steady magnetic field is modulated by an alternating current, and the periodic force acting on a sample placed in the field is translated into a periodic displacement of an elastic strip. The force is measured dynamically in terms of a compensating electrostatic force, and an electronic circuit serves as a null detector. An accuracy of 10^{-4} dyn in the measurement of force is obtained. An attractive feature of the technique is that a small, inexpensive relay electromagnet is employed, which is energized by currents of about 0.1 A to give small fields of about 70 Oe. Sufficient space is available in the pole gap to accommodate large temperature-control devices. Measurements on a paramagnetic sample (ca. 0.1 g) are reported. The method provides a delicate test for the presence of ferromagnetic impurities.

Modifications of the microvibrational methods were developed by several workers; see references given in brackets in [90].

THE FONER OR VIBRATING-SAMPLE TECHNIQUES—SPINNER MAGNETOMETERS

The principle of the Foner magnetometer [91] is shown in Figure 3.40. The sample S vibrating in a field changes its flux and induces in the pickup coils a voltage that is proportional to the magnetic susceptibility of the sample.

It is appropriate to point out at the outset that commercially available vibrating-sample magnetometers are designed for measurements on magnetic materials with very large magnetic susceptibilities such as the ferromagnetic, ferrimagnetic, and some antiferromagnetic and superparamagnetic materials. At the present state of their development, these devices do not seem to be capable of measuring mass diamagnetic susceptibilities ($\chi_m \simeq -1 \times 10^{-7}$) or even feebly paramagnetic susceptibilities ($\chi_m \simeq +10^{-6}$) of materials in which most magnetochemists are interested. In our experience very few modifications of the Foner [91] method are capable of actually measuring diamagnetic susceptibilities. Recent reports by Moss [92] indicate that the use of superconducting magnets has considerably improved this situation. Reference should be made to a 1984 review article by Mulay and Mulay [46].

Generally, most physicists, electrical and electronics engineers, and those investigators who work with magnetic materials of very high intrinsic specific susceptibilities ($\chi_g \simeq 10^{-3}$) think in terms of *magnetic moments* and express the sensitivity of their measuring devices as "a change in the moment of χ emu/g" instead of "a change in χ susceptibility units per gram," with which the chemist is more familiar. Their use of the term *moment* implies a measurement of magnetization, since magnetization, by definition, is the induced moment per

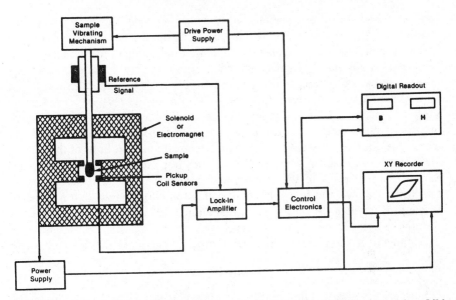

Figure 3.40   Vibrating sample magnetometer block diagram. Reprinted, with permission, from LDJ Electronics, Inc., Troy, MI.

unit volume. The measurement of magnetization I, which varies with the field H (and the temperature) in ferromagnetic and ferrimagnetic materials is naturally more important to the workers in this area than the magnetic susceptibility (χ_g), which is simply a constant defined by I/H for diamagnetic, paramagnetic, and antiferromagnetic materials. A brief description of a vibrating sample magnetometer, which is now commercially available, follows.

Figure 3.40 shows a diagram for a VSM. A rod carrying the sample and a "feedback" coil is vibrated at about 80 Hz. The sample is placed in a field of about 10^4 Oe. The output voltage developed across the sample coil is proportional to the magnetization (magnetic moment) of the sample, whereas that developed across the reference coils is proportional to the direct current in the feedback coil. The circuit is designed to sense the difference between the two voltages (δv); to amplify this difference, which is proportional to the magnetic moment; and, after passing δv through a phase-sensitive detector, to record the difference. The reference voltage can be attenuated in a ratio of $10^5 : 1$; this facilitates measurements of magnetic moments over a range of five orders of magnitude. The following specifications are claimed by the manufacturers. A change in magnetic moment of 5×10^{-4} to 5×10^{-5} emu corresponding to a change in magnetic susceptibility of 5×10^{-8} to 5×10^{-9} cgs units for a 1-g sample in a field of 10^4 Oe can be detected. The absolute accuracy is better than 2%.

In actual practice it is possible (and necessary) to use only a few milligrams of the material (ca. 50 mg) in the powdered form, or, where possible, in the form of a tiny machined sphere (e.g., metals, alloys, some ferrites). Hence, although the sensitivity of the apparatus is stated in terms of a number of units per gram, the actual attainable sensitivity is far less, because small samples must be used. The sample is usually placed rigidly in a small cylinder (2-cm tall and 0.5-cm o.d.) made from nylon, Teflon, or a similar material. For high-temperature work, an appropriate ceramic material (mullite) or boron nitride may be used for the sample holder. The cylinder is then attached to the lower end of the rod, which in turn is connected to a transducer (or a loudspeaker) at the top of the apparatus. Care must be taken that there are no lateral vibrations of the sample and that the sample (in whatever form) does not rattle inside the container. Several modifications of the VSM and comments about such factors as its sensitivity are reviewed by Mulay and Mulay [36–46] along with a number of techniques in which the sample is rotated along a vertical axis. The principle of the above spinner magnetometer is shown in Figure 3.41. A summary of developments in the VSM and rotating sample magnetometers is given in Table 3.12.

PENDULUM MAGNETOMETER

Let us discuss the usefulness and the apparent ease of construction of the *pendulum*-type magnetometers. Figure 3.42 illustrates a horizontal pendulum developed by Weiss and Forrer [115]. It has only one degree of freedom for the sample; the sideways motion is in a direction perpendicular to the magnetic field so that a fairly strong magnetization (as encountered in ferromagnetic and ferrimagnetic materials) can be measured without being disturbed by the attraction of the sample to the magnet's pole piece. Displacement of the arm caused by the force acting on the sample is balanced by adjusting the current in a coil that encloses a small piece of a permanent magnet attached to the arm. The displacement of the arm is sensitively detected by the deflection of a mirror,

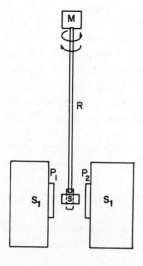

Figure 3.41 Schematic drawing of a *spinning* or a *rotating* coil magnetometer for measuring the remanent magnetization and so on of geological samples. The battery (or ac) operated motor M spins the sample S between two pickup (or inductance) coils P_1 and P_2. These are generally surrounded by magnetic fields S_1 and S_2 (shown as a cross section). The current induced in P_1 and P_2 is measured with a micrometer or a milliammeter. The sample is attached to a *nonmagnetic* (diamagnetic) rod R, so that the anisotropic magnetic properties of the sample S can be measured with respect to a reference material.

Table 3.12 Recent Aspects of Some Vibrating (VSM) and Rotating (RSM) Sample Magnetometers[a]

Topic	Comments
1. RSM to detect weak magnetic moments, rotational hysteresis, anisotropy energy (Flanders [93]).	Paper gives both extensive mathematical analysis of theory and working equations.
2. Compact VSM to fit inside a bore of superconducting magnet (Mangum and Thornton [94]).	Uses a piezoelectric vibrator, simple construction, with a YIG (yttrium iron garnet) standard vibration at ca. 210 Hz.
3. High-sensitivity spinner magnetometers (Morris [95]).	Paleomagnetic applications are described.
4. VSM for microscopic particles of magnetic materials (e.g., $SmCo_5$) (Zilstra [96]).	A thin gold reed is vibrated and is detected by a microscope of high sensitivity ca. 2×10^{-8} erg Oe^{-1}; gives hysteresis curves.
5. Superconducting VSM (Moss [92]).	Biological studies on oxy, deoxy, and met derivatives of hemerythrin.
6. RSM with a standard of known susceptibility and another sample connected to a shaft, which is rotated in a magnetic field with a $v \simeq 55$ Hz (Hudgens [97]).	Measures change in χ_v of a diamagnetic material ($\Delta\chi_v$, ca. 10^{-9} cgs unit).
7. VSM for weakly paramagnetic frozen solutions. The superconducting field coil also acts as detector. Vibration at (20/7) Hz with a 2-cm amplitude (Redfield and Moleski [98]).	Small samples (0.7 mL) with a resolution of $\Delta\chi_v = 10^{-9}$ cgs unit cm^{-3} sensitivity, ca. 2×10^{-7} mol paramagnetics, ca. 1.83 BM for $s = \frac{1}{2}$. Advantages over the SQUID and Faraday balances are discussed.
8. Adaptation of a commercial VSM (e.g., Princeton Applied Research) for use from 0.5 to 1.5 K (Olivera and Foner [99]).	Cerous magnesium nitrate used as a thermometer; constructional details are given; few experimental results are provided.

233

Table 3.12 (continued)

Topic	Comments
9. Calibration of VSM at high temperature (Hines and Moeller [100]).	Use of Gd_2O_3 as a magnetic thermometer.
10. Increasing the sensitivity of VSM by (a) increasing the filling factor (sample volume per effective volume of detection on coils); (b) cooling detection coils (ca. 4 K) to reduce Johnson noise; (c) increasing $1/r^3$ [reduce r from ca. 1.5 to 0.15 cm, where r = distance between sample and detection coils (first-order effects)]. Second-order effects are discussed (Foner [101, 102]). Further improvements suggested; discussion of geometric scaling factors given. (Unfortunately, none of the suggestions appear to have been incorporated in commercial instruments.)	Finite practical limitations exist with regard to most suggestions. A minimum separation r is necessary to accommodate cryostats, etc. Author claims that for 1-g sample at 10 kOe, a $\Delta\chi$ ca. 10^{-13} emu g^{-1} Oe^{-1} may be observed. This is said to be better than the SQUID magnetometer; however, a 1-g sample cannot be inserted into the Foner magnetometer.
11. Auxiliary equipment for VSM; use of gelatin (medication type) capsules to reduce diamagnetic contributions (Drake and Hatfield [103]).	Gelatin dissolves in water; therefore, a new capsule must be used for each measurement; negligible cost is involved.
12. Limitations of using Gd_2O_3 as a temperature thermometer discussed; platinell II thermocouple is recommended (Sill and Drensky [104]).	Temperature increment of the furnace must be slow; ca. 1 min needed for thermal equilibrium.
13. Theory of VSM (and its limitations). Authors discuss the point-dipole approximation, which	Authors tested various configurations of samples and point out that, while point-dipole

in practice is not realized. Report measured deviations from a pure dipole behavior for cylindrical sample for which (diameter: length) ratio is ca. 1.5 (Foiles and McDaniel [105]).

The authors show that:

Foner explicitly recognized the potential problem of sample size and conducted limited tests on paramagnets having irregular shapes; [his] moments agreed to within ±1% and thus the pure dipole approximation (i.e., output signal is directly proportional to volume independent size) was obeyed. However, the maximum linear dimension of these samples was 3 mm and no information was given concerning the detection coil configuration used for these tests.

approximation is fine for paramagnets, ferromagnets, or ferrimagnets, it does not apply to aluminum (Cu is diamagnetic; Na, K, etc, show Pauli paramagnetism). Considerable reduction in the sensitivity of VSM is shown.

14. VSM for high-pressure measurements. A high-pressure cell described by Maple and Wohlleben, designed for use with a Faraday balance was adopted for use with a VSM (Guertin and Foner [106]; also, Cordero-Montalvo [107] and Cordero-Montalvo, Vedam, and Mulay [108]).

Magnetic susceptibility of 0.2-cm^3 sample could be measured. Used $\Delta\chi$ ca. 10^{-7} emu g^{-1} for pressure up to 8 kbar. Good constructional details are given. The system was tested in writers' laboratories and found to work well.

15. Sensitivity of the VSM and force-type (Faraday) magnetometers; Lewis shows that even for idealized geometry with a VSM an apparent sensitivity in susceptibility ($\Delta\chi$ ca. 10^{-13} emu g^{-1} Oe^{-1}) as claimed by Foner [101] is incorrect because of the practical limitations on the size of the sample that can be accommodated (Lewis [109]).

Much of the confusion about sensitivity arose from the fact that one group of workers (e.g., chemists) are interested in the susceptibility measurements χ_g, whereas those working with magnetic materials (e.g, physicists, electrical engineers, and materials scientists) measure magnetization M, which is expressed synonymously as the moment μ. (Note that $M = \mu/v$ and $\chi = M/H$. Sensitivities in both χ and M should be given, and exaggerated claims regarding sensitivity should be avoided by all, especially by the manufacturers.

Table 3.12 (*continued*)

Topic	Comments
16. In his reply to Lewis' letter, Foner [101] states that: (a) The moment sensitivity is unambiguous and is better than that reported to date for any other method (including superconducting devices). (b) The statement that a given moment would correspond to a susceptibility for a 1-g sample at 10^4 G (1 T) permits the reader to be assured about the units (and that the terms *moment* and *susceptibility* have not been confused) and that *mass*, not *volume susceptibility*, is being discussed. (c) By stating the sensitivity in magnetic moment the reader has all that is needed to convert to χ in appropriate units at various fields.	
17. Mathematical analysis of emf induced in a VSM. The authors consider a point dipole in a single-turn coil, instantaneous flux generated, and so on (Bragg and Seehra [110]).	The paper gives a good insight into the theory of VSM using some experimentally observed parameters.
18. Critique of analysis in No. 17 (Guy [111, 112]).	This paper elucidates the limitations of the theory when samples and detector coils of finite size are employed.

19. Frequency-doubling effect in a VSM. The possibility of *frequency-doubling* effect [$A(2\omega_0)$] is considered on theoretical basis (A = amplitude and ω_0 = frequency of vibration) (Guy [113]).

Application of the method to accurate anisotropic measurements are given. Signal: noise ratio is improved by a factor of about 300. (The papers cited in 17 and 18 are very important for understanding the theory of and improvements to VSM.)

20. Foner and Zieba [114] (cf. [46]) describe design analysis of VSM.

A good mathematical analysis of detection coil, sensitivity function, and sample geometry is given.

21. Modifications of VSM using electromagnets or superconducting magnets.

References in Mulay and Mulay [39–46] provide additional information.

---

[a]Additional references are listed by Mulay and Mulay in their 1984 review [46].

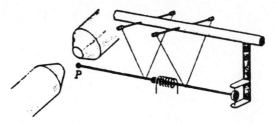

Figure 3.42 Horizontal magnetic pendulum. Reprinted with permission from S. Chikacumi and S. Charap, *Physics of Magnetism*, Wiley, New York, 1964. Copyright © 1964 by John Wiley & Sons, Inc.

which is attached to the arm. The current induced in the coil is a direct measure of the magnetization of the sample. A heating furnace can be introduced easily to surround the sample for measurements above room temperature. In our opinion, it would be difficult, if not impossible, to make measurements below room temperature in the usual cryogenic range (4–300 K) using conventional cryostats.

A vertical pendulum magnetometer, which does not have the above limitations for making measurements in the cryogenic range, is described by Bozorth and Williams [116]. Conventional cryostats, with a sufficiently wide variable temperature access can be used easily; and, needless to say, large furnaces for high-temperature work can be introduced to surround the sample (Figure 3.43). In their vertical system, a stout rod is suspended through an elastic strip of metal and the sample is attached to the other end. A coil C, which surrounds the sample, carries a current to produce a moment (magnetization), which cancels the moment of the specimen. The deflection of the pendulum is detected by a pair of strain gauges G, which are attached on both sides of the elastic metal strip.

The sample and the coil are placed in an inhomogeneous magnetic field produced by tapered magnetic pole pieces. As with the horizontal pendulum, the current in the coil is directly proportional to the moment of the sample. Calibration of the horizontal and vertical pendulum magnetometers can be accomplished by using samples of known saturation magnetizations or magnetic susceptibilities. Recent reviews by Mulay and Mulay [36–46] contain descriptions of other modifications to the pendulum magnetometers.

6.5.5 Variable Field and Other Magnetometers

Brankin and co-workers [117] (cf. [118]) describe an *integrating magnetometer*, which in our opinion should mark a new era in instrumentation. During the past two decades vibrating sample magnetometers (VSM) have been both developed and made available commercially. With the VSM, the vibration or motion of a superconducting cylindrical sample in a magnetic field may affect the penetra-

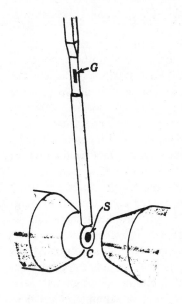

Figure 3.43 Principle of the vertical pendulum method for measuring susceptibility. The sample S and the surrounding coil C are placed in an inhomogeneous field produced by special poles of a magnet. A sensitive transducer G attached to the elastic strip measures the deflection of the pendulum. Reprinted with permission from S. Chikacumi and S. Charap, *Physics of Magnetism*, Wiley, New York, 1964. Copyright © 1964 by John Wiley & Sons, Inc.

tion or expulsion of the magnetic flux. Also, in specific cases, the magnetization near the ends of the superconducting sample may be different from that of the central region.

In their integrating magnetometer, the sample (1–5-mm diameter) is held steady inside a superconducting magnet and is surrounded by appropriate pickup coils. The external field is then swept at about $10\,\mathrm{Oe\,s^{-1}}$ ($10^3\,\mathrm{A\,m^{-1}\,s^{-1}}$), and the voltage across the pickup coils is measured and recorded on an x–y recorder. This voltage is given by

$$V = -(n_1 - n_2)A\,\frac{dM}{dH}\frac{dH}{dt}$$

Here, A is the cross-sectional area of the sample, and dM/dH is the slope of the magnetization curve; this slope defines the magnetic susceptibility χ of the material. The inner pickup coil has $n_1 = 2100$ turns, and the outer coil has $n_2 = 500$ turns; both search coils are approximately 5 mm long.

Since the integrating magnetometer is apparently very easy to construct, it would be worthwhile to investigate its applicability to other materials of special interest to the chemists and materials scientists.

A dynamic magnetometer for the measurement of anisotropy in ferromagnetic materials is described by Gessinger and co-workers [119]. The apparatus can automatically perform a Fourier analysis of the pickup signal by using a lock-in amplifier and sinusoidal reference signals that are synchronized with the pickup signals.

6.5.6   New Superconducting Techniques: The SQUID and Meissner Effect Methods

Several experimenters have adopted the obvious route of using superconducting magnets, which easily give fields H up to approximately 100 kOe in place of the conventional electromagnets (H ca. 20 kOe) in conjunction with the Gouy or the Faraday techniques to obtain substantially large magnetic forces. However, with the advent of superconducting technology over the last two decades, new phenomena based on superconductivity were discussed; and ingenious experimenters have profitably exploited these phenomena (which intrinsically involve measurements of changes in magnetic flux quanta) for making new devices, which range from the measurement of ultracryogenic temperatures (less than 0.2 K) and the detection of gravitational waves to the construction of magnetometers for the measurement of magnetic susceptibilities as small as 10^{-12} emu g^{-1}. These magnetometers are often called *susceptometers, gradiometers*, and so on. In the sections to follow, we describe two typical applications based on the *Meissner effect* and the *Josephson junction*. For a full appreciation of superconductivity, which is becoming a subdiscipline of magnetism, the readers are advised to first read introductory literature, for example, chapters in books by Hall [120] and Kittel [28], and papers by Matisoo [121]. Indeed there are several publications on the alloy and the new oxide superconductivity and its applications that give far more advanced information; however, for our purposes these references are quite adequate.

One of the basic concepts in superconductivity refers to a fundamental unit for magnetic flux; this unit is called the *flux quantum* ϕ_0 given by

$$\phi_0 = (hc/2e) = 2.07 \times 10^{-7}\,\text{G}\cdot\text{cm}^2$$
$$= 2.07 \times 10^{-15}\,\text{Wb}$$

where h, c, and e are, respectively, the Planck constant, the velocity of light, and the charge on the electron. This flux quantum may be regarded as a unit as basic as the Bohr magneton ($eh/4\pi mc$) in magnetism.

THE SQUID SUSCEPTOMETER

The acronym SQUID is for *superconducting quantum interference device*. This device is based on a small junction, consisting of two superconducting strips made of materials such as the Nb–Sn alloy, separated by a thin (100-Å) quartz (SiO$_2$) dielectric barrier. This junction is named after Josephson (cf. [121]), who first showed that supercurrents could "tunnel" (in the quantum mechanical sense) through a thin dielectric barrier without developing a voltage drop (the dc Josephson effect). He further predicted that a critical current exists, above which a voltage would be developed. For finite voltages V the supercurrent in the junction oscillates at a frequency v directly proportional to the applied voltage given by the ac Josephson equation $hv = 2eV$, where the symbols have their usual meanings. It is now customary to use the flux quantum unit ϕ_0 in the literature on SQUID techniques [$\phi_0 = 2.07 \times 10^{-7}\,\mathrm{G\cdot cm^2}$]. The SQUID has been used extensively by many workers since January, 1987, for measuring the magnetic properties of the new oxide superconductors.

Cukauskas and co-workers [122] described in sufficient detail the construction and principles of operating a superconducting *magnetometer*, that is, a *susceptometer* (see Figure 3.44).

The measurement of susceptibility consists of transferring energy E_M of a sample to energy E_R stored by a current flowing through a weakly linked superconducting SC ring. The signal power from the magnetometer is proportional to the energy E_R, and the transfer of energy is accomplished by moving the magnetized sample from a coil L_1 to L_2. The flux change in L_1 induces a current in the superconducting circuit L_1–L_2 and causes a corresponding flux change in another radio frequency coil L_2 and in turn in the SC

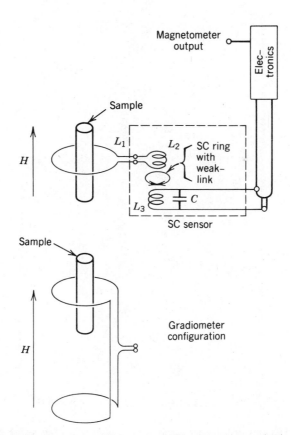

Figure 3.44a Scheme for measuring magnetic susceptibility using a superconducting SC flux transfer circuit L_1–L_2 and a weakly linked SC ring coupled to an rf tank circuit L_3–C. Instead of the single pickup loop L_1, it is often advantageous to use a gradiometer configuration for L_1 as shown in the lower part.

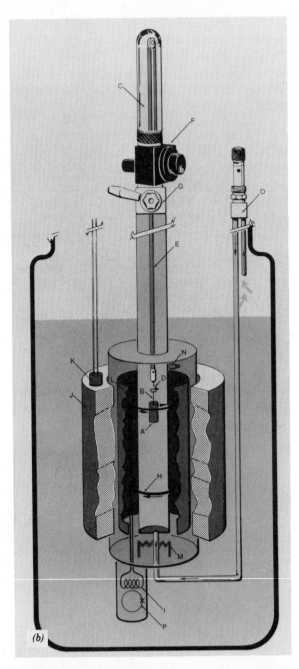

Figure 3.44b A SQUID susceptometer showing constructional details: A, sample; B, cotton thread; D, sample support hook; E, metal tape; F, sample transport mechanism; G, air lock valve; H, pickup coils; I, SQUID input coil; J, superconducting magnet; K, superconducting switch; L, superconducting magnetic shield; M, heater; N, thermometer; O, regulating valve; P, SQUID sensor. Reprinted, courtesy of Biotech Corporation (formerly SHE Corp.), San Diego, CA 92121.

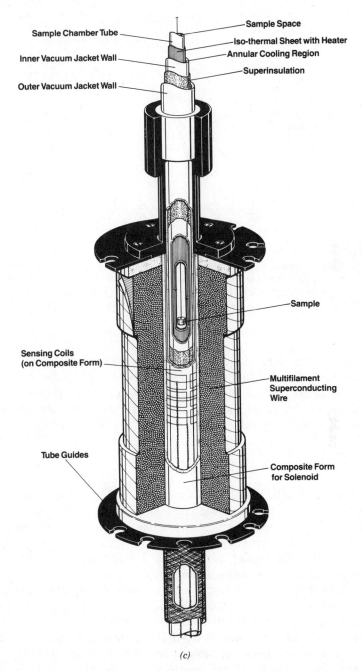

Sample Chamber Tube

Sample Space

Iso-thermal Sheet with Heater

Inner Vacuum Jacket Wall

Annular Cooling Region

Superinsulation

Outer Vacuum Jacket Wall

Sample

Sensing Coils
(on Composite Form)

Multifilament
Superconducting
Wire

Tube Guides

Composite Form
for Solenoid

(c)

Figure 3.44c Constructional details of a SQUID magnetometer made by QUANTUM design, 11578 Sorrento Valley Road, San Diego, CA 92121.

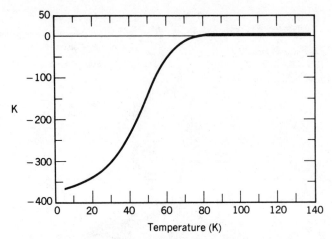

Figure 3.44d Relative change in the volume magnetic susceptibility with temperature of the (new) superconducting compound: $Y_1Ba_2Cu_3O_{7-\delta}$. In the original figure χ was shown in units of μH. This symbol should not be confused for the permeability μ times the field H. It stands for microhenries, which is used in ac susceptibility measurements. The volume susceptibility κ is a dimensionless quantity, although most workers use units such as "cgs" or "emu" per cubic centimeter.

ring (with a weak link, Josephson junction), which is coupled to a L_3 coil-capacitor–radio frequency-tank circuit. In practice, L_1 is replaced by a *split coil* to give a so-called *gradiometer configuration*. With properly matched distances and coupling coefficients of unit, $\frac{1}{2}E_M$ is transferred to the SC ring. The term E_M is related to other measurable parameters by the following relation:

$$E_M = \tfrac{1}{8}\pi B^2 V = 2\pi\chi_v H^2 V$$

where H is the applied field, V is the volume of the sample, and χ_v is its susceptibility. A sketch of a commercially available susceptometer is shown in Figure 3.44b. The volume susceptibility χ_v times the volume V gives the per gram susceptibility σ in electromagnetic units per gram. These susceptometers generally employ thin-walled quartz tubes, 3–6 mm diameter, and the small volumes permit the use of milligram samples.

The paper [122] by Cukauskas and co-workers has several useful and important features. First of all, the bibliography, which consists of 26 references, is more than adequate for the self-education of a novice who may be anxious to enter the newly emerging area of SQUID techniques. Secondly, reference is made to a novel technique, previously described by Day (cf. [121, 122]) to observe especially broadline NMR absorption spectra using essentially the same magnetometer apparatus. Day's technique involves the measurement of magnetization of a sample as the frequency of an applied transverse radio frequency is varied. When the conditions of adiabatic fast passage are satisfied, the total magnetization reverses direction as the applied frequency is swept through the resonant frequency. Thus the microscopic changes in the magnetization of nuclear moments are recorded and not the absorption of the radio frequency power, as is done in the classical NMR spectrometers. The authors cite examples of magnetic resonance absorption for protons (H^1) in water and Delrin (acetyl resin), and for fluorine (F^{19}) in Teflon material. Thus, their SQUID magnetometer constitutes a dual purpose susceptometer and an NMR spectrometer. An additional virtue of their NMR measurement capability is that it automatically provides a means for measuring the actual applied magnetic field from the (superconducting) magnet at the specimen; thus, the NMR spectrometer acts as a gaussmeter.

Obviously, then, the sensitivity of the magnetometer is very high since it is capable of measuring the magnetizations (and hence the susceptibilities) of nuclei in a sample. Their reported measurements exemplify a challenging situation indeed. The total susceptibility (which has been apparently mislabeled as "magnetization" by the authors) for water measured at 4 K and at 100 Oe is stated to be $\chi_T = -0.686 \times 10^{-6}$ cgs units and calculated nuclear paramagnetism $\chi_n = +0.24 \times 10^{-6}$ cgs units. Subtracting this contribution from χ_T yields $\chi_{H_2O} = -0.710 \times 10^{-6}$ cgs, which is approximately the susceptibility for water at room temperature (300 K). The authors stress the high sensitivity attainable with their SQUID magnetometer; for instance, a volume susceptibility change of 10^{-10} cgs units is measurable in a field of 100 Oe; since high stabilized fields up to 30 kOe are available in superconducting magnets, a sensitivity of $\Delta\chi = 10^{-13}$ cgs units is possible and compares well with static techniques of Brill and co-workers [67, 123, 124]. Cukauskas and co-workers [122] note that perhaps the primary feature of their susceptometer is its response time rather than its sensitivity. With the usual 30-MHz operating frequency it should be possible to observe *instantaneous* changes in magnetization at frequencies up to many kilohertz and even to a few megahertz. Thus the susceptometer may be useful for studying fast relaxation rates in photoexcited states or mixing experiments as encountered in biochemical studies (cf. [124]). Applications of the SQUID susceptometer, especially to biosystems are described in Section 8.3.3.

Duret and co-workers [125] describe an ultrahigh-frequency (uhf) superconducting magnetometer that utilizes a new thin-film sensor. Readers interested in the SQUID techniques will find this paper interesting in many ways. With attainment of uhf up to 350-MHz pumping frequency, the use of a particular radio frequency produces an improved signal-to-noise (S/N) ratio, and a sensitivity of $2 \times 10^4 \, \phi/Hz$. Here, ϕ is the flux quantum equal to 2.07×10^{15} Wb $= 2.07 \times 10^7$ G·cm². The thin-film (2000 Å) SQUID is made by vacuum depositing niobium on a fused quartz cylinder (2-mm diameter). It is inexpensive to fabricate and has a higher critical temperature, readily established critical current (0.5–1.5-μA accuracy). The advantages of this thin-film microbridge SQUID are compared with those of a conventional Zimmermann's point contact SQUID (cf. [121]). The paper is well written, and the construction of the entire magnetometer, including vapor deposition synthesis of the SQUID, is detailed.

A magnetic monopole detector utilizing superconducting elements is reported by Alvarez and collaborators [126]. Their rationale for searching the magnetic monopoles (north and south) in the lunar surface was discussed previously. The apparatus consists of a superconducting niobium sensing coil with a core at room temperature; it is shorted by a superconducting mechanical switch and protected against the effects of a variable ambient magnetic field by an adequate shield made of superconducting lead. The authors describe the characteristic features, performance, and sample containers. They found that a search for the monopole in 28 lunar samples with a total weight of 8.4 kg gave negative results. Under Section 3.7 it was noted that there is new evidence for the existence of the magnetic monopole. Additional evidence is given in [45, 46]. Blas Cabrera of Stanford University has conceded that no monopoles can be detected after two years of research (see *Physics Today*, p. 17, April, 1984).

THE MEISSNER EFFECT METHOD

Ingenious techniques for susceptibility measurements based on the unique properties of superconductors and the well-known dc inductance method were developed successfully by Deaver and Gorre [127]. Compared with all the modifications of other methods that range from the force measurements to magnetic resonance, this technique provides much higher accuracy and is quite novel in its theoretical and practical approach. The cost of instrumentation and operation (equivalent to the cost of a good superconducting magnet and liquid helium) are perhaps beyond the reach of the average research laboratory. This would seem to be its only limitation. However, the principle and operation are elegant and allow susceptibility measurements as small as 10^{-10} cgs units/g (units specified by these authors) in a 10^4 Oe field and a wide range of temperatures (2–300 K).

The basic idea here is to measure the total change in flux of two coils (wound in opposition) when the sample is moved from one coil to the other. This situation is reminiscent of the well-known Barnett method (Figure 3.35). The high sensitivity seems to result from the very high fields employed and from the sophistication used in the measurement of the change in flux. This sophistication is based on the Meissner effect in superconductors.

The Meissner effect refers to the complete expulsion of the magnetic field from the interior of a solid superconductor (i.e., the flux B = 0) when it is cooled below its transition temperature; fluxoid quantization comes into play, which means that the only possible values of magnetic flux trapped inside a thick superconducting ring are integral multiples of $hc/2e = 2.07 \times 10^{-7}$ G·cm². This quantization makes it possible to produce a region of zero magnetic field. Now, if a superconducting cylinder is cooled below its transition temperature in an external field, which produces less than half a quantized flux through it, a current is induced in the superconductor to exactly cancel this flux (or fluxoid) and thus to produce the lowest quantized state, that is, zero magnetic field. The current in the cylinder is a direct measure of the flux that existed initially in the cylinder. The magnetic field can thus be measured without access to the volume occupied by the field and with no motion of the superconductor.

In the apparatus, a small sample (0.1 cm^3) is placed in a superconducting coil L_1, which is connected (in opposition) to coil L_2 and then on to coil L_3. Because L_1 and L_2 are wound in opposition, the flux change in the entire assembly—L_1, L_2, and L_3—is twice that which would have resulted from simply removing it from L_1 (this trick of doubling the flux change is well known to those who work with inductance circuits). The voltage induced in another coil, which is coupled to L_2, is proportional to the persistent current resulting from the movement of the sample. Deaver and Gorre [127] discuss appropriate circuits for this purpose and several alternate refinements.

7 PRACTICE: EXPERIMENTAL SETUP FOR MAGNETIC BALANCES AND TORQUEMETERS

7.1 General Laboratory Requirements

The Gouy and Faraday techniques, which depend on accurate measurements of weight changes, naturally require a location free from floor vibrations. Usually any location in a basement, away from elevators and stairways, is preferred. In our laboratory, a frame made from a 2-in.-diameter iron pipe was constructed and was anchored rigidly to the concrete floor. This frame has been very advantageous in supporting a "single-pan" microbalance for refined measurements of diamagnetic susceptibilities with the Gouy technique. Many balance manufacturers (Mettler, Sartorius, etc.) provide vibration-proof tables and designs for masonry supports, and these are quite adequate for most purposes.

The location of a magnetics laboratory should also be free from electrical and magnetic disturbances that may occur around electric generators, air compressors, machine shops, and laboratories using high-voltage equipment such as mass spectrometers and the electron microscopes (transmission or the new scanning type).

7.2 Instrumentation: Salient Features of Literature

A review of various modifications of the Gouy and the Faraday balances, and others reveals that much of the instrumentation is devoted to increasing (1) the magnetic force acting on weakly magnetic materials (feebly paramagnetic or diamagnetic) and (2) the precision and the accuracy of such measurements. These criteria were responsible for the development of such ingenious devices as the well-known Sucksmith balance (which employs an optical system for measuring the force in terms of the distortion of a circular strip of metal). In our opinion the commercial availability of microbalances and analytical balances of high sensitivity with facilities for automatic recording of weights may make it unnecessary to indulge in any further modifications of the basic Faraday and Gouy methods. The educational value of laboratory instrumentation and the immense pleasure that can be derived from such instrumentation cannot be denied. The relatively high cost of commercial instruments and their unavailability in most parts of the world will indeed continue to provide an incentive for further modifications and new instrumentation.

For those interested either in duplicating the instrumentation reported in the literature or in fabricating new instruments for a specific purpose, a careful study of the many excellent books available on instrumentation is recommended. For example, the several volumes entitled *Vacuum Microbalance Techniques* [49] and reviews by Hirsch [128], Thomas and Williams [129], Gordon and Campbell [130] and Banerjea [131] provide a wealth of information on the theory and applications of various balances that can be constructed in the laboratory and on others that are available commercially in many parts of the world. Such study facilitates the selection of a ready-made piece of equipment and/or the choice of materials for constructing one.

Reference should also be made to [35, 41–46] and to literature on supplies, materials, and equipment [14] before purchase or construction. Specific information concerning the many materials, balances, and magnets used in instrumentation follows.

7.3   Materials Used in Instrumentation

7.3.1   Sample Holders: How to Avoid Lateral Motion of Samples

Generally, the long sample tubes required for the Gouy balance and the small "buckets" or "pans" used with the Faraday balance have been fabricated from Pyrex or borosilicate glass. However, many glass compositions contain traces of iron, which are detected in very accurate magnetic measurements at very low temperatures, and their magnetic contributions must be corrected. Hence, fused quartz is selected by many workers; quite often polytetrafluoroethylene (Teflon polymer) or other polymers, such as nylon or Kel F, are also employed. Hurd [132] and Marshall and co-workers [133] made a careful survey of the magnetic susceptibilities of some of these materials. Pyrex glass, which is generally regarded as paramagnetic at all temperatures, shows a transition (to diamagnetism) around 34 K and is strongly temperature-dependent below this temperature. Fused quartz (diamagnetic) is almost independent of temperature. Teflon material (diamagnetic) shows a transition at around 80 K. Reliable experimental data for these materials over a range of temperatures are reported. This information is particularly useful for making accurate corrections for the susceptibility of vessels, and so on, while making precise measurements on samples over a wide temperature range.

LeGrand [134] also studied the magnetic susceptibility of extruded Teflon polymers by the Gouy method. These samples showed apparent paramagnetic susceptibility but became diamagnetic after heating and evacuation. The paramagnetism is reestablished when the sample is kept in contact with air for 1–10 min. Electron paramagnetic resonance (EPR) signals were not observed for the paramagnetic samples.

Sample holders in various shapes for magnetic measurements and sizes, helical springs, and also the quartz fibers employed as suspensions in the Faraday balance or in magnetic anisotropy "torque" apparatus are available from Worden Quartz Products, Inc., which is a subsidiary of the Ruska Instrument Company (6121 Hillcroft Avenue, Houston, TX 77036).

Small sample holders with an internal diameter of, say, 5 mm and a height of 5 mm are made from platinum and gold. Copper is also employed when measurements are restricted to noncorrosive materials such as metals, alloys, and intermetallic compounds. Such metallic sample holders weigh considerably more than quartz holders of comparable size. The use of metallic holders is then limited to relatively high loading capacity (10–100 g) balances, used in conjunction with the Faraday balance. (The Cahn-1000 balance meets these requirements.)

Gold and spec-pure copper are highly diamagnetic, and the net diamagnetism of such sample holders is generally enough to compensate the large moments of a few milligrams of ferromagnetic and ferrimagnetic materials. Thus the movement of the sample holder to one of the magnet poles is prevented in such cases. The magnetic force experienced by the sample is simply calculated by subtracting the negative-repulsive force experienced by the metallic bucket from the total force experienced by the metallic bucket plus sample. The susceptibility (or the moment) is then calculated by using appropriate equations.

The field profile in a superconducting Faraday magnetometer is described by Johannson [135]. Lateral stability is obtained for both paramagnetic and ferromagnetic samples. Experimental aspects and the theory of the magnetometer are described. A small ring, which is made of ferromagnetic nickel wire and surrounds the sample tube (6.4 mm in radius), prevents the instability of strongly ferromagnetic (even single crystal) samples and thus facilitates the centering of the sample at the desired spot in the field.

7.3.2 Suspensions

A Faraday balance suspension made of nylon monofilament line (1-lb test) is recommended by Villa and Nelson [136] because it is shatterproof, diamagnetic, and lightweight (density = 1.5 g/cm^3). This suspension does not change its length with loads up to 500 mg and at liquid nitrogen temperature. The advantages of this nylon suspension over quartz fibers are obvious.

Spraget and Williams [137] describe a novel balance suspension, which also serves as a thermocouple for use with the Cahn RG electrobalance. The setup is used in conjunction with a Faraday susceptibility balance. The thermocouple consists of twisted 0.08-mm enameled wires (0.03% Fe–Au alloy and Cu). The free ends of these wires are connected to the central pivot of the balance and taken along the weighing arm to the suspension point where the wires are glued and the suspension is allowed to hang from that point. The lower thermocouple junction is kept in contact with the sample. The relative susceptibility results for pure tantalum and chromium obtained with this suspension and in fields of 5000 Oe are given. Very good agreement with previous results including an accurate determination of the spin-flop temperature of Cr (123.5 ± 0.5 K) is given. In our opinion, this type of suspension may be adapted as well for the Gouy-type balance.

Zatko and Davis [138] recommend the use of sapphire filaments as hang-down supports for the Faraday and even the Gouy methods. Sapphire (Al_2O_3) filaments are now available commercially (Saphicon Division, Tyco Labs, Milford, NH 03055) and have a high tensile strength. Filaments 0.254 mm (10 mil) in diameter can be bent into circles less than 10 cm in diameter. Thus the sapphire filaments do not break as easily as the quartz filaments, which otherwise pose exasperating experimental problems. The sapphire filaments must be bent into a suspension hook carefully with a minitorch. We have successfully used a stiff tungsten wire for suspension.

7.3.3 Materials and Devices for Removing the Static Charge

Villa and Nelson [136] describe how to use radioactive sources to eliminate static electricity-induced problems, which tend to move the sample bucket (usually made of quartz) close to the surrounding cryogenic chamber (cold finger) and vitiate the sample measurements. These sources are made by evaporating $^{137}CsCl$ in 0.5 N HCl to dryness. One source is attached on the outside of the sample chamber about 1 cm higher than the sample position, and another source is placed inside at the bottom of the sample chamber. The use of a 350-μCi radioactive source of ^{241}Am for removing the electrostatic charge has also been described by earlier workers. An anti-static pistol known as Zerostat (Aldrich Chemical Co., P.O. Box 355, Milwaukee, WI 53201) emits positive charges from a piezoelectric device and is said to neutralize any static negative charges, and has been found to be effective in our laboratory. Another radioactive isotope (Po) and a conductive coating are available from the Cahn Instrument Company (Cerritos, CA 90701). The conductive coating is likely to decompose in high-temperature work.

7.3.4 Materials for Cryogenic and High-Pressure Experiments

Ginsberg [139] reports magnetic susceptibilities of materials that may be used in magnetic instrumentation and cryogenic apparatus. This paper and the erratum [140] concerning the composition of epoxy and polymethane adhesives provide useful information for functions such as the fabrication of sample vessels and the cementing of fiber suspensions. Of particular interest are the names and properties of materials such as phosphor bronze, beryllium–copper (Berylco), and germanium resistor and the sources for obtaining them. The alloys such as Berylco are used in the fabrication of high-pressure vessels for magnetic instrumentation.

7.4 Magnets, Direct Current Sources, and Gaussmeters

The introduction of magnetic resonance techniques since the 1950s has greatly increased the availability of permanent and electromagnets with interchange-able pole faces and variable pole gaps. With the advent of the superconducting technology, several magnets with axial or perpendicular access to the field that are suitable for magnetic measurements are available in the United States, European countries, and Japan. Instrumentation journals such as the *Review of Scientific Instruments*, the *Journal of Physics E*, and the *Electronics Buyers Guide* provide more than adequate listings of manufacturers of magnets. In Asian countries also (such as India) research-type magnets are available.[†]

Permanent magnets are available from the Indiana Steel Company (Valparaiso, Indiana) and Colt Industries (Pittsburgh, PA).

When cost is not a problem, it is indeed advantageous to buy a commercial magnet and a matching dc power supply. However, electromagnets that give moderate fields of about 5000 Oe can be built at low cost. Several publications [141] describe the construction of such magnets. It is even possible to convert a large transformer into a workable electromagnet merely by cutting a pole gap in the core of the transformer at a suitable place. One such design is described by Broersma [142].

A direct current supply obtained from a generator may be used for energizing electromagnets. However, the fluctuations in the average supply are usually so large that they ruin the precision of measurements. Devices for compensating such fluctuations are described [16, 143], but these are cumbersome to build. A battery supply is always advantageous, but this requires constant maintenance and limits prolonged use of the batteries.

Several circuits for stabilized power supplies were described in the literature [143–145]. A tube-type circuit especially adaptable for energizing the electromagnet is described by Figgis and Nyholm [146]. Here again, if the cost of such equipment is inconsequential, it is preferable to buy matching solid-state regulated power supplies from commercial sources. These power supplies provide extreme regulation of current and take into consideration even the most sensitive technique for measuring force; a lower limit of about 3000 Oe is desirable for routine work.

Generally, most magnetic susceptibility measurements are carried out by the so-called relative methods, that is, by comparing the susceptibility of a sample relative to that of a reference material or known susceptibility. These do not need an actual measurement of the field H in the Gouy method or of $H \cdot dH/dX$ in the Faraday method. However, quite often absolute measurements of average susceptibility and of the anisotropic susceptibilities (Section 7.11) require an accurate measurement of the field. While measurement of the field H can be carried out rather easily, that of the gradient dH/dX is more difficult.

The fields can be measured with instruments known as *gaussmeters* or *fluxmeters* or by other sophisticated apparatus. Descriptions of some of the common types of gaussmeters follow.

The Hall effect gaussmeter (gauss is used in place of oersted) consists of a very small probe or a sensing element made of a semiconducting crystal such as indium arsenide. The crystals can be very small (3×5 mm with a thickness of

[†]Manufacturer's addresses in India are available from the Department of Physics, University of Poona, Ganeshkhind, Poona (now spelled as Puné), India, 411007.

less than 0.5 mm). When a small direct current is passed along, say the x axis of the element, and the field H is applied along the z axis, a *Hall* voltage is generated along the y axis, which is proportional to the field H, that is, to the H_z component of the field. This facilitates the measurements of various components of the field. Unfortunately, the measurements are quite dependent on the temperature. The Hall voltages can be measured with great accuracy and permit the measurement of fields as small as 0.001 to about 30 kOe; however, the response becomes nonlinear usually above 10 kOe. With the incremental Hall effect, gaussmeters fields from 10 to 10 kOe can be suppressed to obtain accurate measurements to 1 part in 10,000. The Hall gaussmeter is more than adequate for most magnetochemical work and has been successfully used by us for anisotropy measurements on organometallic compounds. Addresses of manufacturers of Hall-Probe gaussmeters are found in the *Review of Scientific Instruments*, in the *Journal of Physics E*, and in [14].

The rotating coil gaussmeter uses a small flat coil (diameters range from 1 mm to ca. 2 cm) that is mechanically rotated by an electric motor in the field; thus a change in flux occurs, and an alternating voltage is produced that is proportional to the frequency of rotation of the coil and the field H. This voltage is measured by using ac (*lock-in*) amplifiers or after rectification. Fields as small as 1 γ ($\gamma = 10^{-5}$ Oe) can be measured with the sensitive ac techniques. Fields up to 100 kOe are also measured. The accuracy is generally better than 0.01% of the total field. This is adequate for routine magnetochemical work. The rotating coil gaussmeters are available commercially from the Rawson Electrical Instrument Company (Cambridge, MA.).

Nuclear magnetic resonance spectrometers are used to measure fields with great accuracy. In reality, the NMR gaussmeter is a portable NMR spectrometer. It consists of a tiny rf coil (about 0.03 cm^3 in volume) that is connected to the proper rf oscillator, amplifier, oscilloscope, and so on. The coil encloses a sealed sample of water or hydrocarbon, which contains protons H^1. These nuclei are subjected to the resonance experiment; that is, the frequency of the oscillator is gradually varied over a range (0.1–100 MHz) until a sharp resonance peak is observed on the scope. The sharpness depends on the homogeneity of the field. The sample of water usually contains a paramagnetic ion (e.g., Mn^{2+}), which reduces its NMR absorption line width (to ca. 10^{-6} Oe) and gives a sharp peak, as in the high-resolution NMR spectrometer, with which most chemists are familiar. The high-resolution proton resonance spectrometers generally operate over a wide range of frequencies corresponding to a broad scan of fields.

The lower range (ca. 0.1 MHz) in the gaussmeter permits measurement of fields down to approximately 300 Oe. Since the frequency of resonance v is directly proportional to the field ($hv = \gamma H$, where γ is the gyromagnetic ratio) and since the frequency can be measured with great accuracy (1 part in 10^6), fields, large or small, can be measured with the small degree of accuracy. The NMR gaussmeters are available from several commercial sources (see [14]).

With the Faraday balance, it is possible to measure the gradient dH/dX, that is, measure a small increment in the field ΔH over a small (vertical) distance ΔH by placing two small probes at the appropriate positions X_1 and X_2 in the field. These probes can be connected in opposition so that one obtains the difference ΔH between fields H_1 (at X_1) and H_2 (at X_2). Such probe settings can be adapted, especially with the Hall probes, which can be obtained in the form of thin long strips at the ends of which the sensors are attached. The gradient dH/dX or $H \cdot dH/dX$ can also be measured by using the "relative method," that is, by using a reference material of known susceptibility (see Section 7.10.3).

One obvious way to measure a uniform field H without using gaussmeters is to follow the Gouy technique for the susceptibility measurement. When a sample of known volume susceptibility κ is packed in a sample tube of known cross-sectional area A, and the magnetic force on the sample is measured ($g\Delta\omega$, where $\Delta\omega$ is the change in weight of the sample and g' is the gravitational constant), the field may be computed from the following equation:

$$g'\Delta\omega = \tfrac{1}{2}\kappa \cdot A \cdot H^2$$

If the field measured by this technique is then used to compute the susceptibility of another sample, the entire procedure merely corresponds to that of the *relative* method for measuring the susceptibility outlined previously.

7.5   Balances, Automatic Recording Devices, and Helical Springs

Several analytical, semimicrobalances, and microbalances in the single- and double-pan styles are manufactured by the American firms Ainsworth and Sons, Cahn Instrument Company, Perkin-Elmer Company, Fisher Scientific Company, and Testing Equipment Company, by E. Mettler in Switzerland; by Sartorius Werke A. G. in Germany; by Stanton Instruments in England; and by others. Also, a balance using a double-cantilevered beam is made by the Testing Equipment Company England/USA. These manufacturers and several others also make equipment for the automatic recording of weight, and many provide a suspension for attaching to the bottom of the pan that is useful for the Gouy method (Figure 3.45). The criteria for choosing a balance are discussed in Section 7.6.

For studying a continuous change in magnetic susceptibility as a function of temperature of the changes that may occur as a function of time in chemical and physical processes, it is convenient to have an automatic recording device. Robertson and Selwood [147] describe equipment for thermomagnetic analysis, especially of ferromagnetics. It is generally possible to convert any balance, such as the spring, the double-pan, or the single-pan, to automatic recording devices without too much difficulty. As mentioned previously, Hirsch [128], Thomas and Williams [129], Gordon and Campbell [130], Banerjea [131], the volumes that appeared in the *Vacuum Microbalance Techniques* series [49], and articles

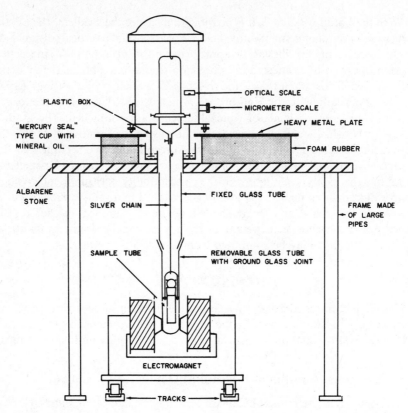

Figure 3.45 Constructional details of the Gouy balance in our laboratory. A frame made of 2-in.-diameter galvanized iron pipe gives support equally good as the concrete pillar shown.

by Mulay and Mulay [35–46] give excellent reviews of several ingenious methods (divided into the deflection and the null methods) adopted by investigators and manufacturers for this purpose.

In the first category, the deflection of the beam or of a pointer attached to the spring is proportional to the change in weight and is converted into a signal, which is then recorded. In the null method the force required to restore the balance to equilibrium is converted into an appropriate signal and then recorded. The deflection methods are quite adequate, considering the magnitude of changes in weight (which can be controlled by adjusting the magnetic field) especially in the Gouy techniques and also that one end of the sample tube always remains in a uniform field. The deflection methods are also easier (and less expensive) to adopt than the null methods. In the Faraday method care must be taken to ensure that the sample remains in the uniform $H \cdot dH/dX$ region.

In the phototube method, the deflection controls the amount of light entering the photocell; the proportional current generated is then amplified and recorded on a potentiometric-type recorder, such as those of Brown, Varian, Leeds and Northrup (Houston), and Texas Instruments Company. A Sartorius phototube device was successfully adapted in our laboratory for measuring and recording the deflections in the Gouy balance. (Apparently, this device is no longer available, but it can be constructed.) In the electronic methods, the deflection is made to cause a proportionate change in the capacitance of the parallel-plate condenser or in the mutual inductance of a coil–plate or coil–coil assembly. These changes are measured with a suitable capacitance or inductance bridge and are converted to a dc signal, which is then displayed on a recorder through conventional electronic circuitry. Methods using novel techniques such as a strain gauge [148] have also been tried. However, a differential transformer or a variable permeance transducer incorporated into a balance yields several advantages (cf. [130]). These devices are commercially available at low cost, and correcting them to the recorder does not involve complicated circuitry.

The deflection method using a helical spring is shown to provide certain advantages (e.g., small *dead space* for adsorption studies and low cost), especially with the Faraday technique. However, for routine work, a null method is always preferred because the Faraday technique demands that the *reference* material and the *sample* be placed at the same position in a uniform $H \cdot dH/dX$ region.

Compared with other techniques of measuring forces, automatically recording changes in torsion is somewhat complicated. Nevertheless, many workers (cf. [130]) and the Sharples Corporation (Bridgeport, PA), have accomplished this by using the principles of an electrobalance as outlined in Section 7.11.

7.6   Requirements of Field Strength and Sensitivity of Balances

Paramagnetic susceptibilities that are very large compared with those of the diamagnetic materials produce appreciable changes in the weight of a sample when a magnetic field is applied. Changes produced by paramagnetic samples of a few tenths of a gram that are placed in a field as low as 3000 Oe may be measured with an analytical balance with a sensitivity of 0.2 mg. However, accurate measurements of diamagnetic susceptibilities on samples of the same size require better accuracy (0.01 mg) in weighing and the use of higher fields. A happy compromise for measurements on paramagnetic, ferromagnetic, and diamagnetic materials in fields ranging from 3000 to 10,000 Oe is to use a semimicrobalance with a loading capacity around 50 g or better. For work on superparamagnetic materials, fields up to 20 kOe are required. The changes in weight may be observed conveniently merely by adjusting the rider or the chain with the conventional double-pan balance or by using the optical (deflection) scale of the single-pan balance. When a variation in the magnetic susceptibility of a system is studied, it is advantageous to restrict weight changes to the range provided by the rider, the chain, to within 1 part in 10^4 or better, which is more than adequate for magnetic susceptibility measurements. However, current regulation may not regulate the field in the pole gap. Devices for regulation of the magnetic field are described in the literature and should be adopted, particularly with homemade electromagnets. Fortunately, electromagnets are usually designed to provide regulated magnetic fields that are quite adequate for magnetic susceptibility measurements.

Figures 3.45 and 3.46 depict a setup that has been operating satisfactorily in our laboratory for some time. This arrangement provides one electromagnetic that is mounted on a movable platform and equipped with interchangeable pole caps for the Gouy and Faraday techniques, and others. It has proven useful for other magnetic work, such as magnetic anisotropy and magnetic resonance studies. In Figure 3.47 is a photograph of the Faraday balance using constant force poles.

Permanent magnets constructed from alloys such as Alnico, Permendur, and the new magnetic materials, which have a high remanence and high coercive force, provide fields as high as 10,000 Oe in a pole gap of 2.5 cm with tapered poles about 5 cm in diameter. Small permanent magnets giving fields of about 2000 Oe may be retrieved from "war surplus" magnetron magnets. The cost of a permanent magnet with reasonable field homogeneity (required broad-line

Figure 3.46 Close-up view of the quartz sample bucket in the field of a 4-in.-pole diameter electromagnet (entire balance shown in Figure 3.52a). This sample is within the region of *constant force*. Reprinted with permission from L. N. Mulay, "Techniques for Measuring Magnetic Susceptibility," in A. W. Weissberger and B. W. Rossiter, Eds., *Physical Methods of Chemistry, Techniques of Chemistry*, Vol. 1, Part 4, Wiley-Interscience, New York, 1972. Copyright © 1972 by John Wiley & Sons, Inc.

Figure 3.47   A micro-Faraday balance in our laboratory, using a Cahn-RG (now called Cahn-2000) improved balance, made by the Cahn Instruments Company, Cerritos, CA. The dc power supply to the 6.5-in. pole electromagnet, vacuum connections are not shown. The balance is mounted on a vertically moving stand to locate the sample position in the constant $H \cdot dH/dX$ region (see schematic in Figure 3.60).

NMR work, etc.) is usually less than the combined cost of a comparable electromagnet and a stabilized power supply for the direct current required to energize it. This and the availability of a steady field over long periods constitute major advantages of a permanent magnet. On the other hand, some mechanical device for moving the permanent magnet over a sufficient distance is required to permit a measurement of force on the sample with and without the field. Permanent magnets, unfortunately, cannot be used for studying easily the susceptibility as a function of field strength, and this property is useful for detecting the presence of ferromagnetic impurities in the sample. However, the field of a permanent magnet can be varied within moderate limits, by passing a current through coils surrounding the magnet. Some variation in the field also may be obtained by using magnetic shunts. Permanent magnets are especially useful when studying changes in magnetic susceptibility that occur as a function of time in certain photochemical and polymerization processes, in processes involving free radicals, and so on.

The choice of field strength depends on the size of the sample, the dimensions of the sample tube, and the sensitivity of the technique for measuring force. Large forces obtained at higher fields can be measured with more precision than small forces; as such no upper limit can be placed on the magnetic field. While studying changes in the magnetic susceptibility of a system, it is important to restrict the changes in weight within the range observable by the particular balance. This is readily done by adjusting the field strength of the electromagnet, the mass of the sample, or both.

The single-pan balance, which has several advantages over the conventional double-pan type, was first adapted for the Gouy technique in 1950 and reported by Mulay [149]. Many single-pan models are equipped with a good damping system (usually air), which facilitates a quick weighing within a few seconds. The enclosed gram and centigram weights are handled by a remote control and thus are protected. An optical deflection system is employed to read the milligrams; the fraction up to 0.01 mg or better may be read from an external Vernier arrangement in the semimicro and micro versions of the single-pan balance. The built-in optical deflection scale incorporates the advantages of the external microscope method of observing deflections used in the sensitive Theorell [150] and Michaelis [151] modifications of the Gouy technique. Apart from these mechanical merits, some single-pan balances (e.g., the Mettler, Cahn 1000 and 2000 models, Sartorius) facilitate weighing under conditions of constant load and, hence, of constant sensitivity. This feature is desirable when studying samples varying in sizes from a few tenths of a gram to several grams.

7.7   Temperature Control

7.7.1   Commercial Sources

There are several manufacturers of cryogenic apparatus in the United States and many appear to produce good apparatus for optical and Mössbauer studies, and so on, at low temperatures. The instrumentation usually involves a direct thermal contact between the sample and a metal block, whose temperature is generally controlled and measured by several well-established techniques. Thus, this type of cryogenic apparatus with continuously variable temperature (1.2 K to room temperature and up) is easily available. Conversely, it is not easy to regulate and measure the temperature of a sample, which is freely suspended from a balance and cannot establish a rigid thermal contact with a metal block. Although a few manufacturers in the United States sell cryostats for use with magnetic balances and with the vibrating-sample magnetometers, the performance of some of the products has not been quite satisfactory in our experience. Considering the high cost of such equipment (in the range of $8000–16,000, counting the hidden cost of essential accessories) and the special needs of different laboratories, it is concluded that construction of temperature-control apparatus is still worth undertaking. The following sections describe various aspects of such instrumentation. The Air Products Company (Allentown, PA) makes reliable cryogenic equipment for susceptibility measurements, as well as for optical, UV, and Mössbauer spectroscopies. These units provide a regulation of 0.1 K over long periods of time.

7.7.2 Measurement and Regulation of Temperature

It is easier to measure, with a thermocouple, the temperature of the atmosphere immediately surrounding the sample tube than to ascertain the temperature of the sample. However, if the sample is allowed to stand at a fixed temperature for about 10–15 min, it quickly reaches thermal equilibrium with its surroundings; this renders direct measurement of the sample temperature unnecessary. A "copper–constantan" or a "platinum–rhodium" thermocouple should be employed, since these are essentially nonmagnetic in nature.

Dail and Knapp [152] describe a novel temperature device and the relevant circuitry for making accurate measurements in conjunction with a Faraday balance. Its unique features include the ability to establish a close physical contact between the sample and the temperature sensor (carbon resistor) and to connect the device to a temperature controller. This allows simultaneous control and measurement of the sample temperature, especially between 4.2 and 80 K. The sample, the sensor, and the lightweight input coils forming a part of the inductive coupling circuit (air-core transformer) are all suspended from the Cahn microbalance. The connection between the sensor and the input coils is made with fine wires, which are attached to the regular (quartz-fiber-type) suspension. The change in resistance of the sensor with temperature is measured externally through an ac-type bridge, a lock-in amplifier, a 2-kHz single source, and so on. The temperature control is provided by a helium Dewar heating coil surrounding the sample and an exchange gas. This concept should prove most useful for other temperature ranges (above 30 K) with the use of other sensors.

It is always desirable to measure the temperature when the magnetic field is turned off, since certain sensors have magnetoresistive properties, which may vitiate the temperature measurement. In this connection, a paper by Belanger [153] describing the behavior of carbon resistors in high magnetic fields is most valuable. Neuringer and Shapiro [154] describe low-temperature thermometry in high magnetic fields that use carbon resistors. Equally useful information on the fabrication of small carbon resistors is given by Booth and Ewald [155], and the use of modified Speer resistors is outlined by Robichaux and Anderson [156].

Several circuits for regulating temperatures were reported in the literature. Many descriptions continue to appear in the *Review of Scientific Instruments*, the *Journal of Scientific Instruments*, and the *Journal of Physics E*. To study a variety of regulators [36–46], Mulay and Mulay may be referenced.

7.7.3   Production of High Temperatures (Approximately 300–1200 K)

Temperatures above room temperature are produced easily with an electric cylindrical furnace, if its coils are wound noninductively to prevent stray magnetic fields. Temperature regulation is accomplished simply with adequate relay mechanisms or by motor-driven rheostats to within 0.1°C of the desired temperature. The heating coils may be wound with a Nichrome resistance wire for use with the Gouy balance. Any distortion of the uniform field by this ferromagnetic material does not seriously affect the measurements. However, the coils should be wound with a wire made of either a platinum or an appropriate nonferromagnetic material that is especially for use with the Faraday balance and that employs a nonuniform field with a constant force, $H \cdot dH/dX$ region. Heating coils made from magnetic materials distort this region considerably and do affect the measurement.

7.7.4   New Trends in Cryogenic Instrumentation

THERMOELECTRIC COOLING

A variety of cryostats are described in instrumentation journals, but not all are designed for classical force (Gouy, Faraday) or electronic (ac or dc induction, NMR, etc.) susceptibility measurements. Some cryostats can be adapted directly or modified for magnetic susceptibility measurements; however, instead of listing them, let us discuss a significant advance in the design of a cryostat that allows it to obtain magnetic susceptibility or anisotropy measurements by either technique. The cryostat was designed for a torque magnetometer and uses thermoelectric cooling modules; its constructional details are given by Birss and Wallis [157]. Mulay and co-workers [3, 5, 6, 31, 35, 36, 82, 85, 118, 158–165] pointed out the possibility of using thermoelectric modules for temperature control, at least over a limited range of temperature. It is gratifying for this "prediction" to become a reality and for temperature to be regulated over a much wider range (77–373 K). Related aspects of temperature control have since been reviewed [36–45] and are included below.

Birss and Wallis [157] use the smallest available standard 15-A Frigistor thermoelectric cooling modules, consisting of four thermocouples. Each thermocouple comprises a pair of conduction elements, and the eight elements are connected electrically in series; however, the elements are arranged so that the hot junctions abut an aluminum plate (9 × 9 mm) and the cold junctions abut a similar plate. These plates are kept about 6 mm apart in thermal, but not electrical, contact with the thermocouple junctions. The maximum heat pumped against zero temperature gradient is 3–5 W, and the ultimate temperature difference between hot and cold junctions is 62°C for zero heat pumped; this assumes that the hot junctions are held at 27°C with the module run at 15 A and 0.4 V. In the temperature control unit there are two cooling modules run in cascade, so that a larger temperature difference can be achieved with a normal water cooling of the hot junction of the second stage. A suggested ratio of the numbers of modules in the first and second states is 1:6.

The versatility of this system, which is sure to please instrumentation lovers, lies in its ability to produce higher temperatures (up to 100°C) by merely reversing the current flow so that the cold junctions now become hot. The temperature range of the unit can be further extended by cooling the hot junctions of the second stage to 77 K with liquid nitrogen. Although the maximum temperature difference between hot and cold junctions is very much reduced at these low temperatures, temperatures between 77 K and room temperature can be obtained readily by using the module with reverse current.

SMALL GAS-COOLED CRYOSTATS
A good review of small gas-cooled cryostats was published by Meyer and Rich [166] and compares the cost and performance of the miniaturized gas refrigeration systems for producing temperatures down to −200°C. The helium compressor cryostats manufactured by the Air Products Company (Allentown, PA) and Lakeshore Cryotronics (Westerville, OH) provide a small cooling head that can be fitted easily between the poles of a magnet. The use of nonmagnetic materials such as copper for the cooling and enclosing the cold end in a thermally insulated jacket do not seem to pose serious problems. However, the continuous *compressing action* does produce undesirable vibrations, which are likely to vitiate the Gouy and Faraday measurements. These effects are now avoided by incorporating a metal bellows between the compressor and the sample chamber.

7.7.5   Fixed Temperatures and Use of Freezing Mixtures

In the low region, temperatures provided by freezing mixtures, low-boiling liquids, and liquified gases [167] are also obtained easily, using a setup as shown in Figure 3.48. The temperature of the sample remains constant as long as the freezing agent is maintained at an adequate level above the sample tube. Reithler [167] describes a cryostat in which different refrigerants are used to obtain constant temperatures.

7.7.6   General Principles for Producing Variable Temperatures in the Cryogenic Range

In general, the production and control of temperatures between 77 K and room temperature are accomplished easily with liquid nitrogen alone, and several references to such cryostats other than the ones cited here may be found in the literature [168]. The use of liquid hydrogen for obtaining temperatures down to 21 K is rather *hazardous*, and one cannot merely substitute liquid hydrogen for liquid nitrogen without introducing adequate safety precautions in the nitrogen cryostat. A design using liquid hydrogen originally developed by Aston and co-workers [169] for NMR experiments appears to be adaptable for the Gouy and Faraday techniques. Scott [170] describes an elaborate cryostat that uses liquid helium, hydrogen, and nitrogen to give temperatures from 1.6 to 300 K. It is particularly appropriate for the Faraday technique.

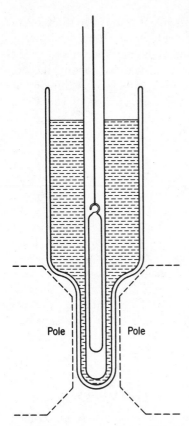

Figure 3.48 Typical Dewar flask for low-temperature susceptibility measurements. Reprinted with permission from P. W. Selwood, *Magnetochemistry*, Wiley-Interscience, New York, 1956. Copyright © 1956 by John Wiley & Sons, Inc.

The choice of a design for a cryostat depends on the specific nature of the magnetic study, the temperature range, the precision of temperature regulation, and, finally, the particular technique used for the susceptibility measurement. With techniques using a magnet, it is important to make the tailpiece of a cryostat as narrow as possible, so that a small pole gap for obtaining high magnetic fields can still be used, allowing enough space for the sample. A Dewar flask with an oval cross section to fit into narrow pole gaps is described by Broersma [75]. A helium Dewar flask with incorporated magnetic pole tips is described by Edwards and co-workers [171].

Special cryostats are required to produce an intermediate temperature that is not furnished by a freezing agent. Such cryostats are constructed according to the following general principles, and a few examples of their typical applications are referenced.

1.   The sample is surrounded by a double-walled metal or glass jacket through which a liquid at the desired temperature is circulated. This provides only a limited temperature range. Low-temperature baths (from −35 to 65°C, with a regulation of 0.01°C) are made by the Wilkens-Anderson Company (Chicago, IL).

2.   A metal block (copper or lead) surrounding the sample is cooled by a cooling agent (e.g., liquid nitrogen) and heated simultaneously by an electric current. Regulation of heat gives a desired temperature somewhere between that of the coolant and the room temperature. A cryostat of this type, as described by Mulay [172] for NMR studies (Figure 3.49), can be used without the rf coil for the Faraday magnetic susceptibility technique, since the temperature gradient over a sample height of about 2 cm is quite negligible.

3.   A coolant such as liquid air is injected into a spiral opening in the metal block; the rate of flow controls the temperature. More sensitive control may be obtained by passing oxygen through a liquid nitrogen condenser and allowing the liquified oxygen to drip into the metal block. The rate of oxygen passing through the condenser controls the rate of cooling. Extreme care should be exercised in using liquid oxygen because it can cause severe fires.

4.   Liquid air or nitrogen is boiled under the sample so that cold vapors arise over a jacket surrounding the sample. Figure 3.50 shows this arrangement (cf. [25]). Several temperature regulators are marketed by manufacturers of high-resolution NMR and EPR spectrometers (e.g., Varian Associates, Palo Alto, CA, supplies many descriptive brochures). The vapors of liquid nitrogen are heated electrically to the desired temperature before they are circulated around the NMR or the EPR sample tube. These regulators are adapted easily for magnetic susceptibility work. In our laboratory a unit made by the Perkin-Elmer Company (Norwalk, CT) has given very good service for the past several years. However, this unit is no longer available and a similar unit made by Varian Associates also gives good results. These use the gas-flow principle.

5.   The sample is surrounded by a double-walled jacket, which in turn is surrounded by a coolant (liquid nitrogen). The space inside the jacket is filled with helium gas, which acts as a heat-transfer medium. Regulation of the pressure of helium gas by a vacuum pump (and some heating) controls the temperature [173]. Figgis and Nyholm [146] use air for heat transfer. A cryostat of this type is manufactured by the Newport Instruments Company, Newport, England. Other descriptions of cryostats that are adaptable for susceptibility work may be found in books on experimental techniques.

6.   For work in the temperature region furnished by liquid helium, principles 1 to 4 are difficult to adapt, since liquid helium boils off very quickly (the specific heat is only 3.6 cal g^{-1} mol^{-1} deg^{-1}). Hence, a new conduction technique, originally developed for nuclear and electron magnetic resonance experiments, may be used for magnetic susceptibility measurements. (Such cryostats are in operation in the Physics Department and Gordon McKay Laboratories of Harvard University.) We have not found descriptions of such cryostats in the literature. One end of a copper tube is immersed to variable extents in liquid helium; thus, different temperatures are produced at the other end, which contains a small sample. This procedure limits the volume over which a uniform temperature can be maintained; however, the technique is promising for the Faraday-type magnetic balance that uses samples of very small volume.

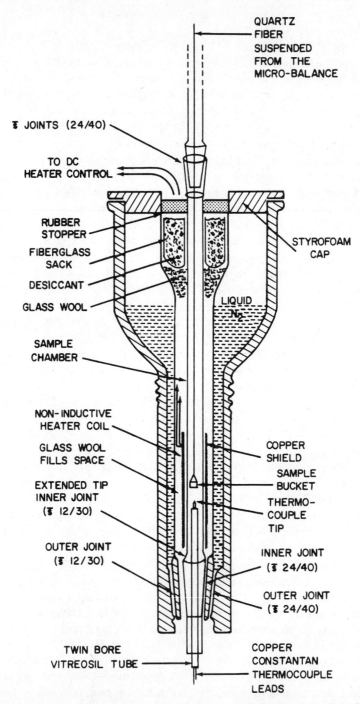

QUARTZ
FIBER
SUSPENDED
FROM THE
MICRO-BALANCE

JOINTS (24/40)

TO DC
HEATER CONTROL

RUBBER
STOPPER

FIBERGLASS
SACK

DESICCANT

GLASS WOOL

STYROFOAM
CAP

LIQUID
N₂

SAMPLE
CHAMBER

NON-INDUCTIVE
HEATER COIL

GLASS WOOL
FILLS SPACE

EXTENDED TIP
INNER JOINT
(§ 12/30)

OUTER JOINT
(§ 12/30)

COPPER
SHIELD

SAMPLE
BUCKET

THERMO-
COUPLE
TIP

INNER JOINT
(§ 24/40)

OUTER JOINT
(§ 24/40)

TWIN BORE
VITREOSIL TUBE

COPPER
CONSTANTAN
THERMOCOUPLE
LEADS

Figure 3.49 A cryostat originally designed for broad-line NMR work can be adapted easily to magnetic susceptibility studies by the Faraday method. The suspended sample bucket shown occupies the NMR coil-and-sample region. A temperature range from approximately 85 to 500 K is obtained easily (Mulay [172]).

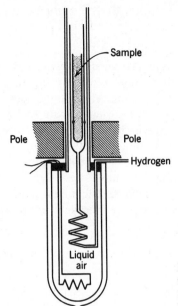

Sample

Pole   Pole

Hydrogen

Liquid
air

Figure 3.50 Apparatus for automatically controlled low-temperature measurements. Reprinted with permission from P. W. Selwood, *Magnetochemistry*, Wiley-Interscience, New York, 1956. Copyright © 1956 by John Wiley & Sons, Inc.

7.   Liquid helium is allowed to "leak down" through a throttle valve and to evaporate gently upward from the bottom of the sample chamber. Control of the throttle valve and of the heating coils surrounding the sample gives a desired temperature. This principle has been adapted in a cryostat for use with a vibrating sample magnetometer. Applicability of this technique to the Gouy and Faraday systems is doubtful. Evaporation of liquid helium directly under the sample may introduce variable buoyancy errors in the measurement of changes in weight. A careful study of the adaptability of this technique would be most desirable.

Birss and co-workers [174] presented a very careful analysis of four different methods that can be used in helium cryostats to obtain temperatures between 4.2 and 77 K. They have employed an unusual method involving heat exchange between the coolant vapor and the specimen. In their cryostat (which can be used over the range 4.2 K and room temperature) the rate of boil-off of liquid helium by way of an electrical heater is controlled by an electronic circuit. The required temperature is obtained by altering the heat exchange between the vapor and the specimen. This cryostat is especially designed to accommodate a very sensitive torque magnetometer. Details of construction and data on the performance of the cryostat in relation to the rate of boil-off of helium and the specimen temperature are given. This apparatus should be very useful for other magnetic techniques, also.

7.7.7   Miscellaneous Methods (Narrow Temperature Range for the Gouy Method)

When working in a narrow range, such as from -10 to $60°C$, where precise temperature measurement is not critical, we have obtained satisfactory measurements in our laboratory by dipping a small thermometer directly into the sample (see Figure 3.51). This procedure eliminates the somewhat cumbersome use of a thermocouple and a potentiometer, which are normally required for large temperature ranges and more accurate measurements. We used this technique with a Gouy balance to study, as a function of acidity and temperature, the changes in the magnetic susceptibility of aqueous solutions containing paramagnetic ions. The tube containing the sample was weighed at room temperature, and the weights used were left unchanged. The tube was then removed, and in an air jacket it was cooled by an adequate freezing agent to a

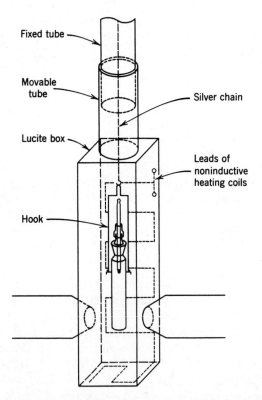

Fixed tube

Movable tube

Silver chain

Lucite box

Leads of noninductive heating coils

Hook

Figure 3.51   Temperature control for varying temperatures over a narrow range, used with our Gouy balance; the Lucite box may be replaced by a glass tube. Reprinted with permission from L. N. Mulay, "Techniques for Measuring Magnetic Susceptibility," in A. W. Weissberger and B. W. Rossiter, Eds., *Physical Methods of Chemistry, Techniques of Chemistry*, Vol. 1, Part 4, Wiley-Interscience, New York, 1972. Copyright © 1972 by John Wiley & Sons, Inc..

desired temperature. Condensed moisture was wiped off the tube, which was then suspended in the Lucite box and weighed quickly with the magnetic field first off and then on. Because of the preliminary adjustment of the weight, these weighings could be made in a few seconds, and the temperature change was found to be less than 1°C.

This method is restricted for use in the temperature range from −10 to 60°C; however, it eliminates the need for widening the pole gap to accommodate a heating conventional unit, which would lower the magnetic field strength and affect the accuracy of the susceptibility measurements.

7.8 Measurements Under Pressure

In Section 1 we discussed the growing interest in the study of magnetic susceptibility, especially in that of solids under high pressure. Part of this interest stems from the geoscientists' attempts to understand the formation of magnetic rocks and minerals in the earth's interior, which is subjected to enormous pressure. Magneticists are interested in elucidating the magnetic interactions of localized and/or itinerant electrons in various solids under pressure. The materials scientist is also concerned with the synthesis of new materials with tailor-made properties by using pressure and temperature as variables. Their success in synthesizing diamonds under very high pressures is well known. The materials scientist investigates magnetic properties as a tool for both characterization and the development of new magnetic materials for technological applications, such as the permanent magnet materials (NdFeB) and superconductors.

Over the past decade a few reports contained outlines about the fabrication of pressure vessels for the investigation of magnetic susceptibility under pressure. Several of these vessels were specially designed for use with the force-type methods, whereas others were fabricated in conjunction with the inductance bridge or the vibrating sample magnetometers. Since these developments are too diverse to review here, they are described briefly in the various sections marked, "Modifications of...," under each instrumentation category, such as the Gouy, the Faraday, the inductance bridge, and the vibrating sample magnetometer. References to the pressure-dependent studies are found in Section 8 and in [43–46].

7.9 Typical Assembly and Procedures for the Gouy Balance

7.9.1 Apparatus

In our laboratory, a Gouy magnetic balance (Figure 3.52a) was assembled from the following units to obtain simplicity in its operation and to permit a maximum flexibility in its application to a variety of research problems. Figure 3.45 shows the arrangement used for enclosing the suspension system to prevent the effects of drafts of air.

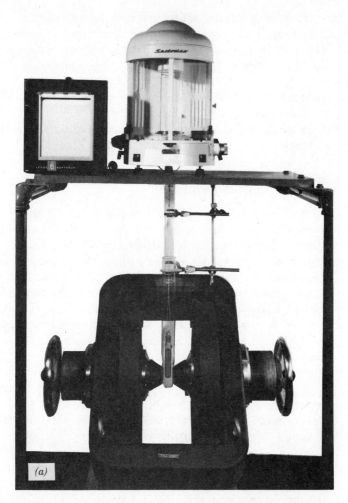

Figure 3.52 (*a*) Gouy balance in our laboratory. A servodevice, consisting of a light-beam and a phototube placed on top of the Sartorius balance, is used for automatic data recording. Reprinted with permission from L. N. Mulay, "Techniques for Measuring Magnetic Susceptibility," in A. W. Weissberger and B. W. Rossiter, Eds., *Physical Methods of Chemistry, Techniques of Chemistry*, Vol. 1, Part 4, Wiley-Interscience, New York, 1972. Copyright © 1972 by John Wiley & Sons, Inc. (*b*) Various Gouy tubes. Simple Gouy tube (left): A, suspension from balance; W, wire loop; C, collar; G, Gouy tube; R, reference mark; and S, specimen. Double-ended Gouy tube (center) to eliminate empty tube correction δ. Double-ended Gouy tube (right) to eliminate δ and solvent correction: S, solution; S′, solvent; and A, air bubble to permit expansion and contraction of solvent.

1.   A Newport type A electromagnet with 4-in.-diameter tapered poles, an adjustable pole gap, and interchangeable straight and Faraday pole caps, giving a maximum field of about 10,000 Oe in a 1.5-cm gap (approximate cost, $4000). A matched Newport type B dc power supply operating from a 110-V, 60-Hz main, giving a maximum direct current of 8 A at 100 V and a regulation of 1 part

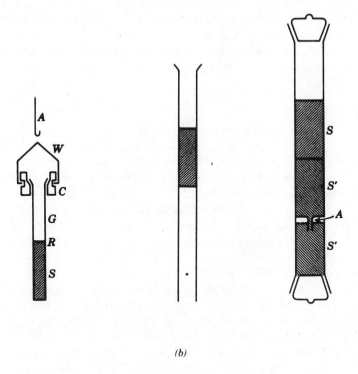

(b)

Figure 3.52 (*continued*)

in 10^4 (approximate cost, $6000). Various sample tubes are shown in Figure 3.52*b*.

In recent work (Figure 3.52*a*) an electromagnet with 6.5-in.-diameter poles tapered to 2.5 in., originally manufactured by Spectromagnetic Company (now part of Varian Associates), and a matching power supply are used. With poles tapered to 1 in., fields of 20,000 Oe in a large pole gap of about 4 cm can be obtained. This has facilitated the inclusion of different cryostats, high-temperature furnaces, and so on, in the 4-cm pole gap. The approximate costs of this magnet and the matching regulated power supply are $4000 and $6000, respectively.

2.   A Sartorius semimicrobalance with a sensitivity of 0.02 mg under normal working conditions. The range on the optical scale is 100 mg; loading capacity is 100 g (approximate cost $2000). A thin silver chain is attached to the bottom of the pan.

3.   A Sartorius photoelectric recording device (approximate cost $1200) and a Leeds and Northrup recorder ($2500). The recording device fits on top of the Sartorius balance and employs a phototube and an amplifier for connecting to a recorder. (Apparently the photoelectric device is no longer available; however, it can be constructed easily.)

4.   A homemade cryostat for regulating temperatures from 77 to 300 K. The arrangement facilitates displacement of air by nitrogen gas, whenever it is required for accurate work. This cryostat is similar to the one used for NMR experiments [172] (Figure 3.49).

7.9.2   Preparation of Samples for the Gouy Method

METALLIC AND CERTAIN NONMETALLIC SOLIDS

Samples of metals, alloys, glass, polymers, and so on, may be obtained in narrow, long, uniform cylinders. The length should be such that the suspended end of the rod lies in a field that is negligible in comparison with the field applied at the lower end. Usually a length between 10 and 15 cm is adequate. The diameter must be such that the rod fitted into the smallest possible gap will move freely inside the cryostat. Rods a few millimeters in diameter are suitable for work over a wide temperature range; however, rods 2–3 cm in diameter may be used with advantage for measurement at room temperature by suspending them directly in the pole gap.

For most purposes, a sample tube made of Pyrex glass is quite adequate. For very accurate measurements, allowance must be made for the temperature dependence of its small paramagnetic susceptibility [132], which arises from certain impurities. The sample containers may be made from rather thin-walled tubing of uniform diameter (available from the Wilmad Glass Company, Buena, NJ) with hooks for suspension. A stirrup made of Pyrex glass, copper, or silver wire may be used to support the tube.

The general considerations for determining the size of the tube are the same as those mentioned previously for rods of metals and alloys. An etched mark may be made at the top of the tube, where the field is known to be negligible. This fixes the volumes of the sample and the reference to be used. For measurements on powdered solids, a tube about 1 cm in diameter and 15 cm in length is convenient; for liquids and solutions, a tube of the same length and about 2 cm in diameter is suitable for work of moderate precision.

For accurate work on solutions a semidifferential method is used (cf. [151, 175, 176]), which employs a tube positioned at the center, where the field is applied (Figure 3.52b). The pure solvent is sealed in the bottom part with a reservoir half filled. This allows the solvent to expand at higher temperatures and, on cooling, prevents a bubble of vapor to form at the septum. The solution is placed in the upper compartment of a double-ended tube (Figure 3.52b, center). The magnetic pull on the two ends of this compensation tube is in opposite directions and is almost canceled if the susceptibility of the contents in the two compartments is the same. Thus by keeping the solvent unchanged in the lower compartment and by varying the concentration of the solution in the upper compartment, changes in weight corresponding to a difference in the susceptibility of the two are obtained. The difference corresponds to the susceptibility of the solute under conditions of magnetic dilution and in the absence of interactions between the solute and the solvent.

The limits of error in the differential method of evaluating a solute susceptibility are considerably lower than in the method employing separate solvent and solution measurements.

POWDERED SOLIDS

Some comments on the packing of powdered solids appear in later sections. Finely ground powder should be used, to be sure that a maximum filling is obtained in a fixed volume. Under normal working conditions, experiments show that a loose packing or a tight ramming of the powder in the tube does not appreciably affect the susceptibility measurements, provided appropriate corrections (descriptions to follow) are made for the presence of air pockets. However, the ramming method may be preferred to the first since it facilitates a maximum filling and minimizes the chances of preferential orientation of the particles in the field. In this method, small and nearly equal portions of the powder are introduced in the tube and are packed after each addition by pounding with the flat end of a ramrod that snuggly fits the tube. If the portions and the number of strokes used in pounding are equalized, it is generally possible to obtain uniform packings of the reference material and sample and a reproducibility of weight to within 1%.

LIQUIDS AND SOLUTIONS

Liquids and solutions do not pose packing problems and may be used in conjunction with sample containers described previously. The sample tube may be sealed for very volatile liquids. A dry box with an atmosphere of nitrogen or special reservoirs and burets in which the flow of liquids is controlled by the pressure of nitrogen gas may be used for filling easily oxidizable samples, such as the solutions of Cr(II) and Fe(II). Sealing the tube is desirable in this case and also with such reactive substances as the free radicals. Work with reactive substances requires calibration of each tube used for individual experiments, but in many cases a tube with a good ground-glass stopper or with a high-vacuum stopcock is quite adequate.

7.9.3   Experimental Procedure

The experimental procedure is based on finding the susceptibility of a sample relative to the known susceptibility of a reference material. A pure liquid, such as benzene, acetone, or distilled water free from dissolved oxygen and carbon dioxide, is recommended as a reference.

MEASUREMENTS ON LIQUIDS AND SOLUTIONS

A Pyrex sample tube of 25-mL capacity and 0.8-cm diameter is convenient for work on pure liquids and solutions; a smaller tube is used if the size of the available sample is small. The stopper, which is lightly greased to prevent evaporation of liquids during weighings, must be handled carefully during the entire operation so that errors in the weight of the tube are not caused by loss of grease or by the acquisition of dust particles.

The empty tube is suspended between the poles of the magnet as shown in Figures 3.45 and 3.46, with its lower tip in the center of the pole gap. It is weighed and the balance beam is left in the released position so that the milligram reading is displayed on the optical scale of the single-pan balance. A suitable current, say 2 A (which, in our equipment, produces a field of 3000 Oe in a pole gap of 3 cm of the electromagnet), is passed for a minute or two, and the weight is noted. A decrease in weight caused by the predominating diamagnetism of glass is readily observed on the optical scale. Care must be taken to avoid prolonged passage of current, which produces heating effects and thereby introduces buoyancy errors in weighing.

The procedure is repeated at other field strengths obtained by passing higher currents. The change in weight ΔW_t corresponding to each amperage is noted. (Since ΔW_t is proportional to the square of the field strength, it should be proportional to the square of the amperage, provided the electromagnet is operated within the linear range of the field-current curves. The satisfactory performances of the balance and the electromagnet is ensured by obtaining a linear plot for ΔW_t versus the square of amperage.)

A reference liquid is enclosed in the sample tube up to a mark (enclosing a volume of about 25 mL) and is weighed as before without the magnetic field. Care is taken to avoid the formation of air bubbles. The sample is weighed in magnetic fields corresponding to the exact amperages used previously for the empty tube. The change in weight ΔW_{t+r}, which corresponds to each setting, is noted. The reproducibility in current settings is facilitated by a sensitive lamp and scale galvanometer connected across a small resistor $(1-10\,\Omega)$ placed in series with the ammeter in the electromagnet circuit.

The tube is dried and is filled to the same mark with the sample. The weighing procedure with the magnetic field off and then on is repeated as previously. This gives the change in weight ΔW_{t+s}, which corresponds to the same fields that were used for the reference.

MEASUREMENTS ON POWDERED SOLIDS

The general procedure followed for measurements on solid particles is the same as that described previously, except that a smaller sample tube (volume 15 mL, i.e., ca. 4 mm) facilitates handling powdered samples of about 0.3 mg. The powder is packed in the tube as described earlier; either a liquid or powdered solid may be used as a reference; such reference materials are listed in Section 7.9.5.

7.9.4   Calculations of Magnetic Susceptibility and Correction for Errors

For these calculations, the following symbols are used: W, weight; ρ, density; κ, volume susceptibility; χ_m, mass susceptibility; ΔW, change in weight on applying the magnetic field; and V, actual volume up to the mark W_r/ρ_r. The subscripts r, s, and t correspond to the reference, sample, and tube, respectively. The susceptibility is $\kappa_{\text{air}} = 0.029 \times 10^{-6}$ cgs units per cubic centimeter at room temperature and pressure. For measurements at a fixed field corresponding to a known current, $\Delta W_r = W_{t+r} - W_t$ and $\Delta W_s = W_{t+s} - W_t$.

If the sample and reference are filled to the same mark in a tube of cross-sectional area A and are subjected to the same magnetic field H under identical conditions of the Gouy experiment, the following equations hold when the tube is surrounded by air (g' is the earth's gravitational constant):

$$\text{force}(s) = g'\Delta W_s = \tfrac{1}{2}AH^2(\kappa_s - \kappa_{air})$$

and

$$\text{force}(r) = g'\Delta W_r = \tfrac{1}{2}AH^2(\kappa_r - \kappa_{air})$$

$$\frac{\Delta W_s}{\Delta W_r} = \frac{\kappa_s - \kappa_{air}}{\kappa_r - \kappa_{air}} = \frac{\kappa_s\rho_s - \kappa_{air}}{\kappa_r\rho_r - \kappa_{air}}$$

Hence,

$$\kappa_s = \left(\frac{\kappa_r\rho_r - 0.029 \times 10^{-6}}{\Delta W_r}\right)\frac{\Delta W_s}{\rho_s} + \frac{0.029 \times 10^{-6}}{\rho_s}$$

or

$$\kappa_s = (\text{tube constant})\frac{\Delta W_s}{\rho_s} + \frac{0.029 \times 10^{-6}}{\rho_s}$$

This equation is used readily in computing the susceptibility of a liquid sample relative to a liquid reference because liquids can be "packed" to a mark in the tube under identical conditions. The evaluation of a constant for a tube and its repeated use in a series of measurements simplify the calculation of susceptibility.

It is almost impossible to powder solids to the same particle size and to pack them uniformly up to a mark in the tube; this introduces variations in the volume of paramagnetic air held in the pockets of the sample. Hence a correction must be made for the susceptibility of air pockets in the powdered solid by including the contribution $\kappa_{air}(1 - W_s/V\cdot\rho_s)$ because of the air enclosed per cubic centimeter of the solid–air mixture.

The susceptibility of a *powdered solid* relative to a *liquid reference* is thus computed from the following relations, which are based on a derivation by French and Harrison [177]

$$\chi_s = \left(\frac{\kappa_r\rho_r - 0.029 \times 10^{-6}}{\Delta W_r}\right)\frac{\Delta W_s}{W_s}\frac{W_r}{\rho_r} + \frac{0.029 \times 10^{-6}}{\rho_s}$$

$$= (\text{tube constant})\frac{\Delta W_s}{W_s}\frac{W_r}{\rho_r} + \frac{0.029 \times 10^{-6}}{\rho_s}$$

For a higher degree of accuracy, another correction must be included, since the volume of the solid packed up to a mark will be less than that of the reference liquid by an amount equal to the volume of the meniscus. In this case

$$V\rho_r = W_r - C$$

where C, the correction for the meniscus of height h cm for a tube of radius between 2 and 4 mm, is given by

$$C = 0.054 \frac{W_s}{h} - 0.0037$$

The final equation is then

$$\chi_s = (\text{tube constant}) \frac{\Delta W_s}{W_s} \frac{(W_r - C)}{\rho_r} + \frac{0.029 \times 10^{-6}}{\rho_s}$$

Sloot and co-workers [178] discuss elegantly the reliability of magnetic susceptibility data obtained with the Gouy balance and focus their attention on large errors caused by the misadjustment of the sample holder. These errors were not suspected by any of the previous experimenters, and it is now difficult to believe that the accuracy in most of the published data was poor, despite claims to the contrary. The authors derive an equation for a correction factor f in terms of angle α, which is the deviation of the sample from the vertical axis, and x, the distance between the lower end of the tube and the center of the magnet. The authors show that even for a small deviation of $\alpha = 1°$ and small values of R, which is used for defining the polar coordinates of positioning, an error of 0 to 2% is introduced in the measurement. For high values of field obtained by minimizing the pole-piece diameter, the positioning of the sample becomes even more important.

7.9.5 Calibration

It is customary to use distilled water as a calibrating agent. However, distilled water contains appreciable amounts of dissolved air, which is paramagnetic. We prefer to employ *conductance water*, which is boiled just before use to remove all air. The susceptibility of water is -0.720×10^{-6} cgs units/g.

Some investigators [179] recommend benzene as a reference ($\chi = -0.7081 \times 10^{-6}$ cgs units/g) but its susceptibility depends on whether it is saturated with air. Several authors [180] have investigated the susceptibility of solutions of nickel chloride. The molar susceptibility of this salt is $4436 \pm 12 \times 10^{-6}$ at 20°C. The variation of the difference in volume susceptibility of a 0.1 M nickel chloride solution and water with temperature, as reported by Michaelis [151] is given in Table 3.13.

Table 3.13 Magnetic Susceptibilities of
0.1 *M* Nickel Chloride Solutions per
Cubic Centimeter[a]

Temperature (°C)	κ
18	0.0446
20	0.0443
22	0.0440
24	0.0437
26	0.0434
28	0.0432

[a]All values are multiplied by 10^{-6}.

Although the susceptibility of a mixture is not always a linear function of concentration, the susceptibility per gram of nickel chloride at 20°C is given by

$$\chi_m = [23.21p - 0.720(1 - p)] \times 10^{-6}$$

where p is the weight fraction of nickel chloride in the solution. The susceptibility of this solution is independent of concentration near 30% $NiCl_2$ by weight. Hence such a solution may be used conveniently for calibration.

Figgis and Lewis [24] recommend MTC {mercury tetrathiocyanatocobaltate(II) $Hg[Co(SCN)_4]$} as an all-around calibrant for solids. It has a *specific* or per gram susceptibility of 16.44×10^{-6} ($\pm 0.5\%$) at 20°C and is said to have exceptionally good packing properties. Curtis [181] suggested tris(ethylenediamine) nickel(II) thiosulfate as a calibrating agent. It has a specific susceptibility of 10.82×10^{-4} ($\pm 0.4\%$) at 20°C.

7.9.6 Modifications of the Gouy Balance

Several modifications of the Gouy magnetic balance have appeared in the literature; a review of important modifications before 1956 is presented by Selwood [25]. The novelty in these modifications lies partly in the use of a permanent magnet in place of an electromagnet and in the techniques employed for suspending the sample and for measuring forces. For instance, Hilal and Fredericks [182] use a "solenoid" arrangement to measure changes in weight of about a few milligrams to a high degree of accuracy not obtainable by the conventional method of swings. The solenoid method is found in some recording balances that are produced by the A.R.A.M. Company in Lyon, France, and by several manufacturers in the United States [14].

There is indeed no limit to implementing ingenious devices in either an ordinary or a susceptibility balance to meet special requirements.

A unique Gouy-type recording apparatus using an Ainsworth model UMD Chainomatic balance was reported [183]. An unusual feature is the addition of a quartz spring that is always kept under tension. The resulting change in transducer output is amplified with a Sanborn carrier preamplifier. A basic advantage of the Ainsworth balance is that it permits a complete evacuation of the balance and the weighing chamber and thus minimizes errors from the buoyancy effects. Advantageously, in this particular Gouy balance setup, a correction for κ_0, the volume susceptibility of the medium surrounding the sample, is avoided; thus,

$$\text{force} = g\Delta W = \tfrac{1}{2}AH^2(\kappa - \kappa_0)$$

where ΔW is the change in weight of sample when the magnetic field H is applied, κ is the volume susceptibility of the sample under investigation, and A is its cross-sectional area.

A Gouy balance suitable for studies of adsorption of (paramagnetic) gases on adsorbents such as silica and alumina, based on the work of Juza and co-workers, is described in the literature [25]. The method is a differential method in that the force of attraction on the sample A is counterpoised by a weight B similar in size to the sample. Any adsorption in chamber A naturally changes the susceptibility of the sample and produces a change in weight in the magnetic field, which may be applied temporarily. The change in weight may be followed by using the electromagnetic force device coupled to the (magnetic) counterpoise in chamber B.

Other modifications developed by Theorell [150] and Theorell and Ehrenberg [184], shown in Figures 3.53 and 3.54, are excellent for studying reactions involving changes in paramagnetism and diamagnetism. These can be followed easily, for they produce significant changes in magnetic forces acting on the sample, which is suspended horizontally from a bifilar suspension. However, with sufficient sensitivity, changes in diamagnetism (e.g., those resulting during polymerization of styrene) also could be followed. A high-field electromagnet is employed for convenience and increased sensitivity. The horizontal sealed tube holds the reaction mixture. The changes in susceptibility, which cause the displacement of the tube, are observed with a micrometer microscope. Selwood [25] calibrated a similar apparatus by filling the tube with a dichromate solution that was gradually undergoing reduction with sucrose. This involved a change from the diamagnetic $(Cr_2O_7)^{2-}$ to the paramagnetic Cr^{3+} ion. A modification by Howland and Calvin [185] (cf. [186]) for work on microgram quantities appears in Figure 3.55.

Instrumentation of a Gouy balance with temperature control is reported by Earnshaw [187] and others [188].

Servochemical networks adapted by Hedgcock and Muir [189] for the Gouy method can be used in conjunction with the Faraday method, also. In the instrument described, a sensitive electrodynamic balance is operated as a null instrument by allowing it to form part of a servomechanical network. The feedback system provides a stiffness of balance movement of 8×10^4 dyn deg^{-1} deflection. The sensitivity is such that changes of 10^{-9} emu g^{-1} are detected in

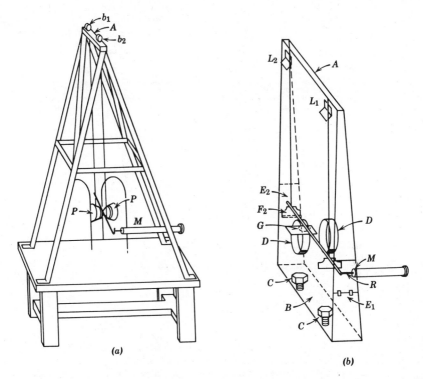

Figure 3.53   (a) Theorell's magnetic deflection balance; (b) constructional details (Theorell [150]).

the susceptibility of large metallic samples having electrical resistance less than $10^{-9} \, \Omega \, cm^{-1}$. The effects of eddy currents in such samples induced by the magnetic field are eliminated. The balance appears to be versatile for measurement on other samples and at different temperatures.

Garaleshko [175] presents a modification of the Gouy method for measuring the magnetic susceptibility of semiconductors and other weakly magnetic substances. The advantage of this method is that several measurements may be taken on a single sample. The use of a special pole tip in the electromagnet of the setup, as well as the use of high-saturating magnetic fields, makes it possible to exclude the effect of ferromagnetic impurities. A detailed description of the apparatus for measurements over a temperature range from 77 to 1000 K is given. The precision of absolute measurements is 1%, and the sensitivity for specific susceptibility is 10^{-6} cgs units.

A striking modification providing tenfold enhancement of the sensitivity of a Gouy balance is described by Henry and Hoyt [190]. The improvement is achieved by enclosing the sample in a double compartment-type Gouy tube and immersing it in a water chamber placed between the poles. This decreases the effective weight of the sample tube (because of the buoyancy effects) and contributes to a higher accuracy in the measurement of changes in weight when

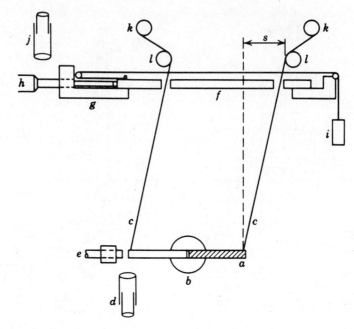

Figure 3.54 Schematic drawing of another sensitive balance, constructed by Theorell and Ehrenberg [184]. This view, illustrating the main principles of the apparatus, is in the direction of the central magnetic field, although the front magnetic pole is not shown. The tube a between the poles b is suspended by two thin wires c and its position relative to the fixed screw e is observed in the microscope d. The upper ends of the wires are kept in position by a score in each of the cylinders (1) and are fixed on the unrolling cylinders. These four cylinders are mounted on the slide f, whose movement in the guide g is controlled by the micrometer h. The slide is maintained firmly against the head of the micrometer by the lead weight i. The displacement s of the slide is read on the micrometer with the aid of a reading glass j.

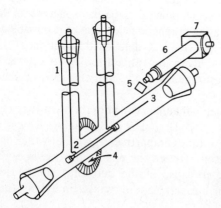

Figure 3.55 Apparatus used by Howland and Calvin [185] to measure the magnetic susceptibility of aqueous ions of the actinide elements: (1) 140-cm glass suspension fiber; (2) thin-wall capillary with control partition; (3) pointer; (4) magnetic pole face (only one shown); (5) mirror; (6) microscope; and (7) filar micrometer.

the field is applied. With a microbalance capable of measuring deflections to 5×10^{-4} cm, a field of 4000 Oe, and sample masses of 0.2 g, the susceptibilities were measured down to 0.1×10^{-6} emu g^{-1}. In our opinion, a major drawback of this technique is that it cannot be used readily for measurements at temperatures other than the room temperature. Although the authors [190] observed no harmful effects from the convection currents at room temperature, measurements at slightly higher temperatures might introduce many sources of error.

Note that the general idea here is similar to that of Chandrasekhar [191] who used floats immersed in oil in his modification of the Sucksmith balance. However, in this case the sample was farther away from the floats and also from the circular spring, which facilitated measurements over a wide temperature range. Dupouy and Haenny [192] describe other interesting immersion techniques.

The construction of a Gouy balance for adsorption measurements of paramagnetic salts between 77 and 250 K is reported by Shiraiwa and co-workers [193].

Until recently, only a few scattered reports on magnetic titrations have appeared in the literature. A systematic study of such titrations and their applications to several systems was made by Heit and Ryan [194], who designed a novel sample-tube titrating assembly for use with the Gouy balance (Figure 3.56). Their review of magnetic titrations and another by Graybill and co-workers [48], mentioned in Section 6.4.1 on the Quincke apparatus, should prove very useful to analytical and inorganic chemists. The magnetic titrations are based on observations of the changes in magnetic susceptibility when a sample is titrated against a reagent. Often the changes in magnetic susceptibility corresponding to conversion of one species of a given magnetic moment into another of a different magnetic moment are quite large and provide an accurate indication of the end point in the titration. The important feature of the new technique [194] is the development of a Gouy-sample assembly that includes a buret and a titration vessel and thereby keeps the total weight of the assembly constant. This assembly is illustrated in Figure 3.56; its use makes the magnetic titrations simple and quick. This arrangement avoids the difficulties involved in the cumbersome procedure of removing the sample tube from the magnetic pole gap, performing an outside titration, reweighing the assembly, and then observing the change in weight corresponding to change in susceptibility.

In [194] the dilution effect and simple methods for correcting it are discussed, and titration curves are interpreted for the possibility of forming different complexes during titration by describing the weight change that occurs when a field versus reagent moles is applied. Complete data and titration curves were provided for reactions of paramagnetic Fe^{2+} and the Ni^{2+} ions that change to diamagnetic species.

An outstanding contribution for the micromeasurement of magnetic susceptibilities using a novel *Cartesian diver* approach was made by Gersonde [195]. His paper sounds impressive in that measurements on a sample

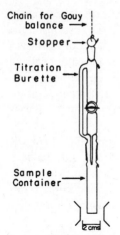

Chain for Gouy
balance

Stopper

Titration
Burette

Sample
Container

2 cms

Figure 3.56 Magnetic titration assembly for the Gouy balance [194].

containing as little as 50 μg of Fe^{3+} ions show change in the molar susceptibility of 130×10^{-9} cgs units for 1 g of Fe^{3+} ions.

Although the operations and principles of the apparatus appear initially to be complicated, the technique should be attractive to microanalysts because it was developed to study magnetic susceptibility changes in microgram quantities of biological materials (e.g., hemoprotein changing to the tertiary structure). The method combines some aspects of the Gouy and the Rabi methods [196] of measuring susceptibility, although the latter method is not stated clearly by the authors. However, the mathematical treatment is quite clear. The equations of the Gouy method are employed inasmuch as a tiny cylindrical (sealed) sample tube moves in and out of a uniform field. The sample tube is suspended by a glass fiber from a Cartesian diver floating in a liquid that surrounds the entire assembly in a larger outer tube.

The principles of the Cartesian diver as a micromanometer for variations of hydrostatic pressure of a liquid is better than the Warburg apparatus by a factor of 1000. This advantage was used cleverly in essentially observing *apparent* displacement of the sample tube. Actually, the displacement of the sample is counterbalanced by *adjusting* the susceptibility of the surrounding float liquid. Rabi used this type of balancing to measure the (anisotropic) susceptibilities of crystals suspended by a fiber and surrounded by an inert solvent. The paramagnetic susceptibility of the surrounding float liquid is controlled by an external *magnetic titration* technique. The molar susceptibility of the sample is given by

$$\chi_{mol} = \chi_g \frac{\Delta C_{sf} VM(1/\sigma)}{m}$$

where V is the volume of the external tube containing the reference liquid, N is the molecular weight of the sample, and m is its amount, respectively; χ_g and C_{sf}

are the specific susceptibility and the concentration of the paramagnetic salt in the float liquid. Because the apparatus is used in a relative manner, changes in C_f (i.e., ΔC_f) are employed to achieve a *magnetic equivalence point* or a *null point*. This simply means that the concentration of the surrounding floating liquid is changed ΔC_f so that the sample tube will not move with the field on or off. (Thus, when the volume susceptibility of the sample equals that of the displaced liquid, there is no magnetic force on the sample.) The factor $(1/\sigma)$ enters as a constant of proportionality between ΔC_r and ΔC_f, where C_r is the concentration of the paramagnetic salt solution in the tube, and the paramagnetic salt and solvent are the same as in the outer float tube. The null point is observed by an optical microscope focused on the Cartesian diver. The apparatus must be free from mechanical vibrations; this is achieved by placing even the heaviest magnet in a sand bath mounted on oscillation elements that seem to imply heavy spring mounts. These precautions are necessary for measuring forces as small as 2×10^{-4} dyn in this floatation balance.

Reference should be made to reviews [36–46, 197–199] and references contained in [199] for other modifications of the Gouy balance.

7.10 Typical Assembly of, and Procedures for, the Faraday Balance

7.10.1 Special Poles for Nonuniform Fields

Since the Faraday method essentially measures the force on a sample suspended in a nonuniform magnetic field, the method can be used to measure the magnetic susceptibility per gram χ_g or the saturation magnetization per gram I_s of a sample; this latter quantity is particularly significant in ferromagnetic and ferrimagnetic materials (see Sections 4.8.3 and 4.8.4). The nonuniform field is produced by employing magnet pole caps of special shapes that are differently designed for the two distinct property measurements. The nomenclature concerning the pole cap shapes used in the measurement of susceptibility and magnetization appears to have caused confusion in the magnetics literature. This confusion can be avoided by simply designating the most useful pole cap shape for susceptibility χ_g measurements as the *constant force* design and that for magnetization I_s measurements as the *constant gradient* design. This distinction follows from the two basic equations stated earlier:

$$F = m \cdot \chi_g H \frac{dH}{dX}$$

If $I_s = \chi_g \cdot H$ represents saturation conditions, then

$$F = m \cdot I_s \cdot \frac{dH}{dX}$$

where F is the force acting on a sample of mass m placed in a nonuniform field H with a gradient dH/dX.

When the pole caps are properly shaped (as shown in Figure 3.25), a nonuniform field with a vertical axis of symmetry X perpendicular to the line y joining the poles is produced such that the product $H \cdot dH/dX$ remains constant over a short distance along X. These pole shapes (Figures 3.26 and 3.26a) (which are labeled the Varian and Heyding and co-workers [200] designs) are called the *constant force designs*, since the force acting on the sample remains constant over an appreciable region surrounding the sample, as shown in Section 7.10.4. Thus two samples of equal mass, suspended within this field region having a constant value of $H \cdot dH/dX$ experience forces proportional to their per gram susceptibilities χ_g; this behavior forms the basis of the relative method of measuring susceptibilities described in Section 7.10.3. Modifications of the constant force pole cap designs, too numerous to list here, were reported in the literature. A summary is given in a brochure published by Varian Associates. I have successfully used the Varian design (Figure 3.26a) with a Varian (spectromagnetic) 6-in.-pole-diameter electromagnet and the Heyding and co-workers [200] design with a Newport 4-in.-pole electromagnet for susceptibility measurements on various diamagnetic, paramagnetic, and antiferromagnetic samples.

For saturation magnetization measurements on ferromagnetic and (magnetically soft) ferrimagnetic materials where magnetic saturation occurs at small magnetic fields, that is, typically 500 Oe, special pole caps are used. These produce a constant gradient dH/dX over a specified distance along the x axis, and the *Sucksmith* design is shown in Figure 3.26. Because the per gram magnetization σ_s measurements are performed in a magnetic field H_s greater than that required for saturation, the product $\chi \cdot H$ equals the saturation magnetization σ_s for values of H greater than H_s. Thus the force F depends on m and dH/dX only, and pole caps with such a design are known as *constant gradient pole caps*. (For measurements on superparamagnetic materials, which are difficult to saturate, high fields in the range 10–20 kOe and low temperatures down to 4 K are desirable.) The Sucksmith design, shown in Figure 3.23, usually has a 0.5-in. region (for 12-in.-diameter pole caps with about a 2-in. pole gap), which may be called the *magic spot*, where even a strongly ferromagnetic sample remains fixed on the plane midway between the magnet poles and is not attracted to either pole. This magic spot behavior is displayed because the region of maximum flux density is located on this plane, which forces a freely suspended sample to remain fixed on this plane. Field mapping procedures similar to those described in Section 7.10.4 allow the magic spot to be located while avoiding the otherwise exasperating problems associated with measurements on strongly ferromagnetic samples.

The *constant gradient* pole caps are used for susceptibility measurements by the relative methods (see Section 7.10.3) if the sample and the reference compound are placed exactly at the same position such that both the field H and the gradient dH/dX acting over their (identical) volumes are also constant. Since these conditions, and especially the positioning of the two compounds at the same spot, are difficult to achieve with a spring balance (which does not involve the *null* principle), the *constant force* pole cap technique is far superior to the *constant gradient* method for susceptibility measurements.

We have successfully used the Sucksmith design [201] and another modification known as the Fereday (not Faraday) design (cf. [25]) in conjunction with a Newport 4-in.-diameter pole electromagnet for measurement of saturation magnetization on various ferromagnetic and ferrimagnetic samples. Designs for other pole shapes are available from Varian Associates.

7.10.2   Coils for Nonuniform Fields

Lewis [202] (see also Cape and Young [203]) made a very significant contribution toward the instrumentation of a Faraday balance by designing coils to provide the dH/dX gradient, which is needed to produce a magnetic force on the sample (Figure 3.26*b*). Traditionally, the gradient is provided by pole tips of different shapes such as those designed by Fereday, Sucksmith, Heyding, and Henry [204]. These have the disadvantage that, in varying the field H, dH/dX is also automatically varied so that the *constant force* region continuously varies with the field H while maintaining a constant pole gap. [Generally the pole gap must be held constant to a *minimum dimension* to accommodate low-temperature cryostats (Dewar flasks) and high-temperature furnaces.] Thus with all designs of pole tips, the region of constant force is large at low values of H and small at high values of H.

The Lewis coils, which are mounted on the flat poles of an electromagnet and energized by an auxiliary dc power supply, can provide different values of the gradient dH/dX, which can be superimposed on the steady field H of the electromagnet. The authors describe the constructional details of the coils, which can handle currents up to 12 A. These provide a constant-force region ($H \cdot dH/dX$) of a few centimeters between the coils with a variation in the gradient of less than $\pm 0.1\%$. This is more than adequate to handle large samples (up to ca. 400 mg) in sample capsules about 0.5 cm in diameter and about 1 cm tall. The authors describe various ways to avoid, for example, the effects of fields (generated by the electromagnet and the coils) on the electrobalance and vibrations of the electrobalance. The concept of using such coils was first proposed by Sucksmith ([201] and later papers).

According to Cape and Young [203], a field gradient dH/dX may also be obtained by using a pair of coils. They use Helmholtz coils (about 7.5-cm radius), canted at an angle of 29° and separated by a distance of 7.5 cm to obtain a *linear field region* (that is, a region of constant force), which is about 4 cm from the axial center of the coils. These coils are placed between the poles of a standard electromagnet (with relatively large poles, at least 15 cm in diameter). As in the Lewis design [202], the coils provide a means of varying the gradient dH/dX independently of the field H produced by the electromagnet. Cape and Young [203] do not give constructional details for the coils or the associated dc power supply. This situation is disenchanting for those wanting to adopt their design.

They do report measurements of magnetization ($\chi \cdot H$) for a gallium-substituted yttrium iron garnet film. The authors designed the coils to obtain especially good control of the force $H \cdot dH/dX$ at very low values of the field H. They contend that because of remanent magnetization in the pole tips the use of special pole tips to obtain field gradients at low values distorts the gradients $H \cdot dH/dX$.

In our opinion, the Lewis design is superior to that of Cape and Young [203] in many ways. It is more versatile, in the sense that it can be used from the very low to the very high fields produced by the electromagnet. In this case the poles can be kept close together to provide the necessary minimum gap for accommodating Dewar flasks, cryostats, and so on. The Cape and Young [203] configuration based on the principle of Helmholtz coils demands that the distance between the coils remain (at all times) equal to the radius of the coils. This would make it impossible to obtain high fields especially at such a large pole separation as 7.5 cm. Cape and Young suggest mounting the coils over the poles of the electromagnet and using superconducting coils to obtain high fields; this approach could be very cumbersome.

Another disadvantage (pointed out by Cape and Young) is that the sample must be located accurately at the same spot relative to the coils. This is not essential in the Lewis design, which provides large (ca. 3 cm) vertical regions of uniform force $H \cdot dH/dX$.

A uniform field in a superconducting magnet may also be produced by winding the superconducting coil on a cone of desired dimensions. One such approach is described by Kikuchi and co-workers [205].

7.10.3   The Relative Method for Measuring Susceptibilities

The magnetic force F_s acting on an isotropic sample of mass m_s and mass susceptibility χ_s in a field H with a gradient dH/dX is given by

$$F_s = H \frac{dH}{dX} m_s \cdot \chi_s = \text{constant} \cdot m_s \cdot \chi_s$$

provided $H \cdot dH/dX$ is constant over the volume of the sample. The force acting on a reference compound placed at the same position as the sample and under identical conditions is given by a similar equation in which the subscript r is used for a reference material with known mass susceptibility χ_r. By combining these equations, the susceptibility of the sample is obtained from

$$\chi_s = \frac{F_s \chi_r m_r}{F_r m_s}$$

Since the forces are measured in terms of the corresponding changes in weight ΔW, χ_s is given by

$$\chi_s = \frac{\Delta W_s \chi_r m_r}{\Delta W_r m_s}$$

Thus in the relative method actual measurements of the field H and gradient dH/dX are avoided. Although the measurement of the field is rather easy, the measurement of the gradient dH/dX is difficult.

For the relative method to be successful, it is essential that the sample and reference be placed at exactly the same position in the field so that according to the theory, $H \cdot dH/dX$ is constant over the volume of the specimen. This is accomplished partly by using very small samples; usually a few milligrams are placed in a small cylindrical container made of quartz (ca. 5 mm in diameter and 6 mm tall). The product $H \cdot dH/dX$ is made constant over a workable region by using the special poles described previously.

Since it is difficult to position the sample and the reference at the same place, especially with the helical-spring and the torsion techniques, the provision of a large region of constant $H \cdot dH/dX$ in the pole gap gives the convenience of placing the specimen within such a region rather than at a particular *spot*. This practical aspect is vital for the helical-spring method; with the *null* method the specimen always remains at the same position. A description of these methods follows.

In addition to the constant-force poles, a number of descriptions of constant-gradient poles are found in the literature. There is some confusion concerning their nomenclature. A distinction between the two types is made only if the magnetic parameters being measured (e.g., the specific susceptibility χ or the specific magnetization defined by $\sigma = \chi \cdot H$) and the procedures for their measurement are clearly stated; these aspects were discussed previously.

Several constant-force and constant-gradient poles are available from Varian Associates, who kindly supplied us their brochure No. 29-J-002-76, which provided additional information on the two types of poles. It will suffice to point out that the region of constant force is determined by the particular value of the field H and the pole gap. For a fixed pole gap this region generally decreases with increasing field. Thus a set of poles gives best overall performance for specific settings of the pole gap and the field. In this context reference should be made to papers by Hill [206] regarding various field profiles and the errors in susceptibility measurements caused by improper positioning of the specimen; an examination of these aspects follows.

In the final calculations of the specific susceptibility, allowance is made for the force acting on the sample holder. Thus, if ΔW_{B+S} is the total change in weight for the *sample plus the bucket*, and ΔW_B is the change in weight for the *bucket only*, the susceptibility χ_s is given by

$$\chi_s = \left(\frac{\chi_r m_r}{\Delta W_{B+r} - \Delta W_B} \right) \frac{\Delta W_{B+S} - \Delta W_B}{m_s} = \frac{K'(\Delta W_{B+S} - \Delta W_B)}{m_s}$$

The instrumentation constant K' is determined by using a number of calibrating or reference compounds, which are listed in Section 7.9.5. Usually the specimen holders are made of quartz or other suitable materials, which are stable at high temperatures. These materials are diamagnetic; thus as in the

Gouy method, the correction ΔW_B is actually added to the ΔW_{B+S} and to the ΔW_{B+r} terms.

7.10.4   Mapping of the Field Profile

The mapping of the field along the vertical direction X, to find a region of constant force $H \cdot dH/dX$ is accomplished by moving the specimen a few millimeters at a time, turning the field on and off, and actually measuring the force (i.e., the change in weight) that is directly proportional to $H \cdot dH/dX$.

With the helical-spring or the deflection balance described in Section 7.10.7, a turning of the top Vernier screw facilitates the raising (or lowering) of the specimen in the field. When an electrobalance is employed with the null method, this is accomplished by mounting the entire balance on a support that can be moved up and down using a Vernier screw. A typical curve for the deflection or the force acting on a specimen (in this case, the empty quartz bucket) as a function of the vertical distance X with respect to some arbitrary point is illustrated in Figure 3.57. Once the flat region is established, the position of the specimen is so adjusted that it remains approximately in the center of the region. Figure 3.46 shows the actual position of the sample between the Heyding poles. Actual measurement of the force, with a small specimen at its different positions, should be made. This gives a far more realistic profile of the field than that obtained from gaussmeters.

Quite often the manufacturers of electromagnets and magnet pole caps provide curves of H^2 as a function of the vertical distance X measured from some arbitrary point. In this case the straight-line regions (i.e., those with a constant slope) are said to correspond to the flat region of constant force. The rationale behind this approach can be seen by rearranging the basic Faraday force equation to give: $H \cdot dH = (F/m\chi)dX$, which on integration gives $H^2 = (2F/m\chi_g)X + 2C$ or $H^2 = (\text{slope})X + C$, where C is a constant of integration. Thus if H^2 is plotted as a function of the vertical distance X for a sample of fixed mass m and mass susceptibility χ_g, a linear region of the resulting curve with a constant slope corresponding to a region of constant force F is obtained.

Several manufacturers find it convenient to supply curves of H^2 versus the distance X simply because it is easy to measure the field H with a small gaussmeter probe placed at different points along the vertical axis X; the gaussmeter probe may consist of a small rotating coil or a Hall effect (semiconductor) device, which produces an electromotive force directly proportional to the field in which it is placed. As mentioned previously, although it is easier to measure the field H at a spot (i.e., over the volume occupied by the probe) than to measure the difference dH between two points, a measurement of the force corresponding to $H \cdot dH/dX$ should be undertaken to map the field profile.

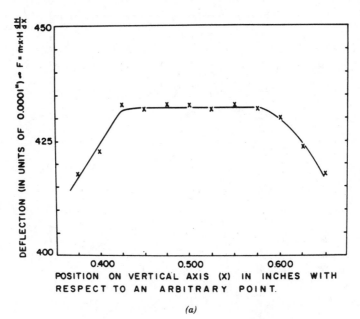

(a)

POSITION ON VERTICAL AXIS (X) IN INCHES WITH
RESPECT TO AN ARBITRARY POINT.

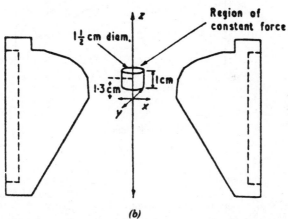

(b)

Figure 3.57 (*a*) Plot of the magnetic force acting on the sample versus the vertical position of the sample with respect to some arbitrary position. The force (measured as a deflection) is proportional to $H \cdot dH/dX$. The flat part represents a *constant force*, between Heyding type poles for a fixed field and pole gap (Martin and Hill [208]). (*b*) Axes of measurement of magnetic force(s) on a sample bucket suspended in a region of constant force [208]. (*c*) Magnetic force (proportional to $H \cdot dH/dX$) acting on a sample as a function of the vertical position (magnet displacement) of the sample. Note that the field profile becomes more peaked at higher fields ($H_{max} = 6.3 \, \text{K} \cdot \text{G}$), that is, at higher magnetizing currents. The field profiles correspond to a 5-cm pole gap (Martin and Hill [208]). Reprinted with permission from L. N. Mulay, "Techniques for Measuring Magnetic Susceptibility," in A. W. Weissberger and B. W. Rossiter, Eds., *Physical Methods of Chemistry, Techniques of Chemistry*, Vol. 1, Part 4, Wiley-Interscience, New York, 1972. Copyright © 1972 by John Wiley & Sons, Inc.

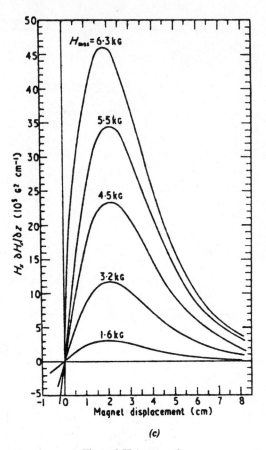

(c)

Figure 3.57 *(continued)*

7.10.5   Experimental Procedure and Calibration

The experimental procedure is the same as in the Gouy method. This involves measurement of (1) the change in weight of the empty quartz bucket, (2) the bucket plus reference, and (3) the bucket plus sample. These measurements are carried out under conditions explained in the preceding sections. The specific susceptibility of the sample is then computed by using

$$\chi_s = \frac{\Delta W_s \chi_r m_r}{\Delta W_r m_s}$$

We used MTC {mercury tetrathiocyanatocobaltate(II), $Hg[Co(SCN)_4]$} extensively as an all-around calibrant for solids. This compound has a *specific susceptibility* of 16.44×10^{-6} ($\pm 0.5\%$) emu g^{-1} at 20°C. Some workers compress calibrating materials into a small pellet to avoid preferential orientation of the microcrystals in the powdered material when the field is turned on. This procedure should be followed especially with $CuSO_4 \cdot 5H_2O$. The mass or per gram susceptibility of MTC is given by

$$\chi = \frac{4.98 \times 10^{-3}}{T + 10}$$

where T is the absolute temperature. This relation is helpful in checking the overall performance of the balance. One should avoid the use of a steel or even nonmagnetic steel and piston for making pellets because these tend to introduce ferromagnetic impurities.

Before making measurements on a sample of unknown susceptibility, it is important to check the overall performance of the balance by measuring the susceptibility of several standards against that of MTC. We found several acetyl acetonates of transition metal ions to be satisfactory for this purpose. Susceptibilities obtained at various temperatures for the Fe(III) and Mn(IV) salts were reported by Jackson [207]. Reference should be made to comments on calibration under Section 7.9.5.

7.10.6   Error Analysis

The following references should be helpful in achieving the highest possible degree of accuracy in susceptibility measurements.

Hill [206] discusses the influence of positioning errors on susceptibility measurements using the Heyding, Taylor, and Hair [200] poles described previously. This careful analysis of errors seldom described by other workers is recommended highly to all experimentalists using this technique. In particular, the influence of the field profile on susceptibility measurements is discussed, taking into account changes in sample size and position. Hill [206] shows that if suitable pole tips are used in conjunction with a movable magnet, measurements of susceptibility can be reproduced to 0.2%. For specimens limited in size to 2 mm, no corrections for dimensions are required to maintain this accuracy. Hill gives excellent graphs and tables for variation of magnetic force as a function of the length of the specimen, the vertical displacement of the sample, and so on.

Calibration procedures for determining regions of constant force at various fields and for different pole gaps for 4-in. Heyding-type pole tips are given by Martin and Hill [208]. Using these pole tips and the method of Senftle and Thorpe, reviewed later, researchers have made absolute susceptibility measurements on standard platinum samples and used them to obtain the field profiles. Martin and Hill discuss valuable methods for measuring the region of constant force $(H \cdot dH/dX)$ and the experimental procedures for obtaining the field profiles (Figures 3.57a and b). Analysis of the data by the Honda–Owen method for obtaining absolute (corrected) susceptibilities is also presented.

Stewart [209] gives a careful analysis of errors sometimes caused if the specimen holder is attracted toward one of the poles and swings out of the vertical axis in a Faraday magnetic balance. Stewart considers various factors such as the gradient forces, image forces arising from the magnetic image of the sample in the pole, and the overall behavior of the Faraday system. He gives equations for predicting the instabilities that occur when the downward magnetic force exceeds a critical value, which is (surprisingly) the same for paramagnetic and ferromagnetic materials, and is independent of the electromagnet current. Lewis [202] suggests that these instabilities can be avoided by using his coils.

Gerritsen and Damon [210] discuss essentially the problems arising in the measurement of small magnetic forces at low temperatures (below 4 K). They used a microbalance that is fundamentally the same as those described in the literature and found that thermolecular flow in the exchange gas may affect considerably the accuracy of the measurement of magnetic forces less than 1 dyn. They suggest that the character of the gas flow limits the degree of equilization of the temperature of the sample and bath. Furthermore, by choosing the dimensions of the suspension and the exchange gas pressure and by shielding the sample adequately from thermal radiation, the interference is reduced to a tolerable amount.

In [195] constructional details for the microbalance and the cryostat are provided; and the effect of exchange gas on (1) the zero reading of the microbalance, (2) the Faraday force on a sample, and (3) the difference between the temperature of the bath and the sample are discussed in great detail. The authors' refinements [210] should be useful when it is necessary to obtain accurate susceptibility measurements in the very low temperature range (ca. 1.28–4.2 K).

7.10.7   Other Assemblies of the Faraday Balance

ASSEMBLY USING A HELICAL SPRING BALANCE

A McBain-type spring balance, familiar to workers studying adsorption, thermogravimetric analysis, and so on, is used in the construction of one type Faraday balance. Generally in the handling of such adsorption balances or similar Faraday-type magnetic balances using quartz helical springs, considerable difficulty is experienced in placing the sample at a given position, where any changes in its weight can be conveniently observed within the span of a measuring microscope. The conventional bellows usually allows positioning over a range as limited as 1 cm. Samples varying widely in weight naturally require a larger range of adjustment both initially and during the course of a measurement. Some unique constructional and operational details of a microbalance were introduced and are described here for studies on magnetic susceptibility, adsorption, and thermogravimetry, and for measuring small changes in susceptibility occurring during these processes [164].

Figure 3.58 is a schematic of the balance in our laboratory. It consists of a helical quartz spring suspended from a bellows assembly, a main sample-adsorption (or thermal-decomposition) chamber, and an inlet gas system.

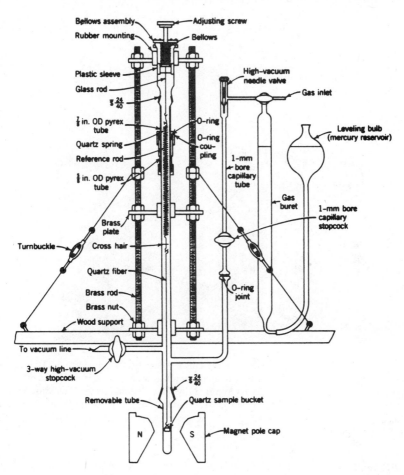

Figure 3.58 Schematic of a quartz helical-spring Faraday balance in our laboratory used for studies on adsorption and thermogravimetric and simultaneous thermomagnetic analysis. This device uses a very small dead-space for adsorption, and so on, studies and is inexpensive. Reprinted with permission from L. N. Mulay, "Techniques for Measuring Magnetic Susceptibility," in A. W. Weissberger and B. W. Rossiter, Eds., *Physical Methods of Chemistry, Techniques of Chemistry*, Vol. 1, Part 4, Wiley-Interscience, New York, 1972. Copyright © 1972 by John Wiley & Sons, Inc.

Figure 3.59 is a photograph of this assembly using a 4-in. electromagnet (cf. Figure 3.46). The cathetometer (measuring microscope) and the vacuum system are excluded from the photograph to bring out the essential details.

A special bellows assembly for sample positioning was constructed from bellows (approximately 2-cm o.d., 7.6 cm long, with a total extension of 5 cm, obtained from Servometer Corporation, NJ); a short section of borosilicate glass pipe (about 2.5-cm i.d., 20 cm long) using a retainer ring, brass bonnet, and a seat (obtained from Kopp Scientific Company, New York, NY); and a vertical screw with a very fine pitch (0.625-cm diameter, 15 cm long, 16 threads per cm). Any

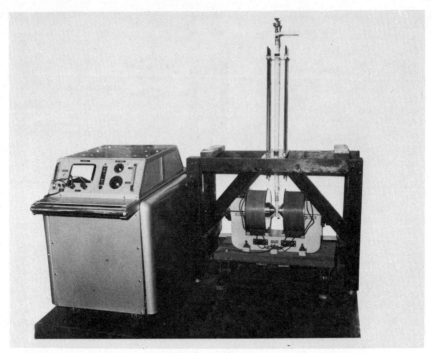

Figure 3.59   The helical spring Faraday balance shown schematically in Figure 3.58. To emphasize the essential features of the apparatus the cathetometer (measuring microscope) and vacuum system are not shown. Reprinted with permission from L. N. Mulay, "Techniques for Measuring Magnetic Susceptibility," in A. W. Weissberger and B. W. Rossiter, Eds., *Physical Methods of Chemistry*, *Techniques of Chemistry*, Vol. 1, Part 4, Wiley-Interscience, New York, 1972. Copyright © 1972 by John Wiley & Sons, Inc.

gyrational motion of the brass cylinder at the base of the bellows assembly during the travel of the vertical screw was avoided by allowing the cylinder to slide through a sleeve machined from a plastic rod to give a snug fitting arrangement similar to a piston and chamber system. The sleeve was lubricated with a small amount of stopcock grease, which provided smooth motion of the bellows. At the base of the bellows, a glass rod (2-mm diameter) with a hook was attached, and a quartz spring and a bucket were suspended from it.

A provision for a coarse positioning of the sample was also introduced by a telescoping arrangement of two concentric glass tubes using two Veeco type C O-ring couplings (about 1.6- and 2.0-cm i.d., Vacuum Electronics Corporation, Plainview, NY), which were soldered together. The entire assembly was vacuum tight to pressures of 10^{-7} torr. When the sample was removed, the spring pulled the fiber up into the chamber, thus making its retrieval very difficult. With this special telescoping arrangement, it was possible to loosen the coupling, lower the fiber to the position at which the sample could be replaced, raise the unit to the desired position, and tighten the coupling. The sample was then accurately positioned by adjusting the vertical screw on the bellows.

The sample was aligned and positioned in the correct region of the magnetic field by observing the position of the cross hair at the bottom of the spring through a microscope (Gaertner Scientific Company, Chicago, IL, $32 \times$ magnification, No. 72905). This microscope was attached to a vertical mechanical stage that could be adjusted over a range of 5 cm with an accuracy of 0.002 mm, using a micrometer drum for measuring deflections of the spring. The sample could be centered in the pole gap in the desired region of the magnetic field by adjusting the four turnbuckles that controlled tilting of the glass chamber and spring assembly. Commercially available cathetometers can be used also for the same purpose. In subsequent work the authors used a Gaertner cathetometer capable of reading displacements to within 0.0001 in.

The quartz bucket was cylindrical in shape, about 0.5 cm in diameter, 0.5 cm tall, and 0.2 g in weight. This was adequate to hold a few milligrams of the sample. The quartz buckets and springs of varying sensitivity were obtained from Ruska Corporation, Worden Quartz Products Division (Houston, TX). A typical spring used for paramagnetic measurements had a sensitivity of 1 mm/2 mg, a maximum load capacity of 1200 mg, an extension of 600 mm, and a helix diameter of 8 mm. A reference rod was provided inside the spring to correct for any changes caused by temperature variations in the length of the spring.

To obtain the best field characteristics, special pole caps were made for the magnet according to the specifications of Heyding, Taylor, and Hair [200]. Under optimum conditions, these pole caps provided almost a 1-cm region of constant force in a pole gap of about 4 cm, with less than 1% variation. This region is found experimentally by placing any sample (the empty quartz bucket may be used as a diamagnetic sample) at different positions about 0.5 mm apart along the vertical axis in the field with respect to any arbitrary starting point and making a plot of deflection observed on application of the field versus the position of the sample (Figure 3.57). A less favorable profile of the field (i.e., a narrow range over which $H \cdot dH/dX$ is constant) is obtained at higher fields. Care must be taken during measurement so that the specimen does not wander from this region.

This balance was specially designed to study the (physical) adsorption of oxygen gas on gamma alumina and to establish the dimerization of oxygen to the O_4 species. The adsorption studies need a relatively small dead space in the balance (ca. 300 cm^3) that is not obtainable in commercial electrobalances, such as the Cahn balance. Mulay and Keys [164] described several applications of the spring balance [165]; these include studies on the thermal decomposition of $CuSO_4 \cdot 5H_2O$.

ASSEMBLY USING A NULL BALANCE

Figure 3.60 is a schematic of a Faraday apparatus. A *null* electrobalance (Cahn RG model now known as R1000) is used. This assembly is rather expensive; however, it can be evacuated to 10^{-8} to 10^{-9} torr and has facilitated studies such as in situ chemisorption.

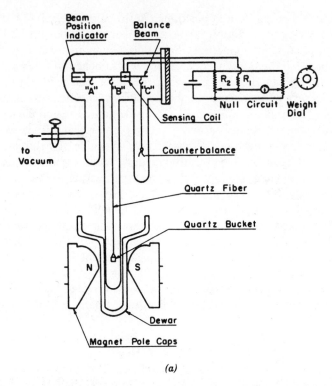

(a)

Figure 3.60a A schematic of a null-type Cahn RG electromicrobalance for in situ magnetic measurements (see Figure 3.47). The Cahn RG (or Cahn 2000 balance) can be replaced by the Cahn-1000 high-load capacity (100 g) sensitivity (few micrograms) in conjunction with a large 12-in. pole-piece electromagnet. Reprinted with permission from L. N. Mulay, "Techniques for Measuring Magnetic Susceptibility," in A. W. Weissberger and B. W. Rossiter, Eds., *Physical Methods of Chemistry, Techniques of Chemistry*, Vol. 1, Part 4, Wiley-Interscience, New York, 1972. Copyright © 1972 by John Wiley & Sons, Inc.

The Cahn balance was mounted on a vertical screw assembly from an old cathetometer (measuring microscope made by Scientific Instruments Company, Cambridge, England). This was in turn mounted on a cross-slide milling table (Sears Roebuck Company, Philadelphia, PA) used for controlling the lateral motion of the two slides at right angles in milling machines, lathes, and so on. Thus the balance had three degrees of freedom, which allowed the sample to be positioned in the most favorable part of the field. The vertical screw assembly may be replaced easily by another single section of the cross slide.

The field profile was mapped in the vertical direction x, as described previously, by controlling the vertical Vernier. Once the midpoint in the region of constant force was established, no further adjustments were necessary. The balance could be evacuated and filled with an inert gas such as helium during the measurements. Very small samples (ca. 10 mg) can be used, and changes in

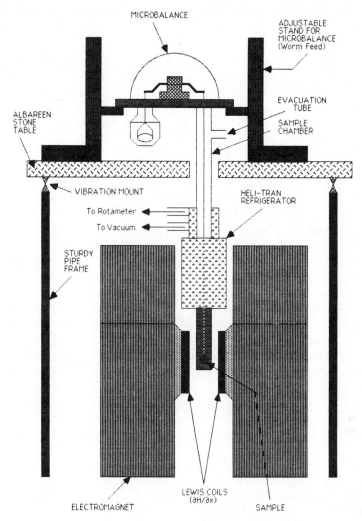

Figure 3.60*b*   A recent (1988) modification of the Faraday balance using the latest instrumentation.

weight with an accuracy of 1 μg can be measured. The balance, which has a load capacity of 2 g, needs frequent calibration—quite often between measurements—to attain a high degree of accuracy.

Griest and Ostertag [211] point out the difficulties encountered in the calibration of a Cahn balance and suggest a simple method of surmounting them. The calibration is generally carried out on the sample loop A, which frequently necessitates breaking the vacuum, separating the hangdown tube, and so on, to insert the calibration weight. The authors describe in detail procedures in which only the readily accessible counterweight (loop C) is used.

The Cahn balance setup was used successfully for studies on superparamagnetism of iron dispersions and for studies on the titanium–oxygen system by Mulay and co-workers [31, 212–217] and for recent studies on Fe–carbon catalysts [198, 218]. The most recent modification, shown in Figure 3.60a, was used for measurements on semiconductors and superconductors.

7.10.8   Modifications of the Faraday Balance

Critical reviews of several modifications of the Faraday balance appear in [2, 25, 27, 219]. Most modifications are novel because of force-measuring techniques used and the pole shapes designed to produce nonuniform fields. In the original Faraday method, a torsion head served to measure the force on a sample placed in a nonuniform field. Here the sample, suspended from a torsion arm, is free to move horizontally. The torsion head is twisted to return the sample to the original place, and the twist is a measure of the force required to just balance the magnetic force at the zero position. Important modifications of this torsion method are named after Curie, Curie–Cheneveau and Wilson, and Oxley. In the Curie–Cheneveau method, a small permanent magnet is moved toward and away from the sample to produce the gradient. An elaborate modification of special significance is described by Cini [220] and by other workers. Some of the microtechniques are illustrated in Figures 3.61 and 3.62 (see [186]).

Sucksmith used a sensitive method in which the sample is suspended from a phosphor bronze ring equipped with an optical system. The displacement of the

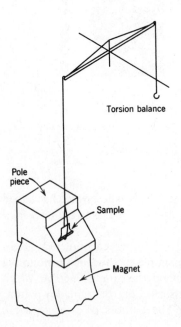

Torsion balance

Pole piece

Sample

Magnet

Figure 3.61 A sensitive magnetic torsion balance for micromeasurements. The sample is placed in a nonuniform field. Reprinted with permission from L. N. Mulay, "Techniques for Measuring Magnetic Susceptibility," in A. W. Weissberger and B. W. Rossiter, Eds., *Physical Methods of Chemistry, Techniques of Chemistry*, Vol. 1, Part 4, Wiley-Interscience, New York, 1972. Copyright © 1972 by John Wiley & Sons, Inc.

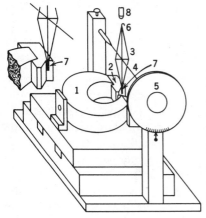

Figure 3.62 A micromagnetic susceptibility apparatus for solids; (1) 5-in. Alnico No. 5 magnet; (2) specially cut pole piece to give uniform dH/dX; (3) quartz frame; (4) torsion fiber; (5) torsion wheel; (6) pointer; (7) sample tube; and (8) microscope. Reprinted with permission from L. N. Mulay, "Techniques for Measuring Magnetic Susceptibility," in A. W. Weissberger and B. W. Rossiter, Eds., *Physical Methods of Chemistry, Techniques of Chemistry*, Vol. 1, Part 4, Wiley-Interscience, New York, 1972. Copyright © 1972 by John Wiley & Sons, Inc.

sample is magnified several hundred times with this system. Chandrasekhar [191] introduced special damping devices in the Sucksmith balance. Other workers replaced the optical system with a flat spiral spring that is linked to a displacement transducer. The voltage produced is recorded or read on an oscillograph. Improvements for handling variable loads (1–80 mg) were incorporated in the Sucksmith balance. Improvements in the Sucksmith balance are described by Pepper and Smith [221]. The optical magnification was doubled by replacing the usual fixed lens by a concave mirror, which with the two plane mirrors, forms a double-pass system. Further magnification may be provided by using four mirrors, instead of the usual two, mounted on the circular spring. The authors [221] also describe a cryostat of the heat leak (type) for use from 85 to 650 K. Four mirrors for the Sucksmith balance were used in my laboratory in early 1964; however, the arrangement was cumbersome.

Milligan and Whitehurst [222] and Jacobsen and Selwood [223] described magnetic balances in which a quartz helical spring is used for measuring small changes in force. In our opinions, as described previously, this is by far the simplest and most sensitive force-measuring device that can be adapted readily for the Faraday technique.

Most of the modifications mentioned thus far use permanent or electromagnets with inclined poles or poles of special design [200] to produce a nonuniform field. Larger poles of different shapes are described by Garber and co-workers [224].

Other sensitive methods were developed by Smith [225], Cini [220], and Pacault [80] and co-workers. In another modification [226] of the Curie–Cheneveau method, a small permanent magnet with straight pole faces is used. The required gradient is obtained by moving the permanent magnet vertically toward or away from the sample. Senftle and co-workers [226] use a quartz helix spring and a measuring microscope for determining the force on the sample. The magnetic susceptibility of a sample relative to that of a standard is simply calculated from

$$\chi = \frac{(d_1 - d_2)(m_2 - m_3)}{(m_1 - m_3)(d_2 - d_3)} \chi_s$$

where d is the deflection (i.e., one-half the difference of maximum-to-minimum deflection observed during the movement of the magnet), m is the mass in grams, χ_s is the susceptibility of the standard (in emu g^{-1}), and the subscripts 1, 2, and 3 refer to the measurements of the sample plus pan, the standard plus pan, and the pan alone, respectively. The chief advantages of the method are (1) submilligram amounts of the sample are required for paramagnetic substances having a susceptibility from 1 to 50 $\times$ 10^{-6} emu g^{-1}, (2) small samples—about 10 mg—are required for weakly paramagnetic and diamagnetic substances to obtain a precision of better than 2%, and (3) only an inexpensive small permanent magnet and a simple experimental setup are needed.

A continuation of work with this equipment by Senftle and Thorpe [227] led to a method [228] of measuring magnetic susceptibility that, according to their claims, is an absolute method that does not require a standard substance. The susceptibility χ is shown to be a function of the area a under the curve of sample displacement versus the distance of the magnet from the sample and of the maximum applied field H_{max}

$$\chi = \frac{2g'a}{h} H_{max}^2$$

where g' is the acceleration caused by gravity and h is the measured static deflection of the helix. In the experimental procedure, the area a of the curve (which resembles the first differential of a Gaussian curve) is measured with a planimeter and H_{max} is measured with a gaussmeter. Candela and Mundy [229] verified the usefulness of this method. Although it is claimed to be *absolute* and to require no standards, a precise measurement of H is needed; whenever this is a requirement, as with the Gouy technique, use of a standard is unnecessary. Thus the volume susceptibility κ (and hence the mass susceptibility $\chi_m = \kappa/$density) can be calculated simply from the basic equation $g'\Delta W = \frac{1}{2}\kappa AH^2$, where the symbols are as stated previously. In the magnetic susceptibility methods a standard with known susceptibility is employed to avoid measuring the magnetic field H and other parameters, such as the area A of the cross section of the sample. This is possible because these factors cancel out when considering the weight changes ratio of the magnetic field of the sample to reference W_s/W_r, which equals the ratio of the susceptibility of the sample to standard χ_s/χ_r.

Candela and Mundy [229] made a comparative study of susceptibilities of powdered samples down to liquid helium temperatures by the Thorpe and Senftle modification and by the Gouy method. The former method requires a small sample, about 3 $\times$ 10^{-2} g smaller than for the Gouy method.

A modification of the Senftle balance is described by Baidakov and co-workers [230]. Its major advantage is that it minimizes the effect of oxygen adsorbed on the solid sample by incorporating a gas-purification system. Measurements can be made on a few milligrams of sample from -100 to $150°C$ either under vacuum up to 10^{-8} torr, or in helium gas up to 1 atm. The apparatus was used for magnetic-susceptibility studies on lead sulfide.

Garber and co-workers [224] provide an elaborate description of a balance designed to give a reproducibility of better than 0.15%; this was particularly used for measurements of the magnetic susceptibility of short cylindrical samples of copper, silver, and gold. Henry and Rogers [231] describe a similar apparatus. A null technique especially useful for paramagnetic materials was developed by Singer [232]. It can be used over a wide range of magnetic field strengths without readjustments, and it automatically detects the presence of ferromagnetic impurities. The balance is semiautomatic, uses small single crystals, and subtracts the contributions from (ferromagnetic) impurities and antiferromagnetic components. Another torque method, which employs a light-beam–mirror–phototube network, originally designed by Penoyer [233] for ferromagnetic anisotropy measurements on crystals is expected to be useful for susceptibility measurements.

An electromagnetic servobalance with a differential transformer was designed to handle large forces (5 g) for ferromagnetic materials [234]. It also readily measures forces as small as 0.02 mg. Butera and co-workers [234] list several references to servobalances.

Special adaptations are described for measurements at very high temperatures [55, 235]. A Faraday magnetic balance for measurement of susceptibilities of highly reactive materials is reported by Kirchmayr and Schindl [236]. The instrument is useful for ferromagnetic or strongly paramagnetic materials in the range 80-1600 K, and an accuracy of $\pm 1\%$ is claimed. The apparatus features temperature regulation for the study of phase changes by magnetic methods.

Mukherjee and Sutradhar [237] describe a modification of a Foex–Forrer-type horizontal balance designed for susceptibility measurements at high temperatures. Obuszko [238] employs an optical method for measuring the magnetic force on a sample in an inhomogeneous field. Small displacements corresponding to from 20 to 400 mg are measured with a linear amplification of 200; the sensitivity is 6.3 dyn mm^{-1}, which can be enhanced. The apparatus is particularly suitable for paramagnetic and ferromagnetic susceptibility measurements.

An apparatus designed by Richardson and Beauxis [239] represents by far the most elaborate and highly automated design for measurement and recording of sample weight, wherein changes in weight correspond to susceptibility measurements, which are obtained by using the Faraday technique during programmed temperature regulation from -196 to $500°C$. It has a commercial vacuum torsion balance (Sartorius Electronic II vacuum microtorsion balance), which gives the advantage of handling a maximum load of 1 g, and it is capable of detecting weight changes in the range 0–20 mg with a sensitivity of 1 μg. As a null instrument, it continuously maintains the sample in the same position by a restoring force; this stipulation is highly desirable for susceptibility measurements, which are reliable only if the sample is maintained in the same region and preferably in the same position in a uniform field gradient. Control circuitry and a power supply are provided to activate the apparatus in a sequence appropriate to the measurement of changes in weight and magnetic susceptibility. Of special significance is the programmed temperature control: from $-196°C$ to room temperature it uses liquid nitrogen and from room temperature to $300°C$ it uses a furnace.

As many as 180 data points (change in weight or magnetic susceptibility versus temperature) were obtained continuously in about 12 h over the entire temperature range. The results were checked by manual observation at certain fixed points. The precision of the *continuous* and *fixed* methods of measuring susceptibility were found to be ± 1 and $\pm 0.5\%$ depending on the stability (0.3%) of the magnetic field. Comparisons are made of magnetic moment and the Weiss constant obtained by the two procedures, and the agreement between the two is quite satisfactory and as good as can be expected in any susceptibility measurements. The technique was employed to study the slow reduction by hydrogen of metal oxides, and a possibility for studying adsorption of certain gases not harmful to the balance components is discussed.

Another automatic vacuum microbalance of the pivot type was built and tested by van Liehr [240]. It can be adapted easily for susceptibility measurements of especially diamagnetic materials and adsorption experiments involving low-surface areas, thermogravimetry, and so on. Its unique features include the handling of high loads up to 2 g, stability over long periods of time, and infrequent calibration. It has a sensitivity of $88\ \mu V\ \mu g^{-1}$ in the most sensitive region. Zero shifts were reduced by a novel frame construction. The main limitation is said to be the precision of the recorder. The balance is automated by an optical beam, a solid-state null indicator, and associated circuitry, which are described in detail. The device features magnetic damping; the oscillation period of the balance is about 11 s.

A Faraday balance that uses a small superconducting magnet is described by Kikuchi and co-workers [205]. The superconducting wire is wound on a cone to produce a large gradient $H \cdot dH/dX = 5.8 \times 10^7\ Oe^2\ cm^{-1}$ with a maximum field of 18 kOe. A Sartorius semimicrobalance with a sensitivity of about 1 g is used. This setup can be used only over a limited temperature range from 1.4 to 4 K, which is obtained by pumping on liquid helium. Although the field obtained with this apparatus is about the same as that obtained by most electromagnets, it illustrates the use of superconducting magnets for susceptibility measurements. The apparatus was used to study the adsorption of copper acetylacetonate and diphenylpicylhydrazyl on silica gel.

Other modifications to this system that also use a microbalance are described by Kolenkow and Zitzewitz [241] and by Soule and co-workers [242]. Soule and co-workers [242] offer a thorough study of magnetic-field gradients between the Davy-type pole pieces, as measured by a silver sphere probe. An electrical solenoid-transducer-type weighing mechanism and a low-temperature cryostat are described. Soule's paper gives insight into resolving the numerous problems associated with obtaining susceptibility measurements by the Faraday method. An ultimate sensitivity of $\pm 0.001 \times 10^{-6}$ emu g^{-1} may be reached with this apparatus; the precision is said to be 0.1%, and the overall error is less than 1%. A crystal anisotropy apparatus (balance) is reported by Neogy and Lal [243].

7.11   Torquemeters for Magnetic Anisotropy

Most of the anisotropy work reported in the literature was on magnetic materials of technological significance. These materials include single crystals, *whiskers* or *thin film* of metals, and alloys. Again in the *nonmetallic* area, relatively more studies were made on *ionic solids*, for example, on *ferrite*-type materials and on salts of rare earths. In general, very few studies are reported on *molecular solids* (i.e., coordination complexes and organometallic compounds). The lack of activity in this area may be attributed to the difficulties in making anisotropy measurements on such materials, which are weakly magnetic (paramagnetic or diamagnetic) as compared to the materials first listed. Usually the *weak* materials can be studied by torsion techniques, which necessitate the handling of very delicate quartz torsion fibers. Furthermore, the anisotropy data can be meaningful only if a complete three-dimensional Fourier analysis of the X-ray data is available. This essentially allows the transformation of the crystal anisotropies (χ_1, χ_2, and χ_3) to the molecular anisotropies (κ_1, κ_2, and κ_3).

An adaptation of the Krishnan *flip-angle* method is described, and a resume of its theoretical aspects was given by Mulay and Fox [163].

The four basic parts of the working apparatus for this method are (1) the torsion head for rotating the crystal, (2) the glass-tube draft shield to prevent disturbing air currents, (3) the electromagnet, and (4) the optical system to permit easy and accurate observation of the crystal. A complete description of this instrumentation, the experimental procedures, and a detailed analysis of error are given by Gordon (cf. [163]). Consequently, only the specific characteristics of the apparatus used in the anisotropy studies on ferrocene and modifications of practical value are included here (Figure 3.63a).

In the original experiments, an air-cooled electromagnet and a stabilized dc power supply were used. High-field Permendur conical pole tips tapering from 10 to 5.8 cm with a 30° angle were also used, and the pole gap was 3.3 cm. The field strengths available for continuous operation under these conditions ranged from 1500 to 9000 Oe. The field strength was measured with a model A gaussmeter with probe FA 21 (Halltest Division, Instrument Systems Corporation, New York, NY). The calibration remained consistent over the entire period of this work. The present setup, which includes a large water-cooled magnet, and details of the motor-driver torsion head appear in Figures 3.63b and c. The entire assembly can be evacuated to prevent moisture condensation at low temperatures.

At first the electrostatic forces between the crystal-fiber system and the wall of the glass draft shield were large enough to be visibly disturbing. Several techniques were tried, such as the introduction of radioactive material on the

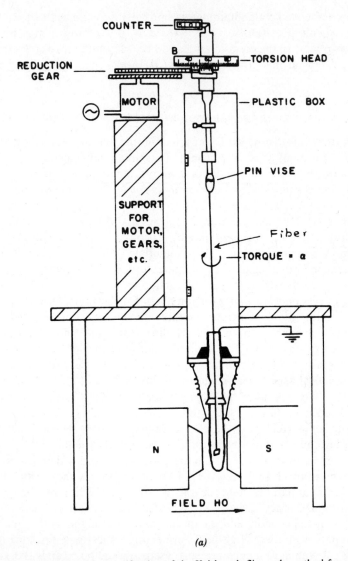

COUNTER

B

TORSION HEAD

REDUCTION
GEAR

MOTOR

PLASTIC BOX

SUPPORT
FOR
MOTOR,
GEARS,
etc.

PIN VISE

Fiber

TORQUE = α

N

S

FIELD HO

(a)

Figure 3.63 (*a*) Schematic of a modification of the Krishnan's flip-angle method for anisotropy, which is used in our laboratory. The quartz fiber is twisted with the help of a motor, reduction gears, and so on. The angle α of the twist is recorded accurately with the counter and the Vernier scale. (*b*) Setup for the flip-angle anisotropy apparatus in our laboratory. This instrument has a large water-cooled electromagnet with 6.5-in.-diameter pole tips tapered to approximately 2.5 in. The plastic shield shown in (*a*) was replaced by a long glass tube, and the torsion head is enclosed in a bell jar to facilitate evacuation of the entire apparatus. Note, (*a*) and (*b*) reprinted with permission from L. N. Mulay, "Techniques for Measuring Magnetic Susceptibility," in A. W. Weissberger and B. W. Rossiter, Eds., *Physical Methods of Chemistry, Techniques of Chemistry*, Vol. 1, Part 4, Wiley-Interscience, New York, 1972. Copyright © 1972 by John Wiley & Sons, Inc. (*c*) Details of the motor-driven torsion head and the counter for the anisotropy apparatus of (*b*).

Figure 3.63 (*continued*)

Figure 3.63 (*continued*)

inner glass wall near the crystal, but none was satisfactory. Finally, a coil of bare copper wire (No. 18 gauge) was forced into the glass draft shield so that it formed a loosely wound spiral pressing against the inside of the entire draft shield. When this copper wire was grounded, the electrostatic disturbances were eliminated completely.

The crystals of ferrocene used in this study were grown by slow evaporation [163]. Only well-formed crystals, free from visible imperfections, were chosen

(5–40 mg). The crystal was carefully aligned on an X-ray goniometer and was then attached with both beeswax and a glue to a short length of glass fiber, called the *orienting fiber*; this in turn was affixed to the fine quartz-fiber system with water-soluble glue (Elmer's Glue-All), which can be thinned to any desired consistency with a little water and dries rather quickly. Moreover, the orienting fiber can be detached easily by dissolving the glue in water. This property was especially advantageous here, since the crystals were not soluble in water. (For water-soluble crystals, an organic glue, e.g., Duco cement, should be used.) The quartz fibers were manufactured by Worden Laboratories, a subsidiary of Ruska Instrument Company (Houston, TX). They ranged in diameter from about 10 to 30 μm and were approximately 20 in. long.

In the following discussion, we see that the anisotropy $\Delta\chi$ corresponding to a particular orientation of a crystal can be measured in terms of the maximum torque required to flip the crystal from its equilibrium position in a fixed field H. The torque α is proportional to H^2. Thus measuring the torque α as a function of H and plotting α versus H^2 gives a straight line, whose slope corresponds to the anisotropy $\Delta\chi$. For fibers with different torsion constants K, a family of parallel lines is obtained. The value of $\Delta\chi$ is determined accurately for each setting by considering the average of slopes of such lines. The torsion constant is determined independently by an *oscillation* method. This method is described in several standard physics textbooks. In the following paragraphs, W is the molecular weight of the crystal and m is its mass. A monoclinic crystal (such as ferrocene) is considered here. The significance of ξ and various symbols is given in Figure 3.64.

For a monoclinic crystal, the relation of the magnetic axes U_1, U_2, and U_3 to the crystalline axes a, b, and c is considered. If the U_3 axis (i.e., the b axis) is along the suspension fiber,

$$\Delta\chi = \chi_1 - \chi_2 = \frac{2WK}{mH^2}\left(\alpha'_{max} - \frac{\pi}{4}\right) = \eta_1$$

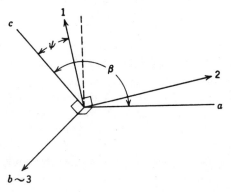

Figure 3.64 Relationship of the crystal axes a, b, c to the principal magnetic axes 1, 2, 3 in ferrocene. Reprinted with permission from L. N. Mulay, *Magnetic Susceptibility*, Interscience, New York, 1966, a reprint monograph based on L. N. Mulay, in I. M. Kolthoff and P. J. Elving, Eds., *Treatise on Analytical Chemistry*, Part 1, Vol. 4, Interscience, New York, 1963, Chap. 38. Copyright © 1966 by John Wiley & Sons, Inc.

If the U_3 axis is perpendicular to the suspension fiber and the a axis of the crystal is along the suspension,

$$\Delta\chi = \chi_1 \cos^2\xi + \chi_2 \sin^2\xi - \chi_3 = \frac{2WK}{mH^2}\left(\alpha''_{max} - \frac{\pi}{4}\right) = \eta_2$$

If both a and b crystalline axes are perpendicular to the suspension, then

$$\Delta\chi = \chi_1 \sin^2\xi + \chi_2 \cos^2\xi - \chi_3 = \frac{2WK}{mH^2}\left(\alpha'''_{max} - \frac{\pi}{4}\right) = \eta_3$$

It is seen readily that the principal anisotropies are given by

$$\chi_1 - \chi_2 = \eta_1$$
$$\chi_1 - \chi_3 = \tfrac{1}{2}(\eta_1 + \eta_2 + \eta_3)$$
$$\chi_2 - \chi_3 = \tfrac{1}{2}(\eta_2 + \eta_3 - \eta_1)$$
$$\cos 2\xi = \frac{\eta_2 - \eta_3}{\eta_1}$$

If the average susceptibility $\bar\chi$ is known from any classical method such as the Faraday, the principal susceptibilities can be determined from the relation

$$\bar\chi = \tfrac{1}{3}(\chi_1 + \chi_2 + \chi_3)$$

coupled with any two of the anisotropies.

As pointed out previously, the principal anisotropies can be transformed into molecular anisotropies κ_1, κ_2, and κ_3 only if the crystal structure is known. Interpretation of such data is given by Mulay and co-workers [6, 163, 244, 245] and in Section 8.3.2 under subsection "Elucidation of Bonding in Organometallics: Magnetic Anisotropy."

Several torquemeters, also called torque magnetometers, have been reviewed [36–46] since the early 1970s. Some of this instrumentation should prove useful for anisotropy work.

7.12   High-Pressure Measurements

In our laboratory a Faraday balance is used to measure magnetic susceptibility under pressures up to 20 kbars (Cordero-Montalvo and co-workers [108] and Cordero-Montalvo [107]). The balance assembly and the high-pressure cell are shown in Figures 3.65 and 3.66a and b, respectively. Our cell design is essentially similar to that of Wohlleben and Maple [246], who mostly followed the constructional details described earlier by Kawai and Sawaoka [247]. The cell is made of diamagnetic Cu–Be alloy (1.5-cm o.d. and about 3 cm tall, weighing

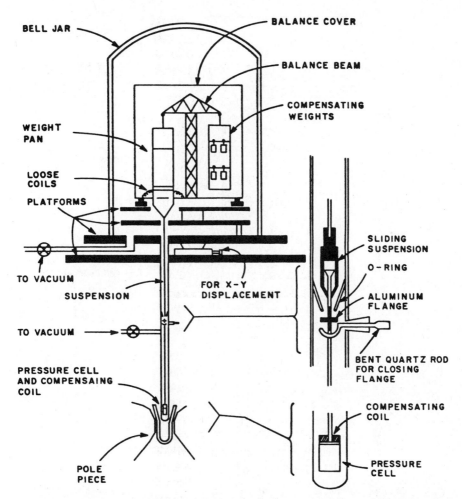

Figure 3.65 Schematic of the Faraday magnetometer constructed in our laboratory for high-pressure susceptibility measurements at room temperature and with refrigerant baths. The special attachment shown at the right side of the figure was designed to permit the removal and replacement of the cell while the upper section of the apparatus remains under partial vacuum.

about 80 g). The diamagnetism of the cell is exactly compensated by wrapping around it an appropriate length of weakly paramagnetic tantalum foil so that the net combination has a zero susceptibility. The large weight (80 g) of the cell necessitated the use of a semimicroanalytical balance, such as the Sartorius balance with a load capacity of approximately 100 g. A few milligrams of the material under study was placed in a tiny Teflon cup (ca. 3-mm i.d., ca. 3 cm tall, with an airtight cap) (Figure 3.66b). The Teflon cup was filled with a mixture of 1:1 n-pentane and isoamyl alcohol. Thus when the pressure cell was subjected to a uniaxial pressure under a pressurizing jack, this pressure was converted to

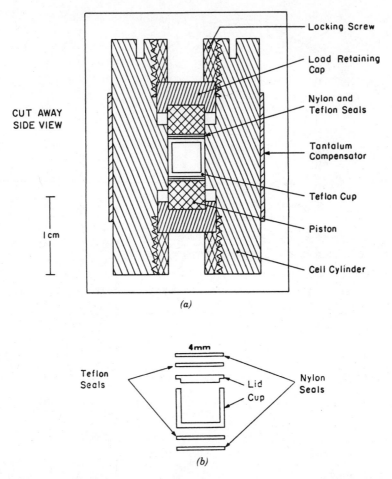

Figure 3.66 (*a*) Schematic of the pressure cell that is capable of reaching 12 kbar, which was constructed for the magnetic susceptibility measurements performed in the present study. (*b*) High-pressure nylon and Teflon seals and Teflon cup and lid used for containing the sample and pressure-transmitting fluid inside the pressure cell.

hydrostatic pressure; and the few milligrams of sample experienced pressure from all directions, which facilitated a study of the change in the susceptibility of the sample under almost ideal hydrostatic conditions. For each pressure measurement the sample was removed from the Faraday balance, and the pressure was adjusted with a pressure jack outside the main apparatus. The entire balance and sample chambers could be evacuated and a small amount of helium gas could be introduced in the system when necessary. As shown in Figure 3.66*a*, an adjustable seal was introduced so that the vacuum in the upper balance bell jar cover could be maintained while the pressure cell was being removed or inserted in the sample chamber. A small coil assembly with its axis parallel to the field was attached to the suspension, through which the leads of the coil were fed and brought outside the balance chamber. This coil was energized by a small direct current so that the magnetic moment (i.e., magnetization) thus produced could oppose exactly the paramagnetic moment of the sample. Since the susceptibility of the pressure cell had been adjusted to zero, the procedure amounted to measuring the current i in the compensating coil, which was directly proportional to the magnetic force on the sample, which in turn was proportional to its mass susceptibility. Hence, $i \propto$ force = $m \cdot \chi \cdot H \cdot dH/dX$.

Thus, in reality the Sartorius balance was used under truly *null* conditions, which enhanced the overall sensitivity of the apparatus; the current i could be measured with great accuracy (1 part in 10^6) with the help of a sensitive Leeds and Northrup potentiometer. The apparatus was used to study valence changes of Sm^{2+} in doped and undoped SmS, which is discussed in Section 8.3.1 under subsection "Insulator–Metal Transitions in Doped and Undoped SmS." A thesis by Cordero-Montalvo [107] and the paper by Wohlleben and Maple [246] give several constructional details of the apparatus and procedures for its operation.

The sensitive Cahn 1000 balance, which has a high load capacity of 100 g and a sensitivity of a few micrograms, facilitates measurements of susceptibility under high pressure. This balance is made by the Cahn Instruments Company (Cerritos, CA).

8   GENERAL APPLICATIONS

8.1   Scope

Applications of the magnetic technique are numerous. Some selected applications of commercial instruments are described for testing magnetic materials and for gas analysis. Selected applications of magnetic measurements are also described in the areas of inorganic, organic, and biological chemistry and materials science, including thermomagnetic analysis; geotype applications are mentioned briefly. With sufficient ingenuity the magnetic technique may be used to solve many other problems than those represented by these examples.

Several studies are related to paramagnetic and ferromagnetic properties, which have many technological applications. Diamagnetism is increasingly applied in such areas as organic chemistry and superconducting charge-transfer complexes. Although new applications are discovered daily for the nuclear and electron paramagnetic resonance and Mössbauer spectroscopy techniques developed since the 1950s, the relatively simple static magnetic techniques continue to be useful in certain respects. These techniques are helpful as exploratory tools and, in several cases, the magnetic studies preceded the more elaborate NMR, EPR, and Mössbauer investigations.

8.2   Applications of Commercial Instruments

Although completely assembled commercial instruments for measuring magnetic susceptibility are now available, equipment for testing magnetic material, such as permeameters, coercimeters, and *B–H* loop tracers, was always readily available over longer periods of time and has proved extremely useful for measuring the technologically relevant properties of magnetic materials, which in turn helped to establish a successful magnetics technology. Instruments for measuring and monitoring the contents of paramagnetic gases such as oxygen have also been used for industrial applications over longer periods of time. Hence, in this section the principles and applications of such commercial instruments are described; in later sections a wider range of susceptibility applications and magnetization apparatus, which can be assembled in the laboratory, or apparatus such as the vibrating sample and the SQUID magnetometers, which have become commercially available, are discussed.

8.2.1   Permeameters, Carbon Analyzers, and Coercimeters

Basically, the permeability of a sample is measured by the magnetic coupling it establishes between two inductance coils. In the low-frequency ac permeator designed by Rogers and co-workers [248], a primary and a secondary solenoid are wound around the sample holder. A small alternating current at 5 Hz is sent through the primary solenoid; and after the induced current is converted to a direct current by a rectifier, it is recorded on a microammeter. A calibration curve is obtained for samples that are of the same size but contain different amounts of carbon. From this the carbon content of the unknown sample, shaped to the same size, is determined.

The classic Isthmus permeameter (cf. [25]), shown in Figure 3.67, is based on the same general principle, except that the sample is first magnetized and then

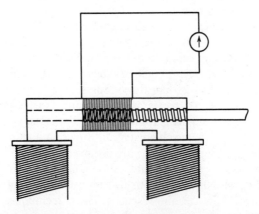

Figure 3.67   The Isthmus permeator. Reprinted with permission from P. W. Selwood, *Magnetochemistry*, Wiley-Interscience, New York, 1956. Copyright © 1956 by John Wiley & Sons, Inc.

withdrawn from the secondary, which is connected to a galvanometer. The results obtained are used extensively to determine the concentration of copper in alloys containing a 1 : 1 ratio of iron and nickel.

Several other variations of such magnetic instruments as the Carbanalyzer [249] and the Carbometer and Siemens ferrometer [250] are described in the literature. Several variations of permeameters, hysteresis apparatus especially designed for the testing of magnetic materials, were reviewed in recent years by Mulay and Mulay [36–46].

In a coercimeter described by Doan [251], the sample is first magnetized in a strong magnetic field and then placed in a variable field, the strength of which is balanced against the residual magnetism of the sample. This is accomplished by moving a secondary coil, which is attached to a galvanometer (Figure 3.68).

Many permeameters, coercimeters, and hysteresis apparatus are available commercially but are too numerous to list here. Information on these is available in *Review of Scientific Instruments*, *Journal of Physics E*, and various buyers' guides published by the electrical and electronics industries in the United States and abroad [14, 15].

8.2.2 Oxygen Analyzers

Among the many physical methods that are adaptable for continuous analysis and recording of oxygen content in a gas sample, the ones used most widely are based on a measurement of oxygen's magnetic susceptibility and on special effects produced in a magnetic field.

Most methods utilize the fact that, among the few paramagnetic gases known, oxygen has a large molar magnetic susceptibility of 3.45×10^{-3} at 20°C and

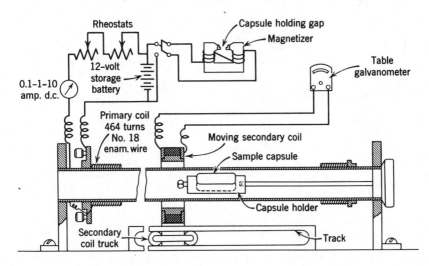

Figure 3.68 A form of coercimeter for determining coercive force, principally of powdered specimens.

standard pressure. All other gases such as nitrogen and carbon dioxide, which are constituents of air, are diamagnetic. This facilitates an analysis of oxygen in air and in other samples containing diamagnetic gases. Magnetic susceptibility data on oxygen were used in conjunction with the EPR technique to study atomic recombination of oxygen. Krongelb and Strandberg [252] found that the surface recombination coefficient for oxygen atoms on quartz is 3.2×10^{-4} per collision, and the second-order volume recombination process is about $3 \times 10^{15} \, cm^6 \, mol^{-2} \, s^{-1}$.

1.   The Hays oxygen analyzer, marketed by the Milton Roy Company (Michigan City, IN). This apparatus is based on the inverse relation between the paramagnetism of oxygen and its temperature. It is similar to an apparatus developed by Klauer and co-workers [253]. Figure 3.69 shows two electrically self-heated, identical nickel coils ($3A$ and $3B$) on the outside of a glass tube, 1. The winding is placed between the poles of a small permanent magnet, 4. The two windings form two legs of a Wheatstone bridge. The sample gas flows from the entrance, 5, to the exit, 6; and oxygen is attracted into the tube, 1, by a magnetic field. The gas is heated by coil $3A$, which decreases its susceptibility. Cooler gas entering at $3A$ pushes the heat gas away from the field, as shown by

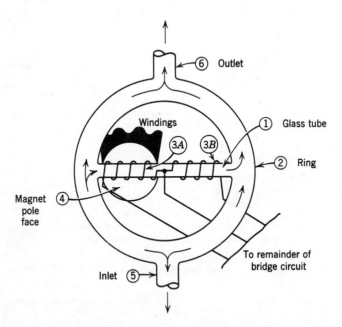

Figure 3.69 Hays Corporation's oxygen analyzer based on the *magnetic wind* principle. Reprinted with permission from L. N. Mulay, *Magnetic Susceptibility*, Interscience, New York, 1966, a reprint monograph based on L. N. Mulay, in I. M. Kolthoff and P. J. Elving, Eds., *Treatise on Analytical Chemistry*, Part 1, Vol. 4, Interscience, New York, 1963, Chap. 38. Copyright © 1966 by John Wiley & Sons, Inc.

the arrow. This flow, often termed the *magnetic wind* cools coil 3*B* and heats coil 3*A*. This difference in temperature changes the resistance of the two coils and produces an unbalance of the Wheatstone bridge, which is converted into a voltage unbalance, amplified, and transmitted to an indicator or a recorder. The unbalance is proportional to the oxygen content of the gas. Thus the instrument is adaptable to an *enclosed* or a *flowing* sample of gas containing oxygen. Several references to a similar oxygen meter and applications are given by Selwood [25] and Krupp [254].

2. The Beckman oxygen analyzer, marketed by the Beckman Company (Fullerton, CA). As shown in Figure 3.70, this analyzer measures directly the volume susceptibility of a gas in an inhomogeneous field and is similar to that employed in the Faraday technique. One end of a test body in the form of a glass dumbbell is placed between the poles of an Alnico permanent magnet. The test body is supported on a silica fiber, which acts as a tension suspension. This fiber measures the torque of the system about the axis of the suspension in terms of the deflection of a smaller mirror attached to the fiber. The torque is proportional to the magnetic field strength, its gradient, and the difference in the volume susceptibilities of the test body and the gas surrounding it. The design is similar to that formulated by Reber and Boeker [56]. A change in the susceptibility of the gas, resulting from a change in its oxygen content, produces a nearly proportional deflection of the mirror [66]. A similar instrument is manufactured by Servomex Instruments, ISA Associates (Langhorne, PA). A

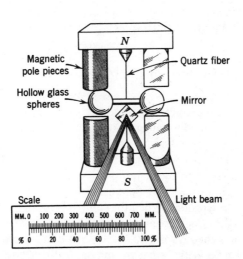

Figure 3.70   Schematic view of Beckman oxygen analyzer measuring system table-model analyzer; N and S are magnet poles. Reprinted with permission from L. N. Mulay, *Magnetic Susceptibility*, Interscience, New York, 1966, a reprint monograph based on L. N. Mulay, in I. M. Kolthoff and P. J. Elving, Eds., *Treatise on Analytical Chemistry*, Part 1, Vol. 4, Interscience, New York, 1963, Chap. 38. Copyright © 1966 by John Wiley & Sons, Inc.

Figure 3.71   The Beckman oxygen analyzer (shown schematically in Figure 3.70).

photograph of the analyzer is shown in Figure 3.71. Mulay and Mulay [36–46] reviewed several patented *gas analyzers*; however, their commercial availability is unknown.

8.2.3 Commercial Instruments for Geophysical Research

THE SPINNER OR ROTATING SAMPLE MAGNETOMETER

Generally, this magnetometer is used to study geological specimens from various locations. Usually mineral samples, bored from rocks, containing ferrimagnetic traces are studied. The sample (usually 1–2 cm in diameter and 2–4 cm tall) is attached to a rod, which in turn is connected to an electric motor. The sample is surrounded by *pickup* probes. Early models used inductance or signal coils for such probes. The Schonsted Instrument Company (Reston, VA 22090) uses *flux gate* probes. When the sample is rotated at a fixed frequency, it generates a signal, of which the amplitude and phase in each plane are measured. This information is then translated into vector data. From such measurements the TRM (thermomagnetic remanent magnetization), and in certain cases, the anisotropic susceptibilities, are measured by comparison with a known standard (Figure 3.71*a*). Basically, the objective is to study the virgin properties of rocks and minerals, which are not subjected to an external field (except that of the earth's magnetic field). In certain cases, a magnetic shield or a set of Helmholtz coils is used to make measurements in a zero field. Mulay and Mulay [38–46] reviewed modifications of such magnetometers and also their calibration techniques.

OTHER INSTRUMENTS

Bison Instruments, Inc. (Minneapolis, MI 55416) supplies several instruments for geophysical research. Their magnetic susceptibility meter is apparently based on an inductance bridge, which is particularly useful for determining approximately the content of magnetite (Fe_3O_4—a ferrimagnetic material) in iron ores. The susceptibility of rocks and other materials in the range from 10^{-5} to 10^{-1} cgs units can be measured. Their instruction manual for Model 3101 contains several useful references to the magnetic measurement techniques applicable to rocks and other materials.

Flux gate magnetometers are available from several manufacturers. Their principles and applications, including the airborne instruments, are described by Dobrin [255]. Reviews by Mulay and Mulay [36–46] contain a few references to such magnetometers.

8.2.4 Alternating Current Inductance Bridge

A mutual inductance bridge for the measurement of ac susceptibility is made by Linear Research, Inc. (5231 Cushman Place, San Diego, CA 92110). Basically, it consists of a four-terminal ac resistance bridge (Model LR-400) with a mutual inductance attachment. The bridge operates at a frequency of 15.9 Hz; mutual inductances from 200 μH (microhenries) to 200 H (henries) can be measured over eight ranges with a resolution of 0.01 μH. Mulay, Cao, and Klemkowsky [76] describe an inductance coil for use with this bridge along with a simple cryostat.

The inductance bridge is particularly suitable for the determination of the relative changes in diamagnetic susceptibility with temperature of superconducting materials. It was shown in Section 3.8 that the volume susceptibility κ is related to the permeability. Thus,

$$\kappa = \mu - 1/4\pi$$

When the Meissner effect is observed, the flux density $\beta = \mu H = 0$ in the superconducting state; that is, no lines of force from an applied field H penetrate the material. Thus μ becomes zero and $\kappa = -1/4\pi \simeq -8 \times 10^{-2}$ cgs units, and κ equals 1 in SI units.

At room temperature κ is small, about -1×10^{-6}. Hence a large change of 10^4 occurs when κ is measured at room temperature and below the critical superconducting temperature. Since about early 1966, several workers have reported the relative change in the ac susceptibility of a new class of oxide materials. Professor Paul W. Chu of the University of Houston reported that an oxide $YBa_2Cu_3O_{9-x}$ becomes superconducting above 95 K, which is well above the temperature of liquid nitrogen, namely, 77 K. In addition to showing that the resistance of this compound becomes zero below the critical temperature, he found that the volume susceptibility shows a relatively large change in the diamagnetic susceptibility $(-\kappa)$ of the material (Figure 3.37b). In his first report the relative change in $-\kappa$ was unfortunately marked in units of microhenries. This symbol represents *microhenries* and not the permeability of μ times the field H. Several papers on the superconductivity of various oxide materials have appeared since January 1987. A few have reported the ac susceptibility appropriately in arbitrary units.

8.3 Selected Applications

8.3.1 Applications to Inorganic Chemistry

PARAMAGNETIC IONS IN SOLUTIONS, GLASSES, AND FUSED SALTS

Wiedemann's additivity law may be used to determine the concentration and oxidation states of paramagnetic ions in solution, in solid mixtures, and in glass. The method is capable of yielding results that are accurate to within 1% or better, depending on the accuracy of the susceptibility measurement. An obvious limitation is that the solvent and the solute should not interact. Further, for paramagnetic ions the system must be magnetically dilute, that is, free from spin–spin interactions between adjacent ions. In recent years, the technique was used profitably to study the nature of such interactions and to derive valuable structural information. A discussion of this aspect follows.

The concentration $p\%$ of a diamagnetic salt in solution may be obtained by measuring the specific susceptibility χ_s^g of the solution and using the following relationship:

$$\chi_s^g = \rho \cdot \frac{\chi_{\text{cation}} + \chi_{\text{anion}}}{\text{mol wt salt}} + (100 - \rho)\chi_{\text{H}_2\text{O}}^g$$

where χ_{cation} and χ_{anion} represent the gram ionic susceptibilities. Some of these values are listed in Table 3.8. The specific susceptibility of water $(\chi_{\text{H}_2\text{O}}^g)$ is taken as -0.720×10^{-6} emu g^{-1}.

For a paramagnetic ion, χ_{cation} is dependent on the temperature and its effective magnetic moment. Often, χ_{cation} is calculated from

$$\chi_{\text{cation}} = \frac{N \cdot \mu_{\text{eff}}^2}{3kT}$$

Some typical values for μ_{eff} are given in Tables 3.8–3.10.

Solids, for instance, aluminum oxide, which support paramagnetic ions on the surface, are analyzed similarly [25]. The specific susceptibility of the particular support used is determined separately.

Remarkably, in some cases the determination of the ratio of paramagnetic ions in two different oxidation states dispersed in colloidal systems such as glass gave results as precise as that of the spectroscopic methods. According to

DeJong (cf. [25]), a glass containing 0.08% total iron in the Fe^{3+} and Fe^{2+} valence states resulted in the following Fe^{3+} percentages: 44.5 to 47% from the ultraviolet spectrum, 44% from the infrared spectrum, and 45.5% from the magnetic susceptibility data, which was treated according to the additivity law, assuming that the magnetic moments for Fe^{3+} and Fe^{2+} are 5.96 and 4.90 Bohr magnetons, respectively. The limited studies in this area were undertaken not only to find the concentration of the paramagnetic ions, but also to unravel structural aspects of glasses containing these ions. Bishay (cf. [25]) recently studied the color and magnetic properties of iron ions in glasses of various types. His paper lists several references to magnetic studies on glass. Iron-containing glasses have been studied by Banerjee [256]. Recent work on fused salts was reviewed by Mulay and Mulay [36–46].

A thesis by Fisher [257] based on the work carried out in our laboratory gives additional references to magnetic studies on glasses containing iron and other transition metal ions.

An interesting application of the magnetic technique to elucidate the coordination geometry of transition metal ions in fused salts is reported by Bailey and co-workers [258], and the results are consistent with spectroscopic interpretations. While the spectroscopy of transition metal ions on fused salts is not new, and the concept of magnetic titrations (see, e.g. [194]) involving changes in the magnetic moment of the species is also not new, the application of the magnetic technique to studies on fused salts is rather unique. The authors investigated the coordination geometry of Ni(II) and Co(II) in various metals and glasses. The salts were placed in a Gouy tube, and the changes in susceptibility were measured as a function of the ratio of metal ion to the ligand. Effective magnetic moments were then obtained. Formation of the $[Ni(CN_4)]^{2-}$ ion in fused KCNS solvent involved changes from a paramagnetic to a diamagnetic species from which an equilibrium constant of approximately $2 \times 10^5 \, M^{-4} \cdot L^{-4}$ was derived. The moments found for the chloride and sulfate glasses are in the ranges typical for the geometries derived from spectroscopic data, namely, tetrahedral and octahedral. The moment in KSCN is also in the expected tetrahedral range. Although the work reported by Bailey and co-workers [258] is somewhat exploratory in nature, it represents a new approach to investigating fused salts.

DETERMINATION OF PURITY AND ANALYSIS OF RARE EARTHS

The shielding of $4f$ electrons in rare earths makes the interactions between their adjacent ions negligible. This corresponds to the ideal situation of *magnetic dilution* and often permits a magnetic analysis of rare earths.

Formerly, the purification of rare earths was based on fractional crystallization. A plot of magnetic susceptibility versus fraction number was quite useful in determining purity. A horizontal plateau in such plots indicated the isolation of a pure rare earth salt.

Although elegant techniques of ion exchange are used currently to separate rare earths, determination of magnetic susceptibility is an excellent criterion for purity and also for *spectroscopic purity*. Actually, the two determinations are quite different. It is known, for instance, that a trace of a ferromagnetic impurity in copper can be detected only by the magnetic technique. Naturally, the choice of the criterion for purity depends on the final use of the material in question.

The magnetic criterion was applied successfully to the diamagnetic lanthanum and lutetium oxides. The analysis of rare earths depends on the application of the additivity law, as discussed previously. It is particularly useful for analysis of binary mixtures of lanthanum oxide (diamagnetic) and gadolinium oxide (paramagnetic). In general, an analysis of a binary mixture is facilitated if the two components have widely different susceptibilities. Accuracy of better than 1% may be achieved by careful control of temperature during susceptibility measurements and by studying the solutions of their soluble salts instead of the solid oxides. Many textbooks [6, 259] list the magnetic susceptibilities of rare earth oxides and related compounds.

POLYMERIZATION OF PARAMAGNETIC IONS IN SOLUTION, PARAMAGNETISM OF GLASSES, AND FUSED SALTS

The magnetic susceptibility method is useful when studying the polymerization of paramagnetic ions in solution. A few typical examples are Fe(III) and Cr(III), studied by Mulay and co-workers [260], and Mo(V), investigated by Sacconi and Cini (cf. [6]).

This technique is particularly useful when studying systems in which large changes in magnetic susceptibility of a solution occur because of the conversion of a species from paramagnetic to diamagnetic (or to one of a very low magnetic moment) or vice versa. Only relatively simple systems with few species of varying magnetic properties can be studied effectively. The magnetic method has another limitation because its applicability is rather doubtful when studying polymerization of diamagnetic ions. The magnetic work by Mulay and Selwood [261] on the dimerization of Fe(III) in solution is illustrated in some detail.

The following pH-dependent equilibrium proposed by Hedstrom [262] was investigated

$$2Fe^{3+} + 2H_2O \underset{}{\overset{k_{22}}{\rightleftharpoons}} \left[Fe \underset{\begin{array}{c}O\\H\end{array}}{\overset{\begin{array}{c}H\\O\end{array}}{\diamond}} Fe \right]^{4+} + 2H^+$$

The magnetic susceptibilities were measured with a Gouy balance for a solution containing 0.061 M Fe(ClO$_4$)$_3$ at a constant ionic strength (3 M NaClO$_4$) at different acidities. Solutions containing other concentrations of iron were also investigated. The perchlorate (*spherical*) anion was chosen because the (ClO$_4$)$^-$ ion has little chelation effect on the cation.

The problem was to find the dimerization constant k_{22}. First, it was necessary to find the specific susceptibility of Fe(ClO$_4$)$_3$ in solution and then, to find the susceptibility per gram atom of total iron in solution from the following data:

P_1 = wt fraction of $Fe(ClO_4)_2$ = 0.0173

P_2 = wt fraction of H_2O = 0.6570

P_3 = wt fraction of $NaClO_4$ = 0.2940

χ_1 = specific susceptibility of $Fe(ClO_4)_2$ (unknown)

χ_2 = specific susceptibility of H_2O = -0.720×10^{-6}

χ_3 = specific susceptibility of $NaClO_4$

$\left(\text{calculated from } \dfrac{\chi_{Na^+} + \chi_{ClO_4^-}}{\text{mol wt } NaClO_4} = -0.3550 \times 10^{-6} \right)$

χ_{sol} = specific susceptibility of the solution
(pH = 0.39, stored for several weeks), measured by the Gouy technique = 0.0856×10^{-6}

According to the additivity law, $\chi_{sol} = \chi_1 P_1 + \chi_2 P_2 + \chi_3 P_3$; substituting the appropriate quantities and solving for χ_1 gave $\chi_1 = 38.4 \times 10^{-6}$. The molar susceptibility χ_m for $Fe(ClO_4)_3$, calculated from $\chi_m = \chi_1 \times$ molecular weight of $Fe(ClO_4)_3$, was $13{,}600 \times 10^{-6}$ cgs units. This result is converted to $\chi_{Fe(t)}$, susceptibility per gram atom (of total) iron in solution, using Pascal's relation; hence,

$$\chi_{Fe(t)} = \chi_{Fe(ClO_4)} - 3\chi_{ClO_4^-}$$

$$\chi_{Fe(t)} = 13{,}600 \times 10^{-4} - 3(-34 \times 10^{-6})$$

$$= 13{,}702 \times 10^{-6} \text{ cgs units}$$

Now,

$$\chi_{Fe(t)} = f \cdot \chi_{Fe^{3+}} + (1 - f)\chi_{dimer}$$

It was assumed that the dimer is diamagnetic and, since diamagnetic susceptibilities are negligible in comparison with paramagnetic susceptibilities (making $\chi_{dimer} \simeq 0$), the following result was obtained:

$$f = \text{wt fraction of } \chi_{Fe^{3+}} \text{ ions} = \frac{\chi_{Fe(t)}}{\chi_{Fe^{3+}}}$$

This assumption was based on the observation [260] that the magnetic moment per gram atom of total iron decreases rapidly to half the original value ($\mu = 5.82$ Bohr magnetons, pH = 0) with increasing concentration of the dimer (at pH = 3), indicating that, to cause such a large increase, the dimeric species must have a negligible magnetic moment (μ ca. 0, that is, diamagnetic). Other arguments supporting this assumption are presented in [261].

At pH = 0, all iron should be formulated as (hexaaquo) Fe^{3+} ions; therefore, a solution of the same composition at pH = 0 gives a value of $14{,}500 \times 10^{-6}$ for

$\chi_{Fe^{3+}}$. Hence, $f = 13,702/14,500 = 0.9448$; and $(1 - f)$, which is the weight fraction of the dimer, $= 0.0552$.

Thus, the procedure involved evaluation of $\chi_{Fe(t)}$ as a function of the pH and the corresponding f value; in all cases $\chi_{Fe^{3+}}$ was taken as 14,500. The dimerization constant k_{22} could then be calculated in each case from the f and $(1 - f)$ values; the average value of 7.5×10^{-2} was obtained from a study of several solutions stored for long periods of time. The method was extended to study the temperature dependence of k_{22} by measuring the susceptibilities of the solution over a range of temperatures from -10 to $60°C$. From this the heat of formation (ΔH) of the dimer was estimated to be about $9.8 \, kcal \, mol^{-1}$.

Brownlow [263] studied the magnetic properties of hydrolyzed species adsorbed on ion-exchange resins and confirmed the existence of species such as the dimer with subnormal magnetic moments, with μ_{eff} less than $5.82 \, \beta$.

Remarkably, another technique, which measures the spin-lattice relaxation times (T_1) from proton magnetic resonance spectra [264], was used to measure the effective magnetic moment of Fe^{3+} ions in solution and indicated that the dimeric ions are almost diamagnetic $(\mu = 0)$ and yielded values between 10^{-2} and 10^{-3} for the dimerization constant k_{22}, which agree well with the magnetic susceptibility data [261]. During the experimental work the spin-lattice relaxation time T_1 for protons in solutions containing the Fe^{3+} and the dimeric species was measured; the reciprocal of T_1 is proportional to μ_{eff}^2 of the paramagnetic ion.

Apart from yielding the same magnitude for the dimerization constant k_{22}, these two independent magnetic methods were successful because they gave results that agreed satisfactorily with an entirely unrelated technique of electrometric titration [262].

Independent evidence for this dimerization was obtained from UV spectrophotometric investigations by Mulay and Selwood [261] on an identical set of ferric perchlorate solutions. These investigations showed absorption peaks at 240 and 335 μm and were assigned to the presence of the dimer $Fe_2(OH)_2^{4+}$ and the (hexaaquo, monomeric) Fe^{3+} ions, respectively. In addition, values for equilibrium constants k_{22} at two temperatures (15 and 51°C) in good agreement with the corresponding values obtained from the magnetic studies were yielded. Mulay and Selwood [261] postulated that the diamagnetism of the dimer was caused by "exchange interactions between adjacent irons which destroys all their paramagnetism." In this manner the observed subnormal moments in colloidal solutions of iron may be attributed to the presence of the diamagnetic dimers in large quantities in equilibrium with the paramagnetic Fe^{3+} ions.

In more exhaustive magnetic and spectrophotometric studies on ferric perchlorate solutions of varying concentration, Mulay and Naylor [197] confirmed these observations and discussed the valence-bond type structure of the diamagnetic dimer (Figure 3.72) and somewhat quantitatively explained the subnormal moment in hydrous ferric oxide (Figure 3.72). In Figure 3.72 each half of the dimer is depicted as a low-spin d^2sp^3 octahedral complex of the Fe^{3+} ion in which one unpaired electron (shown with a dashed arrow) is left on each

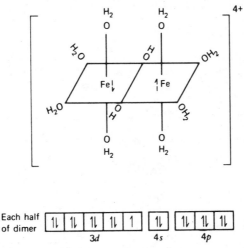

Figure 3.72 The dimeric ion. Bridging between the iron atoms may be through a diol or a dioxo bridge. Reprinted with permission from L. N. Mulay, *Magnetic Susceptibility*, Interscience, New York, 1966, a reprint monograph based on L. N. Mulay, in I. M. Kolthoff and P. J. Elving, Eds., *Treatise on Analytical Chemistry*, Part 1, Vol. 4, Interscience, New York, 1963, Chap. 38. Copyright © 1966 by John Wiley & Sons, Inc.

iron. The dimeric ion is then pictured as two octahedra that share an edge which brings the two irons relatively close and favors an antiferromagnetic coupling, which in the extreme case could yield a diamagnetic dimer. Mulay and Naylor [197] also proposed that species (1) could be antiferromagnetic with a subnormal moment and explained the subnormal moment (ca. 3 BM) in hydrous ferric oxide. In retrospect their dimers represent unique examples of *species in solution* to exhibit antiferromagnetic coupling between neighboring cations, somewhat similar to that observed in magnetically condensed systems in the solid state.

Some workers [265, 266] pursued the concept of a residual moment on the dimer, first proposed by Mulay and Naylor [197]. In their magnetic susceptibility work, Schugar and co-workers [266] considered the possibility of two unpaired spins on the dimer. Ohya and Ono [267] discuss the antiferromagnetism and the superparamagnetism of various (clustered) species in their studies on the Fe^{57} Mössbauer spectroscopy of frozen solutions of ferric perchlorate.

Mulay and Naylor [197] proposed a structure for hydrous ferric oxide in which the diamagnetic (or feebly paramagnetic) dimers are dispersed among the distinctly paramagnetic aquo-Fe^{3+} ions or their first and second hydrolysis products. They showed that at least one aquo-Fe^{3+} ion ($\mu = 5.92$ BM) must combine with a dimeric ion (e.g., $\mu = 0$) to reduce the average moment to the range from 2 to 3 BM. For simplicity, the polymeric structure of hydrous ferric oxide is shown linearly in Figure 3.73. Here, the aquo-Fe^{3+} octahedron shares

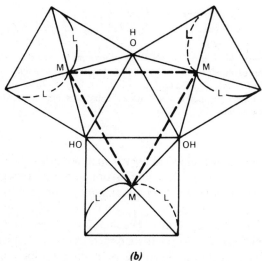

(a)

(b)

Figure 3.73 (a) Proposed structure for hydrous ferric oxide. The bridging may be depicted through ol and/or oxo groups. A three-dimensional spread of the chain is also possible. (b) Plan view of the trinuclear benzoate complex. Here M = Fe(III) and L = $C_6H_5COO^-$.

one corner of the dimeric unit; thus a large distance between the corresponding irons is maintained, which minimizes the magnetic coupling between the two. The authors have shown that a three-dimensional spread of the polymeric chains with or without cross-linking through the appropriate -ol (i.e., the OH^-) or the oxo (O^{2-}) bridges may also be considered. They also considered the possibility of a net spin on the dimer.

Many reports on the polymerization of paramagnetic ions have been written since the publication of [261] in early 1954. Since they are too numerous to be listed here, the reader should refer to [36–46] and to the chapter, "Magnetically Condensed Compounds," in [6] by Mulay and Boudreaux. Some typical examples of such systems follow.

MAGNETICALLY CONDENSED SYSTEMS: METAL CLUSTER COMPLEXES

The early work on the polymerization of paramagnetic ions in solution led to the synthesis of numerous solid polynuclear complexes, that is, coordination complexes containing two or more transition metal ions, which displayed the so-called *intramolecular antiferromagnetism*. These compounds show magnetic exchange interactions between neighboring metal ions of the same or different types; several of these metal-cluster, or *magnetically condensed*, complexes were increasingly useful, particularly in homogeneous catalysis applications.

A typical example of a trinuclear iron complex is the benzoate complex

$$[Fe_3(C_6H_5COO)_5(OH)_3](C_6H_5COO) \cdot H_2O$$

which was studied by Mulay and Ziegenfuss ([245], cf. [6]) along with several other complexes. The proposed structure is shown in Figure 3.73*b*.

Mulay and Ziegenfuss were able to propose this structure based on a study of its susceptibility as a function of temperature (and Fe^{57} Mössbauer spectroscopy) and a careful analysis of the magnetic data based on the Van Vleck–Kambe theory. Readers interested in further elucidation of the structures of such polynuclear metal complexes should refer to the original Mulay and Ziegenfuss paper [245] and to a comprehensive chapter by Hatfield (cf. [6]), who studied exhaustively magnetic properties and X-ray crystal structures of such systems.

EVIDENCE FOR THE FORMATION OF O_4 FROM OXYGEN (O_2) USING THE PHYSICAL ADSORPTION ROUTINE

Since the paramagnetic oxygen molecule has two unpaired electrons ($^3\Sigma$ state), one would expect two such molecules to dimerize to O_4 species through mechanisms involving a magnetic intermolecular coupling, which in the extreme case may *pair off* their spins and possibly give rise to a diamagnetic O_4 species. The concept was proposed by Pauling [64]. For more than 25 years, workers in the Leiden Laboratory, Holland, were engaged in obtaining unequivocal magnetic evidence for such dimerization. Basically, they attempted to control the magnetic dipolar interaction between oxygen molecules in the gas phase by diluting it with noble gases (or *inert* gases in the old sense), like argon, to see whether significant deviations would occur from the Curie–Weiss law, expected for the normal paramagnetic behavior of oxygen. Indications of their early successes were dispelled by the careful work of Knobler [268], who concluded in his 1961 doctoral dissertation that "magnetic susceptibility measurements could give no unequivocal proof for the existence of the dimer."

Mulay and Keys [165] used a novel approach to control the dipolar interaction between oxygen molecules and to develop a new method (independent of seeking deviations from the Curie–Weiss law) to observe directly oxygen dimerization through a large decrease in the paramagnetic susceptibility of oxygen. The technique of physical adsorption of oxygen on the large surface of a powdered solid such as γ alumina (which has the least defect structure) was chosen for controlling the dipolar interaction. A direct observation of a decrease in the oxygen susceptibility seemed possible, because a very large change from the paramagnetism of O_2 to the diamagnetizatism (or reduced paramagnetism) of O_4 was expected for the pressure-dependent equilibrium

$$2O_2 \rightleftharpoons O_4$$

A Faraday microbalance [164] using a quartz helical spring enclosed in a chamber of small dead space (ca. 500 cm³) was especially constructed to measure simultaneously the amount of oxygen adsorbed on γ alumina and the change in its specific magnetic susceptibility χ as a function of pressure. Figure 3.74 shows a typical plot (at a constant temperature of 77 K) of the specific susceptibility of oxygen as a function of pressure. The upper curve shows χ as a function of the amount of oxygen adsorbed per gram of alumina. A family of curves was obtained that show a decrease in the susceptibility of oxygen with increasing adsorption at slightly higher temperatures than and including 77 K.

The large decrease in the susceptibility of oxygen vividly depicts the formation of a polymeric species of oxygen with a lower susceptibility than that of molecular oxygen. The overall decrease in the susceptibility was less at higher temperatures.

From these results Mulay and Keys [165] calculated the equilibrium constants K_x at different temperatures for the above dimerization. The results were treated on the basis that oxygen behaved as a quasiliquid in the adsorbed phase. The K_x values were calculated assuming that all O_4 species exist either in the diamagnetic singlet state or in a paramagnetic triplet state (with half the specific paramagnetic susceptibility of O_2). From linear plots obtained for

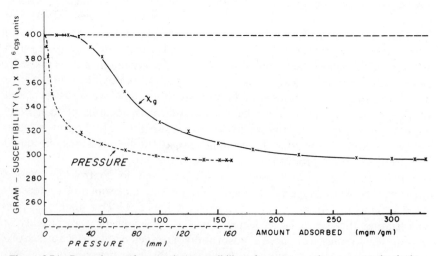

Figure 3.74 Dependence of magnetic susceptibility of oxygen on the amount adsorbed on γ alumina. Dashed curve shows the dependence of susceptibility on pressure of oxygen.

$\log K_x$ versus $1/T$, the heats of dimerization were calculated to be ΔH approximately $-0.5\,\text{kcal mol}^{-1}$ assuming an entirely singlet state, and ΔH approximately $0.7\,\text{kcal mol}^{-1}$ assuming an entirely triplet state for the O_4 species. These small magnitudes for ΔH are indicative of very weak van der Waals-type forces, which appear to hold the two oxygen molecules in the O_4 species.

It was subsequently shown [165] that the physical heat of adsorption of oxygen on γ alumina at very high surface coverages approaches the heat of condensation of oxygen. Thus it is believed that unequivocal evidence was obtained for the dimerization of O_2 and O_4 and for further polymerization of O_2 to $(O_2)_n$-type series in the condensed phase. Wachtel and Wheeler [269] discussed antiferromagnetism in solid oxygen. Key references to some interesting theoretical and experimental work on O_4 appear in [5, 43–45].

DISSOCIATION PHENOMENA

Magnetic susceptibility also has been used to study some dissociations. By taking cryoscopic measurements Bodenstein and co-workers [270] found that chlorine hexoxide in carbon tetrachloride solution was mainly in the form Cl_2O_6. Goodeve and Richardson [271] showed that small amounts of ClO_3 are also present in the liquid. The compound in the gas phase is known to exist entirely as ClO_3, which contains 41 electrons. Therefore, it must contain at least one unpaired electron and show paramagnetism. On the other hand, the dimer Cl_2O_6 contains an even number of electrons and is diamagnetic on the assumption that it is in $^1\Sigma$ state with completely paired spins. The molar magnetic susceptibility of Cl_2O_6 was calculated by adding the experimental atomic values, and that of ClO_3 was calculated by using Van Vleck's formula

$$\chi_m = \frac{4N\beta S(S+1)}{3kT} + N_\alpha$$

where the symbols have their usual meaning. The small additional term N_α represents the underlying diamagnetism, and so on, and can be neglected.

Measurements of magnetic susceptibility of liquid chlorine hexoxide over a temperature range from -40 to $10°C$ showed that the diamagnetism of the liquid was less than that calculated for Cl_2O_6, which indicates a paramagnetic contribution was presumably made by ClO_3; furthermore, the susceptibility was temperature dependent, pointing again toward the presence of a paramagnetic species and an increased dissociation of the dimers. The percentage of ClO_3 by weight varied from 0.721 at $-40°C$ to 1.001 at $10°C$. The dissociation constant varied from 2.54 to 4.91×10^{-3} over the same range. From this the heat of dissociation was shown to be about $1.73 \pm 0.5\,\text{kcal mol}^{-1}$.

STUDY OF OXIDATION–REDUCTION REACTIONS AND DIFFUSION IN SITU

Because the oxidation (or reduction) products of pyrite have a much higher susceptibility than the feebly paramagnetic pyrite (χ_g ca. $0.1 \times 10^{-6}\,\text{cgs g}^{-1}$), they may be removed from coal; hence, it is technologically relevant to study this process. In our laboratory, the kinetics of oxidation was studied [272] by following the changes in the magnetization of the oxidation products. The magnetokinetic approach proved superior to the standard TGA or DTA techniques used by other workers, in that the identification of the products could be carried out by both the magnetic and Mössbauer spectroscopic techniques.

Interestingly, after the slopes of the magnetokinetic curves reach the peak effect (that is, above 100 s and up to the point of approaching constant magnetization), they increase with decreasing particle size. Exploratory runs between 400 and 500°C generally indicated that the oxidation of smaller particles yields larger net magnetization for the reaction product at room temperature.

In Figure 3.75 typical results on the magnetization per gram of the sample are plotted as functions of time during the oxidation of pyrite with three particle-size distributions (250–180, 180–125, and 125–90 μm) at a constant reaction temperature of 450°C. In each case, the magnetokinetic curve shows a peak in magnetization, which occurs during the initial states of pyrite oxidation. These peaks represent a smaller percentage of the total net magnetization as particle size is decreased, and they have been attributed to the formation of γ-Fe_2O_3 on the surfaces of pyrite particles. This is reasonable since a relatively small amount

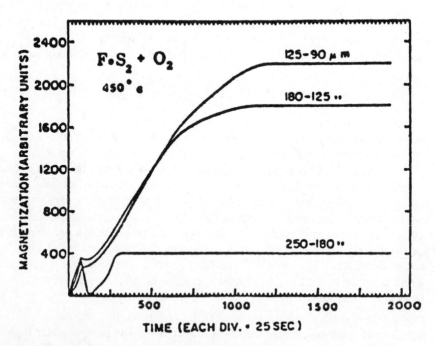

Figure 3.75 Magnetization of pyrite during oxidation at 450°C as a function of time for three particle-size distributions.

of pyrite is expected to have reacted at this point and also the calculated magnetization based on initial weights taken is much too large to be accounted for in terms of the formation of α-Fe_2O_3, which is antiferromagnetic. The decrease in this initial peak is caused by conversion from γ-Fe_2O_3 to α-Fe_2O_3, which starts at about 427°C and is completed at 510°C.

In subsequent work, Marusak and Mulay [273] investigated the possibility of formation of $Fe_2(SO_4)_3$ and the decomposition of $Fe_2(SO_4)_3$ to give iron oxides

$$FeS_2 + \tfrac{11}{4}O_2 \rightarrow \tfrac{1}{2}\alpha\text{-}Fe_2O_3 + 2SO_2$$

$$FeS_2 + \tfrac{11}{4}O_2 \rightarrow \tfrac{1}{2}\gamma\text{-}Fe_2O_3 + 2SO_2$$

To understand the kinetics of the second reaction, which produced initially the most abundant species (γ-Fe_2O_3), plots of $\ln(dM/dT)$ as a function of $1/T$ were obtained; values of dM/dT were taken from the initial slopes of the M versus T curves similar to Figure 3.75 over the temperature range from 425 to 500°C and for a starting particle size of pyrite in the range from 250 to 180 μm. From the $\ln(dM/dT)$ versus $1/T$ curves, an activation energy of approximately 7 kcal mol^{-1} was obtained for the second reaction in reasonable agreement with a value of approximately 11 kcal mol^{-1} reported by earlier workers.

It was concluded that the formation of $Fe_2(SO_4)_3$ is controlled by the starting size of the pyrite used for oxidation, which is in agreement with earlier work. In addition, the starting particle size has a direct effect on the resulting magnetization of the reaction products, which is related to the final particle size. The presence of γ-Fe_2O_3 in the reaction products was not detected in the Mössbauer spectra or in the X-ray diffraction data. This may be attributed to the presence of small quantities of γ-Fe_2O_3 present, the largest concentration being only about 7% by weight as determined by magnetic measurement. The original papers [272, 273] should be consulted for other technical details and for references to relevant work.

A classic application of magnetic susceptibility to a study of oxidation–reduction reactions occurring in a Leclanche dry cell is described by Selwood [25]. The problem was to determine whether the reduction of manganese ion in the pyrolusite, which is employed as a depolarizer in the cell, occurs during the discharge of the cell or during its recovery. A small dry cell was suspended horizontally by two fine copper wires that also acted as conductors for discharging the cell. One end of the cell was placed between the poles of a magnet with a field of about 9000 Oe. Changes in the magnetic susceptibility of manganese during discharge and recovery were expected to produce proportional displacements. These were observed by a micrometer microscope. The deflection during discharge was plotted as a function of the coulombs withdrawn. In the system studied manganese was the only paramagnetic constituent of the cell. Since the magnetic susceptibilities are virtually linear with the oxidation state of manganese in a magnetically dilute system such as the one provided by the disperse structure of pyrolusite, the deflections observed could be interpreted in terms of the change in the oxidation state of manganese in the cell. It was found that the cell contents became more paramagnetic during the discharge of the cell and that manganese reduction occurs only during the discharge.

A detailed investigation of the magnetokinetics of iron atoms diffusion during the antiferromagnetic to ferrimagnetic λ transition in Fe_9S_{10} was undertaken by Marusak and Mulay [274] because of its possible relevance to the removal of pyrite from coal by reducing the feebly paramagnetic pyrite (χ_g ca. 0.1×10^{-6} emu g^{-1}) to a ferrimagnetic species with a much higher magnetization. The antiferromagnetic to ferrimagnetic transition was interesting because it appeared to lend itself to magnetization (σ) techniques whereby the changes $\Delta\sigma$ could be followed as a function of time t at specific points around the λ transition at constant temperature T.

The antiferromagnetic to ferrimagnetic phase transition, referred to as the λ *transition* in the cation-deficient iron sulfide Fe_9S_{10}, has been studied by Marusak and Mulay [274]. Typical results obtained in our laboratory for the magnetization as a function of increasing temperature are shown in Figure 3.76, which depicts the λ-type peak. When the product is cooled, a reversible transition with thermal hysteresis is observed. The magnetic structure of Fe_9S_{10} consists of ferromagnetically coupled planes of iron atoms that are ordered antiferromagnetically along the c axis of the pseudohexagonal NiAs cell. Several theories to describe the λ transition were proposed that differ in minor but significant ways and involve transitions either from one superstructure to

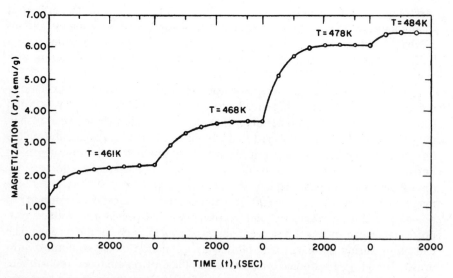

Figure 3.76   Typical magnetokinetic results (per gram σ vs. time t) at various temperatures. Data were obtained in a constant field of 10,000 Oe.

another or to a rearrangement of Fe^{2+} and Fe^{3+} ions. Early workers explained the thermomagnetic behavior of Fe_9S_{10} as follows. Above the β transformation (600 K) both spins and vacancies are disordered. When the temperature is reduced, magnetic order sets in at 600 K; however, because of the vacancy disorder, exact cancellation of the two sublattices occurs, resulting in antiferromagnetism. Around 550 K vacancy ordering occurs in such a way that imperfect magnetic sublattice cancellation and consequent ferrimagnetism result. At a somewhat lower temperature (460 K) ferrimagnetism again reverts to antiferromagnetism. This effect was attributed to the redistribution of the vacancies to a disordered state brought about by an opposite coupling between spin order and vacancy order.

The synthesis of Fe_9S_{10} and the techniques for measuring magnetizations as a function of time and/or temperature are described by Marusak and Mulay [274]. The sample of Fe_9S_{10} was characterized by X-ray diffraction and had the 5 C structure [274].

The magnetization as a function of time for various temperatures about the transition is shown in Figure 3.76. These magnetization growth curves were analyzed to obtain a diffusion coefficient for the iron atoms diffusing into the vacancies and vice versa, by using the well-known concepts and equations, which describe the reaction rate theory as applied to diffusion processes. Finally, Marusak and Mulay [274] obtained a diffusion coefficient of iron atoms along the c axis to be approximately 10^{-18} cm^2 s^{-1} over the temperature range from 440 to 484 K. The original papers discuss details of calculations, possible mechanisms for the diffusion, Mössbauer spectroscopy, and references to other relevant work.

STRUCTURAL ASPECTS OF COORDINATION COMPLEXES

This is by far the largest area in which magnetic susceptibilities and the moments derived therefrom have been used. It is impossible to indicate, even briefly, such applications because of the variety of the complexes and the bonding they involve. Therefore, recent books (cf. Mulay and Boudreaux [6]) and coordination chemistry reviews must be consulted. Some rudimentary concepts are presented here; however, any discussion of the *crystal field theory* presented in [6] is beyond the scope of this chapter.

A major part of magnetochemical studies deals with complexes and compounds of transition elements. Simple measurements reveal the diamagnetic or paramagnetic nature of the material under study. The measurements on solid materials permit a study over a wide range of temperatures. Quite often studies are made on solutions, and the susceptibility of the solute is derived therefrom. Great caution must be exercised in interpreting such data because of: the possibility of dissociation of the material in solution, interaction between the solvent and the solute, and changes in the degree of its magnetic dilution.

In the case of paramagnetic materials, magnetic measurements over a range of temperatures permit evaluation of the Weiss constant and of the magnetic moment per gram atom of the metal ion from the measured molar susceptibility.

In refined calculations the diamagnetic susceptibility of the ligands must be subtracted from the measured molar susceptibility. Diamagnetic corrections for some ligands may be calculated, somewhat approximately, by using the atomic and correction constants of Pascal that are given in Tables 3.2 to 3.5 and 3.7. For simple ions and ligands of low molecular weight, the diamagnetic correction often may be neglected.

Generally the theoretical spin-only formula

$$\mu_{\mathrm{eff}} = \sqrt{n(n + 2)}$$

gives a rough agreement with experiment and allows an estimate of the number of unpaired electrons n in the molecule.

Van Vleck's equations, which take into account the orbital contribution, may be employed to calculate the moments of rare earth ions, in which the unpaired electrons in the $4f$ level are well shielded from the influence of surrounding ions and ligands. Hence, the electrons are free to orient with the applied field. Table 3.9 lists moments for the tripositive rare earth ions. The transition metal ions differ in this respect because the orbital moments of their ions are not free to move and to orient themselves with the applied field. Their magnetic properties are modified by interactions with the surrounding ions and molecules. In certain cases *quenching* of the orbital moment is incomplete and many deviations from the *spin-only* formula are observed. Table 3.8 lists the range of values observed for the magnetic moments of transition metal ions and the moments calculated by the spin-only value.

Pauling's *valence bond* (VB) theory [64] was used to deduce the type of bonding present in simple complex compounds. Generally, Hund's rule is applied; this assumes that electrons fill all the available orbitals before pairing off. Table 3.8 shows that the magnetic moment increases from 0 to 5.9 for Fe^{3+} and Mn^{2+} ($n = 5$) and then decreases to zero. The agreement between theory and experiment is satisfactory up to five unpaired electrons; ions with six, seven, and eight d electrons show somewhat higher magnetic moments than those expected from the spin-only formula. These deviations may be attributed to incomplete quenching of orbital moments. A discussion of spin-orbit coupling is given by Mulay and Boudreaux [6] and in the references therein.

Table 3.14 illustrates the relationship between the magnetic moment and the nature of bonding in complexes of manganese based on the valence-bond theory; these complexes are particularly selected because the oxidation state, coordination number, and chemical bonding of manganese in these complexes vary significantly. Different values of magnetic moment were reported by some workers for the same compound; however, their comparisons with the theoretical moments indicate the presence of the unpaired electrons, shown in the second column of Table 3.14.

A study of coordination compounds has stimulated the most activity in the area of magnetic susceptibility. The earlier interpretations of the results in terms of the valence-bond theory, however, must be treated with caution. Thus, at one

Table 3.14 Magnetic Moments of Some Manganese Complexes[a]

	Unpaired Electrons, n	$\mu = \sqrt{n(n+2)}$	Example	Orbitals Used in Bond Formation				Description of Structure
				3d	4s	4p	4d	
Mn(VII)	0	0.00	MnO_4^{-}[b]	[ ][ ][ ][ ][ ]	[::]	[::][::][::]		sp^3 Tetrahedral
Mn(VI)	1	1.73	MnO_4^{-2}	[↑][ ][ ][ ][ ]	[::]	[::][::][::]		sp^3 Tetrahedral
Mn(II)	1	1.73	$Mn(CN)_6^{-4}$	[↑↓][↑↓][↑][::][::]	[::]	[::][::][::]		"Inner" orbital d^2sp^3 octahedral
Mn(III)	2	2.83	$Mn(CN)_6^{-3}$	[↑↓][↑][↑][::][::]	[::]	[::][::][::]		Same
Mn(II)	3	3.88	$MnPy_2Cl_2$[c]	[↑↓][↑][↑][↑][::]	[::]	[::][::][::]		dsp^2 Square planar
Mn(III)	4	4.90	$Mn(Acac)_3$	[↑][↑][↑][↑][ ]	[::]	[::][::][::]	[::][ ][ ]	"Outer" orbital sp^3d^2 octahedral
Mn(II)	5	5.92	$Mn(dipy)_3 \cdot Br_2$	[↑][↑][↑][↑][↑]	[::]	[::][::][::]	[::][ ][ ]	Same
Mn(II)	5	5.92	$Mn(kojate)_2$	[↑][↑][↑][↑][↑]	[::]	[::][::][::]		sp^3 Tetrahedral

[a]Key: Py, Pyridine; Acac, acetylacetone; kojic acid, 2-hydroxymethyl-5-hydroxy γ-pyrone.

[b]$KMnO_4$ shows a feeble temperature-independent paramagnetism. This does not affect the electronic and structural assignments [27].

[c]Kleinberg, Argensinger, and Griswold (cf. [3]) and Cox and co-workers (cf. [3]) state that this compound is a nonelectrolyte and has a plant structure. However, conflicting evidence is found in the literature [cf. 25]).

time a diamagnetic tetracoordinated Ni(II) complex was interpreted as a square planar complex, whereas a paramagnetic Ni(II) complex was interpreted as a tetrahedral complex.

Ni(II)   ↑↓ ↑↓ ↑↓ ↑↓ ..   .. dsp^2 square planar

Ni(II)   ↑↓ ↑↓ ↑↓ ↑ ↑   ..   . . sp^3 tetrahedral

In terms of the rapidly growing approach of the crystal field theory (see references in [6]), the square planar Ni(II) complexes may be either diamagnetic or paramagnetic, the former occurring with strong-field ligands, and the latter with weak-field ligands. Crystal field theory indicates that tetrahedral Ni(II) complexes should always be paramagnetic.

Actually, the new approach stems from the earlier and pioneering work of Van Vleck [21] and Bethe [275]. As discussed previously, many polynuclear species show subnormal magnetic moments and may even show completely diamagnetic behavior (cf. [264]). This may arise from exchange interactions between electrons on adjacent metal ions or may be caused by π bonding between a metal ion and the bridging group in the polynuclear species [6] and may also be caused by superexchange.

ELECTRONIC TRANSITIONS IN THE MIXED VALENCE OXIDES OF TITANIUM

Mulay and co-workers [31, 276–282] have successfully used magnetic susceptibility and EPR spectroscopy [212–216] to elucidate the phase transitions (such as the semiconductor-to-metal transitions) in oxides of titanium (known as the *Magneli phases*), which can be represented by the general formula Ti_nO_{2n-1}. These authors were able to control and measure the temperature of each sample to within $\pm 0.1°C$ during susceptibility measurements by the Faraday method and thus to pinpoint the temperature T_t of the sharp transition with great accuracy. Figure 3.77 shows a graph of the molar susceptibility as a function of temperature for Ti_3O_5, Ti_4O_7, Ti_5O_9, Ti_8O_{11}, and $Ti_{10}O_{19}$. Oxides of this type are known as the *Magneli phases* and contain Ti^{3+} ($3d^1$, paramagnetic) and Ti^{4+} ($3d^0$, diamagnetic) ions. In early literature, they have been incorrectly referred to as the *nonstoichiometric compounds*—a proper name is *mixed valence* compounds because of the different valences involved. In the low-temperature region the susceptibility was ascribed to the semiconducting behavior because very few itinerant or delocalized electrons are involved, whereas in the high-temperature region, above the transition temperature (T_t), the susceptibility is found to decrease according to the Curie–Weiss law. This behavior was ascribed to the presence of localized ($3d^1$) electrons and to the presence of delocalized electrons.

At higher temperatures the susceptibility appeared to remain constant; this aspect was interpreted in terms of the *Pauli paramagnetism* of the delocalized electrons described under Section 4.6 on magnetic behavior.

From the Curie–Weiss behavior, ($\chi = C/T - \theta$), where C and θ are the Curie and Weiss constants, respectively, an effective magnetic moment μ_{eff}

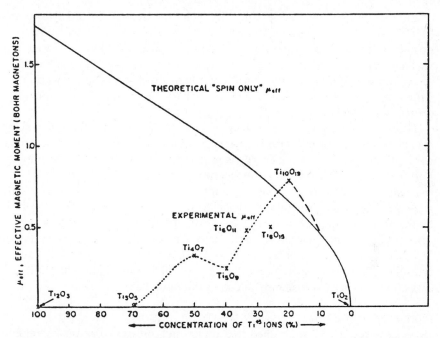

Figure 3.77 Effective magnetic moment as a function of the concentration of Ti^{3+} ions in various oxides. The solid line represents the theoretical *spin-only* values.

$(\mu_{\text{eff}} = 2.832\sqrt{C})$ was calculated. In turn this μ_{eff} gave the number n_p of localized $(3d^1)$ paramagnetic electrons, based on the spin-only formula $[\mu_{\text{eff}} = \sqrt{n(n+2)}]$. From the value n_p and the total number n_T of electrons per mole (calculated from stoichiometry), the number of delocalized electrons n_d was obtained to evaluate certain band parameters, which is explained in [276, 212–216].

Figure 3.77 shows the μ_{eff} per Ti^{3+} ion as a function of the stoichiometry along with a theoretical curve that would be observed if all the $3d^1$ electrons on all Ti^{3+} ions were to remain localized.

Figure 3.77 also shows that in most cases, the observed μ_{eff} per Ti^{3+} ion was less than the *theoretical* value, which suggested the presence of an antiferromagnetic coupling (referred to as *constrained antiferromagnetism*) between clusters of Ti^{3+} ions. The electrons that did not *couple* could be identified with the number n_d of the delocalized electrons. From the n_d values the effective electronic mass m^* could be calculated, which in turn yielded band parameters such as the fermi energy E_F, and so on, on a relative scale as a function of the stoichiometry.

Mulay and co-workers [212–216] also conducted extensive EPR and related magnetic studies, which were correlated with the magnetic parameters, X-ray crystal structure data, Hall mobility, and so on. Most importantly, the EPR studies yielded relative values for the exchange interaction J between the $[Ti^{3+}V_OTi^{4+}]$ and the $[Ti^{3+}-Ti^{3+}]$ clusters, the presence of which was postulated from X-ray structural information. (Here, V_O refers to the oxygen vacancy centers that are present in these oxides.) A plot of the transition temperature T_t and the J values as a function of stoichiometry showed that the values of T_t and J decrease as the stoichiometry varies from Ti_2O_3 and TiO_2 (which is an insulator).

Hence, as the *coupling* (i.e., values of J) increases between the *clusters*, the transition temperature T_t also increases, as one would expect from the simple concept that relatively more energy is required to break the strong homopolar bonds. Mulay and co-workers further proposed that the *constrained antifer-romagnetic coupling* (with fewer available delocalized electrons) is primarily responsible for the semiconducting behavior of titanium oxides in the low-temperature region, whereas the breaking up of such coupling, which is responsible for the creation of more delocalized electrons at higher temper-atures, gives the metallic behavior. Thus, thermomagnetic analysis of the above type (along with EPR spectroscopy, etc.) is a tool for elucidating the electronic transitions in unusual systems such as titanium oxides.

INSULATOR–METAL TRANSITIONS IN DOPED AND UNDOPED SmS

In recent years it was established that under high pressure, drastic crystal and electronic structure changes can be brought about, which then give rise to a change in the valence of the system. Thus the concept of intermediate or fractional valence (such as $Sm^{2.5}$ in which the Sm ion is halfway between the Sm^{2+} and the Sm^{3+}) is invoked. In these situations pressure is treated as a *variable similar to temperature*.

The work of Cordero-Montalvo, Vedam, and Mulay [108] is a typical example; this paper gives references to the pioneering work by others in the high-pressure field. Because space is limited and the system is very complex, only an abstract of their work is presented in this chapter. (Figure 3.78 gives a typical illustration of the transition.)

The compound SmS provides an interesting example, in that this sulfide at normal atmospheric pressure and room temperature is black; however, when pressure is applied (ca. 6 kbar), it changes to a golden yellow color. This has been explained on the basis of a fluctuating valence intermediate between Sm^{2+} and Sm^{3+} with an absence of magnetic ordering even at low temperatures. In [108] is reported measurements of magnetic susceptibility χ for SmS alloyed with GdS, LaS, and YS as a function of pressure P and of composition x of the dopant at 295 K. When entering the intermediate valence (IV) state, a sharp decrease in x similar to that of SmS is obtained. The susceptibility of $Sm_{0.85}Gd_{0.15}$, which retains the high-pressure phase metastably, was measured at 1 atm for several temperatures below ambient; χ deviates from the simple additivity law for SmS and dopant when x is varied. This was attributed to several mechanisms, among which conduction electron enhancement of Sm–Sm exchange interaction and an increase in valence from alloying are the most important. A gradual increase in the valence of Sm with pressure is also observed. It is concluded that the

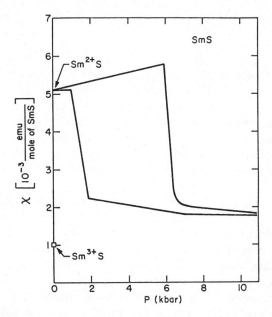

Figure 3.78   Magnetic susceptibility χ_M of SmS as a function of pressure P.

moment of the Gd ion has no role in the transition to the IV state that results from either pressure or alloying and that these two parameters are not equivalent for these alloys.

NEW SUPERCONDUCTING MATERIALS

Professor Bardeen, who is well known for his contributions to superconductivity, astutely remarked in 1956 that it is more fruitful to view a superconductor as an extreme state of diamagnetism (where the susceptibility reaches a maximum value of $-1/4\pi$ below the critical temperature T_c) than as a limiting case of infinite conductivity. Hence, a measurement of diamagnetic susceptibility as a function of temperature has become important in determining T_c and in elucidating the transition to the superconducting state in the old metallic and the new oxide materials.

The yttrium-based and related high-temperature oxide superconductors have attracted considerable attention since early 1986. Bednorz and Muller were awarded the 1988 physics Nobel Prize for their research on the oxide materials. Most researchers studied the diamagnetic susceptibility with the SQUID magnetometer (Section 6.5.6) and the ac susceptibility with an inductance apparatus (Sections 6.5.6 and 8.2.4). References to the pioneering work of Professor Paul Chu of Houston University and to several excellent contributions are found in a conference proceedings, edited by Nelson, Whittingham, and George [283], related proceedings, and in good reviews typically published in *Physics Today* (American Physical Society/American Institute of Physics).

8.3.2   Applications to Organic Chemistry

FREE RADICALS

Many organic and organometallic molecules dissociate into free radicals under conditions of dissolution of solids, pyrolysis, and photolysis. From the magnetic point of view, free radicals are entities containing unpaired electrons and, as such, behave as paramagnetic compounds. The stable free radicals are easily studied by the magnetic susceptibility technique; absence of an orbital contribution to magnetic moment somewhat simplifies the calculation of the magnetic moment by the spin-only formula. A study of free radicals with extremely short lives of about microseconds can be studied very effectively by the technique of EPR.

Free radicals are highly reactive; and, although somewhat stable in the solid state, they are particularly autooxidizable in the dissolved state. Their susceptibility measurements may be made by the Gouy method; in the solid state this procedure is not problematic. However, since the magnetic state of a free radical in solution is usually quite different from that in the solid, measurements for the two states yield copious analytical and structural information about the free radicals. Naturally, their study in solution requires several precautions, including *prevention of* (1) autooxidation by conducting the magnetic measurements in an inert atmosphere, (2) interaction with the solvent wherever possible, and (3) solvent loss by evaporation, a normal precaution whenever working with solutions.

The literature contains many magnetic studies on free radicals in systems such as hexaaryl and hexaalkyl ethane, organometallics, semiquinones, porphyrins, highly conjugated systems, and metal ketyls. Dainton and co-workers [283] have shown that solutions of potassium in ethers, unlike solutions of alkali metals in ammonia and amines [284], are diamagnetic. Many authors, notably, Selwood [25], Walling [84], Hutchinson [285], and Wheland [286], have reviewed the literature. Accounts of the electron paramagnetic resonance studies are given by Ingram [287], Vanngard and Malmstrom [288], and Blois and co-workers [289]. Only one application of the magnetic susceptibility data is outlined here with reference to a classical study of the dissociation of dihexaaryl ethane [25].

The studies of chemical and optical properties and of apparent molecular weights showed that the hexaaryl ethanes probably dissociate into the corresponding triaryl methanes. These methods are in error. In some cases, solutions exhibit color even after a complete disappearance of the free radical; hence, the assumption that only the free radicals are colored and the corresponding intensity calculations give rise to erroneous results. The dissociation constant at 20°C for the hexaphenyl ethane by the colorimetric method is 4.1×10^{-4} in benzene, 19.2×10^{-4} in carbon disulfide, and 1.2×10^{-4} in propionitrile. The heat of dissociation was about 11 kcal/mol. The magnetic method gives 2×10^{-4} for the dissociation constant in benzene (20°C); this is about half the value obtained from the colorimetric method.

The experimental procedure and the nature of calculations follow for determining the dissociation constant of hexaphenyl ethane:

$$\phi_3C\text{—}C\phi_3 \rightleftharpoons 2\phi_3C$$

A known weight of chloromethane is dissolved in benzene and is shaken with silver to form the hexaphenyl ethane. It is assumed that the reaction is complete without any solvent loss. The percentage of ethane (5.49% found in an experiment) is calculated from the stoichiometry. The specific susceptibility of benzene is taken as -0.708×10^{-6}. From the measured specific susceptibility of the solution (-0.700×10^{-6}), and using Wiedemann's law, the specific susceptibility χ_g of ethane is -0.56×10^{-6}. The apparent molar susceptibility χ_m for the partly dissociated ethane is -272×10^{-6} from $\chi_m = \chi_g \times$ molecular weight. Using Pascal's empirical constants, the molar diamagnetic susceptibility for the ethane is calculated as -325×10^{-6}. The apparent diamagnetic molar susceptibility, as measured, is less than the calculated value; the difference is attributed to a paramagnetic contribution caused by the formation of the free radical ϕ_3C from the hexaphenyl ethane. The molar paramagnetism representing partial dissociation is thus $[-272-(-325)] \times 10^{-6} = 53 \times 10^{-6}$. The degree of dissociation α is calculated from the ratio of molar paramagnetic susceptibility (53:2540) obtained for partial dissociation to that expected (2532×10^{-6}) for complete dissociation of the hexaphenyl ethane; thus, α is equal to 2.1%.

The value 2532×10^{-6} is derived in the following way. When completely dissociated, 1 mol of hexaphenyl ethane gives 2 mol of triphenylmethyl free radicals. The molar susceptibility χ'_m corresponding to one unpaired electron is calculated from its magnetic moment $\mu = \sqrt{n(n + 2)} = 1.73$ Bohr magnetons. The simple relation used here is $\mu = 2.84\sqrt{\chi'_m \cdot T}$, where T is the temperature in kelvins; 2.84 is the numerical value of the constant $3k/N\beta^2$; and k, N, and β are the Boltzmann constant, the Avogadro's number, and the Bohr magneton, respectively. When 1.73 is used for μ and 293.16 K for the temperature, χ'_m is 1266×10^{-6}. For 2 mol we obtain 2532×10^{-6}.

The magnetic method is based on certain assumptions. For instance, in using the formula $\mu = 2.84\sqrt{\chi'_m \cdot T}$, it is assumed that the Weiss constant for the triphenylmethyl radical is zero. Although some experimental evidence indicates that the Weiss constant is not too large for such systems, it is often more than negligible. This poses the problems of having to explain any appreciable interactions unexpected in such magnetically dilute systems and of making adequate corrections to the values calculated for χ'_m.

Another inherent difficulty is to determine accurately the diamagnetism of the (paramagnetic) organic free radicals in these solutions. This is commonly done by using Pascal's empirical constants (Tables 3.2–3.5, 3.7, and 3.8) or by measuring the molar susceptibility of the corresponding methane and subtracting from it the atomic susceptibility of hydrogen. However, neither method is capable of allowing for the anomalously large diamagnetism of certain free radicals. This diamagnetism may be as large in magnitude as the paramagnetism of the free radical. It stems from the possibility of motion of electrons over large distances in many resonant structures, which, according to current theories, are said to stabilize the free radicals with respect to ethane. The resonance stabilization may be accompanied by a large increase of molecular anisotropy. These considerations have been used particularly to explain the strange observation [25] that hexa-*p*-biphenylethane and hexa-*p*-*tert*-butylphenyl-ethane were only 70% dissociated and the apparent degree of dissociation was not dependent on temperature.

Despite some shortcomings, the magnetic susceptibility method is very useful, not only in its analytical aspect, but also in testing the theories of free radical stability.

STUDIES ON REACTIONS AND POLYMERIZATION

The oxidation of styrene [290] and the reversible reduction of duroquinone [151] are typical examples of reaction studies *in situ* by the magnetic method. A horizontal Gouy magnetic balance (Figures 3.53 and 3.54) designed by Theorell is quite adequate for such studies.

The work of Farquaharson and Ady [291] is quite representative in polymerization; rather extensive reviews [25] are also available. Recently, Haberditzl (cf. [5]) reviewed this topic thoroughly.

The polymerization of 2,3-dimethylbutadiene is considered. Suppose that $\chi_m M$ and $\chi_m P$ represent the molar susceptibilities of the monomer and a polymer containing n molecules of the monomer; then, according to concepts of additivity,

$$\chi_m P = n\chi_m M + (n - 1)\lambda$$

where $(n - 1)$ corresponds to the new bonds formed between n molecules, and λ is the corresponding correction constant for each bond. The specific susceptibility of the polymer $\chi \cdot P$ is now given by

$$\chi \cdot P = \frac{n\chi_m M + (n - 1)\lambda}{n \times \text{mol wt of monomer}}$$

The polymerization of 2,3-dimethylbutadiene shows a growth curve expected for this inverse relationship between $\chi \cdot P$ and n. In general, by using Pascal's values for the specific bonds involved in polymerization, it is possible to find the degree of polymerization n. The sensitivity of the method decreases rapidly with increases in n because the change in $\chi_m P$ with respect to n varies inversely with n^2, which is seen by differentiating this expression. The specific susceptibilities of 2,3-dimethylbutadiene and its final polymerized product are -0.670×10^{-6} and -0.7305×10^{-6} cgs units, respectively. Figure 3.79 shows the change in diamagnetic susceptibility with time. Although the initial and final products are diamagnetic, a possibility exists for the formation of free radicals, which may

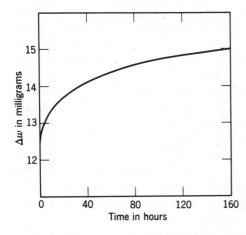

Figure 3.79 Polymerization of 2,3-dimethyl-butadiene; apparent change in weight on application of the magnetic field versus time.

induce the polymerization. Investigations showed that the induction period for the polymerization of 2,3-dimethylbutadiene is about 3 h in the absence of any accelerator. During this period the magnitude of the molar susceptibility of the polymer decreases from -60×10^{-6} to about -42×10^{-6}; the sharp decrease is, therefore, attributed to the formation of paramagnetic free radicals. Simple calculations show that this decrease can be attributed to the formation of 0.7% of species with one unpaired electron. This was verified indirectly by the addition of the same amount of benzoyl peroxide, an accelerator that results in a polymerization with no initial decrease in susceptibility.

A significant new development in organic polymers is reported by Berlin and co-workers [292]. Polymers containing nitrogen and polar groups in a conjugated chain in a macromolecule show a rapid decrease in paramagnetic susceptibility, which falls off rapidly with increasing field. This is said to be similar to a ferromagnetic behavior. The cloud of strongly interacting unpaired electrons is said to unite the whole structure into a single electronic unit. Preliminary experiments show that polyaminoquinones can act as extremely effective catalysts for decomposition of hydrogen peroxide and other oxidation–reduction reactions. Somewhat similar observations are reported by Krause (cf. [3]). The polymers obtained from a reaction of benzidine and chloranil gave polymers with high susceptibility (ca. 1.28×10^6), which indicates a dependence on the applied field. The polyaminoquinones show strong internal hydrogen bonds and form complexes with metals. The susceptibility and further EPR data were used to propose definite structures for the polyaminoquinones. Additional comments on the magnetic properties of (polymeric) nucleic acids are given in later sections.

STRUCTURAL ELUCIDATION OF BONDING IN ORGANICS BASED ON AVERAGE SUSCEPTIBILITIES

Pascal initiated studies on the determination of structures of a multitude of organic compounds, such as the hydrocarbons, alcohols, aldehydes, ketones, and various aromatic systems. He established values for atomic susceptibility constants χ_A and the constitutive correction constants λ for various types of

bonds (e.g., $-C{=}C$, $-C{\equiv}C-$, $-C{\displaystyle{\overset{\textstyle O}{\underset{\textstyle R}{\diagup}}}}$) listed in Tables 3.3–3.5 and 3.7.

Some researchers still find these values, derived empirically from studies on a multitude of organic compounds, useful for quickly obtaining approximate values for the molar susceptibilities of specific organic structures or ligands in coordination compounds. Despite their approximate nature, these values were useful for applying diamagnetic corrections to the measured values of paramagnetic complexes. The diamagnetic corrections can be quite large if the ligands become particularly bulky, that is, of a very high molecular weight. Numerous examples of determining organic structures are given by Bhatnagar and Mathur [2], who summarized the work up to about 1935. In 1955 Hutchinson [285] reviewed work on the determination of whether cyclooctatetracene is a nonplanar, fourfold conjugated system. It was concluded that cyclooctatetracene may be pictured as a conjugated nonaromatic ring. Reference should be made to these early works and to a 1966 monography by Mulay [3], because these sources give insight into the early empirical, yet ingenious, computations that served organic research quite well in its early days. However, despite many attempts, a sound theoretical significance for the constitutive correction constants λ was not found.

In recent years Haberditzl (cf. Chap. 3 in [162]) developed atomic increment systems based on quantum mechanical considerations. In addition, he discussed several applications of diamagnetism, not only to relatively simple types of organic compounds, but also to conformations of complex ring systems, metalated alkyl compounds, charge-transfer complexes, polymerization processes, and the dependence of NMR chemical shifts on the diamagnetism of organic structures. These aspects are too numerous to summarize here even briefly; as such, the reader is referred to Haberditzl's outstanding contributions to applications, to organic systems, and to some inorganic systems.

The use of high-resolution NMR spectroscopy with 100 and much higher MHz frequencies, as well as the sophisticated Fourier transform techniques and *magic angle NMR for solids* have indeed revolutionized organic structural analysis; yet it is important to recognize that a deeper theoretical understanding of the diamagnetism of molecules is helpful, not only in gaining an insight into the theoretical aspects of NMR spectroscopy, but, equally, in exploring other areas such as the charge-transfer complexes (some of which are said to show superconducting properties), polymerization processes, effects of external magnetic fields on enzymes, and biological systems.

ELUCIDATION OF BONDING IN ORGANOMETALLICS: MAGNETIC ANISOTROPY

Most of the anisotropy work reported in the literature (cf. [26]) was on magnetic materials of technological significance. These materials include single crystals, *whiskers* or *thin films* of ferromagnetic metals, and alloys. Again, in the *nonmetallic* (insulators) area, relatively more studies were made on *ionic solids*, for example, on *ferrite*-type materials and on salts of rare earths. In general, very few studies are reported on *molecular solids* in which most chemists are interested (i.e., coordination complexes and organometallic compounds).

Haberditzl (cf. Chap. 3, [162]) gave a comprehensive list of the diamagnetic anisotropies of several aromatic compounds, which elucidated their electronic structures. As a typical example, we summarize the work by Mulay and co-workers [163, 244] on a few metallocenes, which are indeed aromatic. A major part of the diamagnetism of aromatic systems, such as benzene, arises from a circulation of the π electrons in orbits defined by the radii of their rings; and the diamagnetic anisotropy of such molecules, as shown by Pauling [64], provides a measure of the π-electron density in the ring system. Mulay and Fox [163] first studied the diamagnetic anisotropy of single crystals of ferrocene [Cp_2Fe, where Cp is a cyclopentadienyl ring, $(C_5H_5)^-$] to resolve specific problems about its electronic structure. Subsequently, Mulay and co-workers [244] extended the diamagnetic anisotropy measurements to ruthenocene (Cp_2Ru) and to osmocene (Cp_2Os) with a view to (1) elucidating their bonding by appropriately developing their molecular orbital (MO) descriptions, (2) establishing a relative scale of π-electron density in the Cp rings in each of the three metallocenes, and (3) thus quantifying the observed decrease in the electrophilic reactivity in the series $Cp_2Fe > Cp_2Ru > Cp_2Os$.

The systematic examination of the diamagnetic anisotropy of a *homologous* series of metallocenes such as Cp_2Fe, Cp_2Ru, and Cp_2Os led to a unique method of characterizing such molecular solids at the macroscopic and the microscopic levels. Macroscopic characterization is based on establishing precisely the magnitude of the principal susceptibilities (χ_1, χ_2, and χ_3) of a single crystal of metallocene; all parameters have a negative sign, as generally observed in diamagnetic solids. Any changes or imperfections in crystal structure and the inclusion of trace impurities or defects in the lattice are immediately reflected in the anisotropic parameters χ_1, χ_2, and χ_3, which are naturally more sensitive than the trace (or average) susceptibility χ of a polycrystalline solid; χ is equal to $\frac{1}{3}(\chi_1 + \chi_2 + \chi_3)$. The microscopic characterization corresponds to obtaining a complete description of electronic bonding between the metal atom and the ligand. This characterization is accomplished by a further treatment of the (macroscopic) crystal susceptibilities χ_1, χ_2, and χ_3 and various crystal structure parameters (e.g., molecular orientations, number of molecules per unit cell) that yield the orthogonal molecular susceptibility components K_1, K_2, and K_3 defined along appropriate axes within a *single* molecule. The molecular anisotropy ΔK, which is conveniently defined as $\Delta K = K_3 - 1/2(K_1 + K_2)$, then provides insight into the electronic structure of metallocenes. Descriptions of these aspects follow.

Mulay and Fox [163] measured the principal susceptibilities χ_1, χ_2, and χ_3 of a single crystal(s) of ferrocene using Krishnan's flip-angle method, described in

Section 7. Subsequently, Mulay and Withstandly [244][†] measured the principal susceptibilities of single crystals of osmocene and ruthenocene.

By using the X-ray crystal structure data from the literature and Krishnan and Lonsdale's transformations for appropriate crystal structures, Mulay and co-workers calculated the molecular susceptibilities K_1, K_2, and K_3 for each of the above metallocenes. Relevant structural information for these is given in Figure 3.80. The results of their work are summarized in Tables 3.15 and 3.16.

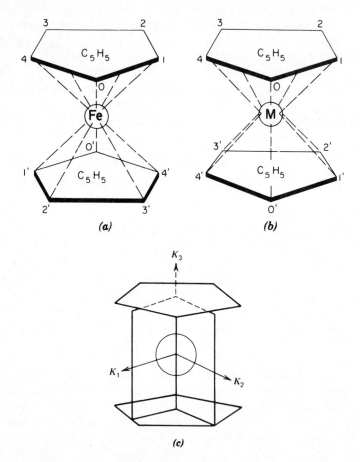

Figure 3.80 (*a*) Staggered structure of ferrocene; (*b*) eclipsed structure of ruthenocene or osmocene; (*c*) molecular axes for the magnetic anisotropy of ferrocene, showing the directions of the principal molecular susceptibilities K_1, K_2 (in the plane of the ring), and K_3 for ferrocene.

[†]A *printing error* appeared in [244]: Values reported for $\chi_a = -137.0$, $\chi_b = -148.0$, and $\chi_c = -165.0$ (all times 10^{-6} emu per mole) correspond to *ruthenocene* and not to *osmocene*. Correct values for osmocene are -183, -189, and -207 (all times 10^{-6}), respectively. The correct molar susceptibilities were since incorporated into subsequent publications, for example, in [5, 6, and 160] and in Table 3.15.

Table 3.15 Summary of Anisotropy Results on Metallocenes

Compound, Crystal Habit, and Space Group	Crystal Anisotropy per mole (-1×10^{-6} emu)	Crystal Susceptibility per mole (-1×10^{-6} emu)	Molecular Susceptibility Components[a] (-1×10^{-6} emu)	Molecular Anisotropy $\Delta K_\perp$ ($\times 10$ emu) $\Delta K_\perp = K_3 - (K_1 + K_2)/2$	π-Electron Density in Each Cp Ring
Ferrocene, $(C_5H_5)_2Fe$, monoclinic, $P2_{1/a}$	a Axis vertical $\Delta\chi = 35.6$; b Axis vertical $\Delta\chi = 39.1$; a and b Axis vertical $\Delta\chi = 10.52$	$\chi_1 = 100.0$, $\chi_2 = 135.0$, $\chi_3 = 139.4$, $\psi = 25°34'$ (ψ is the angle between magnetic axis u_1 and crystal axis c)	$K_1 = 104.1$, $K_2 = 108.6$, $K_3 = 162.3$, $\psi = 25°47'$	56.0	4.6
Ruthenocene $(C_5H_5)_2Ru$, orthorhombic, $Pnma$	a Axis vertical $\Delta\chi = 18.5$; b Axis vertical $\Delta\chi = 26.0$; c Axis vertical $\Delta\chi = 11.3$	$\chi_1 = 137$, $\chi_2 = 148$, $\chi_3 = 165$ (Magnetic and crystal axes coincide; $\psi = 0$)	$K_1 = 148$, $K_1 = 125$, $K_3 = 177$, $\psi = 0$	40.0	3.1
Osmocene, $(C_5H_5)_2Os$, orthorhombic, $Pnma$	a Axis vertical $\Delta\chi = 16.8$; b Axis vertical $\Delta\chi = 24.2$; c Axis vertical $\Delta\chi = 6.4$	$\chi_1 = 183$, $\chi_2 = 189$, $\chi_3 = 207$ (Magnetic and crystal axes coincide; $\psi = 0$)	$K_1 = 189$, $K_2 = 175$, $K_3 = 215$, $\psi = 0$	33.0	2.6

[a]Generally one expects $K_1 = K_2$ because of the symmetry of these molecules and because of the reorientation of the Cp rings about their centrosymmetric axis. The reason for the inequality between K_1 and K_2, especially in ruthenocene and osmocene, is not known and is being investigated by Mulay. The reorientation processes in ferrocene, ruthenocene, and so on in the solid state have been established by Mulay and co-workers [82] using the broad-line proton magnetic resonance technique.

Table 3.16   Comparison of π-Electron Populations in Each Ring in Metallocenes

	Cp_2Fe	Cp_2Ru	Cp_2Os
Diamagnetic aniostropy experiments	4.6	3.1	2.6
MO description (this work)	4.64	3.05	2.75

Table 3.16 clearly shows good agreement between the π-electron density in each Cp ring and that calculated from the molecular orbital descriptions for the three metallocenes.

It is important to note that the π-electron densities decrease in the order ferrocene, ruthenocene, osmocene, which in turn explains the decrease in their electrophilic substitution activity observed empirically by several organometallic chemists. Furthermore, the MO descriptions that postulate a single $d\pi-p\pi$ bond between the Cp ring and the metal atom, which leaves a large π-electron density, are confirmed and rule out the *valence-bond* pictures postulated by early workers. Reference should be made to Mulay and co-workers [160] for further details about these aspects, including, especially, a discussion of the molecular orbital descriptions and the X-ray crystal structure data.

DIAMAGNETIC ANISOTROPY AND CHARACTERIZATION OF GRAPHITES AND CARBONS

Graphite possesses fascinating layer structure (Figure 3.81) and has a large diamagnetic anisotropy. When a single crystal of natural graphite is oriented such that the magnetic vector is perpendicular to the basal plane (aromatic plane) the specific susceptibility $\chi_\perp = -21.75 \times 10^{-6}$ emu g^{-1}. If the magnetic field vector is oriented parallel to the basal plane, the specific susceptibility $\chi_\| = -0.30 \times 10^{-6}$ emu g^{-1}. The ratio $\chi_\perp : \chi_\|$ is a good measure of the diamagnetic anisotropy of the material and for selected natural single crystals [5, 6, 293, 294] is as high as 72. The high diamagnetic anisotropy and the underlying physical phenomena are fairly well known since the pioneering work in 1941 of Ganguli and Krishnan (cf. [5]). They were able to interpret the high susceptibility for a magnetic field oriented perpendicular to the basal plane as being caused by the contribution of π electrons, which are free to move throughout the planes of the benzene-type C_6 rings within the layers of graphite. In a qualitative sense, these free (or quasifree) π electrons are responsible for the high electrical conductivity in graphite along the aromatic or basal plane.

Until recently most studies were made on natural graphite; generally, the specimens from Sri Lanka (Ceylon) were regarded as *perfect* with reference to their crystal structure and desirable physical properties. Commercial processes, based on the pyrolysis of mixtures of aromatic compounds were developed to produce synthetic graphite and graphitic carbons in various grades and forms to meet the numerous applications such materials find in modern technology. For

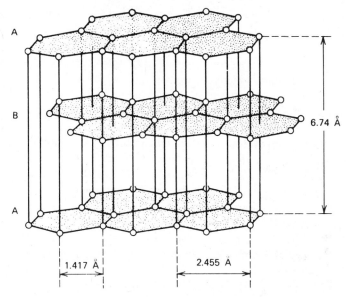

Figure 3.81   Crystal structure of graphite.

instance, graphite is used as a moderator in nuclear reactors; as electrodes for commercial electrolysis and electrical steel making, and in ultrahigh intensity light sources; and as monochromators in neutron diffraction experiments. The evolution of these synthetic materials and the availability of natural graphite from various mineral sources recently prompted a need for developing a tool for their characterization. In view of the pronounced diamagnetic anisotropy of graphite, work was undertaken in our laboratory to characterize various specimens of graphite and graphitic carbons by measuring their diamagnetic properties and by deducing appropriate electronic parameters, such as the degeneracy temperature T_0 and the effective electron mass m^*, which describe in part their band structures.

The magnitude of the $\chi_\perp$ in graphite is so great that the usual closed-shell diamagnetic theory based on the Langevin–Pauling formalism, which is adequate to explain the anisotropy in aromatic compounds, no longer suffices for graphite. Hence, it is proposed that in graphite there is an additional *Landau* diamagnetism caused by the presence of free π electrons. Section 4.6 and [5] and papers cited therein give an adequate background on the free-electron diamagnetism and associated concepts. The following paragraph summarizes the recent band-theory approaches developed for the interpretation of diamagnetism in graphite; later paragraphs outline their applications for deriving useful electronic properties.

Pacault and co-workers (cf. [5]) elaborated on a simple model proposed by Ganguli and Krishnan (cf. [5]) and noted that the semiempirical relation $y = 1 - \exp(-x)$ that was proposed to fit the experimental data can be firmly grounded on existing theories of a two-dimensional electron gas. In their theory this semiempirical relation takes the form to describe the observed susceptibility χ_t at temperature t relative to χ_0 observed at a cryogenic temperature: $\chi_t/\chi_0 = 1 - \exp(-T_0/T)$, where $T_0 = (\nu N'h^2\alpha)/(4\pi mkS)$ is the degeneracy temperature of the two-dimensional electron gas; $\chi_0 = [(\nu\beta^2 N)/(kT_0 S)(1 - \alpha^3/3)$, where N' is the total number of carbon atoms, ν is the number of holes per carbon atom in the band with effective mass $m^* = m/\alpha$, S is the area occupied by the gas, and $\beta = eh/4\pi mc$ is the Bohr magneton. Their theory unjustifiably assumes ν to be temperature independent, implying no thermal excitation of carriers (electrons and holes) from the π band to the conduction band. According to Pacault's (cf. [5]) simple model, $T_0 = \varepsilon_0/k$, where ε_0 is the energy difference from the Fermi level to the top of the valence band, and k is the Boltzmann constant (Figure 3.82).

Smaller values of T_0 signify a *more metallic* behavior, which is considered to be most desirable in several technological applications of graphite. To characterize several graphite samples, their degeneracy temperatures were evaluated from their observed diamagnetisms.

Mulay's approach [294] stemmed from earlier determinations, which showed that specific diamagnetic susceptibility $-\chi$ in carbon materials is proportional to the position of the Fermi level in relation to the edge of the valence band and that oxidation may lower the Fermi level because it produces various structural imperfections that can act as electron traps.

Average and/or anisotropic susceptibilities were measured in my laboratory on appropriate polycrystalline or single-crystal samples depending on their availability. Spectroscopic grade synthetic carbons (AGKSP and L113SP), Graphon samples (carbon black heat treated to 2700°C), and a *SRPG* (stress-recrystallized pyrolytic graphite) sample were studied along with a powdered sample (SP-1) of natural graphite. Most samples were obtained from the Union Carbide Company (Cleveland, OH) and were ascertained to be free from ferromagnetic impurities. In appropriate cases the room-temperature values of the anisotropy ratio ($R = \chi_\perp/\chi_\parallel$) were obtained, and the average susceptibility (χ_{av}) was measured or derived from $\chi_{av} = (\chi_\perp + 2\chi_\parallel)/3$. These values are listed in Table 3.17.

Among the synthetic materials the anisotropy ratio ($R = 38.2$) is highest for the SRPG sample and lowest for the Graphon sample ($R = 1.02$), whereas all samples show smaller differences in their average susceptibilities of $\chi_{av} \simeq -7 \times 10^{-6}$ emu g^{-1}. Thus, the anisotropy ratios R provide a useful characterizational parameter. A discussion follows showing that the average susceptibility χ_{av} of powdered solids is equally useful as a characterization technique provided it is measured over a range of temperatures from the easily accessible temperature of liquid nitrogen (80 K) to about room temperature (300 K).

Plots of the average susceptibility χ_{av} as a function of reciprocal temperature $1/T$ are shown in Figure 3.82a. By fitting an equation, based on Pacault's model

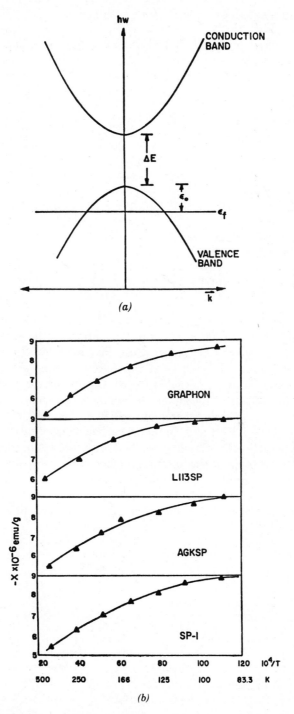

Figure 3.82 (a) Graphitic carbon band as proposed by Marchand and co-workers (cf. [293]). (b) Temperature dependence of diamagnetic susceptibility, polycrystalline graphitic carbons.

Table 3.17 Principal ($\chi_{\parallel}$, $\chi_{\perp}$) and Average (χ_{av}) Diamagnetic Susceptibilities in Units of 10^{-6} emu g^{-1} for Graphites and Graphitic Carbons at Room Temperature

Sample	$-\chi_{\parallel}$	$-\chi_{\perp}$	Anisotropy Ratio, R	$-\chi_{av}{}^{a}$
SP-1 (powder)				6.86
SRPG	0.54	20.60	38.2	(7.23)
AGKSP	8.33	4.60	1.84	(7.09)
AGKSP (powder)				7.10
L113SP	7.06	6.93	1.02	(7.02)
L113SP (powder)				7.01
Graphon carbon black heat				6.25

^{a}Values of χ_{av} in parentheses are calculated from principal susceptibilities.

for the susceptibility of a degenerate two-dimensional electron gas, to the observed curves in this figure, values for the ratio α of the rest mass m to the effective mass m^* of the free carriers (electrons and holes) and the degeneracy temperature T_0 were estimated. These values follow, along with values of v, the number of holes per carbon atom in the band with effective mass m^*.

As seen from Table 3.18 among the synthetic graphites and carbons the stress-recrystallized pyrolytic graphite (SRPG) has the lowest degeneracy temperature ($T_0 \simeq 379$ K) and Graphon carbon black heat the highest with others showing characteristic values that fall somewhere in between. Thus, when applied to magnetic measurements, the parameterizing procedure for reasonable solid-state models yields valuable information on electronic parameters that is characteristic of samples' particle sizes, defect structures, and so on. A paper by Mulay and co-workers [294] and a thesis by Santiago [293] offer further

Table 3.18 Electronic Parameters for Various Graphites and Graphitic Carbon Samples Using Marchand's Model^a

Sample	T_0 (K)	$-\chi_a \times 10^6$	α	v
SP-1	344.2	27.75	296	ca. 10^{-5}
SRPG	379.0	28.00	290	ca. 10^{-5}
AGKSP	389.4	27.95	250	ca. 10^{-5}
L113SP	402.0	27.00	240	ca. 10^{-5}
Graphon carbon black heat	506.0	20.50	188	2×10^{-5}

^{a}Values were obtained by curve fitting.

discussion and significance about these solid-state aspects and parameters such as v and the effective electron mass m^* (or α).

Therefore, natural graphite (SP-1) is still more metallic (with $T_0 = 344.2\,\text{K}$) than the synthetic SRPG sample (with $T_0 \simeq 379.0\,\text{K}$), which suggests the presence of various imperfections (mosaic character, twinning effects) in the synthetic materials. Thus, the best synthetic material available is not as defect free as one of the best (if not *the* best) sources of natural graphite.

Mulay and Boudreaux [5] and Mulay and Mulay [36–45] surveyed new results on graphite, diamond, and various carbons. Carbon and graphitized carbon black are excellent substrates for iron and so on for catalytic work. Papers by Jung and co-workers [218] and Mulay and Prasad-Rao [23] should be read. Mulay and Prasad-Rao showed that with successive doping of carbon atoms, the magnitude of diamagnetic susceptibility $|\chi|$ decreased, thus lowering the Fermi energy deeper into the valence band. When superparamagnetic iron clusters were dispersed on such substrates, they found that the methanation activity resulting from $(CO + H_2)$ increased reasonably on the boron-doped carbon–Fe catalyst, as compared to the undoped C—Fe system. In the first case the increase was attributed to an induced metal-support interaction. From the magnetization versus the ratio [field (H):temp] curves these authors evaluated the average particle size for the Fe clusters by using the Langevin function stated in Section 8.3.5 under subsection "Characterization of Fine Particles: Applications to Catalysis."

ORGANIC CHARGE-TRANSFER COMPLEXES WITH HIGH ELECTRICAL CONDUCTIVITY
Several organic charge-transfer complexes were recently found to show a great increase in electrical conductivity at cryogenic temperatures. These observations had raised the hope of producing organic (polymer) superconductors and have posed challenging problems in the synthesis of such materials and in their characterization. One of the characterizational techniques is based on the measurement of their diamagnetic susceptibilities, which in appropriate situations follow the fluctuations in electrical conductivity with temperature. Since the magnetic susceptibility technique is nondestructive, it has assumed an important role in research on superconductivity. The role of magnetic measurements in such research has been reviewed by Mulay and Boudreaux [5]. In early work charge-transfer complexes formed between TTF and TCNQ shown below and their derivatives were studied as suitable candidates for organic superconductors [210]. These complexes are now considered to be pseudo-superconductors:

TTF   TCNQ

8.3.3   Applications to Biosystems

Applications of magnetic studies to biochemical problems have dealt with structural aspects of components of blood and a study of their reactions. In recent years studies on other biological components, such as ribonucleic acid (RNA) and deoxyribonucleic acid (DNA) were reported; many studies on enzymes and enzymatic reactions were carried out also. A general trend is to use the SQUID magnetometer, which apparently has the highest sensitivity for measuring the small changes in the diamagnetic or feebly paramagnetic components of such biosystems. It is gratifying to note that new instrumentation based on the superconducting quantum interference device (SQUID) is finding increasingly more applications to supplement or to complement the information obtained from nuclear and electron magnetic resonance spectroscopy and from Mössbauer spectroscopy.

The following paragraphs describe some of the *classical work* on biosystems, which was the foundation of the work that began, surprisingly, with Michael Faraday who attempted to measure the effect of a magnetic field on a dead frog and later decided to study the magnetic properties of blood.

STUDY OF COMPONENTS OF BLOOD

This area was pioneered by Pauling [63] with Coryell [65] and by Haurowitz and Kittel [61]. Recently, Pauling (cf. [43]) confirmed the diamagnetism of oxyhemoglobin. This is still a controversial issue. Reference should be made to the work of Cerdonio and co-workers, which has been reviewed by Mulay and Mulay and is cited in [43].

The *heme*, which is an iron–porphyrin complex, is the essential constituent of hemoglobin and related compounds. The terms ferriheme and ferroheme refer to the ferric Fe(III) and ferrous Fe(II) complexes of iron (Figure 3.83).

The general procedure is to study the magnetic susceptibility of such components or their derivatives and to obtain the magnetic moment per gram atom of iron in these compounds. The magnetic criteria outlined previously under coordination compounds (Section 8.3.1) were used to determine the valence state of iron and the bond type in these complexes.

In most cases an octahedral d^2sp^3 bonding is observed. Ferriheme chloride (hemin) has a magnetic moment of about 5.8 Bohr magnetons for iron, which corresponds to the theoretical value 5.92 Bohr magnetons for the $5d$ electrons in Fe(III). The iron in ferrohemoglobin, which is regarded as a conjugated protein containing native globin and the ferroheme, has a moment of about 4.91 Bohr magnetons. This corresponds to the theoretical value of 4.9 Bohr magnetons for the $4d$ unpaired electrons in Fe(II).

The magnetic moments observed for some derivatives of iron protoporphyrin and porphyrin follows, using Rawlinson's scheme, as shown in [25].

A magnetic moment of about 2.83 Bohr magnetons corresponding to two unpaired electrons was not reported in any of the derivatives.

The magnetic moment observed in many derivatives for iron depends on the experimental conditions employed. It is well established that the dimerization

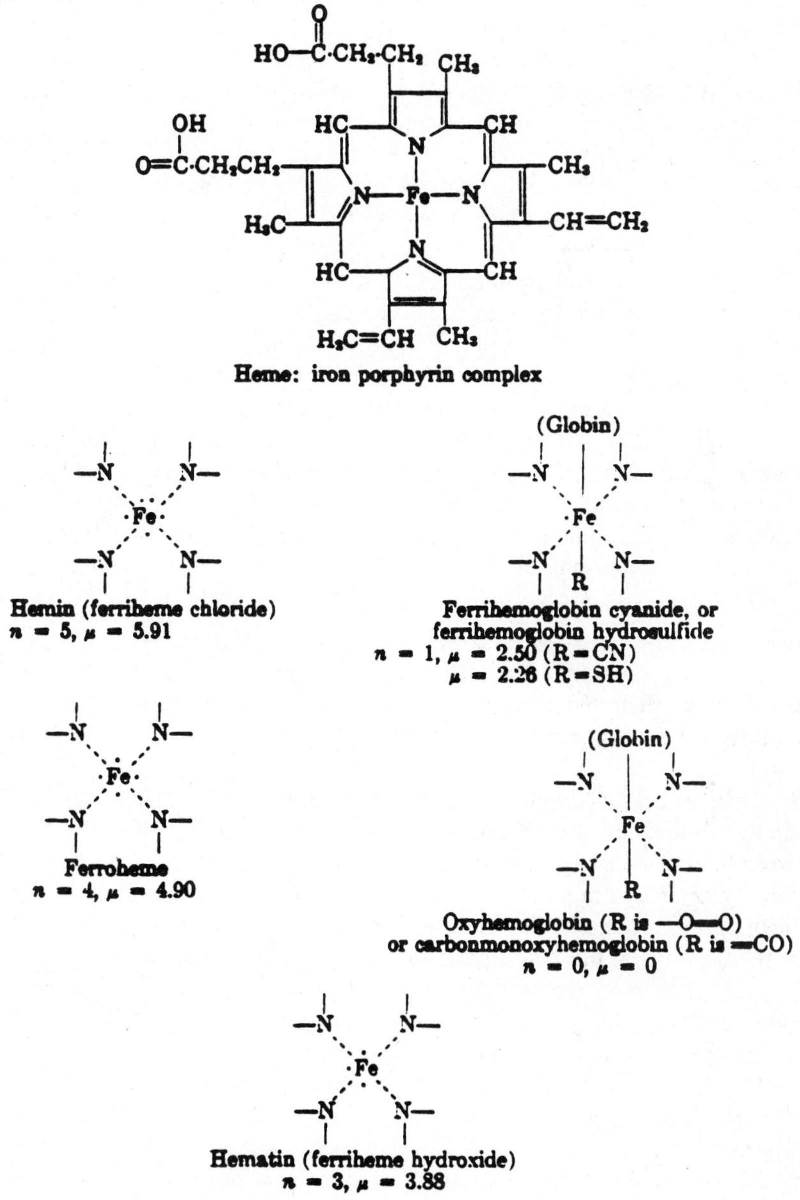

Figure 3.83 Skeletal structures of various iron-containing natural complexes, showing the number n of associated unpaired electrons and the corresponding effective magnetic moments μ. Reprinted with permission from P. W. Selwood, *Magnetochemistry*, Wiley-Interscience, New York, 1956. Copyright © 1956 by John Wiley & Sons, Inc.

and polymerization of Fe(III) ions in solution at moderate acidities are sufficient to depress the average magnetic moment per gram atom of iron [264]. Hence, often the magnetic moments observed, for instance, in hematin, need a cautious interpretation. The effect of pH on the magnetic moment of iron in ferrihemoglobin is discussed under magnetic titrations (Section 8.3.3).

Magnetic moments varying between 2 and 5.89 Bohr magnetons were observed in ferrihemoglobin derivatives containing imidazone, ammonia, azide, and ethanol groups. Some studies are reported on hemochromogens, which are compounds of *ferroheme* and denatured globin (Figure 3.83). Myroglobin, catalase, and other iron-containing compounds were also investigated. The work was reviewed by Selwood [25], Haurowitz and Kittel [61], and others [36–46]. In a classic study, Brill and Williams [124] correlated the absorption spectra of ferric porphyrin complexes with the magnetic moment of iron in these complexes. An analysis of the absorption spectra was used to estimate the amounts of low-spin and high-spin complexes that exist in equilibrium mixtures. Magnetic, oxidation–reduction, and other chemical properties were used to elucidate the nature of groups binding the iron in the hemoproteins, catalase, peroxidase, hemoglobin, and myoglobin. Brill and Williams' [124] work gives an extensive bibliography in this area.

MAGNETOKINETIC STUDIES ON BIOSYSTEMS

Brill and co-workers [67, 123, 124] in their magnetokinetic study of the reaction between ferrimyoglobin and methyl hydrogen peroxide represents a unique application of their magnetic susceptometer, devised for studying kinetics of reactions and flow systems. A brief description of the apparatus is found in Section 6.4.3.

In earlier work, Theorell and co-workers [150, 184] used a Gouy method specially modified for their studies on the red compound formed by the reaction between horse ferrimyoglobin and methyl hydrogen peroxidase. Because of the instability of this compound at the ferrimyoglobin concentration of 650 μm, some of the compound had reverted to ferrimyoglobin during the relatively long time period for making measurements with the Gouy balance. These authors made appropriate corrections for the ferrimyoglobin present, using the spectrophotometric technique, and obtained a tentative value of 3000×10^{-6} cgs units for the paramagnetic susceptibility of the red compound. According to Griffith [295], who introduced other corrections, the revised value is 3300×10^{-6} cgs units.

The flow-system susceptometer has many advantages over the Gouy method. It is capable of handling much smaller concentrations of the paramagnetic material and has a response of a few tenths of a second. In a kinetic study using this apparatus, methyl hydrogen peroxidase was injected into a sample of ferrimyoglobin, and the changes in the magnetic susceptibility were followed on a recorder. The susceptibility of the reaction product was calculated from

$$\chi_{compound} = \chi_{unreacted} - \Delta\chi$$

where

$$\Delta\chi = [10^3 \, \Delta\chi(\text{final})]/[\text{ferrimyoglobin}]$$

and represents the observed change in molar susceptibility. The molar susceptibility of the unreacted ferrimyoglobin existing in two forms was calculated on the basis of additivity relation, using, at 20°C, $\chi_{Fe} = 13{,}980 \times 10^{-6}$ for the brown form and $\chi_{FeOH} = 11{,}330 \times 10^{-6}$ for the red alkaline form.

In the final analysis, Brill and co-workers [123] calculated the magnetic susceptibility of the reaction product to be $(3300 \pm 500)10^{-6}$ cgs units. The rate constants for the reaction between ferrimyoglobin and methyl hydrogen peroxidase were obtained as a function of temperature, using the magnetic and spectrophotometric techniques. Significant differences were observed, particularly at low temperatures, in the results obtained by the two techniques. The authors point out that the magnetically determined rate constants are not expected to obey the Arrhenius relation if two or more magnetic processes contribute to the observed kinetic curve. They considered the possibility of production and disappearance of free radicals in the reaction between ferrimyoglobin and hydrogen peroxide, a theory postulated previously by other workers. Thus the production and disappearance of free radicals affected the time course of magnetic susceptibility, whereas the spectrophotometric method generally followed the overall conversion of ferrimyoglobin to the reaction product. The original paper contains several interesting aspects of this work and discussions about other reactions.

DETERMINATION OF HEMOGLOBIN CONCENTRATION

This method, developed by Taylor and Coryell [296] is quite promising because it gives more accurate results than the conventional gasometric methods based on the determination of carbon monoxide capacity. The hemoglobin solution is reduced to ferrohemoglobin using sodium dithionite ($Na_2S_2O_4$). Ferrohemoglobin, which is paramagnetic, is then saturated with carbon monoxide and converted to carbonmonoxyferrohemoglobin, which is diamagnetic. Therefore, the difference between their susceptibilities represents the paramagnetic contribution of the ferrohemoglobin. The observed difference depends on the concentration of ferrohemoglobin, which in turn corresponds to the original concentration of hemoglobin. A change in molar susceptibility of $12{,}290 \times 10^{-6}$ at 24°C corresponds to one heme; the effective magnetic moment of ferrohemoglobin is taken to be 5.46 Bohr magnetons. In these experiments the diamagnetisms of water, dissolved salts, and proteins cancel out, since the entire change in magnetic properties arises from a change in the state of iron, except for the negligible diamagnetism of added carbon monoxide.

A divided glass tube (18 mm in diameter, 30 cm long) was used to measure the difference between the susceptibility of the hemoglobin solution and of water using the Gouy technique. About 0.3–0.6 g sodium dithionite was used for reducing 30 mL of the solution to ferrohemoglobin. In the actual experiments, the apparent change in weight of the solution on application of the magnetic field is determined. Since this is proportional to the magnetic susceptibility, a calculation of the actual susceptibility is avoided. Represent the change in weight for the ferrohemoglobin by ΔW Hb, for the carbonmonooxyhemoglobin by ΔW HbCO, and for the solution being studied by ΔW; hence, the *molal susceptibility* per heme at 24°C for the solution under investigation is given by

$$\chi M = \frac{(\Delta W - \Delta W\,HbCO)}{(\Delta W\,Hb - \Delta W\,HbCO)} \times 12{,}290 \times 10^{-6}$$

MAGNETIC TITRATIONS

As discussed previously, it is possible to conduct a magnetic study of chemical processes in which large changes in susceptibility arise because of the conversion of a paramagnetic into a diamagnetic species (or to species with relatively small paramagnetism) and vice versa. Such study is particularly facilitated if the susceptibility changes in the reagent and its products are very small. Several magnetic titrations of oxyhemoglobin and ferrihemoglobin with diamagnetic reagents, such as sodium dithionite, sodium hydroxide, and potassium cyanide, were described by Coryell and co-workers [62]. The Gouy technique, which allows handling of large samples (20–30 mL), is very convenient for such titrations.

On reduction with sodium dithionite, oxyhemoglobin, which is diamagnetic, gives hemoglobin, which is paramagnetic. In the experimental procedure, a solution of oxyhemoglobin is introduced in a cylindrical sample tube (ca. 25 mL), which is closed with a rubber cap. Small volumes (ca. 0.5 mL) of a suitable concentration of sodium dithionite solution are injected from a syringe through the rubber cap for each measurement. The magnetic field is applied, and the change in weight ΔW is determined after each addition of the reagent. Appropriate corrections are made for the effect of the added reagent on the observed ΔW values. This correction is determined by performing a separate experiment with the reagent or by using Pascal's constants. It is not necessary to apply corrections for the changes in susceptibility that occur during the oxidation of sodium dithionite, because such changes are diamagnetic, quite small, and fall well within the limits of experimental error allowed by the Gouy method. A curve for the titration of oxyhemoglobin with sodium dithionite is shown in Figure 3.84. The changes in ΔW are quite linear with the volume of the reagent, and a sharp end point is obtained; the curves and the limits of error are comparable to those obtained in the conventional conductometric titrations.

The magnetic titration of ferrihemoglobin is shown in Figure 3.84. The pH was adjusted by the addition of suitable alkali and the same experimental method was used. The reaction was first order in hydroxide ion, and equilibrium constants were calculated. The overall change in the magnetic moment per gram atom of iron during the reaction was calculated. The magnetic moment at the point of inflection, which is somewhat affected by the ionic strength, corresponded to a species with three unpaired electrons.

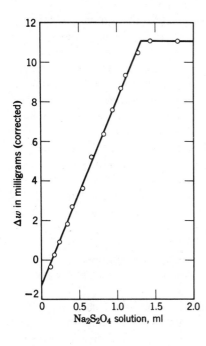

Figure 3.84 Magnetic titration of oxyhemoglobin with sodium dithionite. Reprinted with permission from P. W. Selwood, *Magnetochemistry*, Wiley-Interscience, New York, 1956. Copyright © 1956 by John Wiley & Sons, Inc.

OTHER BIOCHEMICAL AND BIOLOGICAL STUDIES

During the past few years some workers [5] have studied magnetic susceptibilities of carcinogenic compounds in efforts to examine the coexistence of antimitotic and carcinogenic action. They report that the susceptibility of chloramine, triethylenemelamine, urethane, myleran, 6-mercaptopurine, and several others show very high susceptibility. The observed values are higher than the calculated ones and range from 7.5×10^{-6} to 12×10^{-6} cgs units. This is ascribed to a spread of one or more π electrons in these molecules. Much of the work in this area appears to be purely speculative.

Early workers reported work on ESR in nucleic acids. Since then several workers have studied both the electron resonance and magnetic moments of various preparations of nucleic acids. Shulman and co-workers [297] showed that ESR spectra reported previously exhibit a ferromagnetic rather than paramagnetic behavior. This was attributed to trace quantities of iron, which were present in amounts sufficient to explain the observed magnetism, assuming that the magnetic ions exist as concentrated ferromagnetic aggregates rather than as dispersed paramagnetic centers. Recently, this field was reviewed critically [298], and it is suggested that the iron found in nucleoproteins is present in the living cells, although not in the concentrated ferromagnetic form. The latter, an oxide–hydroxide material approximately Fe_3O_4 in composition, may be precipitated during the extraction of nucleic acids or may even form quite slowly in intact, but dried, organisms. These views are supported by the findings of Wacker and Vallee [299].

Blakemore and Frankel [298] have shown that certain aquatic bacteria are magnetotactic and contain tiny internal *compasses* [magnetic materials such as Fe_3O_4 (magnetite) aligned more or less parallel to the long axis of the cell]. The north-seeking bacteria swim in the direction of an applied field and accumulate at the edge of a drop of water. In the absence of an applied field the bacteria swim persistently northward under the influence of the geomagnetic field. The bacteria were named *Aquaspirilum magnetotacticum*. Similarly south-seeking bacteria were collected in the southern hemisphere (near New Zealand). The authors [298] give convincing evidence for the presence of Fe_3O_4 particles by Mössbauer spectroscopy and for the elongated size in the 400–1200 Å (single domain) range by electron microscopy. Experimental setup is described for observing the behavior of bacteria in a water drop exposed to a magnetic field produced by Helmholtz coils. Several papers are listed, which appeared in *Science, Archives of Microbiology, Journal of Bacteriology, Nature (London)*, and so on. Interested readers should refer to the Mulay and Mulay reviews [36, 37] on early work in which authors surveyed *biomagnetism* and described experiments on the effect of magnetic fields on normal and cancer cells and presented a hypothesis for the observed effects (cf. [5]).

Several workers have studied the magnetic properties of *normal* and *diseased* tissues such as those from laboratory animals. Such studies are difficult; interpretations must be made cautiously, because it is difficult to obtain results on *norms* (i.e., *control groups*) for comparison with the diseased materials. Senftle and Thorpe (cf. [37, 38]) reported the magnetic susceptibility of normal liver and transplantable hepatoma tissue. It was shown that the difference is probably caused in large part both by the amount of water held by the cells in the two types of tissue and on adsorbed oxygen. The water in the normal cells appears to be bound, whereas that in the tumor tissues appears to be partially free.

Mulay and Mulay [118] used the helical-spring Faraday balance, described previously, to measure the magnetic susceptibilities of fresh tissues from the *normal* mouse leg and from mouse melanomas (a form of skin cancer) using special quartz containers that could be tightly closed to avoid water evaporation during measurement. Measurements were made at 77, 194, and 294 K. A distinct change in the *relative* moment was found; for the normal leg tissue, S-91 amelanotic melanoma (42 days), and S-91 melanoma (42 days), the values were 0.093, 1.0, and 1.55 (arbitrary units), respectively. This indicated a distinct increase in the paramagnetism of the melanoma. In addition, a study of ESR spectra of Cloudman S-91 and S-91A melanomas and various mouse tissues was correlated with trace-metal analysis and known biochemical reactions. Thus average magnetic moments obtained from temperature-dependence studies showed the following order for tissues: S-91 > S-91A > leg muscle. These results indicated a similar order in the relative concentration of magnetic species (free radicals and/or paramagnetic ions) and were confirmed by ESR. The ESR spectra at different power levels showed that the free radical activity as well as the Cu^{2+} and Fe^{3+} activities attributed to different signals were generally higher in S-91 than in S-91A, which did not show a Fe^{3+} signal. Normal tissues such as liver, spleen, heart, kidney, brain, and leg muscle showed varying amounts of free-radical activity but not signals attributable to paramagnetic ions. Electron spin resonance spectra and trace metal analysis at different stages of S-91 tumor growth indicated a systematic increase in the free-radical activity, total copper and iron concentrations, and paramagnetic ions, presumed to be Cu^{2+} and Fe^{3+}. The S-91A melanoma showed an increase in total copper, but not in free radical and paramagnetic ion (Cu^{2+}) activity after the initial growth. Correlation of these findings with known biochemical reactions suggests that the free-radical activity may be attributed partly to melanine and intermediates of various enzymatic reactions.

Thus, the static magnetometric studies on a great many enzymes complemented and supplemented the information obtained from ESR (this is also known as the EPR technique, where P stands for *paramagnetic*). Many examples on enzymes were critically surveyed by Mulay and Mulay [36–46]. Peterson's recent work is typical and quite comprehensive in nature. In his doctoral thesis [300] Peterson presents investigations of several proteins that contain iron or copper in their active sites, that is, in the small parts of proteins where their functions are performed. The electronic states of the metal ions were investigated by optical and EPR spectroscopies and by the temperature dependence of the magnetic susceptibility to gain information about structure and function of the active sites. A highly sensitive magnetic balance of the Faraday type, used for susceptibility measurements, is described extensively. Studies of the magnetic susceptibility over a wide temperature range are presented for the copper-containing proteins, laccase and ceruloplasmin, and iron-containing two-iron–two-sulfur ferredoxin, and the iron-containing $B2$ subunit of ribonucleotide–reductase. Electron paramagnetic resonance studies of dopamine-β-monooxygenase are also presented. The thesis is based on several papers dealing with magnetic susceptibility studies on (1) laccase and ceruloplasin; (2) two-iron–two-sulfur ferredoxin, which show antiferromagnetism [300], (3) iron center in ribonucleotide–reductase from *Escherichia coli* (*E. coli*) and other nucleotides in which EPR was used as the probe.

8.3.4   Geochemical and Mineralogical Applications

One of the most important recent advances in this area was the development of high-field, high-gradient magnetic separation (HGMS), which was developed in 1971 by researchers at the National Magnet Laboratory (Cambridge, MA). The HGMS utilizes a filamentary matrix of ferromagnetic material (such as stainless steel wool packed into a column) that is placed in a high field, sufficient to saturate the magnetic filaments. The resultant high gradient and high field along the edges of the filaments provide regions for trapping even weakly paramagnetic particles such as CuO. Many potential applications of this trapping were pointed out by researchers at the National Magnet Laboratory and by earlier workers. High-gradient magnetic separation is a well-developed technology based on a vast amount of basic research involving sophisticated mathematical calculations. It is beyond the scope of this chapter to outline these aspects, which were reviewed in detail by Luborsky [301] and in a special issue of the Institute of Electrical and Electronics Engineers, Inc. by Liu [302]. In the following paragraphs are listed several HGMS applications.

The beneficiation of kaolin clays to improve their brightness was accomplished by removal of a few percentage of iron-stained anatase TiO_2, which exists in the form of micron and submicron particles, with a small susceptibility of approximately 10^{-5} emu g^{-1} at a field of approximately 13,000 Oe. Pyrite and other impurities that occur as feebly paramagnetic impurities in coal were also removed. Other components in coal that normally become ash are also paramagnetic and are removed together with the Fe—S. A large proportion of coal contains from 1 to 4% total sulfur of which 75% is paramagnetic and is removable by HGMS. However, it is not possible to remove the organically bound sulfur in coal by any magnetic separation technique.

The vast deposits of taconite that contain very small grains of hematite, goethite, and iron silicates can be beneficiated after grinding the ore to particles less than 30–40 μm. Since the mining industry is exhausting its supplies of iron ore containing high levels of magnetite for steel making, it is proposed that the industry use alternative sources such as the oxidized taconite subjected to HGMS.

The removal of suspended magnetic solids such as iron oxides from steel mill waste water or coagulated ultrafine particle contaminants to form magnetic *seed* particles using HGMS was demonstrated, with the advantage that high speed and large capacities of water can be handled for the removal of very small particles. Other applications include the purification of municipal and industrial waste-water treatment (cf. [44–46]).

High-gradient magnetic separation has biomedical applications also. For example, red blood cells were *filtered* for producing plasma with either very pure or controlled red blood cell populations for biomedical applications. The application is possible because in their deoxygenated state red blood cells are expected to have a susceptibility relative to that of water, that is, 0.3×10^{-6} emu cm^{-3}, which is two orders of magnitude lower than that of CuO. Even so, they have been filtered by HGMS without any observable biological damage.

The list of HGMS applications continues to grow yearly; the SALA Magnetics Corporation (Cambridge, MA) is already producing giant-scale magnetic separators for mineralogical separations, and will supply technical and basic scientific literature on request.

There exist several classical applications of magnetic measurements to geoplanetary and planetary sciences. The examination of the magnetic susceptibilities of several lunar samples was reviewed by Mulay and Mulay [36–45].

Magnetic measurements on samples collected at different depths may be used to locate similar strata below ground, as in regions of oil wells. This is accomplished by plotting magnetic susceptibility as a function of depth.

Variations in susceptibility in different strata are related to similar variations from other walls in adjacent areas. Tectonic structures favorable to accumulation of mineral deposits may be located by this method. Valuable information on the orientation of the earth's magnetic field in previous ages was obtained by measuring the orientation of magnetized impurities at different depths, under the ocean floor, and in sedimentary rocks. Many workers have tried to correlate the magnetic properties of minerals with their geochemical history. Several geochemical and geophysical journals and monographs, published by the American Institute of Mining and Metallurgical Engineers, describe magnetic analysis of minerals in the field, which are helpful for magnetic prospecting. Selected references are given by Selwood [25] and Kauffman [303]. More recent books by Dobrin [255], Nagata [304], and Strangway [305] describe several applications to geophysical prospecting for unraveling the periodic reversal in the earth's magnetic field, its origin, and so on.

8.3.5   Thermomagnetic Analysis

Thermomagnetic analysis has developed into a major tool for (1) the characterization of fine particles (such as the superparamagnetic particles in certain types of catalysts), (2) studies on solid–gas or solid–solid reactions, (3) the identification and/or determination of a ferromagnetic or ferrimagnetic component dispersed in a nonmagnetic (i.e., diamagnetic) matrix, and (4) studies of the structural or electronic phase transitions. The ever-growing ramifications of such analysis are impressive indeed and some of these could be categorized under chemistry, metallurgy, and materials or solid-state science. In our opinion such classification tends to fragment an elegant field of research into too many subtopics. Hence, typical examples of work involving novel approaches, carried out in our laboratory and by others are described briefly without relabeling these applications to fit major disciplines.

CHARACTERIZATION OF FINE PARTICLES: APPLICATIONS TO CATALYSIS
Excellent books by Selwood [22] survey, to about 1975, the magnetic studies on various catalyst materials. Reviews by Mulay and Mulay [36–46] provide selected examples of such studies conducted since the mid-1970s. Research papers or reviews on this subject continue to appear periodically in *Advances in Catalysis* and in journals such as the *Journal of Catalysis, Journal of Surface Science, Journal of Applied Physics, Journal of Applied Catalysis*, and *Journal of Physical Chemistry*.

Readers should first become acquainted with the principles of *superparamagnetism* and *ferromagnetism* discussed in preceding sections (4.8.4 and 4.8.5) and with references cited therein. With this background a summary follows of the work of Mulay and co-workers [161], who characterized in a novel manner two commercial samples of nickel on alumina that were found to contain *both multidomain ferromagnetic* and single-domain *superparamagnetic particles*. The changes in particle profiles (shape and size) during heat treatment as well as some H_2 chemisorption and methanation studies were carried out.

The two commercial catalysts A and B had the following characteristics—
nickel content (wt%): (A) 43 and (B) 67; BET surface area (m^2 g^{-1}): (A) 51 and (B)
117; and average crystallite size (Å): (A) 185 and (B) 74. Magnetization
measurements were performed on a vibrating sample magnetometer as a
function of the field (up to 20 kOe) and over a range of temperature (77–600 K)
to yield especially values for saturation magnetization per gram σ_{sat} of nickel in
the catalyst, the coercive force H_c, and the remanence I_r. Special quartz sample
holders were designed to accommodate in vacuum both the as-received samples
and those received after reaction and/or heat treatment (400–700°C).

The weight fractions x_i of the superparamagnetic (a type) and hence of the
multidomain (c type) particles in the various heat-treated samples were
estimated from the observed curves of the relative magnetization (σ/σ_∞ vs H/T
at 573 K). Typical magnetization curves for catalyst B only are shown in Figure
3.85. An equation for a two-component system was assumed for deriving the
weight fractions of the superparamagnetic (a type) and ferromagnetic (c type)
particles [161]:

$$\frac{\sigma}{\sigma_\infty} = \left[1 - \sum_i x_i \right] + \sum_i x_i L \left[\frac{I_{\text{sp}} v_i H}{kT} \right]$$

The expressions under the summation signs refer to superparamagnetic
particles; L denotes the Langevin function and v_i is the mean volume of the
particles within the x_i fraction. This expression is applicable at high fields such
that saturation magnetization of multidomain particles shows no change and
superparamagnetic particles obey the well-known Langevin function. The
average moment μ_i of the a particle is defined by $\mu_i = I_{\text{sp}} \cdot v_i$ (a magnetization
curve for superparamagnetic particles of 30-Å radius, and at 573 K, using the
Langevin function, is also shown in Figure 3.85 for comparison with the
observed curves).

By assuming an overall mean volume $\bar{v}$ for all superparamagnetic particles, as
in Selwood [25], a nonlinear regression procedure was developed for estimating
the weight fraction x of the a particles and the mean particle diameter of this
fraction based on spherically shaped particles. The regression program accomplished convergence very easily with a conventional hill-climbing subroutine.

The "good" catalyst B showed a superparamagnetic component with a
constant particle size of approximately 27 Å over the entire increasing Htt (heat-
treatment temperature) range with a decrease in weight fraction from 58 to 26%.
On the other hand, catalyst A showed a larger range of particle size (25 to 54 Å)
with a considerable decrease in its relatively lower weight fraction (36 to 12%) of
superparamagnetic particles with increasing Htt. These results were further
supported by the hydrogen chemisorption studies that provided the degree of
dispersion f, which is defined by $f = $ [Ni atoms at the surface]/[total number of
Ni atoms in the bulk per gram catalyst].

In addition to elucidating the particle-size analysis and its correlation with
hydrogen chemisorption studies, Mulay and co-workers [161] performed an

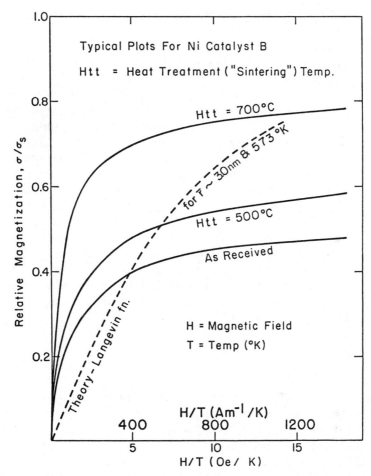

Figure 3.85 Typical magnetization σ/σ_s versus H/T curves for catalyst B for the as-recieved and heat-treated samples. A magnetization curve for superparamagnetic particles (30-Å radius at 573 K) is shown for comparison.

analysis of changes in particle profiles during the heat-treatment procedure by following the changes in the coercive force H_c and the remanence σ_r. These changes were correlated with the relative methanation activity ($CO + 3H_2 \rightarrow CH_4 + H_2O$), which was studied by standard gas chromatography techniques [161]. The reaction was *structure sensitive* to magnetic particle profiles.

Another thermomagnetic application to Ni on Al_2O_3 containing graphite as a binder is described by Mulay and Yamamura [306]. They found that catalysts A and B containing 51 and 37.5% free nickel had almost the same surface area (ca. 148 m^2 g^{-1}) and both had an almost ideal superparamagnetic behavior in their as-received states by showing a good superposition of data points for σ/σ_s

plotted against H/T_m, where T_m is the measuring temperature. The value of σ_s was established both by extending the measurements of σ to a very *low cryogenic* range and also by plotting σ versus $1/H$ and extrapolating $1/H$ to zero, which corresponds to finding σ (saturation) at infinite fields and at temperatures approaching 0 K.

The difference between the two catalysts became clear when they were subjected to heat treatment. Catalyst A continued to show a superparamagnetic behavior at a Htt = 500°C for 4 h. In contrast, catalyst B showed a good superposition only at high values of H/T_m. In the low H/T_m regions three distinct curves A, B, and C were observed for various measuring temperatures T_m (K): (A) 381–502, (B) 351, and (C) 295. For further characterization, they studied the coercive force H_c as a function of T_m for various heat-treated samples.

Since $H_c = 0$ at and above the blocking temperature T_B (which really indicates the collapse of any hysteresis curve at T_B), the T_B values for various heat-treated samples can be calculated. For example, for catalyst B, with a Htt = 500°C for 4 h, a T_B of approximately 400 K can be estimated. The general shape of H_c versus T_m indicated that the *shape anisotropy* and *not the crystalline anisotropy* was more prominent during the heat-treatment procedures. Generally the shape anisotropy contributes more substantially to changes in H_c than the crystalline anisotropy.

Mulay and Yamamura [306] further studied the effects of Htt on the average particle sizes r_{LF} and r_{HF} obtained using the low-field and high-field approximations of the Langevin function, which defined the upper and lower limits of particle sizes. Both from the observed σ/σ_s versus H/T_m curves and appropriate computations, the fractions x and y of r_{LF} and r_{HF} present in the as-received and heat-treated catalysts were obtained, and the average particle size r_{av} value was calculated by taking the appropriate weighted fractions r_{LF} and r_{HF}. Thus, a reasonable idea of particle-size distributions was obtained; for example, $x = 0.72$ for $r_{LF} = 48$ Å, and $y = 0.25$ for $r_{HF} = 32$ Å.

In addition, these workers studied the effects of heat-treatment temperature on the r_{av} and the effects of Htt on H_c at 80 K. The observed changes correlated well with the particle sizes from X-ray diffraction line-broadening studies.

Thus the work of Mulay and co-workers [161] and of Mulay and Yamamura [306] showed that the thermomagnetic analysis of transition metals (e.g., Ni) supported on diamagnetic matrices such as Al_2O_3 yielded not only information on the distribution of (catalytically active) superparamagnetic particles, but also provided information on changes in particle profiles (shape and size) with heat-treatment conditions; and, in appropriate cases [161], good correlations between the chemisorption and/or catalytic activity could be established.

Prior to this work, Mulay and Collins (cf. [161]) showed that a fine (superparamagnetic) dispersion of α-Fe_2O_3 could be produced by using a novel technique of inserting $Fe(CO)_5$ (maximum dimension ca. 9 Å) in a Linde 13 X zeolite cage structure with an aperture of approximately 10 Å. On pyrolyzing the iron pentacarbonyl in air, a fine dispersion of α-Fe_2O_3 was obtained. This was confirmed by *thermomagnetic* analysis at various fields. The initial dispersion had an average particle size of approximately 85 Å, which on annealing up to 900°C increased to approximately 300 Å. The inverse susceptibility $(1/\chi)$ versus T showed a negative Weiss constant θ indicating the antiferromagnetic characteristics of α-Fe_2O_3; this was further confirmed by Mössbauer spectroscopy.

The early (about 1967) work by Mulay and co-workers [158] seemed to open a new area of introducing fine dispersions of Fe or Fe—Co bimetallic clusters *inside* other cage structures of other zeolites, such as the Mobil Oil Corporation ZSM-5 and the corresponding silicalite. This corporation has successfully marketed a new ZSM-5 series of catalysts in recent years for the production of alcohols and synthetic gasoline-range hydrocarbons in the octane range. Their catalyst systems are shape selective, because only certain desirable components can pass through the pore structure of their zeolites.

In a basic study of ZSM-5 impregnated with iron salts or iron carbonyls (e.g., Fe_3CO_{12}), Mulay and co-workers (cf. [307], which contains magnetic studies) investigated the thermomagnetic and Mössbauer characteristics of bifunctional medium-pore (ca. 6 Å) zeolite—iron (and Fe—Co) catalysts used in the synthesis gas conversion. The thermomagnetic properties were particularly helpful in identifying various iron carbides formed during the carbiding of Fe with CO and in the used catalyst. Reference to this work will follow.

Jung, Mulay, and their collaborators [218] have used yet another relatively new support, namely, carbon for dispersing transition metals such as iron. The Monarch Carbolac-1 support with a porous high-surface-area (*amorphous*) structure served as a good support for obtaining fine dispersion of iron with an average particle diameter of approximately 30 Å; this particle size was obtained by using the low-field (LF) and high-field (HF) approximations of the Langevin function, which were well suited to the observed superparamagnetism of this system. This particle size was confirmed by CO chemisorption data. Conversely, iron deposited on a low-surface-area graphitic carbon (V3G) showed the presence of large ferromagnetic particles. The presence of some iron oxides was detected in such systems. From a practical point of view the Carbolac-1/Fe catalyst showed stable hydrogenation activity. Typical thermomagnetic and magnetization curves are shown in Figures 3.86 and 3.87. Several examples of catalyst systems were cited by Mulay and Mulay [41–45].

IDENTIFICATION OF PHASES BY CURIE POINTS

Selwood [22] summarized the classic work up to 1957 especially on the identification of the iron carbides. Mulay and Mulay [36–46] reviewed selected examples of identification of various phases by measurements of Curie (or Néel) temperatures. Here are reported how magnetic balances, vibrating sample magnetometers, and especially certain thermogravimetric analyzers (TGA) are used to determine such transition temperatures and also *new* examples in this area are presented.

An apparatus capable of yielding accurate measurements of magnetization σ as a function of temperature, such as the vibrating sample magnetometer or the

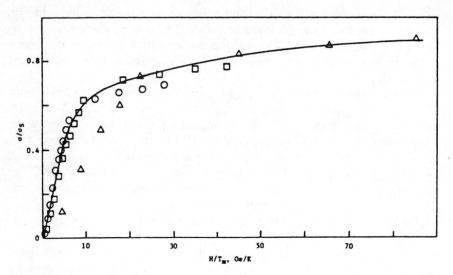

Figure 3.86 Relative magnetization versus field/temperature for 4.5% Fe/V3G (reduced 1 h at 673). Open circles, squares, and triangles show results at 295, 195, and 80 K, respectively. Note that the plots at 295 and 195 K show good superposition.

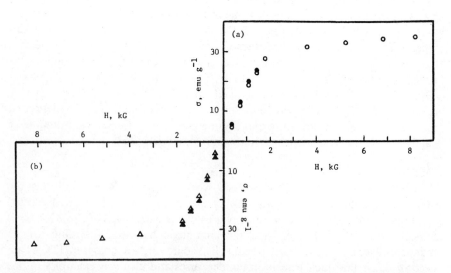

Figure 3.87 Magnetization per gram of iron at 80 K for 5.0% Fe/C-1 (reduced 5 h at 723 K in addition to an 11-h reduction at 673 K): (*a*) initial run measured in a sequence of increasing (○) and decreasing (●) *H*; (*b*) duplicated run measured after reversing the polarity of the magnet in a sequence of increasing (△) and decreasing (▲) *H*.

Faraday balance, can be used to produce σ versus T curves. From the inflection in this curve the Curie temperature T_c, which is a characteristic of the ferromagnetic or ferrimagnetic component, can be determined for identification purposes. Some manufacturers of TGA apparatus, such as the Perkin-Elmer Company (Norwalk, CN), employ a relatively small heating furnace (ca. 0.5 in. o.d., ca. 0.75 in. high) and provide a small permanent magnet, which can surround the furnace containing the sample. If the sample does not decompose (i.e., undergo weight changes) during heating, then the automated TGA apparatus provides magnetization (in some arbitrary units) as a function of temperature. The strength of the field provided by the small magnet with a pole gap of about 3 cm is not specified by the Perkin-Elmer Company (it may be approximated at 300 Oe). However, it appears that such small fields, which do not magnetically saturate the ferromagnetic or ferrimagnetic sample, are still adequate to pinpoint the Curie temperature of various materials either by the inflection in the σ versus T_m curve or through the sharp differential $d\sigma/dT_m$ versus T_m curve; here T_m refers to the measuring temperature.

The TGA manufacturers are interested in providing a temperature calibration for their furnaces by the observation of T_c of standard ferromagnetic samples. The applicability of such TGA apparatus for thermomagnetic research was demonstrated for the determination of T_c of *amorphous* and crystalline phases of metallic glasses such as Fe_{80} and B_{20} [cf. Fyans [308] (Figure 3.88)]. In our opinion, while small fields of a few hundred oersted may be enough for bulk samples, such higher fields of up to several thousand oersteds will be

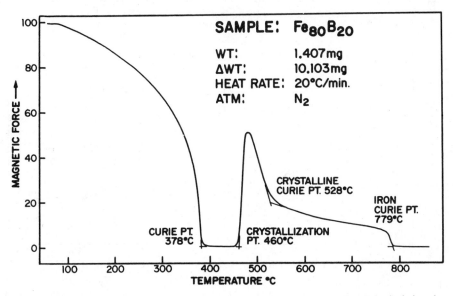

Figure 3.88 Thermomagnetic analysis of a sputtered amorphous film of $Fe_{80}B_{20}$, depicting the Curie point, the crystallization point, and so on.

necessary to saturate magnetically the very small particles of ferromagnetic materials. Most workers have used fields up to 5000 Oe with their magnetometers. The use of sensitive magnetometers for research applications is recommended rather than the use of modified TGA apparatus, which is not designed to give absolute magnetization values. Some of the automated apparatus described by Selwood [22] and by Richardson and Beauxis [239] would be most satisfactory.

Mulay and co-workers [158] used the Curie point T_c determinations to identify species, such as Fe_3C, Fe_3O_4, and Fe, which were formed during the synthesis of glassy carbon. Glassy carbon was formed by the polymerization (and subsequent pyrolysis of) polyfurfuryl alcohol, to which ferrocene dicarboxylic acid (FDA) or vinyl ferrocene (VF) was added (Figure 3.89). The presence of the iron compounds (Fe_3C, Fe_3O_4, etc.) was confirmed by X-ray diffraction and Mössbauer spectroscopic studies.

Mulay and co-workers carried out an extensive thermomagnetic study of the zeolite (ZSM-5) impregnated with Fe or Fe—Co clusters [307]. A summary of their work follows.

The freshly impregnated zeolites indicated the presence of Fe^{3+} species. From an analysis of the paramagnetic susceptibility, a Bohr magneton number of 5.96 was deduced. The magnetization M studies on the reduced samples of ZSM-5 (14.7% Fe) and silicalite (13.6% Fe) indicated that Fe is in the metallic state with 86 and 85% reduction, respectively. The magnetization versus temperature curves for ZSM-5 (11.1% Fe) are shown in Figure 3.90. The carbided sample of ZSM-5 (11.1% Fe) appeared to be the high Curie point form of the Hägg carbide (Fe_5C_2) with $T_c = 540$ K. The used sample of ZSM-5 (11.1% Fe) exhibited a magnetic transition with $T_c = 650$ K, which corresponds

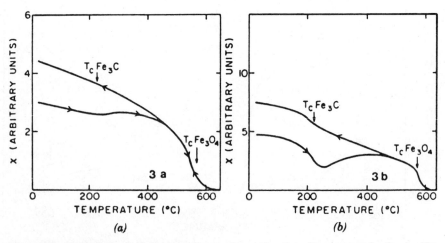

Figure 3.89 Magnetic susceptibility in arbitrary units as a function of temperature for the glasslike carbon polymerized with 10% VF (a) 500°C; (b) 625°C.

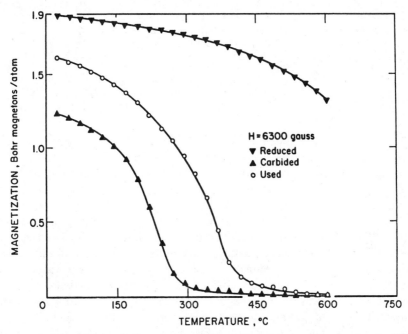

Figure 3.90 Magnetization (Bohr magnetons/iron atom) as a function of temperature for ZSM-5 (11.1% Fe).

to the hcp phase of Fe_2C. The magnetic transition of an iron carbide Cementite (Fe_3C) was masked in the M versus T curve shown in Figure 3.90 since T_c of Fe_3C is about 490 K, below that of the hcp carbide. The hcp phase Fe_2C is stable below 470 K in an atmosphere of synthesis gas. Its presence in the used sample seemed to indicate that it was formed while the catalyst was cooled after the reaction.

The magnetization data (Figures 3.89 and 3.90) on ZSM-5 (5.6% Fe, 4.5% Co) shows that the reduced, carbided, and used samples have large magnetic moments [1.94, 2.04, and 2.61 μ_B/TM (transition metal) atom, respectively, at room temperature] and high Curie points ($> 900°$), which cannot be accounted for on the basis of individual Fe and Co particles. The magnetic data indicate the composition to be that of a Fe—Co alloy, in support of the conclusions reached through the Mössbauer analysis. Hence, one can conclude that the difference in selectivity between ZSM-5 (11.1% Fe) and ZSM-5 (5.6% Fe, 4.5% Co) catalysts arises from the presence of bimetallic TM clusters in the latter, with consequent changes in the average number of $3d$ electrons per TM atom. The M versus T curve (Figure 3.90) of the carbided sample of ZSM-5 (5.6% Fe, 4.5% Co) indicated an irreversible formation of a second phase with a higher moment above a temperature of 450°C.

In addition to this work involving impregnation with Fe and Co salts, Mulay and co-workers (see references in [307]) used thermomagnetic analysis (TMA) to unravel the nature of species in the ZSM system, which was impregnated with $Fe_3(CO)_{12}$.

Figure 3.91 shows the TMA for the carbided and used carbonyl-impregnated samples. The carbided sample displays two phases: χ carbide (Fe_5C_2) and Fe_3O_4. Some samples (not shown in Figure 3.91) contained a small amount of ε carbide.

To estimate the relative amounts of χ carbide and Fe_3O_4, the total magnetization σ was written as $\sigma = \alpha\sigma(Fe_3O_4) + \beta\sigma(\chi Fe_5C_2)$, where α and β are the relative amounts of the two phases. In the carbided sample, $\alpha/\beta = 2.4$; in the used sample, $\alpha/\beta = 3.7$. In samples containing small amounts of ε carbide, its amount was estimated with a similar procedure.

This observation indicated a substantial increase in the Fe_3O_4 phase relative to the χ carbide phase with prolonged use of the carbonyl-impregnated catalyst. This catalyst displayed steady catalytic activity and selectivity at 280°C for a period of 7 days. The percentage aromatics in the liquid hydrocarbons was rather low, about 5%, during the period of steady catalytic activity. This is in contrast to the previously observed behavior of the nitrate-impregnated samples. The carbided sample using nitrate impregnation was essentially single-phase χ carbide. The used samples exhibited, in addition to χ and ε ($Fe_{2.2}C$)

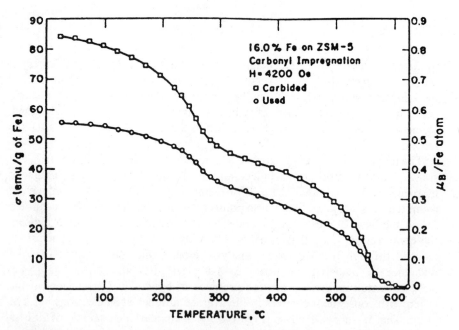

Figure 3.91 Curves of σ versus T for the carbided and used samples ZSM-5 (16% Fe), carbonyl impregnated.

phases, the catalytically inactive θ carbide, Fe_3C [309]. The percentage aromatics in the liquid hydrocarbons was rather high initially, typically about 25%, but increased with time on stream for the nitrate-impregnated catalysts. On heating above 500°C during TMA, the χ and ε carbides were always converted to the more stable and catalytically inactive θ carbide.

In conclusion, their studies have shown (1) the carbonyl-impregnated samples in the as-impregnated samples contain ultrafine γ-Fe_2O_3 of $d = 60$–66 Å, whereas the nitrate-impregnated samples contain α-Fe_2O_3 particles of approximately 100-Å diameter; and (2) the stability displayed by the carbonyl-impregnated catalyst may be influenced by the presence of substantial amounts of Fe_3O_4 and χ-Fe_5C_2 during synthesis gas conversion.

Previously it was shown how the low-field (LF) and high-field (HF) approximations of the Langevin function gave average values for the volume of the superparamagnetic *large* particles $\bar{v}_{LF}$, which saturate easily at low fields, and the volume of *small* particles $\bar{v}_{HF}$, which tend to saturate at high fields. Thus, the limiting values of particle volumes could be determined. This section is concluded by pointing out that it is *now possible* to obtain a *particle-size distribution* from the observed superparamagnetic curve for a particular system. For this purpose one may imagine that (1) n_i particles of volume $\bar{v}_i$ will give a particular curve, (2) n_{ii} particles of volume $\bar{v}_{ii}$ will give another curve, and so on. Finally, the observed curve results from $\Sigma\, n_i v_i$ distribution. By assuming a *log-normal* distribution for the catalyst particles, based on [310], the distribution can be calculated:

$$f(r) = \frac{1}{\sqrt{2\pi} r \ln(S)} \exp -\left(\frac{\ln r - \ln r_0}{\sqrt{2}\ln S}\right)^2$$

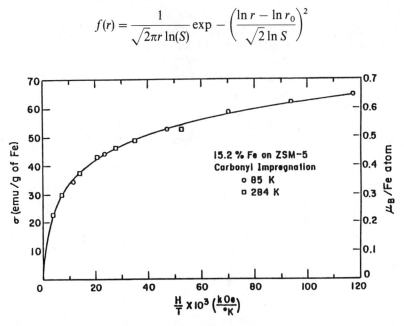

Figure 3.92 Data for σ versus T at 85 and 284 K for sample Z-CO-1 (15% Fe/H-ZSM-5).

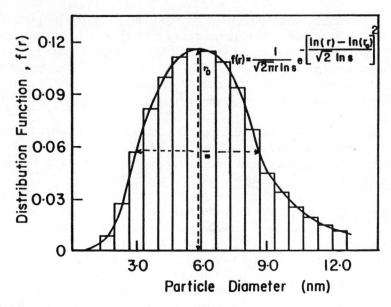

Figure 3.93   Particle-size distribution for sample containing 15% Fe/H-mordenite. In the equation s is the square root of the variance and r_0 is the geometric mean of the distribution.

In this equation, r_0 is the geometric mean of the distribution and S is the square root of the variance of the distribution. By combining these parameters with those of the well-known Langevin function and by using an appropriate computer-fitting program, a particle-size distribution can be obtained. Mulay and co-workers [309] illustrated the applicability of this technique to freshly deposited $\gamma\text{-Fe}_2\text{O}_3$ in ZSM-5 by obtaining a histogram for the particle-size distribution (Figures 3.92 and 3.93). Similarly, such distributions were obtained for the reduced Fe^0 superparamagnetic σ/σ_s versus H/T curves.

Appendix: Numerical Values For Some Important Physical Constants

Electronic charge, e *4.80298×10^{-10} cm$^{3/2}$ g$^{1/2}$ s^{-1} (esu)†
Electron rest mass, m 9.1091×10^{-28} g
Proton rest mass, M_p 1.67252×10^{-24} g

†The items denoted by asterisks are often shown with a minus sign. Other numerical values in a convenient form:

$N\beta$ (Bohr magnetons per mole)	5584.78 erg Oe^{-1} mol^{-1}
$N\beta^2$	0.26073 cm^{-1} erg Oe^{-2} mol^{-1}
$3\kappa/N\beta^2$	7.9971 mol Oe^2 erg^{-1} deg^{-1}
kT	$1.3804 \times 0.5035T$ cm^{-1} mol^{-1}
	(e.g., 208.4 cm^{-1} mol^{-1} at 300 K)

Velocity of light, c	2.997925×10^{10} cm s^{-1}
Planck's constant, h or	6.6256×10^{-27} erg·s
$\hbar$	1.05450×10^{-27} erg·s
Charge-to-mass ratio for	5.27274×10^{17} cm$^{3/2}$ g$^{-1/2}$ s^{-1} (emu);
electron, e/m	1.758796×10^{7} cm$^{1/2}$ g$^{-1/2}$ (esu)
Bohr magneton, μ_B or β	*9.2732×10^{-21} erg Oe^{-1}[†,‡]
	(or 4.66858×10^{-5} cm^{-1} Oe^{-1})
	(also referred to simply as emu)
Nuclear Bohr magneton, μ_{Bn}	5.0505×10^{-24} erg Oe^{-1}
Electron volt, eV	1.60210×10^{-12} erg;
	3.8291×10^{-20} cal
Associated "temperature"	1.16049×10^{4} K
Associated wavenumber	8065.73 cm^{-1}
Boltzmann's constant, k	1.38054×10^{-16} erg K^{-1}
Gas constant, $\mathscr{R}$	8.3143×10^{7} erg K^{-1} mol^{-1}
Avogadro's number, N	6.02252×10^{23} mol^{-1}
(number of atoms per mole)	
Loschmidt's number, N	2.687444×10^{19} cm^{-3}
(number of atoms per cubic centimeter)	
Bohr radius, a_0	0.529167×10^{-8} cm

Acknowledgments

We are indebted to the staff of Professor H. Theorell's magnetic laboratory in Stockholm, Sweden, for a demonstration of their apparatus and techniques. We thank Dr. R. T. Lewis, formerly with the Chevron Research Labs, for helpful discussions concerning his magnetometer and the design of his "gradient coils."

This chapter is based on the following major publications, with regard to selected figures, and so on.

Our reviews that appeared during 1962–1984 in *Analytical Chemistry*, published by the American Chemical Society, Washington, DC; Chapter VII by L. N. Mulay, in *Physical Methods of Chemistry*, Part I, Vol. IV, 1972, Arnold Weissberger and Bryant W. Rossiter, Editors; *Theory and Applications of Molecular Diamagnetism*, 1976; *Theory and Applications of Molecular Paramagnetism*, 1976, L. N. Mulay and E. A. Boudreaux, Eds.; and *Magnetic Susceptibility*, Reprint Monograph, 1963; all published by Wiley, New York.

Grateful acknowledgments to the following were recorded in these publications for permission to reproduce figures, numerical data or quotations: B. I. Bleaney and B. Bleany, S. Broersma, B. N. Figgis, J. Lewis, J. H. Van Vleck, A. Pacault, H. Semat, P. W. Selwood, H. E. White, American Physical Society,

---

‡Some authors use gauss (G) in place of the oersted (Oe).

American Chemical Society, and other scientific journals and organizations mentioned therein.

The benefits derived from other Wiley publications by A. H. Morrish, S. Chikazumi, and S. H. Charap are also recorded here with gratitude.

We sincerely appreciate the encouragement received from our Pennsylvania State University colleagues; namely, Deans Charles L. Hosler (now Vice President for Research and Dean of the Graduate School), John Dutton, Arnulf Muan, and Peter Luckie; and Professors Michael Coleman and Donald Koss.

Professor Stuart Kurtz, who shares a common interest with us regarding magnetic properties of bioorganic materials, continues to provide an additional impetus to our research and writing activities.

References

1. E. C. Stoner, *Magnetism and Atomic Structure*, Methuen, London, 1926; *Magnetism and Matter*, Methuen, London, 1934.

2. S. S. Bhatnagar and K. N. Mathur, *Physical Principles and Applications of Magnetochemistry*, Macmillan, London, 1935.

3. L. N. Mulay, *Magnetic Susceptibility*, Interscience, New York, 1966, a reprint monograph based on L. N. Mulay, in I. M. Kolthoff and P. J. Elving, Eds., *Treatise on Analytical Chemistry*, Part 1, Vol. 4, Interscience, New York, 1963, Chap. 38.

4. D. C. Mattis, *The Theory of Magnetism*, Harper & Row, New York, 1965.

5. L. N. Mulay and E. A. Boudreaux, Eds., *Theory and Applications of Molecular Diamagnetism*, Wiley-Interscience, New York, 1976.

6. L. N. Mulay and E. A. Boudreaux, Eds., *Theory and Applications of Molecular Paramagnetism*, Wiley-Interscience, New York, 1976; H. Semat and H. E. White, *Atomic Age Physics*, Reinhart, New York, 1959.

7. T. I. Quickenden and R. C. Marshall, *J. Chem. Educ.*, **49**, 115 (1972).

8. M. L. McGlashan, *Ann. Rev. Phys. Chem.*, **24**, 51 (1973).

9. N. H. Davies, *Chem. Br.*, **6**, 344 (1970); **7**, 331 (1971).

10. T. P. Abeles and W. G. Bos, *J. Chem. Educ.*, **44**, 441 (1967).

11. L. H. Bennett, C. H. Page, and L. H. Swartzendruber, *Comments on Units in Magnetism*, Special Publication, National Institute of Standards and Technology, Gaithersburg, MD.

12. B. I. Bleany and B. Bleany, *Electricity and Magnetism*, Oxford University Press, London, 1965, and subsequent editions.

13. G. Pass and H. Sutcliffe, *J. Chem. Educ.*, **48**, 181 (1971).

14. *American Laboratory Buyers Guide*, Arlington, MA, 1987.

15. Cahners Computer Center, *United States Industrial Directory*, Denver, CO, 1987.

16. B. Cabrera, *Phys. Rev. Lett.*, **48**(20), 1378 (1982).

17. B. Cabrera, *Sci. Am.*, **246**, 106 (1982).

18. J. E. Crooks, *J. Chem. Educ.*, **56**, 301 (1979).

19. J. E. Wertz and J. R. Bolton, *Electron Spin Resonance*, McGraw-Hill, New York, 1972.

20. S. L. Rock, E. F. Pearson, E. H. Appleman, C. L. Norris, and W. H. Flygare, *J. Chem. Phys.*, **59**, 3940 (1973).

21. J. H. Van Vleck, *Electric and Magnetic Susceptibilities*, Oxford University Press, London, 1932; "The Magnetism of Some Rare Earth Compounds," in H. P. Kallmann, S. A. Korff, and S. G. Roth, Eds., *Physical Sciences*, New York University Press, New York, 1962, Chap. 11.

22. P. W. Selwood, *Chemisorption and Magnetization*, Academic, New York, 1975.

23. M. Prasad-Rao and L. N. Mulay, *J. Chem. Phys.*, **19**, 1051 (1951).

24. B. N. Figgis and J. Lewis, "Magnetochemistry of Complex Compounds," in J. Lewis and R. G. Wilkins, Eds., *Modern Coordination Chemistry*, Interscience, New York, 1960.

25. P. W. Selwood, *Magnetochemistry*, Wiley-Interscience, New York, 1956.

26. A. H. Morrish, *Physical Principles of Magnetism*, Wiley-Interscience, New York, 1965; S. Chikacumi and S. Charap, *Physics of Magnetism*, Wiley, New York, 1964.

27. L. F. Bates, *Modern Magnetism*, Cambridge University Press, London, 1952.

28. C. Kittel, *Solid State Physics*, 2nd ed., Wiley, 1969.

29. A Earnshaw, *Introduction to Magnetochemistry*, Academic, New York, 1968.

30. J. S. Smart, *Effective Field Theories of Magnetism*, Saunders, Philadelphia, PA, 1966.

31. L. N. Mulay and W. J. Danley, *Bull. Am. Phys. Soc.*, **14**(2), 350 (1969); *J. Appl. Phys.*, **41**(3), 877 (1970); D. W. Collins and L. N. Mulay, *IEEE Trans. Magn.*, **4**, 3, 470 (1968); *J. Am. Ceram. Soc.*, **53** (2), 74 (1970).

32. B. D. Cullity, *Introduction to Magnetic Materials*, Addison-Wesley, Reading, MA, 1972.

33. D. W. Collins, J. T. Dehn, and L. N. Mulay, "Mössbauer Spectroscopy and Superparamagnetism," in T. J. Gruverman, Ed., *Mössbauer Methodology*, Vol. III, Plenum, New York, 1966.

34. E. F. Kneller and F. W. Luborsky, *J. Appl. Phys.*, **34**, 656 (1963).

35. L. N. Mulay, *Anal. Chem.*, **34**, 343R (1962); "Techniques for Measuring Magnetic Susceptibility," in A. W. Weissberger and B. W. Rossiter, Eds., *Physical Methods of Chemistry, Techniques of Chemistry*, Vol. 1, Part 4, Wiley-Interscience, New York, 1972.

36. L. N. Mulay and I. L. Mulay, *Anal. Chem.*, **36**, 404R (1964).

37. L. N. Mulay and I. L. Mulay, *Anal. Chem.*, **38**, 501R (1966).

38. L. N. Mulay and I. L. Mulay, *Anal. Chem.*, **40**, 440R (1968).

39. L. N. Mulay and I. L. Mulay, *Anal. Chem.*, **42**, 325R (1970).

40. L. N. Mulay and I. L. Mulay, *Anal. Chem.*, **44**, 324R (1972).

41. L. N. Mulay and I. L. Mulay, *Anal Chem.*, **46**, 490R (1974).

42. L. N. Mulay and I. L. Mulay, *Anal. Chem.*, **48**, 314R (1976).

43. L. N. Mulay and I. L. Mulay, *Anal. Chem.*, **50**, 274R (1978).

44. L. N. Mulay and I. L. Mulay, *Anal. Chem.*, **52**, 199R (1980).

45. L. N. Mulay and I. L. Mulay, *Anal. Chem.*, **54**, 216R (1982).

46. L. N. Mulay and I. L. Mulay, *Anal. Chem.*, **56**, 293R (1984); 1990 and 1992 reviews, in preparation.

47. G. Quincke, *Ann. Phys. Paris*, **24**, 347 (1885); **34**, 401 (1888).

48. G. R. Graybill, J. W. Wrathall, and J. L. Ihrig, *Chem. Instrum.*, **3**, 71 (1971).

49. H. T. Sone, *Philos. Mag.*, **39**, 305 (1920); *Sci. Rept. Series* 1, Tohoku University, **11**, 139 (1922).

50. R. Stossel, *Ann. Phys. Paris*, **10**, 393 (1931).

51. A. Glaser, *Ann. Phys. Paris*, **75**, 459 (1924).

52. F. Bitter, *Phys. Rev.*, **35**, 1572 (1930).

53. G. G. Havens, *Phys. Rev.*, **42**, 337 (1982); **43**, 992 (1933).

54. A. Lallemand, *C. R. Acad. Sci.*, **194**, 1726 (1932).

55. L. Néel, *C. R. Acad. Sci.*, **194**, 1726 (1932).

56. R. K. Reber and G. F. Boeker, *J. Chem. Phys.*, **15**, 508 (1947).

57. V. I. Vaidyanathan, *Indian J. Phys.*, **1**, 183 (1926).

58. A. I. Efimov, dissertation, 1 SPb, 1888, as quoted by Ya G. Dorfman in [219].

59. A. V. Rankine, *Proc. Phys. Soc. London*, **46**, 391 (1934).

60. J. O'M. Bockris and D. F. Parsons, *J. Sci. Instrum.*, **30**, 362 (1963).

61. F. Haurowitz and H. Kittel, *Ber. Bunsenges. Phys. Chem.*, **66B**, 1046 (1963).

62. C. D. Coryell, F. Stitt, and L. Pauling, *J. Am. Chem. Soc.*, **59**, 633 (1937).

63. L. Pauling, "The Electronic Structure of Hemoglobin," in F. J. W. Roughton and J. C. Kendrew, Eds., *Hemoglobin*, Interscience, New York–London, 1949.

64. L. Pauling, *The Nature of the Chemical Bond*, 3rd ed., Cornell University Press, Ithaca, NY, 1960.

65. L. Pauling and C. D. Coryell, *Proc. Natl. Acad. Sci. USA*, **22**, 159, 210 (1936); *J. Phys. Chem.*, **43**, 825 (1939).

66. L. Pauling, R. E. Wood, and J. H. Sturdivant, *J. Am. Chem. Soc.*, **68**, 795 (1946).

67. A. S. Brill, A. Ehrenberg, and H. Den Hartog, *Biochim. Biophys. Acta*, **40**, 313 (1960).

68. R. G. Gordon, *Rev. Sci. Instrum.*, **33**, 729, 1167 (1962).

69. S. J. Gill, G. P. Malone, and M. Downing, *Rev. Sci. Instrum.*, **31**, 1299 (1960).

70. S. P. Yu and A. H. Morrish, *Rev. Sci. Instrum.*, **27**, 9 (1956).

71. R. Stevenson, *Rev. Sci. Instrum.*, **32**, 28 (1961).

72. A. Berkovitz and E. Kneller, Eds., *Magnetism and Metallurgy*, Vol. 1, Academic, New York, 1970; T. R. McGuire and P. J. Flanders, Chap. 4; H. J. Oguey, Chap. 5.

73. S. J. Barnett, *Appl. Phys. Q.*, **23**, 975 (1952).

74. S. Broersma, *Rev. Sci. Instrum.*, **20**, 660 (1949); *Magnetic Measurements on Organic Compounds*, Martinus Nijhoff, The Hague, 1947; *J. Chem. Phys.*, **17**, 873 (1949).

75. S. Broersma, *Rev. Sci. Instrum.*, **24**, 993 (1953).

76. W. A. Norder, *Rev. Sci. Instrum.*, **31**, 849 (1960); L. N. Mulay, W. Cao, and M. Klemkowsky, *Mater. Sci. Eng.*, **100**, L11 (1988).

77. R. A. Ericson, L. D. Roberts, and J. W. T. Dobbs, *Rev. Sci. Instrum.*, **25**, 1178 (1954).

78. J. J. Fritz, R. V. Rao, and S. Seki, *J. Phys. Chem.*, **62**, 703 (1958).

79. H. G. Effemy, D. F. Parsons, and J. O'M. Bockris, *J. Sci. Instrum.*, **32**, 99 (1955).

80. A. Pacault, *Rev. Sci. Acad. Sci. (Paris)*, **84**, 169 (1946); **86**, 38 (1948); *Bull. Soc. Chim. Fr.*, D 371 (1949).

81. J. Joussot-Dubien, *C. R. Acad. Sci.*, **248**, 3165 (1959).

82. L. N. Mulay and A. Attalla, *J. Am. Chem. Soc.*, **85**, 702 (1963).

83. G. Feher and W. D. Knight, *Rev. Sci. Instrum.*, **26**, 293 (1955).

84. C. Walling, *Free Radicals in Solution*, Wiley, New York, 1957.

85. L. N. Mulay and M. Haverbusch, *Rev. Sci. Instrum.*, **35**, 756 (1964).

86. D. C. Douglas and A. Fratiello, *J. Chem. Phys.*, **39**, 3161 (1963).

87. J. L. Deutsch, A. C. Lawson, and M. Poling, *Anal. Chem.*, **40**, 839 (1968).

88. D. F. Evans, *J. Chem. Soc.*, 203 (1959); J. Q. Adams, *Rev. Sci. Instrum.*, **31**, 155 (1954).

89. K. G. Orrel and V. Sik, *Anal. Chem.*, **52**, 567 (1980).

90. Y. L. Yousef, R. K. Gigris, and H. Mikhail, *J. Chem. Phys.*, **23**, 959 (1955); K. Dwight, *J. Appl. Phys.*, **38**, 1505 (1967); R. S. Kaeser, E. Ambler, and J. F. Schooley, *Rev. Sci. Instrum.*, **37**, 173 (1966); J. E. Noakes and A. Arrott, *Rev. Sci. Instrum.*, **39**, 1436 (1968); D. E. Farrell, *Rev. Sci. Instrum.*, **39**, 1452 (1968).

91. S. Foner, *Rev. Sci. Instrum.*, **30**, 548 (1959); S. Foner and E. J. McNiff, Jr., *Rev. Sci. Instrum.*, **39**, 171 (1968).

92. T. H. Moss, *Biochemistry*, **10**, 84 (1971).

93. P. J. Flanders, *Rev. Sci. Instrum.*, **41**, 697 (1970).

94. B. M. Mangum and D. D. Thornton, *Rev. Sci. Instrum.*, **41**, 1964 (1970).

95. P. J. Morris, *J. Phys. E*, **3**, 819 (1970).

96. H. Zilstra, *Rev. Sci. Instrum.*, **41**, 1241 (1970).

97. S. J. Hudgens, *Rev. Sci. Instrum.*, **44**, 579 (1973).

98. A. G. Redfield and C. Moleski, *Rev. Sci. Instrum.*, **43**, 1543 (1972).

99. N. F. Olivera and S. Foner, *Rev. Sci. Instrum.*, **43**, 37 (1972).

100. W. A. Hines and C. W. Moeller, *Rev. Sci. Instrum.*, **44**, 1544 (1973).

101. S. Foner, *Rev. Sci. Instrum.*, **47**, 570 (1975).

102. S. Foner, *Rev. Sci. Instrum.*, **45**, 1181 (1974); **46**, 1426 (1975).

103. R. F. Drake and W. E. Hatfield, *Rev. Sci. Instrum.*, **45**, 1435 (1974).

104. L. R. Sill and S. M. Drensky, *Rev. Sci. Instrum.*, **46**, 221 (1975).

105. C. L. Foiles and T. W. McDaniel, *Rev. Sci. Instrum.*, **215**, 756 (1974).

106. R. P. Guertin and S. Foner, *Rev. Sci. Instrum.*, **45**, 863 (1974).

107. C. Cordero-Montalvo, "Pressure Dependence of Magnetic Susceptibility of Rare Earth Substituted SmS," doctoral dissertation, The Pennsylvania State University, University Park, PA, 1978.

108. C. Cordero-Montalvo, K. Vedam, and L. N. Mulay, "Pressure Dependence of Magnetic Susceptibility of Rare Earth Substituted SmS," in C. W. Chu and J. A. Woolam, Eds., *High Pressure and Low Pressure*, Plenum, New York, 1978.

109. R. T. Lewis, *Rev. Sci. Instrum.*, **47**, 519 (1976).

110. E. E. Bragg and M. S. Seehra, *J. Phys. E*, **7**, 216 (1976).

111. C. N. Guy, *J. Phys. E*, **9**, 433 (1976).

112. C. N. Guy, doctoral dissertation, Cambridge University, England, 1974.

113. C. N. Guy, *J. Phys. E*, **9**, 790 (1976).

114. S. Foner and A. Zieba, *Rev. Sci. Instrum.*, **53**, 1344 (1982); **54**, 137 (1983).

115. P. Weiss and R. Forrer, *Ann. Phys. Paris*, **5**, 153 (1926).

116. R. M. Bozorth and H. J. Williams, *Phys. Rev.*, **103**, 572 (1956).

117. P. R. Brankin, A. R. Eastman, and R. F. Rhodes, *J. Phys. E*, **3**, 312 (1970).

118. L. N. Mulay and I. L. Mulay, *J. Natl. Cancer Inst.*, **39**(4), 735 (1967).

119. H. Gessinger, H. Kronmüller, and R. Bundschuh, *J. Phys. E*, **3**, 468 (1970).

120. H. E. Hall, *Solid State Physics*, Wiley, New York, 1974.

121. J. Matisoo, *Anal. Chem.*, **4**(1), 83A (1969); **41**(2), 139A (1969).

122. E. J. Cukauskas, D. A. Vincent, and B. S. Deaver, *Rev. Sci. Instrum.*, **45**, 1 (1974).

123. A. S. Brill, H. Den Hartog, and V. Legallais, *Rev. Sci. Instrum.*, **29**, 383 (1958).

124. A. S. Brill and R. J. P. Williams, *Biochem. J.*, **78**, 246 (1961).

125. D. Duret, P. Bernard, and D. Zenatti, *Rev. Sci. Instrum.*, **46**, 474 (1975).

126. L. W. Alvarez, H. P. Eberhard, R. R. Ross, and R. D. Watt, *Science*, **167**, 701 (1970).

127. B. S. Deaver and W. S. Gorre, *Rev. Sci. Instrum.*, **38**, 311 (1967).

128. R. K. Hirsch, *J. Chem. Educ.*, **44**, A1023 (1967); **45**, A7 (1968).

129. J. M. Thomas and B. R. Williams, *Q. Rev.*, **20**, 231 (1965).

130. S. Gordon and C. Campbell, *Anal. Chem.*, **32**, 271R (1960).

131. K. S. Banerjea, *Weighing Designs*, Dekker, New York, 1975; A. H. Corwin, "Weighing," in A. Weissberger and B. W. Rossiter, Eds., *Physical Methods of Chemistry*, Vol. 1, Part 4, Wiley-Interscience, New York, 1972, Chap. 1.

132. C. M. Hurd, *Rev. Sci. Instrum.*, **37**, 515 (1966).

133. B. J. Marshall, R. Johnson, D. Follstaetd, and J. Randorff, *Rev. Sci. Instrum.*, **40**, 375 (1969).

134. D. G. LeGrand, *J. Polym. Sci.*, **60**, S71 (1962).

135. T. Johannson, *J. Phys. E*, **9**, 164 (1976).

136. J. F. Villa and H. C. Nelson, *J. Chem. Educ.*, **53**, 28 (1976).

137. H. W. G. Spraget and D. E. G. Williams, *J. Phys. E*, **9**, 317 (1976).

138. D. A. Zatko and G. T. Davis, *Rev. Sci. Instrum.*, **43**, 818 (1972).

139. D. M. Ginsburg, *Rev. Sci. Instrum.*, **41**, 1661 (1970).

140. D. M. Ginsburg, *Rev. Sci. Instrum.*, **42**, 732 (1971).

141. D. Koran, *Laboratory Magnets*, N. V. Phillips Laboratory (Glueilampenfabrieken), Eindhoven, The Netherlands, 1968.

142. S. Broersma, *Am. J. Phys.*, **24**, 500 (1956).

143. W. G. Amey, W. R. Clark, F. M. Kranz, and A. J. Williams, Jr., *Commun. Electron.*, **54-295**, 1 (1954).

144. J. S. Bell and P. G. Wright, *Electron. Eng.*, **32**, 394 (1960).

145. R. B. Block, U.S. Atomic Energy Commission, AECU-4554 (1959); *Nucl. Sci. Abstr.*, **14**, 678 (1960).

146. B. N. Figgis and R. S. Nyholm, *J. Chem. Soc.*, **331** (1959).

147. F. R. Robertson and P. W. Selwood, *Rev. Sci. Instrum.*, **22**, 146 (1951).

148. E. S. Bartlett and D. N. Williams, *Rev. Sci. Instrum.*, **28**, 919 (1957).

149. L. N. Mulay, *Proc. Indian Acad. Sci.*, **34A**, 245 (1951).

150. H. Theorell, *Ark. Kemi Mineral. Geol.*, **A16**(1), 1 (1943).

151. L. Michaelis, "Determination of Magnetic Susceptibilities," in A. Weissberger, Ed., *Techniques of Organic Chemistry*, 2nd ed., Vol. I, Part II, Interscience, New York, 1949.

152. H. W. Dail and G. S. Knapp, *Rev. Sci. Instrum.*, **40**, 1986 (1969).

153. B. C. Belanger, *Rev. Sci. Instrum.*, **40**, 1652 (1969).

154. L. J. Neuringer and Y. Shapiro, *Rev. Sci. Instrum.*, **40**, 1314 (1969).

155. B. L. Booth and A. W. Ewald, *Rev. Sci. Instrum.*, **40**, 1354 (1969).

156. J. W. Robichaux, Jr. and A. C. Anderson, *Rev. Sci. Instrum.*, **40**, 1512 (1969).

157. R. R. Birss and P. M. Wallis, *J. Sci. Instrum.*, **40**, 551 (1963).

158. L. N. Mulay, D. W. Collins, A. W. Thompson, and P. J. Walker, Jr., *J. Organomet. Chem.*, **178**, 217 (1979).

159. L. N. Mulay and J. T. Dehn, *J. Inorg. Nucl. Chem.*, **31**, 3103 (1969).

160. L. N. Mulay and J. T. Dehn, "Magnetic Susceptibility: Characterization and Elucidation of Bonding in Organometallics," in M. Tsutsui, Ed., *Characterization of Organometallic Compounds*, Part 2, Wiley-Interscience, New York, 1970, pp. 439–480.

161. L. N. Mulay, R. C. Everson, O. P. Mahajan, and P. L. Walker, Jr., "Magnetic Characterization of Semiamorphous Nickel Dispersions on Alumina," in R. A. Levy and R. Hawegawa, Eds., *Amorphous Magnetism II*, Plenum, New York, 1977, p. 415.

162. L. N. Mulay, R. C. Everson, and P. L. Walker, Jr., *J. Chem. Tech. Biotech.*, **29**, 1 (1979); *Am. Inst. Phys. Conf. Proc.*, **29**, 536 (1975).

163. L. N. Mulay and M. E. Fox, *J. Chem. Phys.*, **38**, 760 (1963); *Rev. Sci. Instrum.*, **33**, 129 (1982).

164. L. N. Mulay and L. K. Keys, *Anal. Chem.*, **36**, 2383 (1964).

165. L. N. Mulay and L. K. Keys, *J. Am. Chem. Soc.*, **86**, 4489 (1964); **87**, 1192 (1963).

166. J. W. Meyer and A. M. Rich, *Electronics*, **36**(35), 29 (1963).

167. J. C. Reithler, *Bull. Soc. Fr. Mineral. Cristallogr.*, **89**(2), 277 (1966).

168. R. R. Birss and E. W. Lee, *J. Sci. Instrum.*, **37**, 225 (1960).

169. J. G. Aston, B. Bolger, R. Trambarulo, and H. Segall, *J. Chem. Phys.*, **22**, 560 (1954).

170. W. R. Scott, *J. Sci. Instrum.*, **38**, 436 (1961).

171. D. F. Edwards, R. W. Terhue, and V. J. Lazazzera, *Rev. Sci. Instrum.*, **29**, 1049 (1958).

172. L. N. Mulay, *Rev. Sci. Instrum.*, **28**, 279 (1957).

173. A. L. Smith and H. L. Johnston, *Rev. Sci. Instrum.*, **24**, 420 (1953).

174. R. R. Birss, J. I. Gibbs, and P. M. Wallis, *J. Sci. Instrum.*, **1**, 15 (1968).

175. N. P. Garaleshko, *Ukr. Fiz. Zh.*, Ukr. Ed., **7**, 1068 (1962).

176. J. W. Heaton and A. C. Rose-Innes, *J. Sci. Instrum.*, **40**, 369 (1963).

177. C. M. French and D. J. Harrison, *J. Chem. Soc.*, 2538 (1953).

178. J. G. Sloot, C. H. Massen, and J. A. Poulis, *J. Sci. Instrum.*, **2**, 970 (1969).

179. W. R. Angus and W. K. Hill, *Trans. Faraday Soc.*, **40**, 185 (1943).

180. H. R. Nettleton and S. Sugden, *Proc. R. Soc. London*, **A173**, 313 (1939).

181. N. F. Curtis, *J. Chem. Soc.*, 3147 (1961).

182. O. M. Hilal and G. E. Fredericks, *J. Chem. Soc.*, 785 (1954).

183. W. K. Chen, F. B. Koch, and J. M. Sivertson, *Rev. Sci. Instrum.*, **31**, 1157 (1960).

184. H. Theorell and A. Ehrenberg, *Ark. Fys.*, **3**, 299 (1951).

185. J. J. Howland and M. Calvin, U.S. Atomic Energy Commission, AECD-1895, 1948.

186. B. B. Cunningham, *Nucleonics*, **5**, 62 (1949).

187. A. Earnshaw, *Lab. Pract.*, **10**, 157, 294 (1961).

188. M. Simek and O. Navratil, *Chem. Listy*, **53**, 1276 (1959).

189. F. T. Hedgcock and W. B. Muir, *Rev. Sci. Instrum.*, **31**, 390 (1960).

190. P. K. Henry and G. D. Hoyt, *Rev. Sci. Instrum.*, **34**, 446 (1963).

191. B. S. Chandrasekhar, *Rev. Sci. Instrum.*, **27**, 967 (1956).

192. G. Dupouy and C. R. Haenny, *Acad. Sci. (Paris)*, **199**, 781, 843 (1934).

193. Y. Shiraiwa, K. Kikuchi, E. Kishi, H. Nagano, and K. Oshima, *Bull. Soc. Chim. Jpn.*, **35**, 2040 (1962).

194. M. L. Heit and D. E. Ryan, *Anal. Chim. Acta*, **29**, 524 (1963).

195. K. Gersonde, *J. Sci. Instrum.*, **43**, 591 (1966).

196. I. I. Rabi, *Phys. Rev.*, **29**, 174 (1927).

197. L. N. Mulay and M. C. Naylor, "Polymerization of Paramagnetic Fe^{3+} Ions in Solution," in S. Kirschner, Ed., *Advances in the Chemistry of the Coordination Compounds*, Macmillan, New York, 1961.

198. L. N. Mulay and A. V. Prasad-Rao, *IEEE Trans. Magn.*, **19**(5), 1998 (1983).

199. L. N. Mulay, E. G. Rochow, E. O. Stejskal, and N. E. Weliky, *J. Inorg. Nucl. Chem.*, **16**, 23 (1960).

200. R. D. Heyding, J. B. Taylor, and M. L. Hair, *Rev. Sci. Instrum.*, **32**, 161 (1961).

201. W. Sucksmith, *Philos. Mag.*, **8**, 158 (1929).

202. T. R. Lewis, *Rev. Sci. Instrum.*, **42**, 31 (1971).

203. J. A. Cape and R. A. Young, *Rev. Sci. Instrum.*, **42**, 1061 (1971).

204. W. E. Henry, *Phys. Rev.*, **88**(3), 559 (1952).

205. K. Kikuchi, H. W. Bernard, J. J. Fritz, and J. G. Aston, *J. Phys. Chem.*, **69**, 3654 (1954).

206. G. J. Hill, *J. Phys. E*, **1**, Ser. 2, 52 (1968).

207. L. C. Jackson, *Proc. R. Soc. London A*, **140**, 695 (1933); *Proc. Phys. Soc. London*, **47**, 1029 (1935).

208. R. H. Martin and G. J. Hill, *J. Phys. E*, **1**, 1257 (1968).

209. A. M. Stewart, *J. Phys. E*, **2**, 851 (1969).

210. A. N. Gerritsen and D. H. Damon, *Rev. Sci. Instrum.*, **33**, 30 (1962).

211. R. M. Griest and W. Ostertag, *Rev. Sci. Instrum.*, **39**, 1758 (1968).

212. J. F. Houlihan and L. N. Mulay, "EPR Studies on Transitions in Titanium Oxides," in C. D. Graham and J. J. Thyne, Eds., *Proceedings of the 17th Annual Conference on Magnetism and Magnetic Materials*, Vol. 5, Part 1, American Institute of Physics, New York, 1971, p. 316.

213. J. F. Houlihan, W. J. Danley, and L. N. Mulay, *J. Solid State Chem.*, **12**, 265 (1975).

214. J. F. Houlihan and L. N. Mulay, *Mater. Res. Bull.*, **6**, 737 (1971).

215. J. F. Houlihan and L. N. Mulay, *Inorg. Chem.*, **13**, 745 (1974).

216. J. F. Houlihan and L. N. Mulay, *Phys. Status Solidi B*, **61**, 647 (1974).

217. J. F. Houlihan and L. N. Mulay, *Phys. Status Solidi B*, **65**, 513 (1974); *Mater. Res. Bull.*, **11**, 307 (1976).

218. H. J. Jung, A. M. Vannice, L. N. Mulay, R. M. Stansfield, and W. N. Delgas, *J. Catal.*, **76**, 208 (1982).

219. Ya G. Dorfman, *Magnetic Properties and Structure of Matter*, State Publishing House for Technical-Theoretical Literature, Moscow, 1955. English translation, U.S. Atomic Energy Commission, Technical Services, Washington, DC, June, 1961.

220. R. Cini, *Ric. Sci.*, **29**, 506 (1959).

221. A. R. Pepper and J. H. Smith, *J. Sci. Instrum.*, **42**, 328 (1965).

222. W. O. Milligan and H. B. Whitehurst, *Rev. Sci. Instrum.*, **23**, 618 (1952).

223. P. E. Jacobsen and P. W. Selwood, *J. Am. Chem. Soc.*, **76**, 2641 (1954).

224. M. Garber, W. G. Henry, and H. C. Hoeve, *Can. J. Phys.*, **38**, 1595 (1960).

225. E. V. Smith, *J. Sci. Instrum.*, **38**, 466 (1961).

226. F. E. Senftle, M. D. Lee, A. A. Monkewica, J. W. Mayo, and T. Pankey, *Rev. Sci. Instrum.*, **29**, 429 (1958).

227. F. E. Senftle and A. Thorpe, *Nature (London)*, **29**, 410 (1961).

228. P. E. Senftle, M. D. Lee, and A. A. Monkewiz, *Rev. Sci. Instrum.*, **29**, 429 (1958).

229. C. A. Candela and R. E. Mundy, *Rev. Sci. Instrum.*, **32**, 708 (1961); *Inst. Radio. Trans. Instrum.*, **11**, 106 (1962).

230. L. A. Baidakov, L. N. Blinor, Y. V. Zubenko, A. Kazennov, and P. Strakhov, *Vestn. Leningr. Univ.*, *Fiz. Khim.*, **1**, 40 (1966).

231. W. E. Henry and J. L. Rogers, *Philos. Mag.*, **1**, 223, 227 (1956).

232. J. R. Singer, *Rev. Sci. Instrum.*, **30**, 1123 (1959).

233. R. R. Penoyer, *Rev. Sci. Instrum.*, **30**, 711 (1959).

234. R. A. Butera, R. S. Craig, and L. V. Cherry, *Rev. Sci. Instrum.*, **32**, 708 (1961).

235. P. Weiss and G. Foex, *J. Phys. A: Gen. Phys.*, **1**(5), 274 (1911).

236. H. R. Kirchmayr and K. H. Schindl, *Acta Phys. Austriaca*, **22**, 267 (1931).

237. A. K. Mukherjee and N. G. Sutradhar, *Indian J. Phys.*, **37**(12), 616 (1963).

238. Z. Obuszko, *Acta. Polon. Chem.*, **24**(1), 135 (1963).

239. J. T. Richardson and J. O. Beauxis, *Rev. Sci. Instrum.*, **34**, 877 (1963).

240. J. van Liehr, *Rev. Sci. Instrum.*, **39**, 1841 (1968).

241. R. J. Kolenkow and P. W. Zitzewitz, "Microbalance for Magnetic Susceptibility Measurements," in P. M. Waters, Ed., *Vacuum Microbalance Techniques*, Vol. 4, Plenum, New York, 1965.

242. D. E. Soule, C. W. Nezbeda, and A. W. Czanderna, *Rev. Sci. Instrum.*, **35**, 1504 (1964).

243. D. Neogy and R. B. Lal, *J. Sci. Indian Res.*, **21B**, 103 (1962).

244. L. N. Mulay and V. Withstandley, *J. Chem. Phys.*, **43**, 4522 (1965); *Bull. Am. Phys. Soc.*, **11**(1), 41 (1966).

245. L. N. Mulay and G. H. Ziegenfuss, "Exchange Interactions in Trinuclear Fe^{3+} Complexes," in C. D. Graham and G. H. Lander, Eds., *American Institute of Physics Conference Proceedings*, New York, **24**, 213 (1975).

246. D. Wohlleben and M. B. Maple, *Rev. Sci. Instrum.*, **42**, 1573 (1971).

247. N. Kawai and A. Sawaoka, *Rev. Sci. Instrum.*, **38**, 1770 (1967).

248. B. A. Rogers, K. Wentzel, and J. P. Riott, *Trans. Am. Soc. Met.*, **29**, 969 (1941).

249. H. K. Work and H. T. Clark, *Am. Inst. Min. Metall. Pet. Eng. Inst. Met. Div. Spec. Rep. Ser.*, No. 1132 (1939); K. Dwight, *J. Appl. Phys.*, **38**, 1505 (1967); R. S. Kaeser, E. Ambler, and J. F. Schooley, *Rev. Sci. Instrum.*, **37**, 173 (1966); J. E. Noakes and A. Arrott, *Rev. Sci. Instrum.*, **39**, 1436 (1968); D. E. Farrell, *Rev. Sci. Instrum.*, **39**, 1452 (1968).

250. W. S. Thal, *Z. Tech. Phys.*, **15**, 469 (1934).

251. D. J. Doan, *U.S. Bur. Mines Tech. Prog. Rep.*, No. 3268, 91 (1935).

252. S. Krongelb and M. W. P. Strandberg, *J. Chem. Phys.*, **31**, 1196 (1959).

253. F. Klauer, E. Turowski, and V. Wolff, *Angew. Chem.*, **54**, 494 (1941); *Z. Tech. Phys.*, **22**, 223 (1941).

254. H. Krupp, *Z. Angew. Phys.*, **6**, 541 (1954).

255. M. B. Dobrin, *Introduction to Geophysical Prospecting*, McGraw-Hill, New York, 1975, and subsequent editions.

256. B. K. Banerjee, *Indian J. Phys.*, **33**, 201 (1959).

257. W. F. Fisher, "Magnetic Studies on Iron Containing Glasses," Masters dissertation (Geochemistry), The Pennsylvania State University, University Park, PA, 1972.

258. R. A. Bailey, E. N. Balko, and T. Lesniak, *J. Inorg. Nucl. Chem.*, **13(b)**, 1527 (1954).

259. A. Kleber, *Rare Earth Research*, Macmillan, New York, 1961.

260. L. N. Mulay, A. W. Thompson, D. W. Collins, and P. L. Walker, Jr., *J. Organomet. Chem.*, **178**, 217 (1979).

261. L. N. Mulay and P. W. Selwood, *J. Am. Chem. Soc.*, **76**, 6207 (1954); **77**, 2693 (1955).

262. B. O. A. Hedstrom, *Ark. Kemi*, **6**, 1 (1953).

263. C. E. A. Brownlow, *Nature (London)*, **194**, 176 (1962).

264. S. Broersma, *J. Chem. Phys.*, **26**, 1405 (1957); *Rev. Sci. Instrum.*, **34**, 217 (1963).

265. C. Schlenker, R. Buder, M. Schlenker, J. F. Houlihan, and L. N. Mulay, *Phys. Status Solidi B*, **54**, 247 (1972).

266. H. Schugar, C. Walling, R. B. Jones, and H. B. Gray, *J. Am. Chem. Soc.*, **89**, 3712 (1967).

267. T. Ohya and K. Ono, *J. Chem. Phys.*, **57**, 3240 (1972).

268. C. M. Knobler, "On the Existence of Molecular Oxygen Dimer," doctoral dissertation, Leiden University, Drukerig, Pasmans, Gravenhage, Holland, 1961.

269. E. J. Wachtel and R. G. Wheeler, *J. Appl. Phys.*, **42**, 1581 (1971); *Phys. Rev. Lett.*, **24**, 233 (1970).

270. M. Bodenstein, P. Hartek, and E. Padelt, *Z. Anorg. Chem.*, **147**, 233 (1925).

271. C. F. Goodeve and F. D. Richardson, *J. Chem. Soc.*, 294 (1937).

272. L. A. Marusak and L. N. Mulay, *IEEE Trans. Magn.*, **12**(6), 889 (1977).

273. L. A. Marusak and L. N. Mulay, *J. Appl. Phys.*, **50**(11), 7807 (1979).

274. L. A. Marusak and L. N. Mulay, *J. Appl. Phys.*, **50**, 1865 (1979); *Phys. Rev. B*, **21**, 238 (1980).

275. H. Bethe, *Ann. Phys. Paris*, **3**, 133 (1929).

276. W. J. Danley and L. N. Mulay, *Mater. Res. Bull.*, **7**, 739 (1972).

277. L. K. Keys and L. N. Mulay, *Appl. Phys. Lett.*, **9**, 248 (1966).

278. L. K. Keys and L. N. Mulay, *Bull. Am. Phys. Soc.*, **12**, 503 (1967).

279. L. K. Keys and L. N. Mulay, *Jpn. J. Appl. Phys.*, **6**, 122 (1967).

280. L. K. Keys and L. N. Mulay, *J. Appl. Phys.*, **38**, 1445 (1967).

281. L. K. Keys and L. N. Mulay, *J. Phys. Soc. Jpn.*, **23**, 478 (1967).

282. L. K. Keys and L. N. Mulay, *Phys. Rev.*, **154**, 453 (1967).

283. F. S. Dainton, D. M. Wiles, and J. Wright, *J. Chem. Soc.*, 4283 (1960); W. L. Nelson, M. S. Whittingham, and T. F. George, *High Temperature Superconductors* (ACS Symposium Series, 351 and 377), American Chemical Society, Washington, DC, 1987, 1988; J. W. Halley, Ed., *Theories of High Temperature Superconductivity*, Addison-Wesley, Reading, MA, 1988.

284. E. C. Evers, *J. Chem. Educ.*, **38**, 591 (1961).

285. C. A. Hutchinson, "Magnetic Susceptibilities," in E. A. Braude and F. C. Nachod, Eds., *Determination of Organic Structures by Physical Methods*, Academic, New York, 1955.

286. G. W. Wheland, *Advanced Organic Chemistry*, Wiley, New York, 1949.

287. D. J. E. Ingram, *Spectroscopy at Radio and Microwave Frequencies*, Butterworths, London, 1955; *Free Radicals as Studied by Electron Spin Resonance*, Butterworths, London, 1958.

288. T. Vanngard and B. Malmstrom, *Electron Spin Resonance and Biochemistry*, Vol. 37, Nordstet, Stockholm, 1959.

289. M. S. Blois, Jr., H. M. Brown, R. M. Lemmon, R. O. Lindblom, and N. Weissbluth, *Free Radicals in Biological Systems*, Academic, New York, 1961.

290. H. Boardman and P. W. Selwood, *J. Am. Chem. Soc.*, **72**, 1372 (1950).

291. J. Farquaharson and P. Ady, *Nature (London)*, **143**, 1067 (1939).

292. A. A. Berlin, L. A. Blymenfeld, and N. N. Semenov, *Izvest. Akad. Nauk. S.S.R., Otdel. Khim. Nauk.*, 1689 (1959).

293. J. J. Santiago, "Magnetic Studies on Carbon and Graphite," doctoral dissertation, The Pennsylvania State University, University Park, PA, 1972.

294. L. N. Mulay, J. J. Santiago, and P. L. Walker, Jr., *American Institute of Physics Conference Proceedings*, **10**(2), 1520 (1973).

295. J. S. Griffith, *Biochim. Biophys. Acta*, **28**, 439 (1958).

296. D. S. Taylor and C. D. Coryell, *J. Am. Chem. Soc.*, **60**, 1177 (1938).

297. R. G. Shulman, W. M. Walsh, Jr., H. J. Williams, and J. P. Wright, *Biochem. Biophys. Res. Commun.*, **5**, 52 (1961).

298. R. P. Blakemore and J. P. Frankel, *Sci. Am.*, **245**(6), 58 (1981).

299. W. E. C. Wacker and B. L. Vallee, *J. Biochem. Tokyo*, **234**, 3257 (1959).

300. L. Peterson, "Magnetic Studies of Iron and Copper Containing Proteins," doctoral dissertation, Uppsala University, Stockholm, Sweden, 1981; *Biochim. Biophys. Acta*, **18**, 622 (1981); *J. Biol. Chem.*, **21**, 5 (1980).

301. F. E. Luborsky, *Proceedings of Magnetism and Magnetic Materials Conference*, H. C. Wolfe, Ed., American Institute of Physics, New York, 1975.

302. Y. A. Liu, Ed., *Industrial Application of Magnetic Separation*, Institute of Electrical and Electronics Engineers, Inc., Special Issue, 1979.

303. A. R. Kauffman, "Magnetic Methods of Analysis," in W. G. Berl, Ed., *Physical Methods in Chemical Analysis*, Academic, New York, 1951, and editions subsequent to 1961.

304. T. Nagata, *Rock Magnetism*, Maruzen Co. Ltd., Tokyo, Japan, 1961.

305. G. Strangway, *History of the Earth's Magnetic Field*, McGraw-Hill, New York, 1970.

306. L. N. Mulay and H. Yamamura, *J. Appl. Phys.*, **50**(11), 7795 (1980).

307. C. Lo, K. R. P. M. Rao, and L. N. Mulay, *Mössbauer Spectroscopy and Its Chemical Applications*, G. J. Stevens and G. K. Shenoy, Eds. (Advances in Chemistry Series, Vol. 194), American Chemical Society, Washington, DC, 1981, Chap. 27.

308. R. L. Fyans, Special Brochure No. MA-49, Perkin-Elmer Corporation, Instrument Division, Norwalk, CT, March, 1978.

309. R. T. Obermeyer, C. Lo, M. Oskooie-Tabrizi, and L. N. Mulay, *J. Appl. Phys.*, **53**, 2683 (1982).

310. P. J. Desai, personal communication.

Chapter 4

ELECTRON SPIN RESONANCE

Ted M. McKinney and Ira B. Goldberg

1   INTRODUCTION AND GENERAL CONCEPTS

Electron spin resonance (ESR), electron paramagnetic resonance (EPR), and electron magnetic resonance (EMR) are some of the names for a physical measurement technique that requires the presence of a net quantum mechanical angular momentum in the sample under study. Most commonly, this net angular momentum arises from the spin of unpaired electrons associated with the atoms or molecules of the sample, but in some instances it may arise from the net orbital angular momentum of p, d, or f orbitals of gas-phase atoms or molecules.

Net quantum mechanical angular momentum in a sample can also arise solely from the quantized spins of nuclei. This gives rise to the phenomenon of nuclear magnetic resonance (NMR), which is appreciated as a most sophisticated medical diagnostic tool in its incarnation as magnetic resonance imaging [1]. Although there are many formal similarities between the phenomena of NMR and electron resonance, the energies of the reasonantly absorbed quanta are so different and the required instrumentation is so disparate that it is customary to treat NMR separately. Nuclear magnetic resonance is discussed in Chapter 5 of this volume.

Authors have suggested distinct meanings for the terms ESR and EPR [2, 3]. Formally, EPR is the more general term since materials may be paramagnetic without having unpaired electron spins. The more common practice, currently promoted by Chemical Abstracts Service (CAS), is to use ESR as the generic term regardless of the origin of the paramagnetism, although the International Union of Pure and Applied Chemistry (IUPAC) is promoting use of EPR. We use ESR throughout this chapter.

No matter what name is used, the fundamental process observed in a magnetic resonance experiment is the resonant absorption of electromagnetic energy when a sample containing an assemblage of quantized magnetic moments is placed in a magnetic field. Electron spin resonance spectrometers operate at energies corresponding to microwave radiation. Normally the microwave frequency is held constant while the laboratory magnetic field is swept to produce the spectrum. The microwave frequency bands are usually indicated by letter designations of historic origin, and Table 4.1 shows these in relation to the magnetic fields that are routinely required.

Table 4.1 ESR Band Designations[a]

Band Designation	Band Range[b] (GHz)	ESR Frequency[b] (GHz)	Magnetic Field[c]
L band	1–2	1–2	38–75 mT
S band	2–4	2–4	75–150 mT
X band	8–12.4	9–10	320–355 mT
K band	18–26.5	22–24	785–858 mT
K_a band	26.5–40	33–35	1.18–1.2 T
Q band	33–50	34	1.2 T

[a]See Section 7 for a discussion of units used in magnetic resonance.
[b]GHz = gigahertz = 10^9 Hz.
[c]1 T = 10,000 G.

Electron spin resonance is a useful tool in many research fields. Because of the scope of this chapter, only a few examples are given. However, Section 8 lists various textbooks and review articles that are useful to readers seeking specific applications.

There are several different aspects of ESR that make it useful in a variety of studies. The molecular or lattice symmetry of a paramagnetic state is reflected by anisotropy of the g-factor (electron environment), the distribution of an electron delocalized over magnetic nuclei is revealed through hyperfine splittings, and the relative positions of the electrons in molecules or ions containing two or more unpaired electrons are reflected through the dipolar couplings. Numerous dynamic effects are understood through relaxation processes.

As a result this chapter is organized as follows: (1) basic ESR phenomena are described and related to the fundamental relaxation processes, (2) spectroscopic parameters are described, (3) specific cases of normal ESR are described, and (4) multiple resonance techniques and dynamic ESR spectroscopy are reviewed briefly.

1.1   The Dirac Electron

One of the great triumphs of twentieth-century physics was Dirac's relativistic description of the electron [4]. In his treatment, one of the quantization conditions that arose naturally was given the name *spin*. (Another condition, called *charge*, also arose from this treatment, and it implied the existence of the positron long before it was observed experimentally.)

Strictly speaking, electron spin is only meaningful as a quantum mechanical operator, which is used in the mathematical treatments of interactions between electrons and matter and/or radiation. However, it is often convenient to describe the electron trivially as a charged sphere spinning at a fixed rate. Because of the finite mass of the electron there is an angular momentum of magnitude $|\mathbf{S}| = [S(S + 1)]h/2\pi$, where $S = \frac{1}{2}$ is the spin quantum number of the electron. Because of the negative charge on the electron, the spin angular momentum is manifested as a magnetic dipole $\boldsymbol{\mu}$, which is oriented precisely opposite to the direction of the angular momentum vector $\mathbf{S}$ (see Figure 4.1). The magnetic dipole interacts with an external field such that the projection of $\boldsymbol{\mu}$ in the field direction (defined as the z direction in a Cartesian system of coordinates) is

$$\mu_z = -g_e(eh/4\pi m_e c)M_S = -g_e\mu_B M_S \tag{1}$$

where g_e is a fundamental constant of the electron (see Section 4.1.4), e is the electron charge, h is Planck's constant, m_e is the electron rest mass, c is the speed of light, and $M_S = +\frac{1}{2}$ and $M_S = -\frac{1}{2}$ refer to the two quantized spin states of the electron. It is often convenient to refer to these as the $|\alpha_e\rangle$ and $|\beta_e\rangle$ spin states, respectively, or simply as $|\alpha\rangle$ and $|\beta\rangle$ spins when there is no possibility of confusion with nuclear spin states. The quantities in parentheses represent the Bohr magneton, symbolized by μ_B. Figure 4.2 represents the relation between these quantities in the field $\mathbf{B}$. (Local fields in a molecule or crystal can similarly interact with the electron, but discussion of those effects is deferred.)

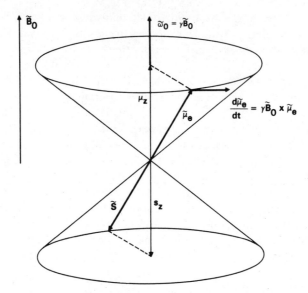

Figure 4.1 Relation between spin angular momentum **S** and spin magnetic moment **μ** for an electron subjected to a magnetic field **B₀**. (Note: a tilde indicates a vector quantity in all figures.)

In this chapter we use the phrase *spin unit* to signify any general entity that possesses quantized electron spins and/or orbital angular momentum. Consequently, when we describe the behavior or properties of a spin unit, the results apply equally well to such diverse structures as a free radical, a transition metal ion, a crystal defect site, an isolated paramagnetic atom, or an isolated electron spin.

1.2 Zeeman Energy Levels

The energies of the $|\alpha_e\rangle$ and $|\beta_e\rangle$ spin states are degenerate in the absence of a magnetic field. The Zeeman splitting is the name given to the lifting of this degeneracy when a sample of atoms or molecules possessing angular momentum is subjected to a magnetic field. For example, a sample of solvated electrons (formed by dissolving sodium metal in liquid ammonia) will have two well-defined electronic Zeeman levels when a magnetic field B is turned on, and the energy separation will increase as the field is made progressively stronger (see Figure 4.2). In addition to the field strength, the energy-level separation depends on two characteristic constants of the electron

$$h\nu = \Delta E = g_e \mu_B B \tag{2}$$

where g_e is called the *electron g-factor* and has a value close to 2 and μ_B is the Bohr magneton $\mu_B = eh/4\pi m_e c$.

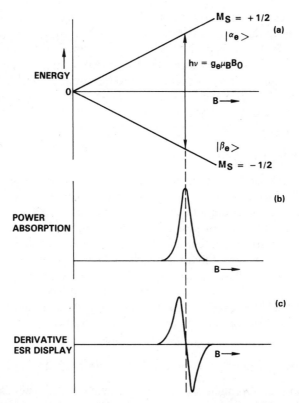

Figure 4.2 (*a*) Zeeman splitting of the energy levels of electron spin states in a magnetic field with varying magnitude B. (*b*) Curve representing energy absorption when an ensemble of unpaired electrons is subjected to microwave radiation to constant frequency ν_0 while the magnetic field B is varied. (*c*) Curve representing the first derivative of curve *b*. This is the way ESR absorptions are normally displayed.

1.3 Larmor Precession

Since the spin magnetic moment always maintains an orientation of $\pm \cos^{-1}(1/\sqrt{3})$ with respect to an applied magnetic field, it can never be colinear with the external field. Consequently it experiences a torque

$$\mathbf{T} = \boldsymbol{\mu} \times \mathbf{B} \qquad (3)$$

so that the equation of motion of the magnetic moment is

$$d\boldsymbol{\mu}/dt = \gamma \mathbf{B} \times \boldsymbol{\mu} \qquad (4)$$

where $\gamma = -g_e e/2m_e c$ is a fundamental constant called the *magnetogyric* (or *gyromagnetic*) *ratio*. (The electron *g*-factor g_e is discussed later.) The preceding equation can be solved to yield the familiar result

$$\omega_0 = \gamma \mathbf{B}_0 \qquad (5)$$

which describes the motion known as *Larmor precession*. The vector ω_0 (corresponding to the angular frequency $2\pi\nu_0$) is defined by the right-hand rule so the precessional motion of the electron spin magnetic moment occurs in the counterclockwise sense (see Figure 4.1). The subscripts in (5) emphasize that the Larmor precession is directly proportional to the strength of the magnetic field. Note the x and y components of the magnetic moment of a given spin are constantly changing and cannot be specified because the x, y, and z components of angular momentum do not commute.

1.4   Free Electron g-Factor

The fundamental unit of electron magnetism, the Bohr magneton, is easily derived on the basis of a loop carrying the current corresponding to the orbital motion of an electron about a nucleus. Thus an electron in a p orbital with one unit ($h/2\pi$) of orbital angular momentum has a magnetic moment of $1\ \mu_B$.

Dirac's relativistic treatment of the electron predicted that the magnetic moment arising from spin should be precisely two Bohr magnetons per unit of angular momentum; that is, $\mu_S = 2.0\ M_S\ \mu_B$. However, careful experimental measurements revealed the true value to be slightly larger, $g_e = 2 \times 1.001159652193 = 2.002319304386$ [5]. This correctional term can be expressed as a power series in the fine structure constant, $\alpha = 2\pi e^2/hc$, and was explained in terms of quantum electrodynamics [6].

1.5   Resonant Absorption by an Ensemble of Spins

Treated quantum mechanically, the component of μ perpendicular to the magnetic field is indeterminate. But when a collection of spins is considered, the bulk magnetization M passes over to the classical limit

$$M = \Sigma\mu \tag{6a}$$

$$M_z = (N_\beta - N_\alpha)\mu_z \tag{6b}$$

$$M_\perp = 0 \tag{6c}$$

with the perpendicular or transverse component being zero because the individual moments are precessing with random phase relations. The net difference $(N_\beta - N_\alpha)$ between populations of the two spin states produces the M_z component, which is aligned with the laboratory field.

When a perturbing field $2\mathbf{B}_1$ is pulsed perpendicular to the laboratory field $\mathbf{B}$, say along the x axis, the net magnetization vector is momentarily tipped away from the laboratory z axis. When the perturbing field is applied periodically, it can be decomposed into two counterrotating, circularly polarized components

$$2\mathbf{B}_1 = \mathbf{i}\,2B_1 \cos \omega t$$

$$= B_1(\mathbf{i} \cos \omega t + \mathbf{j} \sin \omega t) + B_1(\mathbf{i} \cos \omega t - \mathbf{j} \sin \omega t) \tag{7}$$

where the last set of terms, with the minus sign, corresponds to clockwise rotation and can be neglected because it does not interact strongly with the counterclockwise-precessing electron magnetic moments. The field $\mathbf{B}_1 = \mathbf{i} \cos \omega t + \mathbf{j} \sin \omega t$ exerts an additional torque on each of the magnetic moments that comprise $\mathbf{M}$. The composite effective field

$$\mathbf{B}_e = \mathbf{B} + \mathbf{B}_1 \tag{8}$$

acts on the individual moments and causes them to achieve a degree of phase coherence that produces a nonzero transverse component of the bulk magnetization, as indicated in Figure 4.3.

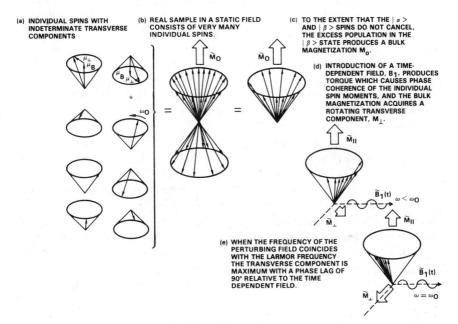

Figure 4.3 Schematic representation of the bulk magnetization $\mathbf{M}_0$ arising in a sample consisting of many unpaired electron spins. A sample comprises an ensemble of individual spins, (a). Note that the transverse component $\mu_\perp$ of any given spin is indeterminate so that when the system is far off-resonance, the transverse components of the individual members of the ensemble are distributed randomly about the precessional cones, (b). The bulk magnetization of the spin system arises from the net excess of electron spins in the $|\beta\rangle$ spin state, (c). In the absence of periodic magnetic disturbance $B_1(t)$ (microwave radiation), the magnetization is stationary along the axis of the external magnetic field. When microwave radiation is applied it produces a torque on the individual moments, imposing a degree of phase coherence among the transverse components of the individual spins, (d). Consequently the bulk magnetization acquires a transverse component that rotates about the magnetic field axis. (The microwave magnetic field also provides torque to bring about transitions between $|\alpha\rangle$ and $|\beta\rangle$ spin states. This introduces enough indeterminacy in the energy of the electron spins to offset the determinancy of the transverse component, and thus conform to the Heisenberg uncertainty principle.) At the resonance frequency, the rotating component of the bulk magnetization $\mathbf{M}_\parallel$ is maximum and 90° out of phase with the microwave radiation field, (e). (At high levels of microwave power, the phase coherence of the transverse components $\mu_\perp$ is so great that the longitudinal magnetization must approach zero to conform to the uncertainty principle. This is the condition known as *saturation*.)

A sample is an ensemble comprising many individual spins, represented as in Figure 4.3a; however, the transverse component of any one spin is indeterminate. When the system is far off-resonance ($|\omega_0 - \omega| \gg 0$), the transverse components of the individual spins that make up the sample ensemble are distributed randomly about the precessional cones as indicated in Figure 4.3b. The bulk magnetization of the spin system arises from the net excess of electron spins in the $|\beta\rangle$ spin state as indicated in Figure 4.3c and in (6b). In the absence of periodic magnetic disturbance (microwave radiation) the magnetization is stationary along the axis of the external magnetic field. When microwave radiation is applied, it produces a torque [corresponding to (3) but arising from the effective field (8)] on the individual moments and imposes a degree of phase coherence among the transverse components of the individual spins, as in Figure 4.3d. This introduces a rotating transverse component into the bulk magnetization. (The microwave magnetic field also provides torque [(3) and (8)] to cause transitions between $|\alpha\rangle$ and $|\beta\rangle$ spin states. This introduces enough indeterminacy in the energy of the electron spins to offset the determinacy of the transverse component and thus to conform with the Heisenberg uncertainty principle.)

At the resonance frequency, the rotating component of the bulk magnetization is maximum and 90° out of phase with the microwave radiation field, as in Figure 4.3e. (At high levels of microwave power, the phase coherence of the transverse components is so great that the longitudinal magnetization must approach zero to preserve agreement with the uncertainty principle. This is the condition known as *saturation*.) When the frequency of the rotating field is the same as the Larmor precession frequency $\omega_0 - \omega = 0$, the periodicity of the perturbing field causes the greatest degree of phase coherence in the spin ensemble. The transverse component of magnetization can induce a current in a coil surrounding the sample (Bloch induction experiment) [7]. (This phenomenon can actually be used as the basis for construction of very narrow band microwave filters [8].) The transverse component of magnetization lags the perturbing field by 90°. The torque also has components in the z direction that are proportional to the magnitude of $\mathbf{B}_1$. The presence of these components is responsible for the transitions between the $|\alpha\rangle$ and $|\beta\rangle$ spin states.

1.5.1 Population of Spin States

The populations of the energy levels corresponding to the $|\alpha\rangle$ and $|\beta\rangle$ spin states are given by the Boltzmann distribution

$$N_\alpha^0/N_\beta^0 = \exp(-\Delta E/kT) \tag{9}$$

In a laboratory field of 340 mT (3400 G), the energy difference is 6.3795×10^{-24} J, corresponding to a microwave frequency of 9.628 GHz. At 20°C, $kT = 4.05 \times 10^{-21}$ J, so the series expansion of the exponential can reasonably be truncated after one term

$$N_\alpha^0/N_\beta^0 \simeq 1 - \Delta E/kT = 0.9984 \tag{10}$$

In principle it is possible to detect this equilibrium net excess of electrons in the lower energy level by suspending a sample from a balance arm and measuring the apparent change in weight when a strong field is applied to the sample (Gouy method) [9, 10]. Since this technique is not very sensitive, it is useful only when large quantities of sample are available, such as for paramagnetic transition metal complexes. Resonance methods of detection are far more sensitive.

1.5.2   Saturation

When the system is subjected to microwave radiation at the resonant frequency, there is a net absorption of energy and the populations change from the equilibrium values. Since the transition probabilities for stimulated absorption and emission are equal ($P_\beta = P_\alpha = P$), this change can be written as

$$dN_\beta/dt = P(N_\alpha - N_\beta) \tag{11a}$$

or

$$= -Pn \tag{11b}$$

where $n = N_\beta - N_\alpha$ is the instantaneous population difference of the two states. The total number of spins in the system $N = N_\alpha + N_\beta$ can be used to reformulate the populations as

$$N_\beta = 1/2(N + n), \qquad N_\alpha = 1/2(N - n) \tag{12}$$

Since N is fixed, (11) can be rewritten as

$$dn/dt = -2Pn \tag{13}$$

which is readily solved as

$$n = n^0 \exp(-2Pt) \tag{14}$$

where n^0 is the equilibrium population difference. This suggests that the population difference is expected to decay to zero if the sample is subjected to the resonance condition for a long interval.

When the oscillating microwave field $\mathbf{B}_1$ has a sufficiently large amplitude and is at the resonance frequency, the two energy levels can approach equal populations with the result that energy absorption from the microwave field decays to nearly zero. This condition is known as *saturation* and forms the basis of pulsed ESR techniques and saturation-transfer spectroscopy.

1.5.3   Longitudinal Relaxation

If the microwave field were turned off at this point, the bulk magnetization would recover its equilibrium value because the transition probabilities for lattice-induced emission and absorption are unequal. The rate constant for this recovery process is symbolized as $1/T_1$, where T_1 has the dimension of time and is called the *spin–lattice* or *longitudinal relaxation time*. The word lattice in this context is a relic from the early days of magnetic resonance when all samples were studied as solids. It is still used to refer to the surroundings of the spin system, be they solid, liquid, or gaseous.

From the Boltzmann distribution (9), it is evident that saturation corresponds to a condition equivalent to a very high temperature. Therefore, the return to the equilibrium magnetization, $M_z^0 = n_{\mu_z}^0$, can be envisioned as thermal equilibration between the spin system and other degrees of freedom in the lattice, such as vibrations or rotations. The mechanism of this equilibration is basically nonradiative; rather, it is stimulated by randomly fluctuating fields present in the sample so that there is an exchange of energy between the spin system and the lattice. In solids this energy transfer may involve vibrations of the entire bulk of the sample (phonon modes), whereas in dilute gases the degrees of freedom into which energy may be transferred are limited to the vibrations, rotations, and translations of discrete molecules.

Restricting attention only to the relaxation process, the rate of change in the population of the lower level in the sample is

$$dN_\beta dt = N_\alpha W_{\alpha\beta} - N_\beta W_{\beta\alpha} \tag{15}$$

where the relaxation transition probabilities for emission $W_{\alpha\beta}$ and absorption $W_{\beta\alpha}$ are unequal. When the sample achieves thermal equilibrium the system is at steady state and the populations attain their equilibrium values so

$$N_\beta^0/N_\alpha^0 = W_{\alpha\beta}/W_{\beta\alpha} \tag{16}$$

Using the expressions of (12), the approach to equilibrium is written as

$$dn/dt = -n(W_{\alpha\beta} + W_{\beta\alpha}) + N(W_{\alpha\beta} - W_{\beta\alpha}) \tag{17a}$$

or

$$= -(n - n^0)/T_1 \tag{17b}$$

by identifying $1/T_1 = W_{\alpha\beta} + W_{\beta\alpha}$ and the equilibrium population difference

$$n^0 = N(W_{\alpha\beta} - W_{\beta\alpha})/(W_{\alpha\beta} + W_{\beta\alpha}) \tag{18}$$

When the results for both excitation and relaxation are combined

$$dn/dt = -2Pn - (n - n^0)/T_1 \tag{19}$$

it is found that in the steady-state condition ($dn/dt = 0$) it is indeed possible to maintain a net population difference between the two spin states when the sample is subjected to continuous radiation at the resonance frequency

$$n = n^0/(1 + 2PT_1) \tag{20}$$

Under such steady-state conditions, the rate of energy absorption is

$$dE/dt = nP\Delta E = n^0 P\Delta E/(1 + 2PT_1) \tag{21}$$

and the sample does not saturate as long as $PT_1 \ll 1$. Consequently, when the microwave field (which determines the magnitude of P) is not too large and T_1 is short (typically 10^{-4}–10^{-9} s), saturation is avoided easily. However, certain experimental methods *require* sample saturation to yield detailed information (see Section 5).

1.5.4   Transverse Relaxation

When the sample is subjected to the resonance condition, the magnetization vector **M** tips away from the laboratory z axis because the transverse components of the individual magnetic moments attain a degree of phase coherence. If the microwave radiation were suddenly turned off, this phase coherence would decay with a rate constant of $1/T_2$, where T_2 is the *spin–spin* or *transverse relaxation time*.

There is no net change in the energy of the spin system as a result of this loss of phase coherence because the transverse components are not quantized. The mechanism of spin–spin relaxation involves the flipping of an $|\alpha\rangle$ to a $|\beta\rangle$ state in concert with the flipping of a $|\beta\rangle$ to an $|\alpha\rangle$ state. Since the angular momentum is completely indeterminate during the transitions, there is no correlation between the orientation of the transverse component of an individual **μ** before and after a transition. Consequently, the spin–spin interactions cause the initially coherent transverse components to become redistributed randomly around the Larmor precession cone.

1.6   The Bloch Equations

It is convenient to view the motion of the magnetization **M** from a frame of coordinates that rotates at the frequency of the microwave radiation. This treatment is developed fully in the standard texts [11–13] and is only summarized briefly here. The rotating frame of coordinates is defined by the unit vectors **i***, **j***, and **k***, which rotate relative to the set of fixed laboratory unit vectors **i**, **j**, and **k** according to

$$\mathbf{i}^* = \mathbf{i} \cos \omega t + \mathbf{j} \sin \omega t \tag{22a}$$

$$\mathbf{j}^* = -\mathbf{i} \sin \omega t + \mathbf{j} \cos \omega t \tag{22b}$$

$$\mathbf{k}^* = \mathbf{k} \tag{22c}$$

The z axes of the two coordinate frames coincide, but the x^*, y^* axes rotate at an angular velocity **ω** ($= \mathbf{k}\omega$) relative to the laboratory axes, Figure 4.4. In the rotating system, the oscillating magnetic field of the perturbing microwave radiation is static, $\mathbf{B}_1 = \mathbf{i}^* B_1$. Moreover, the velocity of rotation diminishes the apparent strength of the laboratory magnet by ω/γ so the net effective field experienced by M in the rotating frame is

$$\mathbf{B}_e = \mathbf{i}^* B_1 + \mathbf{k}^* (B_0 - \omega/\gamma) \tag{23}$$

The apparent decrease in field caused by rotation is clear from an analogy with mechanical tops. A spinning top precessing in a gravitational field appears to be

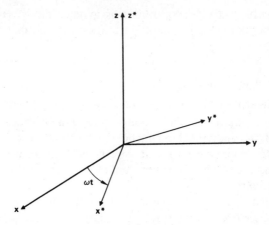

Figure 4.4 Relationship between the laboratory-fixed coordinate system (x, y, z) and the rotating coordinate system (x^*, y^*, z^*). The latter rotates at an angular frequency and sweeps out an angle ωt in time t.

spinning in field-free space when viewed in a frame of coordinates rotating at the precessional velocity.

It is conceptually convenient to develop the Bloch equations for conditions corresponding to a fixed value of the laboratory field $\mathbf{B}_0 (= \mathbf{k}\omega_0/\gamma)$ and to examine the effect of varying the frequency of the perturbing field B_1. However, it is more convenient experimentally to reverse the functionality and to vary the laboratory field B while maintaining the frequency constant. When ω is close enough to ω_0 to produce some degree of phase coherence in the transverse components of the individual μ, the magnetization $\mathbf{M}$ is no longer colinear with the z axis and the transverse components described as M_x and M_y in the laboratory frame are related to U and V in the rotating frame by the transformation (22) and

$$iM_x + jM_y + kM_z = \mathbf{M} = i^*U + j^*V + k^*M_z \qquad (24)$$

The equation of motion for $\mathbf{M}$ is much simpler in the rotating frame:

$$\delta\mathbf{M}/\delta t = \gamma \begin{vmatrix} i^* & j^* & k^* \\ B_1 & 0 & (B_0 - \omega/\gamma) \\ U & V & M_z \end{vmatrix} \qquad (25)$$

The curly derivative is a reminder that this determinant describes the motion in the rotating frame. This equation represents the idealized case of "frictionless" motion of $\mathbf{M}$ in the rotating frame. When the real effects of relaxation are introduced, the rate of change of each component becomes

$$\delta U/\delta t = -V(\omega - \omega_0) - U/T_2 \qquad (26a)$$

$$\delta V/\delta t = U(\omega - \omega_0) - \gamma B_1 M_z - V/T_2 \qquad (26b)$$

$$\delta M_z/\delta t = V_y B_1 - (M_z - M_0)/T_1 \qquad (26c)$$

where the quantity γB_0 is replaced with ω_0 and M_0 is the equilibrium magnetization in the absence of microwave perturbation. This set of equations is called the *phenomenological equations of Bloch* [14] because the effects of relaxation are included empirically without regard for the origin of the relaxation processes. When the microwave field has been applied for many Larmor periods, it is simple to solve the equations for the steady state

$$U = \frac{M_0 \gamma B_1 T_2^2 (\omega_0 - \omega)}{1 + T_2^2 (\omega_0 - \omega)^2 + \gamma^2 B_1^2 T_1 T_2} \tag{27a}$$

$$V = -\frac{M_0 \gamma B_1 T_2}{1 + T_2^2 (\omega_0 - \omega)^2 + \gamma^2 B_1^2 T_1 T_2} \tag{27b}$$

$$M_z = \frac{M_0 [1 + T_2^2 (\omega_0 - \omega)^2]}{1 + T_2^2 (\omega_0 - \omega) + \gamma^2 B_1^2 T_1 T_2} \tag{27c}$$

Thus there are basically two types of magnetic resonance experiments; one maximizes the U signal (*dispersion*) and another maximizes the V signal (*absorption*). Since the U response attains its extrema at nonresonant values of ω, it is more difficult to saturate than absorption and is ideally suited for measurements near liquid helium temperatures where T_1 is long.

The physical quantity actually measured in a magnetic resonance experiment is the susceptibility χ, which is related to the magnetization through

$$\mathbf{M} = \chi \mathbf{B} \tag{28}$$

In the absence of a perturbing microwave field the static susceptibility of a collection of N spins is given by the Curie expression

$$\chi_0 = N g^2 \mu_B^2 [S(S+1)]/3kT \tag{29}$$

For $S = \frac{1}{2}$ systems this becomes

$$\chi_0 = N g^2 \mu_B^2 / 4kT \tag{30}$$

It is desirable to transfer the rotating frame results back into the static laboratory coordinate frame. This produces the components

$$M_x = U \cos \omega t - V \sin \omega t \tag{31a}$$

$$M_y = U \sin \omega t + V \cos \omega t \tag{31b}$$

$$M_z = M_z \tag{31c}$$

Clearly $M_\perp = \sqrt{(M_x^2 + M_y^2)} = \sqrt{(U^2 + V^2)}$ is out of phase with the amplitude of the oscillating microwave field in the laboratory system, and this is conveniently represented by expressing the dynamic susceptibility as a complex number

$$\chi(\omega) = \chi'(\omega) + i\chi''(\omega) \tag{32}$$

although there is nothing "imaginary" about the magnetic susceptibility. The symbol $i = \sqrt{-1}$ is used merely to indicate the phase relation between magnetic dispersion χ' and absorption χ''. The transverse magnetization in the laboratory frame arises from the oscillating microwave magnetic field along the laboratory x axis and produces an x component of magnetization

$$M_x = 2B_1 \cos \omega t [\chi(\omega)] \tag{33}$$

Equation (33) is verified through the series expansions of trigonometric and exponential functions and noting that $\cos \omega t = \text{Re}[\exp(+i\omega t)]$, where Re refers to the real part of the argument. Further use of series† expansion reveals that when the two quantities of (32) are written as complex numbers, M_x is

$$M_x = \text{Re}\{[2B_1 \exp(+i\omega t)][\chi'(\omega) - i\chi''(\omega)]\} \tag{34a}$$

$$= \text{Re}\{2B_1[\cos \omega t + i \sin \omega t][\chi'(\omega) - i\chi''(\omega)]\} \tag{34b}$$

$$= 2B_1\chi'(\omega) \cos \omega t + 2B_1\chi''(\omega) \sin \omega t \tag{34c}$$

Comparison of this result with that of (31a) shows that

$$\chi'(\omega) = U/2B_1 \quad \text{and} \quad \chi''(\omega) = -V/2B_1 \tag{35}$$

When the values of U and V from (27) are introduced along with the substitutions $M_0 = \chi_0 B_0$ and $B_0 = \gamma \omega_0$,

$$\chi'(\omega) = \frac{1}{2} \chi_0 \omega_0 \frac{T_2^2(\omega_0 - \omega)}{1 + T_2^2(\omega_0 - \omega)^2 + \gamma^2 B_1^2 T_1 T_2} \tag{36a}$$

$$\chi''(\omega) = \frac{1}{2} \chi_0 \omega_0 \frac{T_2}{1 + T_2^2(\omega_0 - \omega)^2 + \gamma^2 B_1^2 T_1 T_2} \tag{36b}$$

$$M_z = \frac{\chi_0 \omega_0}{\gamma} \frac{1 + T_2^2(\omega_0 - \omega)}{1 + T_2^2(\omega_0 - \omega)^2 + \gamma_2 B_1^2 T_1 T_2} \tag{36c}$$

Most ESR experiments are performed with detection of the absorption part of the complex susceptibility $\chi''(\omega)$, although many spectrometers are capable of also detecting the dispersion $\chi'(\omega)$. This is particularly useful for studying samples that saturate readily in the absorption mode. The behavior of these two quantities near resonance is shown in Figure 4.5.

1.7   Line Shapes

The spin system absorbs energy from the microwave field because the latter exerts a torque of magnitude VB_1 [see (25)] about the z axis that performs work on the magnetic moments at the rate ωVB_1. Consequently, energy is absorbed at

---

†Identities and series pertinent to such manipulations include:

$$\cos x = \cosh ix = [\exp(ix) + \exp(-ix)]/2$$
$$\exp x = 1 + x + x^2/2! + x^3/3! + \cdots$$
$$\cos x = 1 - x^2/2! + x^4/4! - \cdots = \text{Re}[\exp(ix)]$$
$$\sin x = x - x^3/3! + x^5/5! - \cdots = \text{Im}[\exp(ix)]$$

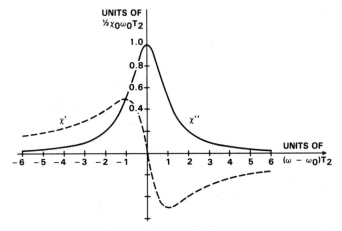

Figure 4.5 Frequency response of the absorption (χ'', ——) and dispersion (χ', ---) components of the complex magnetic susceptibility in a spin system; T_2, the spin–spin relaxation time, is related to the intrinsic line width.

the rate

$$dE/dt = 2\omega B_1^2 \chi''(\omega) \tag{37a}$$

From the value of $\chi''(\omega)$ in (36b)

$$\frac{dE}{dt} = B_1^2 \chi_0 \omega \omega_0 \frac{T_2}{1 + T_2^2(\omega_0 - \omega)^2 + \gamma^2 B_1^2 T_1 T_2} \tag{37b}$$

where χ_0 is the static susceptibility, (29). When B_1 is so small that there is no saturation, this reduces to the form

$$L(\omega) = (T_2/\pi)\{1/[1 + T_2^2(\omega_0 - \omega)^2]\} \tag{38}$$

which is the *Lorentzian line shape* as denoted by $L(\omega)$.

This is cast in more useful form when $1/\gamma T_2 = \Gamma$ is identified as the half-width at half-height (Figure 4.6) of the absorption line. The Lorentzian line is then normalized by expressing the maximum absorption amplitude as $Y_{max}^L = 1/\pi\Gamma$ and expressing the field-dependent amplitude as

$$Y^L(B) = Y_{max}^L \Gamma^2/[\Gamma^2 + (B - B_0)^2] \tag{39}$$

where the functionality $(B - B_0)$ is the difference between the value of the resonant field B_0 and the instantaneous field B when the laboratory magnet is slowly swept while the microwave frequency is held constant. The derivative form, which is usually recorded experimentally, is shown in Table 4.2.

Spectra of organic radicals in nonviscous solution characteristically exhibit this Lorentzian line shape. However, the conditions that prevail in viscous solutions or solid phases or in the presence of unresolved hyperfine structure often cause the resonant field of the individual spin units to occur at slightly

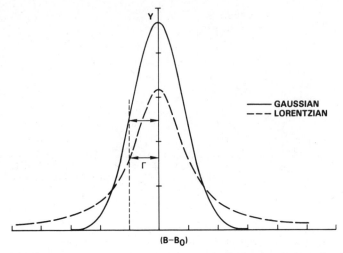

Figure 4.6 Comparison of the Gaussian (——) and Lorentzian (---) absorption line shapes. Curves were calculated on basis of equal half-width at half-height (Γ) for samples containing identical numbers of spins.

different values, depending on the local environment of the spin. This produces an absorption envelope that results from the superposition of many different Lorentzian lines occurring over a statistical distribution of resonant fields, and this situation is represented by the so-called *Gaussian line shape*. In its normalized form this has the functionality

$$Y^G(B) = Y^G_{max} \exp[(-\ln 2)(B - B_0)^2/\Gamma^2] \tag{40}$$

where Γ again represents the half-width at half-height of the absorption envelope and the maximum amplitude is given by

$$Y^G_{max} = (1/\Gamma)[\ln 2/\pi]^{1/2} \tag{41}$$

These line-shape functions apply only to ideal, unsaturated absorptions. If the sample is subjected to an excessively strong microwave field and saturation results, the absorption line loses intensity in the central region and appears to broaden.

The conduction electrons in a metal produce a distinctive *Dysonian line shape* as shown in Figure 4.7. The shape of the ESR absorption for conduction electrons is determined by sample geometry because the asymmetric line shape in Figure 4.7 arises when the skin depth (the distance that the microwave energy penetrates the sample) is small relative to the sample thickness. The ratio of the time necessary for interdiffusion of skin electrons and interior electrons to the spin–spin relaxation time determines the line shape [16, 17]. For thin films conduction electrons exhibit the usual Lorentzian line shape.

Table 4.2 Comparison of Normalized Lorentzian and Gaussian Lines[a]

	Lorentzian Line Shape	Gaussian Line Shape
Equation for normalized absorption	$Y^L(B) = \dfrac{1}{\pi\Gamma} \cdot \dfrac{\Gamma^2}{\Gamma^2 + (B - B_0)^2}$	$Y^G(B) = \left(\dfrac{\ln 2}{\pi}\right)^{1/2} \cdot \dfrac{1}{\Gamma} \exp\left[\dfrac{(-\ln 2)(B - B_0)^2}{\Gamma^2}\right]$
Half-width at half-height	Γ	Γ
Equation for first derivative	$[Y^L(B)]' = \dfrac{1}{\pi\Gamma} \cdot \dfrac{2\Gamma^2(B - B_0)}{[\Gamma^2 + (B - B_0)^2]^2}$	$[Y^G(B)]' = -\dfrac{2(\ln 2)^{3/2}}{\pi^{1/2}\Gamma^3} \cdot (B - B_0) \exp\left[\dfrac{(-\ln 2)(B - B_0)^2}{\Gamma^2}\right]$
Peak-to-peak amplitude	$2[Y^L(B)]'_{max} = \dfrac{3\sqrt{3}}{4\pi} \cdot \dfrac{1}{\Gamma^2}$	$2[Y^G(B)]'_{max} = \left(\dfrac{8}{\pi e}\right)^{1/2} \dfrac{\ln 2}{\Gamma^2}$
Peak-to-peak width	$\Delta B_{pp} = \dfrac{2}{\sqrt{3}}\,\Gamma$	$\Delta B_{pp} = \left(\dfrac{2}{\ln 2}\right)^{1/2}\Gamma$

[a]These expressions for Lorentzian and Gaussian line shapes as well as properties of the second derivatives are found in [15], p. 33, although in somewhat different notation. Reprinted, with permission, from Professor John E. Wertz, Department of Chemistry, University of Minnesota, Minneapolis, MN.

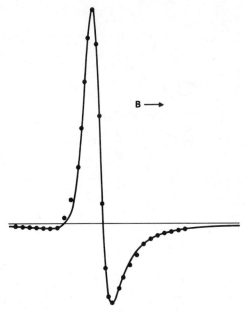

B —→

Figure 4.7 Dysonian line shape observed for a sample of polyacetylene doped with AsF_5 (———). The points correspond to a Dysonian line shape calculated for $T_D \gg T_2$, where T_D is the time required for the electrons to diffuse through the skin depth and T_2 is the spin–spin relaxation time.

1.8   Anisotropic *g*-Tensor

In the first-order formulations of the energy of a quantized system of electron spins, the Hamiltonian is given as

$$\mathcal{H} = g_e \mathbf{B} \cdot \hat{\mathbf{S}} \tag{42}$$

where $\mathbf{B}$ and $\hat{\mathbf{S}}$ are the field and spin operators and the eigenvalues are the Zeeman energy levels of the spin system.

Paramagnetism can arise from either spin or orbital angular momentum. Thus we should recognize that any orbital angular momentum could add to that of an unpaired electron (or vice versa) so that the *g*-factor might be affected. Orbital angular momentum arises from a nonspherically symmetric electron wave function in an atom, ion, or diatomic radical. In polyatomic molecules it often arises from admixture of excited states.

In the case of free atoms or transition metal ions, the orbital angular momentum **L** can couple strongly to spin angular momentum **S** to produce Russell–Saunders coupling

$$\mathbf{J} = \mathbf{L} + \mathbf{S} \tag{43}$$

where **J** is the resultant angular momentum. The magnitude of **J** can assume discrete values

$$J = L + S, \quad L + S - 1, \quad \dots, \quad |L - S| \tag{44}$$

which characterize different electronic states. This is illustrated by oxygen atoms that have $L = 1$ and $S = 1$. In this case, the states $J = 0$ and $J = 1$ are greater in energy by, respectively, 0.028 and 0.020 eV than the $J = 2$ state. The g-factor is given by the Landé interval rule

$$g_J = 1 + \frac{J(J + 1) - L(L + 1) + S(S + 1)}{2J(J + 1)} \tag{45}$$

When the examples of the different states of oxygen atoms are used, $g_{J=2} = g_{J=1} = 1.5$, and $g_{J=0}$ is undefined. In fact, both the $J = 2$ and $J = 1$ states are readily detected; but when $J = 0$, the spin and orbital angular momenta act in opposition so there is zero net angular momentum and the atom is diamagnetic. The Landé rule also shows that for atoms with quantized orbital angular momentum but no spin $J = L$, and the g-factor is $g = 1$.

In the more general case of radicals, radical ions, or transition metal ions in the condensed phase, strong covalent interactions and/or electrostatic interactions quench the orbital angular momentum, causing the g-factors to approach g_e. However, there is usually some residual angular momentum from coupling with excited states. In these cases the g-factor is dependent on the orientation of the paramagnetic molecule with respect to the applied magnetic field. The Hamiltonian is then given by

$$\mathscr{H} = \mathbf{B} \cdot \mathbf{g} \cdot \mathbf{S} \tag{46a}$$

$$= [B_x, B_y, B_z] \begin{bmatrix} g_{xx} & g_{xy} & g_{xz} \\ g_{yx} & g_{yy} & g_{yz} \\ g_{zx} & g_{zy} & g_{zz} \end{bmatrix} \begin{bmatrix} S_x \\ S_y \\ S_z \end{bmatrix} \tag{46b}$$

where $\mathbf{g}$ is a second-rank tensor.

The degree of admixture of excited states can be calculated by perturbation theory. The correction to g, to first order, is given by

$$g_m = g_J + \lambda \sum_{l \neq m} \frac{|\langle l|\hat{L}|m\rangle|^2}{E_l - E_m} \tag{47}$$

where g_m is the g-factor of state m; g_J is the g-factor given by the Landé rule (45); E_l and E_m are, respectively, the energy levels of the excited and ground states (unperturbed); and $\langle l|$ and $|m\rangle$ are the appropriate wave functions. Equation (47), while simple in appearance, is usually difficult to evaluate in most practical cases. (Several examples are discussed in Section 2.2.) Nevertheless, an important functionality is evident from (47). The degree of admixture of angular momentum from the excited state into the ground state is inversely proportional to the energy separation between the states.

It can also be shown that the magnitude of the spin–orbit coupling coefficient λ, and hence the magnitude of the g-shift, is related to the effective atomic number of the nucleus with which the unpaired electron is associated. For heavier elements, there is some shielding of the nucleus by filled electron shells, but λ does increase dramatically with atomic number. It is most readily measured by atomic spectroscopy.

In ESR spectroscopy the g-factor (or components of the g-tensor) is viewed as a parameter that characterizes the spin unit; that is, $\Delta E = g_{eff} \mu_B B_0$, where $g_{eff} = g_e + \Delta g$, and the term *free radical* is usually reserved for systems for which Δg is quite small. The deviation from the free spin value often provides information about the electronic structure of the system under study. Either positive or negative values of Δg are possible. The first-row transition elements, for example, demonstrate both effects. In the $3d^3$ ions, $g \approx 1.98$–1.99, characteristic of incompletely quenched orbital angular momentum. Note that in the Landé formulation, orbital angular momentum acts to decrease g_{eff} to less than g_e. On the other hand Cu(II) ($3d^9$) complexes have anisotropic g-factors in the range $g \approx 2.1$-2.4. This is consistent with the admixture of energetically accessible s orbitals to the extent of leaving behind a positive "hole" in the d orbital. Consequently the change in the algebraic sign of the orbital angular momentum residing in the d orbital produces a positive Δg. Such deviations are particularly dramatic in complexes of the lanthanide elements.

In general, immobilization of an atom or a complex that is subject to spin–orbit coupling reveals different g-factors for different orientations of the electron orbital with respect to the external field. The axes of magnetically dilute single crystals can be rotated relative to the laboratory field to reveal the principal axes of the g-tensor, which, unless it is asymmetric, is found to be diagonal; that is,

$$\mathbf{g}_{diag} = \mathbf{T} \cdot \mathbf{g}_{xtal} = \begin{bmatrix} g_{xx} & 0 & 0 \\ 0 & g_{yy} & 0 \\ 0 & 0 & g_{zz} \end{bmatrix} \tag{48}$$

where $\mathbf{T}$, the diagonalization transform, reduces the tensor to three principal values that are usually not coincident with the crystallographic axes. In many instances it happens that the spin unit has axial symmetry such that $g_{xx} = g_{yy}$, and there are but two principal values, $g_\perp$ and $g_\parallel$ corresponding to perpendicular or parallel orientation relative to the external field. A good discussion of g-anisotropies is given in Chapter 6 of [18], and [19] outlines a general method for the extraction of g-tensor components from single-crystal ESR data.

Even when it is not possible to obtain the sample as a magnetically dilute single crystal it is still possible to determine the principal g-factors from the spectrum of a sample consisting of randomly oriented molecules (powder spectrum). Under these conditions, all orientations of the molecules are equally probable, although the orientation of each spin unit is stationary. The diagrams of Figure 4.8 indicate the idealized absorption curves that would be observed if each individual spin unit exhibited a δ-function line shape (zero line width). It also shows more realistic curves representative of absorptions broadened by normal relaxation processes. The first derivative curves shown below the absorption curves reflect the display mode in which ESR data are most often recorded.

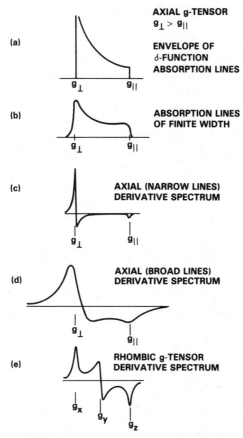

(a)

AXIAL g-TENSOR
$g_\perp > g_\parallel$

**ENVELOPE OF
δ-FUNCTION
ABSORPTION LINES**

(b)

**ABSORPTION LINES
OF FINITE WIDTH**

(c)

**AXIAL (NARROW LINES)
DERIVATIVE SPECTRUM**

(d)

**AXIAL (BROAD LINES)
DERIVATIVE SPECTRUM**

(e)

**RHOMBIC g-TENSOR
DERIVATIVE SPECTRUM**

Figure 4.8 Schematic representations of spectra arising from powders of materials having axial (a)–(d) and anisotropic (e) g-tensors: (a) idealized curve that will arise from a powder sample having an axial g-tensor and infinitely sharp (δ function) line widths, (b) realistic curve resulting for finite width of the individual absorption lines, (c) first-derivative representation of curve b, (d) first-derivative display arising from greater width of the individual absorption lines, (e) first-derivative display of a powder sample having an anisotropic g-tensor $g_x > g_y > g_z$.

When the spin units are allowed to experience rapid tumbling such as occurs in high-pressure gas or in a nonviscous solution, the only experimentally observable quantity is a time-average value related to the trace of the g-tensor

$$\langle g \rangle = \tfrac{1}{3}\mathrm{Tr}\,\mathbf{g} = \tfrac{1}{3}[g_{xx} + g_{yy} + g_{zz}] \tag{49a}$$

or

$$= \tfrac{1}{3}[2g_\perp + g_\parallel] \tag{49b}$$

the latter representing axial magnetic symmetry. In either instance, only a single line corresponding to the value of $\langle g \rangle$ would be observed. This average is called the *isotropic g-factor* even though the parent tensor has anisotropic components.

1.9   Hyperfine Coupling

1.9.1   Fermi Contact Interaction

The information available from ESR is greatly enriched when the electron spin is associated with atoms that possess nuclear magnetic moments. Although the unit of nuclear magnetism (the nuclear magneton)

$$\mu_P = eh/4\pi m_P c \tag{50}$$

(m_P = proton mass) is much smaller than the Bohr magneton by virtue of the mass difference between the proton and electron, nuclei nonetheless experience Zeeman interaction with the magnetic fields present in ESR spectroscopy.

The nuclear Zeeman energy levels are analogous to the $(2S + 1)$ electronic Zeeman levels except that a nucleus of spin I has $(2I + 1)$ possible orientations in a magnetic field. When electron and nuclear spins are associated in the same atom or molecule, there is a mutual interaction of the magnetic moments, and the energy states are further separated into *hyperfine* levels. The situation is shown in Figure 4.9 for the simplest case of the levels of a collection of hydrogen atoms. Note that in the absence of a laboratory field three of the levels are degenerate because the mutually interacting magnetic moments are the only source of magnetic interaction (see, for example [11], Chapter 2). The ESR transitions, which are normally observed at larger values of the laboratory field,

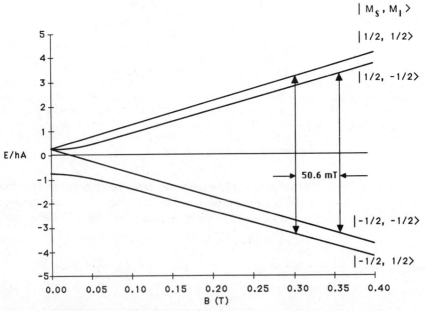

Figure 4.9   Calculated hyperfine energy levels of hydrogen atoms showing curvature near zero field and linear variation at the high-field approximation. The observed ESR transitions correspond to the selection rules $\Delta M_S = \pm 1$ and $\Delta M_I = 0$.

are also indicated in the figure and reflect the ESR selection rules $\Delta M_S = \pm 1$ and $\Delta M_I = 0$. The latter emphasizes the fact that NMR transitions do not normally occur at the field and frequency conditions suitable for ESR. Consequently Figure 4.9 indicates that a collection of hydrogen atoms exhibits two ESR lines that are separated by a quantity called the *hyperfine splitting constant* (hfsc), symbolized by a. Hyperfine splittings arising from interactions between electron and nuclear moments are thus distinguished from *fine structure splittings* arising from mutually interacting electron spins, which are often observed in atomic spectroscopy of excited states, but may also be observed in ESR studies of transition metal ions or triplet states.

For hydrogen atoms, the unpaired electrons are in the ground state $1s$ orbital and consequently have a spherically symmetric probability distribution. The process by which this hyperfine splitting arises was first explained by Fermi [20]. It is caused by the finite probability of contact between the nucleus and an s electron and is represented by a δ function that is nonzero only when the electron probability is finite at the nucleus. For a system comprising one electron and one nucleus the magnetic energy of isotropic interaction is approximately

$$E = hA_N = -\frac{8\pi}{3}|\psi(0)|^2 \mu_e \mu_N \tag{51}$$

where A_N is the isotropic hyperfine coupling constant (in frequency units) of nucleus N and $|\psi(0)|$ is the electronic wave function evaluated at the nucleus ($r = 0$). This so-called *Fermi contact interaction* produces isotropic hyperfine interactions that can provide much information about the electronic structure of ions and radicals.

For atoms in which the unpaired electron resides in a p or d orbital there is, to first order, zero probability of finding the electron at the nucleus. Nonetheless isotropic hfs are observed from such species as Mn(II), which in the ground state is a d^5 ion having five unpaired electrons, one in each of the five d orbitals. This result can be explained in terms of a configuration interaction [21] in which the $3d^5$ configuration is mixed with a configuration having unpaired spin in the $3s$ orbital [as in (47)] thereby producing the isotropic contact hfs. Such configuration interaction is responsible for nearly all observations of isotropic hfs because the unpaired electron is rarely in a pure s orbital.

Appendix 1 lists the hyperfine couplings for selected elements where the calculated values assume that the unpaired electron is confined to a pure atomic orbital (AO).

1.9.2 Anisotropic Dipolar Interactions

When the electron is in an orbital of lower symmetry than the s orbitals there is a different mechanism whereby the electron and nuclear moments can couple. This is the familiar classical situation of two magnetic dipoles interacting mutually through space with the energy [22]

$$E = (\mu_0/4\pi r^3)[\boldsymbol{\mu}_1 \cdot \boldsymbol{\mu}_2 - 3(\boldsymbol{\mu}_1 \cdot \mathbf{r})(\boldsymbol{\mu}_2 \cdot \mathbf{r})/r^2] \tag{52a}$$

where the vectors of (52a) are shown in Figure 4.10. When one electron dipole and one nuclear dipole are separated by the radius vector $\mathbf{r}$, the anisotropic interaction energy reduces to

$$E = [(1 - 3\cos^2\theta)\mu_0/4\pi r^3]\, |\boldsymbol{\mu}_e|\, |\boldsymbol{\mu}_N| \tag{52b}$$

where r is the distance separating the two dipoles and θ is the angle between $\mathbf{r}$ and the external field $\mathbf{B}$.

Consider an axially symmetric (e.g., p or d) orbital containing the unpaired electron. When the axis of this orbital is parallel to the external field ($\theta = 0$ or π), $E = -2\mu_e\mu_N/r^3$; when the orbital is perpendicular to the external field, $E = +1\mu_e\mu_N/r^3$. In a collection of randomly oriented spin units, the probability of the latter orientation is much greater than the former because the $\theta = 0$ and π orientations are represented by the two points corresponding to the poles of a sphere, while the $\theta = \pi/2$ orientations are represented by an infinity of points around the equator. Consequently if the spin units in the powder sample exhibited δ-function line shapes, the envelope of absorption from all of the possible orientations might look like Figure 4.11a. Of course in any real sample the individual lines have finite widths that produce a smoothing effect on the absorption curve, Figure 4.11b. The standard first derivative display, Figure 4.11c, is also included for completeness.

The anisotropic hyperfine splitting can also be represented as traceless tensor

$$\mathbf{T} = \begin{bmatrix} T_{xx} & 0 & 0 \\ 0 & T_{yy} & 0 \\ 0 & 0 & T_{zz} \end{bmatrix} \tag{53}$$

where $\mathrm{Tr}(\mathbf{T}) = [T_{xx} + T_{yy} + T_{zz}] = 0$. In the limit of rapidly reorienting spin units this dipolar part of the hyperfine tensor averages to zero. For the axial

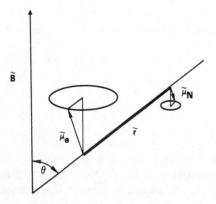

Figure 4.10 Diagram defining the quantities for the dipolar interaction of (52). The vector $\mathbf{r}$, which separates the magnetic moments $\boldsymbol{\mu}_e$ and $\boldsymbol{\mu}_N$, makes an angle θ relative to the applied magnetic field $\mathbf{B}$.

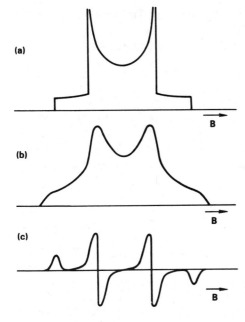

(a)

(b)

(c)

Figure 4.11 (a) Envelope of idealized δ function line shapes arising from dipolar coupling of nuclear and electron spins in a powder sample. (b) Smoothing that results from finite line widths of the individual absorptions. (c) Standard first-derivative display of curve b.

symmetry described previously, $T_\perp = T_{xx} = T_{yy}$ and $T_\parallel = -2T_\perp = T_{zz}$. The terms of **T** that are negative or positive depend on, among other factors, the sign of the nuclear spin magnetic moment.

The diagonal form of the complete hyperfine tensor is given as the sum of the isotropic and anisotropic contributions

$$\mathbf{A} = \begin{bmatrix} T_{xx} + A_0 & 0 & 0 \\ 0 & T_{yy} + A_0 & 0 \\ 0 & 0 & T_{zz} + A_0 \end{bmatrix} = \mathbf{T} + \mathbf{1} \cdot A_0 \qquad (54)$$

where **1** represents the unit tensor (see, for example [11], Chapter 7; [15], Chapter 7; and [17], Chapter 7).

It is recognized readily that the resonance field for any one spin unit fluctuates randomly with time when the sample is not in a rigid medium. In viscous solutions, which impede rapid reorientation of the spin units, the averaging of **T** is incomplete, and this produces spectral line widths that are significantly broader than the intrinsic line width. Anisotropies in the g-tensor likewise introduce motional broadening, but usually the **g** anisotropy is not so great as that of **A**.

Although the principal axes of **g** and **A** are fixed to the molecular framework of the spin unit, there is no a priori reason that requires the two different sets of axes to coincide, although they frequently do. In such cases the spectra obtained in nonviscous solution often exhibit well-behaved, sharp lines. However, when the tensor axes do not coincide, the tumbling of the spin unit even more strongly modulates the resonance field of each individual unit. The net effect is that the lines are excessively broadened, as if there were strong T_2 relaxation processes operating. This problem is addressed in a variety of contexts [23–26].

1.9.3   Hyperfine Splittings From Many Nuclei

When the spin unit contains several magnetic nuclei, the ESR spectrum may appear to be quite complex. If all the nuclei in a set are magnetically equivalent, as for the six hydrogen atoms of the benzene anion radical, it is the net sum of the individual m_I on a given spin unit that determines the nuclear Zeeman energy of that unit. In the benzene anion case, $M_I(=\Sigma m_I)$ can take the value

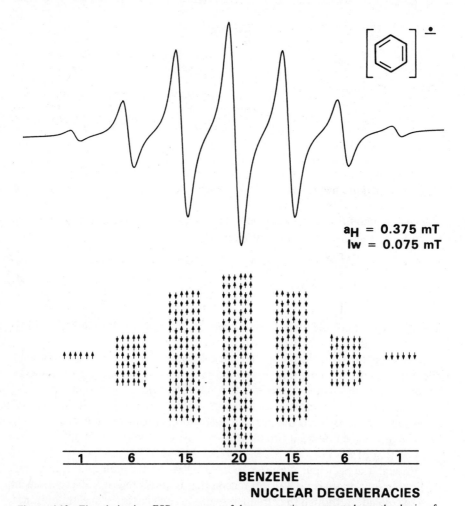

a_H = 0.375 mT
lw = 0.075 mT

BENZENE
NUCLEAR DEGENERACIES

Figure 4.12 First-derivative ESR spectrum of benzene anion computed on the basis of a Lorentzian line shape and the parameters shown. The lower portion shows a schematic representation of the various nuclear spin states present in a sample of benzene anions and indicates the statistical basis for the different intensities of hyperfine lines.

± 3, ± 2, ± 1, or 0. However, taken as an ensemble the statistical probability varies in a way that reflects the number of ways the individual m_I can be combined to yield a given M_I, as indicated in Figure 4.12 for benzene anion radical. With $I = \frac{1}{2}$ nuclei, these probabilities vary with the number of equivalent nuclei as the binomial coefficients, so the benzene anion spectrum consists of seven equally spaced lines having relative intensities of $1:6:15:20:15:6:1$, which are the coefficients of the binomial expansion of $(a + b)^n$, where $n = 6$.

In a radical having n_j different sets of magnetically equivalent nuclei, where a set may contain n_i nuclei of spin I, the total number of lines N_l expected in the isotropic spectrum is

$$N_l = \prod_j (2n_i I + 1)_j \qquad (55)$$

where the product runs over j. Figures 4.13 and 4.14 show the splitting patterns for the anion radicals of biphenylene and naphthalene, respectively. In each case

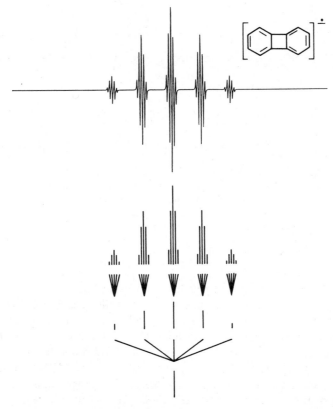

Figure 4.13 Biphenylene anion: first-derivative ESR spectrum computed on the basis of Lorentzian line shape of width 0.05 mT, one set of four equivalent hydrogen atoms with $a_H = 0.2765$ mT and another set of four equivalent hydrogen atoms with $a_H = 0.0216$ mT. The lower portion shows a *stick diagram* reconstruction based on δ-function line shapes. This indicates how the actual spectrum can be rationalized in terms of two successive splittings of the electron resonance into two $1:4:6:4:1$ hyperfine patterns.

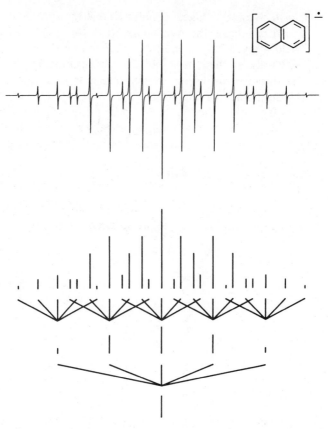

Figure 4.14 Naphthalene anion: first-derivative ESR spectrum computed on the basis of Lorentzian line shape of width 0.05 mT, one set of four equivalent hydrogen atoms with $a_H = 0.484$ mT and another set of four equivalent hydrogen atoms with $a_H = 0.186$ mT. The stick diagram indicates how the spectrum can be rationalized in terms of two successive, overlapping 1:4:6:4:1 hyperfine patterns.

the radical has two sets of four equivalent protons, but the magnitudes of the hfsc cause the spectra to appear quite different. In the case of biphenylene anion the spectrum can be assigned by inspection, whereas the overlapping hyperfine patterns of naphthalene anion require more careful analysis.

For a radical with several different sets of nuclei and nuclear spins—such as Wurster's blue perchlorate (the cation radical of *N,N'*-tetramethyl-*p*-phenylenediamine), which has a set of 4 aromatic hydrogen atoms, another set consisting of the 12 methyl hydrogen atoms, and the set of 2 amino nitrogen atoms—the formula predicts that the spectrum contains a total of $5 \times 13 \times 5 = 325$ lines. However, the intensity ratio between the most and least intense lines can be calculated to be 16,632:1, so it is not possible to observe all these lines in a normal trace.

							Label
			1				NO MAGNETIC NUCLEUS
		1	1	1			1 NUCLEUS, I = 1
	1	1	1				
		1	1	1			
			1	1	1		
	1	2	3	2	1		2 NUCLEI, I = 1
1	1	1					
	2	2	2				
		3	3	3			
			2	2	2		
				1	1	1	
1	3	6	7	6	3	1	3 NUCLEI, I = 1
							ETC.

Figure 4.15 Empirical procedure for predicting intensity ratios for several equivalent $I = 1$ nuclei. The method is easily modified to accommodate equivalent nuclei of any nuclear spin value.

Although the intensity ratios of lines arising from a set of nuclei having I greater than $\frac{1}{2}$ can be written in analytical form [27], it is simple to diagram the buildup in degeneracies (Figure 4.15) because there are seldom more than, say, four equivalent nitrogen atoms in a free radical.

1.9.4 Second-Order Effects

Figure 4.9, the energy-level diagram for hyperfine splitting in hydrogen atoms, indicates nonlinearity in the energy levels at low values of the applied field. This occurs because the applied field is not strong enough to break down the coupling between the electron and nuclear magnetic moments. Stated differently, M_S and M_I are not "good" quantum numbers at low field, and it is the resultant angular momentum $I + S$ rather than S alone that determines the energy levels. (This is discussed further in Section 2.3.1.)

When the applied magnetic field is large and the hyperfine splitting is small, the curvature in the energy levels is not apparent, and the resulting spectra are said to obey the high-field approximation (implying linear separation of the nuclear- and electron-spin energy levels as a function of applied field). However, for a nucleus that exhibits a large hyperfine splitting this approximation breaks down and the line positions are shifted from their first-order positions. The line positions, corrected to second order, are given by

$$B_j = B_0 - aM_j - [a^2/(B_0 - aM_j)][I(I + 1) - M_j^2] \qquad (56)$$

where B_0 is the resonant field in the absence of nuclei, a is the hyperfine splitting constant, and M_j is the spin state of the nucleus responsible for resonance at field

B_j. Note that for positive hfsc the spin states of the resonance lines decrease from $+I$ to $-I$ as the spectrum is swept through increasing fields. The second-order corrections shift each spectral component to lower field positions. Accurate determinations of g-factors should be corrected for such shifts. Likewise, the second-order shifts produce unequal spacing between adjacent hyperfine lines; and this must be considered in accurate determinations of hfsc, particularly when the hfsc is large, as in species such as Mn(II) (see Section 2.4.1). If the hfsc is on the order of B_0, then perturbation theory cannot account for the accurate line positions, and the energy eigenvalues must be determined by solving the spin Hamiltonian.

The second-order effect becomes more complicated when the hyperfine pattern arises from a set of equivalent nuclei. Here, additional lines may be observed because of the breakdown in the approximation that the nuclear spin quantum numbers of the individual nuclei add to produce a unique nuclear spin state. Rather, the lines that arise from superposition of degenerate states in the first-order spectrum are separated into discernible components at second order [28]. Figure 4.16 indicates the second-order shifts predicted for different numbers of equivalent $I = \frac{1}{2}$ nuclei.

The nominal line positions of Figure 4.16 are given by (56), but we must also consider the fact that the spins of a set of n equivalent nuclei combine such that

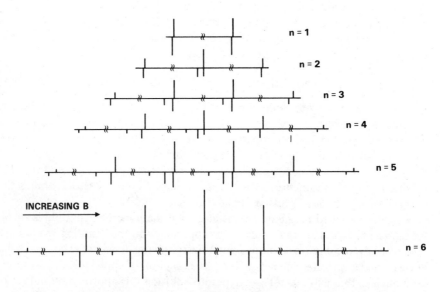

Figure 4.16 Schematic representation of comparison between first- and second-order hyperfine patterns for 1 to 6 equivalent $I = \frac{1}{2}$ nuclei. Since the second-order effects are small, the horizontal axis (magnetic field) is represented with discontinuities between first-order hyperfine components. The standard, binomial coefficient stick diagrams of the first-order patterns are displayed above the field axis while the stick diagrams hanging from the field axis show the corresponding second-order spectra. Note that many components of the second-order patterns are shifted to lower field values and also that when these components are summed their intensities are equal to that of the corresponding first-order line.

the values $I = (I_1 + I_2 + \cdots + I_n)$, $(I_1 + I_2 + \cdots + I_n - 1)$, $\ldots$, M_j contribute differently to the magnetic energy levels. Thus for four equivalent protons, the second-order effect causes the central absorption with a nominal value $M_0 = 0$ actually to be split into three separate lines because the value of I in (56) can be 2, 1, or 0; the M_{+1} and M_{-1} lines each split into only two components because $I = 2$ or 1; while the M_{+2} and M_{-2} lines each have but one component because $I = 2$.

Even if the second-order splittings are not resolved, the interior lines of the multiplet may broaden enough to distort the apparent amplitudes of the lines. Normally the observation of second-order splittings does not reveal additional information about the radical. But when splittings are large enough to reveal third- and fourth-order effects, it may be possible to ascertain the signs of the hyperfine splittings [29].

1.9.5   Mechanism of Hyperfine Splittings

SPIN POLARIZATION

Spin polarization is the name commonly applied to the configuration interaction of the unpaired electron orbital with other accessible orbitals so as to mix s-orbital character into the ground state wave function and thus produce the Fermi contact term responsible for isotropic hyperfine splitting. The mechanism for such configuration interaction is easily treated on the atomic orbital level and can be extended readily to molecular orbitals (MOs) when the unpaired electron is delocalized over several atoms [30–33].

This procedure can be suggested in very qualitative terms by considering a

$\diagdown$
$\dot{C}$—H fragment of an aromatic hydrocarbon radical. The unpaired electron
$\diagup$

resides in a $2p$ orbital centered on carbon; the remaining carbon valence electrons are involved in sp^2 hybrids, one of which forms a σ bond by overlap with the $1s$ orbital of the hydrogen atom. The p orbital is orthogonal to the sp^2 orbitals; hence, according to Hund's rule of maximum multiplicity, the spin of the unpaired electron acts to polarize one of the electrons of the C—H bond into parallel alignment, leaving the other electron preferentially aligned antiparallel in the hydrogen $1s$ orbital, whereby the Fermi contact term arises (see Figure 4.17). This mechanism successfully explained the isotropic splittings in a diverse assortment of free radicals. It is noteworthy that early workers in the field were hard pressed to explain hydrogen hyperfine splittings in aromatic hydrocarbon radicals because the strict confinement of the hydrogen nuclei to the nodal plane of the π orbital of the unpaired electron was predicted to preclude Fermi contact with the nucleus [34].

HYPERCONJUGATION

Replacement of the hydrogen atom by a methyl group in the foregoing scheme reveals the other common mechanism of hyperfine coupling, *hyperconjugation*. Figure 4.18a indicates a simple valence bond (VB) representation of hyperconjugation. The molecular orbital treatment is represented in Figure 4.18b. (In this

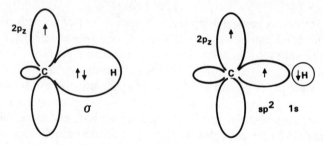

Figure 4.17 Spin polarization in a $\diagup\overset{|}{C}$—H bond fragment. The electron pair in the C—H σ bond is polarized by the unpaired electron in the carbon $2p_z$ orbital. An excess of antiparallel spin is produced in the vicinity of the hydrogen atom.

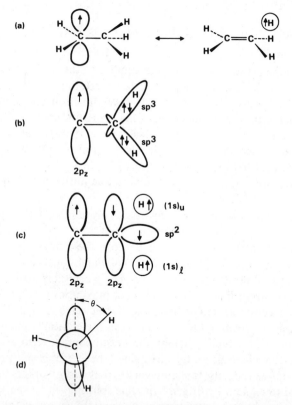

Figure 4.18 Hyperconjugative transfer of unpaired electron spin to a contiguous aliphatic group. (a) Simple valence-bond representation of hyperconjugation in an ethyl radical. (b) Pertinent molecular orbitals that are required for MO description of hyperconjugation. The electron pairs comprising the sp^3 σ bonds can be equally well represented as described in the text and depicted in (c). The hydrogen atoms can be envisioned to participate in two different orbitals—one of σ symmetry, which overlaps with the carbon sp^2 orbital, and the other of π symmetry, which overlaps with the unpaired electron through the intervening carbon $2p_z$ orbital. (d) A Newman projection along the carbon–carbon bond defines the angle θ. The smaller is the value of θ the greater is the efficiency of hyperconjugative spin transmission to an aliphatic hydrogen atom.

416

pictorial treatment one of the methyl protons is eliminated to simplify the diagram.) The sp^3 hybrids that normally describe the C—H bonds of the methyl group (Figure 4.18b) can be reformulated as shown in Figure 4.18c. Here the methyl carbon is considered as an sp^2 hybrid with one of the orbitals pointing to the midpoint of the line joining the two hydrogen atoms. Linear combinations of the $1s$ orbitals of the upper and lower hydrogen atoms yield orbitals of σ and π symmetry

$$\sigma_H = N_\sigma [(1s)_u + (1s)_l]$$

$$\pi_H = N_\pi [(1s)_u - (1s)_l] \tag{57}$$

where N_σ and N_π are normalization constants. This unconventional representation of aliphatic bonds is useful because it illustrates a pathway for delocalization of the unpaired π electron directly into the π_H orbital through the aliphatic carbon "$2p_z$" orbital [35].

If there is twisting about the carbon–carbon bond, the degree of overlap between the p orbitals and the antisymmetric combination of hydrogen orbitals diminishes. This effect is observed experimentally. In radicals where θ, the dihedral angle between the unpaired electron orbital and the projection of the C—H bond, can be estimated it is found that the splitting of hyperconjugatively coupled hydrogen atoms has the form

$$a_{H_\beta} \propto B_0 + B_2 \cos^2\theta \tag{58}$$

where B_0 has a magnitude of about 0.1 mT and represents residual effects of spin polarization, while the larger term $B_2 \approx 5$ mT reflects the sterically variable hyperconjugative contribution. The observation of such steric effects by ESR was the subject of several reviews [36–38].

2 INTERPRETATION OF ELECTRON SPIN RESONANCE PARAMETERS

2.1 Magnitudes of Hyperfine Splitting Constants

2.1.1 Spin Densities and Organic Free Radicals

When the unpaired electron resides in an extended molecular orbital its probability function is distributed over many different atoms. For example the unpaired electron of benzene anion resides in a π orbital, and simple symmetry arguments predict it to have equal probability at each of the six carbon atoms. This is quantified by the concept of spin density ρ_C^π, which is related to the probability of the electron being located at a given carbon atom. For the benzene anion example, $\rho_C^\pi = \frac{1}{6}$ for each carbon atom. (A more rigorous description of benzene anion spin densities follows.)

In less trivial situations, Hückel molecular orbital (HMO) theory is very successful in predicting spin densities in aromatic radicals [39]. Refinements to these calculations, suggested by McLachlan [40] incorporate the effects of electron correlation and produce negative spin densities; that is, a large proportion of $|\alpha\rangle$ spin on one atom often induces a degree of $|\beta\rangle$ spin on an adjacent atom.

The splitting of ring protons a_H in aromatic free radicals is proportional to the spin density on the contiguous carbon atom ρ_C^π and is given quite satisfactorily by the McConnell relation [41, 42],

$$a_H = Q_{CH}^H \rho_C^\pi \tag{59}$$

where the value $Q_{CH}^H \simeq -2.7\,\mathrm{mT}$ applies to a wide range of radicals.

More detailed analysis has shown that cation radicals generally exhibit larger splittings than the corresponding anion radicals. This was attributed to excess charge, where reducing the electron density in a p orbital shrinks the orbital while increasing the electron density expands the orbital [43, 44†]. Since the spin polarization is dependent on the spin density at the nucleus, the value of Q_{CH}^H in (59) is given by [45, 46].

$$Q_{CH}^H = Q(0) + K(1 - q_C^\pi) \tag{60}$$

where q_C^π is the total π electron density in the carbon p orbital. (An sp^2-hybridized carbon atom with one electron in its p orbital is electrically neutral.) The recommended values of the parameters in (60) are $Q(0) = -2.7\,\mathrm{mT}$ and $K = -1.29\,\mathrm{mT}$ [47]. The value of Q_{CH}^H depends on adjacent atoms so that the methyl proton splitting, $a = -2.3\,\mathrm{mT}$, is smaller than predicted from (60).

Hyperconjugative couplings of alkyl substituents are also predicted by an equation that shows proportionality between the coupling and the spin density on the atom adjacent to the pertinent alkyl group [48, 49]

$$a_{H_\beta} = [B_0 + B_2 \cos^2\theta]\rho_C^\pi \tag{61}$$

For nuclei that are involved in the molecular orbital of the unpaired electron the quantification of the hfsc in terms of spin densities is more complicated. Configuration interaction occurs between the delocalized π orbital containing the unpaired electron and the localized σ orbitals terminating at the nucleus responsible for the observed splitting.

For example, consider the four-atom fragment

representing part of an aromatic radical with the unpaired electron in a π orbital delocalized over the molecule. There are several contributions to the hyperfine

†Reference [44] presents an algebraic correction to [43].

splitting exhibited by nucleus Y as shown in Figure 4.19:

1. Spin polarization of the 1s electrons of atom Y by the unpaired electron in the π orbital centered on that atom, symbolized S^Y.
2. Spin polarization of the 2s electrons on atom Y by configuration interaction of the π electron with the σ electrons in the Y—X_i σ bonds, symbolized $Q^Y_{YX_i}$.
3. Spin polarization by the π-spin density residing on the atoms X_i, analogous to the mechanism responsible for the splitting of an aromatic hydrogen atom, symbolized $Q^Y_{X_iY}$.

Each contribution is proportional to the spin density residing on the appropriate atom so that the splitting constant is given by the sum

$$a^Y = \left(S^Y + \sum_i Q^Y_{YX_i} \right) \rho^\pi_Y + \sum_i Q^Y_{X_iY} \, \rho^\pi_{X_i} \tag{62}$$

The theoretical description of such $\sigma-\pi$ interactions was developed in some detail [50]. As an example of the applications of this treatment, consider the nitrogen hfsc exhibited by aromatic nitro groups. There are terms arising from the spin density in the nitrogen orbital as well as from orbitals centered on the

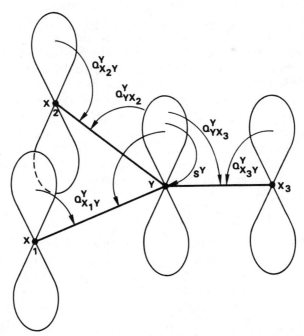

Figure 4.19 Representation of the various $\sigma-\pi$ parameters that are required to calculate the hfsc of atom Y, which hosts an unpaired electron in a $2p_z$ orbital. See text for interpretation of the various configuration interaction terms.

adjacent carbon and the bonded oxygen atoms. This leads to so many adjustable σ–π parameters in the evaluation of terms that it is only practicable to evaluate them by semiempirical methods, which requires correlating the data obtained from many similar radicals [51]. The results of several such studies of semiempirical correlations were noted in reviews [52, p. 41; 53].

The concept of spin density is fundamental to discussions about ESR hyperfine splittings [53]. Although measurement of an ESR spectrum requires a sample consisting of an ensemble of spin units, discussions of spin densities usually model the phenomenon by superimposing the electron probability distribution of a single spin unit on the probability of nuclear spin orientations.

Spin density is a phrase used to quantify the probability of finding electron spin in some particular volume of space. For a given spin unit having one unpaired electron there is unit probability that this electron exists somewhere in the universe

$$\int_0^\infty \int_0^\infty \int_0^\infty \psi\psi^* \, dx \, dy \, dz = \int_0^\infty \psi\psi^* \, dv = 1 \tag{63}$$

where ψ^* is the complex conjugate of the wave function ψ. This trivial result acquires meaning when the volume element is decreased to the dimensions of an atomic or molecular orbital.

For an unpaired electron residing in an atomic $2p_z$ orbital centered on, say, the carbon atom of a planar $CH_3\cdot$ radical, no calculations are required to conclude that $\rho_C^\pi = 1$. In more extended structures, such as a benzene anion radical, the unpaired electron resides in a molecular π orbital that is delocalized over the entire carbon skeleton. The usual approach for calculating the properties of such a delocalized orbital is based on the Hückel approximation that molecular π orbitals can be constructed as linear combinations of atomic p orbitals and that the σ-bond network can be considered invariant. For example, if each C—H fragment of the benzene system is considered as a structure consisting of three planar sp^2-hybrid bonds and an atomic $2p_z$ orbital normal to the plane, the σ-bonding network is composed of overlap of the sp^2 hybrids and the π system is formed by overlap between the six $2p_z$-atomic orbitals (see Figure 4.20). The π orbitals are formed by linear combinations of the ψ_i (the individual atomic orbitals) as

$$\psi_j = c_{j1}\psi_1 + c_{j2}\psi_2 + c_{j3}\psi_3 + c_{j4}\psi_4 + c_{j5}\psi_5 + c_{j6}\psi_6 = \sum c_{ji}\psi_i \tag{64}$$

where ψ_j describes a particular molecular orbital that may contain up to two π electrons. The c_{ji} coefficients describe the fractional contribution of each atomic orbital to the particular j molecular orbital, and the coefficients are normalized within each molecular orbital as

$$\sum_{i=1}^6 (c_{ji})^2 = 1 \tag{65}$$

The specific values of the coefficients arise from use of the variation principle to minimize the energy of each molecular orbital. (Details of the Hückel approach

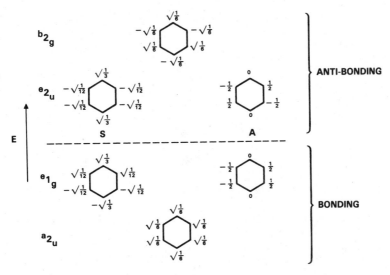

Figure 4.20 π-Molecular orbitals of benzene. The numbers adjacent to carbon atoms represent coefficients of the carbon $2p_z$ orbitals comprising the molecular orbital.

to MO calculations are given in many standard works [39]. A simple summary of many of the basic concepts of MO calculation is found in the appendix to [54].) The results for benzene are summarized in Figure 4.20. In the neutral molecule the π electrons occupy the lowest three molecular orbitals pairwise. The unpaired electron of the anion radical occupies the degenerate antibonding manifold marked S and A for symmetric and antisymmetric orbitals. Note that the average of the squares of the atomic orbital (AO) coefficients in the degenerate manifold is one-sixth at each ring position, and this is identified as the calculated probability for the unpaired electron, that is, the unpaired spin density. The measured hfsc for the splitting of the six equivalent hydrogen atoms in benzene anion led to McConnell's original estimate of $Q \simeq -2.3\,\mathrm{mT}$ in (59). Similar measurements on other highly symmetric radicals (Table 4.3) indicate Q is not a fundamental constant but rather the $\sigma-\pi$ interaction changes with structure, bond strain, and net charge. Moreover the peculiarities of the methyl radical and benzene anion are reflected in the Q values.

The concept of spin density is utilized in ESR studies in many different ways. For example, the results shown in Figure 4.20 can be used to rationalize qualitatively the observation of the observed coupling constants when benzene anion is substituted with electron-withdrawing and electron-donating substituents as shown in Table 4.4. The electron-withdrawing nitrile group stabilizes the symmetric orbital of the degenerate pair by preferential bonding at a site of high electron density. Conversely, the methoxy group stabilizes the antisymmetric orbital because its presence at a nodal position of unpaired electron probability minimizes electron-repulsion interactions [73].

Table 4.3   Hydrogen Hyperfine Splittings in Several Radicals

Radical	Hyperfine Splitting a_H(mT)	Q(mT)[a]	Spin Density	References
H·(1s)		50.7	[b]	[55]
H·(2s)		6.34	[b]	[56]
CH_3·	−2.304	−2.304	1[c]	[57]
C_5H_5·	−0.598	−2.99	$\frac{1}{5}$[c]	[58–60]
$C_6H_6^-$	−0.375	−2.25	$\frac{1}{6}$[c]	[58, 61, 62]
C_7H_7·	−0.391	−2.74	$\frac{1}{7}$[c]	[63–66]
$C_8H_8^-$	−0.321	−2.57	$\frac{1}{8}$[c]	[67–69]

[a]Spin polarization parameter determined from experimental data.
[b]Unpaired electron in pure s orbital.
[c]Determined by symmetry considerations.

Table 4.4   Ring Proton Splittings of Some Substituted Benzene Radicals

Radical	hfsc (mT)	References
$C_6H_6^-$	0.375	[61, 62, 70]
$C_6H_5D^{·-}$	$a_p = 0.341$, $a_m = 0.392$, $a_0 = 0.392$	[71]
$C_6H_5CN^{·-}$	$a_p = 0.842$, $a_m = 0.030$, $a_0 = 0.363$	[72]
$C_6H_5OCH_3^{·-}$	$a_p = 0.064$, $a_m = 0.606$, $a_0 = 0.534$	[73]
$C_6H_5N(CH_3)_2^{·+}$	$a_p = 0.52$, $a_m = 0.08$, $a_0 = 0.17$	[74]
$C_6H_5NO_2^{·-}$	$a_p = 0.418$, $a_m = 0.104$, $a_0 = 0.334$	[75]

The different results for $C_6H_6^-$ and $C_6H_5D^-$ demonstrate how a subtle change in substitution can lift the orbital degeneracy diagrammed in Figure 4.20. The replacement of a proton with a deuteron preferentially stabilizes the antisymmetric orbital as revealed in the unique hfsc of the para proton [71]. Of course, there is considerable thermal population of the symmetric state since the energy difference is quite small.

The seemingly low value for the meta-proton splittings in the benzonitrile anion is observed repeatedly in other systems with similar electron-withdrawing substituents, such as nitrobenzene anions [76]. This is not an anomaly but a demonstration of another effect that arises when a position of zero or low spin density is flanked by sites bearing high spin densities. Evaluation of electron correlation effects shows that the large spin densities can actually induce a net-negative spin density on the neighboring atom. First-order Hückel molecular orbital calculations on nitrobenzene anion suggest zero spin density at the meta positions, meaning that an unpaired electron in an $|\alpha\rangle$ spin state would have zero probability there. But when consideration of the correlation between electrons of opposite spin is introduced into the calculations, it can be shown that the high density of $|\alpha\rangle$ spin at the ortho and para positions actually induces a net probability of $|\beta\rangle$ spin at the meta position, and this is called *negative spin density* in a p^{π} orbital. McLachlan [40] used perturbation theory to devise a method for calculating the contribution of such sites to the hyperfine splitting patterns of radicals.

In a very real sense, many of the approximations made in the course of molecular orbital calculations are so crude there is no reason to anticipate that the results should be as good as they are. Nonetheless, the reciprocity between ESR data and molecular orbital predictions and rationalizations has worked to the benefit of both. Whereas other physical measurements (UV spectrum, dipole moment, ionization potential, etc.) provide only a few points for correlation with molecular orbital predictions, the hfsc that are measured from the ESR spectrum of a complicated radical can yield several data points and provide information that is mapped over the molecule. Such information provides impetus for improvements and refinements in molecular orbital calculation schemes. Conversely, the spin densities calculated for atoms in a complex radical can sometimes give insights into the hfsc in a spectrum that is too complicated to assign on the basis of spectral data alone. However, there is a significant danger in assigning ring proton positions based on molecular orbital calculations alone.

Sales [77] presents an excellent overview of the subject of isotropic hyperfine splittings and computation of unpaired spin densities in organic radicals.

2.1.2 Hyperfine Splitting of Inorganic Radicals and Transition Metal Ions

The mechanisms of hyperfine splittings described in the previous section are also operative in inorganic radicals and transition metal ions. The hfsc observed for such species are useful primarily for determination of the hybridization of the unpaired electron. The A and B parameters (reported in hertz) in Appendix 1 are the hyperfine couplings, which are proportional to the interaction energy between the nucleus and the electron in a particular atom. The isotropic couplings A arise by the Fermi contact mechanism (51) while the anisotropic couplings B result from dipolar interactions (52). For polyatomic species the unpaired electron is in a molecule rather than an atom; consequently, the magnitudes of the hyperfine interactions A_0 and B_0 vary with the peculiarities of bond hybridization. (Conversion of a hyperfine *coupling* in hertz to a hyperfine *splitting*, as measured from a spectrum in tesla, requires knowledge of the g-factor; that is, $a = hA_0/g\mu_B$; see Section 7.4.)

The fractional s and p character of the unpaired electron hybridization can be estimated from the spin densities in the respective orbitals as given by the relations

$$\rho_s = A_0/A \tag{66}$$

and

$$\rho_p = B_0/B \tag{67}$$

In the first approximation, an electron associated with a first-row element is restricted to hybridization having only s and p character; that is, $\rho_s + \rho_p = 1$. Two well-defined examples can be used to illustrate this concept, $CH_3\cdot$ and $CF_3\cdot$. The hyperfine coupling of the methyl radical is $A_0(^{13}C) = 108$ MHz [57]. Based on the value of $A(^{13}C) = 3110$ MHz (in Appendix 1), there is only about 3% s-orbital character. This can be contrasted with the trifluoromethyl radical, where $A_0(^{13}C)$ is 761 MHz [78]. This corresponds to 24.4% s character, almost equivalent to sp^3 hybridization. These results indicate that the methyl radical exists in a planar structure with the unpaired electron in a p orbital and the hydrogen atoms bonded to sp^2 orbitals, while the trifluoromethyl radical has a pyramidal structure with the unpaired electron in an sp^3 orbital with each of the three fluorine atoms bonded to an sp^3 orbital. We can further examine the effect of spin polarization on the hydrogen atoms of $CH_3\cdot$. The hyperfine splitting is about 64 MHz; and since A for hydrogen is about 1410 MHz, the degree of spin polarization is 4.6% in the hydrogen $1s$ orbital.

A second interesting example is the isoelectronic series BF_3^-, $CF_3\cdot$, and NF_3^+ [79]. Table 4.5 lists the hyperfine splittings and spin densities in the central atom and fluorine orbitals. Although the anisotropic hfs of the BF_3^- radical is unknown, the results show that the hybridization seems to shift from sp^3 in the carbon- and boron-centered radicals to a more planar structure in the nitrogen-centered radicals.

Transition metal ions often exhibit hyperfine splitting. Notable examples include copper(II) with four lines ($I = \frac{3}{2}$), manganese(II) with six lines ($I = \frac{5}{2}$), and vanadium(IV) with eight lines ($I = \frac{7}{2}$). In addition, many complexes exhibit *superhyperfine splittings* (shfs). These are hyperfine splittings that arise from ligands. Fluoride often coordinates with transition metal ions and frequently exhibits superhyperfine splittings caused by its large nuclear magnetic moment. Likewise, many nitrogen-containing ligands exhibit superhyperfine splittings.

The principles outlined in Section 1.9 govern the transition metal hyperfine splitting and ligand superhyperfine splitting. However, the d-orbital contribution to hyperfine splitting is not as well defined as are the p-orbital contributions in organic and inorganic radicals. There are three main factors that influence the hyperfine splitting of the transition metal. The first consideration is the degree of mixing of the d orbitals of the ion with the higher energy s and p orbitals. Admixture with s orbitals contributes to the isotropic hyperfine splitting, and admixture with p orbitals contributes to anisotropic hyperfine splitting. Table 4.6 gives calculated values for neutral ions in different configurations [81]. The contribution to the isotropic hyperfine splitting arises when the symmetries of the d orbitals in the transition metal allow coupling to the $4s$ orbital, such as occurs in complexes that exhibit d_{4h} symmetry, where the electron is in an a_{1g}-type orbital [e.g., square planar cobalt(II) complexes]. Contributions to the anisotropic hyperfine splitting arise from admixture with $3p$ or $4p$ orbitals. Assuming a positive nuclear moment, d_{xy}, d_{yz}, d_{xz}, and $d_{x^2-y^2}$ orbitals give a negative contribution to the anisotropic hyperfine splitting, and the d_{z^2} orbital gives a positive contribution. Thus, mixing p character with the

Table 4.5 Comparison of Three Isoelectronic MF_3 Radicals

Radical	$a_M(iso)^a$	$a_F(iso)^a$	ρ_M^s	ρ_M^p	ρ_F^s	ρ_F^p	M—F Bond Hybridization	Reference
$BF_3^{\cdot-}$	428.7	498.8	0.211		0.0104			[80]
$CF_3^{\cdot}$	759.9	404.7	0.244	0.723	0.0084	0.1113	$sp^{3.01}$	[78]
$NF_3^{\cdot+}$	232.8	526.7	0.151	0.957	0.011	0.1116	$sp^{2.47}$	[79]

^{a}Splittings are in megahertz.

425

Table 4.6 Contribution to the Hyperfine Splitting for One Electron in an s or p Orbital (in mT)[a]

Neutral Ion	4s Orbital	Isotropic 3p ($3d^n$)	Anisotropic[b]	
			4p ($3d^{n-1}$ 4p)	4p ($3d^{n-2}$ $4p^2$)
^{47}Ti $\equiv$ ^{49}Ti	−17.57	−14.6	−0.06	−0.21
^{51}V	93.23	82.9	0.24	1.06
^{53}Cr	−22.49	−21.5	−0.05	−0.25
^{55}Mn	109.32	11.2	0.21	1.07
^{57}Fe	16.05	17.2	0.03	0.14
^{59}Co	130.76	147.8	0.21	1.09
^{61}Ni	53.97	64.3	0.09	0.47
^{63}Cu	176.69	220.6	0.09	1.56
^{65}Cu	189.22	236.3	0.10	1.67

[a]Data adapted from [81], with permission.
[b]Given as $2B$, electron configurations in parentheses.

first set of d orbitals results in a decrease of the magnitude of the observed hyperfine splitting, while the reverse is true for the d_{z^2} orbital. The second effect that contributes to the metal splitting is the degree of delocalization onto the ligand. If the probability of the unpaired electron in the ligand orbitals is equal to α^2, then the spin density on the metal is $1 - \alpha^2$. Thus, the greater is the delocalization, the smaller is the metal hyperfine splitting. However, the greater is α, the greater is the superhyperfine splitting of the ligand if the electron is delocalized onto an atom that possesses nuclear spin. It is, however, difficult to find a homologous series with sufficient variation in both the metal and ligand hyperfine splittings to demonstrate the extent of this prediction. The third effect, the charge on the ion, is more difficult to represent. As the charge increases, the electrons are pulled closer to the metal nucleus, thus increasing the superhyperfine splitting. A similar effect, known as the *excess charge effect*, has been observed for the CH-bond segment in aromatic radical ions [43–47, 82]. For example, cations exhibit greater hyperfine splittings than do the corresponding anions.

2.2   Magnitudes of g-Factors

We have not yet considered the magnitude of g-factors other than to point out that for free atoms, the Landé g-factor is given by (45) and that in condensed phases, anisotropy of the molecule-centered g-tensor often produces asymmetric spectra, as shown in Figure 4.8. The seeming paradox of a single-valued Landé factor versus an anisotropic tensor is the result of environmental interactions. The Landé formula pertains to low-pressure gas-phase atoms or ions for which collisional interactions between spin units rarely occur. In other words, the angular momentum of each spin unit is fixed and well defined. But in condensed media, collisional or vibrational interactions with the lattice produce instantaneous variations in the direction of magnetization. This is equivalent to an uncertainty in the net angular momentum of each molecule and therefore produces a broadened spectrum. Consequently, in other than very low-pressure paramagnetic systems, species possessing significant orbital angular momentum are rarely detected. By default, then, it is more usual to detect spin-only paramagnetism for which the g-factor is close to g_e.

If one looked at the periodic table, it would seem that many elements or ions should be paramagnetic species possessing orbital angular momentum. However, when such species partake in bond formation or become coordinated by neighboring atoms, the orbital angular momentum is largely quenched. Since the g-factor of the electron is always g_e, any deviation of the g-factor from g_e represents the contribution of unquenched orbital angular momentum, which arises from orbital angular momentum in excited states, which combines with the electron spin as $L + S$, but where L is no longer a good quantum number. The total magnetic Hamiltonian was given in (46a), but it can be rewritten in a representation that separates the Zeeman ($\mathscr{H}_Z$) and spin–orbit ($\mathscr{H}_{SO}$) contributions

$$\mathscr{H} = \mathscr{H}_Z + \mathscr{H}_{SO} = \mathbf{B} \cdot \mu_B(g_e\hat{\mathbf{S}} + \hat{\mathbf{L}}) + \lambda\hat{\mathbf{L}} \cdot \hat{\mathbf{S}} \tag{68}$$

where λ is the spin–orbit coupling constant. If L were truly equal to zero, as is generally assumed for the first-order ground-state configuration, then no deviations from g_e would be observed.

2.2.1 Deviation of g From g_e

Derivations of the origin of the g-factor for various materials are found in the preceding basic references, such as [3, 12, 15, 18]. To determine the g-tensor, one must solve the Hamiltonian in (68) by second-order perturbation theory and obtain the result in convenient form such as $g\mu_B\mathbf{B} \cdot \hat{\mathbf{S}}$. This procedure is carried out by computing the second-order correction to $\mathscr{H}$, as in [15, Chapter 11].

$$\mathscr{H} = -\sum_n{}' \frac{|\langle G, M_S|(\mu_B\mathbf{B} + \lambda\hat{\mathbf{S}}) \cdot \hat{\mathbf{L}} + g_e\mu_B\mathbf{B} \cdot \hat{\mathbf{S}}|n, M_S'\rangle|^2}{E_n^0 - E_G^0} \tag{69}$$

where G represents the ground state, n represents any electronic or magnetic excited state with spin-state quantum number M_S', E_n^0 and E_G^0 are the energies of the unperturbed states n and G, and the Σ' indicates summation over all excited states with the exception of G. Conventionally, the numerator in (69) is broken down into a series of vector matrix elements, those containing $\hat{\mathbf{L}}$ and those containing $\hat{\mathbf{S}}$

$$\mathscr{H} = -\sum_n{}' \frac{[\langle M_S|(\mu_B\mathbf{B} + \lambda\hat{\mathbf{S}})|M_S'\rangle \cdot \langle G|\hat{\mathbf{L}}|n\rangle][\langle n|\hat{\mathbf{L}}|G\rangle \cdot \langle M_S'|(\mu_B\mathbf{B} + \lambda\hat{\mathbf{S}})|M_S\rangle]}{E_n^0 - E_G^0}$$

$$\tag{70}$$

where the operator S is an effective spin that includes the total angular momentum of the ground state when spin and orbital angular momentum cannot be separated. This distinction is particularly useful for transition metal ions. Upon performing the matrix multiplication implicit in (70), it is possible to factor out terms that correspond to a tensor with the elements

$$\Lambda_{ij} = \sum_n{}' \frac{\langle G|\hat{L}_i|n\rangle\langle n|\hat{L}_j|G\rangle}{E_n^0 - E_G^0} \tag{71}$$

where i and j are the directions with respect to the molecular axes. This tensor depends solely on orbital angular momentum. After appropriate arithmetic manipulations, including addition of the electron spin Zeeman term $\mathbf{B}\cdot\hat{\mathbf{S}}\Gamma$ the spin Hamiltonian becomes

$$\mathcal{H}_S = \mu_B\mathbf{B}\cdot(g_e\cdot\mathbf{1} + 2\lambda\mathbf{\Lambda})\cdot\hat{\mathbf{S}} + \lambda^2\hat{\mathbf{S}}\cdot\mathbf{\Lambda}\cdot\hat{\mathbf{S}} \tag{72}$$

where $\mathbf{1}$ is the unit tensor. The spin Hamiltonian then takes the form

$$\mathcal{H}_S = \mu_B\mathbf{B}\cdot\mathbf{g}\cdot\hat{\mathbf{S}} + \hat{\mathbf{S}}\cdot\mathbf{D}\cdot\hat{\mathbf{S}} \tag{73}$$

The tensors $\mathbf{g}$ and $\mathbf{D}$ are related to $\mathbf{\Lambda}$ as

$$\mathbf{g} = g_e\cdot\mathbf{1} + 2\lambda\mathbf{\Lambda} \tag{74}$$

and

$$\mathbf{D} = \lambda^2\mathbf{\Lambda} \tag{75}$$

The tensor $\mathbf{D}$ includes both spin–orbit dipolar interaction and electron–electron dipolar interaction. The elements of the $\mathbf{D}$ tensor are often rewritten in terms of the zero-field parameters D and E given by

$$D = D_{zz} - (D_{xx} + D_{yy})/2 \tag{76}$$

$$E = (D_{xx} - D_{yy})/2 \tag{77}$$

For species with axial symmetry, that is, D_{3h} or greater, $D_{xx} = D_{yy}$ so that $E = 0$. The tensor $\mathbf{g}$ has been written in different forms, but the expressions of (71) and (74) are particularly useful in emphasizing that it is the addition of orbital angular momentum to the electron spin that contributes to the net interaction of the electron with the applied magnetic field.

The full Hamiltonian, now including electron Zeeman, electron spin–electron spin, electron spin–nuclear spin (hyperfine), and nuclear Zeeman interactions

$$\mathcal{H} = \mu_B\mathbf{B}\cdot\mathbf{g}\cdot\hat{\mathbf{S}} + \hat{\mathbf{S}}\cdot\mathbf{D}\cdot\hat{\mathbf{S}} + h\hat{\mathbf{S}}\cdot\mathbf{A}\cdot\hat{\mathbf{I}} + g_N\mu_N\mathbf{B}\cdot\hat{\mathbf{I}} \tag{78}$$

is not necessarily diagonalized in terms of M_S and M_I, so the terms may not be separable. Some such examples include transition metal ions that exhibit g-factors with large deviations from g_e, species with large spin–spin interactions or species that exhibit large hyperfine splittings. This was discussed in terms of second-order corrections in Section 1.9.4, and it is addressed further in Section 2.3 in the context of gas-phase species.

If the elements of Λ are known, then the g-tensor can be calculated directly. This is done within the principal axis framework of the molecule and is illustrated for O_2^+ in Section 2.2.3.

2.2.2 Organic π-Electron Radicals

The theory of g-factors for conjugated π-electron radicals that exhibit a relatively small deviation of g from the free electron value was derived by Stone [83, 84] using the foregoing basic formalism. The fundamental assumption was that the unpaired electron wave function is completely delocalized over all the p orbitals comprising the conjugated π-electron system, but only the atomic orbitals that contribute to the g-factor need be considered. Thus, the contribution to g from an atom or group of atoms is proportional to the spin density on that atom or group, and the relevant energies such as in (71) are given by the energy of the molecular orbital containing the unpaired electron (see Section 2.1.1) and the energy of the particular atomic orbitals in the group. These calculations provide the following results:

1. The average g-factor of a hydrocarbon radical or radical ion is represented by

$$g = g_e + b + mc \qquad (79)$$

at the Hückel level of approximate molecular orbital calculations. The energy level m of the unpaired electron molecular orbital is given in the Hückel units $E = \alpha + m\beta$; b and c are empirical constants.

2. The anisotropic contributions to the g-factors of a neutral radical are $g_{xx} = g_{yy} = g_e + 3b$, and $g_{zz} = g_e - 3b$.

3. If other atoms are added to the radical or radical ion to extend the conjugation, such as oxygen in semiquinones, or if heteroatoms replace carbon atoms in cyclic structures, the contribution to g is proportional to the spin density on that atom. This latter theme was explored in more detail [85] for semiquinones. The resulting expression for δg_O, the contribution of oxygen to the average g-factor, is

$$\delta g_O = c_O \rho_O / \delta E_{n-\pi^*} \qquad (80)$$

where c_O is an empirical constant for the semiquinone group, ρ_O is the spin density on the oxygen atomic p orbital, and $\delta E_{n-\pi^*}$ is the energy of the $n-\pi^*$ transition. The coefficient $c_O/\delta E_{n-\pi^*}$ was determined for a variety of semiquinones [86] as $8.0 \pm 0.4 \times 10^{-3}$.

Unfortunately, the measurement of g-factors is often neglected. Accurate, absolute measurements require good magnetic field stability and uniformity over the sample. However, measurements relative to a known standard are straightforward. Much information can be gained from g-factors, especially when the molecular structure is of interest. For example, it is possible to determine the spin densities on nuclear-spin-free elements, such as oxygen or sulfur, through the g-factor.

There is considerable experimental evidence for the preceding models. The theory for hydrocarbon radicals and radical ions predicts that the g-factors of radical cations should be smaller than those of radical anions and that a plot of $g-g_e$ versus m should result in a straight line. Values of g-factors for some aromatic hydrocarbon radicals are given in Table 4.7. The empirical constants derived from these data are

$$b = 30.5 \pm 0.4 \times 10^{-5} \qquad c = 16.6 \pm 1.0 \times 10^{-5} \quad \text{for [87]}$$

$$b = 29.8 \pm 0.4 \times 10^{-5} \qquad c = 20.3 \pm 0.7 \times 10^{-5} \quad \text{for [88]}$$

The benzene anion g-factor does not fit these parameters, probably because the molecular orbital energy is large and the approximations used in deriving (79) break down. In addition, the g-factor of this radical was found to be temperature dependent, unlike the other hydrocarbon radical ions [90]. The latter is probably the result of the orbital degeneracy of the benzene anion radical ion [90]. g-Factors of diphenylacetylene radical anions are considerably smaller than the normal hydrocarbon radical ions because of the cylindrical symmetry of the p orbitals about the axis of the sp-bonding orbitals [91, 92].

g-Factors of semiquinones are very solvent dependent because of their strong tendency for hydrogen bonding and electric dipolar (molecular) interactions.

Table 4.7 g-Factors of Aromatic Hydrocarbon Radical Ions and Anions

| | | g-Factor | | | |
| | | Cation | Anion | | |
Hydrocarbon	$|m|$	Radical	Radical	Neutral	References
Benzene	1.000		2.002844		[88]
Biphenyl	0.705		2.002772		[87]
Naphthalene	0.618		2.002737		[87]
			2.002743		[88]
Phenanthrene	0.605		2.002736		[87]
p-Terphenyl	0.593		2.002704		[89]
Pyrene	0.445		2.002717		[87]
			2.002714		[88]
Anthracene	0.414	2.002568	2.002694		[87, 89]
		2.002557	2.002699		[88]
Perylene	0.169	2.002565	2.002689		[87]
		2.002569	2.002657		[88]
Tetracene	0.295	2.002557	2.002671		[87]
		2.002590	2.002672		[88]
Pentacene	0.220	2.002564	2.002663		[87]
		2.002596	2.002676		[88]
Triphenylmethyl	0.000			2.002593	[89]
Perinaphthyl	0.000			2.00265	[88]

Values were measured for semiquinones [85, 86, 88, 93] and protonated semiquinones [94].

The g anisotropy of radical ions is generally small and difficult to evaluate because single-crystal samples cannot be prepared easily and the anisotropic hfs obscures the g anisotropy in frozen solutions. Frequencies greater than approximately 9 GHz, which require proportionately greater magnetic fields to achieve resonance, are used to spread out the spectrum because the span caused by hfs remains unchanged but the g-factor components separate proportionally to the applied field. Several studies were carried out at 35 GHz and low temperature to examine the anisotropy of parabenzosemiquinone and/or its tetrahalogenated analogues [95]. The results agree with expectations; that is, the anisotropy was in the direction predicted by Stone [83, 84], and the spin–orbit coupling of the halogens contributed strongly to the g-factors: p-benzosemiquinone, 2.0047; tetrachloro-, 2.0056; tetrabromo-, 2.0085; and tetraiodo-, 2.0121. Studies of g-factor anisotropy are important, particularly because both the g-factor and hyperfine anisotropies affect the spectrum under conditions of slow molecular tumbling, which is becoming an increasingly significant area of ESR study.

2.2.3 g-Factors of Small Radicals

Electron spin resonance has been used extensively to study atoms and small molecules trapped in matrices, defects in crystals, or radical species generated by ultraviolet or high-energy irradiation. Species such as alkyl radicals, hydrogen atoms, and alkali metals are detected readily. These species are in nondegenerate states and exhibit virtually no orbital angular momentum except that which is added through high-order contributions.

S-STATE ATOMS

Table 4.8 shows the effect of spin–orbit coupling of some S-state atoms. The alkali metal ions are in S states with extremely large separations between the ground state and the next higher orbital possessing angular momentum. As a result, all of the g-factors are very close to the free electron value. Cesium exhibits the greatest spin–orbit coupling and the smallest energy separation between the s and p orbitals and therefore exhibits the greatest deviation from g_e.

DIATOMIC RADICALS IN $^2\Pi$ ELECTRONIC STATES

Many small radicals or atoms trapped in condensed phases are difficult to observe because their electronic states possess significant orbital angular momentum, which induces rapid relaxation of the electron spin. When detection is possible, the sample often must be chilled to liquid nitrogen temperature or lower. In addition these spectra usually exhibit broad lines and g-factors that are significantly different from g_e. In some instances it is possible to observe ESR of species that would exhibit significant angular momentum in their free states because the orbital contribution is largely quenched by electric field effects in condensed systems. Some examples of such mono- and diatomic species include

Table 4.8 *g*-Factors and Hyperfine Coupling of S-State Atoms

Atom	Isotope	g-Factor	Hyperfine Coupling[a]	Spin–Orbit Coupling[b]	References
Hydrogen	1	2.002319	1420.4058		[55, 96]
	2	2.002319	327.3842		[96, 97]
Lithium	6	2.002309[c]	228.208	6	[99]
	7	2.002309[c]	803.512		[99]
Sodium	23	2.002309[c]	171.61	330	[99]
Potassium	39	2.002309[c]	461.723	1140	[98]
Rubidium	85	2.002409	3035.7	4740	[99]
	87	2.002409	6834.687		[99, 100]
Cesium	133	2.002577	9192.8	11500	[99]

[a]Splitting is in megahertz.
[b]In gigahertz.
[c]Deviation from g_e may be experimental uncertainty.

the iodine atom in iodic acid [101]; dioxygenyl, O_2^+ [102]; superoxide, O_2^- [103]; and chlorine oxide [104].

In general the magnetic parameters for such species are highly anisotropic. As a result, ESR detection is possible only for adsorbed atoms and molecules because the adsorption process strongly perturbs the electronic configuration and quenches orbital angular momentum. One cautionary note: Assigning spectra to a molecule containing oxygen is dangerous. Oxygen has no nuclear spin, and its presence or absence must be inferred from secondary evidence. Numerous errors have appeared in the literature.

A relatively simple example of the $^2\Pi_{1/2}$ state can be used to illustrate the application of second-order perturbation theory. The energy-level diagram for diatomic molecules is shown in Figure 4.21, and some radicals composed of first-row elements are shown to the left of the figure. Figure 4.22 shows the network of $2p$ atomic orbitals and the customary coordinate system used to describe such diatomic species. The $2p_x$ and $2p_y$ atomic orbitals combine to form the π_x, π_y, π_x^*, and π_y^* molecular orbitals shown in Figure 4.21b. In the gas phase the π_x, π_y orbitals (or π_x^* and π_y^* orbitals) are degenerate, although spin–orbit coupling breaks the degeneracy by a small amount λ (Figure 4.21c). This degeneracy introduces significant orbital angular momentum about the z axis, which causes fast relaxation of the electron spin so that the species is not detectable by ESR. But in an electric field, such as an orthorhombic field produced by electrostatic charges along the x or y axis (Figure 4.22) in a crystal lattice, the energy levels are split further as indicated in Figure 4.21d. If the splitting is large enough to quench the orbital angular momentum, the species can be observed.

Using the dioxygenyl molecule–ion as a specific example, the energy is split by an amount equal to $(\Delta^2 + \lambda^2)^{1/2}$, where λ is the spin–orbit coupling constant and Δ is the crystal field splitting. In the absence of the crystal field, only the

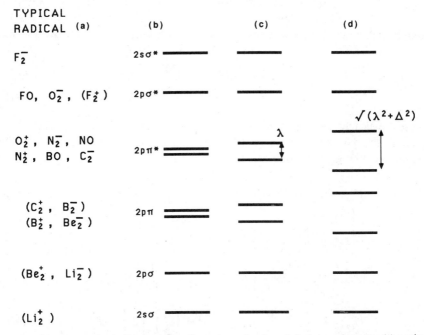

TYPICAL RADICAL (a)	(b)	(c)	(d)

F_2^- — $2s\sigma*$ ▬ ▬ ▬

FO, O_2^-, (F_2^+) — $2p\sigma*$ ▬ ▬ ▬

$\sqrt{(\lambda^2+\Delta^2)}$

O_2^+, N_2^-, NO N_2^+, BO, C_2^- — $2p\pi*$ ▬▬ λ ═ ▬

(C_2^+, B_2^-) (B_2^+, Be_2^-) — $2p\pi$ ▬▬ ▬ ▬

(Be_2^+, Li_2^-) — $2p\sigma$ ▬ ▬ ▬

(Li_2^+) — $2s\sigma$ ▬ ▬ ▬

Figure 4.21 Energy-level diagrams for diatomic molecules of first-row elements: (*a*) various diatomic radicals and their associated charges; those species in parentheses were not observed experimentally; (*b*) the orbital containing the unpaired electron, along with schematic levels that indicate the ordering of molecular orbital energies in the absence of spin–orbit coupling; (*c*) the changes in energy levels when spin–orbit coupling is introduced; (*d*) the further effect of interaction with a crystal field.

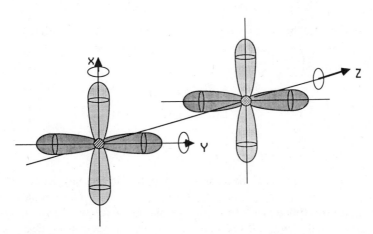

Figure 4.22 Schematic diagram of a diatomic radical illustrating the principal directions. For clarity the $2p_z$ orbitals were not drawn, but the overlap of these two atomic orbitals produces the $2p\sigma$ molecular orbital of Figure 4.21.

diagonal elements of the spin–orbit-coupling tensor $\lambda\,M_L M_S$ exist, so that for a $^2\Pi_{1/2}$ radical the energy levels are split by $\pm 1/2\lambda$. In the presence of a crystal field, the energies of the lower and upper $2p\pi^*$ orbitals, respectively, E_1 and E_2, are

$$E_1 = \tfrac{1}{2}(\Delta^2 + \lambda^2)^{1/2} \tag{81a}$$

$$E_2 = -\tfrac{1}{2}(\Delta^2 + \lambda^2)^{1/2} \tag{81b}$$

These are clearly nondegenerate levels, but the energy separation between them is small, so that excited-state contributions of orbital angular momentum are significant. The ground-state wave functions that enter into these energies are

$$\Psi_{1\alpha}^0 = N_1\left(|1,\tfrac{1}{2}\rangle - \left\{\frac{\lambda}{\Delta} - \left[1 + \left(\frac{\lambda}{\Delta}\right)^2\right]^{1/2}\right\}|-1,\tfrac{1}{2}\rangle\right) \tag{82a}$$

$$\Psi_{1\beta}^0 = N_2\left(|1,-\tfrac{1}{2}\rangle + \left\{\frac{\lambda}{\Delta} + \left[1 + \left(\frac{\lambda}{\Delta}\right)^2\right]^{1/2}\right\}|-1,-\tfrac{1}{2}\rangle\right) \tag{82b}$$

$$\Psi_{2\alpha}^0 = N_2\left(|1,\tfrac{1}{2}\rangle - \left\{\frac{\lambda}{\Delta} + \left[1 + \left(\frac{\lambda}{\Delta}\right)^2\right]^{1/2}\right\}|-1,\tfrac{1}{2}\rangle\right) \tag{82c}$$

$$\Psi_{2\beta}^0 = N_1\left(|1,-\tfrac{1}{2}\rangle - \left\{\frac{\lambda}{\Delta} - \left[1 + \left(\frac{\lambda}{\Delta}\right)^2\right]^{1/2}\right\}|-1,-\tfrac{1}{2}\rangle\right) \tag{82d}$$

where N_1 and N_2 are normalization constants given by

$$N_1 = \left(2\left[1 + \left(\frac{\lambda}{\Delta}\right)^2\right]^{1/2}\left\{\left[1 + \left(\frac{\lambda}{\Delta}\right)^2\right]^{1/2} - \frac{\lambda}{\Delta}\right\}\right)^{-1/2} \tag{83a}$$

$$N_2 = \left(2\left[1 + \left(\frac{\lambda}{\Delta}\right)^2\right]^{1/2}\left\{\left[1 + \left(\frac{\lambda}{\Delta}\right)^2\right]^{1/2} + \frac{\lambda}{\Delta}\right\}\right)^{-1/2} \tag{83b}$$

One can now use the perturbation equations for the g-tensor, with the result that

$$g_z = g_e - 2\left(\frac{\lambda^2}{\lambda^2 + \Delta^2}\right)^{1/2} \tag{84a}$$

$$g_x = g_e\left(\frac{\Delta^2}{\lambda^2 + \Delta^2}\right)^{1/2} - \frac{\lambda}{E}\left[\left(\frac{\Delta^2}{\lambda^2 + \Delta^2}\right)^{1/2} + \left(\frac{\lambda^2}{\lambda^2 + \Delta^2}\right)^{1/2} + 1\right] \tag{84b}$$

$$g_y = g_e\left(\frac{\Delta^2}{\lambda^2 + \Delta^2}\right)^{1/2} + \frac{\lambda}{E}\left[\left(\frac{\Delta^2}{\lambda^2 + \Delta^2}\right)^{1/2} - \left(\frac{\lambda^2}{\lambda^2 + \Delta^2}\right)^{1/2} - 1\right] \tag{84c}$$

A similar treatment can be applied to superoxide ion, but the deviation of g from g_e is exactly opposite to that for dioxygenyl. The simplest way to understand this difference is to recognize that O_2^- consists of pairs of electrons with an extra electron, but O_2^+ consists of pairs of electrons with a hole. This inverts the sign of the energy levels, and consequently inverts the sign of the major part of the deviation from g_e. For example, g-factors for O_2^+ in O_2AsF_6

are $g_x = 2.000$, $g_y = 1.973$, and $g_z = 1.742$ [102]; and for O_2^- impurity in KCl: $g_x = 1.9512$, $g_y = 1.9551$, and $g_z = 2.4359$ [103].

More details on g-factors of small radicals and atoms in condensed phases are found in [105, 106].

2.2.4 σ-Electron Radicals

Most of the radicals that we have discussed are essentially π radicals. But another type of radical is the σ-electron radical, where the unpaired electron is in a σ state. These radicals can then be divided into two subclasses. The first class contains diatomic species such as CN and F_2^-, where the unpaired electron is still predominately in a p-atomic orbital. In the dihalide radical anions, the proportion of s character to the unpaired electron wave function is about 2 to 3%. The g-factors of such species can be calculated much as in the foregoing discussion.

The second class of σ radicals comprises species in which the unpaired electron is localized in a broken σ bond, such as in the phenyl radical ($C_6H_5 \cdot$) or acetylenyl (HC≡C·) radical. This class of radicals is often generated by photolysis or irradiation of molecules that possess bonds that are weak relative to the irradiation energy. For example, the phenyl radical can be generated by photolysis of iodobenzene [107], and the vinyl radical can be generated by 2.8-MeV electron irradiation of liquid ethane [57]. These species typically exhibit g-factors smaller than g_e. Table 4.9 lists several reported σ radicals.

Theoretical models for g-factors of such radicals require extensive molecular orbital calculations because the g-tensor calculations are very dependent on the energy levels, and the degree of hybridization of the unpaired electron is extremely significant.

2.2.5 g-Factors of Transition Metal Ions

The earliest application of ESR involved the study of transition metal ions [114, 115]. Metal ion salts contain a high concentration of paramagnetic material so that great sensitivity was not required to detect signals from single

Table 4.9 g-Factors of Polyatomic σ-Electron Radicals

Radical	Structure	g-Factor	Reference
Acetyl	$RCH_2\dot{C}O$	2.0008	[108]
Carbamoyl	$NH_2\dot{C}O$	2.0017	[109]
Carboxide	$\dot{C}O_2^-$	2.0007	[110]
Nitric oxide	$\dot{N}O_2$	1.9997	[111]
Ethenyl	$CH_2{=}\dot{C}H$	2.0022	[57]
Ethynyl	$HC{≡}\dot{C}$	2.00175	[112]
Phenyl	$C_6H_5\cdot$	2.0021	[113]

crystals or polycrystalline samples. In most cases, when samples of pure paramagnetic solids are studied, hyperfine splittings are not detected because of dipolar and/or exchange broadening, but the g-factor anisotropy is preserved so that one can measure the g-tensor with some accuracy. Since this area is still of great importance in the application of ESR, it is worthwhile to examine the factors that contribute to the g-factors of metal ions. Detailed accounts of the theoretical and experimental aspects of this work are found in many textbooks such as [12, 15, 116–120] and review articles such as [78, 121–123].

Metal ions can be subdivided into groups based on the nature of their magnetic properties. The first case, which is covered here, is the condition where the crystal field in a complex or lattice is strong enough to remove the orbital degeneracy of the ion. Examples include most first-row transition metal ions where the spin–orbit coupling is small relative to the crystal field energies. The spin–orbit coupling is important, however, and these values are given in Appendix 1. It is important to note that the spin–orbit coupling constant of transition metal ions is positive for ions with less than half-filled shells and negative for ions with more than half-filled shells.

The second case, which is analogous to the treatment of gas-phase species, occurs when the crystal field is too weak to break the orbital degeneracy. This is typical of lanthanide and actinide ions in which the spin–orbit coupling is large and the crystal lattice interaction is weaker than it is for transition metal ions because of the larger ionic radii.

In either case, it is convenient to classify the ions according to the spin and orbital angular momentum of the nominally unperturbed ground state. The pertinent symbols are S, spin $(0, \frac{1}{2}, 1, \frac{3}{2}, \ldots)$; L, orbital angular momentum (S for $L = 0$, P for $L = 1$, D for $L = 2$, F for $L = 3$, etc.); and J, total electronic angular momentum $(J = L + S, L + S - 1, \ldots, L - S)$. The so-called *term symbols* are generally arranged as $^{2S+1}L_J$, where the orbital angular momentum is represented by its letter symbol equivalent. For example, Ti^{3+} contains one electron in a $3d$ orbital $(3d^1)$ so the notation for its ground state is $^2D_{3/2}$; and Mn^{2+} contains five unpaired electrons in $3d$ orbitals so its notation is $^6S_{5/2}$. Because there are five d orbitals, each containing an unpaired electron, the state is nondegenerate; therefore, there is no net orbital angular momentum, and $L = 0$.

In the crystal field model, the lattice is represented by point charges. The two basic configurations are the octahedral field, which is shown in Figure 4.23a, and the tetrahedral field, which is shown in Figure 4.23b.

Note that the octahedral field (Figure 4.23a) can be inscribed in a cube, as can the tetrahedral field. However, the tetrahedral field has the effect of opposite point charges acting on the central ion. Thus the energy-level structure is inverted as shown in Figure 4.24. This figure does not show the spin–orbit coupling in the energy-level diagram. In addition, a tetragonal distortion such as shown in Figures 4.23c or d further splits the energy levels. The reader is directed to one of the preceding references [12, 15, 78, 116–123] for the details of this treatment. The crystal field splittings are noted as Δ for the tetrahedral or octahedral fields and as δ for the smaller tetragonal distortion.

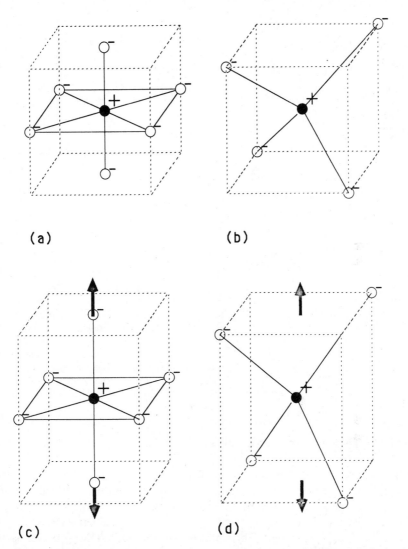

Figure 4.23   Diagram showing the location of ions in complexes subjected to (*a*) octahedral and (*b*) tetrahedral crystal fields; (*c*) shows the effect of tetragonal distortion on the octahedral field, while (*d*) shows the effect of tetragonal distortion on the tetrahedral field.

A useful method of diagramming such energy levels was derived by Tanabe and Sugano [124], and it is found in graphical form in [125]. In some discussions it is customary to express the crystal field splitting Δ in terms of Dq, where D, in this context, is related to the energy of displacing unit electron charge q in the lattice.

As was indicated for the $^2\Pi$-state radicals, the energies and wave functions of the transition metal ions subject to the crystal field perturbation can be

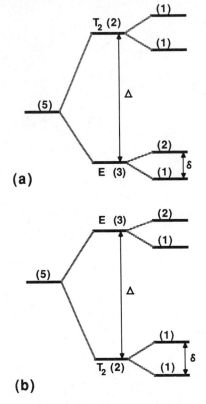

(a)

(b)

Figure 4.24   Splitting of d-electron energy levels in a crystal field. The diagrams apply to $3d^1$ and $3d^6$ ions (a) in an octahedral field and (b) in a tetrahedral field. For $3d^4$ or $3d^9$ ions, (a) represents energy levels for a tetrahedral field, while (b) represents the octahedral field. The numbers in parentheses are the degeneracies of the energy levels. The initial splitting is caused by the crystal field, while the subsequent splittings are caused by tetragonal distortions of the respective fields.

calculated. These in turn are used to compute the Λ matrix needed to calculate the g-tensor. In the following discussion the tetragonal distortion is taken to be in the z direction and is implicitly smaller than the tetrahedral or octahedral field splitting. By convention, the tetragonal splitting is positive for lengthening the distance in the z directions as shown in Figures 4.23c and d. For the following discussion, we deal specifically with axially symmetric cases, so that $g_{zz} = g_\parallel$ and $g_{xx} = g_{yy} = g_\perp$. This method of computing the g-tensor is an approximation. The principal limitation is that the effects of covalency are neglected. This is particularly significant for transition metal ions bound to nitrogen of amine or pyridyl-type ligands. Second, the spin–orbit coupling of the ligand is also neglected. This can be significant when the ion is associated with high atomic weight species such as iodine.

^{2}D STATE IONS: $3d^1$, $3d^7$ (LOW SPIN), AND $3d^9$

$3d^1$.   The ^{2}D state ion has five degenerate states, which split as shown in Figure 4.24a. If a tetrahedral field with a tetragonal distortion is applied, then

$$g_\parallel = g_e \tag{85}$$

$$g_\perp = g_e - 6\lambda/\Delta \tag{86}$$

For an octahedral field with a tetragonal distortion, the spectrum is highly dependent on the magnitude of δ, which contributes most strongly to the g-factor. The effect of such distortions on the g-factor is expressed as

$$g_\parallel = 3 \frac{\delta + \lambda/2}{[(\delta + \lambda/2)^2 + 2\lambda^2]^{1/2}} - 1 \tag{87}$$

$$g_\perp = \frac{\delta + 3\lambda/2}{[(\delta + \lambda/2)^2 + 2\lambda^2]^{1/2}} + 1 \tag{88}$$

These expressions support the observations that spectra of ^2D state ions are very strongly dependent on the magnitude of the tetragonal field. More rigorous forms of (85) and (86) include a smaller additional term of about $-4\lambda/\Delta$, which is added to account for the mixing of the spin–orbit coupling with the ground state. Normally this is not significant.

Octahedral complexes of titanium(III), vanadyl (VO^{2+}), and various vanadium(IV) ions were studied by ESR. Typically they exhibit $g_\parallel$ greater than $g_\perp$, with each parameter having a value smaller than 2. If the tetragonal distortion is too small, the g-factor approaches zero, and spectra are not observed. In intermediate cases, where there is significant residual angular momentum, the spectra exhibit broad lines even at temperatures below 77 K, hence in some instances temperatures below that of liquid helium are required for their observation.

$3d^9$. The most common conformation for $3d^9$ ions, such as copper(II), is octahedral. In this configuration, the state is twofold degenerate, and strong Jahn-Teller distortions broaden the spectrum. For example, the ion $Cu(H_2O)_6^{2+}$ exhibits line widths greater than 20 mT depending on the other ions present. However, the tetragonal field breaks this degeneracy. g-Factors obey the same equations as for the tetrahedral plus tetragonal case for d^1 ions; but since the sign of λ is negative, $g_\parallel$ is greater than g_e. There are many examples of copper(II) complexes in the literature, but the easiest illustration is that of square planar complexes. The g-factors of copper tetraphenylporphyrins vary slightly depending on the solvent or ligand above and below the plane. Typical values for $g_\parallel$ are 2.170–2.193; for $g_\perp$ they are 2.042–2.067 [126].

$3d^7$. Equations (85) and (86) also apply to low-spin $3d^7$ ions in an octahedral field with tetragonal distortion. The low-spin ion has a single spin in a doubly degenerate manifold; $\lambda < 0$, therefore $g_\perp > g_e$, as for $3d^9$ ions. The most familiar ion of this class is Co(II). Cobalt was studied extensively during the past few years because of its potential as a catalyst for liquefaction of coal by hydrogenation.

F-STATE IONS: $3d^2$, $3d^3$, $3d^7$ (HIGH SPIN); AND $3d^8$

Paramagnetic F-state ions contain more than one unpaired electron so the total angular momentum L of the ion in the unperturbed state is 3. These states exhibit sevenfold degeneracy. This is apparent in Figures 4.25a and b. There are two consequences of multiple electron states. First, to obtain multiple spin ions,

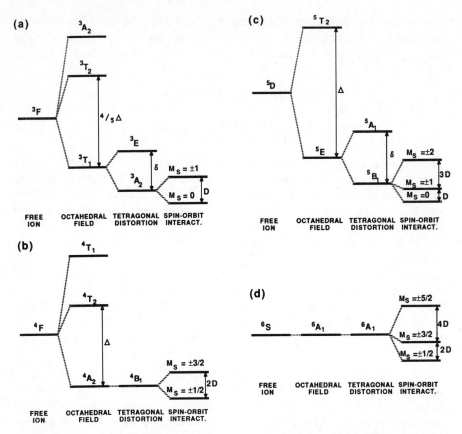

Figure 4.25 Energy-level diagrams for transition metal ions containing more than one unpaired electron in an octahedral crystal field. The levels are initially split by the crystal field with subsequent splittings arising from tetragonal distortions and dipolar interactions. The electronic configurations represented in the figure are (a) d^2, (b) d^3, (c) d^4, and (d) d^5 states.

the orbital angular momentum state in the absence of electrons (see Figure 4.24) must exhibit multiplicity; for example, to obtain a ^3F state that is nondegenerate, there must be two unpaired electrons that occupy equivalent orbital states. If that does not happen, then the ion becomes low spin. Second, with multiple electrons, the dipolar or spin–spin interaction becomes significant.

$3d^8$.   When the ^3F state is split by octahedral and tetragonal fields, it leaves a nondegenerate state. This state has the orbital wave function of $(1/\sqrt{2})\cdot$ $(|2\rangle + |-2\rangle)$, so $L = 0$. The spin–orbit coupling causes mixing with excited states, and the g-factor is given by

$$g_{xx} = g_{yy} = g_{zz} = g_e - 8\lambda/\Delta \tag{89}$$

The zero-field parameter D described in (76) becomes important in many $3d^8$ cases. It can be formulated in terms of the spin–orbit coupling λ and the crystal field splittings δ and Δ as

$$D \approx -4\lambda^2\delta/\Delta^2 \qquad (90)$$

Generally, as the tetragonal distortion of the ion decreases, the spectra become more difficult to observe. The most common $3d^8$ ion is nickel(II). Some early studies [127] give values for Ni(II) in $SrTiO_3$ ($g = 2.204$) and CaO ($g = 2.337$). Often this ion is difficult to detect in solution because modulation by motion of the ligands causes a modulation of D, which in turn causes broadening.

$3d^2$. In an octahedral field, the lowest state is triply degenerate, and Jahn-Teller effects split the ground state and cause rapid relaxation. As a result, most spectra are recorded with sample temperatures near 4 K. In small tetrahedral fields, the electron spin–spin dipolar interaction D is usually extremely large making the spectrum difficult to observe. If the tetrahedral splitting is extremely large, the ion becomes diamagnetic. An important example is that of V(III) in CdS, with $g_{\parallel} = 1.934$ and $g_{\perp} = 1.932$, and $D = 26.5\,GHz$ [128].

The tetrahedral crystal field is the most common for the $3d^2$ ion studies reported in the literature. Because the $3d^2$ ion is equivalent to the $3d^8$ ion with eight holes instead of electrons, the tetrahedral field of $3d^2$ is equivalent to the octahedral field of $3d^8$, so that the g-factors are isotropic and given by (89). Since λ is positive, the g-factor is smaller than g_e. Familiar examples are the MnO_4^{3-} ion in VO_4^{3-} lattice, which exhibits anisotropic g-factors of about 2 [129], and Ti(II) in ZnS, which exhibits an isotropic g-factor of 1.9280 [130].

$3d^3$. The energy-level diagram for d^3 ions in an octahedral field is shown in Figure 4.25b. In an octahedral field, the lowest energy level comprises three states that are energetically equivalent. In conventional d-orbital notation, these are the d_{xy}, d_{yz}, and d_{xz} orbitals. When each contains one unpaired electron, the ground state is nondegenerate. As a result, ESR spectra are readily obtainable, and these species exhibit sharp lines. The wave functions of this state are similar to $3d^8$ ions. Because of octahedral symmetry, the g-factors are isotropic, and

$$g = g_e - 72\lambda/5\Delta \qquad (91)$$

The most commonly studied ion of this class is Cr(III), which is shown in Figure 4.26 in a $SrTiO_3$ host. The Cr^{3+} ion in MgO exhibits a g-factor of 1.9796 and a zero-field splitting of $D = 48.9\,MHz$ (ca. 2 mT) ([131], which contains a tabulation of ESR parameters for various ions in oxide hosts). If the field symmetry is reduced by tetragonal distortion, the g-factor becomes anisotropic and D becomes very large. As a result, only the transition $+\frac{1}{2} \leftrightarrow -\frac{1}{2}$ may be visible in this case.

$3d^7$. There are two possible states of d^7 ions: high spin and low spin. In the high-spin case, the octahedral field creates a degenerate state, and these ions exhibit rapid relaxation so that ESR spectra are difficult to detect. In a tetrahedral configuration, the g-factors and the D tensor are similar to those for the $3d^3$ ion.

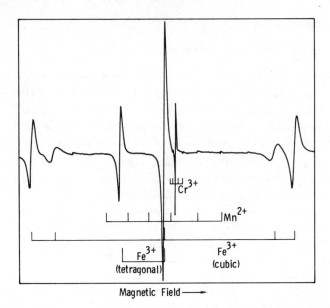

Magnetic Field ⟶

Figure 4.26 Composite spectrum showing ESR absorptions caused by Cr(III) ($3d^3$), Fe(III), and Mn(II) (both $3d^5$) in $SrTiO_3$. One-half of the Fe(III) ions are found in the octahedral site (cubic) and the other half are next to a compensating oxygen vacancy leading to an octahedral site with tetragonal distortion.

Spin-paired, $S = \frac{1}{2}$, ions in which a strong tetragonal splitting causes separation of the lowest state are readily detected by ESR. The Co(II) ion in square planar complexes is typical. Because of the strong influence of the tetragonal distortion, the parameters of these spectra are extremely dependent on the solvent or medium. For example, Co(II) phthalocyanine exhibits $g_{\parallel}$ from 2.007 to 2.029 and $g_{\perp}$ from 2.317 to 2.546 depending on the solvent [132]. Both the cobalt ion and the nitrogen bonded to the cobalt exhibit hyperfine splittings.

S-STATE IONS: $3d^5$, HIGH SPIN

When each of five unpaired electrons occupy one of the five d-electron orbitals, the ion is said to be in the *high-spin configuration*. Since there is only one possible orbital permutation of this configuration, it is in a nondegenerate state, and even in the free ion there is no orbital angular momentum. The term symbol is $^6S_{5/2}$.

Octahedral. In the case of an octahedral crystal field, one would expect g to be isotropic and close to g_e. While this is in fact the case, detailed analysis [15, 118] shows that another term must be added to the crystal field operator that can couple the spin states $|\pm\frac{5}{2}, \mp\frac{3}{2}\rangle$, that is, those that differ by $M_S = \pm 4$. This term is parameterized as a', which is a term in the octahedral crystal field Hamiltonian describing the electrostatic interaction between the nucleus and the outer electrons. Usually a' is small relative to the Zeeman interaction, but it produces a significant alteration of the spectrum as shown in Figure 4.27.

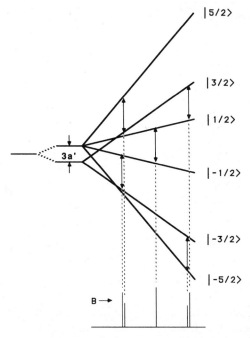

Figure 4.27   Energy-level diagram for a $3d^5$ ion in an octahedral crystal field with the principal axis parallel to B and $g\mu_B \gg a'$; ESR transitions are indicated by double arrows and a representation of the ESR spectrum expected in the high field approximation is shown at the bottom of the diagram.

Instead of one line, five lines are obtained. This is shown for cubic Fe(III) in Figure 4.26, as recorded with the magnetic field parallel to the octahedral axis of the crystal. Solution of the Hamiltonian gives the following energy separations for each transition

$$\Delta E(\pm \tfrac{1}{2}) = a' \pm g\mu_B B \tag{92a}$$

$$\Delta E(\tfrac{5}{2}, -\tfrac{3}{2}) = \tfrac{1}{2}(g\mu_B B - a') \pm [(4\,g\mu_B B)^2 + 16\,a'\,g\mu_B B + 9\,a'^2]^{1/2} \tag{92b}$$

$$\Delta E(\tfrac{3}{2}, -\tfrac{5}{2}) = -\tfrac{1}{2}(g\mu_B B + a') \pm [(4\,g\mu_B B)^2 + 16\,a'\,g\mu_B B + 9\,a'^2]^{1/2} \tag{92c}$$

The same perturbation also applies to Mn(II), but because of the smaller charge on the central ion, it is not always detected. Representative data for these ions are shown in Table 4.10.

Tetragonal Distortion.   Consider two different types of tetragonal distortions. In one a tetrahedral distortion creates a zero-field splitting much larger than the Zeeman energy. Here, only the transitions $|\tfrac{5}{2}, \tfrac{1}{2}\rangle \leftrightarrow |\tfrac{5}{2}, -\tfrac{1}{2}\rangle$ are observed. Evaluation of the energy levels with the field parallel to the tetragonal distortion axis gives the result that $\Delta E = \pm \tfrac{1}{2} g\mu_B B$, which is the conventional spectrum. Evaluation with the field perpendicular to the tetragonal axis gives the result $\Delta E = \pm \tfrac{3}{2} g\mu_B B$. As a result, two lines are observed. The low-field line

Table 4.10 g-Factors and Hyperfine Splittings of d^5 Ions in Various Hosts

Host	Ion	g(iso)	a(mT)	References
CaF_2	Mn^{2+}	1.998	97.8	[133]
CdF_2	Mn^{2+}	2.0026	93	[134]
CdS	Mn^{2+}	2.0016	76	[135]
CdTe	Cr^+	1.9997	12.78	[136]
CdTe	Mn^{2+}	2.010	76	[137]
CdTe	Fe^{3+}	2.0029	10.7	[138]
ZnO	Fe^{3+}	2.0060	9.02	[139]
ZnS	Cr^+	1.9995	13.4	[136, 140]
ZnS	Mn^{2+}	2.0024	64	[141]
ZnS	Fe^{3+}	2.0194	7.8	[142]
ZnSe	Cr^+	2.0018	13.27	[136, 140]
ZnSe	Fe^{3+}	2.0464	6.73	[143]

exhibits a "fictitious" g-factor of 6, and the line at the normal position exhibits a g-factor of 2. The low-field line is dependent on orientation.

If the tetragonal distortion is small, then the entire spectrum is visible. In these cases, as shown in Figure 4.25, the separation of the lines is 2D between the $\pm\frac{1}{2}$ and $\pm\frac{3}{2}$ spin states and 4D between the $\pm\frac{3}{2}$ and $\pm\frac{5}{2}$ spin states. Thus, the lines are equally spaced in the spectrum.

In numerous systems iron resides in highly asymmetric environments, such as in biological compounds and minerals, particularly silicates. Site and crystal symmetries are extremely complex [144]. High-spin iron in a highly distorted tetrahedral site exhibits g-factors near 4. For example, in the thorite lattice ($ThSiO_4$), where the iron occupies a silicon site [145], the g-factors are $g_{xx} = 4.09$, $g_{yy} = 4.38$, and $g_{zz} = 3.04$. Another example is Fe(III) in kaolinite, $Al_2Si_2O_5(OH)_5$, where g is approximately 4, and there are three different sites that are distinguished by different values of the D and E terms of the spin–spin interactions. The different D values are 9.7, 17.6, and > 36 GHz, while the corresponding E values are 2.0, 4.1, and approximately 30 GHz, depending on the site in the mineral [146]. Most of these absorptions appear outside the range of conventional ESR spectrometers.

The observation of g = 4 for iron complexes is often taken as evidence of tetrahedral high-spin iron. While this phenomenon is always observed for this ion, other phenomena can also contribute to the observation of absorption at this field. For example, the orientation-dependent absorption of crystals containing Fe(III) in weak octahedral fields can produce half-field resonance, as can $\Delta M_S = 2$ transitions in multiplet states. Thus, the conclusion that iron is present in a high-spin configuration must be made with extreme caution.

LOW-SPIN d^5 IONS: g-TENSOR

In the presence of a strong octahedral field, the five d electrons are split into two degenerate states. The lower of the two is triply degenerate (symbolized as a T state), and the higher is doubly degenerate (symbolized as an A state) in the absence of any other splitting. However, spin–orbit coupling and other tetragonal fields cause these levels to split. As a result, the electrons are placed in pairs in the lower levels with one unpaired electron. These systems exhibit $S = \frac{1}{2}$, and there is no fine structure. The greater the perturbation, the more easily observed are these spectra.

Because of their high sensitivity to crystal field perturbations, d^5 $S = \frac{1}{2}$ ions exhibit g-factors over a wide range. Theoretical evaluation predicts that $g_{\parallel}$ is less than g_e, and $g_{\perp}$ is less than g_e. g-Factors were observed from less than 1 to about 4.5, depending on the structure of the ion. However, Fe(III) with very strongly bonded ligands in square planar configurations typically exhibits g-factors near g_e and is detectable in solution [147]. Another system, Fe(III) substituted in $K_3Co(CN)_6$, exhibits $g_{\parallel} = 0.915$ and $g_{\perp} = 2.2$ [148].

IRON IN BIOLOGICAL SYSTEMS

Iron is an extremely important component of biological systems. As a result, ESR is used extensively to detect iron in biochemical research. Ehrenberg [149] classified such observations into four major groups:

1. High spin $(S = \frac{5}{2})$ in a square planar (heme) configuration with other ligands or small ions or molecules in at least one of the axial positions (e.g., O_2, CO, H_2, NO, Cl^-, or CN^-). These complexes exhibit $g = 2$ and $g = 6$ by virtue of the large zero-field splitting as described earlier.

2. Low-spin $(S = \frac{1}{2})$ complexes. In certain hemes with various ligands in the axial position, the octahedral field is sufficiently strong to produce spin pairing. In such cases, the complex is highly anisotropic with g_{zz} ranging from 2.5 to 3.2, g_{xx} from 1.2 to 2.0, and g_{yy} from about 2.2 to 2.3.

3. Low spin ferrihemoproteins with g very close to g_e.

4. Thermally balanced high-spin, low-spin systems where the low-spin form is dominant at low temperature, as in the horseradish peroxidases [150]. The literature in this field is extensive, and the reader is directed to the references in the annotated bibliography for additional information.

g-FACTOR–HYPERFINE SPLITTING CORRELATION

In Section 2.1.2, we mentioned the correlation between ligand and transition metal hfs. While this is difficult to demonstrate in a conclusive way, another important correlation exists between the g-factor and hyperfine splittings of the central ion. The correlation was shown for Ti(III), V(IV), and VO^{2+}, which are d^1; for MoO^{3+}, which is d^5; and for Cu(II), which is d^9. Generally speaking the magnitude of the difference between g and g_e decreases monotonically with the magnitude of the hyperfine splitting. This is understood in terms of stronger covalency of the bonding between metal and ligand, which causes a greater energy separation in the ground- and excited-state orbitals. This diminishes the effect of configuration interaction so that the observed g-factor is closer to g_e. Concurrently, the unpaired electron becomes more highly delocalized into the ligand orbitals, decreasing the spin density on the central ion. Figure 4.28 shows examples for various transition metal ion complexes. These plots generally exhibit a considerable amount of scatter. This may arise partially from the effect of inner s electrons on the hyperfine splitting, and from experimental error in the measurement of g-factors.

Since d^5 ions in octahedral fields exhibit g-factors quite near g_e, the foregoing data display is inapplicable. However, the correlation of the hfs of the transition metal ion with the degree of covalency was proposed [136]. One example of this effect is seen in terms of the Mn(II) hyperfine splittings in different lattices [81]. For example, the ^{55}Mn hyperfine splitting decreases as F > Cl > Br > I and O > S > Se > Te when observed in a series of magnesium or cadmium halide lattices or in magnesium, zinc, or cadmium chalcogenide lattices.

2.3   Gas-Phase Radicals

Paramagnetic species in condensed phases are characterized by g-factors that are generally close to 2 because the orbital angular momentum is almost completely quenched by crystal field or environmental effects. However, in the absence of such effects, such as in the gas phase at low pressure, paramagnetic atoms and molecules may have quantized angular momentum from sources other than electron spin. Many of these can be detected by ESR because the angular momentum is a "good" quantum number. By virtue of quantized orbital angular momentum, atoms or molecules may be paramagnetic without spin. Oxygen in the ^{1}D state, and O_2 or NF in the $^1\Delta$ states are primary examples of

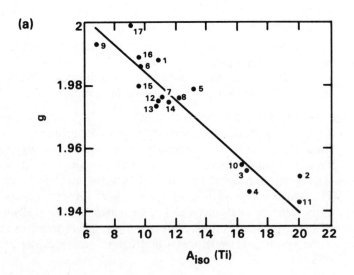

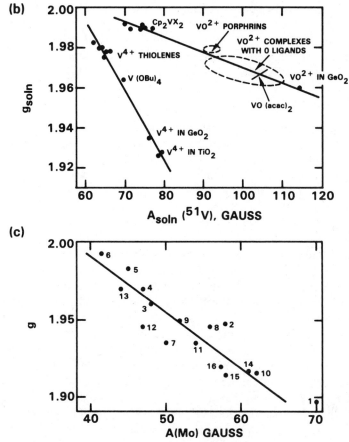

Figure 4.28 (a) Plot of g_{av} versus A_{iso} (Ti) for d^1, Ti(III) ions with a strong axial distortion: (1) Cp_2TiPMe_2, (2) $[Ti(OMe)_4Cl_2]^+$, (3) $(TiOMe)^{2+}$, (4) $(TiF_2)^+$, (5) Cp_2TiCl, (6) $Cp_2Ti(TDT)$, (7) $Cp_2TiCl_2 + AlEt_3$, (8) $Cp_2TiCl_2 + Al(i\text{-}Bu)_3$, (9) $(Cp_2TiH)_2$ (hydrogen-bridged cyclic dimer), (10) $(TiO_2)^+$, (11) Ti^{3+} in CaF_2, (12) $Cp_2TiCl_2 \cdot BCl_2$, (13) $Cp_2TiCl_2 \cdot GaCl_2$, (14) $Cp_2TiCl_2 \cdot AlCl_2$ (15) Cp_2TiPPh_2, (16) $[Cp_2Ti(PPh_2)_2]^-$, and (17) $[Cp_2Ti(PMe_2)_2]^-Na^+$. (b) Plot of g_{av} versus A_{iso} (^{51}V) for a series of vanadium complexes. (c) Plot of g_{av} versus A_{iso} (Mo) for strongly axially symmetric complexes of molybdenum: (1) $(MoOF_5)^{2-}$, (2) $(MoOCl_5)^{2-}$, (3) $(MoOCl_4Br)^{2-}$, (4) $(MoOCl_3Br_2)^{2-}$, (5) $(MoOCl_2Br_3)^{2-}$, (6) $(MoOBr_5)^{2-}$, (7) $[MoO(SCN)_5]^{2-}$, (8) $MoOCl_3$ in EtOH, (9) $MoOCl_3$ in 20% HCl, (10) $Mo_2O_3(SO_4)_2$ in H_2SO_4, (11) $Mo_2O_3(SO_4)_2$ in $H_2SO_4 + KCl$, (12) $trans\text{-}[C_5H_5N][Mo(OCH_3)_2Cl_4]$, (13) $MoOX_5 + S(SH)P(OC_2H_5)_2$, (14) $[MoO(H_2AsO_4)_4]^-$, (15) Mo^{5+} in H_3PO_4, and (16) Mo^{5+} in H_2SO_4. Reprinted, with permission, from [81].

this. The opposite is also true; some atoms may be diamagnetic even though there is spin such as in the oxygen atom $^3P_0$ state where the orbital momentum and spin momenta act in opposition so that there is no net paramagnetism.

In general, ESR is extremely sensitive to paramagnetic atoms and to many diatomic molecules, and often less than 10^9 units per cubic centimeter can be detected. As a result, ESR is used for studies of gas-phase kinetics, even under conditions of very rapid reactions [151].

There are four possible sources of angular momentum in a gas-phase atom or molecule: orbital angular momentum, quantum number L; electron spin, quantum number S; nuclear spin, quantum number I; and for molecules, rotation, quantum number R. To present logically the theory of gas-phase ESR, atoms are discussed first, then diatomic molecules are outlined.

2.3.1   Atoms in the Gas Phase

Gas-phase atoms have been studied by ESR since the late 1940s. Many are easily formed by passing a suitable gas through a microwave discharge operating at 2.45 GHz and applying approximately 100-W power output. A comprehensive review [152] on microwave discharges was published and is helpful to those contemplating research in this field. While many different paramagnetic atoms are generated by this method, it is crucial to flow them rapidly through the ESR cavity so that they are detected before they recombine or react.

The electronic state of the atom is described by the symbol $^{2S+1}L_J$, where L is the letter S, P, D, F,..., which represents the orbital angular momentum as in transition metal ions, and J is the total angular momentum as described in Section 2.3.3. As discussed previously, the g-factor is given by the Landé interval rule (45) to a high degree of accuracy. Deviations from this number are typically small and are accounted by including spin–orbit coupling to excited states. Thus, lighter atoms with fewer electrons exhibit smaller deviations from the Landé g-factor because the spin–orbit-coupling constant is smaller and the energy levels are generally more widely separated.

For cases where $L = 0$, $g = g_e$. Hydrogen and nitrogen are two typical examples. Spectroscopic data for various atoms are given in Table 4.11.

The simplest case is the hydrogen atom where $S = I = \frac{1}{2}$. The energy-level diagram is given in Figure 4.9. The high-field approximation for hydrogen was discussed previously. In general, the high-field approximation is valid when $B \gg 2hAI/g\mu_B$. In practice, if the applied field is larger than the span of the hyperfine splittings by a factor of 5 or more, the wave functions are adequately described by $|S, I, M_S, M_I\rangle$, and the energy levels and accurate wave functions are computed by perturbation theory. Often second- or third-order perturbation theory must be used, but the correction terms are relatively small. Physically, the high-field approximation represents the condition of the nuclear and electron spins being *decoupled*, that is, each interacting independently with the magnetic field because the hyperfine splitting energy is small relative to the electron Zeeman energy. Cases in which the second-order perturbation corrections become large represent the breakdown of this condition.

At low fields, where $B \ll 2hAI/g\mu_B$, the Zeeman term does not cause separation of S and I. Rather, it is better to define a quantum number F, where

$$F = S + I, S + I - 1,\ldots, S - I \tag{93}$$

Since the spins are not decoupled by the magnetic field interaction, the interaction between nuclear and electron spins dominates the Hamiltonian.

Table 4.11   Gas-Phase ESR Data for Atoms of Groups V, VI, and VII

Atom	State	g-Factor	Hyperfine Coupling (MHz)	Quadrupole Coupling (MHz)^a	References
^{14}N	$^4S_{3/2}$		10.4		[153]
^{15}N	^4S	2.00171	14.63		[154]
^{14}N	$^2D_{5/2}$	1.20036	114.69	-0.4	[155]
^{14}N	$^2D_{3/2}$	0.79949	65.3	-0.3	[155]
^{31}P	$^4S_{3/2}$	2.0020	55.15	0.0	[156]
^{75}As	$^4S_{3/2}$	1.9983	66.2	0.13	[156]
^{121}Sb	$^4S_{3/2}$	1.9708	298.6	0.86	[156]
^{123}Sb	$^4S_{3/2}$	1.9708	162.3	1.3	[156]
^{16}O	$^3P_2$	1.500921			[157]
^{16}O	$^3P_1$	1.500986			[157]
^{17}O	$^3P_2$		-218.569	$-10.44\,(-126)$	[158]
^{17}O	$^3P_1$		-4.74	$5.20\,(-92)$	[158]
^{32}S	$^3P_2$	1.500541			[159]
^{32}S	$^3P_1$	1.501029			[159]
^{77}Se	$^3P_2$	1.491414	545.7		[160, 161]
^{77}Se	$^3P_1$	1.50082			[161]
^{123}Te	$^3P_2$	1.461813	-834		[161]
^{125}Te	$^3P_2$	1.461813	$-1,006$		[161]
^{19}F	$^2P_{3/2}$	1.333861	2,009.9	(-446)	[162]
^{19}F	$^2P_{1/2}$	0.6656117	10,244.21		[151, 163]
^{37}Cl	$^2P_{3/2}$	1.3339275	205.288	55.35	[164]
^{35}Cl	$^2P_{3/2}$	1.333924	170.686	43.26	[164, 165]
^{37}Cl	$^2P_{1/2}$				[166]
^{35}Cl	$^2P_{1/2}$		1,037.192		[167]
^{79}Br	$^2P_{3/2}$	1.333921	884.81	-384.878	[164, 168]
^{81}Br	$^2P_{3/2}$	1.333821	953.770	-321.516	[164, 168]
^{79}Br	$^2P_{1/2}$	0.66549	5,332.55		[169]
^{81}Br	$^2P_{1/2}$	0.66566	5,748.15		[169]
^{127}I	$^2P_{3/2}$	1.333995	827.265	1146.36	[170, 171]
^{127}I	$^2P_{1/2}$	0.6654	21,973		[172]

^{a}Parameters in parentheses are off-diagonal matrix elements between J and $J - 1$.

Since gas-phase systems are isotropic

$$\mathscr{H} = g\mu_B \mathbf{B} \cdot \hat{\mathbf{S}} + hA\hat{\mathbf{S}} \cdot \hat{\mathbf{I}} + g_N \mu_N \mathbf{B} \cdot \hat{\mathbf{I}} + \text{higher order} \tag{94}$$

where higher order terms may indicate quadrupole or octopole interactions. In many cases, these higher order terms and the nuclear Zeeman energy $g_N \mu_N B \cdot I$ can be neglected. Often, the nuclear Zeeman terms only produce a shift in the energy levels and do not contribute to the structure of the spectrum.

At $B = 0$, the major term in the Hamiltonian is the hyperfine interaction. The value of F is 1 or 0. For $F = 1$, there are three degenerate states, $M_F = 1, 0,$ and

−1. The term $\hat{\mathbf{S}} \cdot \hat{\mathbf{I}}$ is determined from the operators $\hat{\mathbf{F}}$, $\hat{\mathbf{S}}$, and $\hat{\mathbf{I}}$

$$\hat{\mathbf{S}} \cdot \hat{\mathbf{I}} = (\hat{\mathbf{F}}^2 - \hat{\mathbf{S}}^2 - \hat{\mathbf{I}}^2)/2 \tag{95}$$

where the eigenvalue of $\hat{\mathbf{X}}^2$ is $X \cdot (X + 1)$. Thus, the energy of $F = 1$ is equal to $hA/4$, and the energy of $F = 0$ is $-3hA/4$. The hyperfine interaction is truly a splitting, because the sum of the products of these energies multiplied by the corresponding degeneracy is zero. As shown in Figure 4.9, when a small magnetic field is applied, the energy of $F = 0$ does not change, but the energy of $F = 1$ initially breaks into three levels—one with no change, one negative, and one positive relative to the zero-field value.

The most important case for many atoms and diatomic molecules is the intermediate field. In this region, the Hamiltonian must be solved. It is generally easiest to use the basis set $|S, I, M_S, M_I\rangle$ and to compute the energies and wave functions as linear functions of this set. In standard ESR spectrometers, the field is changed while the frequency is held constant; hence the complete Hamiltonian must be solved for each field. These equations were solved by Breit and Rabi [173], and extensions to different atoms are presented in many of the references in Table 4.11. The procedure is well described in textbooks such as Chapter 2 of [11] and Appendix C of [15]. It is important to remember that the factor $\hat{\mathbf{S}} \cdot \hat{\mathbf{I}}$ can be written

$$\hat{\mathbf{S}} \cdot \hat{\mathbf{I}} = (\hat{S}_x \hat{I}_x + \hat{S}_y \hat{I}_y) + \hat{S}_z \hat{I}_z \tag{96}$$

and the terms $\hat{S}_x$, $\hat{I}_x$, $\hat{S}_y$, and $\hat{I}_y$ couple off-diagonal elements of the matrix. As a result, the energy-level diagram shows curvature of various levels as a function of applied magnetic field.

A similar situation exists for several condensed-phase radicals such as IF_6, where the iodine hfs is very large, $L = 0$ and $S = \frac{1}{2}$. In this case, intermediate field calculations must be carried out. A series of calculations for such species was reported [174, 175]. These are very closely related to the Breit–Rabi equations [173], but apply to more complicated nuclear spin systems.

Oxygen and other Group VI atoms have $P\,(L = 1)$ ground states with two unpaired electrons $(S = 1)$. The major isotopes of oxygen and sulfur possess no nuclear spin. Thus, for these species we only need to consider the coupling of spin and orbital angular momentum. Since neither spin nor orbital angular momentum dominates in the Zeeman energy dependence, these states are best represented by the quantum number J, where the range of possible values was given in (44). As a result, there are three possible states: $J = 2, J = 1$, and $J = 0$. For oxygen, these energies are, respectively, 0, 4750, and 6790 GHz relative to the ground state as shown in Figure 4.29. For the state $J = 1$ and 2, the Landé g-factor is approximately 1.5, but the $J = 0$ state is expected to be diamagnetic. Even though there is orbital angular momentum and electron spin, the angular momenta of the orbit and the spin act in opposition with the consequence that $g = 0$. Thus, the oxygen spectrum is expected to consist of one line that is the sum of the spectra exhibited by the $^3P_2$ and $^3P_1$ states. However, the spectra exhibit six lines as shown in Figure 4.30.

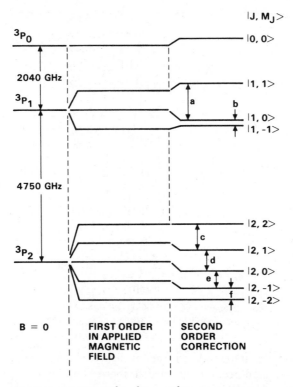

Figure 4.29 Energy-level diagram for the $^3P_2$, $^3P_1$, and $^3P_0$ states of the oxygen atom. The energy levels of the $^3P_1$ and $^3P_0$ states are, respectively, 4750 GHz (158 cm^{-1}) and 6790 GHz (226 cm^{-1}) above the ground $^3P_2$ state. The magnetic Zeeman energy is not drawn on the same scale as the fine structure splittings; also the ESR transitions (↔) are not drawn to scale but represent the sequence that appears in the spectrum of Figure 4.30.

In fact, because of second-order effects described earlier in Section 2.2, the g-factors differ from the calculated Landé g-factor (1.50116, including the relativistic effect of the electron) and are not exactly equal for both states. In addition, rather than one line for each state, there are two lines for the $^3P_1$ and four lines for the $^3P_2$ states, which indicates there is a small perturbation that lifts the degeneracy of each level. Actually, the applied magnetic field tends to decouple L and S [176]. The L and S states of the oxygen atom can be viewed in much the same way as the S and I states of the hydrogen atom (Figure 4.9) except the divergence of the relative Zeeman energies of L and S is so much less rapid that extremely large magnetic fields are required to decouple completely these magnetic states.

For isotopes that exhibit large nuclear hyperfine splittings, a quantum number F is redefined, as

$$F = J + I, \quad J + I - 1, \quad \ldots, \quad J - I \tag{97}$$

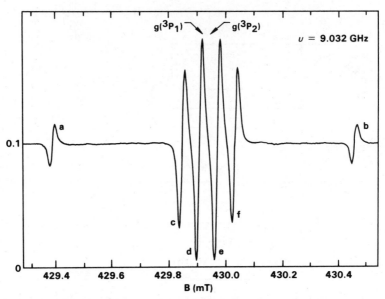

Figure 4.30 Electron spin resonance spectrum of oxygen atoms observed at approximately 9 GHz. The line designations, a–f correspond to the transitions in Figure 4.29.

This may be treated in the same manner as the hydrogen atom described earlier, but if extremely high accuracy is needed, the decoupling of L and S must be considered also. A well-characterized case is the fluorine atom, where $L = 1$, $S = \frac{1}{2}$, and $I = \frac{1}{2}$. Again, there are two P states: $J = \frac{3}{2}$ and $J = \frac{1}{2}$. For the $J = \frac{3}{2}$ state, $F = 2$ or 1; for the $J = \frac{1}{2}$ state, $F = 1$ or 0. The energy levels of the fluorine atom are shown in Figure 4.31. Transitions observed at 9.0 GHz are indicated by solid lines. Some transitions that are not allowed between states $|J, I, M_J, M_I\rangle$ are allowed between states of quantum number $|F, M_F\rangle$. Thus, in the intermediate field case, where I and J are almost decoupled, some weakly allowed transitions occur. These are marked with dotted lines in Figure 4.31.

The $^2P_{1/2}$ state exhibits a 10.2-GHz hyperfine splitting. As a result, at low fields and at very high fields, the transition cannot be observed using a 9-GHz spectrometer. However, at intermediate fields there is a region in which microwave frequencies smaller than the hyperfine splitting can be used.

2.3.2   Diatomic Gas-Phase Radicals

Diatomic and polyatomic radicals in the gas phase exhibit more complicated ESR behavior than atoms because in addition to the angular momentum from the orbit and electron spin, the angular momentum from the rotational level also couples to the total angular momentum. This is discussed in much greater detail by Carrington [177]. For molecules, the orbital angular momentum is designated by Σ for $L = 0$, Π for $L = 1$, Δ for $L = 2$, and so on. As with atoms, the symbol for the electronic state is $^{2S+1}L_J$, where L is now the Greek-letter equivalent and J includes the rotational quantum number R.

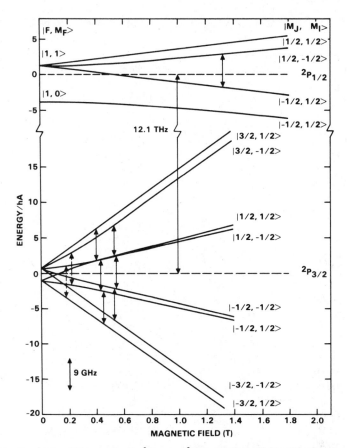

Figure 4.31   Energy-level diagram for the $^2P_{1/2}$ and $^2P_{3/2}$ and P manifolds of the flourine atom. The $^2P_{1/2}$ state is 12.1 THz (404 cm^{-1}) above the $^2P_{3/2}$ ground state. Transitions observed at 9 GHz, ($\leftrightarrow$): strong transitions, (———); weak transitions, (–––).

Diatomic molecules that are detected by ESR include $^1\Delta$, $^2\Pi$, $^3\Sigma$, and $^3\Pi$ states. Rather than discuss these in detail, several examples are given. Table 4.12 lists various diatomic molecules and their reported electronic states.

As in the atomic case of Russell–Saunders coupling between L and S, the various molecular angular momenta may couple and are better described by additional quantum numbers. The pertinent couplings are represented as

$$J = L + S + R \tag{98a}$$

$$= N + S \tag{98b}$$

$$N = R + L \tag{98c}$$

$$F = L + S + I \tag{98d}$$

where international practice introduces a quantity N to represent the sum of rotational and orbital momenta. When $R = 0$, (98) reduces to the atomic case.

Table 4.12 Diatomic Molecules and Electronic States Detected by Electron Spin Resonance

Molecules	States	References	States	References
$^{16}O_2$	$^3\Sigma$	[178–180]	$^1\Delta$	[181, 182]
	$^3\Sigma, v = 1$	[183]		
$^{16}O^{17}O$	$^3\Sigma$	[184]		
$^{32}S^{16}O$	$^3\Sigma, v = 0, 1$	[185–187]	$^1\Delta$	[188]
$^{32}S^{16}O$	$^3\Sigma, v = 0$–6	[189]		
$^{33}S^{16}O$	$^3\Sigma$	[190]	$^1\Delta$	[191]
$^{32}S_2$	$^3\Sigma$	[192]		
$^{76}Se^{16}O$	$^3\Sigma$	[193]	$^1\Delta$	[193]
$^{14}B^{19}F$	$^3\Sigma$	[194]	$^1\Delta$	[194]
$^{16}O^1H$	$^2\Pi_{3/2}, v = 1$–4	[195–197]	$^2\Pi$	[198]
$^{16}O^2H$	$^2\Pi_{3/2}$	[198]		
$^{17}O^1H, {}^{17}O^2H$	$^2\Pi_{3/2}$	[199]		
$^{32}S^1H$	$^2\Pi_{3/2}$	[200–202]		
$^{32}S^2H$	$^2\Pi_{3/2}$	[201]		
$^{33}S^1H$	$^2\Pi_{3/2}$	[190]		
$^{76}Se^1H$	$^2\Pi_{3/2}$	[203, 204]		
$^{76}Se^2H$	$^2\Pi_{3/2}$	[204]		
TeH	$^2\Pi_{3/2}$	[203]		
$^{14}N^{16}O, {}^{15}N^{16}O$	$^2\Pi_{3/2}$	[205]		
$^{14}N^{17}O, {}^{14}N^{18}O$	$^2\Pi_{3/2}$	[206]		
$^{15}N^{17}O, {}^{15}N^{18}O$	$^2\Pi_{3/2}$	[207]		
$^{14}N^{32}S$	$^2\Pi_{3/2}$	[208, 209]		
$^{32}S^{19}F, {}^{76}Se^{19}F$	$^2\Pi_{3/2}$	[210]		
OCl	$^2\Pi_{3/2}$	[211, 212]		
OBr, OI	$^2\Pi_{3/2}$	[213, 214]		
SeF	$^2\Pi_{3/2}$	[214]		

2.3.3 Small Molecules in the Gas Phase

A few important concepts must be explained before elaborating some of the molecular states that were examined by ESR. There are three major coupling schemes that form the basis sets of angular momentum. While none of these is an exact solution of the Hamiltonian, approximate solutions exist to a high degree of accuracy. These schemes were first described by Hund [215], and the three that are of greatest concern are called Hund's cases (a), (c), and (b).

For Hund's case (a), R, L, and S are weakly coupled so that J represents the molecule. Hund's case (c) applies to the situation where L and S are strongly coupled to produce a composite momentum ($L + S$), which is in turn weakly coupled to R so that ($L + S$) and R best represent the molecule. Hund's case (b) represents strong coupling between R and L to form N with subsequent weak coupling to S. Such situations can arise for $L = 0$ or for weak spin–orbit coupling.

SINGLET STATES

Oxygen $^1\Delta_g$ is an example of Hund's case (a) where there is weak coupling between L and R so that the rotational Hamiltonian is given by

$$\mathscr{H}_{rot} = B\mathbf{J}^2 = B(\mathbf{R} - \mathbf{L})^2 \tag{99a}$$

where $S = 0$ and B is the rotational constant. This can be expanded into

$$\mathscr{H}_{rot} = B[R(R + 1) - (L^2 - 2M_R M_L)] + B(L_x^2 + L_y^2) - 2B(R_x L_x + R_y L_y) \tag{99b}$$

The first term establishes the energy level, while the second term is a constant independent of R. The third term splits each rotational level. This effect is called Λ doubling because it couples different L and R states. (Often orbital angular momentum eigenvalues are assigned the value of Λ.) In some cases the separation is small and can be ignored. Often it is significant. In general, it is most significant when B is large and increases with increasing R.

Oxygen ($^1\Delta$) was observed in the $J = 2$ (lowest) and $J = 3$ states. The spectrum for $J = 2$ is shown in Figure 4.32. From first-order theory, the g-factor

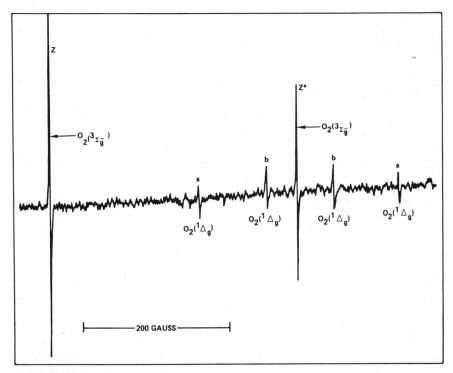

Figure 4.32 Electron spin resonance spectrum showing four lines of O_2 ($^1\Delta_g$) in the $J = 2$ state. Two lines from O_2 in the ground state ($^3\Sigma_g^-$) are also indicated. The intensities of lines designated a and b are in the ratio $2:3$.

is expected to be approximately

$$g_J = L^2/[J(J + 1)] \tag{100}$$

For $J = 2$, 3, and 4 the g_J values are, respectively, $\frac{2}{3}$, $\frac{1}{3}$, and $\frac{1}{5}$. Thus, the required resonance field becomes larger as J increases.

From first-order considerations, the $O_2(^1\Delta)$ state spectrum should consist of one line that is a superposition of transitions between each of the five adjacent M_R levels. However, as shown in Figure 4.32, there are actually four transitions, each separated by 10 mT. This results from second-order mixing of the rotational levels by the magnetic field.

In addition, there is also an interaction of the field with the rotational magnetic moment

$$\mathcal{H} = -g_r\mu_B\mathbf{B}(\mathbf{J} - \mathbf{L}) \tag{101}$$

which causes a shift in the overall spectrum.

Most heteronuclear diatomic molecules exhibit a significant electric dipole moment; as a result these transitions are far more intense than the magnetic transitions of the species described earlier. An example of such a species is NF ($^1\Delta$), which also exhibits hyperfine couplings. In this case, rather than the operator being $\hat{\mathbf{S}}\cdot\hat{\mathbf{I}}$, the hyperfine interaction comes from the orbit and nuclear spin interaction $\hat{\mathbf{L}}\cdot\hat{\mathbf{I}}$. Some of the results are summarized in Table 4.13, and references for the specific species are in Table 4.12.

TRIPLET STATES

By far, the most important triplet-state diatomic molecule that was studied is $O_2(^3\Sigma_g^-)$. Electron spin resonance is one of the few techniques that is used for quantitative determination of this material in the gas phase, especially when it is mixed with other species such as $O_2(^1\Delta)$ and oxygen atoms. Various spectral lines are also used to calibrate the ESR spectrometer sensitivity for quantitative analysis of gas-phase species. The first correct analysis of the oxygen ESR

Table 4.13 Results of ESR Spectra of Molecules in the $^1\Delta$ State

Molecule	State	g_R	g_L	g_r	$A(MHz)$
$^{16}O_2$	$J = 2$	0.666630	0.999866	-1.234×10^{-4}	
	$J = 3$	0.333400		-1.70×10^{-4}	
$^{16}O^{17}O$	$J = 2$	0.66630			^{17}O -424
$^{32}S^{16}O$	$J = 2-4$	0.665	0.99979	-1.6×10^{-4}	
$^{14}N^{19}F$	$J = 2$		1.0001	-1×10^{-4}	^{14}N 109.9
					^{19}F 758.0
					$Q(N)$ 4.1[a]

[a]Quadrupole coupling.

spectra was carried out by Tinkham and Strandberg [178]. The Hamiltonian was derived on the basis of weak coupling between the various angular momentum components. However, both the spin–spin interaction and the spin–rotation interaction are very important. The Hamiltonian is given by

$$\mathscr{H} = \mathscr{H}_{SS} + \mathscr{H}_{rot} + \mathscr{H}_{SO} + \mathscr{H}_{SR} + \mathscr{H}_Z \qquad (102)$$

The specific terms are described below. The spin–spin Hamiltonian is

$$\mathscr{H}_{SS} = \tfrac{2}{3}(\lambda_0 + \lambda_1 |R|)(3S_z - \hat{\mathbf{S}}^2) \qquad (103)$$

This represents the zero-field splitting of the rotational levels by the spin–spin interaction. The rotational Hamiltonian is

$$\mathscr{H}_{rot} = h(B_0 + B_1 |R(R + 1)|)\hat{\mathbf{R}}'^2 \qquad (104)$$

where B_0 and B_1 are, in this context, the rotational constant and the centrifugal distortion coefficient, respectively, and allowance is made in the operator $\hat{\mathbf{R}}'$ that R is not necessarily a good quantum number. The eigenvalues of $\hat{\mathbf{R}}'$ are determined by solving the Hamiltonian matrix. The spin–orbit and spin–rotational terms can be combined as

$$\mathscr{H}_{SO} + \mathscr{H}_{SR} = h\mu\hat{\mathbf{R}} \cdot \hat{\mathbf{S}} \qquad (105)$$

where μ is the spin–rotation coefficient in this context. The Zeeman energy is the usual expression $g\mu_B BS$. Solving these equations is relatively involved. Each level R in the basis set for $^{16}O^{16}O$ can only be odd because the total oxygen ground-state wave function can only be symmetric (Bose statistics). Because $S = 1$ and $L = 0$,

$$J = R, R \pm 1 \qquad (106)$$

Values of the parameters for $^3\Sigma_g^-$ oxygen are given in Table 4.14. The allowed transitions are

$$\langle R, J = R \pm 1, M_J| \rightarrow \langle R, J = R \pm 1, M_J \pm 1| \qquad (\Delta J = 0, \Delta M_J = \pm 1)$$
$$(107a)$$

$$\langle R, J = R \pm 1, M_J| \rightarrow \langle R, J = R \pm 1, M_J \pm 1| \qquad (\Delta J = \pm 2, \Delta M_J = \pm 1)$$
$$(107b)$$

$$\langle R, J = R, M_J = -J| \rightarrow \langle R, J = R, M_J = -J \pm 1| \qquad (\Delta J = 0, \Delta M_J = \pm 1)$$
$$(107c)$$

These are shown for $R = 3$ in Figure 4.33. Each R level is associated with three J levels. As seen in the figure, the applied fields that are normally available are not strong enough to decouple completely the spin and rotation. Within each J level, there are up to $2M_J$ transitions, and there are also transitions allowed between $J = R - 1$ and $J = R + 1$. At about 300 K, the first 20 or so rotational levels are populated. Therefore it is easy to appreciate that the spectrum

Table 4.14   Parameters of Ground-State Molecular Oxygen

g	g-Factor	2.002025
g_r	Rotational g-factor	−0.000126
g_L	Orbital g-factor	−0.002813
B_0	Rotational constant (GHz)	43.100518
B_1	Rotational constant (GHz)	−0.000145
λ_0	Spin–spin coupling (GHz)	59.501342
λ_1	Spin–spin coupling (GHz)	0.000058
μ	Spin–rotation coupling (GHz)	−0.252586

observed at about 9 GHz over the field range from 0.1 to 1.1 T is in fact a very complex pattern consisting of several hundred absorption lines.

DOUBLET STATES

As shown in Table 4.12, many doublet-state species were studied. Of these, the most important is OH· and its isotopically substituted analogues. The Hamiltonian for this molecule must include spin–orbit, spin–rotational, orbit–rotation (including Λ doubling), rotation and centrifugal distortion, and Zeeman interactions. The OH· radicals exhibit large rotational splittings (556 GHz for $^{16}O^1H\cdot$ and 296 GHz for $^{16}O^2H\cdot$). The spin–orbit coupling constant, 4191 GHz, is beyond any possible microwave transition and puts the different levels out of range even at high fields. Because the molecule is so asymmetric, the Λ-doubling term is large and increases as J increases; it is about 1600 MHz at $(L + S) = \frac{3}{2}$ $(R = 0)$. The situation is analogous to the case of the fluorine atom, so that the states are split into $^2\Pi_{3/2}$ and $^2\Pi_{1/2}$. Because of the different perturbations, various states of R, S, and L are mixed.

Detailed discussion is beyond the scope of this chapter, but it is available in [177]. However this molecule does help to illustrate one important concept. Because of the dipole moment, transitions are both electrical and magnetic field allowed. The difference between them is that magnetic transitions are allowed only between states of the same parity while electric field transitions are allowed between states of opposite parity. Each Λ doublet contains one state of each parity. If they are resolved, the electric-field-induced transitions exhibit slightly different spacings. For more information on this, the reader is directed to the appropriate references such as [177, 195–198].

Nitric oxide is also of interest because of its involvement in atmospheric chemistry. In the case of NO, the $^2\Pi_{1/2}$ state is essentially diamagnetic and is about 3600 GHz above the ground state. Only the $^2\Pi_{3/2}$ state is observed. The g-factor is given by

$$g_J = \frac{(L + 2S)(L + S)}{J(J + 1)} \tag{108}$$

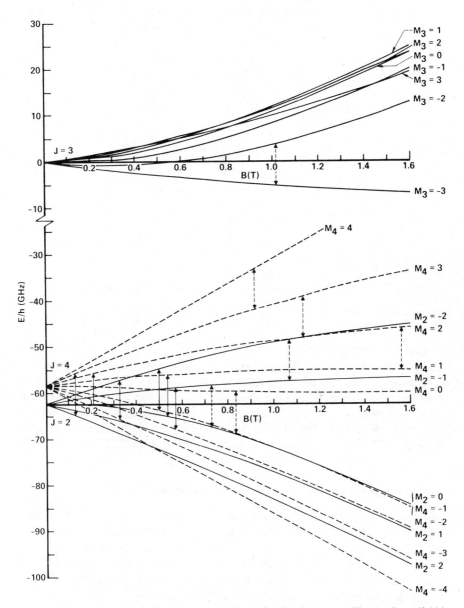

Figure 4.33 Energy-level diagram for the $R = 3$ level of molecular oxygen. The $J = 2$ manifold is shown with the lower solid lines, the $J = 3$ manifold with the upper solid lines, and the $J = 4$ manifold with the broken lines. Electron spin resonance transitions observed at 9 GHz are shown as double arrows; solid vertical arrows represent $\Delta J = 1$ transitions, while broken vertical arrows represent $\Delta J = 0$ transitions; M_J values are shown at the right of the diagram.

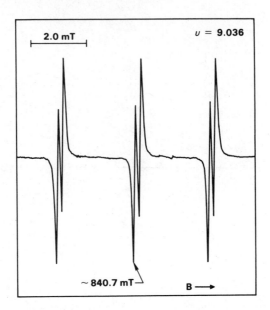

Figure 4.34   High field one-third of the ESR spectrum of NO ($^2\Pi_{3/2}$) in the $J = \frac{3}{2}$ state.

so that the g-factor is 0.8 for $J = \frac{3}{2}$, 0.34 for $J = \frac{5}{2}$, 0.19 for $J = \frac{7}{2}$, and so on. Thus, the resonance fields increase rapidly with J. Again, the primary transitions arise from interaction of the permanent dipole with the electric field of the microwave radiation. The entire spectrum spans about 25 mT at 9 GHz and consists of three sets of six lines, with each set exhibiting an intensity of 3:4:3. In each set, such as the high-field third shown in Figure 4.34, there are three pairs of lines. The hyperfine structure is caused by ^{14}N hfsc of 2.722 mT. The small splitting of about 0.1 mT is caused by Λ doubling.

2.4   Quantitative Measurements by Electron Spin Resonance

Determination of the absolute concentration of spin units in a sample is not a trivial matter. Many of the problems associated with quantitative measurements by ESR were summarized in reviews [216–220]. The difficulties arise from improper processing of the ESR data; instrumental artifacts associated with ESR, particularly over large field scans; and selecting suitable standards for both magnetic field and spin concentration calibration.

2.4.1   Quantitative Measurement Theory

The ESR signal, which is proportional to the microwave susceptibility of the sample, is generally recorded as the *first derivative* of absorption. However it is the *integral* of absorption that is proportional to the concentration of spin units. The principal relationship between the ESR signal and radical concentration for dilute gas-phase species was developed by Westenberg [221, 222], starting from the general definition of power absorption. The power absorption is related to the imaginary part of the susceptibility shown in (109). In general, χ''_{ij} is the frequency dependence of the energy absorption caused by the transition between states i and j, according to

$$\chi''_{ij} = (n_{ij}/h)\mu^2_{ij}f(v - v_0) \tag{109}$$

where n_{ij} is the population difference between states i and j, μ_{ij} is the transition probability, and $f(v - v_0)$ is the line-shape function. Under the conditions that $kT \gg hv$, the concentration of the species is related to the double integral of the first-derivative ESR spectral lines as

$$C = \frac{2kT}{hv_0\mu_B} \times \sum_{ij} \frac{g_{\text{eff}}\mu^2_B Z}{2|\mu_{ij}|^2 \exp(-E_{J,M_J}/kT)} \times \frac{K_I}{P^{1/2}AB_m} \cdot \int_0^\infty \int_0^B S(B')\, dB'\, dB \tag{110}$$

The terms of (110) are arranged in three groups. The first contains fundamental constants and terms that relate to specific experimental conditions; this includes the population difference in the ESR transition of $\Delta E = hv_0$. The second contains terms that describe the magnetic properties of the absorbing species. The third contains instrumental parameters and, finally, the double integral of the ESR signal. The specific parameters are T, the temperature; v_0, the microwave frequency; g_{eff}, the effective g-factor, which is described later; Z, the statistical probability of finding a molecule in the specific electronic and magnetic state (E_{J,M_J}); K_I, an instrumental constant (see following); P, the microwave power; A, the signal amplification; and B_m, the magnetic field modulation amplitude.

Equation (110) is generally applicable to all paramagnetic resonance systems. It was derived for gases [221, 222] and extended to condensed phases [216, 217]. Some results follow. However, the terms in the second group need some explanation. The term g_{eff} is used to correct for the fact that in ESR experiments the magnetic field is scanned while the microwave frequency is held constant. It is given by

$$g_{\text{eff}} = h/\mu_B \cdot dv_0/dB \tag{111}$$

For most species, g_{eff} is merely the resonance condition $hv_0/\mu_B B$. But for species with large hyperfine or zero-field splittings, such as shown for F and O_2 in Figures 4.31 and 4.32, this can deviate significantly from the actual g-factor. Westenberg [220] tabulated these for several gases, and a detailed table of g_{eff} for O_2 was reported [179].

The term Z in (110) is the partition function. This must include nuclear degeneracy, electron and orbit multiplicity, quantized rotation for gases, and thermally accessible electronic states. Usually, only one absorption line is used for analysis. For example, where there are three thermally accessible electronic states, as with $O\,(^3P)$, each must be included in Z. If the four $^3P_2$ lines are broadened such that they cannot be separated, then the summation in (110) must include all four transitions. Similarly, any nuclear-spin degeneracy must be included in Z. When there are many nuclei, the nuclear partition function Z_N is given by

$$Z_N = \prod_k (2I_k + 1)^{n_k} \tag{112}$$

where I_k is the individual nuclear spin of equivalent set k, and n_k is the number of nuclei in that set. As an example, consider that the center line of the spectrum of a set of four equivalent $I = \frac{1}{2}$ nuclei is used for analysis. This line is sixfold degenerate ($1:4:6:4:1$ quintet pattern), while the nuclear partition function Z_N is 16.

For species with $J = S$, such as radicals and transition metals in which the orbital and rotational angular momenta are effectively quenched, the transition probability depends only on the g-factor and spin [217]. When one or more lines of degeneracy D_l are used for analysis, (110) becomes

$$C = \left(\frac{k_B T}{h\nu_0 \mu_B}\right)\left[\frac{1}{g\mu_B \frac{1}{3} S(S + 1)}\right]\left[\frac{\prod_k (2I_k + 1)^{n_k}}{\Sigma_l D_l}\right]\left[\frac{K_I}{P^{1/2} A B_m} \int_0^\infty \int_0^B S(B')\,dB'\,dB\right] \tag{113}$$

This equation is not accurate for very low temperatures or for multiplet-state species with significantly large zero-field splittings where the Boltzmann term must be added to account for population differences.

The foregoing equations are derived on the basis of some important assumptions: (1) a known sample is used to determine the instrumental constant, and the sensitivity over the standard and sample is essentially constant; (2) the Q-factor of the cavity is constant for the sample and standard; (3) the Q-factor remains essentially constant as the resonance condition is swept. The latter condition is generally true for dilute samples, but the use of massive or highly concentrated samples may preclude quantitative measurements [223].

Evaluation of the double integral is most accurately accomplished by computer manipulation of the digitized experimental ESR data. Often it is acceptable to approximate the double integral as the product of the peak-to-peak amplitude of the derivative line and the square of the derivative line width. The validity of this method for making relative estimates of radical concentration is limited to systems that display identical line shapes [224] because the portion of the absorption encompassed by the peak-to-peak width is different for Lorentzian, Gaussian, and even modulation-distorted line shapes [218].

Several experimental factors must be addressed for both absolute and relative concentration measurements. Generally it is straightforward to locate a point sample at the position of maximum spectrometer sensitivity by careful adjustment of the sample position. When a point sample is moved along the axis of a TE_{102} or TE_{01n} cavity, the sensitivity varies as the product of the microwave power density within the cavity and the modulation amplitude, which is also a function of the sample position. For a sample that is positioned along the axis of maximum microwave magnetic field in the cavity (see Section 4.2.3) and has a narrow cross section with respect to the orthogonal axes, the microwave power varies as $\cos^2(\pi z/L)$, where L is the length of the cavity and z is the deviation from the center of the cavity. However, when a sample of finite length is used, it is not satisfactory simply to integrate the sensitivity function over the length of the sample because the field modulation amplitude is generally not uniform over the cavity dimension L. Since this modulation amplitude function is difficult to express analytically, it is easiest to maximize the sensitivity as an empirical function of z.

When a flat sample cell is used in a TE_{102} cavity, as for aqueous or biological tissue samples, the angular orientation of the cell with respect to the microwave field is extremely crucial and difficult to reproduce. It is preferable to leave the cell undisturbed in the cavity and replace liquid samples by draining and refilling whenever possible. Of course, since the microwave loss varies from one solvent system to another, it is necessary that sample and standard be prepared identically; otherwise, the effective power incident on the sample is not reproduced.

Even the sample container can perturb the microwave field and make quantitative measurements unreliable. For instance, when a *dual-sample* TE_{104} cavity is used to record spectra of a sample and a reference material simultaneously, it is necessary to recognize that structures such as variable-temperature Dewar jackets surrounding the sample can concentrate the microwave radiation field and thereby cause the value of the *instrumental constant*, K_I[(110)] to be different for the sample and reference.

Quantitative ESR measurements require exacting experimental technique, and the interested reader is urged to consult the appropriate literature, as in [216–220].

2.4.2 Quantitative Paramagnetic Standards

Two types of standards are used in ESR: one to calibrate the spectrometer sensitivity to determine the concentration of material, and the other to calibrate the magnetic field. The review by Chang [225] lists several materials that are employed as reference standards in ESR spectroscopy, and some are mentioned briefly here.

Numerous solid standards are used. Charred dextrose contains about 5×10^{20} spins/cm^3, has a g-factor of $2.0025922 \pm 2.5 \times 10^{-6}$, and has a moderately narrow line width of 0.06 mT. Sample preparation is critical [226].

Ions of Cr(III) in Al_2O_3 (ruby) are widely used because the ESR properties are thoroughly studied and documented. It is stable and inert to nearly all reagents. Annealed, strainfree samples may be available from the National

Institute of Standards and Technology (formerly, National Bureau of Standards) [227].

1,1-Diphenyl-2-picrylhydrazyl (DPPH) is probably the most widely used ESR standard. In the solid it exhibits one exchange-narrowed line at $g = 2.0036 \pm 0.0003$. The spin concentration decreases with aging so that quantitative measurements require use of freshly crystallized material that is then stored in the cold with the exclusion of air and light. In addition, the method of recrystallization affects the Curie temperature of the material, so it must be used with care [216].

Other quantitative standards include copper sulfate or manganese sulfate. These exhibit broad lines, which can produce different integrated intensities if the field sweep width is changed. The variable degree of hydration affects the metal concentration so the composition must be determined independently. In addition, the Curie temperatures differ from zero.

A wide variety of solutions can be used as standards, but care must be exercised that the sample cell is not moved when standards are replaced by unknowns. Convenient sources of paramagnetic ions are $CuSO_4$, $MnSO_4$, $MnCl_2$, and $CuCl_2$. Potassium peroxylamine disulfonate exhibits narrow lines, and the concentration can be determined optically [228]. The molar absorptivities are 1690 at 248 nm and 2.08 at 545 nm. A solution of 0.05 M carbonate retards decomposition but only for several days.

For gases, only oxygen has been used extensively. This is detailed in [178, 179, 220–222].

2.4.3   Magnetic Field Standards

A variety of magnetic field standards have been used, depending on the span that needs to be calibrated and the degree of accuracy needed.

Extensive work is published on ruby [227]. This material has several advantages, such as small size, which allows for permanent placement of a reference sample inside the cavity. The ESR transitions are orientation dependent, and the absorptions falling at approximately 0.2 and 0.54 T (at X band) conveniently bracket the $g = 2$ region.

In an octahedral lattice Mn(II) provides a convenient reference because the six lines are sharp, ranging from 0.1 to 0.15 mT in width. We have used two reference materials. The first is a sample provided by Miner and co-workers [229]. The material was alleged to be forsterite; however, as shown in Table 4.15 our analysis suggests it is CaO or MgO. The second standard was obtained by grinding and washing calcite from deposits on the beach and then heating it to about 800°C. This was attributed to Mn(II) in CaO and is also shown in Table 4.15. Both data are consistent with results given for CaO, although they differ slightly from the measurements of Rubio and co-workers [230] who reported $a = 8.655 \pm 0.020$ mT and $g = 2.0010 \pm 0.0001$.

Peroxylamine disulfonate, usually used as solutions of the potassium salt, exhibits a nitrogen splitting of 1.305 ± 0.01 mT and $g = 2.00550 \pm 0.00005$ [228, 231]. Freshly prepared material should be standardized by iodometric titration for quantitative measurements.

Table 4.15   Line Separations of Manganese(II) Measured and Calculated to Second Order (in mT)

| M_I | Sample of [229][a] | | CaO | |
	Measured[b]	Calculated[c]	Measured[b]	Calculated[d]
$+\frac{5}{2}$				
	8.234	8.2329	8.188	8.178
$+\frac{3}{2}$				
	8.451	8.4540	8.399	8.399
$+\frac{1}{2}$				
	8.682	8.6750	8.628	8.619
$-\frac{1}{2}$				
	8.990	8.996	8.843	8.853
$-\frac{3}{2}$				
	9.112	9.1171	9.057	9.068
$-\frac{5}{2}$				

[a]In [229] sample was assumed to be forsterite.
[b]Unpublished data, this laboratory.
[c]Calculated from $a = 8.675 \pm 0.005$ mT, $g = 2.00095 \pm 0.00010$, and $\nu = 9.535$ GHz.
[d]Calculated from $a = 8.627 \pm 0.003$ mT, $g = 2.00120 \pm 0.00006$.

2.5   Time-Dependent Processes That Affect Line Widths

In addition to the dipolar (electron–electron and electron–nuclear) interactions that act to broaden ESR lines (Section 1.9.2) there are various other physical and chemical processes that can introduce additional broadening. Often these processes produce unequal line widths within a single spectrum, and such observations can be used to deduce kinetic and thermodynamic information about the system.

Such effects were first observed by NMR spectroscopy, and the analysis developed at that time is extended readily to ESR. For example, the classic paper by Gutowsky and Holm [232] reveals that the methyl groups of N,N-dimethylformamide (DMF) undergo NMR absorption at different magnetic fields because of differences in chemical shielding caused by the dissimilar environments of the groups located cis or trans relative to the carbonyl group. At low temperature, two distinct absorptions are observed, but as the temperature is raised, the absorptions progressively broaden, merge, and eventually coalesce into a single peak because the rotation about the amide C—N bond becomes so rapid that the different methyl environments are averaged.

A similar effect would be observed if a hypothetical radical existed in two different forms, A and B, having different g-factors so that resonance occurs at two different field values separated by δB_0. If the mean lifetimes of the A and B forms are, respectively, τ_A and τ_B, then the mean lifetime of the system is

$$\tau = \tau_A \tau_B / (\tau_A + \tau_B) \tag{114}$$

For equal fractional populations of the A and B forms, as illustrated in Figure 4.35

$$p_A = \tau_A/(\tau_A + \tau_B) \tag{115a}$$

$$p_B = \tau_B/(\tau_A + \tau_B) \tag{115b}$$

and it follows that $\tau_A = \tau_B = 2\tau$.

As exchange occurs, the lines initially decrease in amplitude and also broaden. (The integral of the absorption can be approximated as the product of the peak-to-peak amplitude and the square of the peak-to-peak line width of the derivative display. Consequently, small changes in line width produce large effects on amplitude even though the number of spins is constant.) As the rate increases, the lines broaden more and the separation diminishes from δB_0 to δB_e. This is known as *exchange broadening*. Eventually $\delta B_e \rightarrow 0$ and the lines coalesce. At even higher exchange rates the line sharpens to the *exchange-narrowed* limit. While we are unaware of any real system where a radical

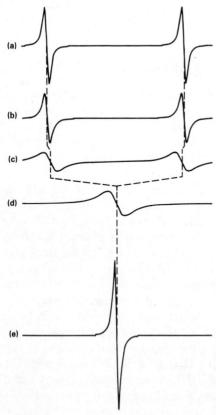

Figure 4.35 Computed spectra for a free-radical system that can exist in two different forms. Spectrum (*a*) represents no exchange, while (*b*) and (*c*) represent increasingly rapid exchange. With sufficiently fast exchange rate, the lines coalesce into the single line shown in spectrum (*d*), which is the *fast exchange limit*. Spectrum (*e*) is an idealized spectrum that will be observed in the limit of infinitely fast exchange. The dashed lines suggest how the line separations decrease as the exchange rate increases.

interconverts between two equally probable g-factors, there are innumerable cases of line sharpening that results from the averaging of $g_{\parallel}$ and $g_{\perp}$ to produce $g_{isotropic}$. The figure indicates the relations between the line width with exchange (Γ), the intrinsic line width in the absence of exchange (Γ_0), the mean lifetimes of the species ($\tau_A = \tau_B = 2\tau$), and the line separations with and without exchange (δB_e and δB_0, respectively) [233]. Note that the exchange-narrowed line occurs at the average position between the values of $B_{0(A)}$ and $B_{0(B)}$ observed in the absence of exchange. If the two forms have unequal probabilities, the exchange-narrowed line will fall at a position given by the weighted mean

$$\langle B \rangle_{av} = (p_A B_A + p_B B_B)/(p_A + p_B) \tag{116}$$

Many different types of processes can occur to broaden ESR lines from their intrinsic limit, Γ_0. Although the nuclear spin system remains unchanged during the interconversion process, hyperfine structure complicates the appearance of the spectrum. Some representative time-dependent processes that can cause changes in the line width include:

1. Alkyl group rotation adjacent to a p or π orbital.
2. Motion of an alkali metal between energetically similar sites of a radical anion.
3. Interconversion of nuclei between axial and equitorial or cis and trans positions.
4. Electron exchange between a radical ion species or between a radical and its diamagnetic precursor.

2.5.1   Exchange Reactions

In concentrated solutions of free radicals, collisions of the type

$$R^{\cdot \alpha} + R^{\cdot \beta} \rightarrow R^{\cdot \beta} + R^{\cdot \alpha} \tag{117}$$

are known as Heisenberg exchanges. Such reactions introduce broadening of hyperfine lines because it is improbable that the nuclear spin configurations of the reactant and product molecules are identical. This can have the effect of shifting the resonance field for each electron spin to some different line of the ESR hyperfine pattern. For example, consider a certain benzene anion radical in which all the nuclear moments are parallel. Figure 4.12 reveals a 63:64 probability that its unpaired electron will be associated with a different set of nuclear spins after it undergoes electron exchange resulting from collision with another benzene anion. But when a different radical that is associated with the central line of the spectrum undergoes such collision, it has a lower 5:16 probability of shifting to another resonance field and so it does not broaden so dramatically at low exchange rates.

Many such electron-exchange reactions have been observed [234, 235], and the second-order rate constants are about 10^9–10^{10} L/(mol·s), characteristic of a diffusion-controlled reaction. This is readily justified because the wave functions of the reactant and product molecules are essentially identical except for the probable differences in nuclear Zeeman energy. The theory of this process has been described in detail in [236].

Implicit in this discussion is a clear warning for experimentalists: To obtain the best possible resolved hyperfine splitting, the radicals should be observed in dilute solutions to preclude exchange broadening.

2.5.2   Electron-Transfer Reactions

When a radical ion sample contains a quantity of the diamagnetic precursor, there is the possibility of interconversion of the type

$$R^{\cdot\pm} + R^0 \rightleftharpoons R^0 + R^{\cdot\pm} \tag{118}$$

Once again, it is improbable that the product radical has precisely the same nuclear spin configuration as the reactant radical so that the electron resonance is expected to shift to a different hyperfine component of the spectrum. Figure 4.36 shows an example of such broadening as the concentration of neutral species is increased [52].

Although there is no net energy change in such reactions, the second-order rate constants may be significantly slower than the diffusion-controlled values characteristic of electron–electron exchange reactions. For example, with the naphthalene anion in tetrahydrofuran (THF), $k_2 = 5.7 \times 10^7$ L/(mol·s) [235]. Benzophenone and benzophenone ketyl interchange with a similar rate constant but display an additional interesting nuance. In the rapid exchange limit (when the concentration of diamagnetic reactant is greater than 2 mol/dm^3) the proton hyperfine structure is obliterated, but there remains a quartet pattern from the sodium counterion ($I = \frac{3}{2}$), which suggests the mechanism involves transfer of sodium atoms (electron plus ion) rather than electron transfer alone [237].

For sodium naphthalenide in a mixed solvent of dimethoxyethane and THF, Hirota [238] reported spectra that suggest a mixture of two different ion-pair species having slightly different sodium hyperfine splittings. Although the different sodium hfs were not resolved, they were inferred from the unequal broadening of the sodium hfs lines as the temperature was changed. This selective broadening of the outer lines of the sodium quartet ($I = \frac{3}{2}$) relative to the inner lines is consistent with the line-width expression [233]

$$\Gamma = \Gamma_0 + \gamma_e \tau p_A p_B (a_1 - a_2)^2 M_I^2 \tag{119}$$

in that the line width of the outer lines ($M_I = \pm\frac{3}{2}$) is greater than that of the inner lines ($M_I = \pm\frac{1}{2}$) when the nucleus can exist in environments characterized by different coupling constants ($a_1 \neq a_2$). (The other symbols in this expression were defined previously.) In another radical system, the tetracyanoethylene anion was sensitive to solvent and the nature of the counterion; the activation energy of the electron-transfer reaction changed by a factor of 2, depending on

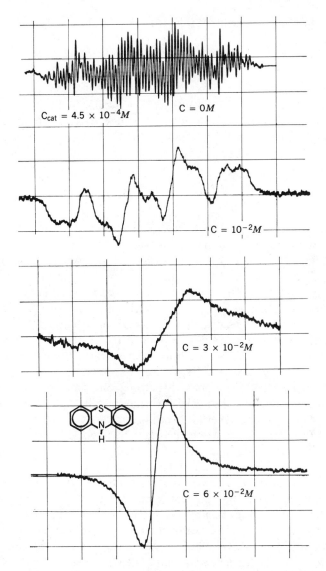

Figure 4.36 Electron spin resonance spectra of phenothiazine cation radicals ($4.5 \times 10^{-4} \, mol/dm^3$) in the presence of increasing concentrations (shown on the figure) of neutral phenothiazine. Reprinted from [52], p. 301, by courtesy of Marcel Dekker, Inc.

the environment [239, 240]. In general, cation radicals exhibit greater electron-transfer rates than anion radicals [241]. This was attributed to reduced repulsion and structural differences.

As shown in Figure 4.36, there are two exchange regions. In the slow exchange region, where hfs are still observed, the broadening varies with the exchange rate and the nuclear spin state of the individual line. In the fast exchange region, the line width ΔB is given by [242]

$$\Delta B = \Delta B_0 + 2.05 \times 10^7 \frac{\Sigma_i d_i B_i^2 / \Sigma_i d_i}{k_{ex} C_R} \tag{120}$$

where ΔB is the Lorentzian line width, ΔB_0 is the line width in the absence of exchange, i represents an individual absorption line, B_i is its resonance-field shift from the center line, d_i is the degeneracy of the line (see Section 1.9.3), k_{ex} is the exchange rate constant for (118), and C_R is the concentration of R in moles per cubic decimeter.

2.5.3 Alternating Line Widths

There is a growing body of examples of free radicals whose spectra show alternately narrow and broad lines. This phenomenon is called the *alternating line width effect* [243] and arises when the radical possesses nuclei that are symmetrically equivalent but whose hfsc are not instantaneously equivalent. Several examples follow. The subject was reviewed in detail [244].

It is necessary first to expand on the earlier concept of equivalent nuclei. Freed and Fraenkel [26] used the phrase *completely equivalent* to characterize nuclei that are symmetrically equivalent and have identical hyperfine splitting constants at any instant. Any interaction that perturbs the hyperfine splitting constants causes each hyperfine splitting constant to change in the same direction. This is called an in-phase hyperfine splitting constant modulation. There are other cases where the nuclei are symmetrically equivalent but physical interactions can cause one hyperfine splitting constant to increase while another undergoes a simultaneous, concerted decrease. Such nuclei are said to be merely *equivalent*, not *completely equivalent*. Interactions that produce this sort of change are said to produce an out-of-phase hyperfine splitting constant modulation. Of course there are yet other cases where nuclei fortuitously exhibit the same hyperfine splitting constant even though they are not related by symmetry. This is called *accidental equivalence*.

To understand processes that can produce out-of-phase modulation of the hyperfine splitting constant, consider the anion radical of dinitrodurene (tetramethyl-3,6-dinitrobenzene). This molecule is so sterically crowded that it is impossible for both nitro groups simultaneously to be coplanar with the benzene ring, a condition that would maximize overlap of the component π electrons and minimize the electronic energy. Rather, when one nitro group moves toward coplanarity, the other one gets squeezed out of the aromatic plane. There is strong evidence that the nitrogen hyperfine splitting constant increases as an aromatic nitro group is forced to twist out of the aromatic plane [76]. Consequently, the motions of the nitro groups of dinitrodurene anion produce instantaneous variations in the nitrogen hyperfine splitting constants that are correlated and out of phase—when one nucleus has a large a_N, the other is small and vice versa. For those radicals in which the nitrogen nuclei have identical nuclear spin states $[(-1, -1), (0, 0), \text{or} (+1, +1)]$, the resonance field is unaffected by the motions. But for those radicals in which the nitrogen spin states are different, the motions produce a shift in the resonance field. Figure 4.37 indicates schematically how these shifts cause the various components of the spectrum to have different line widths. It is noteworthy that the central line of the nitrogen pattern comprises three components—one is sharp and two are quite broad as indicated in Figure 4.37d.

In contrast to the modulation produced by the internal motions of dinitrodurene anion, the spectra of m-dinitrobenzene anion reveal evidence for out-of-phase modulations produced by solvation interactions. When the radicals are generated in nonaqueous solvents containing a small quantity of a polar solvent such as water, there is preferential solvation by the water molecules [245, 246]. But at any given instant, only one nitro group is so solvated, causing a shift in

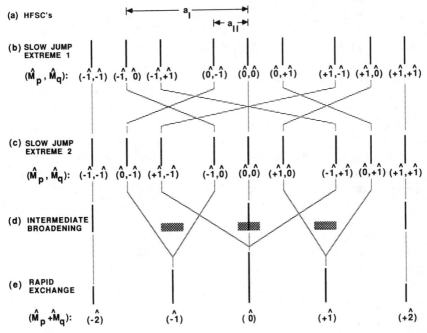

Figure 4.37 Representation of anticorrelated modulation of hfsc, which produces alternating line widths. This diagram applies to a system with two ^{14}N nuclei having the instantaneous splittings shown in (a). In the limit of very slow interconversion the stick diagram of (b) represents the observed spectrum. The *spectral index numbers* (denoted by carets) indicate the nuclear spin states responsible for each line in the spectrum. The nuclear spins follow the selection rule $\Delta M_I = 0$, so the nuclear spin states are invariant when the molecule undergoes some structural change that interconverts the values of the hfsc. Such an interconversion is represented in (c), which has the same appearance as (b) except that the spectral index numbers of some of the lines are altered. At faster rates of interconversion, these lines can become quite broad, (d). At even higher rates of interconversion these lines again sharpen, and in the fast jump limit (e) the spectrum corresponds to that of two equivalent nuclei with a single hfsc that is the average of the two slow-exchange values.

spin density that increases its a_N at the expense of the other. When the other nitro group becomes the solvation site, the a_N values are interchanged, giving rise to the observed line-width alternation in a manner analogous to that shown in Figure 4.37. The effect is accentuated when ethanol is used as the polar solvent [247].

Ion-pairing interactions produced similar anticorrelated hfsc modulations. For example, the pyracene anion displays line-width alternation under selected combinations of solvent and alkalai metal counterion such as reduction with potassium in THF [248]. This was originally attributed to the intramolecular migration of the potassium ion from one edge of the molecule to the other and later to motion between alternate sides of the aromatic plane because of analogous observations on ion pairs of acenaphthene anion [249]. The alternating line widths of the methylene protons indicate that the out-of-phase modulation arises from the association of the potassium ion with the ethylene bridge. However, as the frequency of modulation is increased by changing the sample temperature, the sharp lines of the pattern display quartet structure caused by ^{39}K splittings, which indicates that the modulation occurs by an intramolecular motion rather than by fluctuating associations with random potassium ions.

In a related example, alkali metal ion migration between the two sides of the plane of 1,3,6,8-tetra-t-butyl naphthalene anion was also demonstrated [250].

The foregoing examples can be described in terms of modulation between two different forms. A unique example of modulation between four different forms of the radical is provided by the naphthazarin cation radical [251]. The inner, hydrogen-bonded protons have $a_H = 0.07\,mT$, while the free hydroxyl protons have $a_H = 0.18\,mT$ (see Figure 4.38). The analysis is more complicated than in the dinitrodurene case because there are twice as many configurations. Moreover, in addition to the hyperfine modulation produced by hydrogen bonding, the cis–trans isomerization also modulates the ring proton splittings so that another, secondary alternating line width effect appears in the spectrum.

Three-jump modulations were invoked to explain the line-width alternations caused by hindered rotations of the methyl group in irradiated alanine [252, 253] and of the trifluoromethyl group in anion radicals of 2-trifluoro-methylnitrobenzenes bearing various other substituents at the 4-position [254, 255].

2.5.4 Molecular Motion

Molecular motion, either as tumbling of the paramagnetic spin unit or as restricted motion of the surroundings, can be studied effectively by ESR techniques. Paramagnetic molecules that are chemically compatible with various materials were developed to probe the structure of liquids, emulsions, liquid crystals, and a variety of chemical systems. These materials are called *spin probes* or *spin labels*. Different experimental techniques were also developed that provide data for analysis of the dynamics of such fluids. Two examples are

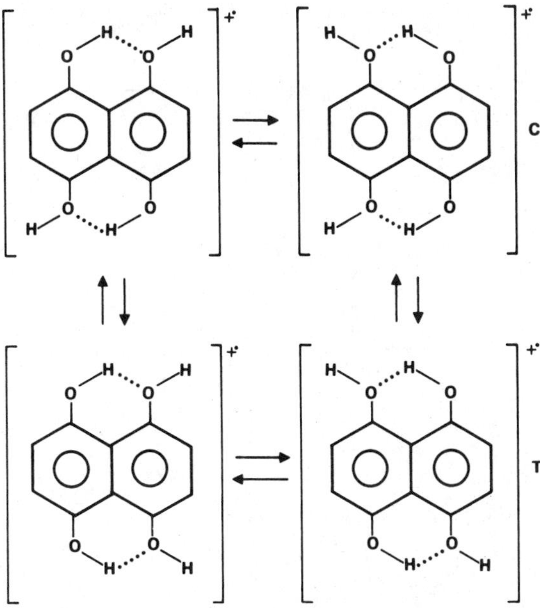

Figure 4.38 Possible conformations of the naphthazarin cation. The two different cis and trans isomers interconvert by the four jumps indicated.

conventional spectroscopy in which the first derivative spectra are recorded, and a saturation transfer technique to be described in Section 5.2. Typically, the analysis of these data is based on spectral simulation, and programs for these purposes are mentioned in Section 6.4.

Theoretical models of the phenomenon of the effects of motion on line shape are extensive. There are contributions to the line widths from g and hyperfine anisotropy, Zeeman and hyperfine interactions, spin–rotation interaction, and in some cases contributions from the nuclear quadrupole interactions. These processes are summarized in Chapters 8 and 9 of [116], and in more detail in [256].

Most of the spectra illustrated in this chapter are those of isotropic systems. In such cases the molecular tumbling is sufficiently rapid to average the various anisotropies such that very sharp absorption lines are observed. This represents the limit of complete averaging. A second extreme limit is that of randomly oriented material in a rigid matrix. Representative spectra are diagrammed in Figure 4.8 for a sample without hyperfine coupling. A more interesting case involves spectra in which both g-factor and hyperfine splittings are observed, such as shown in Figure 4.39c for VO^{2+} in a rigid medium.

Figures 4.39a and b show cases of slightly slowed tumbling, that approach the isotropic limit. For example, if the rotation of the vanadyl ion is rapid, eight lines of equal intensity are expected. However, as the rotation is slowed, the lines begin to broaden. Comparison of these spectra reveals that the relative

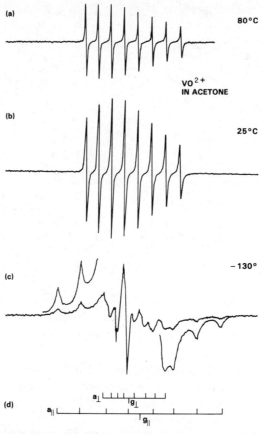

Figure 4.39 Electron spin resonance spectra of vanadyl sulfate in water: (a) spectrum observed for rapidly tumbling molecules at 80°C; (b) spectrum for slower tumbling at 23°C; (c) spectrum for sample immobilized in a frozen matrix at −130°C; (d) stick reconstruction of the parallel and perpendicular components of the spectrum.

amplitudes of the eight lines are more nearly equal at 80°C than at 23°C, because the viscoscity is smaller at the higher temperature, which allows more rapid averaging of the anisotropic components during the time scale of the ESR measurement. (The time scale is essentially the separation of the absorption lines in the anisotropic limit.) The "stick" reconstruction in Figure 4.39d shows that those nuclear spin states with absorptions that shift over the smaller field values on going from perpendicular to parallel orientation have the smaller line widths (greater amplitudes) when the molecule is free to tumble.

In describing the dynamics of spin units in solution, the rotational correlation time τ_c is determined from the ESR spectra. For an isotropic system, the rotational diffusion coefficient D_r is given by

$$D_r = kT/8\pi\eta a \tag{121}$$

where η is the kinematic viscosity and a is the molecular radius. The correlation time τ_c is related to the diffusion coefficient through the correlation function and is given by

$$\tau_c = 4\pi\eta a^3/3kT \tag{122a}$$

$$= a^2/6D_r \tag{122b}$$

As an example of a typical solvent, water molecules have a radius of about 0.35 nm and $\tau_c = 4.5 \times 10^{-11}$ s.

When magnetically anisotropic molecules such as VO^{2+} experience slightly slowed tumbling, it is possible to formulate the line width of each component line as a power series in the nuclear spin state M_I [24, 25, 257–260]

$$1/T_2 = \mathscr{A} + \mathscr{B}m_I + \mathscr{C}m_I^2 + \mathscr{D}m_I^3 + \mathscr{E}m_I^4 \tag{123}$$

where the line width $1/T_2$ is equal to $\sqrt{3/2}$ times the peak-to-peak width of the derivative spectrum. The terms $\mathscr{D}m_I^3$ and $\mathscr{E}m_I^4$ are usually small and are often neglected. The coefficient $\mathscr{A}$ is

$$\mathscr{A} = \frac{1}{T_2(0)} + \frac{2}{15}(\Delta g)^2 \frac{(\mu_B B)^2}{(h/2\pi)^2}\tau_c + \frac{1}{20}(\Delta a)^2 I(I+1)\tau_c \tag{124}$$

where $T_2(0)$ is the relaxation time in the absence of rotational broadening and may be treated as an unknown or estimated from other experimental data, and

$$(\Delta a)^2 = (a_{xx} - a_{iso})^2 + (a_{yy} - a_{iso})^2 + (a_{zz} - a_{iso})^2 \tag{125}$$

and

$$(\Delta g)^2 = (g_{xx} - g_{iso})^2 + (g_{yy} - g_{iso})^2 + (g_{zz} + g_{iso})^2 \tag{126}$$

In (125), (126), and (128) the subscript iso indicates the isotropic parameters. The $\mathscr{B}$ and $\mathscr{C}$ coefficients are expressed as

$$\mathscr{B} = -\frac{4}{15}\frac{\mu_B B}{(h/2\pi)}(\Delta g\,\Delta a)\tau_c \tag{127}$$

where

$$(\Delta g\,\Delta a) = (a_{xx} - a_{iso})(g_{xx} - g_{iso}) + (a_{yy} - a_{iso})(g_{yy} - g_{iso}) + (a_{zz} - a_{iso})(g_{zz} - g_{iso}) \tag{128}$$

and

$$\mathscr{C} = \tfrac{1}{12}(\Delta a)^2\tau_c \tag{129}$$

The vanadyl complexes closely follow the model of axial symmetry. Based on typical values (in megahertz) for such complexes: $a_\parallel = 475$, $a_\perp = 170$, $g_\parallel = 1.945$, and $g_\perp = 1.98$; significant deviations from eight lines of equal intensity occur at values for τ_c longer than about 10^{-11}–10^{-10} s.

While transition metal ion probes are useful for studying solutions, nitroxide spin labels were used as chemical or biological probes in liquids, emulsions, membranes, liquid crystals, and many other systems. Structures of some of the more frequently used spin labels are shown in Figure 4.40. Many applications of spin labels were reviewed and are included in the annotated bibliography. Of particular interest are the series of books edited by Berliner [261]. An example of a spin label applied to the study of a commercially important system is shown in Figure 4.41. Newsprint ink is a mixture of pigment, hydrophobic vehicle, and small amounts of water, and its rheological and graphic properties depend on

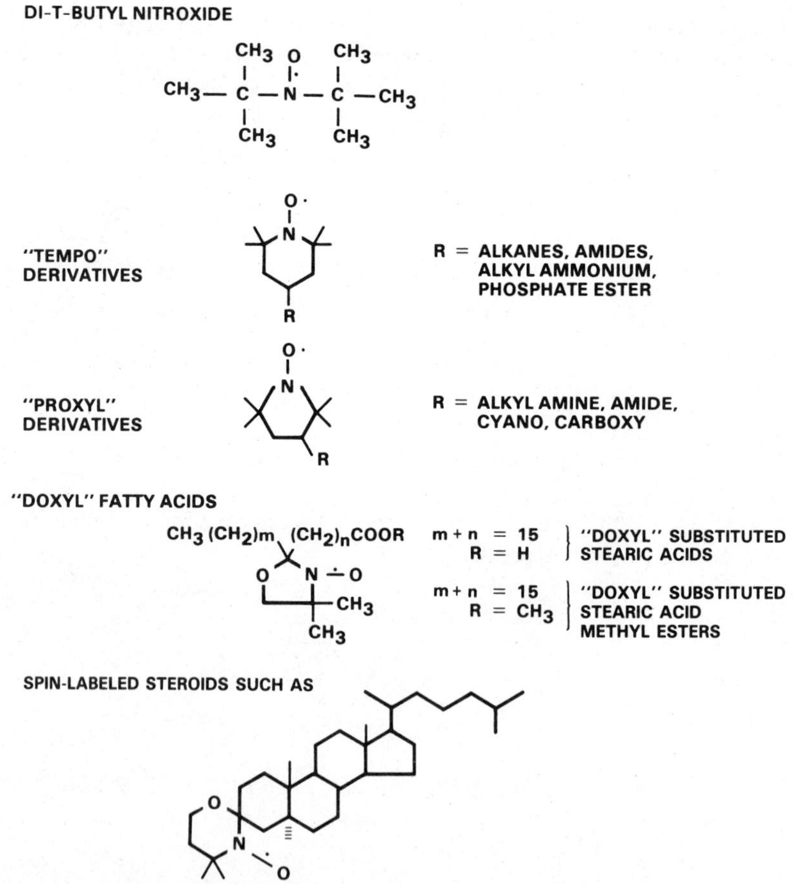

Figure 4.40 Chemical structures of some representative spin labels frequently used in chemical, physical, and biological studies.

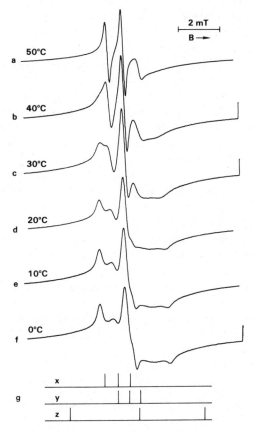

Figure 4.41 (*a–f*) Electron spin resonance spectra of 5-doxyl stearic acid in Flint Aerolith newsprint ink at temperatures between 50 and 0°C. (A stick reconstruction of the components of the anisotropic *g*- and hyperfine tensors is shown at the bottom of the figure.)

the concentration of these components. We have used spin labels to probe the hydrophobic–hydrophilic interface in such mixtures [262].

The spectra of 5-doxyl stearic acid,

in Figure 4.41a–f illustrate a significant change over the range 50–0°C. As the viscosity of the medium increases, the tumbling of the molecule slows markedly. At 50°C, the spectrum appears similar to a slightly impaired tumbling for which averaging is nearly complete. At 0°C, the spectrum appears almost anisotropic, as is indicated by the reconstruction in Figure 4.41g. The doxyl moiety parameters used in the reconstruction were $g_{xx} = 2.0088$, $g_{yy} = 2.0061$, $g_{zz} = 2.0027$, $a_x = 0.589$ mT, $a_y = 0.542$ mT, and $a_z = 3.142$ mT [263].

In early studies the spin labels were described as molecules with axial magnetic symmetry that undergo isotropic tumbling. This is an adequate approximation for the behavior of many chemical systems. But more recent models [263–266] include not only the orthorhombic symmetry, but also nonisotropic rotation where the principal axes of rotational tumbling differ from the magnetic axes. Various computer programs for simulation of such spectra are given in Section 6.4. There are many systems in which it is desirable to investigate slow rotational diffusion. The lower limit at which motion is detected by using nitroxide probes, based on the differentiation of spectra that appear to be representative of rigid matrices, is about 10 μs. However, advanced experimental techniques can extend studies to significantly slower rates (see Section 5.2).

Two pragmatic reviews on the applications of such probes for biochemical applications were published [267, 268]. For theoretical discussions, [261] is most comprehensive.

3   APPLIED ELECTRON SPIN RESONANCE

3.1   Matrix Isolation Techniques

Until the advent of ESR, free radicals were usually considered to be short-lived intermediates whose existence was more often inferred than observed directly. But with the introduction of ESR techniques the presence of free radicals is demonstrated in a wide range of natural materials as well as in a myriad of chemical systems prepared simply for the purpose of producing radicals. Because of improvements in the detection sensitivity of ESR and increasing sophistication in radical production techniques, ever greater effort is expended to demonstrate the existence of ever more transient and elusive unpaired electron species. The techniques of matrix isolation are a case in point.

Unstable free radicals decompose readily and many clever schemes are used to study such transient species. These hinge on several considerations. The radicals can be produced continuously and observed at a steady-state concentration in a flow system. The decomposition rate can be diminished by lowering the temperature of the reaction medium. The reactive species can be surrounded with diluent molecules that are relatively inert to the decomposition pathways of the substrate. Combining these approaches leads to the technique called *matrix isolation* whereby the radicals are separated from each other by some inert substance and studied at low temperatures to inhibit decomposition. This technique was used as early as 1942 [269] in the spectroscopic study of photochemical decomposition of organic compounds as dilute solutions in a rigid glass of EPA (a 5:5:2 mixture of ether:isopentane:alcohol frozen in liquid air).

Adaptation of this methodology to the ESR study of atomic ($H\cdot$, $D\cdot$, and $N\cdot$) and molecular ($CH_3\cdot$ and $HO_2\cdot$) species at liquid helium temperatures was reported in 1958 [270]. In these studies the spin units were prepared by electrodeless electric discharge in a rapidly flowing gas stream and were deposited on a sapphire rod that was in thermal contact with a liquid helium reservoir. Unreacted gas formed the matrix and was typically deposited in 100-fold excess over the concentration of decomposition products. A similar apparatus was used to study samples produced by the vaporization of solids (e.g., CuF_2 vaporized at 900°C) and isolated in matrices of rare gases such as neon or argon, codeposited at liquid helium temperatures. Of course, samples prepared in this fashion are expected to consist of immobilized, randomly oriented spin units, and spectral analysis may be complicated by anisotropic interactions.

Kasai [271] reported on the rare gas matrix isolation of paramagnetic ions formed by photolytic excitation of dilute mixtures of electron donor and acceptor species such as metal atoms and hydrogen iodide. Singly ionized atoms such as Cd^+, Cr^+, and Mn^+ were formed by the reaction

$$M^0 + HI + h\nu \rightarrow M^+ + H\cdot + I^- \tag{130}$$

Molecular radicals were also produced by this technique.

It was found that polycrystalline matrices comprising high symmetry molecules such as sulfur hexafluoride [272], adamantane [273, 274], fluorotrichloromethane [275], neopentane, and tetramethylsilane [276] behave as *rotator solids*. That is, when small, highly symmetric radicals are generated radiolytically in such matrices, considerable motion is possible and well-resolved, almost isotropic spectra are observed frequently even at 77 K. Even when the magnetic anisotropies are averaged incompletely, the resolution of the powder patterns in such matrices is superior to that observed in a rigid glass such as methyltetrahydrofuran.

Utilization of neon as an isolation matrix permits study of elusive cation radicals consisting of several atoms. The ionization energy of neon is great (21.6 eV) as compared to argon (15.5 eV), which permits the study of species having larger electron affinities. Furthermore the spectra tend to be sharp in a neon matrix because essentially all the natural isotopes have zero nuclear spin and there is negligible matrix-radical interaction. Of course the temperature range is limited (2–10 K), which rarely permits the study of temperature-dependent effects such as motional averaging. Much of this work was performed by Knight and co-workers and is the subject of a recent review [277]. Some of the species studied by neon matrix isolation (e.g., CH_4^+, $CH_2D_2^+$, NH_3^+, H_2O^+,

CO^+, N_2^+) relate to upper atmosphere or interstellar chemistry, while others (e.g., AlF^+, SiO^+, AlO^+) provide an experimental foundation for theoretical calculations of electronic structure.

The complementary utilization of ESR matrix-isolation studies along with photon- and electron-pulse excitation of fluid systems was also emphasized [275]. For example, the presumed formation of the radical cation of vinyl-carbazole during a photoinduced polymerization was verified through observation of the spectrum obtained by matrix isolation [278].

3.2   Electron Spin Resonance Imaging

Medical technologists have made great strides in the adaptation of NMR techniques for the visualization of different types of biological tissue in vivo. For this process the sample must be located in a gradient magnetic field rather than the homogeneous field normally employed for spectroscopic studies.

Several groups made preliminary studies of ESR imaging (*zeumatography*) [279]. When a specimen consisting of nonuniformly distributed paramagnetic pockets is placed in a gradient field, the spectrum that results when the laboratory magnet is scanned can be related to the spatial location of the pockets along the gradient field axis.

A simple arrangement for performing such experiments requires only minor modification of existing spectrometers and is capable of spatial resolution of about 1 mm.

It is unlikely that this technique will find wide application in diagnostic medicine; however, it may prove quite useful in selected areas of material science and fundamental studies in the biological sciences.

3.3   Material Science Applications

A symposium [280] on the application of resonance techniques to material science demonstrated the utility of ESR for quantifying and improving performance of technologically important substances such as semiconductors and ceramics. The nature of the ESR signal in amorphous silicon was ascribed to *dangling bonds*, that is, to the unpaired electrons of silicon atoms that have insufficient neighbors to produce the tetrahedral structure of amorphous silicon [281]. The effect of doping acts first to bond with these dangling bonds and to diminish the ESR signal. Further doping causes growth of different ESR signals depending on whether the dopant is *n* or *p* type, and the signals are ascribed to charged weak bonds. Doping also introduces defect sites, which are fundamental to much of modern semiconductor technology. The ability of a dopant atom or hole to capture an electron determines the conducting, insulating, or luminescence properties of the material.

Electron spin resonance is uniquely suited to the study of dopants and defects in semiconductors and insulators. Some examples follow [282, 283].

1.   Doping of gallium arsenide with Cr^{3+} provides sites that can act as either electron traps (conversion to Cr^{2+}) or hole traps (conversion to Cr^{4+} in *p*-type material) [284, 285].

2.   The E' center in irradiated silica, either glassy or in crystalline polymorphic α quartz, was studied extensively and reviewed [286]. Understanding of this site has important implications to the efficiency of optical fibers and to the insulating layers of integrated circuit devices.

3.   Photochromic glasses have the ability to darken on exposure to radiation and reversibly bleach in the dark. One of the simplest systems employs silver chloride as the photochromic material. Addition of low levels of copper sensitizes the photochromic response of oxide glasses, presumably because of the reaction

$$Ag^+ + Cu^+ + hv \rightarrow Ag^0 + Cu^{2+} \tag{131}$$

and this interpretation is supported by ESR studies that show the photolytic darkening to correlate with the concentration of Cu^{2+} [287, 288].

3.4   Temperature-Dependent Spectra

Electron spin resonance spectra have been recorded over a wide range of temperature, from liquid helium cryostats operating at 1.6 K to elevated values of 1500 K [289]. Attainment of such extreme temperatures requires utilization of specialized ESR cavities. An approach to this problem, based on the use of modular cavity design, was described in the review by Berlinger [290].

Extremely low temperatures are often required to slow spin–lattice relaxation processes sufficiently to observe ESR signals of very broad lines. For instance, Nd(III) ions in $La(C_2H_5SO_4)_3 \cdot 9H_2O$ exhibit a spin lattice relaxation time such that $1/T_1 \propto T^9$; therefore, at temperatures above liquid helium the ESR line broadens to unobservability [291]. Recent interest in phase transitions is exemplified in the study of the transition of $SrTiO_3$ from cubic to tetragonal symmetry as the temperature is lowered below 105.1 K [292]. The criticality of phase transitions emphasizes the need for precise control and measurement of temperature. In the study of the $SrTiO_3$ transition, the change in symmetry was probed by doping the crystal with Fe_2O_3 to provide paramagnetic sites. A linear scan of temperature in the region of the critical temperature was applied to permit the relaxation of residual strains in the crystal. A temperature program of 1 K/h was obtained with a resolution of 6×10^{-5} K [293]. The temperature was measured with a calibrated industrial platinum resistance thermometer. Few laboratories have attempted to achieve such precision of measurement and control because of the mistaken notion that temperature has only minor influences on ESR data; however, this is not true for solid-state phase transitions.

Recent work in our laboratory involved the use of a high-pressure, high-temperature coaxial cavity capable of withstanding internal pressure up to 22 MPa and of raising the sample temperature from ambient to 1050 K within 20 s after initiation of heating [294]. This cavity was used in coal liquefaction studies by monitoring the dependence of free radical concentrations in different coals as a function of temperature, hydrogen gas pressure, and the nature and quantity of added hydrogen donor compounds.

The variation in hyperfine splitting constants of organic free radicals as a function of temperature is a fertile field of study. Of course the temperature range used for such studies is rather more limited than the extreme values mentioned earlier. Many studies are limited by the restriction of maintaining the sample in the liquid phase, while the decomposition of the free radicals at high temperatures is also a significant consideration.

Several processes are responsible for the observed variations of isotropic splittings of organic free radicals. One of the earliest explanations for the observed hyperfine splittings of hydrogen atoms in aromatic free radicals invoked out-of-plane vibrations of the C—H bonds [34]. While this was later correctly described as a configuration interaction [30–33] (Section 1.9.5) contribution to aromatic hydrogen couplings, variable temperature studies also revealed small temperature coefficients of such splittings, which are attributed to contributions from thermally populated out-of-plane vibronic states [295]. The hydrogen hfsc is written as a power series, truncated after the second term

$$a_{CH}^{H}(T) = a_0 + a_1 \langle \theta^2(T) \rangle \tag{132}$$

where the term in brackets represents the time average of the out-of-plane angle θ.

Since the sign of Q in the McConnell relation (59) is negative, hydrogen atoms attached to a carbon that bears positive spin density have a negative hfsc value, although the sign of the hfsc cannot be determined from simple first-order spectral effects. However, the contributions from vibrational states, represented by the a_1 term in (132), are positive, so there is a net decrease in the magnitude of the splittings as the temperature is raised. This is observed experimentally. However, there also appears to be a contribution from the spin densities on π atoms adjacent to the atom bearing the hydrogen of interest. When the latter bears a negative spin density and the adjacent atoms bear large positive spin densities, the hfsc has a positive coefficient for the hfsc temperature variation [296]; when this is observed a good case can be made for negative spin density at that site.

When the unpaired electron wave function is nearly degenerate with another wave function, thermal population of the higher state may produce a larger thermal coefficient of the hfsc than vibrational effects if the spin-density distributions of the two wave functions are different. An elegant demonstration of this effect arises when the anion radicals of benzene and benzene-1-d are compared. In unsubstituted benzene anion the wave function of the unpaired electron is doubly degenerate. Replacement of one hydrogen with a deuterium atom lifts this degeneracy [71] and nearly doubles the temperature dependence of the hfsc of the adjacent hydrogen atom. In benzene anion $d(a_H)/dT = -0.140 \times 10^{-6}$ T/K, while in the anion of benzene-1-d the value is -0.274×10^{-6} [297].

For nuclei other than hydrogen, the hfsc are related to the σ–π parameters arising from the interaction of the unpaired electron orbital with lower lying, bonding orbitals of the molecular framework. The hfsc depend on both the spin density on the atom in question and the densities on neighboring atoms (see Section 2.1). When a nucleus such as nitrogen exhibits a temperature-dependent hfsc the variation can arise from either changes in the $\sigma-\pi$ parameters or in the spin densities. Data on cation radicals derived from pyrazine (N,N'-dihydro-1,4-diazine) provide evidence that the $\sigma-\pi$ parameters are temperature dependent in this system [298].

Spin-density variations may occur as the result of torsional motions that act to decouple the molecular π orbital containing the unpaired electron from the p orbital centered on the nucleus responsible for the splitting. An example of this effect was noted in the cation radical of diaminodurene [299]. Some nitroxide radicals also exhibit temperature-dependent nitrogen hfsc that probably occurs because the nitroxide geometry is not well defined [300]; and out-of-plane bending, leading to variation in hfsc, is produced readily by temperature variation.

The earlier description of hyperconjugative coupling to β protons (Section 1.9.5) explained the variation in hfsc with angular orientation relative to the axis of the orbital containing the unpaired electron. The hfsc of β protons are thus expected to exhibit temperature dependence when there is some barrier to free rotation about the bond joining the β carbon to the π system (Figure 4.18). This was first predicted on the basis of data from several nitroalkane anion radicals [49]. The treatment was refined by Fessenden [301] who demonstrated the existence of a rotational barrier about the C_α—C_β bond in alkane and alkene radicals. The temperature variation of the β-proton splitting permitted estimation of the barrier height. Probably the most important aspect of these studies is the demonstration of the validity of the model for angular dependence of the β-proton hfsc. The actual application of this model is useful in explaining the hfsc variations in homologous series of radicals that are presumed to have rotational barriers, regardless of the observation of temperature dependence [36, 37].

In general, methyl groups are assumed to rotate freely so that the time average of the cosine squared term becomes $\langle \cos^2 \theta(t) \rangle = \frac{1}{2}$. When combined with the experimental and theoretical estimates of $B_2 \approx 4.6 - 5.0 \, \text{mT}$, (61) leads to the fortuitous result that the β protons of a methyl group produce the same hfsc as a proton attached at the same site. It is gratifying that it was possible to demonstrate that some methyl group rotations can be frozen out at low temperatures. The $CH_3\dot{C}HCOOH$ radical exhibits three different methyl proton splittings of 120, 76, and 14 MHz at 77 K, while a pattern corresponding to an averaged value of 70 MHz is observed at 300 K [252]. The temperature dependence of ESR splitting constants was the subject of a detailed review [302].

Temperature variation is also important in studies of relaxation processes, kinetics, intramolecular motions, and intermolecular exchange processes, and studies of phase transitions (discussed in Section 3.3).

3.5   Electrochemical Studies

As early as 1930 Michaelis postulated that electrochemical reactions occur in one-electron increments. When ESR began moving from physics to chemistry laboratories in the 1950s it was natural to test this hypothesis. The first report of ESR observation of electrochemically produced free radicals occurred in 1955 and resulted from simply quenching reaction products at liquid nitrogen temperature and transferring the frozen solution to the cavity of an ESR spectrometer [303]. Shortly thereafter, Geske and Maki [304] reported a technique for producing free radicals electrochemically inside the ESR cavity. They called this technique *in situ* (in its original place) *intra muros* (within the walls) *electrochemical generation*, and it offered several distinct advantages over conventional methods for producing free radicals.

Since the radicals are generated directly within the cavity, it is possible to make the ESR observations as soon as the product radical builds up to a satisfactory concentration level and the effects of decay of radicals by subsequent reactions are minimized. Moreover, since electrochemical reactions normally occur in fluid solutions, the resulting free radicals normally exhibit well-behaved isotropic spectra, rather than the anisotropic *powder spectra* that are observed for frozen solutions.

There are many different reagents that have been used to generate free radicals in solution, ranging from oxidation of hydrocarbons in concentrated sulfuric acid to reduction of the same compounds by alkali metals in ethereal solvents. Somewhat less brutal reactions are exemplified by antimony penta-fluoride oxidations and dithionite reductions. But all of the reactions require careful optimization of conditions to produce the desired result. In contrast, an electrochemical electrode can function as an infinitely adjustable redox reagent, depending on the value at which its potential is controlled within the voltage breakdown limits of the solvent–electrolyte system. Consequently, a wide variety of compounds can be converted into free radicals with a single experimental apparatus. The prototype electrochemical ESR cell of Geske and Maki is shown in Figure 4.42, and it still functions well for many present studies. But as more workers expanded the frontiers of the technique, many specialized cells were designed. Several were described in a monograph devoted to the development of electrochemical ESR [54].

Electrochemical techniques of polarography and cyclic voltammetry provide an efficient way to screen chemical compounds as candidates for free-radical formation even if the eventual ESR study does not rely on electrochemical radical generation. The potential at which the candidate compound undergoes oxidation or reduction is a direct measure of the ease of conversion to another species, while the number of electrons per molecule is evaluated readily from the electrochemical data. Another useful piece of information is the degree of electrochemical reversibility of the reaction. For a totally reversible process, the reaction product has a significant lifetime and differs from the parent material only in oxidation state, whereas irreversibility implies that the electrode product

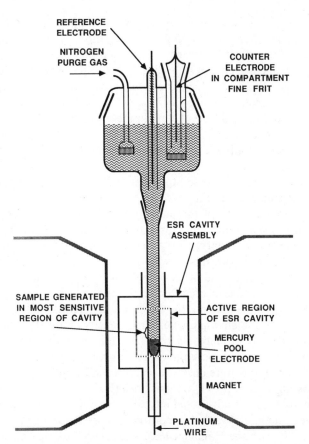

Figure 4.42 Electrochemical cell suitable for in situ, intra muros generation of free radicals for ESR observation. The mercury bead is suitable for cathodic studies, while other sample tubes employ a platinum bead or gauze for anodic studies.

undergoes rapid reaction to some other material. It is sometimes possible to change solvent or electrode material and find a reaction medium that sufficiently stabilizes the product of a "moderately irreversible" reaction to permit ESR observation of the free radical. Or, if the heterogeneous electrode surface facilitates polymerization of the product or other undesirable side reactions, it may still be possible to use the electrochemical data to select a chemical reagent for generating a radical species in a homogeneous medium.

The results of electrochemical–ESR studies were applied in many different ways. In its infancy, the technique was used primarily to produce free radicals for the sake of observing a new species. But its applicability to more fundamental studies was soon apparent. For example, the hyperfine splittings varied with solvent composition for certain classes of free radicals, which led to a clearer understanding of ion–solvent interactions.

Electrochemical reactions are generally carried out in solvents of fairly high dielectric constant containing a high concentration of inert electrolyte to increase the conductivity of the system. Such conditions tend to swamp out any long-lived ion pairs, which can complicate the ESR spectra of radical ions generated in low dielectric solvents. For example, when anion radicals are produced by alkali metal reduction in ethereal solvents it is quite common to observe hyperfine splittings from the alkali metal counterions, which suggests long-lived ion pairs (Section 2.5.2). However, when the anion radicals are produced electrochemically the spectrum only exhibits hfsc from the magnetic nuclei of the anion alone. Although ion association does occur in electrochemical systems, the fluctuations are so rapid that discrete ion-pairs are not revealed in the hyperfine splitting pattern.

Actually, careful analysis of the magnitudes of hfsc for a given radical in different media reveals interesting information. For example, the nitrogen hfsc for nitrobenzene anion is larger in water than acetonitrile, and the value increases monotonically with the mole fraction of water in binary mixtures of the two solvents [305]. The addition of metal ions to acetonitrile solutions of nitrobenzene anion also has an effect similar to the addition of water [306]. Such observations were explained on the basis of preferential solvation (or association) at the site of high unpaired electron density on the free radical [307]. Since different solvents produce different hfsc, mixtures produce intermediate hfsc indicative of the dynamic equilibrium. Further support for this model resulted from low-temperature observation of two differently solvated species of m-nitrophenol anion in water–DMF mixtures [307].

Another early study of substituted nitrobenzene anion radicals generated by electrochemical methods [308] showed a good correlation of redox potential and nitrogen hfsc with the Hammett linear-free-energy $\sigma-\rho$ parameters [309]. Subsequent work on other families of anion, cation, and neutral radicals gave further insights into the effect of substituents on ESR parameters and, more importantly, on the stability of radicals [310].

Studies of electrochemically generated nitroaliphatic anions [311, 312] revealed that the hfsc of the hyperconjugatively coupled β protons showed a strong dependence on the steric bulk of the aliphatic substituent. Similar studies of sterically crowded nitrobenzene anions showed that the nitrogen hfsc increased with crowding caused by twisting of the nitro group out of the aromatic plane. Such observations of steric effects were found in many other systems and were summarized in reviews [37, 38].

The earlier mention of electrochemical reversibility suggests that for many compounds the free-radical electrode reaction product is unstable. Such a situation makes it difficult to observe a high-quality ESR spectrum of the material but does not preclude extracting meaningful kinetic information from such systems. A thin-layer electrochemical cell was devised for this purpose [313]. When it is placed inside the ESR cavity, it is possible to electrogenerate a significant concentration of short-lived radicals and to monitor the decay rate by observing the ESR absorption of a single, selected hyperfine line as a function of time. This particular technique has the advantage that the concentration of electrode product is readily adjusted by selecting the number of coulombs passed through the cell; then the spent solution can be flushed out and replaced with fresh material and the measurement can be repeated without disassembling the apparatus. Kinetic information is extracted by comparing the experimental signal-versus-time curves to simulated curves calculated from knowledge of the current pulse parameters and assumptions about the decay mechanisms and the rate constant [314–316].

A recent series of studies by Albery, Compton, and co-workers [317–326] addressed the topic of kinetic studies of transport and decomposition of radicals produced by electrochemical generation under conditions of steady-state flow of solutions through the ESR cavity. Initial studies involved the use of a cylindrical electrode just above the ESR cavity [317–319]. Later it was found that a "semiannular" or "partial-sector" electrode could be placed inside the cavity without causing too severe degradation of the microwave field; consequently, it was possible to study shorter lived radicals. Interesting modifications of these techniques involved the use of electrodes altered with electroactive polymer coatings [320, 321] and the design of a flat flow cell incorporating a planar electrode suitable for simultaneous spectroelectrochemical–ESR studies [322–325]. A flow system using a cylindrical electrode as part of the central conductor of a coaxial cavity combines advantages of high ESR sensitivity and well-defined hydrodynamic conditions [326].

3.6   Irradiation

A far-reaching application of ESR is in studies of radiation damage. For the most part, these involve energies spanning the range from visible through gamma radiation, but high-intensity infrared, microwave, electron beam, and even ultrasound damage is also investigated by ESR.

3.6.1   Ultraviolet Irradiation Photolysis

The general mechanisms of photolysis of a compound AB by ultraviolet (UV) radiation are

$$AB \rightarrow AB^* \qquad \text{(excited state)} \qquad \text{(133a)}$$

$$\rightarrow A\cdot + B\cdot \qquad \text{(dissociation)} \qquad \text{(133b)}$$

$$\rightarrow A\cdot \cdots B\cdot \qquad \text{(radical pair)} \qquad \text{(133c)}$$

$$\rightarrow AB^{\cdot+} + e^- \qquad \text{(ionization)} \qquad \text{(133d)}$$

Examples of each type of process are revealed in detail by using ESR techniques.

Ultraviolet irradiation can be carried out outside the ESR cavity or within the cavity (in situ). In a simplistic sense, external irradiation is a way to examine stable products, but intermediates that are unstable at ambient temperature may be stable at liquid nitrogen temperatures. In these cases, a simple but effective method involves rapid transfer from the liquid nitrogen irradiation vessel to a cavity containing a Dewar insert that is precooled to the desired temperature.

The in situ method has the advantage that short-lived intermediates generated by photolysis can be observed under steady-state conditions (continuous photolysis), or the decay kinetics can be observed (intermittent photolysis). Microwave cavities (see Section 4.2.3) can be cut with slots that allow about 50% of the incident optical radiation to enter the cavity. Alternatively, the end wall of a rectangular microwave cavity can be fabricated with a small rectangular hole and extended with open waveguide having a cutoff frequency much higher than the resonant frequency of the cavity. Hence, optical or ultraviolet radiation can enter the cavity, but microwaves cannot propagate out. This method is somewhat more effective than using slots because all of the optical radiation can enter the cavity even though the cavity Q factor (Section 4.2.3) might be lower by about 10 to 15%. Both methods were applied extensively to mechanistic studies. This equipment is commercially available from manufacturers of ESR accessories (see Appendix 2).

PHOTOLYTIC PROCESSES

Chemical bond rupture (133b) commonly occurs during or subsequent to UV irradiation, and numerous alkyl and aryl radicals are formed in this manner. Radical pairs are also produced (133c), but they tend to be short-lived if either fragment is reactive toward the solvent; recombination is also an important decay route. Ionization (133d) may occur to form $AB^{\cdot +}$, but such species are generally unstable. Often $AB^{\cdot +}$ decomposes by elimination of positive functional groups, such as hydronium ions, if such are easily formed from the parent ion. Alternatively $AB^{\cdot +}$ can recombine with an ejected electron or capture an electron from solvent. While photolytic processes are a convenient way for producing unstable radicals, the intrinsic instability opens various decomposition pathways so that it is often challenging to discriminate against undesirable side products and generate the desired species selectively.

Photolysis can also be used to generate radicals by an indirect method. For example, peroxides, such as di-t-butyl peroxide are readily decomposed by optical or near-UV irradiation to form alkoxy radicals

$$CH_3C(CH_3)_2O \text{—} O(CH_3)_2CCH_3 \xrightarrow{\;h\nu\;} 2CH_3C(CH_3)_2O\cdot \qquad (134)$$

Alkoxy radicals cannot be detected by ESR, which makes them an ideal reactive intermediate. Such radicals abstract hydrogen from most CH bonds. The CH bond energy varies from about 325 to 440 kJ/(g·mol), while that of the OH bond is about 480 kJ/(g·mol). Thus, in general, alkyl or aryl radicals are formed in the reaction

$$CH_3C(CH_3)_2O\cdot + RH \xrightarrow{\hspace{2cm}} CH_3C(CH_3)_2OH + R\cdot \qquad (135)$$

These reactions are relatively nonspecific, and a variety of products often result. However, direct radical formation by ionizing or dissociative radiation generally leads to an even greater variety of products.

Photolytically produced radicals are often detected under steady-state conditions using conventional spectrometers, continuous irradiation, and flowing solutions of precursor and other reagents. Sometimes more sophisticated approaches are required. These include temporal modulation of the light source (see Section 3.8.2) and use of special sample cells and mixing chambers (Section 3.8.3).

PHOTOEXCITED TRIPLET STATES

Excited triplet states of organic compounds are readily generated in frozen glassy solvents containing about 1 to 5% of the substrate material. The glassy solvent permits light to penetrate with minimal scattering while immobilizing the photoexcited species. Although triplet-state species can be formed in liquid media, they cannot be detected by ESR because the tumbling of the highly anisotropic spin units produces local magnetic field oscillations that induce rapid relaxation, making the lines too broad to observe.

Early ESR studies of organic molecules in the triplet state were accomplished by irradiation of single crystals at low temperatures. With increased instrumental sensitivity it is now often possible to extract the pertinent D and E zero-field parameters (see Section 2.2.1) from the spectra of glassy samples. As an example, consider the spectrum of the naphthalene-d_8 triplet state shown in Figure 4.43. The spectrum exhibits a sharp low-field transition, which corresponds to $\Delta M_S = 2$, and a broad spectrum with sharp singularities in the higher field region, which corresponds to $\Delta M_S = \pm 1$.

The low-field $\Delta M_S = 2$ transition is sufficiently isotropic [327, 328] that small changes in orientation of the spin unit produce only minimal change in the resonance field. Under this condition the two-quantum transition occurs at the field B_{min}, given in terms of the zero-field parameters as

$$B_{min} = \frac{1}{g\mu_B} \left[\frac{h^2 v^2}{4} - \frac{h^2(D^2 + 3E^2)}{3} \right]^{1/2} \tag{136}$$

where D and E are expressed in frequency. The value of $(D^2 + 3E^2)^{1/2}$ is sometimes symbolized as D^*, which should not be confused with a different meaning in excited-state chemistry. The advantage of using this transition analytically is that even though it is first-order "forbidden" (but observable because of the mixing of states), it can be at least an order of magnitude more intense than the $\Delta M_S = \pm 1$ transition because it is more nearly isotropic.

The $\Delta M_S = \pm 1$ transitions in frozen solutions, first reported in [329], exhibit six sharp singularities as shown in Figure 4.43. With the assumption of an isotropic g-factor, which is often adequate for the interpretation of such spectra, these singularities occur at positions of $(hv/g\mu_B) - |D|/g\mu_B$, $(hv/g\mu_B) - (|D| + 3|E|)/2g\mu_B$, $(hv/g\mu_B) - (|D| - 3|E|)/2g\mu_B$, $(hv/g\mu_B) + (|D| - 3|E|)/2g\mu_B$,

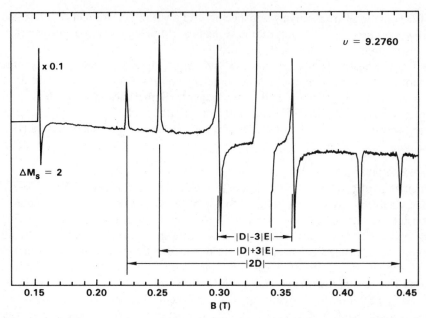

Figure 4.43 Electron spin resonance spectrum of the photoexcited triplet state of naphthalene-d_g at 77 K in ether–pentane–alcohol. The spectrometer gain was decreased by a factor of ten in the region of the $\Delta M_S = 2$ transition to display it on the same scale as the triplet spectrum.

$(h\nu/g\mu_B) + (|D| + 3|E|)/2g\mu_B$, and $(h\nu/g\mu_B) + |D|/g\mu_B$, where D and E are expressed in frequency units. These values may not be exactly equal to the values obtained from single crystals because of environmental effects that distort the principal axes.

Molecules with axes of threefold or greater symmetry exhibit $E = 0$. However, for a molecule of arbitrary symmetry, the results of the Hamiltonian (76), (77) show that $E \leqslant D/3$.

Typical values for D and E of some triplet-state organic species are given in Table 4.16. The D, E, and D^* values can in turn be related to the average distance between electrons, according to (52), where $g\mu_B$ is substituted for $g_N\mu_N$. As a guide, values of D and E are inversely proportional to the third power of the interelectron separation.

Electron spin resonance is also used extensively for kinetic measurements on triplets. Triplet-state lifetimes are determined easily by using a pulsed light source and in some cases signal averaging to obtain a reasonable signal-to-noise (S/N) ratio [331, 332]. Considerable effort was devoted to the studies of porphyrins [333, 334] because of their importance as intermediates or models of intermediates in photochemical processes in biological systems.

Many applications of ESR to studies of photoexcited triplets are described in [330].

Table 4.16 Zero-Field Parameters of Organic Molecules at 77 K in Ethanol–Pentane–Acetone[a]

Molecule	Perprotio		Perdeuterio		
	D (GHz)	E (GHz)	D (GHz)	E (GHz)	D* (GHz)
Benzene					4.78
Naphthalene	3.006	−0.410	2.973	0.426	3.14
Phenanthrene	3.0105	1.3961	3.129	1.399	4.00
Quinoxylene	3.018	0.545			3.13
Quinoline	3.006	0.486			3.20
Dibenzofuran					3.28
Dibenzothiophene					3.43
Diphenyl methylene[b]	12.140	0.575			
Fluorenylidine[b]	12.266	0.8477			

[a]Data derived from [330], Chapter 10.
[b]Ground-state triplet molecules.

3.6.2   Electron-Beam Irradiation

This is a very specialized case of ESR primarily because most researchers do not have access to the linear accelerators or Van de Graaff generators necessary to create sufficiently intense or energetic electron beams; and as a result, this subject is not discussed at length. In the simplest type of experiment, an energetic electron beam is passed through holes (ca. 2.5-cm diameter) drilled through the magnet and holes (ca. 7-mm diameter) drilled through both sides of the microwave cavity so that it passes through the sample. Such electron beams are pulsed, and the repetition rate can be varied up to about $1000\,s^{-1}$. Liquid samples are maintained at low temperatures and flowed through the cavity [335, 336].

Electrons can interact with the sample in two ways: (1) if the electron passes close to the nucleus, it is retarded by the nuclear charge and emits electromagnetic radiation (bremsstrahlung); (2) the electron can collide with the medium and can excite, ionize, or attach to the molecule. The proportion of energy dissipated by bremsstrahlung as compared to that by collisional interactions increases with the energy of the beam. Below approximately 0.1 MeV, bremsstrahlung radiation is negligible, but it becomes the dominant mode above 10 MeV. In addition, the rate of bremsstrahlung is proportional to the square of the charge of the nuclei in the medium, so that it is dominant for heavy atoms. This irradiation method is important in studies of chemically induced electron polarization discussed under kinetics (Section 3.8.3).

It is important to note that a "poor researcher's" electron beam irradiation is formed by electron discharge, such as from a Tesla coil or a high-voltage arc. Such apparatus is used to form radicals in the gas phase that may be trapped subsequently [337].

3.6.3 Ionizing Irradiation

Important sources of radiation include X and γ rays. Because of the hazards associated with this radiation, the irradiation is done in solid samples, usually at liquid nitrogen or lower temperatures to minimize migration of atoms or molecular fragments formed by the radiation. In addition, it is advisable to use lead foil to shield any portions of the sample holder that will be in the spectrometer cavity but do not actually contain sample. Since ionizing radiation produces paramagnetic sites in glass or quartz, such shielding diminishes the amplitude of spurious signals.

The most common γ-radiation source is ^{60}Co with energies of 1.33 and 1.17 MeV at rates of 10^5 to 2×10^7 R/h. The total doses for sample preparation are typically from 10^6 to 10^8 rad. (One roentgen is the amount of radiant energy that must be absorbed by $1\,cm^3$ of dry air to produce ions carrying 0.3333×10^{-9} C of charge. Since the absorption of radiation varies with the target medium, other units must be used to define dosage. One rad is the unit of radiation required to impart an energy of 100 ergs/cm^3 [336].) Rarely, however, is the actual dosage measured in ESR experimentation, although it is possible to carry out actinometry to determine this value.

Typical X-ray energies used in ESR are in the range from 100 to 200 keV at dose rates of $5 \times 10^4 - 1.2 \times 10^6$ R/h. Total dosages are comparable to that of $^{60}Co\,\gamma$ radiation.

Gamma- and X-ray photons having energies in the 100-KeV to about 10-MeV range interact with matter primarily by the mechanism that produces the Compton effect [338]. When such a photon collides with an atom it loses a small fraction of its original energy and momentum by causing ejection of an electron from one of the outer shells. This process can be repeated many times; thus each photon can generate a large number of *secondary* electrons. The major differences between X and γ rays are that γ rays are more energetic but exhibit a smaller absorption per unit pathlength. Low-energy X rays are typically absorbed after passing through a few millimeters of liquid, while ^{60}Co radiation penetrates 30–40 cm.

Numerous studies were carried out using ionizing radiation on many different substances. Work on semiconductors was reviewed [339–341]. Extensive work was reported in a wide variety of inorganic single crystal and polycrystalline materials. A good review of this subject is by Wertheim and co-workers [342], although the references are no longer current.

Examples of radiation-induced defect sites that are commonly encountered are F centers, which are electrons substituted for anion vacancies; V_K centers, which are cation vacancies shared by two neighboring ions so that species such as F_2^- in LiF can form; H centers, which are similar species but which occupy one anion site; V_F centers, which are similar to V_K centers but in the vicinity of a cation vacancy; V_t centers, which are trihalide anion sites adjacent to cation and anion vacancies. Each site can be determined by the hyperfine patterns that relate to the type and number of nuclei, the g-factors that relate to the crystal field effects, and the site symmetry that can best be analyzed in single crystal media. In addition, many impurity sites that are normally not detected are observed after radiation damage because they exhibit different electron capture or ionization properties than the normal lattice. Such is the A (for acceptor) center where, for example, a divalent ion is replaced by a monovalent halogen. Since this site is metastable, electrons generated by irradiation are captured by the cation.

By far, most irradiation experiments were carried out on organic or nontransition metal inorganic species. Here, too, many different sites may form. Typical pathways are shown below. Radicals can form directly by electron attachment or electron trapping

$$XY_n + e^- \rightarrow XY_n^- \tag{137}$$

where XY_n^- is paramagnetic. An example is BF_3, which adds one electron to become the anion [80]. A second mechanism is electron attachment followed by dissociation

$$AB_m + e^- \rightarrow AB_m^- \tag{138a}$$

$$AB_m^- \rightarrow AB_{(m-1)} + B^- \tag{138b}$$

This can yield a variety of products. Typically, $AB_{(m-1)}$ is paramagnetic. For example, SF_6 under γ radiation forms paramagnetic SF_5 and diamagnetic F^- [343], and NF_4^+ forms NF_3^+ and F^- [79].

3.6.4 Effects of Ultrasound

While not an electromagnetic phenomenon, ultrasound was recently reintroduced as a technique to initiate chemical reactions. This is evidenced by the new consortium, "Sonochemistry Development Club," at the Atomic Energy Authority Harwell Laboratory in England. Ultrasound induces unique reactions in solids, liquids, and gases [344]. The mechanism is not well defined, but the effect in liquids is thought to be the result of cavitation [345], which causes a bubblelike void in the liquid. The collapse of this void may allow small volumes of the liquid to attain instantaneous temperatures from 800 to 2450 K.

Some important work [346, 347] showed that hydrogen atoms and hydroxyl radicals may form in ultrasonically stimulated aqueous solutions. These species were detected by spin trapping. Another report [348] demonstrated that radicals are generated in various explosives such as trinitrotoluene when subjected to ultrasound at temperatures well below the temperature at which thermally generated radicals begin to form.

Paramagnetic defects created by ultrasound in solid ZnS were also reported [349].

3.7   Coal

Because of the "energy shortage" of the mid-1970s, coal is a subject of extensive study by a variety of techniques, including ESR. Over the years, raw coals and coals altered by various chemical and physical treatments were studied, and still there is much uncertainty about the origin of the ESR signal and the reference state to which it should be compared. There are numerous reviews, but the most comprehensive are those of Retcofsky [350, 351].

The conventional method of preparing coal prior to recording its ESR spectrum is to grind it in a nitrogen atmosphere or under water, followed by vacuum drying at about 110°C [350, 351]. Although the spectra strongly depend on the rank and source of the coal, a typical spectrum exhibits two major lines that are superimposed, as shown in Figure 4.44. The absorptions are generally assumed to arise from organic radicals, although this is not conclusive.

The sharp line is attributed to fusinite materials (derived from charred organic matter), while the broader lines are attributed to excinite, vitrinite, and inertinite [350, 351]. If the sample is not evacuated, then the interaction of the

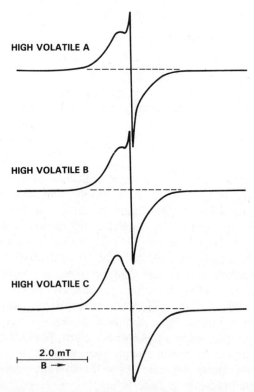

Figure 4.44   Typical ESR spectra observed for three different coal samples, where the spectrum is a superposition of sharp and broad components.

spins with adsorbed oxygen broadens these lines, and they are not resolved. Studies of several coals that were meticulously extracted from deep inside a bed, without exposure to the atmosphere, showed that oxygen causes an irreversible increase in the free-radical content [352]. Likewise, drying of conventionally prepared coals [353–355], even at temperatures well below 110°C [355], also causes an irreversible increase in ESR signal intensity. Mechanical grinding [356] may also cause such an enhancement. In the study of [356] air was not excluded, but in a separate study [357] it was shown that ball-milling caused methyl methacrylate to copolymerize with coal.

Coal is an extremely heterogeneous material, and the foregoing examples of irreversible transformations suggest that geological samples should not be assumed to be at chemical equilibrium. Recently [358], different macerals (organic equivalents of minerals) were separated; and the excinite, vitrinite, and inertinite components were shown to exhibit different g-factors, line widths, radical concentrations, and saturation behavior.

Despite these complexities, ESR parameters were correlated with the chemical properties of a wide range of coals [350]. The free-radical concentration C_r per gram of coal is approximately related to the rank of the coal (R = percentage carbon) according to (139), although the standard deviation is around 50%.

$$\log C_r = 3.2 \times 10^{-3} R + 16.7 \tag{139}$$

The peak-to-peak line width ΔB_{pp} (in mT) is related to the hydrogen content as

$$\Delta B_{pp} \sim 0.15 \, (\%H) - 0.12 \tag{140}$$

which indicates that a large portion of the observed line width arises from unresolved hfsc. Comparison of coals hydrogenated with D_2 gas shows a significantly smaller line width than the same coals heated in H_2 gas [359]. Line widths measured at 35 GHz are greater than those measured at 9 GHz because the different g-factors produce greater separation of the line components at the higher field [360].

g-Factors also correlate with elemental analyses. The most abundant element in coal, with the exception of carbon and hydrogen, is oxygen. Sulfur, though present in low concentration, has a large spin–orbit coupling that can influence the g-factor as shown by the approximate correlation

$$g \sim 2.0026 + 7.1 \times 10^{-5} [(\%O)\lambda_O + (\%S)\lambda_S] \tag{141}$$

where λ_O and λ_S are the spin–orbit couplings of oxygen and sulfur (in reciprocal centimeters, see Appendix 1) and the coefficient 7.1×10^{-5} cm is approximately $1/\Delta E$ observed for semiquinones in (80) [86]. The value 2.0026 is also close to that predicted for a neutral aromatic radical [83, 84].

As mentioned earlier, virtually every possible chemical treatment (solvents, Lewis acids, Lewis bases, oxidants, reductants, electrochemical) and physical treatment (X and γ irradiation, thermal, mechanical) of coal has been examined with ESR. In view of the interest in coal liquefaction and coal gasification, special cavities [294, 361, 362] were designed for in situ measurements under high pressure and elevated temperature to observe the time dependence of transient radical formation and decay. These yielded important mechanistic information, particularly with regard to the time dependence of radical concentration. With the growing interest in graphitic carbon as a structural material and continuing interest in coal as a fuel source, this technique is significant.

The utilization of ENDOR (electron nuclear double resonance) [363–365] and pulsed ESR [365–367] was also undertaken recently to learn more about the fundamental nature of coal-related radicals. This work is not yet conclusive, but promises to provide very interesting and important information in coal research.

Most papers on ESR of coal appeared in *Fuel, ACS Symposia Series*, and *Preprints, ACS Division of Fuel Chemistry*. The interested reader should consult the reviews mentioned [350, 351] and recent years' issues of these periodicals.

Electron spin resonance was suggested for application in the fuel industry as an on-line monitor for the quality of coal liquid products [368] and as a well-logging device in secondary oil recovery operations [369]. It was also utilized to detect the presence of vanadium compounds in feedstocks that would otherwise deactivate the cracking catalysts in refinery operations [370].

3.8   Kinetics

Electron spin resonance is used extensively in studies of kinetics of radical and radical ion reactions. The technique is so sensitive key intermediates of reactions can be monitored. In chemical kinetics, three general methods of study are employed: flow reactors, real-time measurements, and indirect methods. The application of ESR to these techniques is described below.

3.8.1   Flow Reactors

Flow reactors were developed to monitor steady-state concentrations. This allows the use of very long instrumental time constants to obtain good S/N ratios so that absolute or relative concentrations can be determined accurately. Such systems were applied successfully to liquid- and gas-phase systems. For a first- or pseudo-first-order reaction the logarithm of the ESR signal is plotted against the flow velocity at a constant point of injection of the reactants into the flow reactor or against distance at constant flow velocity in the reactor. For example, when the point of injection x is varied at constant flow velocity v, the pseudo-first-order rate coefficient k can be evaluated from

$$\log S = A + kx/v \tag{142}$$

where S is the signal intensity and A is an arbitrary constant.

For higher order reactions there is an additional aspect that is considered when using ESR as the detector in kinetic studies, namely, the nonuniform sensitivity over a sample of finite dimensions. If one considers a point sample, such as a small fleck of DPPH, then the sensitivity is measured as a function of position within the cavity. It is found that the sensitivity as a function of the length z and the x and y positions within the cavity is the product of the square of the microwave magnetic field intensity $B_1^2(x, y, z)$ times the modulation field intensity $B_m(x, y, z)$ in the absence of modulation broadening. The ESR signal S is then given by the integral of this function

$$S = \iiint_{sample} B_1^2(x, y, z)B_m(x, y, z) \, dx \, dy \, dz \tag{143}$$

Values of $B_1^2(x, y, z)$ are determined easily from the solutions to the microwave field equations [371–373], but values of the modulation field are more difficult to determine. Because the modulation is applied parallel to the laboratory magnetic field, cylindrical cavities do not exhibit cylindrically symmetric sensitivity. This asymmetry is insignificant when the sample tube is small with respect to the cavity diameter; however, for large-diameter tubes, such as those used in gas-phase kinetics, it may be important.

For concentration detection in studies of first-order reactions, it is easily shown that the sensitivity profile does not affect the slope of the logarithmic plot, which is the important experimental parameter. However, if the reaction is other than first order, the cavity sensitivity function may become an important factor in the data analysis. Hence, numerical solutions based on assumptions about the reaction scheme may be required to determine kinetic parameters.

Much creativity was demonstrated in experimental designs for flow reactors. Several reviews of the use of ESR in gas-phase kinetics were published [374, 375]. The original apparatus described by Westenberg and de Haas [221] was subsequently modified for specific needs. One example of a flow reactor used in this laboratory for the study of the recombination of atoms on metal surfaces is shown in Figure 4.45. In this reactor the reacting atoms are generated

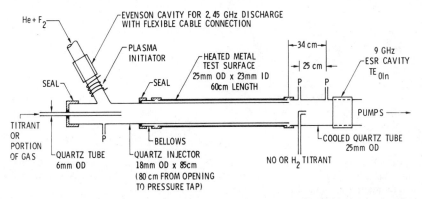

Figure 4.45 Gas-phase flow reactor for the study of atom recombinations on metal surfaces.

by passing the gas and a carrier through a microwave discharge, as discussed in Section 2.3. To study recombination, the point at which the gas contacts the surface is varied by moving the injector. When the homogeneous reaction is of interest, the microwave discharge may be fixed, and the reacting gas is injected through a tube much smaller in diameter than the flow reactor. In this case the position of the injector tube is varied, and the flow velocity is maintained approximately constant [376]. An example of the effectiveness of the method for detecting low concentrations is shown in Figure 4.46. If a computer is available to acquire and integrate the signal, a great deal of sensitivity enhancement can be realized.

Solutions are also studied in much the same manner, but in this case it is easier to vary the flow rate than to vary the point at which the reaction is initiated. The sample can be flowed through a tube a few millimeters in diameter if it is not lossy, or through a capillary or a flat cell if dielectric losses are a problem. Several examples of flat cells are shown in Figure 4.47.

A recent step in flow-reactor design was the application of a high-pressure liquid chromatography pump to impel liquid through a heated capillary tube in the ESR cavity [377]. Using this method, steady-state liquid flows of a few cubic centimeters per minute could be maintained at conditions up to 700°C and 20 MPa. In this way solvents such as benzene or light alkanes can be used. The uniqueness of the system is that the buildup and decay of radicals under thermolytic conditions can be studied easily. This was applied to the pyrolysis of bibenzyl to form toluene [378]

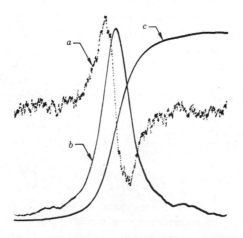

a First derivative output of ESR
b First integral of ESR signal
c Second integral of ESR signal

Figure 4.46 Display of the ESR signal (a) of a sample of flourine atoms at a concentration of about 10^9 atoms/cm^3, (b) integral of the spectrum, and (c) second integral. Such signal processing is required for accurate quantitative measurements.

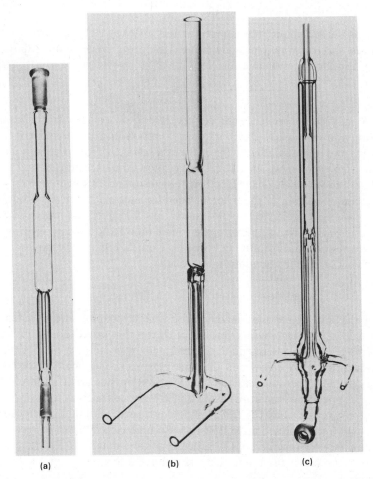

(a)   (b)   (c)

Figure 4.47 Flat cells used in ESR spectroscopy: (*a*) standard flat cell suitable for study of aqueous samples; (*b*) cell with mixing vortex just below the flat, observation portion of the cell; rapidly flowing reactive solutions are mixed intimately just prior to ESR observation; (*c*) mixing cell similar to (*b*) but with the entire assembly enclosed in a Dewar tube to provide control of the sample temperature by thermal contact with a gas stream.

$$C_6H_5-CH_2-CH_2-C_6H_5 \rightarrow 2C_6H_5-CH_2 \cdot \qquad (144)$$

$$C_6H_5-CH_2 \cdot + C_6H_5-CH_2-CH_2-C_6H_5 \rightarrow$$
$$C_6H_5-CH_3 + C_6H_5-\dot{C}H-CH_2-C_6H_5 \qquad (145)$$

A side benefit is that high-resolution spectra are obtained from this procedure.

A detailed review of different techniques in steady-state measurements was published [379], and it contains many useful techniques and ideas for implementing such experiments.

3.8.2   Modulation Techniques

An interesting technique that is more or less a steady-state approach is called *modulation spectroscopy*. This was first used in optical spectroscopy [380] and is ideal for examining reversible reactions

$$A \rightleftharpoons B \tag{146}$$

where B may be an excited state, isomer, dimer, or excimer. In this case, the excitation source, typically illumination, is varied in a periodic manner. The amplitude and phase relative to the excitation source is monitored as a function of excitation frequency, and these data allow the mechanism and kinetic parameters to be determined. Electron spin resonance is directly amenable to this technique because of the inherent need for modulation. If the modulation required for the kinetic study is less than a few kilohertz, the field modulation can be used as a carrier. Otherwise, the source modulation is used directly. Very few such applications are found in ESR [381], partly because the mathematics of solving for the kinetic equations are difficult.

3.8.3   Real-Time Measurements

One of the most important methods for kinetic determinations is real-time measurement. This can reveal transient processes that occur too fast for steady-state detection. The intermediate is generated by a suitable method directly inside the microwave cavity and the time dependence is then recorded. (If the kinetics are slow, then the material can be mixed outside and transferred to the cavity.) Stopped-flow apparatus using an ESR flat cell such as shown in Figure 4.47b or c was described [382, 383]. These are used to obtain time-dependent kinetic data for a representative ESR line. Complete spectra can be recorded by using auxiliary field-sweep coils to generate a rapid magnetic field ramp. When the same ramp is also used to trigger a signal-averaging computer, the S/N ratio is enhanced.

A more common real-time method employs generation by pulsed optical or electron beam radiation. Flash lamps [384, 385] and lasers are ideal for this purpose [331, 386]. Often a higher magnetic modulation frequency (see Section 4) is needed for kinetic measurements if reaction times fall below about 0.1 ms. Significant effort was devoted to modifying spectrometers for 1 to 2-MHz detection [387]. (These references serve as examples and are by no means exhaustive.)

One of the most dramatic results observed by time-resolved ESR is that molecules often dissociate into radicals that are spin polarized

$$A-B \rightarrow A(\cdot\uparrow) + B(\cdot\downarrow) \tag{147}$$

or transfer atoms with spin-polarized electrons

$$R + SH \rightarrow RH(\cdot\uparrow) + S(\cdot\uparrow) \tag{148}$$

where $(\cdot\uparrow)$ and $(\cdot\downarrow)$ indicate, respectively, α- and β-spin polarization. This creates a transient overpopulation of one spin state, typically from 10 to 100-ns duration, and results in very strongly enhanced emission (excessive α spin) or absorption (excessive β spin). This effect is known as *chemically induced dynamic electron polarization (CIDEP)*. Because of the short lifetimes, high-frequency field modulation must be used, or an unmodulated signal must be taken directly from the detector or detector preamplifier. This effect and the nuclear analogue CIDNP are the subject of intense study [388–390] because they reveal the dynamics of the electrons within the molecule.

Two mechanisms are invoked for this effect. In the first, the *radical-pair mechanism*, the dissociated molecular fragments remain together and interact part of the time as singlet states and part of the time as triplet states. The spin densities then vary over different regions of the pair and depend on the nuclear spin states. Thus when the fragments finally separate, different ESR transitions may appear to have opposite polarizations.

The second mechanism, the *triplet mechanism* occurs directly through the triplet state of the precursor. If the fragments separate quickly, then the spin polarization of all transitions are equal and of the same sign. An example of this behavior is the photolysis of *p*-benzoquinone (PBQ) in isopropyl alcohol [391].

$$PBQ \xrightarrow{\;h\nu\;} {}^3PBQ \tag{149}$$

$${}^3PBQ(\cdot\uparrow\cdot\uparrow) + (CH_3)_2CHOH \longrightarrow {}^2PBQH(\cdot\uparrow) + (CH_3)_2C(\cdot\uparrow)OH \tag{150}$$

$$(CH_3)_2C(\cdot\uparrow)OH + PBQ \longrightarrow {}^2PBQH(\cdot\uparrow) + (CH_3)_2C{=}O \tag{151}$$

Detailed explanation of this effect is beyond the scope of this chapter, and the reader is directed to the literature, particularly [389, 390].

3.8.4   Spin Trapping

An important class of reactions provides evidence of free-radical intermediates in systems where the parent radical is too short-lived to permit direct observation. The utility of these so-called *spin-trapping reactions* was recognized independently by three different groups [392–394] and each reviewed the field [395–397]. Usually the spin trap is either a nitroso compound (R—N=O) or a nitrone, often phenyl *N*-tert-butylnitrone [$(CH_3)_3C$—$N(O){=}CHC_6H_5$], and each reacts with unstable free radicals to produce longer lived nitroxide radicals as

$$R_a{-}N{=}O + R_b{\cdot} \rightarrow R_a - \dot{N}(O){-}R_b \tag{152}$$

$$C_6H_5CH{=}N(O)C(CH_3)_3 + R{\cdot} \rightarrow C_6H_5CHR{-}\dot{N}(O){-}C(CH_3)_3 \tag{153}$$

The nitroxide radicals are called *spin adducts*, and since these products are fairly long-lived, the reactions are useful for monitoring the production rate of unstable free-radical intermediates by a technique known as *spin counting* (that is, monitoring the concentration of a spin adduct as a function of time).

Janzen and co-workers [398–400] have reported kinetic data for several reactions such as

$$(CH_3)_3CO\cdot + (CH_3)_3CN{=}O \rightarrow (CH_3)_3CO\dot{N}(O)(CH_3)_3 \qquad (154)$$

$$C_6H_5\cdot + C_6H_5CH{=}N(O)C(CH_3)_3 \rightarrow (C_6H_5)_2CH\dot{N}(O)C(CH_3)_3 \qquad (155)$$

$$C_6H_5CO_2\cdot + C_6H_5CH{=}N(O)C(CH_3)_3 \rightarrow C_6H_5CH(O_2CC_6H_5)\dot{N}(O)C(CH_3)_3$$
$$(156)$$

with second-order rate constants of 1.5×10^6, 1.2×10^7, and 4×10^7 L/(mol·s), respectively.

Spin-trapping reactions have wide utility in studies of mechanism and kinetics; the interested reader should consult the reviews cited previously [395–397]. One noteworthy warning results from recent work by Ram and Stanbury [401]. In their study of Ir(IV) and Fe(III) oxidation of azide ion to the azidyl ($N_3\cdot$) radical, the highly complex kinetics were simplified to pseudo-first order, and the rates were accelerated in the presence of spin traps such as phenyl *N*-tert-butylnitrone. Clearly, care must be exercised that the measuring technique does not alter the course of the reaction of interest.

4   ELECTRON SPIN RESONANCE INSTRUMENTATION

A very rudimentary instrument for carrying out ESR spectroscopy at microwave frequencies is indicated in Figure 4.48, along with a more familiar spectrophotometer to suggest the corresponding functions of several of the components. The source of the microwave radiation is a klystron. The microwaves are conducted to the sample through a waveguide. The sample, contained in the sample tube, is held in a microwave cavity between the poles of a magnet. The detector is typically a diode that produces a direct-current (dc) output related to the level of incident microwave power. The direct current from the crystal is displayed on a recorder or oscilloscope. While it is possible to operate in a mode analogous to visible spectrometers by holding the magnetic field fixed and varying the microwave frequency, in practice it is much easier to hold the klystron frequency fixed and vary the magnetic field because the cavity is a fixed frequency device that acts as a high-Q-tuned circuit.

Although this configuration for an ESR spectrometer is reasonable, it is rarely used because of its poor sensitivity. Various improvements involving modulation of the magnetic field and lock-in (phase-sensitive) or superheterodyne detection give much higher sensitivity and are used in most commercial instruments.

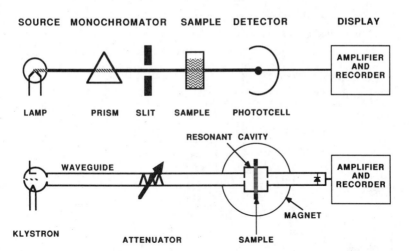

Figure 4.48 Diagram compares the analagous components of an optical spectrophotometer and a rudimentary ESR spectrometer.

4.1 Basic Spectrometer

Since the radiation used in ESR spectroscopy is in the microwave ranges shown in Table 4.1, an understanding of ESR instrumentation requires some knowledge of the operation of microwave components. A brief description of these and other spectrometer components and the design of a typical commercial ESR spectrometer is given here. Detailed discussions of the design and operating techniques of ESR spectrometers are found in the general references, particularly in the books by Poole [371], Alger [402], Wilmshurst [403], and Ingram [404]. A basic spectrometer as shown in Figure 4.49 contains many different components. The radiation source is a klystron tube, which is connected by waveguide or coaxial lines to the various isolators, attenuators, phase shifters and couplers that are needed to condition the radiation and present it to the sample located in a resonant cavity. The microwave radiation is maintained at a fixed frequency by an automatic frequency control (AFC) feedback circuit. The sample cavity is usually connected to one arm of a balanced hybrid T or circulator bridge. Absorption of energy by the sample causes a change in the microwave energy reflected to the diode detector. The magnetic field at the sample is produced by superimposing a sinusoidal modulation field on a much more intense field produced by the large laboratory electromagnet. The latter field is swept slowly (several seconds to several hours) through the field values that encompass the ESR spectrum. The modulation field is obtained by applying an oscillating voltage (at a frequency typically between 35 Hz and 100 KHz) to a set of auxiliary coils, which are often mounted in the walls of the ESR cavity parallel to the pole caps of the laboratory magnet. This compound field causes the amount of microwave energy that the sample absorbs to be modulated at the

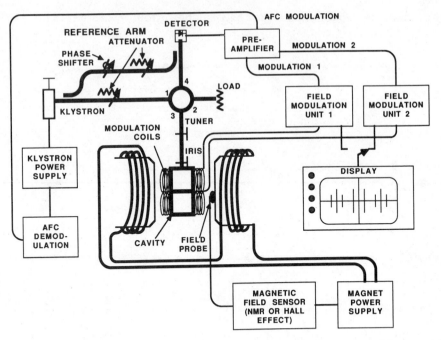

Figure 4.49 Block diagram of a basic ESR spectrometer employing reference arm and dual sample cavity.

same frequency. The resulting alternating-current (ac) signal that appears at the microwave detector crystal is subjected to several stages of amplification, followed by demodulation in a phase-sensitive detector. The resulting direct-current signal is then presented to a recorder. An important consequence of this detection scheme is that the demodulated signal is a close approximation to the first derivative of the absorbed power with respect to the magnetic field applied by the laboratory magnet (provided that the modulation amplitude is small in comparison to the ESR line width).

4.2 Spectrometer Components

4.2.1 Klystrons

The source of radiation is a klystron. These tubes are available for frequencies from 2.5 to 220 GHz, although most analytical work is performed in the X-band (3-cm) region at about 9 GHz. A klystron can be tuned over about ±3% of its central frequency by varying the dimension of the resonant cavity inside the tube. The output frequency is also a function of the resonator and reflector voltages applied to the klystron by the power supply. It is usually stabilized against temperature variations by immersion in an oil bath or by forced air or water cooling. A feedback AFC circuit, utilizing a portion of the signal reflected from the sample cavity, stabilizes the klystron frequency to about 1 to 10 ppm.

The output power of typical klystrons used in ESR spectrometers is about 150 to 1000 mW. Recently, solid-state Gunn diodes were also used for ESR sources. These devices are available with noise levels that compare favorably with klystron below about 8 GHz, but they are limited in power to about 100 mW.

4.2.2 Waveguides, Attenuators, Isolators

The microwave radiation is conveyed to the bridge, sample, and detector by a waveguide, which is a hollow, rectangular tube of copper or brass often plated with silver or iridium for better electrical conductivity. The dimensions of the waveguide depend on the wavelength of the microwave radiation employed. For X band, the dimensions are 0.9×0.4 in. for a rectangular waveguide. The microwave power propagated down the waveguide is decreased by inserting a variable attenuator. This device can be either a piece of resistive material that is inserted into the waveguide or an adjustable polarizer that is rotated relative to a second stationary polarizer. This latter is known as a *rotary vane attenuator* and produces extremely small phase shifts as compared to other types of attenuators. The power at the sample may thus be attenuated by a maximum factor of 10^2 to 10^6 (20–60 dB). This adjustment is important in preventing saturation. Reflection of microwave power back to the klystron is prevented by an isolator, which is a strip of ferrite material that passes microwaves in only one direction. This helps to protect the klystron and to stabilize its frequency.

4.2.3 Cavities

In most ESR measurements the sample is contained in a resonant cavity (Figure 4.50). This is essentially a piece of waveguide one or more half-wavelengths long in which a standing wave is set up. The cavity is analogous to a tuned circuit (e.g., a parallel R-L-C combination) used at lower frequencies. A measure of the quality of the cavity, which directly affects spectrometer sensitivity, is its Q-value or Q-factor, defined as

$$Q = \frac{\text{energy stored in cavity}}{\text{energy lost per cycle}} \tag{157}$$

The standing wave pattern in the cavity has orthogonal magnetic and electric field components

The cavity Q-factors and frequencies are very sensitive to geometry; and it is worthwhile considering several different geometries that are widely used; namely, the rectangular TE_{102} cavity and the cylindrical TE_{011} and TM_{110} cavities. Cavities are designated as the TE_{lmn} (transverse electric field) or TM_{lmn} (transverse magnetic field) mode with subscripts l, m, and n. For TE modes these subscripts designate the number of magnetic half-wavelengths in a given direction. For the rectangular cavity, these designate the numbers, respectively, in the height (a), width (b), and length (L) directions. For the cylindrical cavity, these subscripts designate the number of magnetic half-wavelengths along,

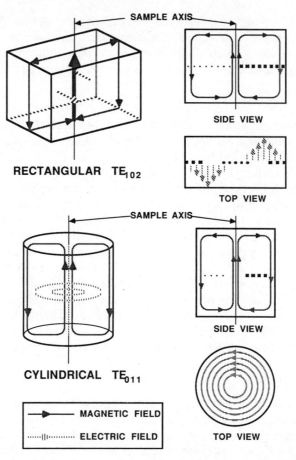

Figure 4.50 Diagrams showing the geometries and the electric and magnetic components of the microwave standing-field patterns in a rectangular TE_{102} and a cylindrical TE_{011} cavity.

respectively, the circumference, the radius (a), and the length (L). For TM modes the subscripts designate the number of half-wavelengths of electric field in the corresponding directions. An excellent discussion of cavities, magnetic and electric field patterns, Q-factors, and frequency is given in [371–373].

A rectangular cavity is shown in Figure 4.51 along with the waveguide connections and impedance matching devices used on spectrometers manufactured by Varian Associates, Inc. Typically, the Q of such a cavity without a sample (the *unloaded Q*) is about 7000.

Since the component of the microwave radiation that interacts with the paramagnetic sample is the radio frequency magnetic field, the sample is placed in the cavity at the location where this field is greatest. However, the radio frequency electric field also interacts with the electric dipoles in the sample; and if the sample has a high dielectric constant (i.e., is *lossy*), the Q of the cavity may

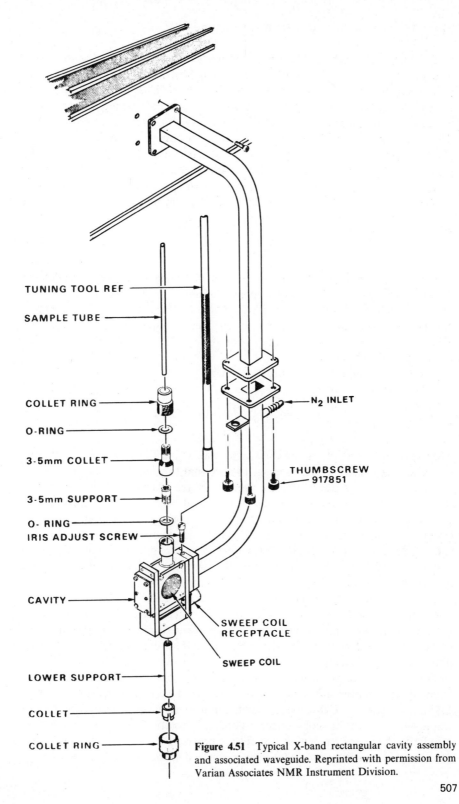

TUNING TOOL REF

SAMPLE TUBE

COLLET RING

O-RING

3-5mm COLLET

3-5mm SUPPORT

O-RING

IRIS ADJUST SCREW

CAVITY

LOWER SUPPORT

COLLET

COLLET RING

N₂ INLET

THUMBSCREW
917851

SWEEP COIL
RECEPTACLE

SWEEP COIL

Figure 4.51 Typical X-band rectangular cavity assembly and associated waveguide. Reprinted with permission from Varian Associates NMR Instrument Division.

507

be decreased significantly. Therefore, the sample is usually located in the cavity in a position of maximum radio frequency magnetic field and the minimum radio frequency electric field. Rectangular flat cells such as shown in Figure 4.47 with a thickness of about 0.25 mm and sample volume of 0.05 mL often are used for aqueous samples (which are particularly lossy) in rectangular cavities. For the same reason, biological tissue samples are confined to planar geometry by placing them between flat, demountable quartz plates. With such a cell, the loaded Q of the cavity is about 2250. Tubing of 3 to 5-mm i.d. with sample volumes of about 0.15 to 0.5 mL can be used with samples that are not lossy.

The frequency f of a TE_{10n}-mode cavity is given by

$$f = \frac{c}{2}\left[\frac{1}{a^2} + \left(\frac{n}{L}\right)^2\right]^{1/2} \tag{158}$$

where c is the speed of light and the other symbols have the meanings defined earlier. The TE_{104}-mode cavity is twice the length and twice the number of magnetic half-wavelengths as the TE_{102} mode. If both cavities have the same a and b dimensions, they are resonant at the same frequency. For example, the TE_{102} cavity of Figure 4.51 has dimensions of $a = 22.7$ and $b = 10.1$ mm and is resonant at about 9.5 GHz.

The Q factor of a TE_{10n} cavity is given by

$$Q = \left(\frac{c}{f\delta}\right)\left(\frac{abL}{2}\right)\left(\frac{1}{a^2} + \frac{n^2}{L^2}\right)^{3/2}\bigg/\left[\frac{L}{a^2}(a + 2b) + \frac{n^2a}{L^2}(L + 2b)\right] \tag{159}$$

where δ is the skin depth of the cavity material. Table 4.17 gives the skin depths for various materials. It is important to note that the smaller is the skin depth, the higher is the Q; therefore, the more sensitive is the spectrometer. These theoretical calculations are developed for an ideal cavity that is a sealed box. In practice, the frequency is slightly lower than one predicts because of the openings required for the sample and the microwave coupling iris. The Q-factor is rarely greater than 70% of the theoretical value because of the presence of the openings, resistive junctions between the various cavity components, and imperfections in the walls. Since the Q-factor depends strongly on efficient electrical conductivity in the cavity walls, the electrical contact between the sidewalls, end plates, and the frame of the cavity is very critical to attaining high Q. If material spills in the cavity, it is important to clean it immediately to prevent deterioration of performance. Scratches that intersect the mode current paths on the walls also degrade the Q.

Other variations on the basic cavity design include rotatable cavities for studying anisotropic effects in single-crystal and solid-sample studies. Dual cavities, consisting of two TE_{102} cavities joined to form a TE_{104} cavity, are employed for simultaneous spectroscopic observation of a sample and a standard. These are particularly useful for precise g-value determinations and quantitative measurements. Slots are machined into the walls of the cavity

Table 4.17 Resistivities and Skin Depths of Common Metals[a]

Metal	Conductivity, σ $(\Omega \cdot m)^{-1}(\times 10^{-7})$	Skin Depth $(\delta)^{b}$ 100 kHz (mm)	9.6 GHz (μm)
Silver	6.13	0.203	0.656
Copper	5.80	0.209	0.674
Gold	4.10	0.249	0.802
Aluminum	3.54	0.267	0.863
Rhodium	1.98	0.357	1.153
Irridium	1.89	0.366	1.18
Tungsten	1.79	0.380	1.226
Molybdenum	1.75	0.383	1.237
Brass	1.57	0.404	1.306
Platinum	1.00	0.503	1.624
Palladium	0.91	0.578	1.704
Titanium	0.232	1.04	3.37
Magnetic Metals (δ For Electrical Component Only)			
Nickel	1.28	0.444	1.435
Iron	1.03	0.495	1.60
Cobalt	1.6	0.397	1.28

[a]These values are either values for the element at room temperature or values that depend on the alloy and seem to be appropriate for cavity calculations.
[b]$\delta = (\pi\mu_0\sigma f)^{-1/2}$.

(parallel to the current direction in the walls) without degrading appreciably the cavity Q-factor. Observation of species generated by irradiation of the sample with ultraviolet or visible light are thus carried out. Cells are also available for electrogeneration of paramagnetic species directly in the cavity and for chemically producing them by flow-mixing reagent streams near the sensitive portion of the cavity. Transverse electric field TE_{101}-mode microwave cavities are available for very lossy samples. This cavity geometry provides a twofold increase in the S/N ratio over TE_{102}-mode cavities when used with aqueous samples [405].

Cylindrical TE_{011} cavities exhibit high Q-factors, typically from 12,000 to 18,000 at X band. In this mode, the microwave electric field is strongest midway between the cylindrical walls and the cavity axis. Consequently, cylindrical sample tubes that are small enough to avoid encroachment of the electric dipoles into this region are appropriate sample holders. High loss materials such as aqueous samples can be examined only in capillary tubing. The cylindrical geometry makes it possible to rotate the magnet around the cavity to study orientation dependence when high angular precision is required.

The TE_{01n}-mode cavities have n half-wavelengths in the cylindrical height (L) direction. Therefore, for $n = 1$, the most sensitive position is the center of the cavity. For $n = 2$ the most sensitive position is displaced from the cavity center by about $L/4$. So-called X-band *wide-access* cavities, which can accommodate a 1-in.-diameter sample tube for gas-phase work, are often called TE_{10n} because they are tunable from $n = 1$ to 4 (when empty). However, when a quartz tube is placed in the cavity, the lowest usable frequency is the TE_{012} mode; and when a large-diameter tube is needed, only quartz can be used because Pyrex glass is too lossy.

The resonant frequency of a TE_{01n} cavity is given by

$$f = \frac{c}{2}\left[\left(\frac{3.8317}{\pi r}\right)^2 + \left(\frac{n}{L}\right)^2\right]^{1/2} \tag{160}$$

which is the same functional dependence as the TE_{102} rectangular cavities. However, since the dominant term may be the first term under the square root, depending on design, the frequency may change only by a few hundred megahertz for increments of n. The theoretical Q-factor of the cavity is given by

$$Q = \left(\frac{c}{f\delta}\right)\frac{3.8317}{2\pi}\left[1 + \left(\frac{n\pi}{2 \times 3.8317}\right)^2\left(\frac{a}{L}\right)^2\right]\left[\frac{1}{1 + \left(\frac{n\pi}{2 \times 3.8317}\right)^2\left(\frac{a}{L}\right)^3}\right] \tag{161}$$

In this cavity, the current lines are primarily circumferential, so that thin slits can be cut in the walls or arcs cut in the end plates without severe degradation of Q. In addition, the junction between the end plates and the cylindrical walls is not critical. However, the body of the cavity must be cut from a single piece.

A third design that is particularly suitable for lossy samples is the TM_{110} mode cavity [406]. These cylindrical cavities are shortened in the height direction. For example, an empty cavity resonant at about 9.4 GHz is about 39-mm diameter and only 19-mm thick. The axis of the cavity is parallel to the applied magnetic field, and the microwave magnetic field pattern appears similar to that of the TE_{102} cavity in Figure 4.50 except the magnetic field pattern resembles a pair of cyclotron "dees" rather than rectangles with rounded corners. The flat sample is located in a plane coinciding with a diameter of the cylinder. Because of the nature of the cavity, a 16–18-mm-wide flat cell of about 0.3-mm thickness offers about a 6:1 enhancement of S/N ratio when compared to the TE_{102} cavity. This, in part, results from a greater filling factor and a lesser intrusion of the sample into microwave electric field region.

Commercially available cavities for X-band spectrometers include TE_{102}, TE_{103} (designed for optical irradiation), TE_{104} (dual sample), TE_{011}, TE_{01n} (wide access), and TM_{110} (aqueous sample cavity) models. While the specific choice depends on the details and constraints of the experiment, some general recommendations can be made. For general purpose work, the TE_{102}-mode cavity is best. Slightly lower S/N ratio is obtained with the TE_{104}-mode cavity, but it is nonetheless ideal for simultaneous recording of sample and reference spectra if the field homogeneity is adequate. For large-volume, lossy samples, the TM_{110}-mode cavity is generally the best choice. For small samples, with low-to-moderate loss, the TE_{011} cavity offers best performance. This is rivaled by the split-ring cavity described below. Finally, for gas-phase samples, the TE_{01n} cavity is best; but for small magnets with a limited gap opening, often the TE_{102}- or TE_{011}-mode cavities can provide adequate performance for a wide variety of useful experimentation.

The *split-ring* or *loop-gap* cavity was recently applied to magnetic resonance [407]. The cavity consists of an outer *guard* of the approximate dimensions of a *TE* mode cavity of the same frequency and an inner loop or ring that is open at one or more places. The electric field is maximum within this gap, while the magnetic field is maximum within the center of the loop. The frequency and Q-factor depend on the diameter, thickness, and length of the loop; the width of the gap or gaps; and the dimensions of the guard. Theoretical expressions are available for several geometries [408]. Although several cavities were marketed, these are generally designed by trial and error for specific needs. They are used at frequencies between 200 MHz and 35 GHz. Maximum Q-factors are typically about 1000. The major advantage of these cavities is that they exhibit an extremely large filling factor and good spatial separation of the microwave electric and magnetic fields. Separation was further improved by adding an internal guard between the electric and magnetic field regions [409].

Microwave cavities were also used for in vivo ESR studies. These are large reentrant cavities that are designed for use at about 1 GHz. At this frequency the dielectric loss caused by water is minimal. Two examples were reported recently [410, 411]. These cavities are modeled as a two-ridge waveguide [412, 413] or as large split-ring cavities as described previously. As an example of performance, 10^{-6} mol of a TEMPOL spin label was detected in a 0.025-kg mouse placed in the cavity [411].

While microwave cavities are usually modeled as closed structures without openings, provision must be made for coupling the resonant structure to the microwave propagation network. Frequently, irises or slots of various sizes are used. For example, the resonant cavity can be attached to the waveguide with a standard-sized flange and coupled by an iris. Matching of waveguide elements (which is analogous to impedance matching in conventional circuits) is accomplished by screws or stubs, which can be positioned in the waveguide (Figure 4.49, slide screw tuner) or across the coupling iris (Figure 4.51, iris adjustment screw). Microwave antennas can be used to couple waveguide to coaxial cable. These, in turn, interface to the cavity through coupling loops that excite either the microwave magnetic or electric field in the cavity. Magnetic mode coupling loops are the most common and are easily adjusted. Dipoles that excite the electric field are much more difficult to adjust and are seldom used.

4.2.4 Detectors

Detectors used for microwave frequencies of 1 GHz and above are usually rectifying diodes analogous to the standard diodes that are encountered in a power circuit, but they operate at higher frequencies. The incident microwave power is converted to rectified direct-current output, which is then filtered by the detector circuit. In general, detectors are constructed of a small piece of semiconductor material such as silicon or gallium arsenide fabricated in a P–N junction. The semiconductor is placed on a ground plane, and electrical contact to the semiconductor is made through a thin filament ending in a small needle to minimize capacitance in the circuit, hence the name *point-contact diode*. Since circuit capacitance is the primary limitation on the useful frequency range, there are numerous proprietary designs that optimize performance in specific frequency regions.

Figure 4.52 shows the output voltage of a typical detector diode as a function of incident power. As indicated on the figure there are two operating regions of a detector: the square law and the linear regions. The square-law region occurs at incident powers less than about 10 μW. At these powers, the output power P_{OUT} is proportional to the square power of the input power P_{IN}. As a result, the output voltage V_{OUT} is linear with respect to the input power through the output resistance R_{OUT} as shown in

$$P_{OUT} = V_{OUT}^2/R_{OUT} \propto P_{IN}^2 \tag{162}$$

$$V_{OUT} \propto P_{IN} \tag{163}$$

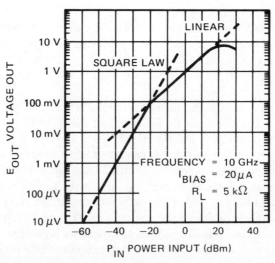

Figure 4.52 Voltage out versus incident power for a typical detector in the X-band frequency range. Reprinted with permission from FEI Microwave, Inc. (formerly, TRW Microwave), 825 Stewart Drive, Sunnyvale, CA 94086.

and in Figure 4.52. In the linear region, the output power is linear with the input power

$$P_{OUT} = V_{OUT}^2/R_{OUT} \propto P_{IN} \tag{164}$$

so that

$$V_{OUT} \propto P_{IN}^{1/2} \tag{165}$$

Typically ESR experiments are carried out with the detector biased well into the linear region to obtain optimum sensitivity. This is accomplished by a *reference arm* as shown in Figure 4.49. Usually the bias power P_{BIAS} is about 0.1–1 mW so that the power absorbed by the sample P_{ABS} is only a small perturbation to the total power incident on the detector. For all practical purposes then the detector output is a linear function of the ESR absorption. [That is, $V_{OUT} \propto P_{IN}^{1/2} \propto (P_{BIAS} - P_{ABS})^{1/2}$. Since $P_{BIAS} \gg P_{ABS}$, the change in output voltage with ESR absorption $\delta V_{OUT}/\delta P_{ABS}$ is proportional to $\frac{1}{2} P_{ABS}$.] Because the power absorbed by an unsaturated sample is proportional to B_1 [see (27b) and (37b)], the ESR signal is in turn proportional to $P_{IN}^{1/2}$.

The reference arm shown in Figure 4.49 also contains a phase shifter, which is used to match the phase of the bias microwave power to that of the power reflected from the cavity in order to discriminate against the ESR dispersion signal, which differs from the absorption signal by 90°.

Detectors are subject to noise that results from thermal excitation of electrons into the conduction band of the semiconductor diode. Such thermal noise is called *flicker noise* or $1/f$ *noise* because it decreases linearly with the reciprocal of the frequency. Eventually it falls below the frequency-independent noise at a value called the *noise corner*. Until around 1970 the available silicon detectors exhibited noise corners of about 100 kHz. For this reason, 100-kHz field modulation became standard in ESR. The disadvantage of 100-kHz modulation is that it creates 3.6-μT sidebands on the spectral lines. Newer Schottky barrier detectors exhibit noise corners of about 10 kHz.

Before Schottky barrier diodes became available, investigations of narrow ESR lines required use of a technique called *superheterodyne detection* to avoid sideband distortion of line shapes. Superheterodyne systems utilize two klystrons, one tuned to the microwave cavity and the second shifted from 30 to 60 MHz from the first. These signals are fed into a mixer that encodes the ESR signal on the difference frequency, which can be amplified readily. However, the newer detectors made this method obsolete except for very low temperature work where extremely low microwave power is required to avoid saturation.

The detector sensitivity is usually reported in terms of a noise figure (NF) which is usually expressed in decibels

$$NF = 10 \log_{10} \frac{\text{noise power from detector}}{\text{thermal noise from equivalent resistor}} \tag{166}$$

The value for NF usually increases with frequency. While this number is very meaningful, it is generally not as convenient as the tangential sensitivity (P_{TSS}) given by

$$P_{TSS} = \frac{3.22\sqrt{(B\,NF_D R_V)} \times 10^{-7}}{K} \qquad (167)$$

where (167) expresses P_{TSS} in milliwatts but it is usually reported in decibels relative to one milliwatt (dBm); B is the detection bandwidth in hertz; NF_D, the noise figure of the detector used in the measurement, is expressed as a ratio; R_V is the effective resistance in ohms; and K is the detector sensitivity in volts per watt (typically 300–2000) and for small signals can be considered the derivative or the slope of the curve in Figure 4.52. Representative performance figures are that P_{TSS} is the power required to give an 8-dB (6.3) S/N noise ratio, and B is 2 MHz. Equation (167) illustrates the effect of various parameters. A typical value of P_{TSS} for a good detector is about -53 dBm (5 nW). For continuous ESR measurements, B is about 1 to 10 Hz, so that it is evident the actual sensitivity is about 5 pW.

Figure 4.53 shows performance characteristics of several detector diodes. The silicon point-contact diode is typical of detectors used before about 1970. Optimal noise figures are about 7.0 to 8.0 dB from about 100 kHz to 12 GHz. Below 100 kHz the flicker noise dominates. This detector can be operated as high as about 40 GHz, depending on design, but performance falls off rapidly beyond about 20 GHz. The Schottky barrier diode exhibits noise figures as low as 5.5 or 6.0 dB and can perform to much higher frequencies than the point-contact diode. Some detectors operate as high as about 80 GHz depending on design. The flicker noise limit is about 10 kHz, so that it allows for much lower

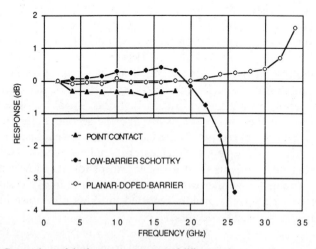

Figure 4.53 Comparison of the frequency response of different microwave detectors; based on data reprinted, with permission, from Hewlett Packard Company.

field modulation frequency than the point-contact diode. The convention of 100-kHz field modulation arose because of the noise characteristics of the silicon point-contact diode. However, in many studies of atoms in the gas phase or organic radicals in liquids, the 100-kHz modulation sidebands (3.6 μT) distort the line shape. The Schottky barrier diode makes it possible to use lower frequency modulation for such narrow lines, although 100 kHz is still commonly used for less exacting studies. There are also detectors that make use of newer technology, such as the Hewlett-Packard Company Planar-Doped-Barrier diode. This product is not yet applied generally in ESR spectrometers.

Most recently, solid-state, low-noise microwave amplifiers are available for frequencies up to 40 GHz. High-quality gallium arsenide field-effect transistor (GaAs FET) amplifiers with noise figures as low as 2.3 dB in the X-band region are available. These are expensive and require careful installation and sophisticated circuit protection to prevent damage. They provide no advantage for routine applications, but they are indispensable for detecting very low-intensity signals and in certain pulse experiments.

4.2.5   Magic-*T* and Circulator Bridges

Rather than employ a detection technique that requires the observation of a small decrease in a large base signal, which prevents very high amplification, a bridge arrangement is generally used. This permits the observation of a small microwave signal and is analogous to the resistance-bridge arrangement used in gas chromatography. Microwave bridges (which are analogous to impedance bridges in conventional circuits) can be of the *magic-T* (or *180°-hybrid-T*) or *circulator* variety. These are shown in Figure 4.54. In the hybrid-*T*, power from the klystron and attenuator entering arm 3 divides equally between arms 1 and 2, which terminate in a resonant cavity and a matched load, respectively. If the impedance of arms 1 and 2 are the same, no power will enter arm 4 to the detector. Under these conditions the bridge is *balanced*. If the impedance of arm 1 changes, say because the Q of the resonant cavity changes when ESR absorption by the sample occurs, the bridge becomes unbalanced, and the difference between the power reflected from arms 1 and 2 enters arm 4 and is detected as the ESR signal.

Circulators behave in a way similar to hybrid-*T* bridges, but the numbering convention is different. Microwave power enters port 1 but is divided between the clockwise and counterclockwise paths. Since each path must complete a loop, twice as much power is incident on the cavity. As before, the difference between the power reflected from the cavity arm and the load arm, at ports 2 and 4, enters port 3, which houses the detector. This bridge geometry (whether waveguide, ferrite, or strip line) has an advantage over the hybrid-*T* arrangement because it presents twice as much power to the detector and to the cavity; thus the S/N ratio is doubled with a circulator element as the microwave bridge. A magic-*T* bridge is used in the Varian V-4502 and E-3 spectrometers, while a circulator is employed in the Varian *E* line, series and in spectrometers

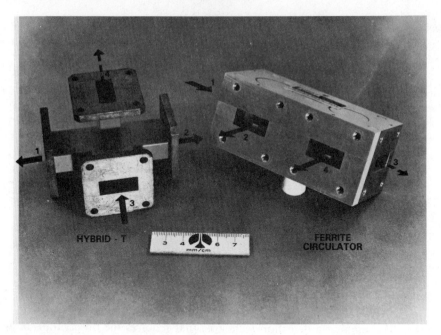

Figure 4.54 Comparison of hybrid-T (or magic-T) bridge element and a microwave circulator.

manufactured by Bruker Analytische Messtechnik (previously marketed in the USA by IBM Instruments, Inc., a former subsidiary of International Business Machines Corporation).

4.2.6   Magnets

An electromagnet capable of producing fields of at least 500 mT is required for X-band ESR. The homogeneity of the field for solution studies should be about 5 μT for studies of organic radicals and 0.1 to 0.2 mT for most transition metals over the ESR sample region. The homogeneity and stability requirements are much less critical than those for high-resolution NMR studies; therefore the stabilizing systems, shimming coils, and sample spinning techniques used in NMR are not necessary. Thus, magnets with 4 or 6-in. pole-piece diameters are employed in ESR spectrometers, although 9-in. or larger magnets offer wider gaps and greater field capability than smaller magnets. The ESR spectrum is recorded by slowly varying the magnetic field through the resonance condition by sweeping the current supplied to the magnet by the power supply; this sweep can be accomplished with a variable-speed motor drive or with analogue or digital ramp circuits. Both the magnet and the power supply may require water cooling. Generally the magnetic field is regulated by using a feedback circuit to sense changes in the magnetic field and correct for these changes. This approach involves the use of a field-strength sensor (e.g., a Hall probe or rotating coil magnetometer), which generates a signal that is proportional to the field.

4.2.7 Modulation Coils

The modulation of the signal at a frequency consistent with a good S/N ratio in the crystal detector is accomplished by imposing a small sinusoidal variation on the laboratory magnetic field. Peak-to-peak modulation amplitudes from 0.005 to 4 mT are commonly used. This variation is produced by supplying an alternating-current signal to modulation coils mounted parallel to the pole faces of the laboratory magnet. For low-frequency modulation (400 Hz or lower) the coils are mounted outside the cavity or on the magnet pole pieces. Higher modulation frequencies (1 KHz or higher) cannot penetrate metal effectively (see Table 4.17), and either the modulation coils must be mounted inside the resonant cavity or cavities constructed of a nonmetallic material (e.g., quartz, plastic, or ceramic) with a thin silver plating must be employed.

At high-modulation amplitudes (e.g., larger than 1 mT), heating of the cavity can occur. This causes dimensional changes in the cavity and may result in baseline drift.

4.3 Sensitivity

The sensitivity of a spectrometer, in terms of the smallest detectable concentration of a given species, depends on spectral characteristics (e.g., line width, relaxation times) and sample characteristics (e.g., solvent). The absorption signal depends on the magnetic susceptibility of the sample, which, in turn, is a function of the number of unpaired electrons or spins contained in the cavity. In general, the minimum number of detectable spins N_{min} is proportional to several variables [371, 414]

$$N_{min} \propto \frac{\Delta v}{Q_0 \eta v_0} \left(\frac{\Delta f k T}{2 P_0} \right)^{1/2} \tag{168}$$

where Δv is the line width of the absorption, Q_0 is the Q of the unloaded cavity, η is the filling factor described in (169), v_0 is the resonance frequency of the absorption, Δf is the bandwidth of the response of the spectrometer, P_0 is the incident power, k is the Boltzmann constant, and T is the detector temperature in degrees kelvin. The filling factor measures the effectiveness of coupling the sample to the microwave magnetic field and is expressed as the integral of the power density over the sample divided by the integral of the power density over the volume of the cavity

$$\eta = \int_{V_s} B_1^2 \, dV_s \Big/ \int_{V_c} B_1^2 \, dV_c \tag{169}$$

where B_1 is the radio frequency magnetic field, V_s is the sample volume, and V_c is the cavity volume.

To maximize the sensitivity of the spectrometer, several parameters in (168) must be optimized. For example, the incident power can be increased, assuming that the absorption is not saturated, and the extent of filtering can be increased (i.e., the Δf decreased). Often it is possible to use a higher microwave frequency (and thus a higher B) to increase the population difference between the two energy levels. For a given quantity of sample, where the geometry is kept the same (except for the cavity dimensions, which are inversely related to the frequency), it can be shown that the sensitivity is proportional to $v_0^{7/2}$. In practice, however, the cavity Q decreases at higher frequencies, and klystrons and detectors are noisier. Theoretically, a 100-fold increase in sensitivity is predicted on increasing the frequency from 9 to 35 GHz; in fact, only a tenfold gain is usually realized. On the other hand, if the sample size is unlimited, the sensitivity per unit volume of sample actually decreases at higher frequencies. If the same geometry is maintained, the minimum detectable concentration is theoretically proportional to $v_0^{-1/2}$.

Finally, consider the filling factor and the cavity Q. For a given sample the best S/N is obtained when the cavity Q is reduced to two-thirds of that of the unloaded cavity by introducing the sample. This puts some constraint on the filling factor for a given sample with a minimum reduction of Q. As a general rule, at X-band frequencies, rectangular cavities provide better performance for lossy samples, while cylindrical cavities show better sensitivity for small-volume or solid samples. For X-band spectrometers concentrations of about $10^{-9} M$ can probably be determined in samples with very low dielectric losses. For aqueous solutions, $10^{-7} M$ probably represents a realistic estimate of the lower limit of detection. Of course, signal-averaging techniques can improve the S/N ratio and increase the effective sensitivity, assuming that the magnetic field and spectrometer parameters are maintained constant over the time interval necessary to accumulate the signals. Signal averaging is probably more useful in the detection and identification of species than in their quantitative determination.

5   ADVANCED EXPERIMENTAL TECHNIQUES

5.1   Multiple Resonance Techniques

Many combinations of multiple resonance experiments were carried out with electron spin as one of the multiple resonance conditions. Among the most common are electron nuclear double resonance (ENDOR); electron electron double resonance (ELDOR); and electron nuclear TRIPLE resonance, often called ENDOR TRIPLE or TRIPLE resonance. Several recent monographs are dedicated to these subjects. References [415, 416] provide detailed descriptions of ENDOR and ELDOR, while [416] also includes TRIPLE resonance. Since this active research area can command an entire chapter, this section of our chapter merely provides an introduction to these techniques.

5.1.1   Electron Nuclear Double Resonance

The ENDOR technique is a method by which the NMR spectrum of a paramagnetic species is recorded, but the use of normal NMR techniques is precluded because of the enhanced relaxation rates of nuclei in the vicinity of unpaired electrons. As a result, special techniques are needed to record the spectrum. The main advantages of ENDOR are its high resolution, which permits resolution of overlapping lines, and its ability to allow identification of the specific atoms that contribute to the ESR spectrum. Its main disadvantages are relatively low sensitivity, high cost, and experimental difficulties. Electron nuclear double resonance modules are available as accessories to conventional spectrometers (see Appendix 2).

Figure 4.55 provides a basic understanding of the ENDOR experiment. Inevitably, the magnitude of the nuclear Zeeman energy E_N

$$E_N = -g_N \mu_N B M_I \tag{170}$$

is smaller than that of the electron Zeeman energy E_E

$$E_E = g \mu_B B M_S \tag{171}$$

because of the large differences in the magnetic moments between nuclei and

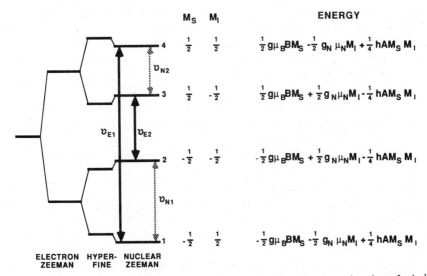

Figure 4.55 Energy-level diagram for ENDOR spectroscopy of an electron and nucleus of spin $\frac{1}{2}$. The diagram shows the electron Zeeman, hyperfine and nuclear Zeeman energies, and the two possible electron spin (v_{E1} and v_{E2}) and nuclear spin (v_{N1} and v_{N2}) transitions.

electrons. In general, the hyperfine energy E_H

$$E_H = hAM_S M_I \tag{172}$$

is larger than the nuclear Zeeman energy. For example, at a magnetic field of 330 mT, the electron splitting is about 9248 MHz, and the proton Zeeman splitting is 14.05 MHz. This value corresponds to a nuclear Zeeman splitting of about 0.05 mT. In organic radicals, α-proton hyperfine splittings can be as large as 2.5 mT.

The possible transitions are shown in Figure 4.55. The allowed electron transitions ν_{E1} and ν_{E2} do not alter the nuclear Zeeman energy because the nuclear spin states are stationary for these transitions; nor do the allowed nuclear transitions ν_{N1} and ν_{N2} alter the electron Zeeman energy because the electron spin states are likewise stationary at these frequencies. However, because of fast nuclear relaxation and small population differences among the nuclear spin states, normal NMR detection does not provide sufficient sensitivity. First, the NMR lines are too broad, and second, the bandwidth of the NMR spectrometer is not sufficient to cover the frequency range spanned by the hyperfine frequency range. To overcome this difficulty, experiments are carried out at constant field by partially saturating the ESR transition with microwave energy, that is, transition ν_{E1}, and scanning the radio frequency (rf) range. In this example, the saturating energy pumps some population of state 1 to state 4. This increases the population difference between states 3 and 4 and between states 1 and 2 by the same amount, which creates a population difference far greater than the equilibrium thermal population difference. When the NMR frequency is reached, the population difference created by ESR saturation is reduced, and a change in the absorption of the microwave energy occurs as manifested by a change in the power incident on the microwave detector. Because saturation can only create equal populations of states, use of excessive power when pumping the ESR transition can cause the absorption of energy to be less pronounced than if the transition is only partially saturated. Other reasons for this relate to relaxation rates and instrumental features beyond the scope of this chapter, and the reader is directed to [415, 416]. It is noteworthy that the intensity of the ENDOR transition depends on the various relaxation rates between nuclear and electron spin states as well as between opposite nuclear and electron spin states (cross relaxation). However, the ENDOR signal (which is only a few percent of the optimal ESR signal) is enhanced by approximately 10 over the NMR transition [417].

The frequencies of the NMR transitions shown in Figure 4.55 can be computed. The frequency ν_{N1} for the transition of state 1 to state 2 is given by $(E_2 - E_1)/h$, where the subscript refers to the state shown in Figure 4.55. Specifically,

$$\nu_{N1} = g_N \mu_N B - \tfrac{1}{2} hA \tag{173}$$

and

$$v_{N2} = g_N \mu_N B + \tfrac{1}{2}hA \qquad (174)$$

Thus, the ENDOR spectrum for the system depicted in Figure 4.55 consists of a pair of lines centered at the frequency corresponding to the nuclear Zeeman transition and separated by the hyperfine splitting.

It is easily appreciated by examining diagrams such as Figure 4.55 and the lower portions of Figures 4.13 and 4.14 that a nucleus of any spin or a set of equivalent nuclei gives rise to only two lines. Thus, the biphenylene and naphthalene anions shown, respectively, in the upper portions of Figures 4.13 and 4.14 should each produce four ENDOR lines. Assuming the microwave resonance frequency to be 9248 MHz as previously, the spectrum of biphenylene anion (hfsc 0.223 and 0.019 mT) consists of four lines located at 7.80, 13.52, 14.58, and 20.3 MHz. The spectrum of the naphthalene anion (hfsc 0.490 and 0.183 mT) consists of three lines at 8.93, 19.17, and 27.78 MHz. The line predicted at 0.32 MHz may not be observed because the bandwidth of the ENDOR spectrometer may be too limited to permit measurements at such a low frequency. The high resolution of the NMR transitions combined with the great decrease in the number of ENDOR lines as compared to the ESR spectrum provide the reason for the popularity of ENDOR in the study of organic radicals [418]. An example is shown in Figure 4.56, taken from [92]. Analysis of the ESR spectrum was virtually impossible without the aid of ENDOR. Note also that pairs of lines in the spectrum are symmetrically displaced on either side of the proton resonance frequency v_H as in the preceding description.

The typical ENDOR spectrometer consists of several additional components beyond those shown in Figure 4.49. First, a broadband high power radio frequency source is needed to sweep the nuclear resonance frequency range. This usually consists of a frequency synthesizer that can be swept through the necessary frequency range and a power amplifier to deliver the required intensity. These components constitute the major cost of the unit. For proton resonance applications, a frequency range from 5 to 30 MHz is usually sufficient if the low-frequency lines are not needed. However, for multinuclear applications where the hyperfine couplings can be large, a range from 0.5 to 100 MHz or more is desirable. Since broadband circuitry must be used, the Q-factor of the system is very low. Therefore, high power is needed; typically this is 100–300 W, although even greater power may be required in some applications. The ENDOR cavity must also have provision for the radio frequency sweep coils. The radio frequency magnetic field must be normal to both the microwave and static magnetic fields. A typical geometry is shown in Figure 4.57, where the sample is placed in a cylindrical cavity, and the radio frequency coils are placed on a quartz support, such as a variable temperature Dewar insert. In some spectrometers, an appropriate matching network is needed to couple the radio frequency power into the radio frequency coils within the cavity. Newer units have broadband capability, and minimal matching is required. Older spec-

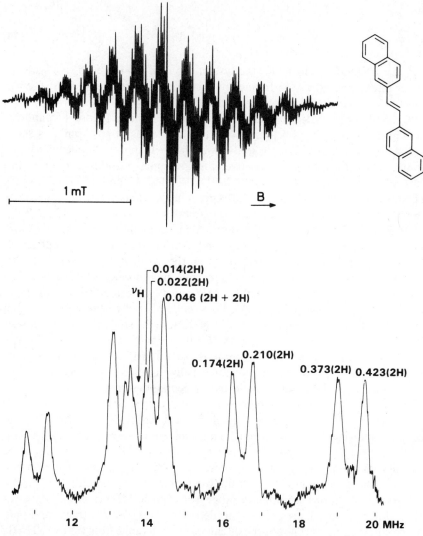

Figure 4.56 Comparison of ESR (upper) and ENDOR (lower) spectra of 2,2′-dinapthylethylene anion radical. Reprinted, with permission, from F. Gerson, I. B. Goldberg, and T. M. McKinney, *Magn. Reson. Chem.*, **26**, 319 (1988). Copyright © 1988 by John Wiley & Sons, Ltd.

trometers used a matching circuit that was tuned to the frequency by a variable capacitor synchronized to the frequency sweep. As the tuning circuit ages, mismatches can introduce spurious signals that are misinterpreted easily as part of the spectrum. In many experiments it is also advisable to use a field frequency lock to keep the ESR absorption frequency constant.

From an experimental point of view, selecting optimum parameters for an ENDOR experiment is quite difficult. The nuclear relaxation rates may be quite

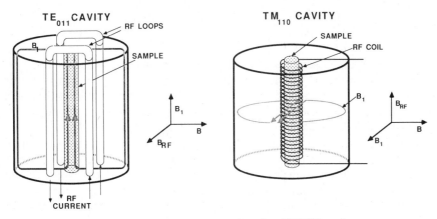

Figure 4.57 Schematic representation of an ENDOR cavity.

different for different sets of equivalent nuclei, especially when there may be exchange. In addition, the relaxation rates are affected by temperature, solvent characteristics (especially viscoscity and polarity), dissolved impurities (especially oxygen, which promotes rapid electron relaxation), and concentration of paramagnetic species [415, 416, 419, 420].

Electron nuclear double resonance was applied to many problems [419], including analysis of spectra of organic radicals in solution and resolution of overlapping lines, liquid crystalline and frozen solutions, single crystals, powders, and molecular processes. The application to organic hydrocarbon radicals in solution was briefly described earlier. This is by far the most extensive use of ENDOR. However, many molecules or radicals of practical interest contain magnetic nuclei other than protons. As a result, it is desirable to apply ENDOR to heteronuclear systems. Since the intensity of the ENDOR signal can vary so extensively, Plato and co-workers [420] carried out a theoretical analysis of the signal intensity that would be expected for various nuclei as a function of the microwave, radio frequency, and modulation magnetic fields. The analysis was based on the theories developed by Freed and co-workers [421–425]. They found, both experimentally and theoretically, that the key parameters in determining the sensitivity were the trace of the square of the hyperfine tensor $\mathrm{Tr}\,|\mathbf{A}^2|$, quadrupole coupling, rotational correlation time (including temperature), and radical concentration. Interestingly, higher temperatures are needed for optimal detection when there is a large quadrupolar contribution. The nuclei studied included ^{1}H, ^{2}H, ^{10,11}B, ^{13}C (natural abundance), ^{14}N, ^{15}N, ^{17}O, ^{19}F, ^{27}Al, ^{29}Si, ^{31}P, ^{33}S, ^{35,37}Cl, and the alkali metals. Based on known parameters of radicals and current instrumentation, it was concluded that there is little prospect of detecting naturally abundant ^{13}C or ^{10,11}B, ^{27}Al, ^{29}Si, ^{33}S, and ^{35,37}Cl in liquids. The estimated relative intensities of lines among the different elements and between the high- and low-frequency ENDOR lines agreed reasonably well with their theoretical estimates. However, since that report naturally abundant ^{13}C [426], $^{10,11}B$ [427], ^{27}Al [417], and ^{29}Si [417, 426] were detected, but the parameters of the radicals were different from those used in the calculations.

Liquid crystalline materials, which can be aligned in a magnetic field, were also used as solvents to obtain some information on the paramagnetic anisotropy of certain molecules [428]. The bulk alignment also creates a preferential alignment of a solute molecule. This alignment is not as complete as it would be for a single crystal or frozen liquid, and it is described by an order matrix. The action of the alignment is often sufficient to remove accidental magnetic equivalences of nuclei within molecules [429] or it can facilitate detection of quadrupolar couplings [430]. The ENDOR lines tend to be broader in liquid crystal media than in liquids, which makes detection more difficult. The problem is even greater for powder samples. Because of the complications caused by multiple relaxation, the observation of ENDOR is far more difficult than that of ESR in such media.

Additional complications arise in powder or frozen liquids because the entire ESR transition is not saturated. As a result, a combination of elaborate modulation and multiple resonance techniques is used [417], often combined with computer data acquisition and spectrum simulations [431, 432]. Electron resonance techniques are described further in another volume of this series, *Determination of Crystalline and Amorphous Solids*.

A significant use of ENDOR is to characterize multielectron-spin organic radicals. Here one must distinguish between an assembly of doublet, *polyradical*, and multiplet states. The distinctions are based on the dipolar coupling between electrons and on the exchange interactions. Two unpaired electrons on the same molecule can be configured as a doublet–doublet, diradical, or triplet. In a multiplet state, there is no detectable ESR signal in solution (see Section 3.6) because the strong dipolar interaction causes rapid relaxation in the rotating molecule. Such species are not accessible to ENDOR. An assemblage of doublet states where the dipolar and exchange interactions are very small is analogous to radicals as described previously. The intermediate case of polyradicals is most advantageously analyzed by ENDOR because the ESR spectra can be very complex [433]. Here the exchange energy is much greater than the hyperfine energy, but it is generally smaller than the Zeeman energy. Observation by ENDOR requires that the dipolar coupling be small, typically less than about 100 MHz.

If an energy-level diagram similar to Figure 4.55 is constructed for a diradical, where $S = 1$, where the radical moieties are identical, three ENDOR lines are expected for each set of nuclei. The ENDOR frequencies are $v_N - hA$, v_N, and $v_N + hA$. In general,

$$v = v_N - |S|hA, \qquad v_N - |S|hA + 1, \qquad \ldots, \qquad v_N + |S|hA \qquad (175)$$

An interesting study that reflects this behavior is the ESR and ENDOR of a tetragalvinol [434] observed through four successive oxidation states. As the number of unpaired electrons increases, the ESR lines broaden and lose resolution. However, the ENDOR spectra remain well resolved.

A technique called *ENDOR–induced electron spin resonance* [435] was developed for resolving overlapping ESR spectra caused by several components. In this method, the ESR desaturation induced by a radio frequency field is recorded. However, the radio frequency field is tuned such that it excites only one of the paramagnetic species present while the magnetic field is swept through the range of ESR transitions. The deconvoluted spectrum is the difference between the ordinary ENDOR spectrum under partial saturation and that recorded in the presence of the resonant radio frequency field. For this reason, it appears similar to an absorption spectrum. The technique requires modifications to the standard ENDOR spectrometer so that the NMR frequency can remain tuned to the proper transition, but it can be very useful in many practical studies. One such example was in the determination of deuterio positions in partially deuterated phenylenyl radical. Since the deuterio species exhibit different spectra from the protiated species, ENDOR-induced spectra of each of the components can be determined [436].

A technique applied to solid systems called *angle selected ENDOR spectroscopy* [437, 438] is used to determine the position and Fermi contact interaction of protons in complex ligands relative to the transition metal ion. The example used to develop the method was the bis(2,4-pentanedionato)cupric complex. In powdered samples, the resonance is selected to a certain point that corresponds to a singularity in the spectrum, and the ENDOR is recorded. If this is repeated at several singularities, data similar to those obtained for single crystals are obtained. This allows the nuclear spin states to be followed as a function of position in the spectra. Since the data ultimately mimic those obtained for single crystals, the spatial location of the protons with respect to the transition metal can be evaluated. The measurement requires extensive averaging and analysis, but it can provide much useful information on the structure of complexes without preparing a large single crystal.

Electron nuclear double resonance is also used to study intramolecular processes by line-width variations, as are ESR (Section 2.5) and NMR. The time scale of ENDOR is similar to that of ESR, 10^{-9}–10^{-4} s. Much the same processes are detected in ENDOR as in ESR, but the simplifying features of ENDOR allow the line-width changes to be more easily resolved in complex systems. One such example is the rotation of phenyl groups in N,N-diphenyl-N'-picrylhydrazyl radical [439]. Several applications to cation exchange in ion pairs, rotation of aryl groups, ring inversions and torsional oscillations of alkyl groups were cited in [417]. However, in using ENDOR, the simplification of the spectrum is offset by the complexity of the overall relaxation processes. As a result, line widths must be estimated from samples that do not show the dynamic effect being studied, or from the same sample at higher or lower temperatures where the dynamic effect does not occur.

5.1.2   Electron Nuclear TRIPLE Resonance

Electron nuclear TRIPLE resonance is an extension of ENDOR that was principally developed to study solutions of radicals. In these measurements the ESR transition is partially saturated as in ENDOR, but two different NMR transitions are stimulated. Hence three radiation frequencies are used, leading to the name TRIPLE resonance. There are two different irradiation schemes that can be used. The first, called *special TRIPLE* resonance [440], is applicable to one set of equivalent nuclei. The second, called *general TRIPLE* resonance [441], is applicable to two or more sets of equivalent nuclei. The latter technique was first applied to solids at low temperature [442] and given the name double ENDOR.

It is convenient to understand ENDOR and special TRIPLE resonance [440] by using Figures 4.55 and 4.58. Figure 4.58 is an elaboration of Figure 4.55 where all relaxation processes are shown as broken lines and the energy levels are staggered for convenience. As an example of the ENDOR experiment consider the case in which transitions v_{E1} and v_{N2} are used. Also, the nuclear

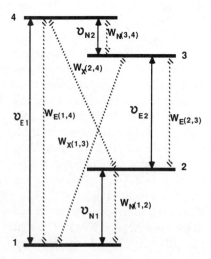

Figure 4.58 Energy-level diagram similar to that of Figure 4.55 but with energy levels staggered to show TRIPLE resonance. Diagram also shows relaxation of electron spin W_E, nuclear spin W_N, and cross relaxation W_X.

relaxations $W_N(1,2)$ and $W_N(3,4)$ are slow with respect to the electron relaxations $W_E(2,3)$ and $W_E(1,4)$. Typically, $W_E \gg W_N \gg W_X$. As a result, the sensitivity of the ENDOR experiment is limited by the rate at which state 4, which is pumped by transitions ν_{E1} and ν_{N2}, is depopulated. This is called a *bottleneck*. By pumping state 2 by ν_{N1}, this bottleneck can be removed. In a practical sense, the ratio of the intensities of the ENDOR transition to that of the ESR transition, (I_{ENDOR}/I_{ESR}) is approximately 0.1. By relieving the bottleneck, the corresponding ratio $(I_{SPECIAL}/I_{ESR})$ is approximately 1. Although this is never attained, the special TRIPLE lines exhibit the square of the Lorentzian line shape, and are therefore sharper.

In this experiment, as in ENDOR, the magnetic field and the microwave transition are held constant. One radio frequency is swept positive and the other negative from the nuclear Zeeman transition frequency. The absolute value of the difference between the radio frequency and the proton Zeeman frequency is plotted on the abscissa. Since both nuclear transitions are attained simultaneously, the special TRIPLE spectrum looks like the high-frequency half of the ENDOR spectrum with the nuclear resonance at zero frequency. Because the bottleneck is removed, the amplitude of the signal is proportional to the number of equivalent nuclei. This feature and the greater sensitivity are the major advantages of special TRIPLE over ENDOR.

In the general TRIPLE resonance experiment, the magnetic field, the microwave transition, and one radio frequency transition are held constant. The remaining radio frequency is modulated and swept over the desired spectral region. The resulting spectrum appears similar to the ENDOR spectrum with several important exceptions. When the swept oscillator reaches the pump frequency, the absorption at that point is very low in intensity because this transition is saturated. The counterpart, however, is enhanced because this condition corresponds to the special TRIPLE experiment. In addition, the relative intensities of the various lines of the general TRIPLE resonance spectrum change. Using the pyrene radical anion as an example, the various intensities can be compared. This anion exhibits hyperfine couplings of -0.475, -0.208, and $+0.109$ mT. The values were determined by ESR [443], and the signs are confirmed by NMR [444]. The ENDOR spectrum, Figure 4.59a, exhibits six lines corresponding to the three hyperfine splittings of the radical. Each set of equivalent nuclei is responsible for a pair of lines symmetrically disposed about the proton Zeeman transition; that is, lines 1 and 6 correspond to the 13.3-MHz splitting, lines 2 and 5 correspond to the 5.8-MHz splitting, and lines 3 and 4 correspond to the 3.1-MHz splitting. Important information can be extracted from comparison of relative intensities of the low- and high-frequency lines I_L/I_H of the unpumped pairs. In Figures 4.59b, c, and d, when a high-frequency line is pumped by the constant radio frequency source, the low-field counterpart is enhanced. In Figure 4.59b the ratio I_L/I_H for lines 2 and 5 is greater than unity, but it is less for lines 3 and 4. In Figure 4.59c the ratio for lines 1 and 6 is greater than unity, while it is less for lines 3 and 4. In Figure 4.59d the ratios for lines 1 and 6 and for lines 2 and 5 are both less than unity. These

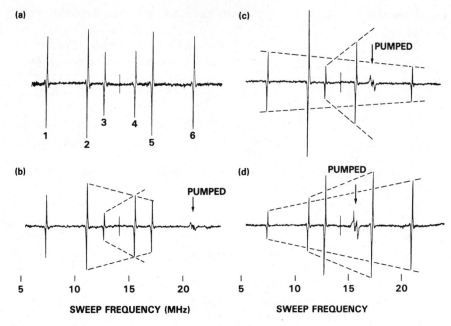

Figure 4.59 Spectra of the pyrene anion: (*a*) ENDOR spectrum; (*b*) general TRIPLE resonance spectrum with line 6 pumped; (*c*) general TRIPLE resonance spectrum with line 5 pumped; (*d*) general TRIPLE resonance spectrum with line 4 pumped. TRIPLE spectra recorded at half gain as ENDOR. Dotted lines show relative intensity of unpumped high-field and low-field line pairs. Reprinted from [441] with permission.

results show that the signs of the two larger splittings differ from that of the smallest splitting. It is also important to note that when the low-field lines are pumped, mirror image spectra are obtained; that is, pumping low-frequency lines enhances the corresponding high-frequency component.

Usually, the relative intensities are compared to the ENDOR spectra, and a parameter V is defined as the ratio of the signal amplitudes of a given transition in the general TRIPLE spectrum with respect to the same transition in the ENDOR spectrum. In this representation the relative signs of the splittings of radicals with only two equivalent sets of protons can be determined. References [441] and Chapter 14 of [416] offer a complete discussion.

Thus general TRIPLE resonance is useful for determining the relative signs of hyperfine splittings. This in turn is important for positional assignments of splittings in complicated radicals.

The major drawbacks to these experiments are (1) they are difficult. The complexity described for ENDOR is compounded because yet another frequency source must be controlled; and (2) they are extremely expensive because two very stable high-power, radio frequency sources are required. Therefore, the use of TRIPLE resonance is currently not widespread.

5.1.3   Electron Electron Double Resonance

The ELDOR technique [415, 416, 445] is similar to ENDOR in that two irradiation frequencies are used, but in ELDOR both frequencies are used to irradiate the electron spin transitions. This can be understood in part by Figures 4.55 and 4.59, where v_{E1} is used in a way similar to ENDOR in that it is tuned to an ESR transition and serves as a probe frequency to monitor the change in absorption, but it is of sufficiently low intensity that there is no significant saturation. The second irradiating microwave frequency is of sufficiently high power to cause saturation. For example, in Figure 4.55, if v_{E1} is the probe frequency, v_{E2} is the saturating radiation. As v_{E1} is scanned, it will reach the second transition. In Figure 4.59, when the second transition is reached, the population of state 1 is depleted such that the nuclei relax to cause a reduction of the ESR signal intensity. Thus, when the microwave frequency difference between the two sources is equal to the hyperfine splitting, an ELDOR signal is observed. This is given by

$$v_{E1} - v_{E2} = A \tag{176}$$

However, rapid nuclear relaxation can be induced through electron–nuclear dipolar coupling because of the electron spin flip. Thus transitions called *forbidden–allowed* (because the electron transition is allowed but the nuclear transition is forbidden) are observed. These occur at

$$v_{E1} - v_{E2} = A/2 - v_N, \ A/2 + v_N \tag{177}$$

where v_N is the nuclear Zeeman transition frequency. At high temperatures and high spin concentrations, only allowed transitions (176) are observed. However, at temperatures below about 20 K and at low spin concentrations the forbidden–allowed transitions may actually dominate the spectrum. The change from low to high concentration is defined by the point at which intramolecular relaxation begins to dominate over the intermolecular dipolar coupling.

The primary use of ELDOR is in the study of relaxation processes. Biological systems containing iron were studied [446] because ELDOR allows detection of large couplings, and because favorable relaxation conditions are brought about by the saturating radiation. Another use for ELDOR is in studies of methyl group rotational tunneling [447].

Electron electron double resonance spectrometers are available commercially or have been home-built as described in Chapter 2 of [416]. The detection system is essentially a standard microwave bridge. However, the pumping magnetic field must be orthogonal to the observation magnetic field. Therefore a pump source must be designed in conjunction with the microwave cavity. For example, if a high-Q cavity is used for the pump, an overall cavity must be constructed such that it is a bimodal design. This can consist of two cross-coupled TE_{102} mode cavities or a TM_{110}-mode cavity. In either case, the high Q

precludes a broad scan unless the scanning method includes some means of altering the frequency of the pump mode without affecting the observation mode. Rotating paddles and screws were used to accomplish this goal. If such a high-Q cavity is used, then the microwave power required can be less than 5–10 W. An easier alternative method employs a microwave helix as the pump. This is a low-Q device and is easily tuned for very broad scans; however, high-power sources from 100 to 200 W are needed, and they are extremely expensive.

5.2 Saturation Transfer Spectroscopy

To extend the use of ESR to the regime of slow rotational diffusion, the technique of saturation transfer (ST) was developed [448]. A review of the applications is given in [449]; overviews, applications, and detailed theoretical treatments are found in [261, 450–452]. This is an extremely active field since it provides a tool to explore biological membranes, emulsions, and interfaces. Understandably, new theoretical work, reviews, and applications appear often.

Normal ESR measurements are capable of detecting molecular motions with a rotational correlation time τ_c of about 10^{-7}–10^{-10} s, although other effects (such as exchange or intramolecular reorientations) can expand the ESR time-scale window to about 2×10^{-6}–5×10^{-11} s. The primary causes of line broadening were discussed in Section 2.5.4, but in this section we concentrate on bulk molecular rotations such as those manifested in the ESR spectra of Figure 4.41. At temperatures below about 20°C the ESR spectra do not change, and analysis of the data suggests the rotational correlation time in this system is about 2×10^{-7} s. Saturation transfer spectroscopy is a method developed to explore the effects of molecular motion in complex systems on time scales slower than those accessible by normal ESR techniques.

The following discussion is restricted to nitroxide spin labels such as in Figure 4.40 because they have received the greatest experimental and theoretical study and analysis. It is convenient to think of the spectra in Figure 4.41 as comprising very many *spin packets*, that is, groups of molecules that have the same orientation and nitrogen nuclear spin state at a given instant, similar to the situation for solid powder samples. However, in nonrigid samples the spin packets are able to reorient at some characteristic rate. This means the resonance condition for a given spin-labeled biomolecule can *diffuse* among a continuum of spin packets.

In Section 1.5 we pointed out that T_1 is a measure of the rate at which the bulk magnetization of a spin system relaxes back to thermal equilibrium and that T_2 is a measure of the rate of loss of phase coherence in a spin system subjected to resonant microwave energy. Figure 4.3 illustrates these concepts for a system that comprises only one spin packet. But in a real system, such as a nitroxide spin-label system undergoing slow rotation, the rotational motion can provide additional relaxation pathways. Ultimately when such a sample is irradiated with enough microwave power to partially saturate a given spin packet at resonance, the molecular motion can carry the saturation (and phase coherence) into a neighboring packet. This occurs if the molecules rotate enough to change their resonance field by about the width of a spin packet $1/\gamma T_2$ in a time comparable to T_1.

Saturation transfer ESR is based on the observation that the cut-off value of the standard ESR time-scale window occurs at approximately the same value as the spectrometer detector modulation f_M (expressed as angular frequency ω_M)

$$2\pi f_M = \omega_M \sim 1/\tau_c \tag{178}$$

In standard ESR only signals that are in phase with the modulation frequency are observed because $\gamma B_M \omega_M \ll 1/T_1 \tau_c$, and the electron spin system magnetization is always in equilibrium with the compound laboratory magnetic field $\mathbf{B} + \mathbf{B}_M$. But at the condition $\gamma B_M \omega_M \sim 1/T_1 \tau_c$ signals can be observed that are both in phase and out of phase with the modulation because the electron cannot "keep up" with the changing magnetic environment caused by the rotational diffusion of the radical in the magnetic field. This condition is produced by use of either a large modulation amplitude or a high modulation frequency. This phenomenon has some similarity to the conditions of rapid (versus slow) passage described in the early magnetic resonance literature [14, 453] where the magnetic field was swept rapidly through resonance. The modulated magnetic field superimposed on the static or slowly swept field is analogous to the rapidly scanned field in the earlier treatments. However, ST is a measure of the spectral diffusion of the bulk magnetization (M_z) as compared with phase coherence (M_x and M_y) measured in fast passage.

As implied in the title of the method, the signal must be partially saturated so there is coherence to the spins. The saturation process was shown in Figure 4.3. To examine the spin packet, one must achieve a magnetization that appears similar to Figure 4.3e. When this occurs, the transfer of saturation as a function of reorientation of the molecule is measured by relating the time required for molecular motion τ_c to the modulation frequency. In the absence of saturation, few of the spins are correlated, so no transfer process is observed. At the other extreme, complete saturation is also ineffective because the bulk magnetization is destroyed.

Saturation transfer has proven so important in biological studies that some discussion about the experimental procedure, display of the ESR data, and information that is obtained is in order. Native biological systems contain very few radicals. As a result, a paramagnetic probe is needed to generate the ESR spectrum. Some examples are shown in Figure 4.40. The ability to analyze the time-dependent phenomena depends, in part, on the extent to which the anisotropic splittings and g-factors are known. This is implicit in the discussion in Section 2.5.4. The magnetic parameters of many commonly used spin probes were given in [261], and new probes, which exhibit compatibility with specific systems, continue to be synthesized.

Because the signal amplitude and to a large extent the shape of the spectra are related to the modulation amplitude, the modulation system must be completely characterized. The modulation system includes the effect of the modulation coils on the distribution of modulation amplitudes within the sample, phase stability of the lock-in amplifier, and modulation amplitude as a function of frequency. These parameters are determined by appropriate measurements, but it is important to recognize that the modulation amplitude varies spatially within the microwave cavity. From an experimental point of view, it is best to use relatively small samples that occupy less than about half of the height of the cavity when the geometric center of the sample is coincident with the center of the microwave cavity. Then the microwave power and modulation amplitudes are essentially constant over the sample [454]. The microwave magnetic field intensity distribution within the cavity is also important because it affects the extent to which the sample is saturated. Field distributions within many different cavities are given in [371–373], and the effects of sample dielectric properties on the microwave field distribution were also investigated [455].

Saturation transfer data have been displayed in many ways. A convention has evolved that calls U and V, respectively, the dispersion and absorption modes of the spectrometer (see Section 1.6). In addition, there is a subscript 1 or 2, which indicates the harmonic of the detected modulation frequency (first or second derivative displays). An unprimed symbol indicates the detection is *in phase* with the modulation amplitude; primed symbols indicate the detection is 90° *out of phase* with the modulation amplitude. Detection modes U_1, U_1', U_2, U_2', V_1, V_1', V_2, and V_2' were considered. Most commonly reported are spectra U_1' and V_2'. As one may expect, the in-phase spectra generally contain rather less ST information, so they are harder to analyze; and the dispersion mode U', which contains abundant ST information, suffers from instrumental noise problems [456].

Regardless of the display mode chosen, ST spectra are extremely sensitive to different instrumental parameters. In this context, the phase stability of the spectrometer system is acutely important because the signal amplitude of the in-phase component is often significantly more intense than the out-of-phase portion. As a result, a small phase error in deconvoluting the modulated signal can result in a large distortion in the appearance of a primed spectrum.

Combination of the facts that the sample size should be as small as possible and that the out-of-phase signal amplitude is weak, puts a constraint on the paramagnetic concentration. If the concentration is made too high, dipolar and exchange broadening can mask some of the information in the spectra.

Finally, for all but a few special cases, the ST spectra cannot be calculated in closed form. As a result, simulations are used extensively. These are also described in detail [261, 451] with additional extensive references therein. Simulations for isotropic, axially symmetric, and three-dimensional anisotropic rotational diffusion were reported. Since there are so many parameters that can influence the appearance of saturation transfer spectra, generally it is better to simulate the conditions in-house rather than to depend on reproducing reported conditions.

5.3   Pulsed Electron Spin Resonance Spectroscopy

Pulsed ESR spectroscopy would also require an entire volume to describe fully its theory, measurement system, and experimental techniques and applications. The only volume of this kind to date is [457], but several instructive review articles are also available [458–460].

The concept of pulsed magnetic resonance was understood to be implicit in the Bloch equations [461] and was applied to protons in 1950 [462, 463] when a radio frequency pulse of large $|\mathbf{B}_1|$ was applied to an ensemble of protons, and the free induction decay (fid) was observed. This is akin to the situation in Figure 4.3e, which shows a highly coherent spin system resulting from stimulation by radiation at the Larmor frequency. When this radiation ceases, the coherence disappears at a rate related to T_2, and a detector tuned to detect a signal oscillating in the laboratory xy plane will respond as in Figure 4.60c.

Pulsed ESR experiments can be divided into two main categories. Saturation recovery ESR (SRESR) is used principally to study relaxation times, while Fourier transform ESR (FTESR) is used to elucidate details of the spectrum. In either case, one or more intense microwave B_1 pulses are applied to the sample as bursts of some definite duration, and then the ensuing behavior of the spin system is monitored. However, there are several distinctions between the two experimental techniques.

It is useful first to define the qualitative words *short, long, small*, and *large* or *high* in terms of their effect on saturation of a spin system. *Short* and *long* refer to the duration of a pulse relative to the spin–lattice (T_1) or spin–spin (T_2) relaxation times, depending on the particular experiment. *Small* refers to a magnitude of B_1 that causes insignificant perturbation of the spin system from thermal equilibrium as shown in Figure 4.3c. *Large* or *high* refers to a pulse of such intensity that the magnetization of the spin system is transferred to the transverse direction and M_z approaches zero because the α and β states are equally populated (saturation). (Figure 4.3e indicates the phase coherence relationship but is not intended to suggest saturation because there is still a significant M_z component in this drawing.) Pulse experiments are more easily performed in NMR than in ESR because of the lower frequencies and longer relaxation times in the former. The time scales in pulsed NMR are about milliseconds to microseconds, depending on relaxation time, while in ESR the time scale is in the 10-ns to 1-μs range. Thus it was not until the late 1960s that electronic circuitry was fast enough to carry out pulsed ESR; and even so, low temperatures were required to increase the relaxation times sufficiently to permit observation [464–467].

In the SRESR experiment, the response of the spin system to a saturating pulse is monitored. Following the pulse, Figure 4.60a, low-intensity microwave radiation is used to probe the sample continuously, as in a normal ESR spectrometer. When a high B_1 pulse is applied to the sample, the normal ESR signal (which is proportional to the M_z component of magnetization) decreases because the magnetization is shifted to the xy plane. After the pulse, the signal

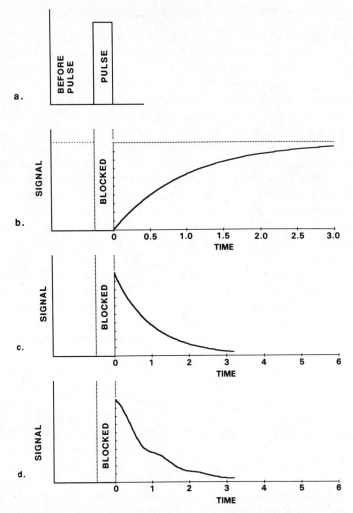

Figure 4.60 Response of an electron to single microwave pulses: (*a*) ideal pulse form; (*b*) saturation recovery signal; (*c*) free induction decay signal with spectrometer set to resonance at pulse frequency; (*d*) free induction decay signal with spectrometer set slightly off resonance.

begins to recover its equilibrium value at a rate proportional to $1/T_1$ (see Figure 4.60*b*). This experiment is independent of the phase of the high-intensity B_1 pulse.

On the other hand, the FTESR experiment detects the free induction decay of the spin system, Figure 4.60*c*. In this experiment, the pulse is designed to bunch the transverse magnetic moments to produce a strong transverse magnetization, as suggested in Figure 4.3*e*. Since this transverse magnetization rotates in the laboratory frame at the Larmor frequency, it can be detected readily as a signal oscillating in the plane normal to the direction of the applied pulse. No additional microwave probe radiation is required. Of course, the transverse magnetization signal is zero in the absence of a perturbing pulse. The time scale of this experiment is about T_2. This experiment is extremely sensitive to the pulse phase and it is possible to distinguish various features of the spin system by using different pulse sequences.

There are several complicating features of the SRESR experiment. Foremost, while it is relatively easy to apply enough power to saturate significantly the ESR transition, it is extremely difficult to apply this pulse quickly enough to avoid phase coherence in the spin system (as in Figure 4.3e). Consequently, in the SR experiment there is always some fid component present. This is especially true if T_1 and T_2 are similar in magnitude, which often occurs in ESR studies. As a result, a spectrometer designed for SR studies must also be capable of monitoring the fid [468, 469] to compensate the SR signal properly.

A block diagram of a pulse ESR system is shown in Figure 4.61. This can be used for both SR- and FTESR experiments. The heavy lines depict the microwave circuit used for pulsing, while the shaded lines indicate the circuit used for measuring the continuous ESR signal. The circles represent three- and four-port circulators. Positive intrinsic negative diodes (PIN diodes), which act as rapidly adjustable attenuators, are used as gates either to control the amplitude of the pulse or to modulate the phase of the pulse relative to an audiofrequency reference signal. An additional PIN diode network is needed to isolate the ESR detector during each high-power pulse and thus to protect it

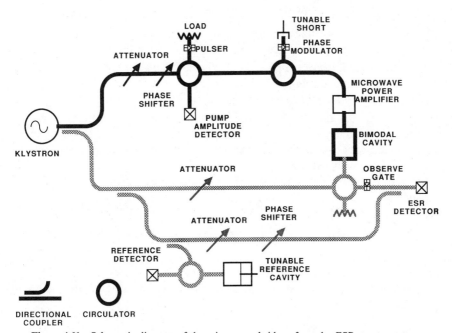

Figure 4.61 Schematic diagram of the microwave bridge of a pulse ESR spectrometer.

from fatal overload. This design concept was refined in more recent instruments [470], but the working principles are identical. Conventional ESR is detected by opening the observation gate during the "off" side of the pulse duty cycle. If the bimodal cavity were replaced with a conventional cavity and the pulse network removed, a spectrometer equivalent to that shown in Figure 4.49 results.

Three procedures have been used to separate the fid signal from the SR signal:

1.   The pump power is modulated 180° at a low audiofrequency. This alternately inverts the phase of the fid signal so that successive cycles cause cancellation of this component and the SR signal remains.

2.   The observe gate is modulated at an audiofrequency. In this case, the SR signal is modulated and the fid signal is not modulated. Therefore, detection in phase with the gate modulation results in an undistorted SR signal. However, this method may damage the detector by high-power pulses.

3.   The reference phase of the pumping power is set such that the fid dispersion (27a) is detected. The magnetic field is then set to the absorption maximum of the line without magnetic field modulation (which would produce the first derivative). The detection arm is then set for absorption (90° phase difference from dispersion). Under this condition, both the steady-state and transient dispersion signals are nulled and only the SR signal remains. This is by far the most versatile approach because the converse settings can be used to detect the fid signal with exclusion of the SR signal.

There are several limitations imposed on the experimental equipment for performing pulse experiments.

1.   The minimum time required to digitize the spectral information. If the signal relaxes quickly, it may not be possible to record the transient response.

2.   The magnitude of the pulse field. The faster are the relaxation processes, the greater are the pumping power requirements. Also, greater pumping power is required to produce fid in systems with large hyperfine splitting.

3.   The limit of the *ringdown time* of the microwave cavity, that is, the time required for the microwave power stored in the resonant cavity to dissipate. As described in Section 4.2.3, the cavity Q is the ratio of power stored-to-power lost per cycle. Qualitatively, the ringdown time is about the cavity Q divided by 2π times the resonant frequency v_0. In a more practical sense, if the time dependence of the magnetic field pulse has a Gaussian profile, the minimum pulse time t_P is [464]

$$t_P = 4Q/(2\pi v_0) \tag{179}$$

However, lower Q-factors reduce the value of B_1, that can be applied to the sample. The approximate value of B_1 at the radio frequency magnetic maximum of the cavity is given by

$$B_1 \sim 2[PQ/(2\pi v_0 c^2 \varepsilon_0 V_C)]^{1/2} \tag{180}$$

where V_C is the cavity volume and P is the incident microwave power. Thus, a balance is needed between both the cavity Q and incident microwave power.

4.   The requirement for mutual isolation of both modes (orthogonality) of the bimodal cavity as described in Section 5.1.3.

The theoretical foundations of pulsed ESR are developed fully in [457], and it is beyond the scope of this chapter to outline the mathematics. Nevertheless, the most simple result of the transient response is given by [457]

$$\frac{m_y(t)}{m_{z\infty}} = y_0 \exp\left(-\frac{t}{T_2}\right) + \gamma B_1 T_2 \left\{ \left[\frac{(z_0 - 1)}{(1 + \gamma^2 B_1^2 T_1 T_2)}\right] \exp\left(-\frac{t}{T_1} - \gamma^2 B_1^2 T_2 t\right)\right.$$

$$\left. + \frac{1}{1 + \gamma^2 B_1^2 T_1 T_2} - z_0 \exp\left(-\frac{t}{T_2}\right)\right\} \tag{181}$$

This equation represents the condition of a homogeneous line on resonance, where B_1 and B_2 are, respectively, the microwave observing field and the pump field. The observing field is along the x direction and the pump field is along the y direction (laboratory frame) of Figure 4.4. (The bimodal cavity is specifically designed to permit this configuration.) It is straightforward to examine the spin behavior under certain limiting conditions. The first term $y_0 \exp(-t/T_2)$ is the fid. The sign of y_0 is inverted when the phase of the pump signal is changed by $180°$. The term $\exp(-t/T_1 - \gamma^2 B_1^2 T_2 t)$ represents the saturation recovery signal in the limit of $B_1 \to 0$.

By changing the field to an off-resonance condition, the first term has an oscillating behavior

$$m_y(t)/m_{z\infty} = [y_0 \cos(\delta\omega t) - x_0 \sin(\delta\omega t)] \exp(-t/T_2) + \text{other terms} \tag{182}$$

where $\delta\omega$ is the deviation of the radial frequency from resonance. An example is illustrated in Figure 4.60d. If the species under study has multiple lines, the spectrometer may be set to resonance at only one line. As a result, the other lines will cause a modulation of the fid signal. This can be used to determine the spectra by Fourier transform techniques, and it is described in detail in [457]. However, a pictorial view is given for NMR in [471]. Several sources of Fourier transform theory and applications are given in [472–474].

While the fid signal can be used to reconstruct the spectrum, various multiple pulse sequences have also been used to elucidate various features of the spectrum. A commonly used two-pulse sequence is the $90°$-τ-$180°$ sequence, where τ, the delay time between pulses, is varied systematically. Figure 4.62a illustrates this sequence. The emission *echo* arises from rephasing of the transverse magnetization, because the different hyperfine components have different Larmor frequencies, and after an appropriate interval the spin system again acquires phase coherence. As τ is varied, if there is hyperfine structure present, the echo pattern traces an envelope. Consider for example a system with

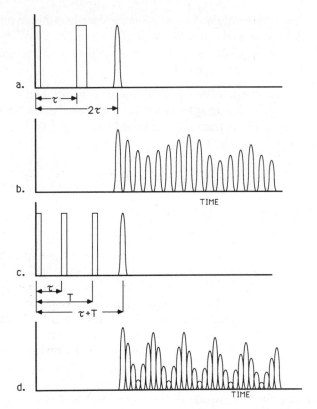

Figure 4.62 Schematic of the two-pulse and three-pulse spin–echo sequences: (*a*) the $90°-\tau-180°$ sequence gives an echo at time τ after the second pulse; (*b*) by carrying out a series of the $90°-\tau-180°$ sequences, incrementally increasing τ each time, the envelope of the spin–echo pulses is modulated at the nuclear hyperfine frequencies and their sums and differences; (*c*) the $90°-\tau-90°-T-90°$ sequence also gives an echo at time τ after the third pulse; (*d*) the $90°-\tau-90°-T-90°$ sequence is used to eliminate the sum and difference modulation frequencies so that only modulation at the nuclear hyperfine frequencies remains. This is carried out by fixing time τ and incrementally increasing time T.

only an electron and a proton in a magnetic field. The magnetic field and the electron field add vectorially to produce a resultant field about which the proton magnetic moment precesses. When the electrons are flipped by an intense resonant microwave pulse, the direction of electron magnetization changes from $+S_z$ to $-S_z$. This reverses part of the field at the proton and alters the effective field. If the pulse duration is short with respect to the Larmor precession frequency of the proton, some protons cannot alter their direction of spins to follow the new effective field. Only those protons that are near the static field direction can follow the new effective field. This creates a branching of two sets of proton spins, which cause a modulation. When the second pulse is applied, the electron spins are flipped back and may create an echo. The echo results from a modulation of the electron precession frequency by the proton precession frequency. If τ is varied in a series of experiments, the echo that is observed consists of a series of maxima and minima, which is essentially a modulation. The two-pulse echo is modulated by ω_{N1}, ω_{N2}, $|\omega_{N1} - \omega_{N2}|$, and $\omega_{N1} + \omega_{N2}$, where ω_{N1} and ω_{N2} refer to the nuclear precession frequencies.

A three-pulse sequence, $90°\text{-}\tau\text{-}90°\text{-}T\text{-}90°$, that leaves only the nuclear precession frequencies without the sums and differences was developed. Here, τ is usually kept constant at the value that creates a minimum echo in the two-pulse sequence, and T is varied systematically. These results are more easily interpreted, and the experiment is analogous to ENDOR.

Applications of electron spin echo spectroscopy to chemical and biochemical problems are described in [457, 458, 460, 475, 476].

5.4   Future Directions of Electron Spin Resonance

Based on *Chemical Abstracts Selects* and other published review articles and books a few extrapolations can be made about future ESR research. If there are no major dislocations in the apparent direction of scientific research, such predictions may be realistic. During the past decade, ESR changed from a self-contained field of research to a tool used to approach a wide variety of scientific and technological problems. Such applications will surely continue. However, ESR is neither as matured as other methods of instrumental analysis, nor is the demand for ESR as widespread; therefore, the equipment needed to address a given problem is not always available. Improvements and new devices for ESR will continue to be developed in those few laboratories where sufficient expertise in electronic and microwave instrumentation resides. Such advances will benefit from developments in high-speed electronics and microwave instrumentation, which are burgeoning because of the demands from the fields of data processing, communications, and defense. Development of measurement techniques at low frequencies will also be important. Samples containing a significant amount of water, such as in biological systems, create unique experimental problems because the optimum range for study is about 500 MHz to about 2 GHz, where the dielectric loss characteristics are lessened. However, intrinsic sensitivity falls off at such relatively low ESR frequencies, hence, larger samples and more efficient resonant structures are required. On the other hand, the need to observe large zero-field and hyperfine splittings in technologically interesting materials will require development of ESR techniques applicable to higher frequencies (greater than 50 GHz). Such high frequencies will probably require use of high-field, pulsed magnets. Time-resolved ESR studies, and pulsed ESR in particular are proliferating and are limited only by available instrumentation. Improvements in the temporal resolution will make it possible to measure rapid kinetics and perhaps pulsed ESR will even replace conventional ESR as has happened in NMR.

During the past few years the ESR section of *Chemical Abstracts Selects* was dominated by various studies of biological and biochemical processes, reactions, membrane structures, and configurations of proteins and enzymes. It is likely that such endeavors will continue. However, the perturbations introduced into the local environment of a native biological or biochemical system by the addition of a spin probe, whether nitroxide or transition metal ion, was never fully resolved, and it is anticipated that much creativity will continue to be displayed in this field.

Since it is not one of the mainstream instrumental methods, very few analytical laboratories routinely use ESR. However, it gained limited acceptance as an analytical tool for special problems because there are certain problems that only ESR can address. One such field is that of semiconductors, particularly the III–V types, and to a lesser extent silicon, where only ESR can satisfactorily diagnose certain dopant or impurity sites. This subject is not reviewed in this chapter but several general references are given in Section 8.8.

The use of ESR in studying interfaces, including membranes and emulsions as in the foregoing description, and also adsorption, electrochemistry, and gas–solid reactions, will also benefit from advances in techniques and instrumentation. Several groups are making strides in the development of time-resolved Fourier transform ESR spectrometers (so-called *2-D FTESR*) for studies of extremely rapid chemical reactions and molecular motions. Similar techniques were used in NMR for the past decade, but only now is it becoming technologically feasible to apply them to ESR.

Applications to structure determinations of organic, inorganic, and transition metal paramagnets, and to diagnosing reaction chemistry of organic, inorganic, and polymer systems will continue and expand. The recent application of ESR to probing the components of high-temperature superconductors is open to considerable debate. Several areas of controversy arise because of the difficulty of preparing pure ceramic materials, and perhaps only the impurities are detected. Furthermore, microwave radiation should not penetrate far into a superconductor; therefore, only those states on or near the surface should be observable below the transition temperature. Along such lines, observations at lower frequencies and fields may be more useful for this type of investigation.

Electron spin resonance zeumatography (Section 3.2) is a relatively new idea, prompted largely by the extension of NMR to magnetic resonance imaging (MRI) medical diagnostics. There are many potential applications of this technique both in fundamental research and in technologically interesting applications. Biological systems, thin films, semiconductors, and any other systems that contain paramagnetic components are possible candidates for imaging investigations. Furthermore the isotropy and rates of diffusion could be evaluated by doping such systems with paramagnetic probes. Because the required equipment is relatively expensive, even compared with MRI, it will doubtless remain principally an experimental method for years to come. But if significant applications to diagnostic or lifesaving medical procedures are discovered, the question of expense will become less burdensome.

6   COMPUTATION IN ELECTRON SPIN RESONANCE

The use of computers in ESR falls into three major classifications:

1.   Off-line analysis including simulations and data reduction.
2.   Data acquisition and control of continuous experiments, including on-line interactive programming.
3.   Data acquisition and control of pulse-type experiments.

Most ESR software is reported in the literature. A review by Vancamp and Heiss [477] summarizes the material available up to about 1980.

Often software is available from the authors of the particular program, but this procedure is not archivally reliable. In addition, most software is ancillary to the subject of research publication and is thus not generally accessible through abstracting services. This makes access to such information difficult for the research community and for other users of applied ESR. To bridge the void, several archives have maintained at least an index of relevant programs. The first is the Quantum Chemistry Program Exchange† (QCPE, see Section 6.4 for specific program numbers, names, and applications) where off-line programs are available for numerous simulations. These programs are available on tape, in card form, or as listings. Very often the programs need to be modified to run on a given computer, but the documentation is usually thorough. Before it was disbanded, IBM Instruments began an index of programs for all phases of ESR. This is continued at California State University at Los Angeles‡ (CSULA, see Section 6.4 for specific program identifiers and applications). Listings are not maintained.

6.1   Off-Line Programs

Off-line programs are generally used for data analysis after spectra are recorded. These are outlined in Section 6.4 with references to specific programs where possible. In many cases, published programs or calculations that form the basis of programs are directly applicable; however, unique cases exist where calculations must be modified. Section 6.4 divides ESR simulation programs into different classes. In principle, the most general programs accommodate a less general case as a specific subset. Each term in the Hamiltonian that is required in the spectrum simulation must be implemented by computation. Therefore, a consequence of generality is that more program space and increased computation time are required. Isotropic and single-crystal spectra have received the

---

†Quantum Chemistry Program Exchange, Department of Chemistry, Indiana University, Bloomington, IN 47405, USA.
‡Dr. Richard Keys, California State University at Los Angeles, Department of Chemistry, Los Angeles, CA 90032, USA.

most attention by far. These two classes of spectra are simulated by computing the line positions and the degeneracy of each line, and a line-shape function is then convoluted onto the stick spectrum with an intensity equal to the appropriate degeneracy factor. In general, single-crystal spectra with coparallel g- and hyperfine tensors appear similar. The advantage of using the appropriate programs for the particular simulation is that programs for single-crystal spectra allow the magnetic axes to be related to the crystal axes, and calculations are carried out over each orientation of the crystal with respect to the magnetic field without the experimenter determining each new g-factor and set of hyperfine couplings for each orientation. Programs that incorporate nonparallel g- and hyperfine tensors are unique and cannot be used readily to reproduce isotropic spectra.

Simulations of spectra that represent species with random orientation are very time-consuming. These simulations are based on the assumption that the spectrum is the convolution of the probability that the magnetic axes of a species are oriented in a given direction onto the orientation-dependent spectrum.

Figure 4.63 shows the process for simulating spectra representing a system that contains an ensemble of species in which the magnetic axes are randomly oriented with respect to the magnetic field. Figure 4.63a illustrates ortho-rhombic symmetry. To include all possible orientations, the spectrum must be simulated over one-eighth of a sphere. If, as usual, the line widths are assumed to be independent of orientation, then the line width need not be added until the end of the simulation, and essentially the simulation of each orientation is a stick figure. The magnetic field axis is then divided into boxes, and each time a simulation fits into the box representing the field region between $B(i)$ and $B(i + 1)$, a count of one is added to the box. This is repeated for many orientations; the line width is then convoluted onto the stick spectrum, and the

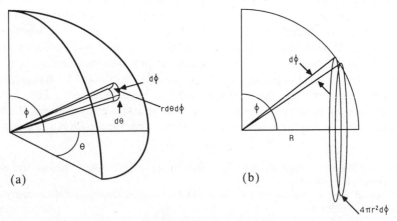

(a)   (b)

Figure 4.63 Sectors utilized for computation of ESR spectra in samples consisting of randomly oriented radicals: (a) orthorhombic symmetry; (b) axial symmetry.

spectrum is plotted. When there is an orientation-dependent line width, a full spectrum must be computed for each orientation, and the resulting intensities must be added to each box. This is obviously more time-consuming than when a constant line width is assumed.

Spectra of axially symmetric species are somewhat simpler, as shown in Figure 4.63b. Only one-quarter of a circle needs to be calculated. In this case, the stick figure intensity added to a box is related to the specific orientation as shown in

$$I_{B_0}^{B_0 + \delta B} = \sum_i D_i g_\perp^2 \left[\frac{g_\parallel^2}{g^2} + 1 \right] \delta(B - B_0, \delta B) \cdot 4\pi \sin \phi \, \Delta\phi \tag{183}$$

where $I_{B_0}^{B_0 + \delta B}$ is the intensity of the simulated spectrum in the range of field B_0 to $B_0 + \delta B$; D_i is the nuclear and electron spin multiplicity of transition i; g is given by $(g_\perp^2 \sin^2\phi + g_\parallel^2 \cos^2\phi)^{1/2}$; δ is a *modified delta function*, which is unity between the field range of B_0 and $B_0 + \delta B$ of the simulation; and $\Delta\phi$ is the increment of the calculated polar angle. For example, for the computation where the species is oriented with the unique axis parallel to the magnetic field, the intensity is given the value of 1. When an orthogonal axis is oriented parallel to the field, the intensity is assigned the value 4π. Since all other possible orientations must fall between these two extremes, only computations over a 90° arc need be executed.

The number of orientations required to simulate an arbitrary case is difficult to estimate. As the number of spectral computations increases, the computer time increases. However, if the number of orientations is too small, then the simulated spectrum becomes too discontinuous. This discontinuous appearance is equivalent to recording the spectrum of a limited number of crystallites so that the condition for a statistical distribution of orientations is not fulfilled. Several programs use random-number or Monte Carlo methods to reduce the computation time and eliminate the discontinuities.

Time-dependent spectra are more difficult to compute and in general are beyond the scope of this review. Many papers report computations that represent specific cases, and each has unique sets of assumptions. To our knowledge, no reviews cover the specific range of applicability of all methods. The two general methods are the *density matrix formulation* [478] and the *modified Bloch equations* [479]. The latter method is used most effectively for systems that undergo exchange and group reorientation as described in Section 6.4, under intramolecular effects, where the exchange can be represented as a jump between one magnetic state and another, as in the rotational reorientation of a methoxy group.

During the period when ESR was directed toward high-resolution spectroscopy to elucidate molecular structure, several programs were designed to increase its reliability by fitting the spectral data to a simulation by least-squares analyses (see Section 6.4). Often the analogue spectrum was digitized separately, so that the discrepancies between experimental and computed spectra were compared directly by the computer program. More recent adaptations of this method utilize data that are stored directly from the spectrometer in digital form. This technique is diminishing in importance because the current research trends are swinging away from purely structural studies. In addition, the magnetic field scan capabilities and sensitivity of modern spectrometers have increased both the S/N ratio and the fidelity of the recorded spectrum to a true representation, so that direct comparison of theoretical and experimental data is virtually as reliable, within experimental uncertainty, as the least-squares methods used earlier.

A second application of the computer to ESR spectroscopy is to improve the S/N ratio or the format of spectral presentation so that the spectra are interpreted more easily. Aside from straightforward signal averaging or a smoothing method, such as sliding mean methods or binomial curve fitting that are standard techniques, several unique applications are reported. Among the oldest methods used to provide enhanced resolution is the addition of the third derivative to the first derivative [480, 481]. The third derivative exhibits a major peak with two smaller ones on either side. Summing the first and third derivatives enhances the center and subtracts interferences from the tails of the derivative. The result is to sharpen the absorption even though the third derivative is relatively noisy. This can be accomplished by an analog summation of first and third derivatives (detection at three times the fundamental modulation frequency) or by taking the third derivative from the digitized spectrum, smoothing, and adding to the raw data.

Digital smoothing using various techniques is also employed. A method was proposed to reconstruct a spectrum that was recorded at a time constant so great as to cause distortion [482]. Methods developed for Fourier transform NMR spectroscopy and reviewed by Ernst [483] were recently employed [484] for spectrum manipulation. Resolution enhancement was achieved by computing the Fourier transform, operating on the spectrum to convert the Lorentzian line shape to Gaussian, and transforming back to the field-dependent spectrum. Smoothing was accomplished by operating on the spectrum by a filter function, $[1 - a\exp(-bt)]$, and transforming back to the field-dependent spectrum.

A technique called *cepstral analysis* [485] was reported. Essentially the logarithm of the Fourier transform is computed, and the spectrum is transformed back to a field-dependent function. In principle, only one pair of lines remains in the spectrum, which is high in intensity relative to other lines from the same set of nuclei. Autocorrelation of a spectrum was also used to enhance the spectrum S/N ratio and to reduce the number of lines [486]. Mathematically, either direct cross-correlation methods or FT methods are used. The process reinforces the hyperfine splittings to a greater extent than the noise increases. A related method of autocorrelation of a series of isotopically substituted radicals was used to determine coupling constant assignments [487].

6.2 Interactive Computer Control

There are many routine control features needed in ESR: control of microwave power, magnetic field (step size and initial and final or center value and scan range), sweep time, delay before scan, modulation amplitude, signal amplification, response time constant, temperature, and excitation method (e.g., electrochemical, photolytic). Similar sets of control features are needed for ENDOR. Some or all of these features may be important for a specific application. Numerous programs and interfaces for interactive control of ESR are reported in the literature or are commercially available. Very often, if there are specific applications or unique equipment, special programs and interfaces must be constructed. Generally, software available from a manufacturer is dedicated to their computer and their spectrometer, and most frequently the software covers only routine operations such as recording data, signal averaging, spectrum simulations, and other common operations.

There are many older or customized spectrometers that are not readily interfaced to computers as well as many personal computers for which no commercial ESR software is available. Nevertheless, numerous interfaces were reported for diverse equipment produced by Varian, Bruker (also marketed by IBM Instruments, a former subsidiary of International Business Machines Corporation), and O. S. Walker Company, Inc. (formerly Magnion and also Strand Laboratories) magnets; Varian V-4502 series spectrometers and consoles; Japan Electron Optics Laboratories Limited (JEOL) equipment; and others. In addition some software is published for Nicolet Instrument Corporation, Apple Computer, Inc., IBM-compatible, and Hewlett-Packard computers. Also software for diverse specialized tasks is reported in the literature, usually in *Review of Scientific Instruments, Journal of Physics E*, and *Journal of Magnetic Resonance*. The repositories of QCPE and CSULA (see Section 6) are invaluable.

6.3   Data Acquisition and Control of Pulsed Electron Spin Resonance

At this time, pulse spectrometers are not available commercially, although Bruker has announced that they have developed one. Consequently, all software and interfaces are accessible only in the literature. References to this information are found in [459] and in those programs archived by QCPE and CSULA (see Section 6.4).

6.4   Computer Programs for Electron Spin Resonance

Untold hours of effort have been devoted to designing, coding, and debugging programs for ESR spectroscopy. Programs that accurately simulate ESR spectra are almost mandatory to support proposed assignments of complicated spectra such as that shown in Figure 4.56. Even so, caution is required because seemingly small discrepancies between experimental and computed spectra may stem from quite large errors in the assumed values of the hsfc.

Different computational approaches are required for isotropic and anisotropic spectra. Even more demanding are the intermediate situations that arise with electron exchange reactions or restricted molecular motions, which occur with slow molecular tumbling or intramolecular reorientations.

Molecular orbital calculations also play a crucial role in ESR spectroscopy since the spectra are often used to map electron densities over an entire molecule

Table 4.18 Outline and References to Calculation and Simulation Programs for Electron Spin Resonance

Calculation or Simulation	Repository Codes[a]	Additional References
Isotropic (e.g., Free Rotation in Solution or Solid and Gas Phase)		
First-order hyperfine only	QCPE	No. 160 (ESSP2) No. 128 (ATLAS) No. 83 (ESRSPEC)	[488]
Including second- and higher order hyperfine	QCPE CSULA	No. 210 (ESRSPEC2) SIM2, SIM7	[233, 489]
Hyperfine splitting of the order of the Zeeman term	QCPE CSULA	No. 311 (ESRJOB, ESRCALC) SIM20	
Gas-phase diatomic species Rotation, spin and orbit moments Rotation and electron dipole	CSULA	SIM20	
Rigid Media		
Single crystal Parallel *g*-factor and hyperfine tensors, first order only	CSULA	SIM1, SIM14	[490]
Axial *g*-factor and hyperfine tensors, second order	CSULA	SIM18	[491, 492]
Nonaxial *g*-factor and hyperfine tensors	QCPE	No. 295 (FIBRE)	

	Source	Program	References
Nonparallel g-factor and hyperfine tensors	CSULA	SIM4	[493]
Multiple unpaired electrons	CSULA	SIM7	
Random Media			
g-Factor only, $S = \frac{1}{2}$	CSULA		
Parallel g-factor and hyperfine tensors, $S = \frac{1}{2}$	QCPE	No. 295 (FIBRE)	[494–496]
	CSULA	No. 265 (SIM14A)	
	CSULA	SIM15	
Nonparallel g-factor and hyperfine tensors, $S = \frac{1}{2}$	CSULA	SIM8, SIM10, SIM18	[79, 179, 493]
Multiplet spin states ($S > \frac{1}{2}$) with spin–spin coupling	CSULA	SIM4, SIM8	[497, 498]
	QCPE	No. 192 (FIELDS)	
ENDOR of powder samples of low-symmetry molecules			[431]
Time-Dependent Effects			
Restricted molecular rotation			
Somewhat fast rotation			[499]
Somewhat slow rotation			[500, 501]
Intermediate rotation			[502]
Saturation transfer			[418, 456, 503]
Intermolecular electron exchange	QCPE	No. 209 (ESREXN)	

Table 4.18 *(continued)*

Calculation or Simulation	Respository Codes[a]		Additional References
Intramolecular effects			
Alternating line-width effect			[244]
Restricted group rotation	CSULA	SIM5	[49, 504]
Counterion or proton migration			[505]
Least-Squares Fit			
To resonance fields	QCPE	No. 243 (PARA)	[506–508[b]]
To simulation	QCPE	No. 197 (ESRCON)	[509]
Correlation analysis	CSULA	SIM12, SIM13	[510]
Molecular Orbital Calculations			
Organic radicals	QCPE	Many listings	
Transition metals—crystal field	CSULA	SIM9, SIM22	

[a]QCPE, Quantum Chemistry Program Exchange, Department of Chemistry, Indiana University. CSULA, Dr. Richard Keys, Department of Chemistry, California State University at Los Angeles.
[b]Most general case of anisotropic noncoincident g and A tensors.

and consequently yield more detailed information (because of the multiple hfsc observed) than other spectroscopic or physical methods that produce only a datum representative of the molecule. Consequently ESR is often credited with having helped stimulate progress in the field of molecular orbital calculations.

Much computational effort is accessible through the QCPE and CSULA repositories noted earlier. Table 4.18 denotes some of the major areas of ESR calculations and simulations. While the references are not exhaustive they should nonetheless be helpful in preventing unnecessary duplication of other workers' efforts.

7   CONVERSION FACTORS AND UNITS IN MAGNETIC RESONANCE SPECTROSCOPY

More than any other field, magnetic resonance has been affected by the change in units from cgs (centimeter-gram-second) to SI (Systéme International). This change has two consequences: first, and most important, it causes difficulty in reading the literature because often an understanding of the origin of the unit is required to understand the convention; second, because of this difficulty, many magnetic resonance data are presented in terms of the cgs units, and although SI units are required [511, 512], journal editors have been inconsistent in demanding their use. This laxity is also perpetuated because even new ESR equipment is calibrated in terms of archaic units.

This section bridges the gap between the *old* and the *new* systems and aids in interpreting older literature. Whether cgs, SI, or other units are preferred by the experimenter is not as important as having all data reported in a standard manner to provide efficient communication among workers.

In Table 4.19 are the most recent values of fundamental constants [5, 513, 514] frequently used in magnetic resonance. Table 4.20 gives standard units and their names. More details on various unit systems can be found in [515–518].

7.1   Spectroscopic Units

Accepted SI units include the kilogram, meter, and second, and other units derived from them. A major change is that the joule rather than the erg or calorie is the standard unit of energy. However, frequency is retained for spectroscopy so that energy is given by

$$E = hv \tag{184}$$

where the Planck constant is given in Table 4.19. Very often reciprocal centimeters and occasionally degrees kelvin or calories are used in magnetic resonance to represent energy levels. Conversion factors between these terms and acceptable SI units are given in Table 4.21.

The unit of electron volts is also retained in the SI system, but this is intended principally for high-energy physics. However, since this unit is also used in magnetic resonance, it is included in Table 4.21.

Table 4.19 Fundamental Constants Frequently Used in Magnetic Resonance

Constant	Symbol	Value in SI Units[a,b]	Value in cgsm Units[c]
Speed of light in vacuum	c	2.99792458×10^8 m/s	$2.99792458 \times 10^{10}$ cm/s
Permeability of vacuum	μ_0	$12.56637061 \times 10^{-7}$ N/A^2	1.0
Permittivity of vacuum	ε_0	$8.85418782 \times 10^{-12}$ F/m^d	1.0
Planck constant	h	$6.6260755(40) \times 10^{-34}$ J·s	$6.6260755 \times 10^{-27}$ erg·s
Electron charge	e	$1.60217733(49) \times 10^{-19}$ C	4.803250×10^{-10} esu
Electron mass	m_e	$9.1093897(54) \times 10^{-31}$ kg	$9.1093897 \times 10^{-28}$ g
Proton mass	m_P	$1.6726231(10) \times 10^{-27}$ kg	$1.6726231 \times 10^{-24}$ g
Electron magnetic moment	μ_e	$9.2847701(31) \times 10^{-24}$ J/T	$9.2847701 \times 10^{-21}$ erg/G
Electron g-factor	$g_e^{\,e}$	$2.002319304386(20)$	2.0023193044
Bohr magneton	μ_B	$9.2740154(31) \times 10^{-24}$ J/T	$9.2740154 \times 10^{-21}$ erg/G
Proton (unshielded) magnetic moment	μ_P	$1.41060761(47) \times 10^{-26}$ J/T	$1.4107076 \times 10^{-23}$ erg/G
Proton (H$_2$O, 25°C) magnetic moment	$\mu_{P'}$	$1.41057138(47) \times 10^{-26}$ J/T	$1.41057138 \times 10^{-24}$ erg/G
Nuclear magneton	μ_N	$5.0507866(17) \times 10^{-27}$ J/T	$5.0507866 \times 10^{-24}$ erg/G
Proton g-factor	g_P	$5.585564(17)$	5.585564
Avagadro constant	N_A	$6.0221367(36) \times 10^{23}$ mol^{-1}	6.0221367×10^{23} mol^{-1}
Molar gas constant	$\mathscr{R}$	$8.314510(70)$ J/(mol·K)	8.31451×10^7 erg/(mol·K)
Boltzmann constant	k	$1.380658(12) \times 10^{-23}$ J/K	1.380658×10^{-16} erg/K

[a] Given in primary SI units, except N = newton, J = joule, K = degree, T = tesla, mol = N_A molecules.

[b] Uncertainty of the last two digits are given in parentheses; c, μ_0, and ε_0 are exact.

[c] Uncertainty in cgs not evaluated, but should be equivalent to the SI units.

[d] F = farads = coulombs per volt.

[e] $g_e = 2(\mu_e/\mu_B)$.

Table 4.20 Standard Units

Quantity	SI		cgsm		
	Unit	Dimensions	Unit	Dimensions	Conversion
Length, l	Meter, m	L	Centimeter, cm	L	1 m = 100 cm
Mass, m	Kilogram, kg	M	Gram, g	M	1 kg = 1000 g
Time, t	Second, s	T	Second, s	T	
Force, **F**	Newton, N	$M\,L/T^2$	Dyne, dyn	$M\,L/T^2$	1 N = 10^5 dyn
Energy, E	Joule, J	$M\,L^2/T^2$	Erg	$M\,L^2/T^2$	1 J = 10^7 erg
Frequency, ν	Hertz, Hz	$1/T$	Hertz, Hz	$1/T$	
Current, I	Ampere, A	I	Abampere	$M^{1/2}L^{1/2}/T$	1 A = 0.1 abampere
Charge, q	Coulomb, C	I T	Abcoulomb	$M^{1/2}\,L^{1/2}$	1 C = 0.1 abcoulomb
Voltage, V	Volt, V	$V = J/I\,T$			
Magnetic pole	Ampere meter, Am	I L	Not named	$M^{1/2}\,L^{3/2}/T$	
Magnetic moment	Ampere square meter, Am^2	I L^2	Ergs per gauss, ergs/G	$M^{1/2}\,L^{5/2}/T$	
Magnetization	Ampere per meter, A/m	I/L	Gauss, G	$M^{1/2}/L^{1/2}\,T$	
Inductance	Henry, H	ML^2/I^2T^2			
Magnetic flux	Weber, Wb	ML^2/IT^2			
Magnetic field	Weber per meter, Wb/m^a	I/L	Oersted, Oe	$M^{1/2}/L^{1/2}\,T$	1 A/m = $(1000/4\pi)$ Oe
Magnetic flux density	Tesla, T^b	M/LIT^2	Gauss, G	$M^{1/2}/L^{1/2}\,T$	1 T = 10^4 G

^{a}Also known as ampere turns per meter.
^{b}Also known as webers per square meter.

Table 4.21 Conversion Factors Between Acceptable SI Units and Other Common Units[a]

Unit	Joules	Hertz	Electronvolts
1 J =	1.0	$1.50918897 \times 10^{33}$	6.2415063×10^{18}
1 erg =	1.0×10^{-7}	$1.50918897 \times 10^{26}$	6.2415063×10^{11}
1 Hz =	$6.6260755 \times 10^{-34}$	1.0	$4.1356692 \times 10^{-15}$
1 ev =	$1.6021773 \times 10^{-19}$	$2.41798836 \times 10^{14}$	1.0
1 cm^{-1} =	$1.9864474 \times 10^{-23}$	$2.99792458 \times 10^{10}$	$1.23984244 \times 10^{-4}$
1 K =	1.380658×10^{-23}	2.803674×10^{10}	8.6173856×10^{-5}

[a]Thumbnail conversions: $1 \text{ cm}^{-1} \simeq 300 \text{ GHz} \simeq 1.44 \text{ K} \simeq 1.99 \times 10^{-23} \text{ J} \simeq 1.24 \times 10^{-4} \text{ eV}$. For 1 K, $(kT/hc) \simeq 0.695 \text{ cm}^{-1}$.

7.2 Magnetic Units

Magnetic units are complicated by the fact that different systems were derived using slightly different considerations [515–517]. However, dimensionality must be preserved in all calculations, so that results are directly interconvertible. To understand each system of units the basic conventions must be given. The five common unit systems in magnetism are mksa (meter-kilogram-second-ampere), SI, cgsm (centimeter-gram-second-magnetic), cgse (centimeter-gram-second-electrostatic), and Gaussian. Magnetic parameters in the SI and mksa systems are almost equivalent. The cgsm system, also called the electromagnetic system, is the cgs system, except that electric current is defined by the force between two conductors with current flow. The cgse, or electrostatic, system defines electrical charge by the force acting between two electrical charges. The Gaussian system is *symmetrical* in that cgsm magnetic units and cgse electrical units are incorporated. The cgse system is less common in magnetics and is not described here.

7.2.1 Centimeter-Gram-Second-Magnetic System

The cgsm system was defined on the basis of Ampere's law for the force **F** acting between two equal lengths l of conductors carrying currents I_1 and I_2, separated by distance r. This is shown in (185), which is traditionally written in differential vector notation, but for the purposes of illustrating units, a shorthand notation is used.

$$F = k_m \mu \frac{(I_1 l_1)(I_2 l_2)}{r^2} \tag{185}$$

In this system $k_m = 1$, and the permeability μ is assigned the value of unity for a vacuum. When F is 1 dyn, r is 1 cm, and $I_1 = I_2$, the current is *defined* as 1 abampere. Therefore, the dimensions of abamperes are expressed in cgsm units as $1 \text{ abamp} = 1 \text{ dyn}^{1/2} = 1 (g^{1/2} \cdot cm^{1/2})/s$.

Equation (185) is analogous to Coulomb's law, which defines the electric charge. Since the charge is an electric pole, we draw the analogy that the current multiplied by length is equivalent to a magnetic *pole*

$$q_m = I \cdot l \equiv \text{abampere cm} \tag{186}$$

This quantity is a mathematical fiction, but shows that the magnetism is derived from current flow. By rewriting (185) in terms of poles

$$F = \mu_0 \frac{q_{m1} q_{m2}}{r^2} \tag{187}$$

and substituting $k_m = 1$ and μ_0 for μ, we define the magnetic field H as the limit of F as q_{m2} approaches 0

$$H = \lim_{q_{m2} \to 0} \frac{F}{q_{m2}} \tag{188}$$

Thus, the magnetic field intensity H equivalent to 1 Oe corresponds to a force of 1 dyn per unit pole. The dimensions of magnetic field therefore are represented as abampere/cm, $dyn^{1/2}/cm$, or $g^{1/2}/(cm^{1/2} \cdot s)$. The magnetic flux, or magnetic flux density in gauss, is related to the magnetic field intensity by

$$B = \mu H \tag{189}$$

where μ (dimensionless) represents the permeability of the medium in which the field H is applied. It is important to note that B and H are equivalent in cgsm units.

In magnetic measurements we are interested primarily in the effects of materials placed in static or microwave magnetic fields. Thus, we must introduce the magnetization of the medium, which is denoted as M and is equal to the number of magnetic moments per unit volume.

A magnetic moment m is conceptually equivalent to a dipole moment in which two equal and opposite charges are placed apart at distance r. The value of m is then

$$m = (q_m l)r \equiv \text{abampere cm}^2 \tag{190}$$

The energy of the dipole, in vacuum, in a magnetic field is therefore equal to the vector scalar product of $\mathbf{m}$ and $\mathbf{H}$ ($E = \mathbf{m} \cdot \mathbf{H}$). Thus the dimensions of $\mathbf{m}$ are ergs per gauss. By combining the dimensions from (188) and (190), energy ($E = \mathbf{m} \cdot \mathbf{H}$) in ergs is equivalent to the square of abampere multiplied by centimeter.

The magnetization of a sample is the number n of magnetic moments per unit volume

$$M = nm/v \equiv \text{abampere/cm}$$
$$= g^{1/2}/cm^{1/2} \cdot s \tag{191}$$

This is dimensionally equivalent to the gauss, as shown following (188). The magnetization is often reported in units of emu (electromagnetic units), which are equivalent to the gauss.

From Gauss' law, which states that the surface integral of a field is equivalent to 4π times the poles encircled by the surface integral, we obtain an expression for the magnetic flux density

$$B = H + 4\pi M \tag{192}$$

Those unit systems in which the factor 4π appears in the equation are called nonrational. (Surprisingly, the term *irrational* is not applied to magnetic units.) From (189) and (192), we obtain an expression for μ

$$\mu = 1 + 4\pi M/H \tag{193}$$

The ratio of M/H is called the *magnetic susceptibility* χ, which in cgsm units is dimensionless. For paramagnetic materials, $M = \chi H$, so that $\mu = 1 + 4\pi\chi$.

7.2.2   Gaussian Units

The Gaussian system was an attempt to combine the magnetic cgsm units and the electrical cgse units to form a consistent set. Unfortunately this led to inconsistency in reporting results. In the Gaussian and cgse systems, the electrical charge is defined in terms of Coulomb's law

$$F = k_e \varepsilon \frac{q_1 q_2}{r^2} \tag{194}$$

where q_1 and q_2 are the electrical charges and r is the separation between them. The constants $k_e = 1$ and ε are chosen to be dimensionless. For a vacuum, $\varepsilon = \varepsilon_0 = 1$. Thus, the unit of charge, called the *statcoulomb*, is given as 1 statcoul = 1 $\text{dyn}^{1/2} \cdot \text{cm} = 1\,(\text{g}^{1/2} \cdot \text{cm}^{3/2})/\text{s}$.

With reference to cgsm units, the abampere can be multiplied by time to define the unit of charge called the *abcoulomb*, and the statcoulomb can be divided by time to define the statampere. Dimensional analysis shows that the statcoulomb and abcoulomb differ by a factor of centimeters per second, and the statampere differs from the abampere by the same factor. It is possible to show that these differ by the speed of light c, which is 2.99792×10^{10} cm/s. Substituting cgse units into (185), with definitions of k_e, k_m, μ_0, and ε_0 each equal to unity, we find that

$$F = \frac{(q_1/t \cdot l_1)(q_2/t \cdot l_2)}{c^2 r^2} \tag{195}$$

Thus, numerically 1 abampere $\equiv c$ statamperes, and 1 abcoulomb $\equiv c$ statcoulombs.

Following the parallelism of cgsm we now define the equivalent of the magnetic pole, $q_m = ql/tc$ in terms of statcoulombs: This factor has the dimensions of $(g^{1/2} \cdot cm^{3/2})/s$ as in cgsm above. From this definition, the units of magnetic field are equivalent in both cgsm and Gaussian units.

7.2.3   Systéme International and Meter-Kilogram-Second-Ampere Units

The mksa system (A ≡ ampere) is similar to the SI system. In the mksa system, the ampere is defined, and from this quantity, the coulomb as a unit of charge is derived as an ampere second.

Prior to about 1948, the ampere was defined as the amount of electric current required to deposit 0.00111800 g of silver from a salt solution in 1 s [519]. Cells were built to a specific design and placed in series with the experiment for which current was to be determined. Copper was often used as a secondary standard to minimize the cost of such experiments. This was the unit of current used with Ampere's law to determine the magnetic flux in the mksa system. Because of the uncertainties inherent in the early measuring techniques, the ampere determined in this way (international-ampere) is 0.999835 of the ampere presently defined by magnetic phenomena (absolute ampere—not abampere, which is the cgsm analogue).

The basis of the absolute ampere (A) is Ampere's law, which is similar in form to (185) for the force between two parallel wires

$$F = k_M \mu \frac{(I_1 l_1)(I_2 l_2)}{r^2} \tag{196}$$

The value of μ for a vacuum μ_0 is set equal to $4\pi \times 10^{-7}$. The value of k_M is set equal to $1/2\pi$. Current is defined as 1 A when l_1, l_2, and r are each 1 m, and the force between the wires is 2×10^{-7} N $[1 \text{ N} = 1(\text{kg} \cdot \text{m})/\text{s}^2]$. The factor of 2 arises because the current in each wire contributes to the force. A similar practical definition of the ampere is the current that, when flowing in the same direction through each of two identical coaxial current loops of radius r and separation a, where $a \ll r$, in a vacuum, produces a mutual attractive force of $\{F = [6(\mu^2 \cdot r^4)/a^4] \cdot 10^7 \text{ N}\}$. This is a consequence of Ampere's law.

A coefficient of 4π is introduced to make the mksa and SI systems *rational*. Dimensional analysis shows that $\mu_0 = 4\pi \times 10^{-7}$ N/A^2. These units are equivalent to henrys per meter, H/m. The unit 1 H = 1 V/A = 1 J/A^2. As a result of these units, μ_0 is often called the *inductance of free space*. (The SI unit H for henry should not be confused with the symbol **H** for magnetic field.) The magnetic flux can now be defined. Equation (196) is rewritten

$$F = I_2 l_2 B \tag{197}$$

where B is defined in vacuum as

$$B = \mu_0 \frac{I_1 l_1}{r^2} \tag{198}$$

The units of B are weber per square meter or tesla, which is equivalent to Newtons per milliampere (N/mA). Coulomb's law is used to convert these units into units of kilograms, meters, and seconds. The mksa or SI equivalent of (194) is

$$F = \frac{q_1 q_2}{4\pi\varepsilon_0 r^2} \tag{199}$$

where $\varepsilon_0 = 10^7/(4\pi c^2) \approx [10^{-9}/(36\pi)]\, \text{s}^2/\text{m}^2$. Using (199), we find that the dimensions of the coulomb are given as $1\,\text{C} = 1\,\text{N}^{1/2}\,\text{s} \equiv 1\,\text{kg}^{1/2}\cdot\text{m}^{1/2}$. Substituting $1\,\text{A} = 1\,\text{C/s} \equiv 1\,\text{N}^{1/2}$ into $1\,\text{T} = 1\,\text{Wb/m}^2 = 1\,\text{N}/(\text{m}\cdot\text{A})$, the magnetic flux density has the dimensions $(\text{kg}^{1/2}\cdot\text{m}^{1/2})/\text{s}$. This is dimensionally equivalent to the cgsm unit of gauss.

The magnetic field **H** is the effect of current and is defined as in (199) without μ_0

$$\mathbf{H} = \frac{I l}{r^2} \tag{200}$$

and has the dimensions of amperes per meter (A/m). For a medium with permeability μ

$$\mathbf{B} = \mu \mathbf{H} \tag{201}$$

Because of the inconvenient magnitude and units of μ, the permeability of a material is often reported in terms of the relative permeability κ

$$\kappa = \frac{\mu}{\mu_0} \tag{202}$$

where κ is a dimensionless term and corresponds to the cgsm value of μ.

To describe the magnetic properties of a material we again define the magnetic pole as $q_m = I l$, and the magnetic moment as $m = I l r$ as in (190). The dimensions of the magnetic moment are ampere square meter $(\text{A}\cdot\text{m}^2)$, analogous to the dimensions in the cgsm system. The magnetization is also defined as in (191), but the dimensions are amperes per meter. In the SI or mksa systems, H and M are in the same units and are additive. Thus,

$$B = \mu_0(H + M) \tag{203}$$

The major difference between the cgsm (and Gaussian) system and the SI (and mksa) system is that in the latter, the coefficient 4π is included in μ_0, so that it is not needed to multiply M.

Equating (201) and (203) yields

$$\mu = \mu_0(1 + M/H) = \mu_0(1 + \chi_m) \tag{204}$$

and from (202),

$$\kappa = (1 + \chi_m) \tag{205}$$

7.3   Use of Units

Magnetic nomenclature is often incorrectly interchanged among unit systems, or even within a given unit system. For example, often the relative permeability is improperly reported as μ in SI systems, or B is called magnetic field instead of magnetic flux density in any of the unit systems. The alert reader will have recognized such inconsistencies in this chapter; however, they conform to current usage in ESR spectroscopy. The SI system has the advantage that when units are used correctly, the physics behind the magnetic phenomena often becomes transparent.

Since the situation with regard to the imprecise usage of magnetic units is not likely to be rectified in the near future, the reader should be aware of these problems, and should not assume that values for reported parameters are correct without first analyzing the results.

7.4   Conversion Between Energetic and Magnetic Molecular Parameters

Parameters such as hyperfine, orbital, or zero-field splittings are often reported as either magnetic units (i.e., gauss, tesla) or energy-related units (i.e., reciprocal centimeter or frequency). Conversion among energy units can be accomplished with the aid of Table 4.21. Conversion from magnetic to energy units is somewhat more difficult because of the g-factor dependence.

When magnetic units are used, the relationship to energy or frequency units is given by

$$E_x = hv_x = g\mu_B M_x \tag{206}$$

where M_x is the specific magnetic parameter given in units of tesla or gauss, and E_x and v_x are the corresponding energy or frequency units. For example, to convert first-order hyperfine splitting a from tesla to frequency units (A),

$$A = g\mu_B a/h \tag{207}$$

This can be referenced to the free electron g-factor

$$A = 0.28026 \frac{\text{GHz}}{\text{T}} \frac{g}{g_\text{e}} a \tag{208}$$

where the coefficient is the magnetogyric ratio.

8   BIBLIOGRAPHIC MATERIALS BY CATEGORY

8.1   Basic Textbooks and General Treatments

Alger, R. S., *Electron Paramagnetic Resonance: Techniques and Applications*, Wiley-Interscience, New York, 1967.

Assenheim, H. M., *Introduction to Electron Spin Resonance*, Plenum, New York, 1967.

Atherton, N. M., *Electron Spin Resonance*, Hilger and Watts, London, 1975.

Carrington, A., and A. D. McLachlan, *Introduction to Magnetic Resonance*, Harper & Row, New York, 1967.

Coogan, C. K., N. S. Ham, S. N. Stuart, J. R. Pilbrow, and G. V. H. Wilson, *Magnetic Resonance*, Plenum, New York, 1970.

Geschwind, S., Ed., *Electron Paramagnetic Resonance*, Plenum, New York, 1972.

Ingram, D. J. E., *Free Radicals as Studied by Electron Spin Resonance*, Butterworths London, 1958; *Spectroscopy at Radio and Microwave Frequencies*, Plenum, New York, 1967; *Biological and Biochemical Applications of Electron Spin Resonance*, Hilger, London, 1969.

McLauchlan, K. A., *Magnetic Resonance*, Clarendon, Oxford, 1972 (elementary presentation of basic material).

McMillan, J. A., *Electron Paramagnetism*, Reinhold, New York, 1968.

Pake, G. E., *Paramagnetic Resonance: An Introductory Monograph*, Benjamin, New York, 1962 (a classic text).

Skobel'tsyn, D. V., Ed., *Quantum Electronics and Paramagnetic Resonance*, Plenum, New York, 1971.

Slichter, C. P., *Principles of Magnetic Resonance With Examples From Solid State Physics*, Harper & Row, New York, 1963 (general treatment of magnetic resonance directed toward NMR but extremely useful).

Symons, M. C. R., *Chemical and Biochemical Aspects of Electron-Spin Resonance Spectroscopy*, Wiley, New York, 1978 (introductory-level book stressing chemistry rather than physics).

Wertz, J. E., and J. R. Bolton, *Electron Spin Resonance, Elementary Theory and Practical Applications*, McGraw-Hill, New York, 1972 (comprehensive, textbook treatment of ESR).

8.2   Electron Spin Resonance Theory

Dixon, W. T., *Theory and Interpretation of Magnetic Resonance Spectra*, Plenum, New York, 1972.

Farach, H. A., and C. P. Poole, "Solving the Spin Hamiltonian for the ESR of Irradiated Single Crystals," *Adv. Magn. Reson.*, **5**, 229 (1973).

Harriman, J. E., *Theoretical Foundations of Electron Spin Resonance*, Academic, New York, 1978.

Misra, S. K., "Evaluation of Spin Hamiltonian Parameters From ESR Data of Single Crystals," *Magn. Reson. Rev.*, **10**, 285 (1986) (provides up-to-date summary of numerical calculations; notation is difficult to follow).

Poole, Jr., C. P., and H. A. Farach, *The Theory of Magnetic Resonance*, Wiley-Interscience, New York, 1972.

Rudowicz, C., "Concept of Spin Hamiltonian, Forms of Zero Field Splitting and Electronic Zeeman Hamiltonian and Relations Between Parameters Used in EPR. A Critical Review," *J. Magn. Reson.*, **13**, 1 (1987) (outlines the theoretical foundations of the concept of spin Hamiltonian, systemizes the operators, Hamiltonian forms, axis systems and parameters of transition metal ion ESR studies, and points out pitfalls and errors in the ESR Hamiltonian literature).

Zhidomirov, G. M., Ya. S. Lebedev, S. N. Dobryakov, N. Ya. Shteinshneider, A. K. Chirkov, and V. A. Gubanov, *Interpretation of Complex ESR Spectra*, US Department of Commerce, NTIS, Springfield, VA, 1985.

8.3 Instrumentation

Feher, G., "Sensitivity Considerations in Microwave Paramagnetic Resonance Absorption Techniques," *Bell Syst. Tech. J.*, **36**, 449 (1957) (frequently referenced and extremely pertinent discussion of sensitivity).

Poole, Jr., C. P., *Electron Spin Resonance: A Comprehensive Treatise of Experimental Techniques*, 2nd ed., Wiley-Interscience, New York, 1983.

Wilmshurst, T. H., *Electron Spin Resonance Spectrometers*, Hilger, London, 1967.

8.4 Organic Radicals and Radical Ions

Carrington, A., "Electron Spin Resonance Spectra of Aromatic Radicals and Radical-Ions," *Q. Rev. Chem. Soc.*, **17**, 67 (1963) (summarizes historically important early work).

Geske, D. H., "Conformation and Structure as Studied by Electron Spin Resonance Spectroscopy," *Prog. Phys. Org. Chem.*, **4**, 125 (1967) (steric effects as revealed in hyperfine splittings).

Kaiser, E. T., and L. Kevan, Eds., *Radical Ions*, Interscience, New York, 1968 (Chaps. 1–8, especially).

Sales, K. D., "The Theory of Isotropic Hyperfine Splitting Constants for Organic Free Radicals," *Adv. Free Radical Chem.*, **3**, 139 (1969) (contains parameters for splittings of σ and π organic radicals).

Solodovnikov, S. P., and A. I. Prokof'ev, "The Electronic Structure of Radical Anions," *Russ. Chem. Rev.*, **39**, 591 (1970).

Williams, F., and E. D. Sprague, "Novel Radical Anions and Hydrogen Atom Tunneling in the Solid State," *Acc. Chem. Res.*, **15**, 408 (1982).

8.5 Inorganic Radicals

Atkins, P. W., and M. C. R. Symons, *The Structure of Inorganic Radicals*, Elsevier, New York, 1967.

Symons, M. C. R., *Chemical and Biochemical Aspects of Electron Spin Resonance Spectroscopy*, Van Nostrand Reinhold, New York, 1978.

8.6   Transition Metal Ions and Complexes

Abragam, A., and B. Bleaney, *Electron Paramagnetic Resonance of Transition Ions*, Clarendon, Oxford, 1970 (very fine text on all aspects of transition metal ESR).

Al'tshuler, S. A., and B. M. Kozyrev, *Electron Paramagnetic Resonance in Compounds of Transition Elements*, Wiley, New York and Keter Publishing House, Jerusalem, Israel, 1974 (one of the most detailed treatments of this subject).

Buckmaster, H. A., and D. B. Delay, "Electron Paramagnetic Resonance of Transition-Metal, Lanthanide and Actinide Ions in Solids," *Magn. Reson. Rev.*, **6**, 139 (1980) (comprehensive review paper).

Carlin, R. L., *Magnetochemistry*, Springer-Verlag, New York, 1986 (wide-ranging monograph on metallic and rare earth elements; includes chapter on ferromagnetism and antiferromagnetism).

Goodman, B. A., and J. B. Raynor, *Adv. Inorg. Chem. Radiochem.*, **13**, 235 (1970) (a very comprehensive review).

Kokoszka, G. F., and G. Gordon, *Tech. Inorg. Chem.*, **7**, 151 (1968).

Konig, E., "Electron Paramagnetic Resonance," in H. A. O. Hill and P. Day, Eds., *Physical Methods in Advanced Inorganic Chemistry*, Wiley, New York, 1968, Chap. 7.

Kuska, H. A., and M. T. Rogers, "Electron Spin Resonance of First Row Transition Metal Complexes," in E. T. Kaiser and L. Kevan, Eds., *Radical Ions*, Wiley, New York, 1968, Chap. 13.

Low, W., *Paramagnetic Resonance in Solids*, Academic, New York, 1960 (one of the most cited reviews).

McGarvey, B. R., "Electron Spin Resonance of Transition-Metal Complexes," *Transition Met. Chem.*, **3**, 89 (1966).

Orton, J. W., *Electron Paramagnetic Resonance*, Gordon & Breach, New York, 1970 (an extremely readable summary of theory and experiment).

Sorin, L. A., and M. V. Vlasova, *Electron Spin Resonance of Paramagnetic Crystals*, Plenum, New York, 1973.

Yen, T. F., Ed., *Electron Spin Resonance of Metal Complexes*, Plenum, New York, 1969.

8.7   Biological Systems

Berliner, L. J., and J. Reuben, Eds., *Biological Magnetic Resonance*, Vols. 1–7, Plenum, New York, 1978–1987 (articles therein not referenced here).

Chapman, D., and J. A. Hayward, "New Biophysical Techniques and Their Application to the Study of Membranes," *Biochem. J.*, **228**, 281 (1985) (general review including the role of ESR and its limitations).

Cohen, J. S., *Magnetic Resonance in Biology*, Vols. 1, 2, Wiley-Interscience, New York, 1980, 1982.

Feher, G., Ed., *Electron Paramagnetic Resonance with Applications to Selected Problems in Biology*, Gordon & Breach, New York, 1970.

Fung, L. W. M., and M. E. Johnson, "Recent Developments in Spin Label EPR Methods for Biomembrane Studies," *Curr. Top. Bioenerg.*, **13**, 107 (1984).

Herak, J. N., and K. J. Adamic, Eds., *Magnetic Resonance in Chemistry and Biology*, Dekker, New York (1975).

Hoff, A. J., "Applications of ESR in Photosynthesis," *Phys. Rep.*, **54**, 75 (1979).

Knowles, P. F., D. Marsh, and H. W. Rattle, *Magnetic Resonance of Biomolecules: An Introduction to the Theory and Practice of NMR and ESR in Biological Systems*, Wiley-Interscience, New York, 1976.

Swartz, H. M., J. R. Bolton, and D. C. Borg, Eds., *Biological Applications of Electron Spin Resonance*, Wiley-Interscience, New York, 1972.

8.8   Special Applications

Benedek, G. B., *Magnetic Resonance at High Pressure*, Interscience, New York, 1963.

Buchachenko, A. L., *Stable Radicals*, Plenum, New York, 1965.

Cubitt, J. M., and C. V. Burek, *A Bibliography of Electron Spin Resonance Applications in the Earth Sciences*, GEO Abstracts, England, 1980.

Chang, T.-T., "The Calibration Methods and the Reference Materials in ESR Spectroscopy," *Magn. Reson. Rev.*, **9**, 65 (1984).

DeVries, K., and D. Roylance, *Encycl. Polym. Sci. Eng.*, **5**, 687 (1986) (applications of ESR to polymer studies; particular emphasis on mechanical damage).

Eaton, S. S., and G. R. Eaton, "Signal Area Measurements in ESR," *Bull. Magn. Reson.*, **1**, 130 (1980).

Fraissard, J. P., and H. A. Resing, Eds., *Magnetic Resonance in Colloid and Interface Science*, Reidel, Hingham, MA, 1980.

Goldberg, I. B., and A. J. Bard, "Electron Spin Resonance Spectroscopy," in P. J. Elving, M. M. Bursey, and I. M. Kolthoff, Eds., *Treatise on Analytical Chemistry*, Vol. 10, Part 1, 2nd ed., 1983 (application of ESR in analytical chemistry).

Iwasaki, M., "ESR of Irradiated Organic Fluorine Compounds," *Fluorine Chem. Rev.*, **5**, 1 (1971).

Kaufmann, E. N., and G. K. Shenoy, Eds., *Nuclear and Electron Resonance Spectroscopies Applied to Materials Science*, Elsevier, New York, 1981 (covers various approaches to material diagnostics ranging from the fundamental to the pragmatic).

Lancaster, G., *Electron Spin Resonance in Semiconductors*, Plenum, New York, 1967.

Lewis, I. C., and L. S. Singer, "ESR and the Mechanism of Carbonization," *Chem. Phys. Carbon*, **17**, 1 (1981).

Lunsford, J. H., "ESR of Adsorbed Species," *Catal. Rev.*, **8**, 135 (1973).

McKinney, T. M., "Electron Spin Resonance and Electrochemistry," in A. J. Bard, Ed., *Electroanalytical Chemistry*, Vol. 10, Dekker, New York, 1977, pp. 97–278.

McWhinnie, W. R., "ESR and NMR Applied to Minerals," in F. J. Berry and D. J. Vaughan, Eds., *Chemical Bonding: Spectroscopy in Mineral Chemistry*, Chapman & Hall, London, 1985, pp. 209–249.

Mims, W. B., *The Linear Electric Field Effect in Paramagnetic Resonance*, Clarendon, Oxford, 1976.

Petrakis, L., and J. P. Faissard, Eds., *Magnetic Resonance: Introduction, Advanced Topics and Applications to Fossil Energy*, Reidel, Holland, 1984.

Poole, C. P., and H. A. Farach, "Electron Spin Resonance of Minerals: Part I Nonsilicates," *Magn. Reson. Rev.*, **4**, 137 (1977).

Poole, C. P., H. A. Farach, and T. P. Bishop, "Electron Spin Resonance of Minerals: Part II Silicates," *Magn. Reson. Rev.*, **4**, 2256 (1978).

Ranby, B., and J. F. Rabek, *ESR Spectroscopy in Polymer Research*, Springer-Verlag, New York, 1977.

Spaeth, J.-M., "Spectroscopic Studies of Defects in Ionic and Semi-Ionic Solids," *NATO ASI Ser., Ser. B*, **147**, 205 (1986) (technique oriented).

Thomann, H., and L. R. Dalton, "ENDOR Studies of Polyacetylene," in T. A. Skotheim, Ed., *Handbook of Conducting Polymers*, Vol. 2, Dekker, New York, 1986, pp. 1127–1156.

Volkov, S. V., "Spectroscopy of Molten Salts," *Rev. Chim. Miner.*, **15**, 59 (1978) (unique review not covered elsewhere).

Wertheim, G. K., A. Hausmann, and W. Sander, *The Electronic Structure of Point Defects*, North-Holland, Amsterdam, 1971.

8.9   Dynamic Effects

Banci, L., I. Bertini, and C. Luchinat, "Electron Relaxation," *Magn. Reson. Rev.*, **11**, 1 (1986).

Dalton, L. R., B. H. Robinson, L. A. Dalton, and P. Coffey, "Saturation Transfer Spectroscopy," *Adv. Magn. Reson.*, **8**, 149 (1976).

Freed, J. H., and J. B. Pedersen, "The Theory of Chemically Induced Dynamic Spin Polarization," *Adv. Magn. Reson.*, **8**, 2 (1976).

Hyde, J. S., "Saturation-Transfer Spectroscopy," *Methods in Enzymology*, **49**, 480 (1978) (qualitative description of the process).

Lepley, A. R., Ed., *Chemically Induced Magnetic Polarization*, Wiley, New York, 1973.

Molin, Y. N., K. M. Salikhov, and K. I. Zamaraev, *Spin Exchange: Principles and Applications in Chemistry and Biology*, Springer-Verlag, New York, 1980.

Muus, L. T., and P. W. Atkins, Eds., *Electron Spin Relaxation in Liquids*, Plenum, New York, 1972.

Muus, L. T., P. W. Atkins, K. A. McLauchlan, and J. B. Pedersen, Eds., *Chemically Induced Magnetic Polarization*, Reidel, Holland, 1977.

Poole, C. P., and H. F. Farach, *Relaxation in Magnetic Resonance:Dielectric and Mossbauer Applications*, Academic, New York, 1971.

Standley, K. J., and R. A. Vaughan, *Electron Spin Relaxation Phenomena in Solids*, Plenum, New York, 1969 (qualitative description of relaxation processes).

Sullivan, P. D., and J. R. Bolton, "The Alternating Linewidth Effect," *Adv. Magn. Reson.*, **4**, 39 (1972).

Sullivan, P. D., and E. M. Menger, "Temperature Dependent Splitting Constants in the ESR Spectra of Organic Free Radicals," *Adv. Magn. Reson.*, **9**, 1 (1977).

Weger, M., "Passage Effects in Paramagnetic Resonance Experiments," *Bell Syst. Tech. J.*, **39**, 1013 (1960) (an often overlooked reference, particularly useful in saturation transfer spectroscopy).

8.10   Double Resonance

Dorio, M. M., "Electron–Electron Double Resonance," *Magn. Reson. Rev.*, **4**, 105 (1977).

Dorio, M. M., and J. H. Freed, Eds., *Multiple Electron Resonance Spectroscopy*, Plenum, New York, 1979.

Eachus, R. S., and M. T. Olm, "Electron Double Resonance Spectroscopy," *Science*, **230**, 268 (1985).

Kevan, L., and L. D. Kispert, *Electron Spin Double Resonance Spectroscopy*, Wiley-Interscience, New York, 1976.

Kurreck, H., B. Kirste, and W. Lubitz, "ENDOR Spectroscopy—A Promising Technique for Investigating the Structure of Organic Radicals," *Angew. Chem. Int. Ed. Engl.*, **23**, 173 (1984) [original *Angew Chem.*, **96**, 171 (1984)] (includes ENDOR and ENDOR TRIPLE resonance, and details of ^{2}H, ^{13}C, and ^{15}N nuclei).

8.11 Pulsed Electron Spin Resonance

Dikanov, S. A., and Yu. D. Tsuetsov, "Structural Applications of the Electron Spin Echo Method," *J. Struct. Chem. USSR*, **26**, 7676 (1985) (translation).

Hore, P. J., "Time Domain ESR," *Primary Photoprocesses in Biology and Medicine* (1984): *Bressanone, Italy, NATO ASI Ser.*, Ser. A, **85**, 131 (1985).

Narayana, P. A., and L. Kevan, "Fourier Transform Analysis of Electron Spin Echo Modulation Spectroscopy," *Magn. Reson. Rev.*, **7**, 239 (1983).

Norris, J. R., M. C. Thurnauer, and M. K. Bowman, "Electron Spin Echo Spectroscopy and the Study of Biological Structure and Function," *Adv. Biol. Med. Phys.*, **17**, 365 (1980) (useful introductory reference to basic theory and practical applications to biological systems).

Stillman, A. E., and R. N. Schwartz, "Study of Dynamical Processes in Liquids by Electron Spin Echo Spectroscopy," *J. Phys. Chem.*, **85**, 3031 (1981).

Weissman, S. I., "Recent Developments in EPR: Transient Methods," *Ann. Rev. Phys. Chem.*, **33**, 301 (1982).

8.12 Journals and Periodicals

Waugh, J. S., Ed., *Advances in Magnetic Resonance*, Vols. 1–11, Academic, New York, 1965–1983 (collection of monographs published at irregular intervals).

Analytical Chemistry, issue *Analytical Reviews*, section "Electron Spin Resonance" (appears as issue No. 5—alternate even-numbered years).

Annual Reports, Chemical Society, London (contains annual review of ESR literature).

Annual Review of Physical Chemistry, Palo Alto, CA (contains ESR review section reflecting the interests of the author).

Bulletin of Magnetic Resonance, International Society of Magnetic Resonance, Columbia, SC (quarterly review journal).

Electron Spin Resonance, Specialist Periodical Reports, Vols. 1–10, Chemical Society, London (literature reviews published periodically at 18-month intervals).

Faraday Discussions of the Chemical Society, **78** (1984) (this issue is entirely devoted to radicals in condensed phases).

Journal of Magnetic Resonance, Academic, New York (periodical journal).

Magnetic Resonance Reviews, Gordon & Breach, New York (quarterly periodical of review articles).

Organic Magnetic Resonance (periodical on magnetic resonance applied to organic chemistry—mostly NMR).

8.13   Data Compilations

Landolt–Bornstein Numerical Data and Functional Relationships in Science and Technology, Springer-Verlag, New York. Group II, Atomic and Molecular Physics:
Vol. 1, *Magnetic Properties of Free Radicals*, 1965.
Vol. 2, *Magnetic Properties of Coordination and Organometallic Transition Metal Ions*, 1966.

Vol. 8, *Magnetic Properties of Coordination and Organometallic Transition Metal Compounds*, Supplement 1 (1964–1968) to Group II, Vol. 2, published 1976.

Vol. 9, *Magnetic Properties of Free Radicals*, Supplement and extension to Group II, Vol. 1.

Part a, *Atoms, Inorganic Radicals and Radicals in Metal Complexes*, 1977.
Part b, *Organic C-Centered Radicals*, 1977.
Part c1, *Organic N-Centered Radicals and Nitroxide Radicals*, 1979.
Part c2, *Organic O-, P-, S-, Se-, Si-, Ge, Sn-, Pb, As-, and Sb-Centered Radicals*, 1979.
Part d1, *Organic Anion Radicals*, 1980.
Part d2, *Organic Cation Radicals and Polyradicals and Index of Substances for Group II, Vol. I, and Group II, Vol. 9*, 1980.

Vol. 10, *Magnetic Properties of Coordination and Organometallic Transition Metal Compounds*, Supplement 2 (1969, 1970) to Group II, Vol. 1, published 1979.

Vol. 11, *Magnetic Properties of Coordination and Organometallic Transition Metal Compounds*, Supplement 3 (1971, 1972) to Group II, Vol. 2, published 1981.

Vol. 12, *Magnetic Properties of Coordination and Organometallic Transition Metal Compounds*, Supplement 4 (1973, 1974) to Group II, Vol. 2, subvolume b, *Electron Paramagnetic Resonance*, published 1984.

Vol. 17, *Magnetic Properties of Free Radicals*, Supplement and extension to Group II, Vol. 9, Subvolume a, *Inorganic Radicals, Radical Ions and Radicals in Metal Complexes*, 1987; Subvolume b, *Nonconjugated Carbon Radicals*, 1987.

(This series has by far the greatest available information density on number of data per kilogram basis.)

Bielski, B. H. J., and J. M. Gebecki, *Atlas of Electron Spin Resonance Spectra*, Academic, New York, 1967 (numerous errors; check data against original references).

Hershenson, H. M., *Nuclear Magnetic Resonance and Electron Spin Resonance: Index for 1958–1963*, Academic, New York, 1965.

Lebdev, Y. S., D. M. Chernikova, N. N. Tikhomirova, and V. V. Volvodskii, *Atlas of Electron Spin Resonance Spectra Theoretically Calculated Multicomponent Symmetrical Spectra*, Plenum, New York, 1963 (not terribly useful in view of wide availability of in-house computational facilities).

APPENDIX 1   HYPERFINE AND SPIN–ORBIT COUPLING CONSTANTS OF ATOMS AND SPIN–ORBIT COUPLING CONSTANTS OF IONS IN COMMON STATES

Nucleus	Spin	Electronic State	Hyperfine Coupling Isotropic $A(MHz)^{a,b}$	Anisotropic $B(MHz)^{c,d}$	Spin–Orbit Coupling $\lambda(cm^{-1})^{e,f}$	Comments
^1H	1/2	$1s^1$	1,420			
^1H	1		218			
^6Li	1	$2s^1$	152		0.34	
^7Li	3/2		401			
^9Be	3/2	$2s^2$	-358		1	
^{10}B	3	$2s^22p^1$	672	17.8	11	
^{11}B	3/2		2,020	53.1		
^{13}C	1/2	$2s^22p^2$	3,110	90.8	28	
^{14}N	1	$2s^22p^3$	1,540	47.8	76	
^{15}N	1/2		2,160	-67.1		
^{17}O	5/2	$2s^32p^4$	$-4,628$	-144	152	
^{19}F	1/2	$2s^22p^5$	47,910	1,515	272	
^{23}Na	3/2	$3s^1$	866		11	
^{27}Al	3/2	$3s^23p^1$	2,746	59	75	
^{29}Si	1/2	$3s^23p^2$	$-3,381$	-86.6	142	
^{31}P	1/2	$3s^23p^3$	10,178	287	231	
^{33}S	1/2	$3s^23p^4$	2,715	78	382	
^{35}Cl	3/2	$3s^23p^5$	4,664	137	586	
^{37}Cl	3/2		3,880	117		
^{39}K	3/2	$4s^1$	231		38.5	
^{45}Sc	7/2	$3d^14s^2$	1,833	-105.8	67	
$^{45}Sc^{2+}$	7/2	$3d^1$		-116.8	80	
$^{47,49}Ti$	5/2, 7/2	$3d^4$	-492	34.6	111	
$^{47,49}Ti^{3+}$	5/2, 7/2	$3d^1$		43.7	60	
^{51}V	7/2	$3d^34s^2$	2,613	211.8	158	
$^{51}V^{4+}$	7/2	$3d^1$		297	110	
^{53}Cr	3/2	$3d^54s^1$	-630	58	223	
$^{53}Cr^{3+}$	3/2	$3d^3$		68.3	77	(high spin)
$^{53}Cr^{5+}$	3/2	$3d^1$		85.9	380	
^{55}Mn	5/2	$3d^44s^2$	3,063	-314	0	
$^{55}Mn^{2+}$	5/2	$3d^5$		-321	0	(high spin)
^{57}Fe	1/2	$3d^64s^2$	450	-49.1	-185	
$^{57}Fe^{2+}$	1/2	$3d^6$		-50.1	-200	
$^{57}Fe^{3+}$	1/2	$3d^5$		-56.4	0	(high spin)
^{59}Co	7/2	$3d^74s^2$	3,666	-429	-170	
$^{59}Co^{2+}$	7/2	$3d^7$		-436	-172	(high spin)
^{61}Ni	3/2	$3d^84s^2$	1,512	-189	-320	
$^{61}Ni^{2+}$	3/2	$3d^8$		-193	-315	
$^{61}Ni^{3+}$	3/2	$3d^7$		-212	-238	
^{63}Cu	3/2	$3d^{10}4s^1$	4,952	-606		
^{65}Cu	3/2	$3d^{10}4s^1$	5,305	-706		
$^{63}Cu^{2+}$	3/2	$3d^9$		-667	-830	

Appendix 1 (*continued*)

Nucleus	Spin	Electronic State	Hyperfine Coupling Isotropic $A(MHz)^{a,b}$	Anisotropic $B(MHz)^{c,d}$	Spin–Orbit Coupling $\lambda(cm^{-1})^{e,f}$	Comments
$^{65}Cu^{2+}$	3/2	$3d^9$		−701		
^{65}Zn	5/2	$4s^2$	1,251			
^{75}As	3/2	$4s^24p^3$	9,582	255	1,550	
^{77}Se	3/2	$4s^24p^4$	13,468	376	1,668	
^{79}Br	3/2	$4s^24p^5$	21,738	646	2,460	
^{79}Br	3/2		23,432	696		
^{81}Rb	5/2	$5s^1$	1,012		158.4	
^{87}Rb	3/2		3,417			
^{95}Mo	5/2	$4d^35s^1$	−3,528		116	
^{97}Mo	5/2	$4d^35s^1$	−3,601			
$^{95}Mo^{3+}$	5/2	$4d^3$		104.7	232	
$^{97}Mo^{3+}$	5/2	$4d^3$		95.5		
^{107}Ag	1/2	$4d^{10}5s^1$	−3,520		1,790	
^{109}Ag	1/2	$4d^{10}5s^1$	−4,044			
$^{107}Ag^{2+}$	1/2	$4d^9$		110		
$^{109}Ag^{2+}$	1/2	$4d^9$		126		
^{127}I	5/2	$5s^25p^5$	26,000	8,180	5,060	
^{133}Cs	7/2	$6s^1$	2,298		364	

[a]Corresponds to unpaired electron in an s orbital with the principal quantum number of the lowest unfilled s orbital
[b]Data taken from [15] and [81].
[c]For p orbitals, $B = \frac{2}{5}h^{-1}g_N\mu_N g_e\mu_B\langle r^{-3}\rangle$. The principal values are −1, −1, or +2 times the value quoted. The value quoted is for d orbitals $m_l = \pm2$, $B = -\frac{4}{7}h^{-1}g_N\mu_N g_e\mu_B\langle r^{-3}\rangle$. The value of $\langle 3\cos^2\theta - 1\rangle$ is $-\frac{4}{7}$ for $m_l = \pm2$, $+\frac{2}{7}$ for $m_l = \pm1$, and $\frac{4}{7}$ for $m_l = 0$. The principal values are also −1, −1, and +2 times the value quoted.
[d]Data for p orbitals taken from [15]; data for d orbitals taken from [81].
[e]Data taken from [520–522].
[f]The value cited is λ rather than ζ. The relationship between terms is $\lambda = \pm\zeta/2S$. For a d^n orbital, the sign is positive if n is less than 5, negative if n is greater than 5, and zero if $n = 5$.

APPENDIX 2 SOURCES OF EQUIPMENT AND MATERIALS FOR ELECTRON SPIN RESONANCE EXPERIMENTATION

This alphabetized list contains companies that supply equipment, accessories, or consumables specifically intended for ESR experimentation. Many other companies distribute items that have been or can be adapted for ESR, but they are too numerous to mention. The list is complete as of August, 1987.

Material	Supplier
Cryogenic temperature controllers	APD Cryogenics P.O. Box 2802 Allentown, PA 18105 Telephone: (215)-481-3100

Material	Supplier
Spin labels and spin traps	Aldrich Chemical Company, Inc. 940 W. St. Paul Ave. Milwaukee, WI 53232 Telephone: (414)-273-3850
ESR spectrometers (frequencies of 1, 4, 9, 25, and 35 GHz), pulsed ESR, ENDOR, ENDOR TRIPLE	Bruker Analytische Messtechnik Silberstreifen, D-7512 Rheinstetten-Fo. Federal Republic of Germany Telephone: 0721/51610
Cavities, temperature controllers; accessories: glassware (photochemical and electrochemical); magnets: 9, 10, 12, and 15 in.	Bruker Instruments, USA Manning Park Billerica, MA 01821 Telephone: (508)-667-9580
ESR spectrometers and magnets: (sales in Asia and Australia)	JEOL Ltd. 1418 Nakagami, Aikahima, Tokyo 196 Japan
Microwave accessories for ESR	Micro-Now Instrument Company, Inc. 8260 N. Elmwood Street Skokie, IL 60076 Telephone: (312)-677–4700
Spin labels and spin traps	Molecular Probes, Inc. 24740 Lawrence Road Junction City, OR 97448 Telephone: (503)-998–6254
Cryogenic temperature controllers	Oxford Instruments of North America 3A Alfred Circle Bedford, MA 01730 Telephone: (617)-275–4350
Flash lamps and photochemical apparatus	Photochemical Research Associates 45 Meg Drive London, Ontario, Canada N6E 2V2 Telephone: (519)-686–2950
Conventional electro-magnets: 4, 5, 7, 9.5, 10, 12, and 15 in.	Walker Scientific Company, Inc. Rockdale Street Worcester, MA 01606 Telephone: (617)-852–3674

Material	Supplier
Quartz items for ESR: flat cells, sample tubes, Dewar flasks, custom capabilities; recording supplies	Wilmad Glass Company, Inc. Route 40 & Oak Road Buena, NJ 08310 Telephone: (609)-697–3000

References

1. P. Mansfield and P. G. Morris, *NMR Imaging in Biomedicine*, Academic, New York, 1982.

2. D. J. E. Ingram, *Free Radicals as Studied by Electron Spin Resonance*, Butterworths, London, 1958, p. 16.

3. J. E. Harriman, *Theoretical Foundations of Electron Spin Resonance*, Academic, New York, 1978, p. 2.

4. P. A. M. Dirac, *Proc. R. Soc. London*, **A117**, 610 (1927).

5. E. R. Cohen and B. N. Taylor, "The 1986 Adjustment of the Fundamental Constants," *International Council of Scientific Unions, CODATA Bulletin*, No. 63, Pergamon, New York, 1986.

6. M. J. Levine and J. Wright, *Phys. Rev. Lett.*, **26**, 1351 (1971).

7. F. Bloch, W. W. Hansen, and M. Packard, *Phys. Rev.*, **70**, 474 (1946).

8. F. E. Reisch, R. W. Grant, M. D. Lind, G. P. Espinosa, and I. B. Goldberg, *IEEE Trans. Magn.*, **11**, 1256 (1975).

9. L. G. Gouy, *C. R. Acad. Seances Paris*, **109**, 935 (1889).

10. J. A. McMillan, *Electron Paramagnetism*, Reinhold, New York, 1968, Chap. 6.

11. A. Carrington and A. D. McLachlan, *Introduction to Magnetic Resonance*, Harper & Row, New York, 1967, pp. 172–182; reprinted, Wiley, New York, 1979.

12. G. E. Pake, *Paramagnetic Resonance*, W. A. Benjamin, New York, 1962, pp. 21–23; G. E. Pake and T. L. Eastle, *The Physical Principles of Electron Paramagnetic Resonance*, 2nd ed., Benjamin, Reading, MA., 1973, pp. 25–27.

13. C. P. Slichter, *Principles of Magnetic Resonance with Examples from Solid State Physics*, Harper & Row, New York, 1963; 2nd rev. ed., Springer-Verlag, New York, 1980.

14. F. Bloch, *Phys. Rev.*, **70**, 460 (1946).

15. J. E. Wertz and J. R. Bolton, *Electron Spin Resonance Elementary Theory and Practical Applications*, McGraw-Hill, New York, 1972; reprinted, Chapman & Hall, New York, 1986.

16. G. Feher and A. F. Kip, *Phys. Rev.*, **98**, 337 (1955).

17. F. J. Dyson, *Phys. Rev.*, **98**, 349 (1955).

18. C. P. Poole, Jr., and H. A. Farach, *The Theory of Magnetic Resonance*, Wiley-Interscience, New York, 1972; 2nd ed., 1987.

19. M. P. Byrn and C. E. Strouse, *J. Magn. Reson.*, **53**, 32 (1983).

20. E. Fermi, *Z. Phys.*, **60** 320 (1930).

21. A. Abragam and M. H. L. Pryce, *Proc. R. Soc. London*, **205A**, 135 (1951).

22. K. A. McLauchlan, *Magnetic Resonance*, Clarendon, Oxford, 1972, p. 85.

23. H. M. McConnell, *J. Chem. Phys.*, **25**, 709 (1956).

24. D. Kivelson, *J. Chem. Phys.*, **27**, 1087 (1957).

25. D. Kivelson, *J. Chem. Phys.*, **33**, 1094 (1960).

26. J. H. Freed and G. K. Fraenkel, *J. Chem. Phys.*, **39**, 326 (1963).

27. C. I. Perrin and J. E. Sanderson, *Mol. Phys.*, **14**, 395 (1968).

28. R. W. Fessenden, *J. Chem. Phys.*, **37**, 747 (1962).

29. R. W. Fessenden and R. H. Schuler, *J. Chem. Phys.*, **43**, 2704 (1965).

30. H. M. McConnell and D. B. Chesnut, *J. Chem. Phys.*, **28**, 107 (1958).

31. R. Bersohn, *J. Chem. Phys.*, **24**, 1066 (1956).

32. S. I. Weissman, *J. Chem. Phys.*, **25**, 890 (1956).

33. A. D. McLachlan, H. H. Dearman, and R. Lefebvre, *J. Chem. Phys.*, **33**, 65 (1960).

34. S. I. Weissman, J. Townsend, D. E. Paul, and G. E. Pake, *J. Chem. Phys.*, **21**, 2227 (1953).

35. J. P. Colpa and E. de Boer, *Mol. Phys.*, **7**, 333 (1963–64).

36. E. G. Janzen, "Stereochemistry of Nitroxides," in E. Eliel and N. Allinger, Eds., *Topics in Stereochemistry*, Vol. 6, Wiley, New York, 1971, p. 177.

37. D. H. Geske, "Conformation and Structure as Studied by Electron Spin Resonance Spectroscopy," in A. Streitweiser, Jr. and R. W. Taft, Eds., *Progress in Physical Organic Chemistry*, Vol. 4, Interscience, New York, 1967, p. 125.

38. G. A. Russell, "Semidione Radical Anions," in E. T. Kaiser and L. Kevan, Eds., *Radical Ions*, Interscience, New York, 1968, p. 87.

39. A. Streitweiser, Jr., *Molecular Orbital Theory for Organic Chemists*, Wiley, New York, 1961.

40. A. D. McLachlan, *Mol. Phys.*, **3**, 233 (1960).

41. H. M. McConnell, *J. Chem. Phys.*, **24**, 764 (1956).

42. H. M. McConnell, *J. Chem. Phys.*, **24**, 632 (1956).

43. J. P. Colpa and J. R. Bolton, *Mol. Phys.*, **6**, 273 (1963).

44. J. R. Bolton, *J. Chem. Phys.*, **43**, 309 (1965).

45. T. C. Sayetta and J. D. Memory, *J. Chem. Phys.*, **40**, 2748 (1964).

46. J. R. Bolton, *J. Phys. Chem.*, **71**, 3099 (1967).

47. J. R. Bolton, *J. Phys. Chem.*, **71**, 3702 (1967).

48. M. C. R. Symons, *J. Chem. Soc.*, **1959**, 277.

49. E. W. Stone and A. H. Maki, *J. Chem. Phys.*, **37**, 1326 (1962).

50. M. Karplus and G. K. Fraenkel, *J. Chem. Phys.*, **35**, 1312 (1961).

51. P. H. Rieger and G. K. Fraenkel, *J. Chem. Phys.*, **39**, 609 (1963).

52. A. J. Bard and I. B. Goldberg, in J. N. Herak and K. J. Adamic, Eds., *Magnetic Resonance in Chemistry and Biology*, Dekker, New York, 1975, p. 41; p. 301.

53. J. R. Bolton, "Electron Spin Densities," in E. T. Kaiser and L. Kevan, Eds., *Radical Ions*, Interscience, New York, 1968, p. 1.

54. T. M. McKinney, "Electron Spin Resonance and Electrochemistry," in A. J. Bard, Ed., *Electroanalytical Chemistry*, Vol. 10, Dekker, New York, 1977, p. 97.

55. J. P. Wittke and R. H. Dicke, *Phys. Rev.*, **103**, 620 (1956).

56. J. W. Heberle, H. A. Reich, and P. Kusch, *Phys. Rev.*, **101**, 612 (1956).

57. R. W. Fessenden and R. H. Schuler, *J. Chem. Phys.*, **39**, 2147 (1963).

58. R. W. Fessenden and S. Ogawa, *J. Am. Chem. Soc.*, **86**, 3591 (1964).

59. S. Ohnishi and I. Nitta, *J. Chem. Phys.*, **39**, 2848 (1963).

60. P. J. Zandstra, *J. Chem. Phys.*, **40**, 612 (1964).

61. T. R. Tuttle, Jr., and S. I. Weissman, *J. Am. Chem. Soc.*, **80**, 5342 (1958).

62. J. R. Bolton, *Mol. Phys.*, **6**, 219 (1963).

63. D. E. Wood and H. M. McConnell, *J. Chem. Phys.*, **37**, 1150 (1962).

64. S. Arai, S. Shida, K. Yamaguchi, and Z. Kuri, *J. Chem. Phys.*, **37**, 1885 (1962).

65. A. Carrington and I. C. P. Smith, *Mol. Phys.*, **7**, 99 (1963–64).

66. G. Vincow, M. L. Morrell, W. V. Volland, H. J. Dauben, Jr., and F. R. Hunter, *J. Am. Chem. Soc.*, **87**, 3527 (1965).

67. T. J. Katz and H. L. Strauss, *J. Chem. Phys.*, **32**, 1873 (1960).

68. H. L. Strauss, T. J. Katz, and G. K. Fraenkel, *J. Am. Chem. Soc.*, **85**, 2360 (1963).

69. R. D. Allendoerfer and P. H. Rieger, *J. Am. Chem. Soc.*, **87**, 2336 (1965).

70. M. T. Jones, *J. Am. Chem. Soc.*, **80**, 174 (1960).

71. R. G. Lawler, J. R. Bolton, G. K. Fraenkel, and T. H. Brown, *J. Am. Chem. Soc.*, **86**, 520 (1964).

72. A. Carrington and P. F. Todd, *Mol. Phys.*, **6**, 161 (1963).

73. K. W. Bowers, "Orbital Degeneracy in Benzene and Substituent Effects," in E. T. Kaiser and L. Kevan, Eds., *Radical Ions*, Interscience, New York, 1968, p. 211.

74. B. G. Pobedinskii, A. L. Buchenko, and M. B. Nieman, *Russ. J. Phys. Chem. (Eng. Trans.)*, **42**, 748 (1968).

75. G. L. Swartz and W. M. Gulick, Jr., *Mol. Phys.*, **30**, 871 (1975).

76. D. H. Geske, J. L. Ragle, M. A. Bambenek, and A. L. Balch, *J. Am. Chem. Soc.*, **86**, 987 (1964).

77. K. D. Sales, *Adv. Free-Radical Chem.*, **3**, 139 (1969).

78. M. T. Rogers and L. D. Kispert, *J. Chem. Phys.*, **46**, 3193 (1967).

79. A. M. Maurice, R. L. Belford, I. B. Goldberg, and K. O. Christe, *J. Am. Chem. Soc.*, **105**, 3799 (1983).

80. R. L. Hudson and F. Williams, *J. Chem. Phys.*, **65**, 3381 (1976).

81. B. A. Goodman and J. B. Raynor, *Adv. Inorg. Chem. Radiochem.*, **13**, 135 (1970).

82. I. C. Lewis and L. S. Singer, *J. Chem. Phys.*, **43**, 2712 (1965).

83. A. J. Stone, *Mol. Phys.*, **6**, 509 (1963); **7**, 311 (1963).

84. A. J. Stone, *Proc. R. Soc., London Sect. A*, **A271**, 424 (1963).

85. D. C. McCain and D. W. Hayden, *J. Magn. Reson.*, **12**, 312 (1973).

86. B. S. Prabhananda, *J. Chem. Phys.*, **79**, 5752 (1983).

87. K. Möbius, *Z. Naturforsch.*, **20a**, 1102 (1965).

88. B. G. Segal, M. Kaplan, and G. K. Fraenkel, *J. Chem. Phys.*, **43**, 4191 (1965), with correction by R. Allendorfer, *J. Chem. Phys.*, **55**, 3615 (1971).

89. K. Möbius and M. Plato, *Z. Naturforsch.*, **24a**, 1078 (1969).

90. M. T. Jones, T. C. Kuechler, and S. Metz, *J. Magn. Reson.*, **10**, 149 (1973).

91. I. B. Goldberg and A. J. Bard, *Chem. Phys. Lett.*, **7**, 139 (1970).

92. F. Gerson, I. B. Goldberg, and T. M. McKinney, *Magn. Reson. Chem.*, **26**, 319 (1988).

93. T. Yonezawa, T. Kawamura, M. Ushio, and Y. Nakao, *Bull. Chem. Soc. Jpn.*, **43**, 1022 (1970).

94. C. C. Felix and B. S. Prabhananda, *J. Chem. Phys.*, **80**, 3078 (1984).

95. B. S. Prabhananda, C. C. Felix, J. S. Hyde, and A. Walvekar, *J. Chem. Phys.*, **83**, 6121 (1985).

96. A. G. Prodell and P. Kusch, *Phys. Rev.*, **88**, 184 (1952).

97. R. Beringer and A. M. Heald, *Phys. Rev.*, **95**, 1474 (1954).

98. P. Franken and S. Koenig, *Phys. Rev.*, **88**, 199 (1952).

99. P. Kusch and H. Taub, *Phys. Rev.*, **75**, 1477 (1949).

100. L. Essen, E. G. Hope, and D. Sutcliffe, *Nature (London)*, **189**, 298 (1961).

101. C. E. Bailey, *J. Chem. Phys.*, **59**, 1599 (1973).

102. I. B. Goldberg, K. O. Christe, and R. D. Wilson, *Inorg. Chem.*, **14**, 152 (1975).

103. W. Känzig and M. H. Cohen, *Phys. Rev. Lett.*, **3**, 509 (1959).

104. M. Trainer, M. Helten, and D. Knapska, *J. Chem. Phys.*, **79**, 3648 (1983).

105. P. W. Atkins and M. C. R. Symons, *The Structure of Inorganic Radicals*, Elsevier, New York, 1967.

106. M. Symons, *Chemical and Biochemical Aspects of Electron Spin Resonance Spectroscopy*, Van Nostrand Reinhold, New York, 1978.

107. V. A. Tolkachev, I. I. Chkheidze, and N. Ya. Buben, *Zh. Strukt. Khim.*, **3**, 709 (1962); *Chem. Abs.* **58**, 6351d (1960).

108. R. C. McCalley and A. L. Kwiram, *J. Am. Chem. Soc.*, **92**, 1441 (1970).

109. T. Yonezawa, I. Noda, and T. Kawamura, *Bull. Chem. Soc. Jpn.*, **41**, 766 (1968).

110. S. A. Marshall, A. R. Reinberg, R. A. Serway, and J. A. Hodges, *Mol. Phys.*, **8**, 225 (1964).

111. G. R. Bird, J. C. Baird, A. W. Jache, J. A. Hodgeson, R. F. Curl, Jr., A. C. Kunkle, J. W. Bransford, J. Restrup-Anderson, and J. Rosenthal, *J. Chem. Phys.*, **40**, 3378 (1963).

112. F. J. Adrian and V. A. Bowers, *Chem. Phys. Lett.*, **41**, 517 (1976).

113. P. H. Kasai, E. Hedaya, and E. B. Whipple, *J. Am. Chem. Soc.*, **91**, 4364 (1969).

114. B. Bleaney and K. W. H. Stevens, *Rep. Prog. Phys.*, **16**, 108 (1953).

115. K. D. Bowers and J. Owen, *Rep. Prog. Phys.*, **18**, 304 (1955).

116. N. M. Atherton, *Electron Spin Resonance, Theory and Applications*, Halsted, New York, 1973.

117. A. Abragam and B. Bleaney, *Electron Paramagnetic Resonance of Transition Ions*, Clarendon, Oxford 1970; reprinted by Dover, New York, 1986.

118. J. W. Orton, *Electron Paramagnetic Resonance: An Introduction to Transition Group Ions in Crystals*, Iliffe, London, 1968.

119. L. A. Sorin and M. V. Vlasova, *Electron Spin Resonance of Paramagnetic Crystals*, Plenum, New York, 1973.

120. W. Low and R. S. Rubins, *Paramagnetic Resonance in Solids*, Academic, New York, 1960.

121. W. Low, *Solid State Phys.*, Supplement 2 (1960).

122. H. A. Kuska and M. T. Rogers, "Electron Spin Resonance of First Row Transition Metal Complex Ions," in E. T. Kaiser and L. Kevan, Eds., *Radical Ions*, Wiley, New York, 1968, p. 579.

123. B. R. McGarvey, "Electron Spin Resonance of Transition-Metal Complexes," in R. L. Carlin, Ed., *Transition Metal Chemistry*, Vol. 3, Dekker, New York, 1966, p. 89.

124. Y. Tanabe and S. Sugano, *J. Phys. Soc. Jpn.*, **9**, 753 (1954).

125. J. S. Berkes, "Energy Level Diagrams for Transition Metal Ions in Cubic Crystal Fields," Monograph No. 2, Materials Science and Engineering Department, The Pennsylvania State University, University Park, PA 1968.

126. J. M. Assour, *J. Chem. Phys.*, **43**, 2477 (1965).

127. P. Auzins, J. W. Orton, and J. E. Wertz, "Electron Spin Resonance Studies of Impurities in II–IV Compounds," in W. Low, Ed., *Proceedings, 1st International Conference on Paramagnetic Resonance, Jerusalem*, 1962, Academic, New York, 1963, Vol. 1, p. 90.

128. F. S. Ham and G. W. Ludwig, "Paramagnetic Properties of Iron Group Ions in Tetrahedral Coordination," in W. Low, Ed., *Proceedings 1st International Conference on Paramagnetic Resonance, Jerusalem*, 1962, Vol. 1, Academic, New York, 1963, p. 130.

129. A. Carrington and M. C. R. Symons, *Chem. Rev.*, **63**, 443 (1963).

130. J. Schneider and A. Räuber, *Phys. Lett.*, **21**, 380 (1966).

131. W. Low and E. L. Offenbacher, *Solid State Phys.*, **17**, 135 (1965).

132. J. M. Assour, *J. Am. Chem. Soc.*, **87**, 4701 (1965).

133. J. M. Baker, B. Bleaney, and W. Hayes, *Proc. R. Soc. London Sect. A*, **A247**, 141 (1958).

134. T. P. P. Hall, W. Hayes, and F. I. B. Williams, *Proc. Phys. Soc. London*, **78**, 883 (1961).

135. P. B. Dorain, *Phys. Rev.*, **112**, 1058 (1958).

136. R. S. Title, *Phys. Rev.*, **131**, 623 (1963); **133**, A1613 (1964).

137. G. F. Dionne, *Phys. Rev.*, **137**, A743 (1965).

138. L. C. Kravitz and W. W. Piper, *Phys. Rev.*, **146**, 322 (1966).

139. W. M. Walsh, Jr., and L. M. Rupp, Jr., *Phys. Rev.*, **126**, 952 (1962).

140. T. L. Estle and W. C. Holton, *Phys. Rev.*, **150**, 159 (1966).

141. J. Schneider, S. R. Sircar, and A. Räuber, *Z. Naturforsch.*, **18a**, 980 (1963).

142. A. Räuber and J. Schneider, *Z. Naturforsch.*, **17a**, 266 (1962).

143. J. Dieleman, *Philips Res. Rep.*, **20**, 206 (1965).

144. C. P. Poole, Jr., H. A. Farach, and T. P. Bishop, *Magn. Reson. Rev.*, **4**, 225 (1978).

145. R. Hubin, *Bull. Soc. Fr. Mineral. Cristallogr.*, **97**, 417 (1974); *Chem. Abs.*, **83**, 100852m (1975).

146. R. E. Meads and P. J. Malden, *Clay Miner.*, **10**, 313 (1975).

147. N. G. Connelly, J. A. McCleverty, and C. J. Winscom, *Nature (London)*, **216**, 999 (1967).

148. B. Bleaney and M. C. M. O'Brien, *Proc. Phys. Soc. London Sect. B*, **B69**, 1216 (1956).

149. A. Ehrenberg, *Ark. Kemi*, **19**, 119 (1962); *Chem. Abs.*, **57**, 5485c (1962).

150. L. S. Meriwether, S. D. Robinson, and G. Wilkinson, *J. Chem. Soc. A*, 1488 (1966).

151. I. B. Goldberg, H. R. Crowe, and D. Pilipovich, *Chem. Phys. Lett.*, **33**, 347 (1975).

152. A. T. Zander and G. M. Heiftje, *Appl. Spectrosc.*, **35**, 357 (1981).

153. M. A. Heald and R. Beringer, *Phys. Rev.*, **96**, 645 (1954).

154. C. J. Ultee, *J. Chem. Phys.*, **43**, 1080 (1965).

155. H. E. Radford and K. M. Evenson, *Phys. Rev.*, **168**, 70 (1968).

156. W. G. Zijlstra, J. M. Henrichs, J. H. M. Mooy, and J. D. W. Van Voorst, *Chem. Phys. Lett.*, **7**, 553 (1970).

157. H. E. Radford and V. W. Hughes, *Phys. Rev.*, **114**, 1274 (1959).

158. J. S. M. Harvey, *Proc. R. Soc. London Sect. A*, **A285**, 581 (1965).

159. R. L. Brown, *J. Chem. Phys.*, **44**, 2827 (1966).

160. W. G. Zijlstra, J. M. Henrichs, and J. D. W. Van Voorst, *Chem. Phys. Lett.*, **13**, 325 (1972).

161. P. Tiedemann, D. Mihelcic, and R. N. Schindler, *Ber. Bunsenges. Phys. Chem.*, **75**, 751 (1971).

162. H. E. Radford, V. W. Hughes, and V. Beltran-Lopez, *Phys. Rev.*, **123**, 153 (1961).

163. J. S. M. Harvey, *Phys. Rev.*, **181**, 1 (1969).

164. J. S. M. Harvey, R. A. Kamper, and K. R. Lea, *Proc. Phys. Soc. London Sect. B*, **B76**, 979 (1960).

165. V. Beltran-Lopez and H. G. Robinson, *Phys. Rev.*, **123**, 161 (1961).

166. L. Davis, Jr., B. T. Feld, C. W. Zabel, and J. R. Zacharias, *Phys. Rev.*, **76**, 1076 (1949).

167. A. Carrington, D. H. Levy, and T. A. Miller, *J. Chem. Phys.*, **45**, 4093 (1966).

168. J. G. King and G. Jaccarino, *Phys. Rev.*, **94**, 1610 (1954).

169. P. B. Davies, B. A. Thrush, A. J. Stone, and F. D. Wayne, *Chem. Phys. Lett.*, **17**, 19 (1972).

170. K. D. Bowers, R. A. Kamper, and C. D. Lustig, *Proc. Phys. Soc. London Sect. B*, **B70**, 1176, 1957.

171. V. Jaccarino, J. G. King, R. A. Satten, and H. H. Stroke, *Phys. Rev.*, **94**, 1798 (1954).

172. H. V. Lilenfeld, R. J. Richardson, and F. E. Hovis, *J. Chem. Phys.*, **74**, 2129 (1981).

173. G. Breit and I. I. Rabi, *Phys. Rev.*, **38**, 2082 (1931).

174. A. R. Boate, J. R. Morton, and K. F. Preston, *J. Magn. Reson.*, **24**, 259 (1976).

175. A. Hasegawa, M. Hayashi, C. M. L. Kerr, and F. Williams, *J. Magn. Reson.*, **48**, 192 (1982).

176. A. Abragam and J. H. Van Vleck, *Phys. Rev.*, **92**, 1448 (1953).

177. A. Carrington, *Microwave Spectroscopy of Free Radicals*, Academic, New York, 1974.

178. M. Tinkham and M. W. P. Strandberg, *Phys. Rev.*, **97**, 951 (1955).

179. I. B. Goldberg and H. O. Laeger, *J. Phys. Chem.*, **84**, 3040 (1980).

180. R. Tischer, *Z. Naturforsch.*, **22a**, 1711 (1967).

181. A. M. Falick, B. H. Mahan, and R. J. Myers, *J. Chem. Phys.*, **42**, 1837 (1965).

182. T. A. Miller, *J. Chem. Phys.*, **54**, 330 (1971).

183. T. J. Cook, B. R. Zegarski, W. H. Breckenridge, and T. A. Miller, *J. Chem. Phys.*, **58**, 1548 (1973).

184. C. A. Arrington, Jr., A. M. Falick, and R. J. Myers, *J. Chem. Phys.*, **55**, 909 (1971).

185. J. M. Daniels and P. B. Dorain, *J. Chem. Phys.*, **45**, 26 (1966).

186. A. Carrington, D. H. Levy, and T. A. Miller, *Proc. R. Soc. London Sect. A*, **A298**, 340 (1967).

187. H. Uehara, *Bull. Chem. Soc. Jpn.*, **43**, 886 (1969).

188. A. Carrington, D. H. Levy, and T. A. Miller, *Proc. R. Soc. London Sect. A*, **A293**, 108 (1966).

189. P. B. Davies, F. D. Wayne, and A. J. Stone, *Mol. Phys.*, **28**, 1409 (1974).

190. A. Carrington, D. H. Levy, and T. A. Miller, *Mol. Phys.*, **13**, 401 (1967).

191. T. A. Miller, *J. Chem. Phys.*, **54**, 1658 (1971).

192. F. D. Wayne, P. B. Davies, and B. A. Thrush, *Mol. Phys.*, **28**, 989 (1974).

193. A. Carrington, G. N. Currie, D. H. Levy, and T. A. Miller, *Mol. Phys.*, **17**, 535 (1969).

194. A. H. Curran, R. G. MacDonald, A. J. Stone, and B. A. Thrush, *Chem. Phys. Lett.*, **8**, 451 (1971); *Proc. R. Soc. London Sect. A*, **A332**, 355 (1973).

195. H. E. Radford, *Phys. Rev.*, **122**, 114 (1961).

196. P. N. Clough, A. H. Curran, and B. A. Thrush, *Chem. Phys. Lett.*, **7**, 86 (1970); *Proc. R. Soc. London Sect. A*, **A323**, 541 (1971).

197. J. M. Brown, M. Kaise, C. M. L. Kerr, and D. J. Milton, *Mol. Phys.*, **36**, 553 (1978).

198. H. E. Radford, *Phys. Rev.*, **126**, 1035 (1962).

199. A. Carrington and N. J. D. Lucas, *Proc. R. Soc. London Sect. A*, **A314**, 567 (1970).

200. H. E. Radford and M. Linzer, *Phys. Rev. Lett.*, **10**, 443 (1963).

201. C. C. McDonald, *J. Chem. Phys.*, **39**, 2587 (1963).

202. M. Tanimoto and H. Uehara, *Mol. Phys.*, **25**, 1193 (1973).

203. H. E. Radford, *J. Chem. Phys.*, **40**, 2732 (1964).

204. A. Carrington, G. N. Currie, and N. J. D. Lucas, *Proc. R. Soc. London Sect. A*, **A315**, 355 (1970).

205. R. L. Brown and H. E. Radford, *Phys. Rev.*, **147**, 6 (1966).

206. N. A. Ashford, F. H. Jarke, and I. J. Solomon, *J. Chem. Phys.*, **57**, 3867 (1972).

207. F. H. Jarke, N. A. Ashford, and I. J. Solomon, *J. Chem. Phys.*, **64**, 3097 (1976).

208. A. Carrington, B. J. Howard, D. H. Levy, and J. C. Robertson, *Mol. Phys.*, **15**, 187 (1968).

209. H. Uehara and Y. Morino, *Mol. Phys.*, **17**, 239 (1969).

210. A. Carrington, G. N. Currie, T. A. Miller, and D. H. Levy, *J. Chem. Phys.*, **50**, 2726 (1969).

211. A. Carrington, P. N. Dyer, and D. H. Levy, *J. Chem. Phys.*, **47**, 1756 (1967).

212. H. Uehara, M. Tanimoto, and Y. Morino, *Mol. Phys.*, **22**, 799 (1971).

213. A. Carrington, P. N. Dyer, and D. H. Levy, *J. Chem. Phys.*, **52**, 309 (1970).

214. J. M. Brown, C. R. Byfleet, B. J. Howard, and D. K. Russell, *Mol. Phys.*, **23**, 457 (1972).

215. F. Hund, *Linenspektren und Periodisches System der Elemente*, Springer, Berlin, 1927.

216. I. B. Goldberg, *Electron Spin Reson.*, **6**, 1 (1978–1979).

217. I. B. Goldberg and A. J. Bard, "Electron-Spin-Resonance Spectroscopy," in P. J. Elving, M. M. Bursey, and I. M. Kolthoff, Eds., *Treatise on Analytical Chemistry*, 2nd ed., Part 1, Vol. 10, Wiley, New York, 1983, p. 226.

218. M. L. Randolph, "Quantitative Considerations in Electron Spin Resonance Studies of Biological Materials," in H. M. Schwartz, J. R. Bolton, and D. C. Borg, Eds., *Biological Applications of Electron Spin Resonance*, Wiley, New York, 1972, p. 119.

219. S. S. Eaton and G. R. Eaton, *Bull Magn. Reson.*, **1**, 130 (1980).

220. A. A. Westenberg, *Prog. React. Kinet.*, **7**, 23 (1973).

221. A. A. Westenberg and N. de Haas, *J. Chem. Phys.*, **40**, 3087 (1964).

222. A. A. Westenberg, *J. Chem. Phys.*, **43**, 1544 (1965).

223. I. B. Goldberg and H. R. Crowe, *Anal. Chem.*, **49**, 1353 (1977).

224. D. B. Chesnut, *J. Magn. Reson.*, **25**, 373 (1977).

225. T.-T. Chang, *Magn. Reson. Rev.*, **9**, 65 (1984).

226. R. H. Hoskins and R. C. Pastor, *J. Appl. Phys.*, **31**, 1506 (1960).

227. T.-T. Chang, D. Foster, and A. H. Kahn, *Natl. Bur. Stand. J. Res.*, **83**, 133 (1978).

228. M. T. Jones, *J. Chem. Phys.*, **38**, 2892 (1963).

229. G. K. Miner, T. P. Graham, and G. T. Johnston, *Rev. Sci. Instrum.*, **43**, 1297 (1972).

230. J. Rubio, E. Munoz, J. Boldu, Y. Chen, and M. M. Abraham, *J. Chem. Phys.*, **70**, 633 (1979).

231. J. Q. Adams and J. R. Thomas, *J. Chem. Phys.*, **39**, 1904 (1963).

232. H. S. Gutowsky and C. H. Holm, *J. Chem. Phys.*, **25**, 1228 (1956).

233. G. K. Fraenkel, *J. Phys. Chem.*, **71**, 139 (1967).

234. J. P. Lloyd and G. E. Pake, *Phys. Rev.*, **94**, 579 (1954).

235. T. A. Miller and R. N. Adams, *J. Am. Chem. Soc.*, **88**, 5713 (1966).

236. M. P. Eastman, R. G. Kooser, M. R. Das, and J. H. Freed, *J. Chem. Phys.*, **51**, 2690 (1969).

237. S. I. Weissman, *Z. Elektrochem.*, **64**, 47 (1960).

238. N. Hirota, *J. Phys. Chem.*, **71**, 127 (1967).

239. M. Ogasawara, H. Takaoka, and K. Hayashi, *Bull. Chem. Soc. Jpn.*, **46**, 35 (1973).

240. M. T. Watts, M. L. Lu, and M. P. Eastman, *J. Phys. Chem.*, **77**, 625 (1973).

241. B. A. Kowert, L. Marcoux, and A. J. Bard, *J. Am. Chem. Soc.*, **94**, 5538 (1972).

242. R. Chang and C. S. Johnson, Jr., *J. Am. Chem. Soc.*, **88**, 2338 (1966).

243. J. R. Bolton and A. Carrington, *Mol. Phys.*, **5**, 161 (1962).

244. P. D. Sullivan and J. R. Bolton, *Adv. Magn. Reson.*, **4**, 39 (1970).

245. J. H. Freed, P. H. Rieger, and G. K. Fraenkel, *J. Chem. Phys.*, **37**, 1881 (1962).

246. J. H. Freed and G. K. Fraenkel, *J. Chem. Phys.*, **41**, 699 (1964).

247. C. J. W. Gutch and W. A. Waters, *Chem. Commun.*, **1966**, 39.

248. E. de Boer and E. L. Mackor, *J. Am. Chem. Soc.*, **86**, 1513 (1964).

249. M. Iwaizumi, M. Suzuki, T. Isobe, and H. Azumi, *Bull. Chem. Soc. Jpn.*, **40**, 1325 (1967); *Bull. Chem. Soc. Jpn.*, **40**, 2754 (1967).

250. I. B. Goldberg and H. R. Crowe, *J. Phys. Chem.*, **80**, 2603 (1976).

251. J. R. Bolton, A. Carrington, and P. F. Todd, *Mol. Phys.*, **6**, 169 (1963).

252. A. Horsfield, J. R. Morton, and D. H. Whiffen, *Mol. Phys.*, **4**, 425 (1961).

253. I. Miyagawa and K. Itoh, *J. Chem. Phys.*, **36**, 2157 (1962).

254. E. G. Janzen and J. L. Gerlock, *J. Am. Chem. Soc.*, **89**, 4902 (1967).

255. J. W. Rogers and W. H. Watson, *J. Phys. Chem.*, **72**, 68 (1968).

256. C. P. Poole and H. A. Farach, *Relaxation in Magnetic Resonance: Dielectric and Mossbauer Applications*, Academic, New York, 1971.

257. D. Kivelson, *J. Chem. Phys.*, **45**, 1324 (1966).

258. R. N. Rogers and G. E. Pake, *J. Chem. Phys.*, **33**, 1107 (1960).

259. R. Wilson and D. Kivelson, *J. Chem. Phys.*, **44**, 154, 4440 (1966).

260. P. W. Atkins and D. Kivelson, *J. Chem. Phys.*, **44**, 169 (1966).

261. L. J. Berliner, Ed., *Spin Labeling—Theory and Applications*, Academic, New York, 1976, Vol. 2, 1979.

262. I. B. Goldberg, M. Cher, T. A. Fadner, and B. E. Blom, *Proc. TAGA*, 14, 1986.

263. E. Meirovitch and J. H. Freed, *J. Phys. Chem.*, **84**, 3281 (1980).

264. E. Meirovitch and M. S. Broido, *J. Phys. Chem.*, **88**, 4316 (1984).

265. J. H. Freed, "The Theory of Slow Tumbling ESR for Nitroxides," in L. J. Berliner, Ed., in *Spin Labeling—Theory and Applications*, Academic, New York, 1976, pp. 53–132.

266. M. S. Broido and E. Meriovitch, *J. Phys. Chem.*, **87**, 1635 (1983).

267. L. W.-M. Fong and M. E. Johnson, *Curr. Top. Bioenerg.*, **13**, 107 (1984).

268. D. Chapman and J. A. Hayward, *Biochem. J.*, **228**, 281 (1985).

269. G. N. Lewis and D. Lipkin, *J. Am. Chem. Soc.*, **64**, 2801 (1942).

270. C. K. Jen, S. N. Foner, E. L. Cochran, and V. A. Bowers, *Phys. Rev.*, **112**, 1169 (1958).

271. P. H. Kasai, *Acc. Chem. Res.*, **4**, 329 (1971).

272. R. W. Fessenden and R. H. Schuler, "Electron Spin Resonance of Radiation-Produced Radicals," in M. Burton and J. L. Magee, Eds., *Advances in Radiation Chemistry*, Vol. 2, Wiley, New York, 1970, p. 1.

273. M. B. Yim and D. E. Wood, *J. Am. Chem. Soc.*, **97**, 1004 (1975).

274. M. B. Yim and D. E. Wood, *J. Am. Chem. Soc.*, **98**, 2053 (1976).

275. T. Shida, E. Haselbach, and T. Bally, *Acc. Chem. Res.*, **17**, 180 (1984).

276. A. Hasegawa, M. Shiotani, and F. Williams, *Faraday Discuss. Chem. Soc. London*, **63**, 157.

277. L. B. Knight, Jr., *Acc. Chem. Res.*, **19**, 313 (1986).

278. Y. Shirota, K. Kawai, N. Yamamoto, K. Tada, T. Shida, H. Mikawa, and H. Tsubomura, *Bull. Chem. Soc. Jpn.*, **45**, 2683 (1972).

279. S. S. Eaton and G. R. Eaton, *J. Magn. Reson.*, **59**, 474 (1984) and references therein.

280. E. N. Kauffman and G. K. Shenoy, Eds. *Nuclear and Electron Resonance Spectroscopies Applied to Material Science*, Elsevier, New York, 1981.

281. D. K. Biegelson, "Electron Spin Resonance Studies of Amorphous Silicon," in E. N. Kauffman and G. K. Shenoy, Eds., *Nuclear and Electron Resonance Spectroscopies Applied to Material Science*, Elsevier, New York, 1981, pp. 85–94.

282. T. A. Kennedy, "Defects in III–V Semiconductors Studied Through EPR," in E. N. Kauffman and G. K. Shenoy, Eds., *Nuclear and Electron Resonance Spectroscopies Applied to Material Science*, Elsevier, New York, 1981, p. 95.

283. D. L. Griscom, "ESR Studies in Amorphous Insulators," in E. N. Kauffman and G. K. Shenoy, Eds., *Nuclear and Electron Resonance Spectroscopies Applied to Material Science*, Elsevier, New York, 1981, p. 103.

284. G. H. Stauss, J. J. Krebs, S. H. Lee, and E. M. Swiggard, *Phys. Rev. B*, **22**, 3141 (1980).

285. U. Kaufmann and J. Schneider, *Appl. Phys. Lett.*, **36**, 747 (1980).

286. D. L. Griscom, "Defects and Impurities in α-Quartz and Fused Silica," in S. T. Pantelides, Ed., *The Physics of SiO_2 and Its Interfaces*, Pergamon, New York, 1978, pp. 232–252.

287. C. L. Marquardt, *Appl. Phys. Lett.*, **28**, 209 (1976).

288. C. L. Marquardt, J. F. Guiliani, and G. Gliemeroth, *J. Appl. Phys.*, **48**, 3669 (1977).

289. J. Shinar and V. Jaccarino, *Phys. Rev. B*, **27**, 4034 (1983).

290. W. Berlinger, *Magn. Reson. Rev.*, **10**, 45 (1985).

291. P. L. Scott, "Spin-Lattice Relaxation of Some Rare Earth Salts," in W. Low, Ed., *Paramagnetic Resonance*, Academic, New York, 1963, p. 399.

292. K. A. Müller and A. Rigamonti, Eds., *Local Properties at Phase Transitions*, North-Holland, Amsterdam, 1976, p. 187.

293. M. D'Iorio, W. Berlinger, and K. A. Müller, *Phase Transitions*, **4**, 31 (1983).

294. I. B. Goldberg and T. M. McKinney, *Rev. Sci. Instrum.*, **55**, 1104 (1984).

295. A. H. Reddoch, C. L. Dodson, and D. H. Paskovich, *J. Chem. Phys.*, **52**, 2318 (1970).

296. P. D. Sullivan and N. A. Brette, *J. Phys. Chem.*, **79**, 474 (1975).

297. R. G. Lawler and G. K. Fraenkel, *J. Chem. Phys.*, **49**, 1126 (1968).

298. M. R. Das and G. K. Fraenkel, *J. Chem. Phys.*, **42**, 792 (1965).

299. A. T. Bullock and C. B. Howard, *J. Chem. Soc., London, Faraday Trans. II*, **71**, 1008 (1975).

300. Y. Ellinger, R. Subra, A. Rassat, J. Douady, and G. Berthier, *J. Am. Chem. Soc.*, **97**, 476 (1975).

301. R. W. Fessenden, *J. Chim. Phys.*, **61**, 1570 (1964).

302. P. D. Sullivan and E. M. Menger, *Adv. Magn. Res.*, **9**, 1 (1977).

303. D. E. G. Austen, P. H. Given, D. J. E. Ingram, and M. E. Peover, *Nature (London)*, **182**, 1784 (1958).

304. D. H. Geske and A. H. Maki, *J. Am. Chem. Soc.*, **82**, 2671 (1960).

305. J. Q. Chambers, III, T. Layloff, and R. N. Adams, *J. Phys. Chem.*, **68**, 661 (1964).

306. T. Kitagawa, T. Layloff, and R. N. Adams, *Anal. Chem.*, **36**, 925 (1964).

307. J. Gendell, J. H. Freed, and G. K. Fraenkel, *J. Chem. Phys.*, **37**, 2832 (1962).

308. A. H. Maki and D. H. Geske, *J. Am. Chem. Soc.*, **83**, 1852 (1961).

309. L. P. Hammett, *Physical Organic Chemistry: Reaction Rates, Equilibria and Mechanisms*, McGraw-Hill, New York, 1940, p. 168; 2nd ed., 1970.

310. E. G. Janzen, *Acc. Chem. Res.*, **2**, 279 (1969).

311. L. H. Piette, P. Ludwig, and R. N. Adams, *J. Am. Chem. Soc.*, **83**, 3909 (1961).

312. L. H. Piette, P. Ludwig, and R. N. Adams, *J. Am. Chem. Soc.*, **84**, 4212 (1962).

313. I. B. Goldberg and A. J. Bard, *J. Phys. Chem.*, **75**, 3281 (1975).

314. I. B. Goldberg, A. J. Bard, and S. W. Feldberg, *J. Phys. Chem.*, **76**, 2550 (1972).

315. I. B. Goldberg and A. J. Bard, *J. Electroanal. Chem. Interfacial Electrochem.*, **38**, 313 (1972).

316. I. B. Goldberg and A. J. Bard, *J. Phys. Chem.*, **78**, 290 (1974).

317. W. J. Albery, B. A. Coles, and A. M. Couper, *J. Electroanal. Chem.*, **65**, 901 (1975).

318. W. J. Albery, A. T. Chadwick, B. A. Coles, and N. A. Hampson, *J. Electroanal. Chem.*, **75**, 229 (1977).

319. W. J. Albery, R. G. Compton, A. T. Chadwick, B. A. Coles, and A. J. Lenkait, *J. Chem. Soc. Faraday Trans. I*, **76**, 1391 (1980).

320. W. J. Albery, R. G. Compton, and C. C. Jones, *J. Am. Chem. Soc.*, **106**, 469 (1984).

321. W. J. Albery and C. C. Jones, *Faraday Discuss. Chem. Soc.*, **78**, 193 (1984).

322. B. A. Coles and R. G. Compton, *J. Electroanal. Chem.*, **144**, 87 (1983).

323. R. G. Compton, D. J. Page, and G. R. Sealy, *J. Electroanal. Chem.*, **161**, 129 (1984).

324. R. G. Compton, D. J. Page, and G. R. Sealy, *J. Electroanal. Chem.*, **163**, 65 (1984).

325. R. G. Compton, P. J. Daly, P. R. Unwin, and A. M. Waller, *J. Electroanal. Chem.*, **191**, 15 (1985).

326. R. G. Compton and A. M. Waller, *J. Electroanal. Chem.*, **195**, 289 (1985).

327. J. H. van der Waals and M. S. de Groot, *Mol. Phys.*, **2**, 333 (1959).

328. M. S. de Groot and J. H. van der Waals, *Mol. Phys.*, **3**, 190 (1960).

329. W. A. Yager, E. Wasserman, and R. M. R. Cramer, *J. Chem. Phys.*, **37**, 1148 (1962).

330. S. P. McGlynn, T. Azumi, and M. Kinoshita, *Molecular Spectroscopy of the Triplet State*, Prentice-Hall, Englewood Cliffs, NJ, 1969.

331. J. T. Warden and J. R. Bolton, *Rev. Sci. Instrum.*, **47**, 201 (1976).

332. B. Smaller, E. C. Avery, and J. R. Remko, *J. Chem. Phys.*, **55**, 2414 (1971).

333. L. D. Kispert, J. Joseph, C. Lin, and J. R. Norris, *Am. Chem. Soc. Symp. Ser.*, **321**, 129 (1986).

334. H. van Willigen, T. K. Chandrashekar, U. Das, and M. H. Ebersole, *Am. Chem. Soc. Symp. Ser.*, **321**, 140 (1986).

335. R. W. Fessenden, *J. Phys. Chem.*, **68**, 1508 (1964).

336. B. Smaller, J. R. Remko, and E. C. Avery, *J. Chem. Phys.*, **48**, 5174 (1968).

337. R. Mangiaracina and S. Mrozowski, *Proceedings of the Fifth Conference on Carbon, University of Pennsylvania*, 1961, Pergamon, New York, 1962, p. 89.

338. M. R. Wehr and J. A. Richards, *Physics of the Atom*, Addison-Wesley, Reading, MA, 1960, pp. 178–181.

339. V. K. Bashenov, *Phys. Status Solidi A*, **10**, 9 (1972).

340. G. D. Watkins, "Vacancies and Interstitials in Semiconductors," in J. E. Whitehouse, Ed., *Radiation Damage and Defects in Semiconductors, Proceedings of the International Conference*, Institute of Physics, London, 1973, pp. 228–237.

341. T. A. Kennedy, *Magn. Reson. Rev.*, **7**, 41 (1981).

342. G. K. Wertheim, A. Hausmann, and W. Sander, *The Electronic Structure of Point Defects as Determined by Mossbauer Spectroscopy and by Spin Resonance*, North-Holland, Amsterdam, 1971.

343. A. Hasegawa and F. Williams, *Chem. Phys. Lett.*, **45**, 275 (1977).

344. W. T. Richards and A. L. Loomis, *J. Am. Chem. Soc.*, **49**, 3086 (1927).

345. P. Boudjouk, *J. Chem. Ed.*, **63**, 427 (1986).

346. K. Makino, M. M. Massoba, and P. Reisz, *J. Am. Chem. Soc.*, **104**, 3537 (1982).

347. K. Makino, M. M. Massoba, and P. Reisz, *J. Phys. Chem.*, **87**, 1369 (1983).

348. T. M. McKinney and I. B. Goldberg, *J. Phys. Chem.*, in press.

349. B. M. Gorelov, O. A. Korotchenkov, I. V. Ostrovskii, and M. K. Sheinkman, *Pis'ma Zh. Tekh. Fiz.*, **11**, 1315 (1985); *Chem. Abs.*, **104**, 5412k (1986).

350. H. L. Retcofsky, *Coal Sci.*, **1**, 43 (1982).

351. A. G. Sharkey and J. T. McCartney, "Physical Properties of Coal and its Products," in M. A. Elliott, Ed., *Chemistry of Coal Utilization*, Wiley, New York, 1981, Sect. 4.13 ("Electron Spin Resonance," contributed by H. L. Retcofsky), Chap. 4, pp. 241–252.

352. H. Ohuchi, M. Shiotani, and J. Sohma, *Fuel*, **48**, 187 (1969).

353. S. W. Dack, M. D. Hobday, T. D. Smith, and J. R. Pilbrow, *Fuel*, **63**, 1510 (1983).

354. S. W. Dack, M. D. Hobday, T. D. Smith, and J. R. Pilbrow, *Fuel*, **63**, 39 (1984).

355. I. B. Goldberg, T. M. McKinney, K. E. Chung, and R. Galli, *Fuel*, **65**, 241 (1986).

356. V. V. Lebedev, T. M. Khrenkova, and N. L. Goldenko, *Khim. Tverd. Topl.*, **12**, 144 (1978); *Chem. Abs.*, **90**, 89801r (1979).

357. M. Sakaguchi, T. Yokono, Y. Sanada, J. Shoma, and H. Kashiwabara, *Fuel*, **60**, 136 (1981).

358. B. G. Silbernagel, L. A. Gebhard, G. R. Dyrkacz, and C. A. A. Bloomquist, *Fuel*, **65**, 558 (1986).

359. R. P. Skowronski, J. J. Ratto, I. B. Goldberg, and L. L. Heredy, *Fuel*, **63**, 440 (1984).

360. I. B. Goldberg, unpublished results.

361. L. Petrakis and D. W. Grandy, *Free Radicals in Coals and Synthetic Fuels*, Elsevier, Amsterdam, 1983.

362. S. Shimokawa, E. Yamada, and K. Makino, *Bull. Chem. Soc. Jpn.*, **56**, 412 (1983).

363. H. L. Retcofsky, H. R. Hough, M. M. Maguire, and R. B. Clarkson, *Adv. Chem. Ser.*, **192**, 37 (1981).

364. L. S. Singer and I. C. Lewis, *Carbon*, **22**, 487 (1984).

365. H. Thomann, I. B. Goldberg, C. Chiu, and L. R. Dalton, "Electron Spin Echo ENDOR Studies of Coals," in L. Petrakis and J. P. Fraissare, Eds., *Magnetic Resonance, Introduction, Advanced Topics and Applications to Fossil Energy*, Reidel, Dordrecht, Holland, 1984, p. 629.

366. S. Schlick, M. Narayana, and L. Kevan, *Fuel*, **65**, 873 (1986).

367. D. C. Doetschman and D. Mustafi, *Fuel*, **65**, 684 (1986).

368. L. S. Singer, U.S. Patent 4,455,527, June 19, 1984.

369. S. W. Nicksic and G. W. Starke, U.S. Patent, 4,560,663, December 24, 1985.

370. F. E. Dickson and L. Petrakis, *Anal. Chem.*, **46**, 1129 (1974).

371. C. P. Poole, Jr., *Electron Spin Resonance, A Comprehensive Treatise on Experimental Techniques*, 2nd ed., Wiley, New York, 1983.

372. I. G. Wilson, C. W. Schramm, and J. P. Kinzer, *Bell Syst. Tech. J.*, **25**, 408 (1946).

373. J. P. Kinzer and I. G. Wilson, *Bell Syst. Tech. J.*, **26**, 410 (1947).

374.   A. A. Westenberg, *Johns-Hopkins Appl. Phys. Lab. Tech. Digest*, **6**, 9 (1966).

375.   A. B. Nalbandyan, *Russ. Chem. Rev.*, **35**, 243 (1966).

376.   I. B. Goldberg, *J. Phys. Chem.*, **84**, 3199 (1980).

377.   R. Livingston and H. Zeldes, *Rev. Sci. Instrum.*, **52**, 1352 (1981).

378.   R. Livingston, H. Zeldes, and M. S. Conradi, *J. Am. Chem. Soc.*, **101**, 4312 (1979).

379.   Ya. S. Lebedev, *Russ. Chem. Rev.*, **37**, 402 (1968).

380.   L. F. Phillips, *Modulation Techniques in Chemical Kinetics*, Pergamon, Oxford, England, 1973.

381.   H. H. Günthard, *Ber. Bunsenges. Phys. Chem.*, **78**, 1110 (1974).

382.   J. C. Kertesz and W. Wolf, *J. Phys. E*, **6**, 1009 (1973); *Am. Lab.*, **7**, 101 (1975).

383.   S. A. Jacobs, G. W. Kramer, R. E. Santini, and D. W. Margerum, *Anal. Chim. Acta*, **157**, 117 (1984).

384.   D. C. Doetschman, *Rev. Sci. Instrum.*, **43**, 143 (1972).

385.   R. W. Yip, *Rev. Sci. Instrum.*, **40**, 1035 (1969).

386.   D. Beckert and K. Mehler, *Exp. Tech. Phys.*, **31**, 397 (1983); *Chem. Abs.*, **100**, 112092w (1984).

387.   G. E. Smith, R. E. Blankenship, and M. P. Klein, *Rev. Sci. Instrum.*, **48**, 282 (1977).

388.   A. R. Lepley and G. L. Closs, *Chemically Induced Magnetic Polarization*, Wiley, New York, 1973.

389.   L. T. Muus, P. W. Atkins, K. A. McLauchlan, and J. B. Pedersen, *Chemically Induced Magnetic Polarization*, Reidel, Dordrecht, Holland, 1977.

390.   A. D. Trifunac and R. G. Lawler, *Magn. Reson. Rev.*, **7**, 147 (1982).

391.   A. D. Trifunac, M. C. Thurnauer, and J. R. Norris, *Chem. Phys. Lett.*, **57**, 471 (1978).

392.   E. G. Janzen and B. J. Blackburn, *J. Am. Chem. Soc.*, **90**, 5909 (1968).

393.   S. Forshult, C. Lagercrantz, and K. Torssell, *Acta. Chem. Scand.*, **23**, 522 (1969).

394.   G. R. Chalfont, M. J. Perkins, and A. Horsfield, *J. Am. Chem. Soc.*, **90**, 7141 (1968).

395.   M. J. Perkins, "The Trapping of Free Radicals by Diamagnetic Scavengers," in *Essays on Free-Radical Chemistry*, Special Publication No. 24, The Chemical Society, Burlington House, London, 1970, Chap. 5, p. 97.

396.   E. G. Janzen, *Acc. Chem. Res.*, **4**, 31 (1971).

397.   C. Lagercrantz, *J. Phys. Chem.*, **75**, 3466 (1971).

398.   E. G. Janzen and C. A. Evans, *J. Am. Chem. Soc.*, **95**, 8205 (1973).

399.   E. G. Janzen and C. A. Evans, *J. Am. Chem. Soc.*, **97**, 205 (1975).

400.   E. G. Janzen, D. E. Nutter, Jr., and C. A. Evans, *J. Phys. Chem.*, **79**, 1983 (1975).

401.   M. S. Ram and Stanbury, *J. Phys. Chem.*, **90**, 3691 (1986).

402.   R. S. Alger, *Electron Paramagnetic Resonance Techniques and Applications*, Wiley-Interscience, New York, 1968.

403.   T. H. Wilmshurst, *Electron Spin Resonance Spectrometers*, Hilger, London, 1967.

404.   D. J. E. Ingram, *Spectroscopy at Radio and Microwave Frequencies*, 2nd ed., Plenum, New York, 1967.

405.   D. Leniart, *Varian Instrum. Applic.*, **10**, 8 (1976).

406.   J. S. Hyde, *Rev. Sci. Instrum.*, **43**, 629 (1972).

407. J. S. Hyde and W. Froncisz, *Electron Spin Reson.*, **10A**, 175 (1986).

408. M. Mehdizadeh, T. K. Ishii, J. S. Hyde, and W. Froncisz, *IEEE Trans. Microwave Theory Tech.*, **31**, 1059 (1983).

409. M. Ono, T. Ogata, K. Hsieh, M. Suzuki, E. Yoshida, and H. Kameda, *Chem. Lett.*, **1986**, 491.

410. A. Sotgiu and G. Gaultieri, *J. Phys. E*, **18**, 899 (1985).

411. A. Sotgui, *J. Magn. Reson.*, **65**, 206 (1985).

412. D. Dasgupta and P. K. Saha, *IEEE Trans. Microwave Theory Tech.*, **29**, 47 (1981).

413. D. Dasgupta and P. K. Saha, *IEEE Trans. Microwave Theory Tech.*, **31**, 938 (1983).

414. G. Feher, *Bell Syst. Tech. J.*, **36**, 449 (1957).

415. L. Kevan and L. D. Kispert, *Electron Spin Double Resonance Spectroscopy*, Wiley-Interscience, New York, 1976.

416. M. M. Dorio and J. Freed, Eds., *Multiple Electron Resonance Spectroscopy*, Plenum, New York, 1979.

417. H. Kurreck, B. Kriste, and W. Lubitz, *Angew. Chem., Int. Ed. Engl.*, **23**, 173 (1984).

418. K. Möbius, *Ber. Bunsenges., Phys. Chem.*, **78**, 1116 (1974).

419. R. S. Eachus and M. T. Olm, *Science*, **230**, 268 (1985).

420. M. Plato, W. Lubitz, and K. Möbius, *J. Phys. Chem.*, **85**, 1202 (1981).

421. J. H. Freed, *J. Chem. Phys.*, **43**, 2312 (1965).

422. J. H. Freed, *J. Phys. Chem.*, **71**, 38 (1967).

423. J. H. Freed, D. S. Leniart, and J. S. Hyde, *J. Chem. Phys.*, **47**, 2762 (1967).

424. J. H. Freed, D. S. Leniart, and H. D. Connor, *J. Chem. Phys.*, **58**, 3089 (1973).

425. J. H. Freed, *J. Phys. Chem.*, **78**, 1155 (1974).

426. H. Bock, B. Heirholzer, H. Kurreck, and W. Lubitz, *Angew, Chim. Int. Ed. Engl.*, **22**, 787 (1983).

427. W. Kaim and W. Lubitz, *Angew. Chim. Int. Ed. Engl.*, **22**, 892 (1983).

428. H. R. Falle and G. R. Luckhurst, *J. Magn. Reson.*, **3**, 161 (1970).

429. B. Kirste, H. Kurreck, H.-J. Fey, Ch. Hass, and G. Schlömp, *J. Am. Chem. Soc.*, **101**, 7457 (1979).

430. K. P. Dinse, K. Möbius, M. Plato, R. Biehl, and H. Haustein, *Chem. Phys. Lett.*, **14**, 196 (1972).

431. L. R. Dalton and A. L. Kwiram, *J. Chem. Phys.*, **57**, 1132 (1972).

432. B. M. Hoffman, J. Martinsen, and R. A. Venters, *J. Magn. Reson.*, **59**, 110 (1984).

433. H. van Willigen, M. Plato, K. Möbius, K. P. Dinse, H. Kurreck, and J. Reusch, *Mol. Phys.*, **30**, 1359 (1975).

434. B. Kirste, W. Harrer, and H. Kurreck, *Angew. Chem. Int. Ed.*, **20**, 873 (1981).

435. B. H. Robinson, L. A. Dalton, A. H. Beth, and L. R. Dalton, *Chem. Phys.*, **18**, 321 (1976).

436. C. Hass, B. Kirste, H. Kurreck, and G. Schlomp, *J. Am. Chem. Soc.*, **105**, 7375 (1983).

437. G. C. Hurst, T. A. Henderson, and R. W. Kreilick, *J. Am. Chem. Soc.*, **107**, 7294 (1985).

438. T. A. Henderson, G. C. Hurst, and R. W. Kreilick, *J. Am. Chem. Soc.*, **107**, 7299 (1985).

439. R. Biehl, K. Möbius, S. E. O'Connor, R. I. Walter, and H. Zimmerman, *J. Phys. Chem.*, **83**, 3449 (1979).

440. K. P. Dinse, R. Biehl, and K. Möbius, *J. Chem. Phys.*, **61**, 4335 (1974).

441. R. Biehl, M. Plato, and K. Möbius, *J. Chem. Phys.*, **63**, 3515 (1975).

442. R. J. Cook and D. H. Wiffen, *Proc. Phys. Soc. London*, **84**, 845 (1964).

443. G. J. Hoijtink, J. Townsend, and S. I. Weissman, *J. Chem. Phys.*, **34**, 507 (1961).

444. M. E. Anderson, G. E. Pake, and T. R. Tuttle, *J. Chem. Phys.*, **33**, 1581 (1960).

445. J. S. Hyde, J. C. W. Chien, and J. H. Freed, *J. Chem. Phys.*, **48**, 4211 (1968).

446. K. Schepler, "Electron Spin–Spin Interactions in Three Biological Systems," Thesis, University of Michigan, Ann Arbor, MI, 1975.

447. C. Mottley, L. D. Kispert, and S. Clough, *J. Chem. Phys.*, **63**, 4405 (1975).

448. J. S. Hyde and L. Dalton, *Chem. Phys. Lett.*, **16**, 568 (1972).

449. J. S. Hyde, *Methods Enzymol.*, **49**, 480 (1978).

450. L. J. Berliner and J. Reuben, Eds., *Biological Magnetic Resonance*, Vols. 1–7, Plenum, New York, 1978–1987.

451. L. R. Dalton, Ed., *EPR and Advanced EPR Studies of Biological Systems*, CRC Press, Boca Raton, FL, 1985.

452. L. W.-M. Fung and M. E. Johnson, *Curr. Top. Bioenerg.*, **13**, 107 (1984).

453. M. Weger, *Bell Syst. Tech. J.*, **39**, 1013 (1960).

454. P. Fajer and D. Marsh, *J. Magn. Reson.*, **49**, 212 (1982).

455. M. Delmelle, *J. Magn. Reson.*, **51**, 245 (1983).

456. D. D. Thomas, L. R. Dalton, and J. S. Hyde, *J. Chem. Phys.*, **65**, 3006 (1976).

457. L. Kevan, and R. N. Schwartz, Eds., *Time Domain Electron Spin Resonance*, Wiley-Interscience, New York, 1979.

458. J. R. Norris, M. C. Thurnauer, and M. K. Bowman, *Adv. Biol. Med. Phys.*, **17**, 365 (1980).

459. P. J. Hore, *NATO ASI Ser.*, Ser. A, **85**, 131 (1985).

460. S. A. Dikanov and Yu. D. Tsvetkov, *J. Struct. Chem. USSR*, **26**, 766 (1985).

461. H. C. Torrey, *Phys. Rev.*, **76**, 1059 (1949).

462. E. L. Hahn, *Phys. Rev.*, **77**, 297 (1950).

463. E. L. Hahn, *Phys. Rev.*, **80**, 580 (1950).

464. W. B. Mims, *Rev. Sci. Instrum.*, **36**, 1472 (1965).

465. A. M. Raitsimring, Yu. D. Tsvetkov, G. M. Zhidomirov, V. E. Khemlinskii, and A. G. Semenov, *Dokl. Phys. Chem.*, **172**, 101 (1967).

466. I. M. Brown, *J. Chem. Phys.*, **52**, 3836 (1970).

467. I. M. Brown and D. J. Sloop, *Rev. Sci. Instrum.*, **41**, 1774 (1970).

468. M. Huisjen and J. S. Hyde, *J. Chem. Phys.*, **60**, 1682 (1974).

469. M. Huisjen and J. S. Hyde, *Rev. Sci. Instrum.*, **45**, 669 (1974).

470. R. W. Quine, G. R. Eaton, and S. S. Eaton, *Rev. Sci. Instrum.*, **58**, 1709 (1987).

471. T. C. Farrar and E. D. Becker, *Pulse and Fourier Transform NMR*, Academic, New York, 1971.

472. R. N. Bracewell, *The Fourier Transform and Its Applications*, 2nd ed., McGraw-Hill, New York, 1986.

473. L. Glasser, *J. Chem. Educ.*, **64**, A228, A260, and A306 (1987).

474. E. O. Brigham, *The Fast Fourier Transform*, Prentice Hall, Englewood Cliffs, NJ, 1974.

475. P. A. Narayana and L. Kevan, *Magn. Reson. Rev.*, **7**, 239 (1983).

476. H. Thomann, L. R. Dalton, and L. A. Dalton, *Biol. Magn. Reson.*, **6**, 143 (1984).

477. H. L. Vancamp and A. H. Heiss, *Magn. Reson. Rev.*, **7**, 1 (1987).

478. A. G. Redfield, *Adv. Magn. Reson.*, **1**, 1 (1965). A. G. Redfield, *IBM J. Res. Dev.*, **1**, 19 (1957).

479. F. Bloch, *Phys. Rev.*, **70**, 460 (1946).

480. L. C. Allen, H. M. Gladney, and S. H. Glarum, *J. Chem. Phys.*, **40**, 3135 (1964).

481. S. H. Glarum, *Rev. Sci. Instrum.*, **36**, 771 (1965).

482. B. Mohos, M. Zobrist, A. von Zelewsky, and G. Galambos, *Anal. Chem.*, **48**, 231 (1976).

483. R. R. Ernst, *Adv. Magn. Reson.*, **2**, 1 (1966)

484. K. Roth and B. Kirste, *J. Magn. Reson.*, **63**, 360 (1985).

485. D. W. Kirmse, *J. Magn. Reson.*, **11**, 1 (1973).

486. K. D. Bieber and T. E. Gough, *J. Magn. Reson.*, **21**, 285 (1976).

487. R. A. Jackson and C. J. Rhodes, *J. Chem. Soc. Chem. Commun.*, 1278 (1984).

488. E. W. Stone and A. H. Maki, *J. Chem. Phys.*, **38**, 1999 (1963).

489. Y. Siderer and Z. Luz, *J. Magn. Reson.*, **37**, 449 (1980).

490. B. H. Robinson and L. R. Dalton, *Chem. Phys.*, **36**, 207 (1979).

491. J. R. Pilbrow and M. W. Winfield, *Mol. Phys.*, **25**, 1073 (1973).

492. J. E. Harriman, *J. Chem. Phys.*, **65**, 2288 (1976).

493. L. K. White and R. L. Belford, *J. Am. Chem. Soc.*, **98**, 4428 (1976).

494. C. Daul, C. W. Schlapfer, B. Mohos, J. Ammeter, and E. Gamp, *Comput. Phys. Commun.*, **21**, 385 (1981).

495. D. L. Griscom, P. C. Taylor, D. A. Ware, and P. J. Bray, *J. Chem. Phys.*, **48**, 5158 (1968).

496. P. C. Taylor and P. J. Bray, *J. Magn. Reson.*, **2**, 305 (1970).

497. W. V. Sweeney, D. Coucouvanis, and R. E. Coffman, *J. Chem. Phys.*, **59**, 369 (1973).

498. V. Beltran-Lopez and J. Jimenez M, *J. Magn. Reson.*, **48**, 302 (1982).

499. A. Baram, Z. Luz, and S. Alexander, *J. Chem. Phys.*, **64**, 4321 (1976).

500. S. Lee and D. P. Ames, *J. Chem. Phys.*, **80**, 1766 (1984).

501. S. A. Goldman, G. V. Bruno, C. F. Polnaszek, and J. H. Freed, *J. Chem. Phys.*, **56**, 716 (1972).

502. P. Coffey, B. H. Robinson, and L. R. Dalton, *Chem. Phys. Lett.*, **35**, 360 (1975).

503. P. Coffey, B. H. Robinson, and L. R. Dalton, *Mol. Phys.*, **31**, 1703 (1976).

504. L. Kevan and S. Schlick, *J. Phys. Chem.*, **90**, 1998 (1986).

505. J. Heinzer, *J. Magn. Reson.*, **13**, 124 (1974); *Mol. Phys.*, **22**, 167 (1971).

506. O. Dracka, *J. Magn. Reson.*, **65**, 187 (1985).

507. H. A. Buckmaster, R. Chatterjee, J. C. Dering, D. J. I. Fry, Y. H. Shing, J. D. Skirrow, and B. Venkatesan, *J. Magn. Reson.*, **4**, 113 (1971).

508. S. K. Misra, *Physica B and C*, **124**, 53 (1984).

509. A. G. Motten, D. R. Duling, and J. Schreiber, *J. Magn. Reson.*, **71**, 34 (1987).

510. A. Motten and J. Schreiber, *J. Magn. Reson.*, **67**, 42 (1986).

511. J. S. Dodd, Ed., *The ACS Style Guide, A Manual for Authors and Editors*, American Chemical Society, Washington, DC, 1986.

512. A. Waldron and P. Judd, Eds., *Physical Review Style and Notation Guide*, Bulletin of the American Physical Society, **28**(6), 1983.

513. E. R. Cohen and B. N. Taylor, *J. Res. Natl. Bur. Stand.*, **92**, 85 (1987).

514. E. R. Cohen and B. N. Taylor, *Rev. Mod. Phys.*, **59**, 1121 (1987).

515. B. Bleaney, "Magnetism," in *Encyclopedia Britannica—Macropedia*, Vol. 11, Encyclopedia Britannica, Inc., Chicago, IL, 1981.

516. H. E. Duckworth, *Electricity and Magnetism*, Holt, Reinhart, & Winston, New York, 1960.

517. E. R. Cohen and P. Giacomo, *Physica*, **146A**, 1 (1987).

518. W. T. Scott, *The Physics of Electricity and Magnetism*, Wiley, New York, 1966; 2nd ed., R. E. Krieger, Huntington, New York, 1977.

519. A. J. Allmand, *The Principles of Applied Electrochemistry*, Longman's Green and Co., New York, 1912, pp. 6 and 32; 2nd ed., 1924.

520. C. E. Moore, *Atomic Energy Levels*, Circular 467, Vols. I–III, United States National Bureau of Standards, Gaithersburg, MD, 1945.

521. T. M. Dunn, *Trans. Faraday Soc.*, **57**, 1441 (1961).

522. D. S. McClure, *J. Chem. Phys.*, **17**, 905 (1949).

Chapter **5**

NEW NUCLEAR MAGNETIC RESONANCE EXPERIMENTS IN LIQUIDS

George A. Gray and James N. Shoolery

1 INTRODUCTION

Periodic absences from one's area of speciality to study in another field usually offer stimulating opportunities for new ideas. Imagine then the plights of users, practitioners, or experts in nuclear magnetic resonance (NMR) spectroscopy who decided five years ago to spend extended periods of time considering, say, the finer aspects of mass spectroscopy or X-ray diffraction. On returning from their sabbaticals, such scientists are poised to take new approaches and to challenge previously intractable problems. However, unless they have as-similated, along with their newly acquired skills, the fast-moving events that have occurred in NMR during their absences, they will be astonished by the rapid development of NMR and its associated capabilities, and their knowledge will be seriously outdated.

The last few years have been the most exciting and productive in the short history of NMR. The previous developments of homogeneous and higher field magnets, multinuclear capabilities, and the introduction of pulsed and Fourier methods were spaced over a much longer time, and it took years for new experiments to gain widespread attention and use. New experiments were tied to new hardware, and the generation of new approaches generally relied on the development of new instrumentation. Because of the "home-built" and spe-cialized nature of technique development, rapid deployment of new experiments was impossible.

Within the last five years particularly, instruments have been introduced that have accelerated the development of new experiments and also made their widespread use possible. The result has been an explosion in the number of strategies for and applicable approaches to NMR. This chapter gives an overview of these developments and provides general classifications by which advantages and drawbacks may be understood. On modern instruments the instrument time actually consumed by one experiment is shrinking, while the time required to design and execute the spectroscopic approach intelligently is growing. Fortunately, the information gleaned from the effort is growing at a much faster rate.

The topic *new experiments* is specifically intended to cover new ways of generating spectroscopic information. In quantum mechanics it is couched in the terms of new Hamiltonians; that is, the techniques are not specifically oriented toward one nucleus or type of sample, but rely on fundamentally new methods of perturbing the nuclear spin system. The resulting signals are detected in the traditional manner. Hence, the focus in *new experiments* is on the events *prior to* signal acquisition. There is nothing unique about pulses of rf (radio frequency) in this classification. Continuous wave methods can and have been implemented. However, the instrumental convenience and the ability to affect "instantaneously" nuclear spins gives high-power, pulsed NMR a decided advantage. Thus, the term *pulse sequence* has come to symbolize the recipe for "preparing" a nuclear spin system. Variation of placement, timing, and phases of these pulses, on one or more nuclei, provide the richness and diversity on which these rapid developments rest.

2   THE ACTION OF NUCLEAR SPINS DURING A PULSE SEQUENCE

It is useful to review some of the basic terms and characteristics of most, if not all, new experiments.

2.1   The Rotating Frame of Reference

In the rotating frame of reference the basic precessional nature of the nuclear spin is factored out, leaving the spin at equilibrium and oriented along the magnetic field direction (Figure 5.1*a*).

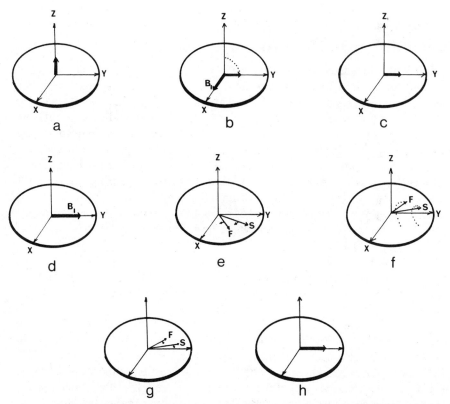

Figure 5.1   The action of a refocusing pulse (*a*) equilibrium magnetization lies along the magnetic field direction (*z* axis); (*b*) an rf field (**B**$_1$) applied along the *x* axis in the rotating frame causes precession of the net magnetization in the *zy* plane; (*c*) the rf field is turned on just long enough to leave the magnetization at right angles to the main magnetic field direction—a 90° pulse; (*d*) the rf field can be phase shifted by 90° with respect to the initial direction; (*e*) off-resonant spins will precess at a rate given by their chemical shifts; the faster precessing nuclei will gain in phase with respect to the slower precessing nuclei; (*f*) a 180° pulse along the *y* axis will reverse the precession an equivalent amount; (*g*) the faster precessing spins are now "behind" the slower precessing spins and will gain in phase during the next precession period and refocus precisely (*h*).

2.2   The 90° Pulse

In the rotating frame, an rf field can be viewed as stationary and, as with any magnetic field acting on a spin, causes the spin to precess about it at right angles (Figure 5.1b). The rf field ($\mathbf{B}_1$) is present only during the pulse. The nuclear spin (symbolized by a vector) will precess by 90° during the time termed the 90° *pulse width* (PW_{90}). As the rf field is switched off, the spin will be aligned at right angles to the static magnetic field (Figure 5.1c) and will generate a current in a receiver coil. Pulses of 180, 270, and 360° refer to correspondingly larger pulse widths and correspondingly different final spin orientations. The actual length of time to perform the 90° pulse is a function of the input rf power, coil efficiency, and magnetic field strength. This value is a measure of the strength of the $\mathbf{B}_1$ rf field by the relationship

$$\gamma B_1 / 2\pi = 1/(4 * PW_{90})$$

2.3   Refocusing 180° Pulses

Often, immediate detection following a 90° pulse is not desired. Other pulses and specific time delays frequently are needed to generate the appropriate data. Once the various spins have been placed at right angles to the static $\mathbf{B}_0$ field, the $x-y$ components, whose frequencies are proportional to their displacement from the position of the rf carrier frequency, of this magnetization begin to precess about $\mathbf{B}_0$. Spins deshielded from the carrier position will precess in a direction different from those shielded with respect to the carrier. This dephasing is frequency dependent and is corrected for in the normal "phasing" done by the operator on the transformed spectrum [this type of correction is usually appropriate only for very small periods of dephasing before data acquisition (0–30 μs)]. In the pulse sequences following, both much longer periods and 180°-refocusing pulses are used.

The 180° pulse often employs a 90° *phase shift* of the rf field (not to be confused with the term 90° *pulse*). This is symbolized in Figure 5.1d as a $\mathbf{B}_1$ vector at right angles to the $\mathbf{B}_1$ field used to place the magnetization in the $x-y$ plane.

Many NMR experiments utilize the elementary sequence: 90°–t–180°–t–. To examine the effect of this portion, consider a simple system of two independent spin vectors (Figure 5.1e), one precessing at a faster rate (F). Application of a 180° refocusing pulse along the y axis forces $\mathbf{B}_1$ precession of the individual spin vectors to new positions within the $x-y$ plane (Figure 5.1f), after which normal $\mathbf{B}_0$ precession continues (Figure 5.1g). At time 2t following the beginning of the sequence the spins have returned aligned along the y axis, since precisely the same distance is to be regained by the F spin as it had achieved before the 180° pulse. This type of refocusing eliminates phase correction that is necessary because of the dephasing of spins during the pulse sequence period 2t and refocuses any spreading on one type of spin vector that is caused by magnetic field inhomogeneity. If detection is initiated after time 2t, a signal is present,

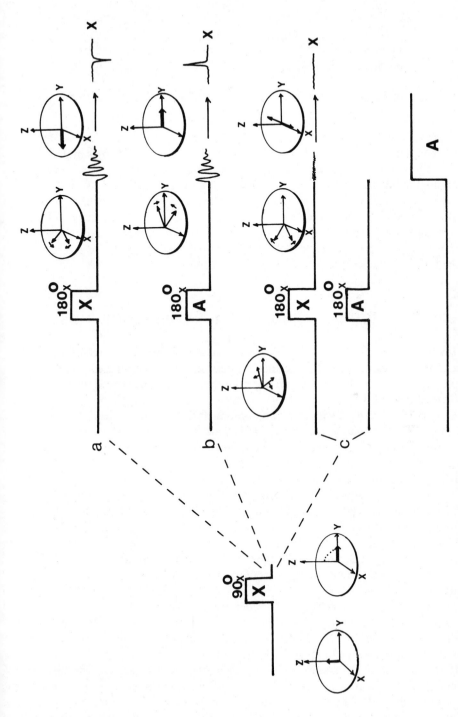

Figure 5.2 The action of two refocusing pulses on an AX spin system: (a) a simple 180° pulse on the x spin along the x axis produces an inverted signal; (b) a 180° pulse on the A spin interchanges the labels, which causes a reversal of the sense of precession and leads to an upright signal; (c) simultaneous 180° pulses lead to divergence of the spin vectors such that they are antiphase. Decoupling of the A spin causes a nulling of the X resonance.

which is normally referred to as an echo or a spin echo. The amplitude of the echo at the start of the free induction decay (fid) is governed only by natural T_2 (spin–spin relaxation time), not by the degradation normally expected from $\mathbf{B}_0$ inhomogeneities present during $2t$.

2.4   Simultaneous A- and X-Refocusing Pulses in an AX-Coupled System

This combination of pulses is used frequently in pulse sequences. The same desirable features just described apply to both A and X nuclei when "simultaneous" 180° pulses are applied to both types of nuclei (closely spaced pulses have essentially the same effect as simultaneous pulses). In addition, the labels on the spin vectors are interchanged (Figure 5.2) since the spins are J coupled. Thus, instead of refocusing after $2t$ the spins continue to diverge. After $1/J$ they will focus along the $-y$ axis, and after $2/J$ they will refocus along the $+y$ axis. Note that after $1/2J$ the spins are aligned along the $\pm x$ axis. These aspects are central for the following polarization transfer pulse sequences.

3   SPECTRAL EDITING USING J-MODULATED SPIN ECHOS

The favorable properties of the 180° refocusing pulse have been exploited in two main efforts: (1) acquisition of more spectral information by causing spectra to depend on both J and the number of coupled nuclei, and (2), discussed later, use of polarization-transfer methods.

Of course, the simplest way to make homonuclear or heteronuclear spectra depend on J is to run a *coupled* spectrum. Frequently, an intermediate level of information, such as a knowledge of the number of attached protons for the various carbon atoms in a ^{13}C NMR spectrum, is required. Because of the low sensitivity of ^{13}C and the severe overlap in a coupled spectrum, this approach is not very universal. For many years single-frequency, off-resonance decoupling (SFORD) was the method of choice [1]. This technique yields partially collapsed multiplets in the X spectrum for XH-coupled spin systems. The multiplets are broad because of uncollapsed long-range couplings and are difficult to interpret in complex molecules. The desirable features of broadband-decoupled spectra are compelling, particularly for small samples or congested spectra.

Fortunately, ^{13}C spin–echo pulse sequences have been developed [2–17] that allow broadband decoupling and carbon multiplicity selection (the number of attached protons). Patterned after an experiment that was introduced for studying proton–proton couplings [18], the central theme is the amplitude and phase modulation of, for example, ^{13}C signals arising from heteronuclear spin coupling. This modulation describes the action of ^{13}C magnetization generated by a pulse under the condition of coupling to another spin; therefore, the decoupler must be turned off for some time. The effect may be obtained in either

(1) gated (interrupted) decoupling or (2) the application of simultaneous proton and ^{13}C 180° pulses. The former has been implemented more widely although it requires a refocusing period twice as long as the latter—a consideration to remember in short T_2 (e.g., polymer) situations [10]. The former pulse sequence is noted in Figure 5.3 along with the effects of varying t within the pulse sequence. Three useful characteristics of this approach are obvious: (1) quarternary carbon atoms are unmodulated, (2) all protonated carbon atoms are nulled at $t = 1/2J$, and (3) quaternaries and methylene carbon atoms have an opposite phase to that of methyl and methine carbon atoms at $t = 1/J$.

The investigator thus has the option of searching for quaternaries that might be masked by protonated carbon atoms either by setting $t = 1/2J$ or by sorting by $CH–CH_3$ and $CH_2–C$ with $t = 1/J$. It is also possible to combine normal spin–echo and modulated spin–echo spectra to arrive at edited subspectra containing only carbon atoms of one type [6–14].

In situations of limited instrument time and/or small samples it is more critical to complete the separation into C, CH, CH_2, and CH_3 types unequivocally within the minimum time. A straightforward J-modulated spin–echo experiment with $t = 1/J$ performs the first task of obtaining a $C–CH_2$ and $CH–CH_3$ separation. Combinations at other values of t have also been proposed [6, 7] but these usually rely on at least three separate measurements. Pei and Freeman [10] have taken advantage of the greater sensitivity of the CH null at $t = 1/2J$ to propose adding just one more measurement to the above $t = 1/J$

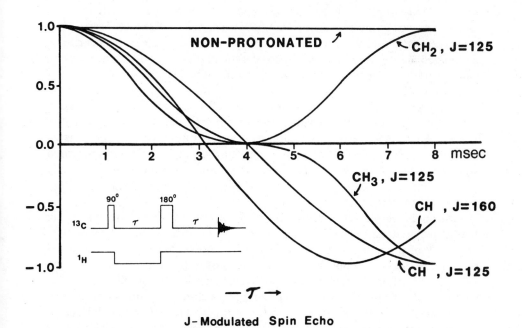

Figure 5.3 Signal amplitudes in J-modulated spin echoes as a function of J_{CH}.

value. In analyzing the residual signals of CH_2 and CH_3 at $t = 1/2J$ for values of J ranging from 165 to 125 Hz, they found that these carbon atoms contribute negligible CH_2 (4%) and CH_3 (1%) signals for $t = 1/2J$ when J is set to 145 Hz. However, CH signals (ca. 21%) are usually positive for aliphatic CH and negative for aromatic CH for $t = 1/2J$, thus clearly separating CH from CH_3 and clearly identifying the quaternary carbon atoms. The method will be less useful for oxy-bound CH groups where J values range from 140 to 150 Hz and thus give a much better null.

Jakobsen and colleagues [11] approached the $CH-CH_3$ separation problem by noting the steeper variation of intensity of the CH_3 signal as a function of t. At $t = 0.152/J$, CH and CH_3 have the maximum intensity *difference* (0.58 and 0.19 relative to 1.0 at $t = 0$); by using this value of t along with $t = 1/J$ reasonable assignments are obtained with only two spectra.

Patt and Shoolery [8] addressed the problem of optimizing sensitivity in this experiment in their APT (attached proton test). The $90-t-180-t$ sequence has the effect of saturating longer T_1 nuclei and therefore requires longer equilibrium delays. In normal experiments it is customary to use an intermediate pulse angle that is set relative to the T_1 and repetition rate of the experiment. By incorporating a second 180° pulse just prior to acquisition they show that the z magnetization remaining after an initial sub-90° pulse is restored to the z axis by the last 180° pulse, thereby allowing faster repetition of the experiment.

Interesting extensions of the J-modulated spin–echo technique have been published [19–21] for application to deuterated molecules. Here, gating of a deuterium decoupler replaces gating of the proton decoupler. All types of deuterated carbon null at a t of $1/3J$, and CD_2 and CD carbon atoms have opposite phases at $1/2J$. Because of the former fact deuterated solvents can be eliminated from ^{13}C spectra, and the general elimination of deuterated carbon peaks is possible in ordinary spectra. Again, CD_2 and CD_3 signals are much less sensitive to the null position (or value of J to obtain a null) than CD signals.

Polarization transfer techniques (following) offer similar solvent-suppression capabilities. However, they usually eliminate nonprotonated carbon atoms also. If these techniques are used in dilute solutions, any residual monoprotonated solvent species is capable of generating signals, possibly obscuring desired signals. In this case the J-modulated spin–echo technique for deuterated systems has the advantage of nulling *all* deuterated signals, even those from partially deuterated solvents.

Instead of gating off a deuterium decoupler to achieve J-modulation in deuterated systems, Doddrell and colleagues [22] developed the alternate approach of providing a 90° deuterium pulse midway through the refocusing period. This pulse is equivalent to the 180° proton pulse in the "proton flip" version [4] of the J-modulated spin–echo experiment for X—H coupled systems. They proposed that phase cycling be used to eliminate nondeuterated rather than deuterated resonances in ^{13}C spectra. This technique could be rather useful in mechanistic studies of deuteration, particularly in cases where strong nondeuterated resonances could obscure small changes associated with deuteration.

The delayed detection aspects of the spin–echo technique can be used to eliminate protonated carbon atoms from a ^{13}C spectrum. In one case [14], no J-modulated period is used. Rather, the proton decoupler is placed off-resonance and is broadband-modulated as usual. The distance off-resonance is enough to severely broaden the resonances of protonated carbon atoms without significantly affecting the line width of nonprotonated carbon atoms. By delaying acquisition for a short time following the initial 90° pulse the signals for the protonated carbon atoms decay rapidly and are not detected (Figure 5.4). The quaternary carbon atoms are detected easily and phasing the spectrum is made simpler by inserting a refocusing 180° pulse midway between the 90° pulse and acquisition. Cookson and Smith [14] applied this technique to petroleum mixtures and were able to detect clearly the quaternary aromatic carbon atoms. Extension of this method to nonprotonated carbon atoms in macromolecules is direct and straightforward and allows the direct detection of nonprotonated sites, which, for example, are hidden under a large protonated carbon envelope. Rutar [23] uses an ordinary spin–echo sequence with decoupling during acquisition but applies simultaneously a 90° pulse on the protons with a refocusing 180° pulse on the carbons. This procedure also cancels protonated ^{13}C signals but with less sensitivity to interpulse spacing (J dependence) than the ordinary J-modulated spin echo.

Beloeil and colleagues [12] introduced a new type of spin–echo sequence that produces either C–CH$_2$ or CH–CH$_3$ ^{13}C subspectra. It involves an initial 45° pulse followed by t–180°–t sequence and a final monitoring 45° pulse that is either in phase or 180° out of phase with respect to the original 45° pulse. The proton decoupler is gated off during the second $t = 1/J$ period.

Nakashima and Rabenstein [13] used ^{13}C spin–echo techniques to separate signals from isotopically enriched compounds into singly and doubly enriched subsets. Under continuous proton decoupling, normal and spin–echo spectra are obtained. In the latter, lines from carbon–carbon coupled patterns are 180° out of phase with respect to signals coming from ^{13}C–^{12}C pairs. Addition of the two types of spectra give spectra consisting only of doubly enriched species.

Spin–echo techniques have also been described for indirect detection of labeled species in proton spectra [24–29]. Simple proton spin–echo spectra can be obtained with and without a ^{13}C 180° pulse that is coincident with the proton 180° pulse. In the latter case, the doublet arising from a proton coupled to the ^{13}C is out of phase by 180° from the singlet in the ^{12}C—H species. Other protons in the molecule have signals unaffected by the ^{13}C 180° pulse so that a simple difference between the two types of spectra is sufficient to cancel the main spectrum, revealing the CH pattern. ^{13}C decoupling during acquisition gives a single resonance.

This technique can be used to monitor the amplitude of a resonance as a function of time in an isotope exchange (equilibration) reaction. The decided advantage here is the enhanced sensitivity of the proton NMR experiment, relative to that obtained by direct detection. Of course, this technique has widespread potential application, including monitoring of specific sites, chemical dynamics, and exchange reactions. It need not be limited to CH systems but

GASOIL WITH IRON TRIFLUOROACETYLACETONATE
SPIN ECHO BROAD BAND OFF-RESONANCE
DECOUPLING (SEBBORD)

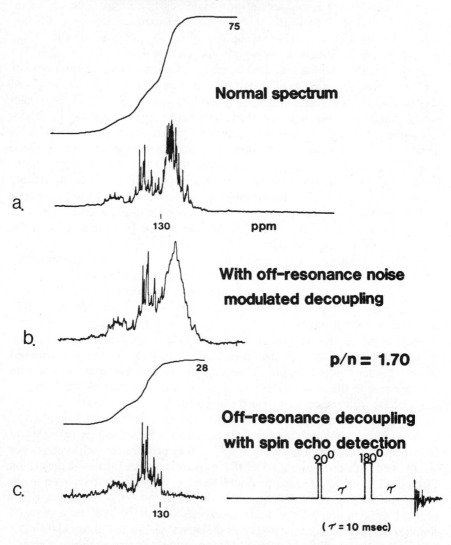

Figure 5.4 Use of spin–echo broadband off-resonance decoupling: (*a*) normal ^{13}C spectrum and integral of the aromatic carbon atoms in a gas oil petroleum fraction; (*b*) effect of off-resonance noise-modulated decoupling, showing broadening of the protonated aromatic carbons; (*c*) effect of delaying the data acquisition 20 ms by the use of a spin–echo pulse sequence; the incoherently broadened ^{13}C signals from the protonated carbon atoms become dephased and only the nonprotonated carbon atoms are detected and integrated, allowing a calculation of the ratio of protonated to nonprotonated aromatic carbon atoms in this complex mixture.

594

can be used in any situation in which there is heteronuclear J coupling; for example, one might study the local environment surrounding a metal atom within a catalyst by examining protons coupled to the metal.

Similarly, Bolton [30] described a technique for selecting only singlets and center lines of triplets within a proton spectrum. The key to the approach is the summing of data from a series of different spin–echo experiments, each with different spin–echo delays. Modulated multiplet components add destructively, while the unmodulated singlets add directly. In macromolecules this procedure can simplify the proton spectrum enormously, and information can be extracted from what was previously an undecipherable pattern.

Spin–echo multiplet selection [18] is normally adjusted to place odd- and even-order multiplets antiphase (for a specific value of J). A spread of J values will normally produce nonideal spin–echo spectra containing phase and multiplet anomalies. Spin–echo multiplet selection can be improved by the addition of Z filters [31]. Here, a $90°-t_z-90°$ pair of pulses is inserted immediately before the acquisition of the data. The process can be thought of as placing the magnetization back along the z axis by the first of the two pulses; the undesired components of magnetization can then be transformed into oscillatory magnetizations that can be suppressed by coadding several spectra for different t_z. The final $90°$ pulse converts the "stored" z magnetization into observable transverse magnetization. With proper phase cycling, this sequence produces time-averaged spin–echo multiplets that are significantly reduced in phase and multiplet distortions.

Finally, spin–echo techniques were applied to the problem of chemical exchange-broadened lanthanide-shifted proton spectra at high magnetic fields [32]. When an ordinary CPMG (Carr–Purcell–Meiboom–Gill) pulse train of refocusing pulses was used, the exchange-broadened resonances decayed away rapidly during the spin–echo pulse train while the narrower resonances retained their signal amplitude, thus a method for controlled separation of sharp and broad signals was provided.

EXAMPLE 1   QUATERNARY CARBON CHARACTERIZATION USING THE ATTACHED PROTON TEST

An example of the use of the APT sequence as an aid to assigning the ^{13}C spectrum of a triterpene derivative obtained is shown in Figure 5E.1. The normal ^{13}C spectrum of the derivative showed 31 lines, including the carboxyl carbonyl at 182 ppm and a new carbonyl group at 160 ppm. The APT spectrum obtained with an 8-ms delay revealed 12 peaks with reversed phase, representing a total of 12 methyls and CH groups. The proton spectrum showed seven methyl peaks, leaving five CH groups. Seventeen CH and quaternary carbon atoms are present, and the APT spectrum with a 4-ms delay exhibits six peaks that are from quaternaries. The CH group at 75 ppm and the carbonyl at 160 ppm are small negative peaks in the 4-ms APT spectrum because they have larger coupling constants than 125 Hz, which gives a null at 4 ms.

The number of CH_3, CH_2, CH, and quaternary carbon atoms has not changed in the derivative, but one new carbonyl has been added. Furthermore, the added carbonyl has a

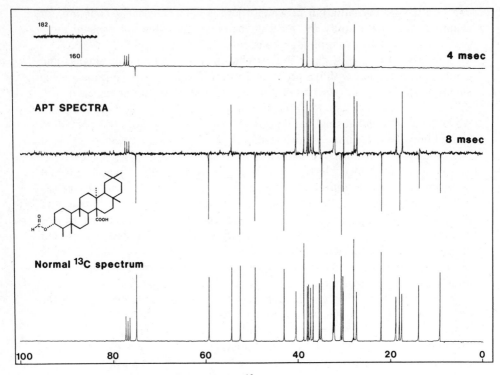

Figure 5E.1   Normal and APT ^{13}C spectra of a triterpene derivative.

proton attached to it. Since the carboxyl carbonyl is still present at 182 ppm and the CH at 75 ppm indicates an oxygen substituent, the only possibility is that the hydroxyl has been converted to a formyl ester.

4   POLARIZATION TRANSFER METHODS

Those familiar with the routine acquisition of ^{13}C NMR spectra are aware of the consequences of the nuclear Overhauser effect (NOE). Saturation of protons, or more accurately, equalizing of populations in the various spin states available to the protons, has the effect of increasing the net ^{13}C magnetization of those carbon atoms relaxed by the protons by a factor of up to three times the equilibrium magnetization. Most analytical or survey ^{13}C spectra are obtained with continuous broadband proton decoupling and any resultant NOE. Characteristics of this mode of operation are (1) the possibility of variable NOE, (2) a repetition rate governed by ^{13}C T_1, and (3) the detection of both protonated and nonprotonated carbon atoms. The first aspect makes quantitation difficult. The second affects net sensitivity, and the third has the prospect of undesirable signals in certain situations. Spectral editing is possible by using *J*-

modulated spin–echo or multiple-quantum methods; if desired, NOE may be retained.

A fundamentally different approach to signal excitation is present in polarization-transfer methods. These methods rely on the existence of a resolvable J coupling between two nuclei, one of which (normally the proton) serves as a polarization source for the other. The earliest of these types of experiments was SPI (selective population inversion) [33] in which low-power selective pulses are applied to a specific X satellite in the proton spectrum for an X—H system. If the resultant population inversion is followed by detection, an enhanced multiplet in the X spectrum is produced. A basic improvement that removes the need for selective positioning of the proton frequency was the INEPT (insensitive nucleus excitation by polarization transfer) technique introduced by Morris and Freeman [34]. This technique uses strong non-selective pulses and gives general sensitivity enhancement.

The pulse sequence has a basic polarization transfer portion (Figure 5.5) that produces a net inversion of one of the proton spin states. Following the occurrence of an X-nucleus 90° pulse there is enhanced magnetization in the X multiplet. The signal enhancement is proportional to the ratio of the magnetogyric ratios of the two nuclei involved; that is, for X—H experiments, factors of 4 and 10 are realized for ^{13}C and ^{15}N, respectively. The repetition rate of the experiment is dictated by the T_1 of the polarization source nucleus. Since this is typically the proton, often more signal per unit time can be obtained than by NOE. Since the relaxation mechanism for the X nucleus is not involved, variable or negative NOE is not a problem, as can be the case for ^{15}N and ^{29}Si. One drawback is the effect of short T_2 values on the sensitivity improvement. Since the sequence requires delays of about $1/J$ before detection (XH coupled systems with J values from 10 to 100 Hz require delays of from 100 to 10 ms), short-proton T_2 values or X-nucleus T_2 values in decoupled polarization-transfer experiments can lower expected gains radically. For example, if the protons have a T_2 of 30 ms (line width in the proton spectra of ca. 10 Hz), two-thirds of the proton magnetization will decay away in 30 ms. This would be the same time necessary to attain polarization transfer for a system having an X—H J coupling of 16 Hz. For typical one-bond couplings of 40–250 Hz for ^{15}N and ^{13}C, this loss is less important because of the $1/J$ delay dependence. Although NOE falls off as field strength and molecular size increase, it still may have a net advantage in some situations because of T_2.

While the first reported INEPT experiment dwelt on a generation of enhanced *coupled* spectra, later reports [35–38] extended the basic sequence to allow broadband decoupling; hence, the goal of general sensitivity enhancement in a more widely useful form was reached. Spectral editing is a side benefit that is made possible through the proper selection of a refocusing period. No quaternary carbon atoms are usually observed because delays are typically set for J values that are appropriate for protonated X nuclei. XH-only spectra can be obtained easily, as well as spectra in which XH_2 is inverted with respect to XH and XH_3. The combination of these two spectra then allows direct

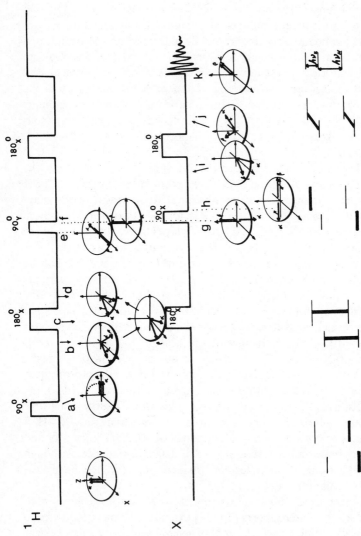

Figure 5.5 The INEPT pulse sequence: (a) proton magnetization is put into the transverse plane by a 90° pulse; (b) with an AX spin system, the two magnetization components, corresponding to protons with ^{13}C spin alpha and beta, diverge for a long enough period to bring them 90° out of phase (1/4J); (c) a refocusing 180° pulse reverses the spin vector labels; (e) after a total time of 1/2J the vectors are antiparallel; (f) a 90° phase-shifted ^1H pulse places the spin vectors antiparallel along the z axis, and a spin inversion is performed on one of the spin states, as indicated in the energy-level diagram; (g) this inversion causes a similar antiparallel orientation of the X nuclei coupled to the protons; (h) a 90° pulse on X places this antiparallel magnetization in the xy plane; (i) the spins are allowed to precess for a time 1/4J and simultaneous 180° pulses on X and ^1H cause alignment of the vectors at (k). At this time broadband decoupling can be applied.

598

identification of XH, XH_2, and XH_3 signals in even the most complex molecules in a manner similar to that in the preceding description of the J-modulated spin–echo experiments.

EXAMPLE 2   SENSITIVITY ENHANCEMENT FOR ^{15}N USING INSENSITIVE NUCLEUS EXCITATION BY POLARIZATION TRANSFER

Insensitive nuclear excitation by polarization transfer can have very dramatic results for low-sensitivity nuclei. The natural abundance ^{15}N spectrum of 0.056 M angiotensin II is examined in several ways in Figure 5E.2. Clearly, it is time-consuming to obtain an NOE-enhanced spectrum, although it does allow recording of signals from nitrogens having labile protons. The INEPT spectra are an order of magnitude greater in sensitivity, and this additional performance allows spectral editing in a fraction of the time. A longer run with less line-broadening shows the close proximity of several of the ^{15}N resonances. The edited spectra are obtained by simply setting a multiplicity parameter that selects a proper delay based on J_{NH} in the refocused INEPT pulse sequence [G. A. Gray, unpublished].

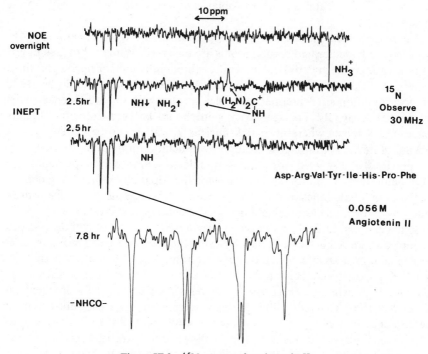

Figure 5E.2 ^{15}N spectra of angiotensin II.

Several useful variations of the basic INEPT experiment have been reported [39–41]. Morris [39] proposed using INEPT-generated ^{13}C signals to determine T_1 values of attached protons. In this sequence a 180° pulse on the protons followed by a variable delay is placed in front of the ordinary refocused INEPT pulse train. An increased delay results in a set of spectra exhibiting ordinary inversion-recovery behavior, and the data is analyzed in the ordinary way. Since the time-dependence observed is that arising from the partial return to equilibrium of the *protons* after the initial proton 180° pulse, the corresponding T_1 values are those of the proton(s) attached to the observed ^{13}C. Since proton spectra can be very overlapped and difficult to analyze, this method offers a way to use the dispersion of ^{13}C to obtain proton-relaxation information.

Spin–lattice relaxation of low-gamma nuclei has been studied using INEPT methods by adding a 90° pulse on the low-gamma observe nucleus just at the time when detection would ordinarily begin [40]. This places the transferred polarization along the $+z$ or $-z$ axis (depending on its phase of the 90° pulse). A variable delay followed by a monitoring 90° pulse determines the extent of relaxation toward equilibrium, and thus measures the observe nucleus T_1 with the advantage of polarization transfer for higher sensitivity (e.g., for ^{15}N and ^{29}Si) and shorter equilibration time (governed by proton T_1). Of course, the method could be extended to systems of other nuclei, for example, phosphorus-bound metals in catalysts where polarization transfer is by way of the phosphorus.

An analogous case is present in the ^{13}C-deuterium INEPT experiment [41–42]. Here, only deuterated carbon atoms are observed. While the magnetogyric ratio factor should give only a theoretical three-fifths enhancement, the short deuterium T_1 allows very rapid accumulation. The technique should be useful for determination of deuterated X-nuclei spectra without interferences from protonated X signals. The sensitivity gain should be very dramatic for fully deuterated X nuclei where normal X-nucleus T_1 values can be extremely long because of inefficient dipolar relaxation from deuterium or other mechanisms. As previously discussed for the J-modulated spin–echo experiment for ^{13}C—deuterium [17], this experiment offers the capability of detecting deuterated sites at small concentration in, for example, studies of the mechanism of deuteration (or a reaction using deuterated reactants, e.g., oxidants).

Spin–echo and INEPT methods can be reversed to detect only those protons J coupled to an X nucleus [43, 44]. By using the pulse sequence in Figure 5.6a we can obtain a spin–echo difference spectrum. The 180° pulse on the ^{13}C interchanges the spin labels on the proton vectors while in alternate scans the labels are left unchanged (Figure 5.7). The resulting pair of spectra are identical except that the ^{13}C satellite lines in one spectrum are inverted. Subtraction of the two spectra yields just the satellite lines and provides a sensitive way to examine satellite resonances. Suppression of the primary proton spectrum must be excellent if natural abundance satellites are to be observed, hence a stringent requirement on spectrometer stability and reproducibility. Suppression factors of 50 [44] and 30 [43] have been obtained by using this technique. An improved suppression factor of 800 has been obtained by the reversed INEPT [43] technique. In this technique (Figure 5.8) proton signals are generated from ^{13}C magnetization, which is transferred to the protons for detection. Proper phase cycling can greatly reduce nontransferred magnetization.

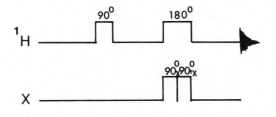

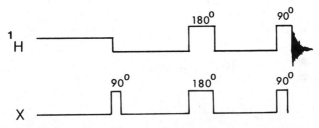

Figure 5.6   Pulse sequences for spin–echo difference and reversed INEPT.

The INEPT technique offers several advantages; it also has some undesirable properties (1) there are, at times, lines and antiphase components missing from coupled spectra; (2) there is a fairly strong J dependence in cases where signals should null, which leads to residual unwanted signals; and (3) there are variable refocused INEPT delay periods, depending on the multiplicity desired, that lead to possible variation in intensity from variations in T_2 among the resonances. Doddrell and colleagues [45, 46] developed a distortionless enhancement by polarization transfer (DEPT) pulse sequence (Figure 5.9) that addresses these problems to a large extent. The DEPT spectra of coupled X nuclei have the normal binomial distribution of intensities; the J dependence for XH selectivity is better than with INEPT, and the multiplicity selection relies on a variable proton flip angle rather than on variable delays, thus the T_2 dependence is factored out. By choosing $\theta = 45$, 90, and 135° (Figure 5.10) in separate spectra, spectral editing is possible through proper combinations of these spectra to give XH, XH_2, and XH_3 subspectra. The DEPT spectra are more sensitive overall to spin relaxation during the pulse train since the relevant delay is of magnitude $3/2J$ while INEPT typically ranges from $1/J$ to $5/4J$. The DEPT equivalent of inverse INEPT [43] has also been discussed [47–49].

Some basic improvements in INEPT and DEPT for *coupled* spectra have been reported. The former (INEPT$^+$) can generate undistorted multiplets by the addition of a 90° pulse on the protons just prior to detection *in the refocused INEPT* version of the experiment [50, 51]. Similarly, DEPT may be improved by adding (on alternate transients) a 180° pulse on the protons just before

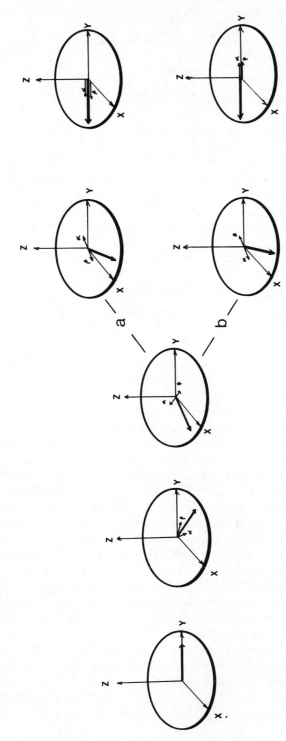

Figure 5.7 Spin–echo difference. After a 90° pulse on the protons, spin precession proceeds with rates determined by relative offsets from the transmitter frequency. Protons not coupled to an adjacent ^{13}C form the bulk of the magnetization and are symbolized by the large vector. Those protons adjacent to a ^{13}C have either a faster (by $+J/2$) or slower (by $-J/2$) precession frequency. These are denoted as alpha or beta vectors. At a time $1/2J$ these vectors are antiparallel. A 180° pulse on the protons rotates the magnetizations about the x axis. In (a) the two 90° pulses on X mutually cancel, but in (b) they perform a 180° pulse on X, reversing the proton spin labels. In either case the spin vectors align at a time $1/J$, either opposed to or aligned with the bulk magnetization. Subtraction of the two data sets cancels the bulk signal leaving signals arising only from protons coupled to X.

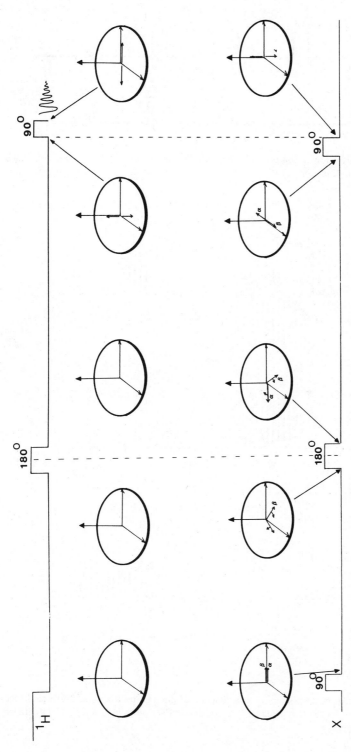

Figure 5.8 Reversed INEPT. After a 90° pulse on X the spin vectors dephase at a rate determined by J and the offset from the X-transmitter frequency. The difference in rates is just J_{XH}. At a point where this divergence is 90°, a 180° pulse pair is used to refocus and interchange spin labels. Further divergence for a time $1/4J$ leaves the vectors antiparallel at which time an X 90° pulse restores the vectors to the z axis. The spin inversion of one of the spin states causes a parallel inversion in the coupled protons. This is sampled by a 90° pulse on the protons. The phases of the first X pulse and last proton pulse can be cycled to eliminate any proton signals coming from protons not coupled to X.

603

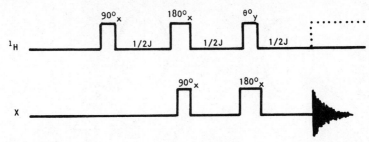

Figure 5.9 The DEPT (distortionless excitation by polarization transfer) pulse sequence, which features a variable angle proton pulse for multiplicity selection while it retains a constant total time of $3/2J$ for the sequence.

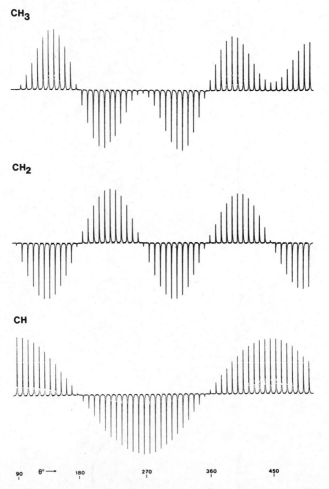

Figure 5.10 Variation of peak intensity as a function of the final proton pulse flip angle in the DEPT pulse sequence for CH, CH_2, and CH_3 carbon atoms. Each plot shows actual DEPT data for individual carbon types.

604

detection of the coupled fid to reduce phase anomalies (DEPT$^+$) [51]. Somewhat better correction to DEPT can be made by adding several more 180° pulses (DEPT^{++}); however, the total length of the pulse train will then be increased by a period $1/2J$ [51].

Further development and analysis [52, 53] of the DEPT sequence has resulted in a more generalized universal polarization transfer sequence (UPT) [54, 55]. The DEPT sequence will not transfer polarization to a spin-one nucleus, whereas the INEPT sequence can transfer from an arbitrary number of spin-$\frac{1}{2}$ nuclei to a nucleus of aribtrary spin quantum number. The UPT pulse sequence has variable angle X and H nucleus pulses and gives equivalent enhancement to that of INEPT.

EXAMPLE 3   USE OF THE DISTORTIONLESS ENHANCEMENT BY POLARIZATION SEQUENCE TO ASSIGN THE SPECTRA OF TWO ISOMERS

A natural product with the molecular formula $C_{11}H_{20}O_2$ can be shown by conventional ^1H and ^{13}C NMR techniques to occur as a 3:1 mixture of two isomers. Selective decoupling of the region from 1.0 to 2.0 ppm resulted in a sharp singlet from the quaternary carbon bearing the hydroxyl. Long-range coupling of the type $^3J_{CCH}$ exhibits† a Karplus-type angular dependence that will result in very small couplings for dihedral angles near 90° and observable couplings for angles near 0 or 180°. Under the conditions of the experiment, only the coupling to the CHOH ring proton will be observable, and its absence indicates a dihedral angle near 90°. A molecular model shows that strong hydrogen bonding between the hydroxyls can occur if the OH and the *gem*-dimethyl substituent are cis and that the forces associated with the formation of such a bond will stabilize a nonplanar configuration of the carbocyclic 5-ring, which will have a 90° dihedral angle.

The ^{13}C spectrum of this compound (Figure 5E.3) exhibits isomeric shifts of the lines for most carbon atoms, and these shifts range from 0 to 0.5 ppm for all but three of the carbon atoms. The assignment of the spectrum must be made before it is possible to confirm which carbon atoms exhibit shifts larger than 0.5 ppm and then to determine their magnitude. The DEPT spectra obtained with the proton pulses set at 135, 90, and 45° are shown in Figures 5E.4a, b, and c, respectively. Since DEPT is a polarization transfer technique, the two quaternary carbon atoms and the three solvent lines disappear. Figure 5E.4b identifies the vinyl CH, the CHOH, and the remaining CH as the lines at 145.6, 81.7, and 50.1 ppm in the major isomer, and 142.8, 81.7, and 50.1 ppm in the minor isomer. Figure 5E.4a exhibits inverted CH$_2$ peaks and confirms the assignment of the vinyl CH$_2$ and the two ring CH$_2$ carbon atoms at 111.6, 32.7, and 22.3 for the major isomer and 115.0, 32.4, and 21.8 ppm for the minor isomer. The remaining positive lines in Figure 5E.4a establish the shifts of the gem methyls and the lone methyl as 29.6, 29.3, and 21.1 ppm in the major isomer, and 29.7, 29.2, and 24.4 ppm in the minor isomer.

Three carbon atoms show a large isomeric shift: vinyl CH, vinyl CH$_2$, and the lone methyl. These shifts are -2.8, $+3.4$, and $+3.3$ ppm, respectively, and progress from the major to the minor isomer. These observations are consistent with an assignment of the major isomer as 1a and the minor isomer as 1b.

Placement of the CH$_3$ cis to the hydroxyl as in 1a generates a steric compression shift

---

† F. W. Wehrli and T. Wirthlin, *Interpretation of Carbon-13 NMR Spectra*, Heyden, London, 1978, p. 56, contains a discussion and references.

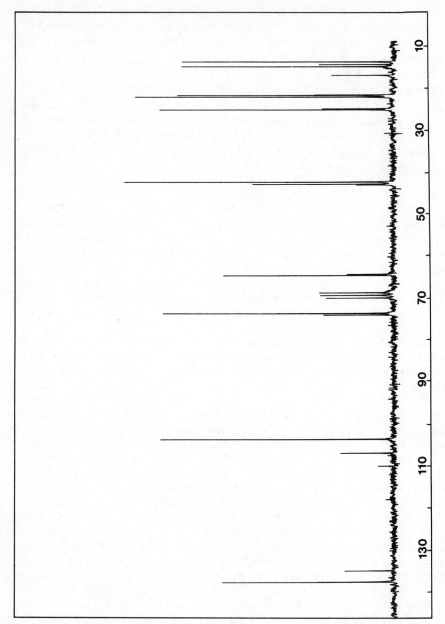

Figure 5E.3 Normal ^{13}C spectrum of unknown compound.

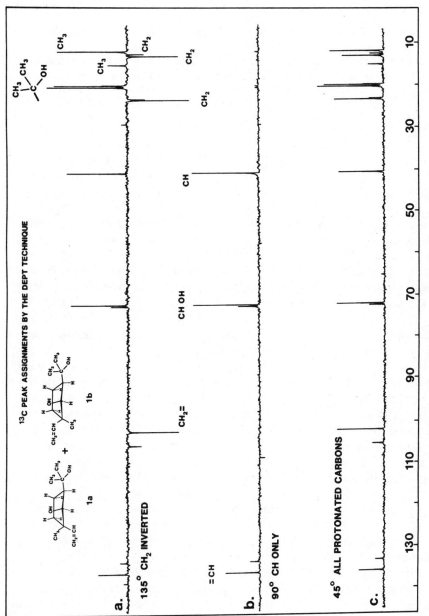

Figure 5E.4 The DEPT ^{13}C spectra of unknown compound.

of the methyl ^{13}C resonance to a smaller chemical shift value, while simultaneously relieving a similar steric compression shift of the vinyl CH in 1b. The isomerization is thus shown to affect only the vinyl and lone methyl carbon atoms strongly and establishes the stereochemistry of the isomers.

The DEPT sequence is the preferred sequence for assignment of mixtures containing small and large peaks because of its insensitivity to errors in the estimate of J_{CH} or variations of J_{CH} for different carbon atoms. Without a considerable increase in the time devoted to data acquisition, small peaks can easily lose enough amplitude in the APT experiment to make detection difficult.

Rutar has discussed [56] a method for selective polarization transfer (SEPT). The sequence permits "editing" of *coupled* spectra [57]. By varying the phase of the last proton 90° pulse, spectra result in which methylenes are in or out of phase with CH and CH_3. Addition of the two spectra gives only methine- and methyl-coupled patterns, while subtraction produces only coupled methylene patterns. The multiplets have the usual INEPT intensity distribution. Similar results can be obtained by using DEPT for protonated and nonprotonated carbon atoms [58, 59]. Rutar has extended this sequence to obtain decoupled spectra. This sequence permits elimination of CH_2 resonances by combination of two spectra. This could have an application in polymer systems with a large CH_2 resonance, for example, linear polyethylene, thus revealing CH and CH_3 resonances close to the strong signal. In the same manner, combination of decoupled DEPT spectra for $\theta = 45$ and 135° should produce equivalent data. For spectrometers in which pulse imperfections are a more serious source of error than phase missettings, the DEPT version may have a disadvantage. However, the larger number of pulses in Rutar's sequence may negate any advantage gained by using only a phase shift to generate the two subspectra.

These sequences have involved nonselective excitation. In certain cases it may be useful to transfer polarization from only one nucleus to another to help simplify and/or interpret complex spectra or to correlate chemical shifts of coupled heteronuclei [60–63]. Brandeau and Canet [60, 61] developed techniques that use a semiselective 90° pulse on the resonance of the A nucleus in an AX spin-coupled system. This is followed by a 90° phase-shifted 90° pulse applied to the A nuclei and then a nonselective 90° pulse to monitor the X nucleus. The semiselective 90° pulse is generated by a pulse train of small angle pulses [64] and has the strength of about $J/2$ (in hertz), which is sufficient to affect both lines in the A spectrum, tipping them in opposite directions in the rotating frame. As in INEPT, refocusing pulses allow broadband decoupling when the X nucleus is observed. By sequential application of this technique to different A nuclei in the heteronuclear case [60], for example, in the proton carbon case, it is possible to derive chemical shift correlations. Of course, the method is superior to simple selective decoupling because the other X nuclei are not excited and thus do not generate a large and overlapping background signal. Carbon–carbon connectivities can be determined by applying the pulse train to one ^{13}C resonance and observing all possible CC couplings to the selected carbon.

A somewhat analogous method for the heteronuclear case was reported by Shaka and Freeman [65], but in a reverse sense. A pulse train selectively excites a single on-resonant ^{13}C resonance. A period of free-precession of the proton-coupled ^{13}C spins then follows. Simultaneous proton and carbon nonselective 90° pulses accomplish polarization transfer *only to those protons spin coupled to the selectively excited carbon atoms*. After another period of free precession of the proton vectors, the proton signal is detected under broadband decoupling of the chosen carbon. This type of heteronuclear shift correlation benefits from the intrinsically higher proton receptivity, although it is governed by the T_1 of the carbon.

Although INEPT has been applied primarily to protonated nuclei, it is possible to excite nonprotonated nuclei, with concomitant sensitivity advantages, by selecting interpulse delays that are appropriate for a long-range XH coupling. Bax and colleagues [66] have suggested a modification of the INEPT experiment to correct two major problems present for small XH coupling. First, homonuclear multiple-quantum coherence is generated by the second 90° pulse in the INEPT sequence because the relatively long interpulse spacing allows significant dephasing of the protons. This coherence cannot be transformed into X-transverse magnetization. Second, the $1/J$ delay after magnetization transfer is relatively long for small J, and thus allows couplings to other protons to modulate the X magnetization. This spreads the X multiplets and reduces the efficiency of the refocusing period with subsequent signal loss. By using selective low-power proton pulses [66] only the desired protons are affected, no multiple-quantum coherence is generated, and the X-magnetization vectors are not affected by couplings to protons other than those that experienced the selective pulses. Of course, the technique also directly correlates the chemical shifts of the two nuclei, and signal assignment is thus permitted.

Polarization transfer can also be accomplished in a fundamentally different manner. This is by way of the extension to liquids of the cross-polarization techniques normally used in solids. This J-cross polarization [67, 68] technique must be capable of dynamic ramping of the decoupler amplitude to accommodate variations that occur in J for proper Hartmann–Hahn [69] matching and thus polarization transfer. Therefore, it is more difficult to add to standard spectrometers and consequently has had little growth in applications.

5 TWO-DIMENSIONAL NUCLEAR MAGNETIC RESONANCE— BASIC PRINCIPLES

One of the most dramatic developments in NMR has been the invention, exploration, and expansion of two-dimensional NMR (2-D NMR). Two-dimensional NMR is an extension of ordinary NMR. The basic principle was invented by Jeener [70] and covers essentially all two-dimensional experiments in NMR, as well as other areas. The general 2-D NMR experiment is characterized by up to four time periods:

Preparation, ..., Evolution (t_1), ..., Mixing (t_{mix}), ..., Detection (t_2)

Preparation time is necessary to bring the system to a known state; for example, this might include adequate time to bring the spins to an equilibrium magnetization followed by a 90° pulse to place the magnetization in the transverse plane. During the *evolution* period, the spins are allowed to evolve within a specific environment that may be different from any environment in the other time periods. This evolution time is usually incremented through a range of values, each experiment resulting in a separate fid that is recorded for a different evolution time followed by the same total detection period. The *mixing* period may be as short as the length of a pulse (appropriate for transverse magnetization or spin population changes within a spin system) or longer when longitudinal magnetization is to be exchanged.

The result is a collection of fid's, the number of which equals the number of different values of the evolution time period. All fid's are then transformed in the usual way, resulting in a collection of spectra (Figure 5.11); the familiar inversion-recovery T_1 experiment is an example of this stage of the process. A new collection of fid's is assembled by taking data points from each spectrum at given frequency values.

For example, the values of spectral intensity in each spectrum found at a frequency f are taken and arranged in a data table. This is repeated for every f value in the spectrum, resulting in N interferrograms, where N is the number of points in the original spectrum. These are then Fourier transformed, resulting in N spectra. These N spectra, presented on a two-dimensional plot, are now characterized by *two* frequency axes. One frequency (F_2) is always that of the normal observe spectrum since it results from the FT (Fourier transform) of

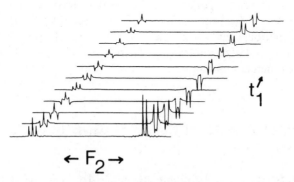

Figure 5.11 Variation of peak amplitude as a function of evolution time. Each spectrum represents the Fourier transform of one fid. The spectra differ only in the length of the evolution time, which is zero for spectrum 1 and is incremented by a constant amount for subsequent spectra. The phase or amplitude of each peak is modulated as a function of the evolution time, depending on the particular pulse sequence used. The first part of two-dimensional data processing generates this type of data.

normal fid's. The significance of the remaining (F_1) frequency is determined by the pulse sequence used.

There have been major advances şince around 1980 in both the number and variety of the pulse sequences used to perturb the spins and also in the computing power and data-processing techniques employed.

All two-dimensional experiments use essentially the same data-reduction process. Several years ago the data-acquisition phase of two-dimensional experiments was usually the shortest, with double transformation requiring up to several hours followed by, again, comparable plotting time. Off-line data processing was often used for faster processing, but this severely limited any widespread use of two-dimensional methods. Two-dimensional processing capabilities became generally available on commercial spectrometers after 1979, and with the introduction of flexible pulse programmers the use of two-dimensional methods has grown quickly. Very recently, new computer hardware, including array processors, has revolutionized two-dimensional data processing, allowing complete data reduction, including a full two-dimensional display in less than 5 s (for a typical 512 × 512 matrix). This speed, coupled with responsive color graphics, has permitted direct, on-line data processing and has removed the formidable time barrier to full use of 2-D NMR spectroscopy.

6 *J*-RESOLVED TWO-DIMENSIONAL NUCLEAR MAGNETIC RESONANCE

The key to the nature of the F_1 domain lies in the environment in which the spins precess *during the evolution period* t_1. This environment may include the interactions of spin coupling, the chemical shift (which governs the rate of precession relative to the transmitter frequency as zero), and magnet inhomogeneity (which causes identical nuclei to have different precession rates). Refocusing pulses, as seen before, can remove some, or all, of these interactions. After the 90° "preparation" pulse has occurred, off-resonance spin vectors will begin to dephase or precess. Use of a strong nonselective 180° refocusing pulse on the observe nucleus midway through the evolution period reverses the dephasing, which arises from the off-resonance or chemical shift effect, for all observe spins. Thus, for every value of t_1 each spin vector is returned to the same orientation as it possessed after the first 90° pulse; that is, chemical shift effects are removed. The resultant two-dimensional spectrum may be thought to be rather simple, as indeed it is for noncoupled nuclei—singlets are centered at coordinates $(F_1 = 0, F_2 = $ chemical shift value); that is, singlets are unaffected by the presence of the evolution period except for T_2-related reduction of intensities. As a function of the evolution period their intensities are described by zero-frequency decays; hence, $F_1 = 0$ for each singlet.

As the name implies, *J*-resolved 2-D NMR is useful for extracting spin-coupling information. To understand this, recall the action of the refocusing pulse in the spin–echo pulse sequence for the A spins in an AX-spin system (Figure 5.12).

The effect of the X-nucleus 180° pulse is the interchange of the spin labels, which allows further divergence of the spin components. Otherwise, with no X pulse the components will always refocus identically at $t = t_1$. In this case no t_1-dependent modulation of the fid is generated, and only T_2 information is probed. However, with the X pulse *J-dependent* modulation is introduced, because, depending on J, the vectors will be at varying phase separations as a function of the time allowed for diverging. This information is coded into the phase of the A-nucleus signals at the start of t_2, the data accumulation period. A full experiment, then, explores this dependence by repetition for the entire array of regularly spaced t_1 periods from zero to some maximum t_1.

The two major categories of this experiment are homonuclear and heteronuclear *J*-resolved 2-D NMR. In the former, only one 180° pulse is necessary since the A 180° pulse serves for both A and X. For the latter, it is also customary to decouple the X nucleus from A during t_2 for sensitivity improvement, rather than to have the AX-coupling information present in both frequency dimensions. Equivalent information for the above "proton flip" experiment may be obtained in the heteronuclear case simply by turning on the decoupler in place of the 180° pulse (Figure 5.13) half-way through the evolution period. Since the spin component divergence is unsymmetrical with respect to the 180° A pulse,

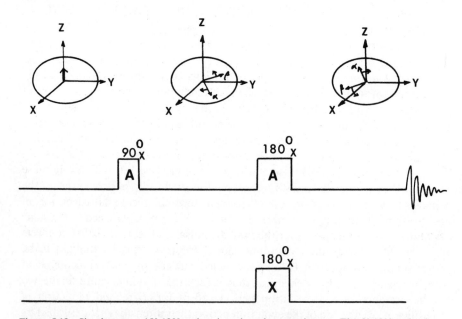

Figure 5.12 Simultaneous AX-180° pulses in spin–echo experiments. The X-180° pulse interchanges the direction of precession of the A-spin, *J*-component vectors, retaining the divergence rather than causing a refocusing. Since this divergence is caused by J_{AX}, the coupling information is encoded as phase modulation of the detected A signals.

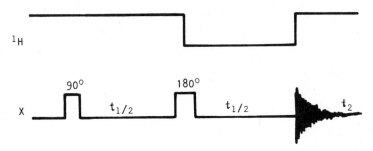

Figure 5.13 Heteronuclear two-dimensional J-pulse sequence. If the decoupler is gated off during one-half the evolution time, refocusing of magnetization vector dephasing arising from J_{XH} is prevented. This J information therefore determines the phases of the resulting broadband decoupled X signal collected during t_2.

the divergence is not refocused, and the J-coupling dependence is encoded into the phase of the decoupled A spin.

The heteronuclear J-resolved two-dimensional experiment produces completely separate information in both domains: A-nucleus chemical shift in F_2 and AX coupling in F_1. This is particularly valuable, for example, in CH cases since complete coupling patterns can be extracted for each ^{13}C *without overlap of adjacent patterns.* Information can be extracted easily, in contrast to the direct observation that is required of highly overlapped coupled one-dimensional spectra (Figure 5.14).

The homonuclear J-resolved two-dimensional spectrum is different in that AX coupling is active in *both* t_1 and t_2; this is reflected in the existence of spin multiplets in both dimensions. The two-dimensional spectrum of a homonuclear AX system can be represented in contours of intensity (looking "top down," Figure 5.15).

Each spin pattern is present, although no one slice perpendicular to F_2 contains a whole pattern. A computer process known as *tilting, rotating,* or *shearing* can be used to realign the spin patterns (45° tilt) perpendicular to F_2 [71]. This permits display and plotting of "slices" of the two-dimensional data— an F_1 spectrum at some specified F_2 value. These slices are the full coupled spin patterns (Figure 5.16). The spin–echo nature of the experiment produces narrow lines, and thus very highly resolved multiplets. In highly congested, but first-order, proton spectra this technique can be of immense value.

Second-order effects do produce additional (combination) lines in the spectrum that are not interpreted easily. These fall at positions that do not tilt onto an F_2 position of a true chemical shift. They can complicate *projected* spectra where the F_1 domain is collapsed to leave a chemical-shift-only spectrum. These *projections* are useful in giving single "one-line-per-nucleus" spectra (Figure 5.16).

Homonuclear J-resolved two-dimensional data are normally projected from a data presentation in absolute-value or magnitude form since a *phase-twist* [72] is present in phased data. The long tails from the absolute value line shape arise

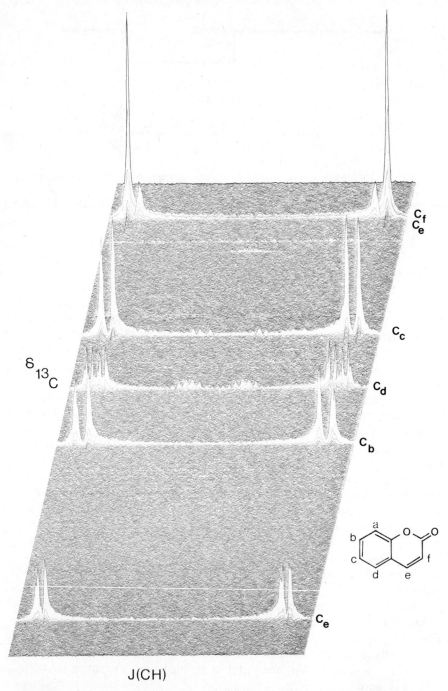

Figure 5.14 Heteronuclear two-dimensional J spectrum of coumarin. These data are characterized by the ^{13}C chemical shift along F_2 and J_{CH} along F_1. The large splittings result from one-bond couplings, while the fine structure arises from long-range J_{CCH} and J_{CCCH} couplings. Differences in these long-range patterns reveal information concerning the number of nearby protons.

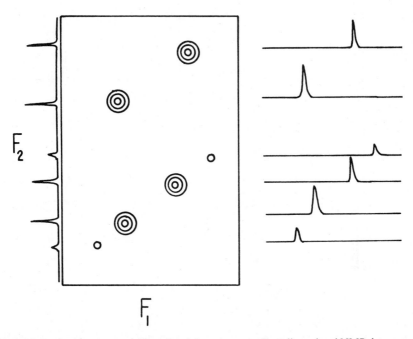

Figure 5.15 Homonuclear two-dimensional J-contour map. Two-dimensional NMR data are most often presented as topographical maps wherein increasing peak intensity is represented by increasing numbers of contours. This "top-down" approach is useful for making direct measurements. HOM2DJ data, as obtained from the two-dimensional transform process are "mixed" in the sense that F_2 has both shift and coupling information present. Slices at any particular value of F_2 at best contain only one line of a multiplet, here illustrated for the case of a doublet and quartet.

from the dispersion mode component and can give very unsatisfactory line shapes in projected data. Bax and colleagues [73] have overcome this problem by shaping the fid's and/or interferrograms by a mathematical function that forces symmetry and generates what appears to be a full echo, hence, the term *pseudoecho processing*. Display of the transform of this data removes the dispersion mode characteristics, and absolute value spectra can have the same line shapes as those in the absorption mode. Of course, signal is degraded at either end of the pseudoecho processed data, and sensitivity is thus reduced significantly.

For macromolecules the normal $90°-t-180°-t$ acquire pulse sequence produces considerable loss of sensitivity because of the irreversible loss of signal through the short T_2 values. Macura and Brown [74] proposed that data acquisition begin immediately following the 180° pulse.

Polarization transfer may be used in heteronuclear J-resolved 2-D NMR [75, 76] for signal enhancement, shorter equilibration periods, and elimination of nonprotonated resonances—the same features as in INEPT or DEPT. In this experiment the evolution time is the refocusing period rather than the $1/4J$, $1/2J$, and $3/4J$ values normally used for refocused INEPT or the variable theta used

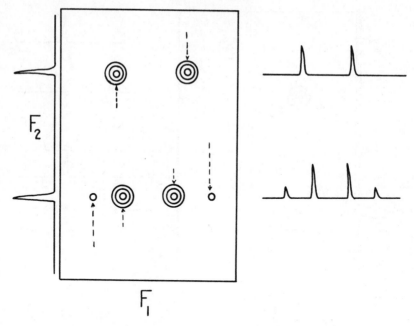

Figure 5.16 Homonuclear two-dimensional J-data rotation. The "mixed" F_2 data in homonuclear two-dimensional J spectra can be rearranged into pure J versus observe chemical shift axes by a 45° *rotation* or *tilt* operation. Slices perpendicular to F_2 at the chemical shift values now give pure J spectra. Projections onto the F_2 axis give single lines at the observe nucleus chemical shifts.

for DEPT. As usual, the polarization transfer part of the sequence includes a $1/2J$ period; thus, the two-dimensional sequence can be tailored to specific types (or J values) of XH-coupled pairs, for example, certain long-range couplings for quaternary carbon atoms whose normal T_1 values may be too long to permit ordinary heteronuclear two-dimensional J experiments.

Bax and Freeman [77] introduced an elegant method for generating specific heteronuclear two-dimensional J data. Their pulse sequence is identical to that of the normal *proton-flip* except that a weak proton 180° pulse is applied to *one* proton selectively in place of the strong nonselective 180° pulse normally used midway through the evolution period (Figure 5.17). This pulse inverts the inner (long-range coupling) satellites of a proton while leaving the outer (one-bond coupling) satellites unaffected. Thus only long-range couplings to this *one* proton modulate the observe nucleus spin–echo; therefore, there is only *one* such modulation for any observe nucleus resonance. Since only long-range couplings are involved in the F_1 dimension, good digitization is possible. The experiment probes the local environment of *one* proton only and thus is especially valuable in structural studies where extensive long-range XH couplings can produce not only extremely overlapped coupled one-dimensional spectra, but also heavily overlapped heteronuclear two-dimensional J spectra. The ability to identify which carbon atoms are in the neighborhood of a specific proton will

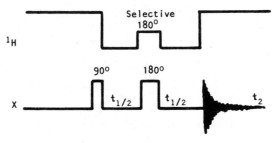

Figure 5.17 Selective heteronuclear two-dimensional J-pulse sequence. The low-power $180°$ proton pulse selectively inverts protons over a very narrow band, for example, 25 Hz. Carbon-13 atoms are either bound to the selected proton or more than one-bond removed. In the former case the large J_{CH} splitting puts both proton spin states outside of the effective bandwidth. If the ^{13}C is more than a few bonds away from the selected proton, no coupling is present. In both situations there is no modulation of the detected ^{13}C signal in t_2 as a function of t_1. However, if ^{13}C atoms two and three bonds away are coupled to the inverted proton their ^{13}C intensities are modulated as a function of t_1, and the two-dimensional data show only long-range heteronuclear couplings.

undoubtedly be valuable in spectral analysis of congested spectra and unknown structures. Figure 5.18 illustrates the selective nature of this experiment.

Bax [78] and Rutar [79, 80] separately proposed a novel class of heteronuclear J-resolved two-dimensional experiments that would allow a choice between one-bond or long-range two-dimensional J spectra.

The preparation phase may consist of a period to allow establishment of NOE, followed by an X-90° pulse [78], or polarization transfer, such as in DEPT [79, 80]. The X-nucleus pulse sequence is the same as in the normal heteronuclear two-dimensional J experiment. However, the proton-180° pulse is replaced by the element $90°(x)-1/2J-180°(x)-1/2J-90°(\pm x)$, where J is the value of the direct XH-coupling constant (Figure 5.19). As in the normal two-dimensional J-pulse sequence, *only protons inverted by this new element* will modulate the X-nucleus signal and therefore give rise to multiplets in the final two-dimensional spectrum. The fascinating property of this new sequence is that in one phase combination *only* long-range XH couplings appear; however, when the other phase combination is selected, only *direct* one-bond XH couplings are observed. In either case small F_1 ranges from 20 to 40 Hz can be used for good digitization, and extensive folding can be permitted (and later corrected) in F_1 for one-bond couplings for good precision.

For the case of $90°(+x)$ as the last proton pulse, protons experiencing a large one-bond coupling to carbon will be inverted, while remote protons, or those that are only weakly coupled to a carbon, will be restored to normal z magnetization; and there will be, therefore, a clean separation based on J. Reversing the phase of the last 90° proton pulse reverses the sense of polarization so that the remote protons experience the population inversion, and the bonded protons are restored to the z axis.

It is possible to obtain proton–proton coupling constants from 2-D NMR spectra *indirectly*, that is, through the resonance of another observe nucleus.

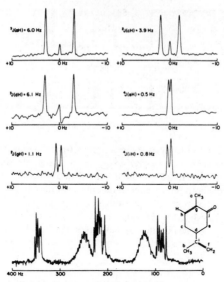

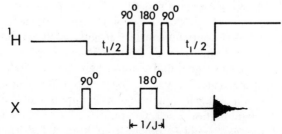

Figure 5.18 Selective heteronuclear two-dimensional J example. The normal coupled ^{13}C spectrum of carvone is complex and difficult to interpret. However, very detailed long-range CH couplings for specific CH pairs can be obtained using the *selective heteronuclear two-dimensional J experiment*. Here only proton H_h is inverted, and F_1 traces for carbon atoms a, c, d, e, g, and i are plotted showing the specific CH_h couplings. The experiment can be performed in turn for other protons to provide other specific pairs. Good digitization in F_1 is possible since the couplings are typically less than 10 Hz. Reprinted, with permission, from A. Bax and R. Freeman, *J. Am. Chem. Soc.*, **104**, 1099 (1982). Copyright (1982) American Chemical Society.

Figure 5.19 *J*-discriminated heteronuclear two-dimensional J pulse sequence. The phase of the last proton pulse determines whether one-bond or long-range XH couplings modulate the ^{13}C intensities during t_2. This permits very fine digitization for long-range couplings since small spectral widths in F_1 can be selected. Adapted, with permission from Varian Associates, Palo Alto, CA [78, 79].

Bodenhausen [81] proposed inserting simultaneous proton and observe nucleus-180° pulses midway through the evolution period in a magnetization transfer two-dimensional experiment proposed by Maudsley and Ernst [82]. This collapsed the proton chemical shift in F_1, leaving only $J(HH)$ and $J(XH)$ contributing to the J spectra in F_1. For highly congested proton spectra, this technique allows the HH multiplets to be spread by the dispersion of the X nucleus (such as ^{13}C or ^{31}P) rather than by the proton shift, as with ordinary homonuclear two-dimensional J spectroscopy. Analogous approaches based on INEPT [83] or DEPT [84, 85] that result only in HH multiplets in F_1, and thereby suppress the XH interaction, have also been described. These approaches have considerable potential as methods that can be used to extract stereochemical information for molecules whose normal proton spectrum is far too overlapped to be interpreted.

The problem of *phase twist* in homonuclear two-dimensional J spectra has been addressed in two treatments. Shaka and colleagues [86] described a computational method that corrects for the phase twist by factoring out the theoretical response and replacing it with an absorption mode representation. True absorption-mode homonuclear two-dimensional J data were obtained by Williamson [87] by using the same basic pulse sequence; however, in a separate set of data the pulse sequence is augmented by another 90° pulse just prior to acquisition. Data from both experiments are combined for every value of t_2 for each value of t_1, generating quadrature (complex) data points. The second set is for an effective reversed precession during t_1, and coaddition of the two experiments converts phase modulations into amplitude modulation, in a way similar to that in which quadrature information is generated in shift-correlation experiments. Pure absorption-mode spectra without phase twist can be produced, and higher resolution data in complex molecules would thus be possible.

Keeler [88] published a method that uses heteronuclear two-dimensional J spectra for assigning proton multiplicity in ^{13}C two-dimensional J spectra. Since each F_1 trace represents the CH-coupled multiplet, this information is, in principle, available after only a few t_1 values. Unfortunately, the truncation of a small t_1 data set leads to "sinc" wiggles in the F_1 slices, and the desired information is obscured. This system has been automated [88], and low-resolution J spectra that are useful for multiplicity characterizations are produced that are independent of relaxation, J values, and pulse calibrations assumptions, all of which are required when using the preceding spectral editing techniques.

7   CHEMICAL SHIFT CORRELATION TWO-DIMENSIONAL NUCLEAR MAGNETIC RESONANCE

7.1   Homonuclear Shift Correlation

Probably the most widely used 2-D NMR experiments are those that relate the chemical shifts of different nuclei; that is, F_1 and F_2 represent chemical shift axes, and the class includes homonuclear and heteronuclear categories. Again, the environment during the evolution time dictates the interpretation of the new information. The simplest example of this experiment is the homonuclear chemical-shift correlation [89–91], also known in one form as COSY (correlation spectroscopy). The most widely used pulse sequence is rather simple: $90°-t_1-90°$ followed by acquisition. The resulting peaks in the two-dimensional plot fall into three categories: axial, diagonal, and off-diagonal. The axial peaks lie along $F_1 = 0$ and result from the longitudinal magnetization sampled by the last pulse. The diagonal peaks are caused by the magnetizations that remain associated with the same spins before and after the second 90° pulse; that is, the magnetization has the same frequency during t_1 and t_2. The off-diagonal peaks confirm the fact that the magnetization present at one frequency in t_1 is transferred to another frequency after the second 90° *mixing* pulse. This is only possible if the two nuclei at the two frequencies share the same spin system; that is, they are J coupled. This last feature forms the basis of the widespread use of homonuclear shift-correlated 2-D NMR since it gives equivalent information to that provided by spin-decoupling experiments. The spread of information in two dimensions actually makes the interpretation easier. Proper phase cycling [91] removes the axial peaks, which simplifies the data.

Variations on the basic experiment include the $90°-t_1-45°$ acquire experiment [90]. This version narrows the patterns on the diagonal, thus permitting the analysis of correlations of nuclei closely spaced in chemical shift.

While the vast majority of homonuclear two-dimensional shift correlation has been in proton NMR, other studies have appeared applied to ^{31}P in cellular phosphates [92], ^{11}B in polyhedral boranes [93, 94], and ^{183}W in heteropolytungstates [95]. Other applications such as establishing C—C linkages in enriched metabolites have been demonstrated [96] in experiments using specifically labeled nutrients for biosynthesis.

EXAMPLE 4   THE USE OF HOMONUCLEAR SHIFT CORRELATION IN SPECTRAL ASSIGNMENT

The multiplets observed in proton NMR spectra result primarily from spin–spin coupling between protons on the same or neighboring carbon atoms and can be used to confirm the presence of chemical bonds between various moieties in a complex structure. Conversely, coupling patterns in known or proposed structures can be predicted, and the coupling patterns may be used to assign protons by matching the predictions. In complex spectra, typical for molecules of interest to modern organic chemists, it is first necessary to determine the protons that are mutually coupled. Until the advent of 2-D NMR, this information was obtained through a series of spin-decoupling experiments—at best a time-consuming procedure that was frequently subject to confusion in highly overlapped spectral regions. By detecting nuclear precession modulated at the chemical shift frequencies of *both* spin-coupled nuclei, two-dimensional correlation spectroscopy (COSY) can demonstrate coupling between protons by displaying off-diagonal peaks in a square data array. This procedure greatly reduces the overlap of significant data. Furthermore, all spin–spin coupling data are obtained in a single experiment and displayed in a single plot. The following example shows how a spectrum with several very similar multiplets can be assigned to correspond with a proposed structure.

This structure was proposed for the sample:

In Figure 5E.5 the methyl protons fall at 1.2 ppm; and, as shown by the dotted lines, correlate strongly to the multiplet at 4.05 ppm and confirm that it is H_5. Similarly, as expected, the multiplets at 1.85 and 2.1 ppm show a strong correlation for H_{2a} and H_{2e}, respectively. Both H_{2a} and H_{2e} show correlations to the multiplet at 4.5 ppm, which must be either H_1 or H_3. This cannot be determined from the chemical shift of the multiplet, but can be determined by noting that H_1 should show additional correlations to the H_1 protons. This is not observed for the 4.5 ppm multiplet, which is therefore assigned as H_3. Confirmation for this assignment comes from the observation of a correlation between H_{2a} and the multiplet at 4.25 ppm, which is H_1. Since both H_1 and H_{2e} are equatorial protons and the ring is surely distorted by the steric forces between the large substituents, the coupling between H_1 and H_{2e} is essentially zero and gives no correlation in the two-dimensional plot and no observed splitting in the multiplet arising from H_{2e}.

Further confirmation for H_1 is found in the correlation to *both* protons at 2.45 and 2.7 ppm. These can only be the H_1 protons; as predicted, they show a strong correlation.

The correlations between the $H_{1'}$ protons and the multiplet at 5.7 ppm confirm the latter to be $H_{2'}$. The assignment of $H_{3'}$ to the multiplet at 6.2 ppm follows from the strong correlation between the multiplets at 5.7 and 6.2 ppm. The *magnitude* of the coupling $J_{2'3'}$ can be determined easily from the line spacing of the multiplets and is found to be about 15 Hz, which confirms the existence of the molecule as the E isomer (2' and 3' trans), and not the Z isomer (2' and 3' cis).

The multiplet from $H_{4'}$ is at 6.35 ppm and its correlation peaks to $H_{3'}$ are not resolved from the diagonal peaks for $H_{3'}$ and $H_{4'}$, the square pattern on the diagonal does indicate this coupling. The correlations to $H_{5'}$ (cis) and $H_{5'}$ (trans) are clearly visible opposite the multiplets at 5.08 and 5.2 pp.

The multiplet at 4.5 ppm, previously confirmed as H_3, shows a very weak correlation to the sharp peak at 5.4 ppm. The coupling is unresolved in the spectrum, but does appear in the two-dimensional plot, confirming the peak at 5.4 ppm as H_4. Since H_4 shows small coupling to H_3 and negligible coupling to H_5, it is certainly equatorial.

A strong correlation appears on the two-dimensional plot between H_3 and the multiplet at 6.4 pp. This cannot be coupling between H_3 and H_4, but will be expected between H_3 and the NH proton. The NH is broadened, either by exchange or by the effect of the ^{14}N quadrupole moment, but this does not interfere with the observation of the coupling in this case.

The aromatic protons all fall in the multiplet at 8.3 ppm and show no correlations to any other protons.

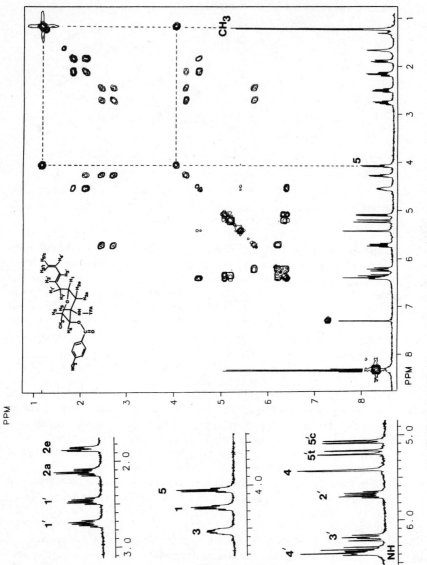

Figure 5E.5 COSY and one-dimensional spectra of unknown compound.

While this spectrum could be assigned without the two-dimensional data, the straightforwardness and ease of interpretation are very appealing. In more complex spectra, the two-dimensional data permit assignments that otherwise would be impossible.

The normal homonuclear shift-correlation two-dimensional pulse sequences generate spectra that have positive and negative components within spin multiplets. To avoid cancellation of phased data in closely spaced multiplets and to make contour and stacked plotting practical, spectra may be presented in the absolute value mode. Pseudoecho [73] processing reduces the baseline distortions (at the expense of sensitivity) and is used routinely along with triangular symmetrization [97] to reduce F_1 *noise* and to suppress artifacts. F_1 noise results from spectrometer instabilities and is manifested as extra noise in slices in F_1 perpendicular to F_2.

Absorption mode spectra are desirable for good line shapes and better resolution. Braunschweiler and Ernst [98] obtained phased COSY data by introducing a rapid series of pulses *during the mix period* and adding the resultant data together for a selection of mix times for each value of t_2. The pulse sequence that occurs during the mix time can be as simple as repetitive 180° pulses of either constant or phase-alternated phases and is designed to factor out chemical shifts, leaving only the isotropic coupling. An additional feature of the technique is that all pairs of nuclei present in a spin system show cross peaks, even if they are not directly J coupled, because of multiple-step, relayed-coherence transfer generated by the pulsing during the mix period. Essentially, the 180° pulses propagate the magnetization throughout the individual spin systems. Of course, isolated spin systems would not show mutual cross peaks. Because of the *total* nature of this correlation, the technique is referred to as TOCSY (total correlation spectroscopy).

The utility of homonuclear shift-correlation 2-D NMR has been demonstrated in probing the existence of very small J couplings [99, 100]. As pointed out by Bax and Freeman [99], introduction of a short delay after each of the pulses in the 90°–t_1–90° sequence emphasizes the off-diagonal peaks that have arisen from small J values. Batta and Liptak [100] exploited this capability to establish interglycosidic linkages in oligosaccharides by way of the four-bond $J(HCOCH)$ coupling (estimated to be lower than 0.2 Hz). Post-pulse delays of 0.4 s were needed here; hence, the technique is useful only for relatively long T_2 protons since a minimum of 0.8 s elapses between the first pulse and data acquisition.

A review of two-dimensional techniques as applied to proton NMR of proteins has been published. It collects and discusses the many experimental techniques developed in the study of proteins by the groups of Ernst and Wuthrich [101].

7.2 Heteronuclear Shift Correlation

The second major form of shift-correlation 2-D NMR is heteronuclear two-dimensional shift correlation. The experiment as proposed by Maudsley, Ernst, and co-workers [102] can be visualized as generating proton magnetization, for example, by a 90° pulse, and then letting the magnetization precess for a time t_1. The extent of precession (in the rotating frame) is proportional to the distance off-resonance from the proton transmitter frequency and therefore its phase is *proton-shift* dependent. Simultaneous 90° pulses in the proton and X observe nucleus transfer magnetization to X—just as in the INEPT experiment. The phase of the X-nucleus magnetization is hence coded with the chemical shift of the attached proton. The experiment is carried out for the full range of t_1 values to produce the two-dimensional spectrum. Since the X-nucleus shift is present in F_2, the XH pair has both chemical shifts identified by extrapolation to the individual axes from the XH peak in the two-dimensional data.

Net magnetization is not transferred to the X nucleus; and a two-dimensional experiment of this kind shows the XH couplings in both dimensions, in addition to the H and X chemical shifts. This sequence can be modified (Figure 5.20) by introducing an X-nucleus 180° pulse midway during the t_1 period to remove the effect of XH splittings in the t_1 period, and thus in F_1, and small delays prior to and after the simultaneous 90° pulses can be inserted to prevent the cancellation of out-of-phase signals (the delays allow antiphase components to focus). Broadband decoupling of ^{1}H during acquisition removes the XH coupling from the F_2 dimension. Proper phase cycling [103] can remove the axial ($F_1 = 0$) peaks, which have no intrinsic interest, and also permits the proton transmitter (the decoupler) to be centered in the proton spectrum with pseudoquadrature detection in F_1. This allows finer digitization for finite data tables.

Ammann and colleagues [104] illustrated how two-dimensional methods can be used to assign protons and carbon atoms in lupane, a C_{30} triterpene containing only carbon and hydrogen. Heteronuclear *J*-resolved 2-D NMR was used to assign the number of protons to each carbon, and heteronuclear chemical shift correlation (HETCOR) allowed assignment of the associated proton shifts and many of the homonuclear HH coupling constants. Ikura and

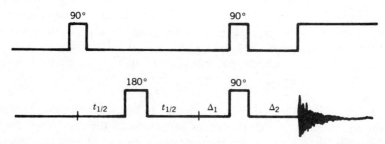

Figure 5.20 Heteronuclear chemical shift correlation. The precession of proton spins following the first 90° pulse allows the proton shift (offset from the decoupler) to determine the phase of the X-nucleus magnetization obtained from the simultaneous X and ^{1}H 90° pulse. The 180° observe pulse decouples the observe nucleus from the protons during t_1. The two fixed delays are set to allow multiplet components to coalesce, first before magnetization transfer and second, before broadband decoupling.

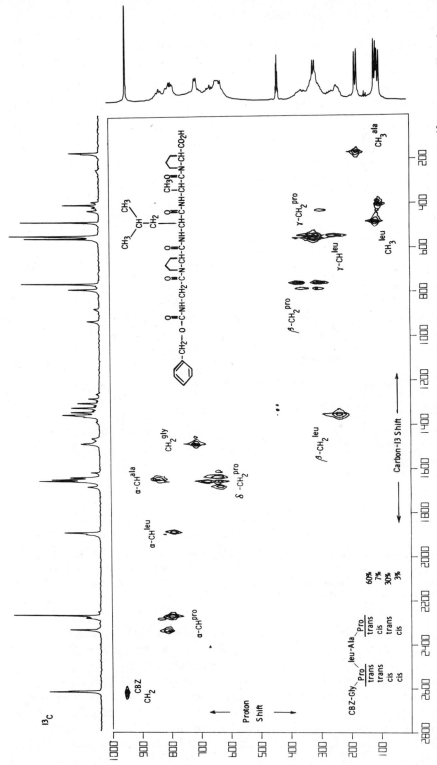

Figure 5.21 Heteronuclear chemical shift correlation, $^{13}C/^1H$. The good additivity of ^{13}C chemical shifts allow direct assignment of the ^{13}C spectrum. Then, for example, the α protons may be assigned directly knowing the assignments of the attached carbon atoms. Reprinted from G. A. Gray, *Org. Magn. Res.*, **21**, 111 (1983) by permission of John Wiley & Sons, Ltd.

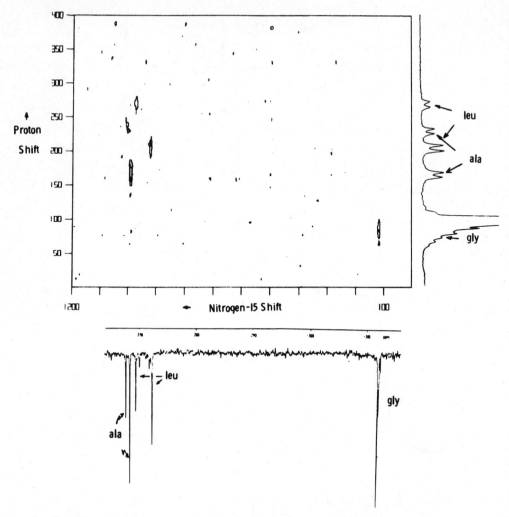

Figure 5.22 Heteronuclear chemical shift correlation, $^{15}N/^{1}H$. Polypeptide ^{15}N spectra are characterized by closely spaced ^{15}N resonances for the α-nitrogen atoms. Knowledge of α-proton assignments permits direct assignment of the ^{15}N signals. Reprinted from G. A. Gray, *Org. Magn. Res.*, **21**, 111 (1983) by permission of John Wiley & Sons, Ltd.

Hikichi [105] used the same techniques on *d*-biotin. A mixture of allylnickel complexes was also studied by Benn [106], and Morris and Hall [107] examined a series of carbohydrates. Larger molecules are, of course, feasible. Chan and Markley [108] assigned histidine-, tyrosine-, and phenylalanine-protonated carbon resonances in uniformly enriched (20%) oxidized ferridoxin by heteronuclear two-dimensional shift-correlation NMR. Polypeptide resonances have also been assigned [109, 110] (Figure 5.21). In these systems H—H homonuclear, H—C heteronuclear, and N—H heteronuclear two-dimensional correlation experiments have allowed direct confirmation of assignments and connectivities, in particular the NH [109–111] nitrogen atoms and protons (Figure 5.22). Shift-correlation experiments for H—D [112] and H—F [113] have also been reported.

EXAMPLE 5   USE OF HETERONUCLEAR SHIFT CORRELATION FOR ISOMER IDENTIFICATION

Structures **1** and **2** differ only in the arrangement of the *gem*-dimethyl groups and cannot be distinguished from one another by analysis of their ^{1}H and ^{13}C NMR spectra. Long-range CH spin–spin couplings to the quaternary carbon *c* will differ in a predictable and unequivocal way but are too complex to resolve and analyze because of the many protons on the nearby carbon atoms. The two-dimensional heteronuclear correlation experiment offers an effective way out of this dilemma. The pulse sequence causes modulation of carbon magnetization at the proton chemical shift frequency of any protons coupled to it as the evolution time is incremented. Magnetization is transferred from the protons to the carbon by the spin coupling, and the fixed delays are set to $1/2J_{CH}$ and $1/3J_{CH}$, respectively. If $J_{CH} = 7$ Hz, the two-bond and three-bond couplings will control the magnetization transfer. Standard two-dimensional data processing then allows a contour plot, as in Figure 5E.6, to be displayed, which shows correlations at the chemical shift coordinates of coupled C—H pairs.

The left side of the plot shows clearly that carbon *c* is coupled to all *four* of the gem methyls and therefore must lie *between* the *gem*-dimethyl groups; hence, structure **1** is confirmed. Coupling through 4 bonds is not observable, so structure **2** is excluded.

Methyl groups 1 and 4 correlate with carbon *a*, while 2 and 3 correlate with carbon *b*. Methyl groups 1 and 2 are equatorial and lie over the plane of the aromatic indole ring, which results in an upfield shift relative to the axial methyl 3 and 4.

The CH$_2$ carbon also shows a correlation to the methyls 1 and 2 on the adjacent carbon *a*, but not to the methyls on carbon *b*.

1

or

2

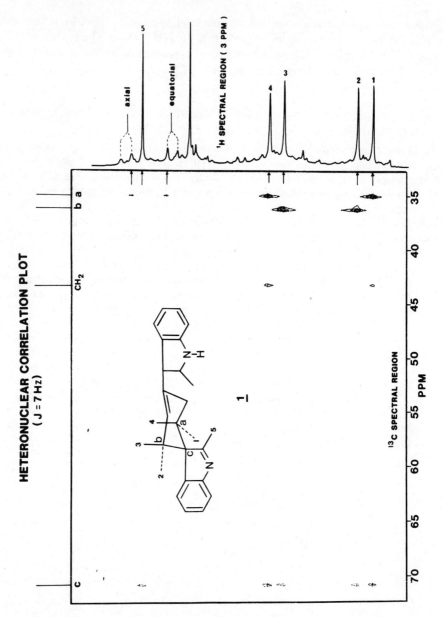

Figure 5E.6   Long-range heteronuclear ^{13}C/^{1}H chemical shift correlation spectrum of **1**.

Bolton and Bodenhausen [114–117] used ^{31}P—^{1}H two-dimensional shift-correlation techniques to obtain HH couplings in cellular phosphates. The phosphorus serves to pick out proton subspectra containing conformationally relevant couplings. Pardi and colleagues [118] obtained H—H and P—H two-dimensional chemical shift-correlation spectra from the backbone of oligonucleotides. The latter spectra contain F_1 multiplets for each ^{31}P chemical shift value. These correspond to the ^{31}P-coupled protons on the two sugars linked together by the phosphate, thus directly establishing the linkage between the sugars. Two-dimensional shift correlations for B—H have been reported in polyhedral boranes where the technique was useful for resolving heavily overlapped ^{11}B signals [119].

Heteronuclear chemical shift correlation techniques can be used to infer spin–lattice relaxation times of the protons attached to the observe nucleus [120]. This is accomplished by saturation of the protons and observe nucleus followed by a variable time t (saturation recovery) during which the observe nucleus is continued at saturation (by repeated pulsing) and the attached proton is remagnetized. This process is followed by the normal H—X two-dimensional shift-correlation experiment. The t dependence of the two-dimensional peak intensities is then used to extract ^{1}H T_1 values by exponential analysis.

The preceding techniques have relied on a single-coherence (magnetization) transfer to establish shift correlation. This usually involves X-nucleus detection with its concomitant lower sensitivity (relative to that of the proton). Bodenhausen and Ruben [121] proposed a two-step process whereby magnetization is transferred from H to X; it evolves for a time as X magnetization and is then reconverted by a reverse-magnetization transfer to the protons and detected with high sensitivity. Proper phase cycling ensured that only these transferred signals time-averaged coherently. The method would require X-nucleus decoupling during t_2 to obtain highest sensitivity. The disadvantages of the technique are the many pulses (10) required, which expose the spectral quality to pulse imperfections, and the T_2 attenuation of magnetization throughout the several delays. The same indirect technique was used to obtain the ^{199}Hg chemical shift in organomercurial phosphates [122].

The DEPT sequence has been modified [123, 124] to allow heteronuclear two-dimensional shift correlation. Use of different theta pulses can give phase-sensitive CH, CH$_2$, and CH$_3$ separate two-dimensional spectra, although a simple J-modulated spin–echo one-dimensional spectrum combined with a normal proton–carbon two-dimensional shift-correlation experiment usually provides the same information.

EXAMPLE 6   USE OF DECOUPLED MULTIPLET HETERONUCLEAR CHEMICAL SHIFT CORRELATION FOR SENSITIVITY ENHANCEMENT

As an illustration of the value of decoupled multiplets in heteronuclear chemical shift 2-D NMR, the ^{13}C-^{1}H spectra in Figure 5E.7 were run using identical conditions, except for the pulse sequence. The spectrum on the right (menthol in CDCl$_3$) resulted from the normal HETCOR pulse sequence, while the one on the left utilized the triplet of pulses midway through the evolution time [125]. The effect is most dramatic for a heavily split proton such as that near the center of the plot. Nonequivalent methylenes do, however, retain their geminal H—H splitting. A factor of 4–25 in time savings may be obtained; thus, the amount of material needed for the analysis is reduced.

The preceding heteronuclear shift correlation techniques produce spectra that have H—H spin multiplets in the F_1 dimension. These can be invaluable in certain situations where the pattern is obscured in the one-dimensional

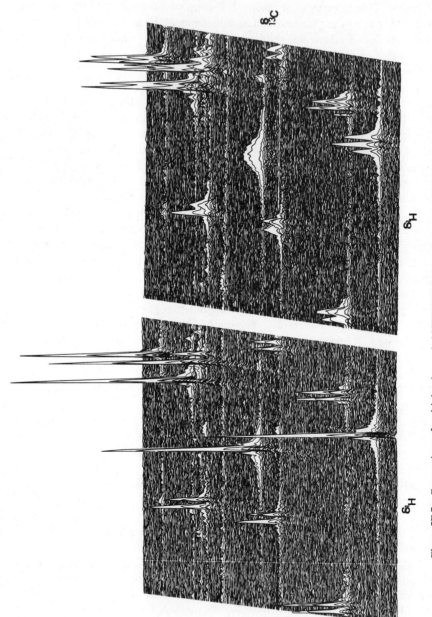

Figure 5E.7   Comparison of multiplet-decoupled HETCOR (left) with normal HETCOR spectrum.

spectrum. At other times sensitivity considerations would argue for collapsing these multiplets, thereby gaining at least a factor of 2 in intensity. In other cases, highly congested two-dimensional spectra may have overlapped H—C correlations. Bax [125] developed a method for collapsing these multiplets that essentially replaces the single X-nucleus 180° pulse at the midpoint of the evolution time with the element $90°(x, H)-1/2J-180°(x, H)180°(x, X)-1/2J-90°(-x, H)$ (Figure 5.23). The resultant two-dimensional shift correlation spectra are characterized by single peaks for X—H bonds. Slices in F_1 show proton singlets at the appropriate chemical shift. Of course, a projection onto the F_1 axis would give a proton "stick" spectrum.

These correlation methods rely on single-quantum coherence or magnetization transfer. Bolton [126] modified the process to detect only zero-quantum coherence transfer from protons to the observe nucleus. The frequency of this modulation is the frequency offset of the proton minus the frequency offset of the observe nucleus, both relative to their respective transmitters. The advantages of zero-quantum spectroscopy are the weaker dependence of F_1 multiplets on field inhomogeneity and the absence of heteronuclear XH coupling in the F_1 domain when only one proton is coupled to the observe nucleus. Experimental P—H zero-quantum correlation two-dimensional data show well-resolved HH coupling patterns for cytidine-3'-phosphate and phosphothreonine.

EXAMPLE 7   USE OF LONG-RANGE HETERONUCLEAR CHEMICAL SHIFT CORRELATION

Heteronuclear chemical shift correlation 2-D NMR can give valuable information concerning molecular neighborhoods. Here, the utility of this sequence is illustrated for cholesterol acetate in $CDCl_3$. In Figure 5E.8 the two-dimensional spectrum on the right is a normal HETCOR two-dimensional shift correlation spectrum with the proton one-dimensional spectrum inset for convenience. A portion of the long-range heteronuclear two-dimensional spectrum for $J = 7\,Hz$ is shown on the left. Note that the methyl protons at 0.62 ppm are correlated to four nearby carbon atoms. The other methyl protons show similar long-range correlations. These correlations provide very strong structural constraints. In combination with homonuclear two-dimensional shift correlation (COSY) spectra, these techniques permit *connectivities* to be established and local structural fragments to be constructed, even for completely unknown materials.

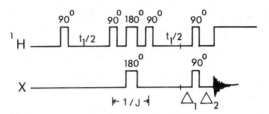

Figure 5.23 Multiplet-decoupled heteronuclear shift correlation. The normal heteronuclear shift correlation sequence (Figure 5.20) is modified by placing the 90°–t–180°–t–90° triplet coincident with the X-nucleus 180° pulse. This has the effect of removing HH couplings in the F_1 dimension. Reprinted, with permission from Varian Associates, Palo Alto, CA [125].

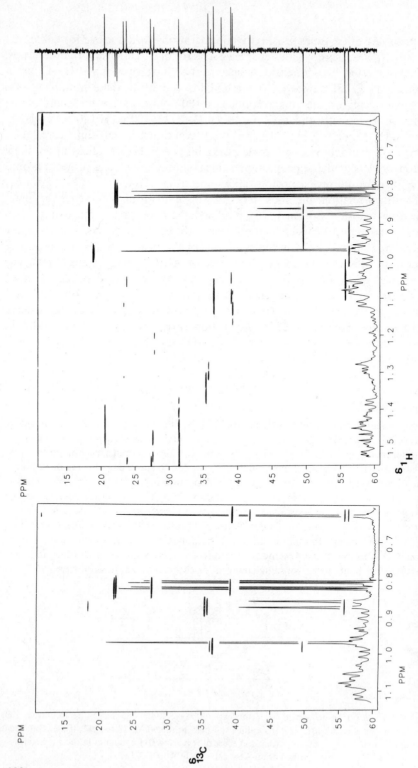

Figure 5E.8 Comparison of normal HETCOR (right) and long-range heteronuclear shift correlation (left) data for cholesterol acetate.

8   RELAYED COHERENCE TRANSFER TWO-DIMENSIONAL SHIFT CORRELATION

The foregoing shift correlation techniques are useful in relating *pairs* of nuclei. Any extension capable of relating at least three nuclei permits establishment of *connectivity* patterns that are invaluable in establishing structural assignments. For example, if in a propyl group the methyl carbon has been assigned, C—H shift correlation permits assignment of the methyl protons. Any technique that permits assignment of the adjacent methylene protons can then permit assignment of the methylene carbon by normal heteronuclear shift correlation. The chain may then be extended to the next methylene, and so on. The key is the connection between the protons and how to achieve it. Once this is made, all carbon atoms and protons, in principle, can be assigned and a molecular framework described.

This goal has been realized by the reports of Eich, Bodenhausen, and Ernst [127], and by Bolton and Bodenhausen [128] who have reported homonuclear and heteronuclear two-dimensional techniques for relaying coherence transfer between spins in a coupled network. This technique allows the observe X nucleus to produce correlations not only with bonded nuclei, but also with next-nearest and more remote neighboring protons. When this technique is combined with normal H—X shift correlation 2-D NMR, the desired connectivities can be made.

In the homonuclear case [127], the second $90°$ pulse of a homonuclear two-dimensional shift correlation $90°-t_1-90°$ pulse sequence is replaced with a $90°-t-180°-t-90°$ sequence. The first pulse of this triplet sequence may be thought of as mixing coherences or magnetizations within the spin network. These coherences then evolve (with the $180°$ pulse to refocus multiplets). Finally, the last pulse monitors the extent of distribution of these magnetizations among the various nuclei in the spin system. Incrementation of t_1 leads to the normal two-dimensional experiment. Any new correlations that appear, relative to the normal COSY experiment, indicate relayed coherence, or neighboring *non-coupled* spins.

The heteronuclear case [128] uses the same basic sequence as the homonuclear case except for the additional 180 and $90°$ observe-nucleus pulses. In this way the information concerning proton–proton couplings is transferred to the X-observe nucleus. For example, a specific carbon, could have two or three peaks in the RELAY two-dimensional spectrum, one from its bonded proton and the other one or two from protons two bonds away. Since in most cases only vicinal proton–proton couplings are resolved, only protons on the neighboring atom provide relayed intensity, which thus simplifies spectral analysis. Bolton [129] applied the sequence in the form of a 16-transient phase cycle to the C—H case for pentanol. One main virtue of the technique is that, as opposed to multiple-quantum two-dimensional connectivity experiments such as INADEQUATE [130], the RELAY experiment obtains signals from molecules in which there is only *one* ^{13}C. Apart from factors such as proton T_2 and relayed coherence-transfer efficiencies, the RELAY experiment should have a substantial sensitivity advantage. However, practical experience indicates that the heteronuclear RELAY two-dimensional experiment is significantly less sensitive than the ordinary C—H chemical shift correlation experiment [96, 131].

Bax reported [131] an improvement in the heteronuclear RELAY two-dimensional experiment where the number of transients per cycle is reduced to four and the sensitivity is increased by 40%. This can allow older spectrometers with limited phase-cycling capability to use RELAY 2-D NMR techniques.

Wagner [132] applied homonuclear RELAY to proteins, relying on the correlations of the NH proton to the β proton of the amino acid in cases of α-proton degeneracy. He reports the use of a $90°-t_1-90°-t'-180°-t'-90°-t''-180°-t''-90°-t_2$ multiple RELAY experiment, which gives NH correlations up to gamma protons.

Delsuc and colleagues [133] used RELAY in a form of H–X–H rather than the customary H–H–X. When this technique was applied as a test case to diethyl phosphite, cross peaks were observed between the methylene and phosphite protons, indicating a mutual coupling to the phosphorus (no cross peaks were noted when the phosphorus pulses were omitted). This can serve to match neighboring units within an oligonucleotide once one proton is assigned.

Kogler and colleagues [134] reported the use of *purging* pulses prior to the RELAY experiment to remove the intensities normally obtained by the directly bound proton. This *J filtering* is effective for one-bond J values from 125 to 225 Hz with suppression factors of up to 99%. The data is simplified and interpretations are easier for the more congested spectra.

Heteronuclear RELAY has also been applied to triglucoses by Bigler and colleagues [135] where they were able to assign all carbon resonances as well as prove the carbon connectivities in compounds whose proton spectra are exceedingly complex and difficult to assign or interpret.

9 MAGNETIZATION TRANSFER TWO-DIMENSIONAL NUCLEAR MAGNETIC RESONANCE

One of the more exciting aspects of 2-D NMR is the class of experiment that can probe intramolecular interactions, proximity of nuclei, and chemical exchange. In some of the previously mentioned two-dimensional experiments *transverse-*(x, y) magnetization transfer occurs; however, in some other experiments longitudinal (z) magnetization is transferred. The nuclear Overhauser effect is one example of population redistribution by way of an *incoherent* process. Other types include saturation transfer experiments where the mechanism is chemical exchange.

Alternate ways to study chemical exchange or NOE have included line shape analysis at or near coalescence and selective saturation or inversion with subsequent magnetization propagation throughout the spin system. The former technique can be very difficult for broad coalescence peaks and requires certain temperatures to be established. The latter is very useful for a limited number of lines but becomes time-consuming for more than a few lines and very difficult for closely spaced lines. The two-dimensional method requires no particular special conditions and therefore is attractive for studies of chemical and biochemical dynamics. This class of experiments was proposed by Ernst and colleagues [136, 137] and elaborated on by Macura and Ernst [138]. The basic pulse sequence is $90°-t_1-90°-t_{mix}-90°$. The second $90°$ pulse restores transverse magnetizations created by the first $90°$ pulse to the z axis. The magnitude and direction of the resultant z-axis magnetization is a function of the precession of the spins during t_1. Some components will be positive, some negative, and some zero. These components will oscillate as a function of t_1. The resultant z magnetizations will be in a nonequilibrium state and spin–lattice relaxation processes will facilitate attainment of equilibrium. Since proton relaxation is primarily dipolar and occurs by means of other protons, this nonequilibrium magnetization will then be redistributed by mutual spin flips to other protons. The final $90°$ pulse monitors the extent of magnetization transfer. The two-dimensional experiment samples all degrees of magnetization transfer as it increments t_1.

The two-dimensional shift correlation spectrum thus produced is character-ized by the usual diagonal peaks, which result from magnetization that remains at the same frequency in t_1 and t_2. Phase cycling can remove the axial peaks, just as in homonuclear shift correlation two-dimensional NMR. The off-diagonal peaks are the result of magnetization that has been transferred in a spin–lattice relaxation sense from one type of spin to another. This may be a result of true chemical exchange occurring during the mix time, or simply from spins close enough in space to provide mutual dipolar relaxation. This sequence has been used to indicate NOEs (in which case the experiment is termed NOESY) in polypeptides and proteins [139, 140]. It is particularly effective in macro-molecules where there is much slower molecular tumbling and more favorable NOEs. By varying t_{mix} the rates of magnetization transfer can be probed and thus the proximity of protons for NOE, or chemical kinetics for true exchange is measurable [141].

One of the serious problems with the preceding pulse sequence is that coherence transfer by means of J coupling can occur through zero- or multiple-quantum processes from the last two pulses in the sequence. These can produce so-called J cross peaks in the two-dimensional exchange spectrum [142]. This complication can make interpretation very unreliable or require unnecessary repetition of the experiment as a function of t_{mix}. Macura and colleagues [143] proposed a technique that takes advantage of the difference in behavior of the *incoherent* nature of magnetization transfer and the interfering zero- and multiple-quantum *coherent* processes. The latter two oscillate in time during t_{mix} while the NOE or chemical-exchange cross peaks grow and decay exponentially. Since the latter decays away slowly, a variation of the mix period through a range of values during the two-dimensional experiment will force the oscillatory coherences to appear at positions displaced from their normal positions along the F_1 direction, and the symmetry usually associated with the homonuclear correlation spectrum will be destroyed. Symmetrization about the diagonal will retain only symmetric data, thus eliminating J-correlation cross peaks. Choice of proper phase cycling [142] and use of a homospoil pulse in the mix period are also useful techniques to reduce undesirable zero- and multiple-quantum coherence transfers.

In some situations it may be desirable to keep a fixed mixing time. Macura and colleagues [144] modified the basic $90°-t_1-90°-t_{mix}-90°$ sequence by inserting a 180° pulse in the t_{mix} period. Its position is a function of t_1 and occurs at a time $[t_{mix} + kt_1]/2$. As t_1 increases, the 180° pulse moves from halfway through the mix period to some time later. Zero- and multiple-quantum J-coherence transfer peaks are shifted again in F_1, and symmetrization eliminates them efficiently.

Some of the more interesting problems of chemical exchange arise in biochemical areas, particularly proton exchange in water. Schwartz and Cutnell [145] used the two-dimensional NOE approach to study NH exchange with solvent protons in water. Here the large solvent resonance makes the normal *hard-pulse* version inapplicable because of the 100,000:1 dynamic range between solute and solvent protons. These workers have replaced the final monitoring 90° pulse with a *long-soft* pulse that produces a null in excitation at the water resonance. Because the last pulse suppresses the water signal, cross peaks corresponding to NH to water proton-magnetization transfer do not appear. Of course, this means that symmetrization techniques will fail if used here. Since the cross peaks for *intermolecular* chemical exchange cannot arise from J correlations, the lack of symmetrization is not as critical as in NOE two-dimensional studies where spin-coupled nuclei may also participate in cross relaxation.

Two-dimensional ^{31}P magnetization transfer techniques have been applied to the study of metabolism in rats [146, 147], in particular the enzyme-catalyzed exchange of the terminal phosphate in ATP with creatine phosphate. The same techniques allowed confirmation of isomer interconversion in a pentapeptide that exists in the form of several isomers because of the presence of two prolines [110]. Here, NH protons from alanine and leucine appear in the form of major (60%) and minor (30%) isomers, the remainder having nonresolved shifts. The two-dimensional NOE spectrum revealed cross peaks between major and minor leucine NH protons, as well as cross peaks for the alanine major–minor isomer interconversion (see Figure 5E.9).

EXAMPLE 8   USE OF TWO-DIMENSIONAL NUCLEAR MAGNETIC RESONANCE TO STUDY CHEMICAL EXCHANGE

Magnetization transfer 2-D NMR has been used primarily to establish NOEs between protons for structural and/or stereochemical studies. In other cases it can be used to demonstrate magnetization transfer by chemical exchange. Here, cross peaks between NH protons in a polypeptide do not indicate the presence of NOE since the relevant cross peaks connect signals from protons in different molecules!

In Figure 5E.9, the four downfield doublets are alanine and leucine NHs from the major and minor isomers. The full plot shows both shift correlation (lower left triangle) and magnetization transfer two-dimensional (upper right triangle). The inset box shows the latter data for the NH region. Note that the cross peaks are for alanine major–minor

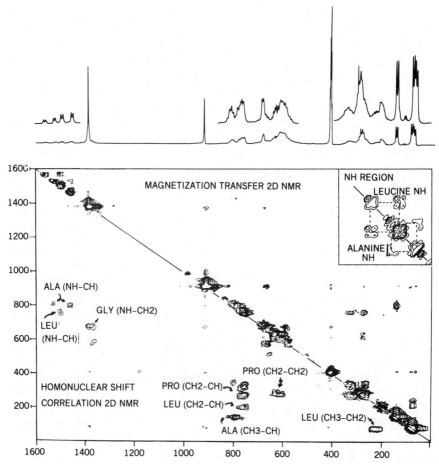

Figure 5E.9   Homonuclear shift correlation (lower triangle) and magnetization transfer correlation (upper triangle) two-dimensional data for a polypeptide. The inset box shows the full magnetization transfer data for the NH region. The cross peaks show chemical exchange in the form of isomer interconversion.

and leucine major–minor pairs. The cross peaks demonstrate the dynamic interconversion of one isomer into the other, in equilibrium, during the mixing period. Since the experiment requires no prior choices, except for the mix time, based on the molecule studied, this experiment has great value in confirming or determining the presence of chemical exchange.

Huang and colleagues [148] used these techniques to explore ^{13}C chemical-exchange networks. Complications from J correlations are absent from these natural-abundance spectra. Normal $90°-t_1-90°-t_{mix}-90°$ sequences were used in the presence of proton decoupling (for NOE, sensitivity, and simplicity). Refocused INEPT was also used in place of the first $90°$ pulse to prepare the initial magnetization and to gain additional sensitivity. This extension suggests future flexibility and selectivity since the INEPT portion of the sequence can be tailored for specific J values (long range or direct), thus allowing very precise control over the site from which magnetization can evolve. Huang and colleagues [148] applied two-dimensional strategies to the classic exchange problems of ring-motion in decalin, bond shift in bullvalene, and solvation shell exchange in aluminum complexes. The longer relaxation times of carbon-13 can actually be advantageous in these studies since they permit longer mix times for slower exchange processes. The other major advantage of the two-dimensional technique is that it can be performed on a very slowly exchanging system whose lines are still narrow.

Rinaldi and Baldwin [149] described a new heteronuclear version of the two-dimensional magnetization transfer experiment. The $90°-t-90°-t_{mix}$ portion of the normal sequence is retained, with an X-(observe) nucleus 180° pulse midway through t_1 to decouple the X and H spins. Cross relaxation between X and H occurs during the mix period followed by a 90° X pulse, and the magnetization is sampled under full proton decoupling. This signal is modulated lated according to the amount of cross relaxation involved during the mix period. Cross peaks in the two-dimensional spectrum reflect the existence of this cross relaxation. Typically, this is dipolar in nature. Note that J coupling is not necessary between the correlated nuclei. Existence of cross peaks in sterochemically rigid systems, particularly for quaternary carbon atoms, can permit local neighborhoods to be mapped if dipolar relaxation is present. Rinaldi and Baldwin [149] point out the value of such an experiment for probing metal in binding sites in complexes with biological or organic ligands.

As molecular systems become more complex and as the chemical shift difference between mutually relaxed or exchanging spins becomes smaller, the presence of the large *diagonal* peaks in the two-dimensional plot becomes more of a problem. Bodenhausen and Ernst [150] developed difference two-dimensional experiments that cancel the diagonal ridge and leave the cross peaks unaffected. The two-dimensional experiment is first performed in the normal manner (A), then with zero mix time (B), and finally in one of two possible ways: (C) all nuclei are saturated by a train of 90° pulses followed first by a mix period and then by (B), or (D) all nuclei are inverted by a 180° pulse followed first by a mix period and then by (B). The saturation-recovery difference technique requires combination of signals as $A − B + C$, which gives a vanishing diagonal. The inversion-recovery difference technique combines the data as $A − B/2 + D/2$, and the diagonal is again nulled. Normally, eight-transient phase cycling and incremented mix times [142] are necessary to eliminate J cross peaks; therefore, if a sufficiently complex pulse programmer is also available, these techniques may be combined into one pulse sequence to produce directly a spectrum without a diagonal.

Another combination experiment is a combined COSY–NOESY pulse sequence in which the normal homonuclear shift correlation acquisition is followed by a third 90° pulse and a separate acquisition [151, 152]. In this, the first acquisition period serves as the mixing period. This places a minimum on the length of the mix period and could rule out the use of the preceding techniques [142, 143] to discriminate against J-correlation peaks. If the mix period is short enough, however, an additional delay can be placed prior to the last pulse and used as above to discriminate against J cross peaks in the NOESY portion of the data. The technique offers an economy of time since the full NOESY data set can be acquired with only a small percentage of extra time.

Finally, the need to maximize efficiency in determination of exchange rates has led Bodenhausen and Ernst [153] to invent the *Accordian* approach. The basic pulse sequence is $90-t_1-90-kt_1-90$, where k is a constant ranging from 10 to 100. Thus, the evolution time and mix times both expand in a regular fashion. Information on rates is present in the *line shapes* of the cross peaks. The phase-sensitive F_1 line shape of a cross peak can be subjected to direct line-shape analysis as a superposition of Lorentzian lines, each producing a rate constant that will determine its width. Alternately, the line shape can be subjected to a reverse Fourier transformation for the same purpose. Another technique is to take sums and differences of two F_1 slices at the two chemical shifts for exchanging spins. These linear combinations produce pure Lorentzians. The line widths at half-height of the signals in the sum are direct measures of the relaxation rate. The corresponding difference line widths can be used to calculate directly the chemical exchange rate constants. Generally this last method can be used most readily in spectrometers that can accommodate the normal add–substract software.

10 MULTIPLE-QUANTUM ONE- AND TWO-DIMENSIONAL METHODS

Aside from a relatively small group of those interested in the spin physics of systems subjected to a train of radio frequency pulses, few others thought much about multiple-quantum processes in NMR prior to 1980 [154]. The field seemed esoteric and unrelated to *real* chemical problems.

Things have indeed changed! One of the most exciting *general* techniques that has emerged since then is double-quantum coherence NMR as applied to ^{13}C. Bax and colleagues [130, 155] defined an elegant technique for establishing carbon–carbon connectivity information that, in its one-dimensional version, allowed the suppression of monolabeled ^{13}C resonances while retaining the signals from dilabeled molecules (they term this technique INADEQUATE, for incredible natural abundance double-quantum experiment). By matching equal values of J_{CC} throughout the spectrum, connectivities *within the molecular backbone* can be established. The basic pulse sequence of $90°-t-180°-t-90°-T-90°$ uses the first two 90° pulses to generate multiple-quantum coherence. This evolves for a short time and is reconverted back into single-quantum magnetization by the last 90° pulse. Only these signals survive the 32 transient phase cycling in the experiment. Suppression factors of up to 1000 are realized, revealing even long-range CC coupling normally hidden by the large mono-labeled uncoupled centerband. The experiment is very demanding on both sensitivity and spectrometer stability since the signals are 0.01% in natural abundance and are further split by CC coupling. Still, the information is direct and can unequivocally link carbon atoms. Automated least-squares analyses [156] can reduce the time required for matching all possible pairs of equal couplings. These can also produce isotope shifts that may have structural utility. The method does require an estimate of average CC couplings. This can be obviated by using a two-dimensional technique that treats the period between the first two 90° pulses as the evolution period in a two-dimensional experiment [157]. Slices in the resultant F_1 dimension at the carbon chemical shifts in F_2 give J spectra with inhomogeneities handled just as in the analogous J-resolved homonuclear and heteronuclear experiments described earlier. Phase-sensitive plots give very highly resolved multiplets with couplings measured to ± 0.03 Hz or better.

These methods still rely on establishing a carbon–carbon connectivity based on identical J_{CC} values. In molecules where there are many similar couplings this method can fail. Fortunately, extension of the pulse sequence to sample *directly* the double-quantum coherence can remove this problem. Bax and colleagues [158] introduced this method, which uses the same pulse sequence but treats the period between the last two 90° pulses as the evolution period. The double-quantum coherence precesses during this period with a frequency characteristic of the *sum* of the chemical shifts of coupled CC pairs. Thus F_1 becomes a double-quantum frequency axis in the two-dimensional data. Both carbon atoms carry the modulation in t_1 and thus have intensity at the same F_1 value. This coincidence of intensity at a common value of F_1 is direct proof of the existence of a coupling between the carbons. The size of the coupling can readily differentiate between bonded carbons and long-range coupled carbon atoms (the period between the first two 90° is normally set to produce optimum double-quantum coherence for bonded, strongly coupled carbon atoms). The data is most conveniently plotted as a F_1/F_2 contour intensity plot. The F_1 signals comprise two doublets symmetrically disposed about a line of slope 2 in frequency units. This is of great help in assigning peaks in noisy spectra since the constraints are several and the F_2 values are already available from the one-dimensional spectrum. The method was further improved to permit quadrature detection in both domains [159]. Half the data were obtained in the normal manner, but a novel 45° (z) composite pulse was inserted in the evolution period for the other half of the data. This pulse induces a 90° phase shift in the double-quantum coherence, which permits quadrature detection. Mareci and Freeman [160] achieved quadrature detection in F_1 by a simpler method based on selective detection of the coherence-transfer echo. The only change is to increase the last pulse in the INADEQUATE sequence to a 135° pulse. Antiecho responses are greatly attenuated and identical spectral widths can be used both in F_1 and F_2.

The powerful property of this general experiment is the ability to trace out the connectivity pattern in the molecular backbone (apart from heteroatom interruptions). Connectivity can be established by beginning at some F_2 chemical shift value, finding any intensities along F_1 at this F_2 value, and pairing these intensities with any corresponding intensities sharing the same F_1 value. This establishes all connectivities of the carbon at this value of F_2. Repeating the process for the now-identified carbon atoms establishes their connectivities. In this manner the full connectivity pattern can be analyzed.

The inherent low sensitivity of this experiment can be helped prior to the INADEQUATE part of the sequence by combining polarization transfer techniques such as INEPT [161]. These experiments do show an increase in observed sensitivity, but most of the pulses do place severe demands on spectrometer stability and relaxation properties of both protons and carbons.

Turner proposed a somewhat different approach to detecting carbon–carbon connectivities [162–164]. Instead of the preceding $90°-t-180°-t-90°-t_1-90°-t_2$ pulse sequence, Turner proposes a Hahn echo method that involves a $90°-t-180°-t-90°-90°-t_1/2-90°-t_1/2-t_2$ pulse sequence. The resulting spectrum is similar in format to the homonuclear shift correlation spectrum with a principal diagonal (no diagonal peaks, however—these are suppressed by phase cycling) about which correlated (bonded) carbon atoms are symmetrically displaced. Turner [163, 164] analyzed sensitivity considerations and compared the various two-dimensional methods of obtaining carbon–carbon connectivity information; however, several of his conclusions were challenged by Bax and Mareci [165].

The one-dimensional version of INADEQUATE has been used to establish connectivities in patchoulol (a 14-carbon polycyclic alcohol) without the use of model compounds [166]. It has also been used to examine all glycolysis metabolites [167] and stereochemistry of simple sugars [168] and conformation analysis by means of long-range CC coupling constants in n-octanol [169].

The detection of carbon–carbon connectivities by using a two-dimensional method has been reported for monensin sodium salt [170], a trimer of biacetyl [171], a biosynthetically ^{13}C-labeled riboflavin [172], and an organochromium complex [173].

Schmitt and Gunther [174] modified the original INADEQUATE one-dimensional pulse sequence to allow selection of the double-quantum evolution time to suppress one doublet selectively. This need arises from the problem of having two carbons of very similar shift, coupled to two other carbons with similar couplings. When the transmitter is placed in the center of one of the AX doublets, a double-quantum frequency difference between the centers of both doublets results. Since the two pairs are 90° out of phase, an INADEQUATE double-quantum frequency evolution period equal to one quarter the double-quantum frequency allows us to discriminate between the pairs; however, this is possible only if the assignment is correct. When pairs are matched with observed spectra, assignments can be verified, as they were for C-6 and C-7 of 2-fluorobiphenylene [174].

Levitt and Ernst [175] applied the methods of composite [176] (or sandwich) pulses to the INADEQUATE experiment. They found this modification to be essential for optimum center peak suppression, principally to correct for off-resonance effects. They also cover the same correction to the DEPT and other arbitrary sequences.

EXAMPLE 9   USE OF TWO-DIMENSIONAL INCREDIBLE NATURAL
ABUNDANCE DOUBLE-QUANTUM EXPERIMENT FOR STRUCTURAL
ANALYSIS

The INADEQUATE pulse sequence utilizes double-quantum coherence effects to detect the ^{13}C signals from pairs of coupled ^{13}C nuclei in natural abundance, while it suppresses the signals from the ^{13}C nuclei in molecules with only one ^{13}C. The timing of the sequence determines the coupling constant J_{CC} for which optimum signals are obtained; and if this is set for J_{CC} in the ranges from 35–70 Hz, only molecules with *directly bonded* pairs of ^{13}C nuclei are observed. When all of these signals fall on a one-dimensional frequency axes, it is nearly impossible to sort them out; however, by introducing an incremental delay between the creation of double-quantum coherence and its reconversion to observable precessing nuclear magnetization, it is possible to produce a two-dimensional display that greatly simplifies this task. The establishment of some (or in favorable cases, all) of the carbon–carbon bonds in a molecule provides very powerful constraints on the possible types of structures and is of great value to the chemist concerned with the elucidation of unknown structures of complex natural products. This application is illustrated by the following problem.

A new compound named marmelerin, isolated from *Croton sonderianus*, was subjected to mass spectral and NMR analysis, which yielded a molecular formula $C_1{}^5H_1{}^8O$. Although ^1H and ^{13}C NMR data allowed the structure to be narrowed down to two choices, I or II

a decision could not be reached on the strength of the available data. It was clear, however, that if the carbon–carbon connectivities could be established by the two-dimensional double-quantum coherence experiment (CCC2D), the correct choice could be made, and all other features of the structure would be confirmed.

The only problem encountered was the long T_1 values associated with the nonprotonated carbons, which fell in the 20–30-s range. A few milligrams of chromium acetyl acetonate were added to the solution of 250 mg marmelerin in 1.0 mL $CDCL_3$ to reduce the T_1 values into the 1–2-s range. An overnight acquisition of data was then sufficient to yield the connectivity plot shown in Figure 5E.10.

Interpretation of the connectivity plot is straightforward, based on the rule that a carbon–carbon bond is established when peaks in the two-dimensional plot are found that meet the following three conditions (1) there are two doublets with the same double-quantum frequency, that is, falling on the same horizontal level in the plot; (2) both doublets are centered around frequencies of the single lines corresponding to two carbon atoms in the standard broadband-decoupled spectrum; (3) the pairs of doublets with the same double-quantum frequency are also symmetrically located about one diagonal line through the entire data set. The peaks are simply numbered in sequence (in this case, 1–15) and their multiplicities, determined by an APT or DEPT experiment, are noted to specify the number of protons bonded to each carbon atom.

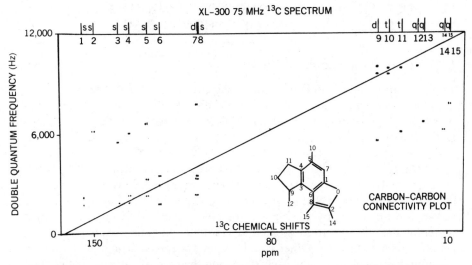

Figure 5E.10 Carbon–carbon connectivity plot obtained from a double-quantum two-dimensional experiment. Carbon atoms at different chemical shifts that have common pairs of doublets at the same value of the vertical (double-quantum) axis are bonded.

Starting with the unique sp^3 carbon with one proton (the CH carbon numbered 9), the connectivities 9–10, 10–11, 11–4, 4–3, and 3–9 are apparent and lead to the moiety

The connectivities 4–5, 5–7, 7–1, 1–6, and 6–8 expand this moiety to

Carbon atoms 1 through 8 are clearly sp^2 carbon atoms, 1 and 2 have chemical shifts typical of aromatic carbon atoms bonded to oxygen, and 7 and 8 show the upfield shift characteristic of oxygen in the beta position. This gives the skeleton

The methyl groups are found to display the connectivities 12–9, 13–5, 14–2, 15–8 leading to structure **I** even though the connectivities 3–6 and 2–8 are not observed directly.

11   MULTIPLE-QUANTUM ONE- AND TWO-DIMENSIONAL METHODS: SPECTRAL SIMPLIFICATION AND EDITING

11.1   One-Dimensional Methods

The generation of multiple-quantum coherence and its subsequent indirect detection has been an active area in chemical physics. Rather than attempting to cover this topic comprehensively, in this section we describe some of the methods that have been introduced and may be of general use in some situations (see [154] for a general review).

Several reports have been published that propose the utilization of multiple-quantum coherence for spectral simplification and editing. Hore and co-workers [177–179] applied the INADEQUATE pulse sequence to proton spectra. They accomplish this [177] by placing the observe nucleus transmitter in the exact center of the AX spectrum so that the double-quantum frequency for the coupled AX pair is zero. Other coupled spin systems will now have double-quantum frequencies that will oscillate during the period between the two final 90° pulses. By coadding a number of spectra generated by variation through a range of double-quantum evolution times, those coupled multiplets with nonzero double-quantum frequencies will destructively add, leaving only the AX spin system spectrum. A general knowledge of the chemical shifts of A, X, and $J(AX)$ is necessary, although this may be refined by a trial spectrum. Hore and colleagues [177] term this experiment DOUBTFUL (double-quantum transitions for finding unresolved lines). The same technique was applied to a 14-base-pair nucleotide [178]. In this case different spectra were generated from a coupled aromatic AX system by systematically varying the transmitter frequency to bring different AX pairs into the zero double-quantum frequency condition. All eight cytosine H-5, H-6 pairings were found. The extension of the method to triple-quantum coherence was demonstrated [179] for the globular peptide *E. coli lac* repressor headpiece. This experiment picks out isolated three-spin systems in proton NMR spectra that are severely overlapped. By placing the transmitter at the average chemical shift of the ABX spins, it is possible for only these signals to survive the coadding.

Sorensen and colleagues [180] addressed the problem of uniform excitation of multiple-quantum coherence with the goal of producing multiple-quantum filtered spectra. The normal INADEQUATE sequence of pulses is augmented by an additional $t–180–t$ segment, hence mirroring the first part of the sequence. The pulse sequence is thus symmetrical about t_1. By coadding spectra using a range of t values, excellent spin filtering is obtained and the resultant spectra are able to be phased and have proper amplitudes.

An experiment has been reported that can produce "edited" spectra by carbon type with low dependence on the spread of J values [181, 182]. In its initial version it uses the same basic pulse sequence as the "proton-flip" version of heteronuclear 2-D J NMR (Figure 5.24). After a delay for relaxation and NOE build-up, an X-nucleus 90° (x) pulse places magnetization in the transverse

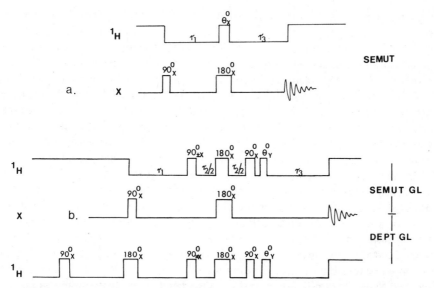

Figure 5.24   SEMUT pulse sequences (a) collection of four separate fid's for $\theta = 0, 180, 60$, and $120°$ allow spectral editing of ^{13}C data into C, CH, CH$_2$, and CH$_3$ spectra with very low dependence on J; (b) SEMUT GL improves the purity of spectral editing; in the DEPT GL version, it allows sensitivity enhancement by means of polarization transfer. Adapted, with permission, from Varian Associates, Palo Alto, CA [181, 182].

plane. It evolves with the decoupler off for a time $2t$ with a $180°$ (x) refocusing pulse at time t along with a theta (x) proton pulse. The decoupler is turned on for broadband irradiation at time $2t$, and the X-nucleus fid is obtained. Theta values of 0, 180, 60, and $120°$ (Figure 5.25) are used to provide four spectra. These may be combined to form CH$_3$–CH and CH$_2$–C subspectra or the pure C–CH–CH$_2$–CH$_3$ subspectra. A spread in J values leads to mixing of one subspectrum *down* only into the next, not *up*; that is, J *cross-talk* occurs from a CH$_n$ to a CH$_{(n-i)}$ only and not to CH$_{(n+i)}$, and is greatest for CH$_{(n-1)}$. The degree of

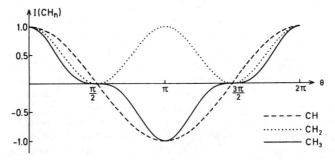

Figure 5.25   Flip-angle dependence of ^{13}C intensities in SEMUT. The fixed delay is set to $1/2J$. By combining data sets for different theta values spectral editing is possible.

separation is shown to be equivalent to that obtained by DEPT subspectral editing. This technique, called SEMUT (subspectral editing using a multiple-quantum trap), has the advantage over DEPT of using fewer pulses and no phase shifting. It is also shorter and therefore less susceptible to cases involving short-proton or X-nucleus T_2 values. It also detects the quaternary X-nucleus signals. It can be at a disadvantage with respect to DEPT or INEPT in those cases of long X-nucleus T_1 and short-proton T_1, since the latter determines the repetition rate of the polarization transfer experiments. The collection and subsequent sorting of SEMUT raw data can be automated easily, and direct edited spectral presentation for ease of interpretation is then possible.

In the later SEMUT GL version [182] a proton $(90°-t_2-180°-t_2-90°)$ *purging sandwich* is inserted after the first t period. The X-nucleus 180° pulse is centered with respect to this *sandwich*. Following this segment the regular theta pulse on the protons occurs with delay t, as in the normal SEMUT sequence. Proper adjustment of the sequence can reduce the level of J cross talk to less than 2% of the proper edited peaks. Sorensen and colleagues [182] also proposed extension of the "purging sandwich" to DEPT for better edited spectra. Note the familial relationship to the heteronuclear two-dimensional J- and shift-correlation pulse sequences of Bax [78, 125] and Rutar [79], which use similar pulse sandwiches.

Other one-dimensional editing techniques have been proposed by Bulsing and colleagues [183] that require the use of other than the four-quadrature phases. They replace the theta pulse in DEPT with a $90°(x)90°(q)$ pulse pair. Variation of the q phase and proper phase cycling can produce DEPT spectra with reduced J cross talk.

Sorensen and Ernst [184] reported use of a sign-labeled polarization-transfer version of INADEQUATE that can permit determination of signs of CC coupling constants with the advantages of sensitivity that accompanies polarization transfer.

EXAMPLE 10   SPECTRAL EDITING OF CHOLESTEROL ACETATE WITH THE SUBSPECTRAL EDITING USING A MULTIPLE-QUANTUM TRAP PULSE SEQUENCE

The very simple SEMUT sequence does not require a phase-shift capability in the proton decoupler channel and, therefore, can be implemented readily on older FT NMR spectrometers. When spectral editing with low cross talk between C, CH, CH_2, and CH_3 subspectra are the objective, SEMUT is superior to APT by virtue of its lower dependence on the value of J_{CH}.

Oxygen substituents such as hydroxyl or acetoxyl groups tend to raise J_{CH} about 10%, from around 128 to 140 Hz. In the APT sequence, which inverts CH carbon signals when the decoupler downtime is set to $1/J$, the same degree of inversion is not achieved for the CH and CH_3 groups bonded to carbon as for those bonded to oxygen.

In this example, τ was set at $1/2J$ for $J = 130$ Hz. Spectra were run with the proton flip angle of 0 and 180°. The 0° pulse gave a normal spectrum (trace *a*) with all carbon peaks positive, while the 180° pulse inverted the CH and CH_3 carbon atoms (trace *b*). When spectra a and b are added, CH and CH_3 are canceled, and only C and CH_2 are left; however, when b is subtracted from a, only CH and CH_3 are left. Figure 5E.11 shows the

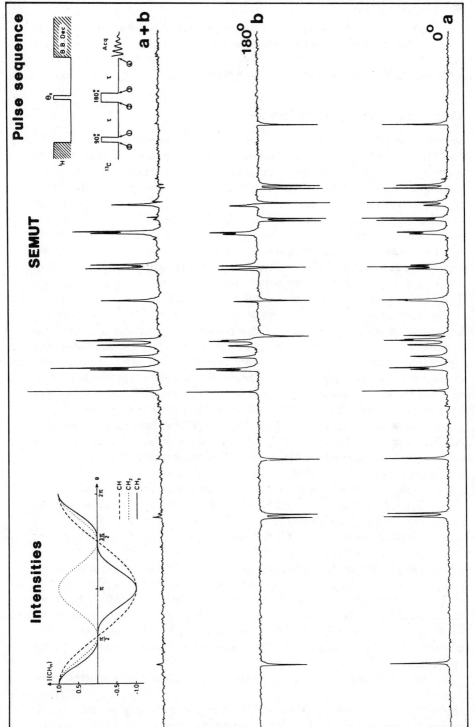

Figure 5E.11 SEMUT ^{13}C spectra of cholesterol acetate.

0 and 180° spectra and their sum. All CH and CH_3 peaks cancel equally well. With $J_{CH} = 140\,Hz$, the peak at 74 ppm would have exceeded its maximum negative value in trace b if the APT were used and would be recorded as a positive peak in the sum spectrum, a + b, which might confuse its assignment.

11.2   Two-Dimensional Methods

Several two-dimensional multiple-quantum techniques have been introduced [185–189]. Multiple-quantum filtering is the use of multiple-quantum coherence (MQC) to simplify homonuclear two-dimensional chemical shift-correlated NMR spectra [185]. Proper selection of the *order* of the MQC present between the evolution and detection periods can eliminate singlets and higher order multiplets if desired. The pulse sequence $90°-t_1-90°-90°-t_2$ has phase cycling programmed for the desired order of MQC that is present between the last two 90° pulses. Singlets cannot produce MQC and thus are canceled. Diagonal peaks are also greatly attenuated. Higher order filtering can allow selection of certain spin systems with other overlapping resonances canceling. This can be of great value in macromolecular spectra.

The same approach has also been separately reported [186] and applied to four-quantum filtering. All peaks with a homonuclear two-dimensional shift-correlated spectrum involving two- and three-spin systems are eliminated. For both cases, phase shifters capable of intermediate values are necessary. One-dimensional spectra using multiple-quantum filtering were reported in both as well, and the same pulse sequences with fixed time were used. Lower order spin patterns are eliminated from the resultant one-dimensional spectrum.

The two-dimensional version of INADEQUATE has been applied to proton NMR. This is complicated by the presence of three or more coupled spins as opposed to the simpler AX or AB case for ^{13}C in natural abundance. Mareci and Freeman [187] showed that the two-dimensional spectrum can be simplified by using a value between 90 and 180° for the flip angle of the last pulse. As in INADEQUATE, coupled protons have intensity at the same double-quantum frequency, symmetrically displaced about a diagonal of slope 2. There are no strong diagonal peaks as in COSY spectra to complicate proton–proton correlations for protons of similar shift.

Boyd and colleagues [188] applied the technique to proteins, with particular emphasis on the characterization of tyrosine, tryptophane, and phenylalanine residues. Kessler and colleagues [189] used spin filtering to obtain a homonuclear two-dimensional J spectrum in which the singlets were suppressed. Intensity distortions and difficulties associated with variation of excitation efficiency and J were avoided by using the symmetric excitation/recursion cycles of Sorensen and colleagues [180].

Bolton [190] used zero-quantum coherence to obtain heteronuclear two-dimensional shift-correlation spectra. The pulse sequence, $90°(H)-t-180°(H)180°(X)-t-90°(H)90°(X)-t_1-90°(H)$, has phase cycling so that the X signals are modulated only by the zero-quantum coherence as a function of t_1.

Proton–proton couplings are present in the resultant two-dimensional spectra and slices perpendicular to F_2 can give proton spectra of protons coupled to X. Line widths are predicted to be only 60% of those obtained in the conventional heteronuclear two-dimensional shift correlation experiment, and this is useful for extracting the proton–proton fine structure.

Bax and colleagues [191–193] developed a technique for obtaining heteronuclear shift correlations by way of zero-quantum and double-quantum coherence where, for higher sensitivity, detection is on the protons. The first part of the sequence converts all longitudinal proton magnetization coupled to an X spin into zero- and double-quantum coherences. These are converted back into transverse magnetization after an evolution period. Fewer pulses are involved than the double INEPT technique of Bodenhausen and Reuben [121], and proper phase cycling can allow use of the technique in protonated solvents. Because of the higher sensitivity t, ^{15}N—H shift correlations were obtained on 18-mM solutions in 5-mm tubes in 3 h.

EXAMPLE 11 ^{15}N SENSITIVITY ENHANCEMENT USING MULTIPLE-QUANTUM TWO-DIMENSIONAL NUCLEAR MAGNETIC RESONANCE

Indirect detection of rare-spin nuclei has been used for many years to overcome the intrinsically low sensitivity of nuclei such as ^{15}N. This is especially critical and difficult for biologically relevant problems where exchangeable protons are often of interest and the solvent of choice is H_2O. Indirect detection usually involves additions of many separate transients, and the only surviving signals are those from protons coupled to the rare spin. Spin–echo differences or reversed INEPT or DEPT allow detection of the protons coupled to the rare spin but do not give information of the rare-spin chemical shift.

An elegant use of multiple-quantum 2-D NMR for addressing these problems has been reported by Bax and colleagues [191, 192] and Griffey and colleagues [193]. Their pulse sequence converts all proton magnetization of protons coupled directly to the rare-spin into equal amounts of heteronuclear zero- and double-quantum coherence. These coherences evolve during t_1 at frequencies given by the difference and sum of the two chemical shifts as measured with respect to their transmitters. Therefore, a two-dimensional experiment encodes the rare-spin chemical shift value and reveals it in the F_1 dimension.

This experiment has been used to examine ^{15}N resonances of 15N-enriched uridine in 1 mM t-RNA(met) in 90% H_2O/10% D_2O.† The data in Figure 5E.12 were obtained in 9.6 h. The imino protons have resonances far downfield in the proton spectrum, here spanning over about 7 ppm. The two-dimensional contour plot gives the double-quantum form of the data. The F_1 projection can be used to obtain the ^{15}N chemical shift for each ^{15}N, as well as to provide the linkage, and therefore the assignment, to their bonded protons.

Bolton also extended multiple-quantum techniques to the RELAY experiment [194]. This overcomes the problem encountered when neighboring and remote protons are degenerate or nearly so.

†A. Zens and D. Foxall, Varian Associates, Palo Alto, CA (unpublished). Sixty-four fid's were taken; 2048 transients each with a repetition rate of 0.264 s on an XL-400 using a 5-mm tube. The sample used was courtesy of D. Poulter, University of Utah, Salt Lake City, UT.

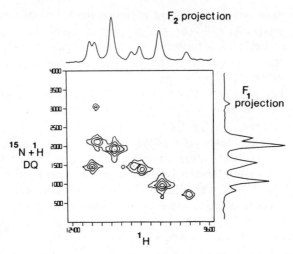

Figure 5E.12 Multiple-quantum two-dimensional spectrum of ^{15}N-enriched uridine in $1\,mM$ t-RNA. Each peak corresponds to a different ^{15}N in the polynucleotide.

Similar information has been obtained by Sorensen and Ernst [195] using *pseudomultiple-quantum* techniques. After an initial 90° and halfway through the evolution time, another 90° pulse on the protons generates sum and difference frequencies for the coupled protons (these may later be sorted out by proper phase cycling into pseudozero- and double-quantum data). The resulting magnetizations are then transferred by a DEPT cycle to the directly connected carbon atoms. No fixed delay to rephase homonuclear couplings is necessary. Signals for both neighboring and remote protons are present for each carbon, as in the basic RELAY experiment.

Several theoretical treatments of multiple-quantum NMR have appeared recently [196–198].

12   SUPPRESSION OF UNWANTED NUCLEAR MAGNETIC RESONANCE SIGNALS

One of the most common problems faced by the NMR spectroscopist is the presence of unwanted resonances in the NMR spectrum. These may arise from solvent or from the analytical sample. In extreme cases; for example, when water is used as a solvent for proton NMR, the solvent may totally determine the operational approach.

The tactics used in these situations can be divided into two main groups: those suitable when a single strong resonance is present; and those applicable to the remaining, more general cases when one (or more) strong resonance(s) appears. Several avenues have been explored for the former. Suppression factors of several hundred have been achieved by simply saturating the unwanted resonance prior to the monitoring pulse. This works extremely well when there is no exchange of magnetization between the irradiated spins and those that are to be observed. The WEFT (water-eliminated Fourier transform) technique [199] minimizes solvent resonances by using the differential in T_1 between small solvent molecules and the usually larger solute molecules. After an initial 180° pulse, all spins are inverted. During a delay period the solute nuclei repolarize (assuming they have much shorter T_1 values). When the solvent nuclei go through zero z magnetization, a 90° pulse is applied, and only the equilibrium magnetization of the solute nuclei is detected.

More sophisticated techniques have been developed that do not generate strong solvent signals [200, 201]. Redfield's 2-1-4 weak-pulse sequence [200] generates an observing pulse with an excitation function that falls off precipitously at an offset frequency coincident with the solvent resonance. Analytical peak amplitudes are highly nonlinear and have large phase corrections. The effect is felt even several parts per million away from the null point, and the technique is still practical for reducing strong signals even though these may cover a wide region, for example, 600–900 Hz in protons. Since the response function maximizes at the transmitter offset, nulls at the solvent, and falls off steeply on the other side of the solvent, lines on the other side of the solvent are attenuated.

Plateau and Gueron [201] developed a sequence that substantially reduces the baseline distortions normally accompanying the Redfield sequence and also allows examination of the whole spectrum. The sequence, $90°(Y)-t-90°(-Y)$, termed *jump and return* cancels the solvent (single) line to first order and provides strictly constant phase across the spectrum, with the exception of a 180° phase change at the solvent frequency. The sequence uses strong 90° pulses, and the transmitter must be set at the solvent frequency. The technique has an elegance of simplicity and achieves essentially the same suppression as Redfield's technique: The first pulse brings the z magnetization into the transverse plane. Off-resonance spins precess in opposite directions at a rate governed by their displacement from the transmitter (in the rotating frame of reference). After a short time t of about 125 μs, the second 90° pulse, phase shifted by 180°, restores the solvent magnetization back to the $+z$ axis. As it is on-resonance, the solvent magnetization undergoes *no precession* during t. Spins that have precessed during t will generate a signal during the acquisition period immediately following the second 90° pulse. The setting of the 90° pulse width is not exceedingly critical since its effect is reversed by the pairing of antiphase pulses. Only the sine amplitude distortions remain, which in principle can be corrected empirically. Water suppression by a factor of 500 has been achieved [210]. Plateau and Gueron also propose a modified version that gives second-order cancellations at the expense of small phase shifts. Other approaches have used a $45°-t-45°$ [202] and $90°(X)-t-90°(X)$ [203] 1–1, which allows sequences that require placing the transmitter away from the water.

Plateau and colleagues [204] extended their jump-return sequence to a 1-2-1 version. The spectrum amplitude versus frequency profile has a cosine-squared dependence, which gives second-order suppression like that of the Redfield method, but uses strong pulses. In this case the solvent nuclei are off-resonance, with the transmitter placed in the region of interest. The t delays in this case are enough for the water protons to precess exactly 180° (ca. 250 μs) while the on-resonance protons do not precess. Essentially, the first pulse creates a nonzero XY magnetization, which (for water) precesses just far enough before the second pulse to be returned to the z axis. The second half repeats the process, further rotating the on-resonant protons but again returning the water protons to the z axis. The authors indicate the extension to higher order binomial distributions of 1-3-3-1 and 1-4-6-4-1 to give third- and fourth-power dependencies.

Sklenar and Starcuk [205] also described the 1-2-1 sequence, and Turner [206] pointed out the binomial distribution properties of those sequences.

Hore [207] examined the family of binomial coefficient-based solvent suppression pulse sequences and recommended the 1-3-3-1 type as the best compromise between efficiency and ease of setting up. This version has produced suppression factors of more than 1000.

A common property of spectra obtained from the Redfield 2-1-4 or other 2-1-4 sequences is a baseline roll: Plateau and colleagues [204] analyze this as a consequence of imposing a frequency-dependent phase correction on the spectrum. The primary effect is on the remaining solvent peak. While the center of the strong residual signal is correctly phased, the wings of the line become non-Lorentzian and cause a baseline undulation. They discuss a procedure to minimize this whereby a simulated Lorentzian line matching that of the solvent line is subtracted prior to phase correction. This removes the unwanted undulation. The procedure bears some similarity, as they discuss, to a data-shifting procedure described by Roth and colleagues [208]. This technique eliminates the solvent line including wings without the need to simulate a spectrum of the residual solvent signal; however, increased spectral amplitude distortion is a side effect.

A wide variety of solvent-elimination and other experiments is possible using selective excitation of one group of spins. For example, the act of saturation of one resonance is a form of selective excitation, in this case, saturation. In 1978 Morris and Freeman [209] developed a more general method of selective excitation. This technique uses a series of small flip-angle pulses separated by a specific delay. The cumulative effect of these pulses is to tip only one magnetization vector toward the XY plane. All other spins have a non-cumulative effect exerted on them and thus remain essentially unperturbed. If applied to solvent suppression, this sequence would be performed to tip the solvent resonance 90°, followed by a nonspecific *hard* 90° pulse applied to all spins. The solvent magnetization is rotated to the $-z$ axis by this pulse, while the previously unexcited analytical spins are rotated into the XY plane and are detected. Phase cycling in all permutations can further attenuate the averaged solvent signal. This series of pulses was named *DANTE* (delays alternating with nutations for tailored excitation).

Haasnoot [210] used the selectivity of the DANTE pulse sequence to handle the problem of similar relaxation times for solvent and solute in the WEFT experiment. Instead of a nonselective 180° pulse, a DANTE pulse train is used to invert the solvent magnetization selectively. After a suitable delay timed to coincide with the solvent magnetization crossing through zero z magnetization, a strong, nonselective 90° pulse is used to sample the spectrum of the previously unaffected solute spins. This technique is suitable for homonuclear NOE experiments since selective spin saturation can be performed during the period between the DANTE inversion pulse and the monitoring 90° pulse, whereas in ordinary WEFT all spins are initially perturbed from equilibrium and are changing in z magnetization for at least part of the time between the two pulses.

Solvent-suppression techniques are just as important for 2-D NMR. The ready extension of simple solvent saturation is the technique of solvent irradiation during the equilibration time *and* during the evolution period. Block–Siegert shifts can be significant in F_1 in COSY and NOESY experiments because of the saturation during t_1. Hosur and colleagues [211] proposed a two-dimensional experiment to calibrate the decoupler strength and the consequent numerical correction on the two-dimensional data. Wider and colleagues [212] compared several saturation methods and recommended that for proteins saturation be absent during both t_1 and t_2.

Long pulses, such as Redfield 2-1-4, 1-3-3-1, or DANTE, can also be used in place of the 90° pulses normally encountered in two-dimensional pulse sequences; observation of J correlation or NOE in water is thus possible [145, 213, 214].

EXAMPLE 12   SOLVENT-SUPPRESSION METHODS

Several solvent-suppression methods exist. The simplest method uses a deuterated solvent that reduces the solvent resonance 100-fold. The remaining solvent magnetization can be saturated using a decoupler (Figure 5E.13a).

In systems where the rate of proton exchange is slow, it is possible to observe labile protons such as NHs in a protein. In these cases, such as the 20-mM bovine pancreatic trypsin inhibitor solution in 90% H_2O illustrated in Figure 5E.13b, simple presaturation not only adequately eliminates the H_2O resonance, but also retains the signals from the NH protons. In the fully general case it is not always possible to use these techniques because many NH protons become saturated when they undergo rapid exchange with solvent protons. For example, the study of lysozyme requires a fully protonated solvent to retain all the NH signals. Additionally, many pulse sequences require 90° pulses; this places impractical dynamic range demands on the hardware.

The ideal experiment must provide excitation to those parts of the spectrum that are of interest but must not excite the strong solvent signal—the 1-3-3-1 pulse sequence is effective for this and is also relatively "forgiving" of instrumental nonidealities [207]. As illustrated in Figure 5E.13c, a 10-mM lysozyme solution with 90% H_2O/10% D_2O as solvent, gives a 1-3-3-1 spectrum in only 64 transients. The residual H_2O signal indicates a suppression factor of about 1000. The sequence produces a distortion of relative intensities in the spectrum but does allow examination of labile protons. The 1-3-3-1 pulse train can be used to provide 90° pulses in two-dimensional sequences such as

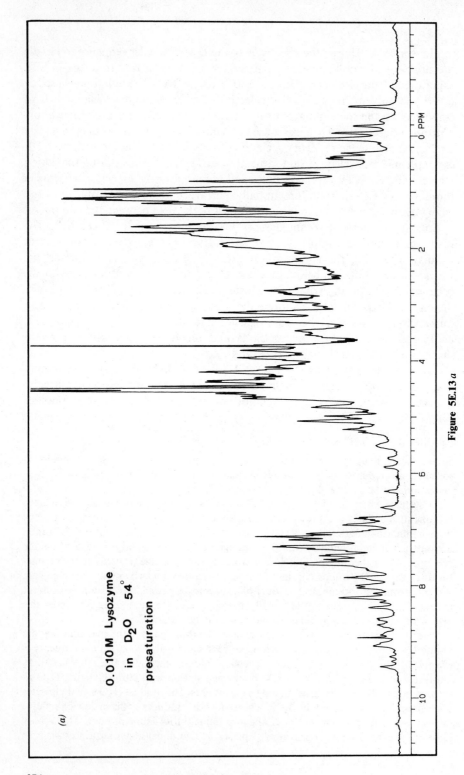

0.010M Lysozyme
in D$_2$O   54°
presaturation

(a)

Figure 5E.13 a

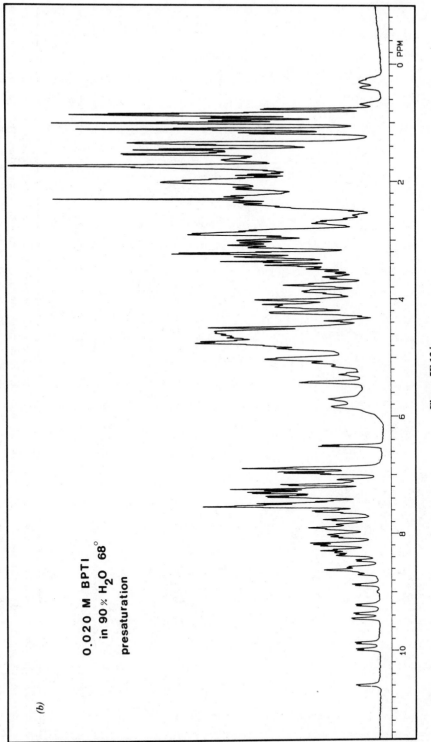

0.020 M BPTI
in 90 % H_2O 68°
presaturation

(b)

Figure 5E.13 *b*

655

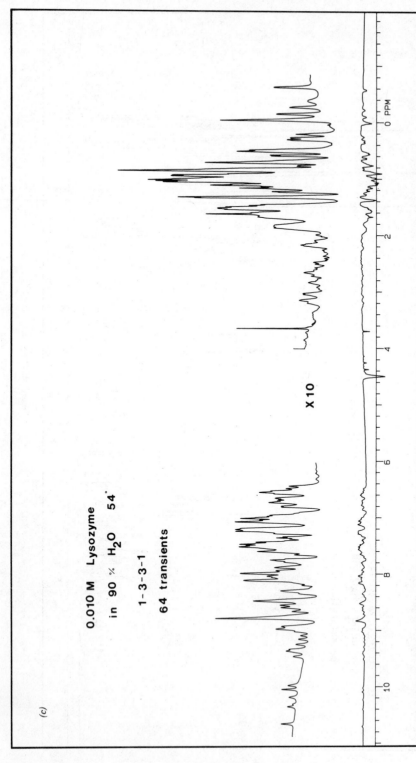

Figure 5E.13 (a) Lysozyme proton spectrum with presaturation for suppression of the residual solvent signal. (b) Bovine pancreatic trypsin inhibitor proton spectrum with presaturation for water suppression. In this case NH resonances do not exchange with solvent and are thus visible in the 9 to 11 ppm region. (c) NH protons in lysozyme do exchange with solvent; thus, the 1-3-3-1 pulse sequence is used to monitor the NH region.

COSY and NOESY. In fact, the double-quantum heteronuclear two-dimensional experiment described in Example 11 (a proton observe experiment in 90% H_2O) used 1-3-3-1 pulses to excite the protons.

13   CONCLUSIONS

Although by no means encyclopedic in coverage, the foregoing material does give a fair representation of the exciting progress made in high-resolution NMR during the recent past (through early 1984). This excitement is reflected in the growing interest in and utilization of NMR by an ever-widening group of users. It is with great anticipation that we look forward to the expanding use of these new techniques.

Acknowledgments

We thank Varian Associates for permission to use the figures appearing in this chapter, with the exception of Figure 5.18 (reprinted with permission from the *Journal of the American Chemical Society*) and Figures 5.21 and 5.22 (reprinted from *Organic Magnetic Resonance* by permission of John Wiley & Sons, Ltd.). All spectra were obtained on Varian XL-series superconducting NMR spectrometers.

References

1.   R. R. Ernst, *J. Chem. Phys.*, **45**, 3845 (1966); H. J. Reich, M. Juntelat, M. T. Messe, F. J. Weigert, and J. D. Roberts, *J. Am. Chem. Soc.*, **91**, 7445 (1969).

2.   F. A. L. Anet, N. Jaffer, and J. Strouse, *Poster H-10*, 21st Experimental NMR Conference, Tallahassee, FL, 1980.

3.   D. L. Rabenstein and T. K. Nakashima, *Anal. Chem.*, **51**, 1465A (1979).

4.   M. H. Levitt and R. Freeman, *J. Magn. Reson.*, **39**, 533 (1980).

5.   C. LeCocq and J.-Y. Lallemand, *J. Chem. Soc. Chem. Commun.*, 150 (1981).

6.   D. J. Cookson and B. E. Smith, *Org. Magn. Reson.*, **16**, 111 (1981).

7.   D. W. Brown, T. K. Nakashima, and D. L. Rabenstein, *J. Magn. Reson.*, **45**, 302 (1981).

8.   S. L. Patt and J. N. Shoolery, *J. Magn. Reson.*, **46**, 535 (1982).

9.   D. J. Cookson and B. E. Smith, *Fuel*, **62**, 34 (1983).

10.   F.-K. Pei and R. Freeman, *J. Magn. Reson.*, **48**, 318 (1982).

11.   H. J. Jakobsen, O. W. Sorensen, W. S. Brey, and P. Kanyha, *J. Magn. Reson.*, **48**, 328 (1982).

12.   J.-C. Beloeil, C. LeCocq, and J.-Y. Lallemand, *Org. Magn. Reson.*, **19**, 112 (1982).

13.   T. K. Nakashima and D. L. Rabenstein, *J. Magn. Reson.*, **47**, 339 (1982).

14.   D. J. Cookson and B. E. Smith, *Fuel*, **62**, 39 (1983).

15.   M. R. Bendall, D. M. Doddrell, and D. T. Pegg, *J. Am. Chem. Soc.*, **103**, 4603 (1981).

16. M. R. Bendall, D. T. Pegg, D. M. Doddrell, S. R. Johns, and R. Willing, *J. Chem. Soc. Chem. Commun.*, 1138 (1982).

17. M. R. Bendall, D. T. Pegg, D. M. Doddrell, and D. H. Williams, *J. Org. Chem.*, **47**, 3021 (1982).

18. I. D. Campbell, C. M. Dobson, R. J. P. Williams, and P. E. Wright, *FEBS Lett.*, **57**, 96 (1975).

19. P. Schmitt, J. R. Wesener, and H. Gunther, *J. Magn. Reson.*, **52**, 511 (1983).

20. J. R. Wesener, P. Schmitt, and H. Gunther, *J. Am. Chem. Soc.*, **106**, 10 (1984).

21. J. R. Wesener and H. Gunther, *Org. Magn. Reson.*, **21**, 433 (1983).

22. D. M. Doddrell, J. Staunton, and E. D. Lave, *J. Magn. Reson.*, **52**, 523 (1983).

23. V. Rutar, *J. Magn. Reson.*, **53**, 235 (1983).

24. K. M. Brindle, J. Boyd, I. D. Campbell, R. Porteous, and N. Soffe, *Biochem. Biophys. Res. Commun.*, **109**, 864 (1982).

25. R. J. Simpson, K. M. Brindle, F. F. Brown, I. D. Campbell, and D. L. Foxall, *Biochem. J.*, **202**, 573 (1982).

26. K. M. Brindle, F. F. Brown, I. D. Campbell, D. L. Foxall, and F. J. Simpson, *Biochem. J.*, **202**, 589 (1982).

27. K. M. Brindle, R. Porteous, and I. D. Campbell, *J. Magn. Reson.*, **56**, 543 (1984).

28. T. Ogino, Y. Arata, and S. Fujiwara, *Biochem.*, **19**, 3684 (1980).

29. K. M. Brindle, I. D. Campbell, and R. J. Simpson, *Biochem. Soc. Trans.*, **11**, 280 (1983).

30. P. H. Bolton, *J. Magn. Reson.*, **45**, 418 (1981).

31. O. W. Sorensen, M. Rance, and R. R. Ernst, *J. Magn. Reson.*, **56**, 527 (1984).

32. J. M. Bulsing, J. K. M. Sanders, and L. D. Hall, *J. Chem. Soc. Chem. Commun.*, 1201 (1981).

33. K. G. R. Pachler and P. L. Wessels, *J. Magn. Reson.*, **28**, 53 (1977); H. J. Jakobsen and W. S. Brey, *J. Am. Chem. Soc.*, **101**, 760 (1979).

34. G. A. Morris and R. Freeman, *J. Am. Chem. Soc.*, **101**, 760 (1979).

35. G. A. Morris, *J. Am. Chem. Soc.*, **102**, 428 (1980).

36. D. P. Burum and R. R. Ernst, *J. Magn. Reson.*, **39**, 163 (1980).

37. D. M. Doddrell and D. T. Pegg, *J. Am. Chem. Soc.*, **102**, 6388 (1980).

38. P. H. Bolton, *J. Magn. Reson.*, **41**, 287 (1980).

39. G. A. Morris, *J. Magn. Reson.*, **41**, 185 (1980).

40. J. Kowalewski and G. A. Morris, *J. Magn. Reson.*, **47**, 331 (1982).

41. P. L. Rinaldi and N. J. Baldwin, *J. Am. Chem. Soc.*, **104**, 5791 (1982).

42. P. L. Rinaldi and N. J. Baldwin, *J. Am. Chem. Soc.*, **105**, 7523 (1983).

43. R. Freeman, T. H. Mareci, and G. A. Morris, *J. Magn. Reson.*, **42**, 341 (1981).

44. M. R. Bendall, D. T. Pegg, D. M. Doddrell, and J. Field, *J. Am. Chem. Soc.*, **103**, 934 (1981).

45. D. M. Doddrell, D. T. Pegg, and M. R. Bendall, *J. Magn. Reson.*, **48**, 323 (1982).

46. D. T. Pegg, D. M. Doddrell, and M. R. Bendall, *J. Chem. Phys.*, **77**, 2745 (1982).

47. M. R. Bendall, D. T. Pegg, D. M. Doddrell, and J. Field, *J. Magn. Reson.*, **51**, 520 (1983).

48. D. M. Doddrell, D. G. Reid, and D. H. Williams, *J. Magn. Reson.*, **56**, 279 (1984).

49. W. M. Brooks, M. G. Irving, S. J. Simpson, and D. M. Doddrell, *J. Magn. Reson.*, **56**, 521 (1984).

50. D. T. Pegg, D. M. Doddrell, and M. R. Bendall, *J. Magn. Reson.*, **51**, 264 (1983).

51. O. W. Sorensen and R. R. Ernst, *J. Magn. Reson.*, **51**, 477 (1983).

52. M. R. Bendall, D. T. Pegg, and D. M. Doddrell, *J. Magn. Reson.*, **53**, 81 (1983).

53. D. J. Pegg and M. R. Bendall, *J. Magn. Reson.*, **53**, 229 (1983).

54. M. R. Bendall and D. T. Pegg, *J. Magn. Reson.*, **52**, 164 (1983).

55. M. R. Bendall, D. T. Pegg, G. M. Tyburn, and C. Brevard, *J. Magn. Reson.*, **55**, 322 (1983).

56. V. Rutar, *J. Magn. Reson.*, **53**, 49 (1983).

57. V. Rutar, *J. Magn. Reson.*, **53**, 135 (1983).

58. M. R. Bendall and D. T. Pegg, *J. Magn. Reson.*, **53**, 272 (1983).

59. D. M. Doddrell, W. Brooks, and J. Field, *J. Am. Chem. Soc.*, **105**, 6973 (1983).

60. J. Brandeau and D. Canet, *J. Magn. Reson.*, **47**, 419 (1982).

61. J. Brandeau and D. Canet, *J. Magn. Reson.*, **47**, 159 (1982).

62. H. J. Jakobsen, P. J. Kanyha, and W. Brey, *J. Magn. Reson.*, **54**, 134 (1983).

63. D. M. Doddrell, W. Brooks, J. Field, and R. Lynden-Bell, *J. Am. Chem. Soc.*, **105**, 6973 (1983).

64. G. A. Morris and R. Freeman, *J. Magn. Reson.*, **29**, 433 (1978).

65. A. J. Shaka and R. Freeman, *J. Magn. Reson.*, **50**, 502 (1982).

66. A. Bax, C.-H. Niu, and D. Live, *J. Am. Chem. Soc.*, **106**, 1150 (1984).

67. R. D. Bertrand, W. B. Mouiz, A. N. Garroway, and G. C. Chingas, *J. Am. Chem. Soc.*, **100**, 5227 (1978); *J. Magn. Reson.*, **35**, 283 (1979); *J. Am. Chem. Soc.*, **101**, 4058 (1979); *J. Am. Chem. Soc.*, **102**, 2526 (1980); *J. Chem. Phys.*, **74**, 127 (1981).

68. K. J. Packer and K. M. Wright, *J. Magn. Reson.*, **41**, 168 (1980).

69. S. R. Hartmann and E. L. Hahn, *Phys. Rev.*, **128**, 2042 (1962).

70. J. Jeener, Lecture, Ampere International Summer School, Basko Polje, Yugoslavia, 1971.

71. W. P. Aue, J. Karhan, and R. R. Ernst, *J. Chem. Phys.*, **64**, 4226 (1976).

72. G. Bodenhausen and R. Freeman, *J. Magn. Reson.*, **28**, 471 (1977).

73. A. Bax, R. Freeman, and G. A. Morris, *J. Magn. Reson.*, **43**, 333 (1981).

74. S. Macura and L. R. Brown, *J. Magn. Reson.*, **53**, 529 (1983).

75. D. M. Thomas, M. R. Bendall, D. T. Pegg, D. M. Doddrell, and J. Field, *J. Magn. Reson.*, **42**, 298 (1981).

76. V. Rutar and T. C. Wong, *J. Magn. Reson.*, **53**, 495 (1983).

77. A. Bax and R. Freeman, *J. Am. Chem. Soc.*, **104**, 1099 (1982).

78. A. Bax, *J. Magn. Reson.*, **52**, 330 (1983).

79. V. Rutar, *J. Am. Chem. Soc.*, **105**, 4095 (1983).

80. V. Rutar, *J. Magn. Reson.*, **56**, 87 (1984).

81. G. Bodenhausen, *J. Magn. Reson.*, **39**, 175 (1980).

82. A. A. Maudsley and R. R. Ernst, *Chem. Phys. Lett.*, **50**, 368 (1971).

83. G. A. Morris, *J. Magn. Reson.*, **44**, 277 (1981).

84. D. T. Pegg and M. R. Bendall, *J. Magn. Reson.*, **55**, 114 (1983).

85. V. Rutar, *J. Magn. Reson.*, **56**, 413 (1984).

86. A. J. Shaka, J. Keeler, and R. Freeman, *J. Magn. Reson.*, **56**, 294 (1984).

87. M. P. Williamson, *J. Magn. Reson.*, **55**, 471 (1983).

88. J. Keeler, *J. Magn. Reson.*, **56**, 463 (1984).

89. W. P. Aue, E. Bartholdi, and R. R. Ernst, *J. Chem. Phys.*, **64**, 2229 (1976).

90. N. Nagayama, A. Kumar, K. Wuthrich, and R. R. Ernst, *J. Magn. Reson.*, **40**, 321 (1980).

91. A. Bax, R. Freeman, and G. Morris, *J. Magn. Reson.*, **42**, 164 (1981).

92. J. M. VanDivender and W. C. Hutton, *J. Magn. Reson.*, **48**, 272 (1982).

93. T. L. Venable, W. C. Hutton, and R. N. Grimes, *J. Am. Chem. Soc.*, **104**, 4716 (1982).

94. T. L. Venable, W. C. Hutton, and R. N. Grimes, *J. Am. Chem. Soc.*, **106**, 29 (1984).

95. C. Brevard, R. Schimpf, G. Tourne, and C. M. Tourne, *J. Am. Chem. Soc.*, **105**, 7059 (1983).

96. G. A. Gray, unpublished work.

97. R. Bauman, A. Kumar, R. R. Ernst, and K. Wuthrich, *J. Magn. Reson.*, **44**, 761 (1981); R. Bauman, G. Wider, R. R. Ernst, and K. Wuthrich, *J. Magn. Reson.*, **44**, 402 (1981).

98. L. Braunschweiler and R. R. Ernst, *J. Magn. Reson.*, **53**, 521 (1983).

99. A. Bax and R. Freeman, *J. Magn. Reson.*, **44**, 542 (1981).

100. J. C. Stephens, J. L. Roark, D. G. Lynn, and J. L. Riopel, *J. Am. Chem. Soc.*, **105**, 1669 (1983); G. Batta and A. Liptak, *J. Am. Chem. Soc.*, **106**, 248 (1984).

101. G. Wider, S. Macura, A. Kumar, R. R. Ernst, and K. Wuthrich, *J. Magn. Reson.*, **56**, 207 (1984).

102. A. A. Maudsley, A. Kumar, and R. R. Ernst, *J. Magn. Reson.*, **28**, 303 (1977); A. A. Maudsley, L. Muller, and R. R. Ernst, *J. Magn. Reson.*, **28**, 463 (1977); G. Bodenhausen and R. Freeman, *J. Am. Chem. Soc.*, **100**, 320 (1978); R. Freeman and G. Morris, *J. Chem. Soc. Chem. Commun.*, 684 (1978).

103. A. D. Bax and G. Morris, *J. Magn. Reson.*, **42**, 501 (1981).

104. W. Ammann, R. Richarz, T. Wirthlin, and D. Wendisch, *Org. Magn. Reson.*, **20**, 260 (1982).

105. M. Ikura and K. Hikichi, *Org. Magn. Reson.*, **20**, 266 (1982).

106. R. Benn, *Z. Naturforsch.*, **37b**, 1054 (1982).

107. G. A. Morris and L. D. Hall, *J. Am. Chem. Soc.*, **103**, 4703 (1981).

108. T.-M. Chan and J. L. Markley, *J. Am. Chem. Soc.*, **104**, 4010 (1982).

109. H. Kessler, W. Hehlein, and R. Schuck, *J. Am. Chem. Soc.*, **104**, 4534 (1982).

110. G. A. Gray, *Org. Magn. Reson.*, **21**, 111 (1983).

111. G. E. Hawkes, L. Lian, and E. W. Randall, *J. Magn. Reson.*, **56**, 539 (1984).

112. H. Bleich, S. Gould, P. Pitner, and J. Wilde, *J. Magn. Reson.*, **56**, 515 (1984).

113. J. T. Gerig, *Macromolecules*, **16**, 1797 (1983).

114. P. H. Bolton and G. Bodenhausen, *J. Magn. Reson.*, **43**, 339 (1982).

115. P. H. Bolton, *J. Magn. Reson.*, **45**, 239 (1981).

116.   P. H. Bolton, *J. Magn. Reson.*, **46**, 91 (1982).

117.   P. H. Bolton and G. Bodenhausen, *J. Magn. Reson.*, **46**, 306 (1982).

118.   A. Pardi, R. Walker, H. Rapoport, G. Wider, and K. Wuthrich, *J. Am. Chem. Soc.*, **105**, 1652 (1983).

119.   D. C. Finster, W. C. Hutton, and R. N. Grimes, *J. Am. Chem. Soc.*, **102**, 400 (1980).

120.   A. G. Avent and R. Freeman, *J. Magn. Reson.*, **39**, 169 (1980).

121.   G. Bodenhausen and D. J. Ruben, *Chem. Phys. Lett.*, **69**, 185 (1980).

122.   D. A. Vidnsek, M. F. Roberts, and G. Bodenhausen, *J. Am. Chem. Soc.*, **104**, 5452 (1982).

123.   M. R. Bendall and D. T. Pegg, *J. Magn. Reson.*, **53**, 144 (1983).

124.   M. H. Levitt, O. W. Sorensen, and R. R. Ernst, *Chem. Phys. Lett.*, **94**, 540 (1983).

125.   A. Bax, *J. Magn. Reson.*, **53**, 517 (1983).

126.   P. H. Bolton, *J. Magn. Reson.*, **52**, 326 (1983).

127.   G. Eich, G. Bodenhausen, and R. R. Ernst, *J. Am. Chem. Soc.*, **104**, 3731 (1982).

128.   P. H. Bolton and G. Bodenhausen, *Chem. Phys. Lett.*, **89**, 139 (1982).

129.   P. H. Bolton, *J. Magn. Reson.*, **48**, 336 (1982).

130.   A. Bax, R. Freeman, and S. P. Kempsell, *J. Am. Chem. Soc.*, **102**, 4849 (1980).

131.   A. Bax, *J. Magn. Reson.*, **53**, 149 (1983).

132.   G. Wagner, *J. Magn. Reson.*, **55**, 151 (1983).

133.   M. A. Delsuc, E. Guittet, N. Trotin, and J. Y. Lallemand, *J. Magn. Reson.*, **56**, 163 (1984).

134.   H. Kogler, O. W. Sorensen, G. Bodenhausen, and R. R. Ernst, *J. Magn. Reson.*, **55**, 157 (1983).

135.   P. Bigler, W. Ammann, and R. Richarz, *Org. Magn. Reson.*, **22**, 109 (1984).

136.   J. Jeener, B. H. Meier, P. Bachmann, and R. R. Ernst, *J. Chem. Phys.*, **71**, 4546 (1979).

137.   B. H. Meier and R. R. Ernst, *J. Am. Chem. Soc.*, **101**, 6441 (1979).

138.   S. Macura and R. R. Ernst, *Mol. Phys.*, **41**, 95 (1980).

139.   A. Kumar, R. R. Ernst, and K. Wuthrich, *Biochem. Biophys. Res. Commun.*, **95**, 1 (1980).

140.   C. Bosch, A. Kumar, R. Baumann, R. R. Ernst, and K. Wuthrich, *J. Magn. Reson.*, **42**, 159 (1981).

141.   A. Kumar, G. Wagner, R. R. Ernst, and K. Wuthrich, *J. Am. Chem. Soc.*, **103**, 3654 (1981).

142.   S. Macura, Y. Huang, D. Suter, and R. R. Ernst, *J. Magn. Reson.*, **43**, 259 (1981).

143.   S. Macura, K. Wuthrich, and R. R. Ernst, *J. Magn. Reson.*, **46**, 219 (1982).

144.   S. Macura, K. Wuthrich, and R. R. Ernst, *J. Magn. Reson.*, **47**, 351 (1982).

145.   A. L. Schwartz and J. D. Cutnell, *J. Magn. Reson.*, **53**, 398 (1983).

146.   P. B. Garlick and C. J. Turner, *J. Magn. Reson.*, **51**, 536 (1983).

147.   R. S. Balaban, H. L. Kantor, and J. L. Ferretti, *J. Biol. Chem.*, **258**, 12787 (1983).

148.   Y. Huang, S. Macura, and R. R. Ernst, *J. Am. Chem. Soc.*, **103**, 5327 (1981).

149.   P. L. Rinaldi and N. J. Baldwin, *J. Am. Chem. Soc.*, **105**, 5167 (1983).

150.   G. Bodenhausen and R. R. Ernst, *Mol. Phys.*, **47**, 319 (1982).

151. C. A. Haasnoot, F. J. M. Van DeVen, and C. W. Hilbers, *J. Magn. Reson.*, **56**, 343 (1984).

152. A. Z. Gurevich, I. L. Barsukov, A. S. Arseniev, and V. F. Bystrov, *J. Magn. Reson.*, **56**, 471 (1984).

153. G. Bodenhausen and R. R. Ernst, *J. Magn. Reson.*, **45**, 367 (1981); *J. Am. Chem. Soc.*, **104**, 1304 (1982).

154. G. Bodenhausen, *Prog. Nucl. Magn. Reson. Spectrosc.*, **14**, 137 (1980).

155. A. Bax, R. Freeman, and S. P. Kempsell, *J. Am. Chem. Soc.*, **102**, 4849 (1980); *J. Magn. Reson.*, **41**, 349 (1980).

156. R. Richarz, W. Ammann, and T. Wirthlin, *J. Magn. Reson.*, **45**, 270 (1981).

157. A. Bax, R. Freeman, and S. P. Kempsell, *J. Magn. Reson.*, **41**, 349 (1980).

158. A. Bax, R. Freeman, and T. A. Frienkel, *J. Am. Chem. Soc.*, **103**, 1202 (1981).

159. A. Bax, R. Freeman, T. A. Frienkiel, and M. H. Levitt, *J. Magn. Reson.*, **43**, 478 (1981).

160. T. H. Mareci and R. Freeman, *J. Magn. Reson.*, **48**, 158 (1982).

161. O. Sorensen, R. Freeman, T. Frenkiel, T. Mareci, and R. Schuck, *J. Magn. Reson.*, **46**, 180 (1982).

162. D. L. Turner, *Mol. Phys.*, **44**, 1051 (1981).

163. D. L. Turner, *J. Magn. Reson.*, **49**, 175 (1982).

164. D. L. Turner, *J. Magn. Reson.*, **53**, 259 (1983).

165. A. Bax and T. H. Mareci, *J. Magn. Reson.*, **53**, 360 (1983).

166. A. Neszmelyi and G. Lukacs, *J. Chem. Soc. Chem. Commun.*, 999 (1981).

167. N. E. Mackenzie, R. L. Baxter, A. I. Scott, and P. E. Fagerness, *J. Chem. Soc. Chem. Commun.*, 145 (1982).

168. A. Neszmelyi and G. Lukacs, *J. Am. Chem. Soc.*, **104**, 5342 (1982).

169. M. A. Phillipi, R. J. Wiserma, J. R. Braivard, and R. E. London, *J. Am. Chem. Soc.*, **104**, 7333 (1982).

170. J. A. Robinson and D. L. Turner, *J. Chem. Soc.*, 150 (1982).

171. J. Hudec and D. L. Turner, *J. Chem. Soc. Perkin Trans.* 2, 951 (1982).

172. P. J. Keller, Q. L. Van, A. Backer, J. F. Kozlowski, and H. G. Floss, *J. Am. Chem. Soc.*, **105**, 2505 (1983).

173. R. Benn, *Angew. Chemie*, **94**, 633 (1982).

174. P. Schmitt and H. Gunther, *J. Magn. Reson.*, **52**, 497 (1983).

175. M. H. Levitt and R. R. Ernst, *Mol. Phys.*, **50**, 1109 (1983).

176. M. H. Levitt, *J. Magn. Reson.*, **33**, 473 (1979); S. P. Kempsell and M. H. Levitt, *J. Magn. Reson.*, **38**, 453 (1980); M. H. Levitt and R. Freeman, *J. Magn. Reson.*, **43**, 65 (1981); M. H. Levitt, *J. Magn. Reson.*, **48**, 234 (1982); M. H. Levitt, *J. Magn. Reson.*, **50**, 95 (1982).

177. P. J. Hore, E. R. P. Zuiderweg, K. Nicolay, K. Dijkstra, and R. Kaptein, *J. Am. Chem. Soc.*, **104**, 4286 (1982).

178. P. J. Hore, R. M. Scheek, A. Volbeda, and R. Kaptein, *J. Magn. Reson.*, **50**, 328 (1982).

179. P. J. Hore, R. M. Scheek, and R. Kaptein, *J. Magn. Reson.*, **52**, 339 (1983).

180. O. W. Sorensen, M. H. Levitt, and R. R. Ernst, *J. Magn. Reson.*, **55**, 104 (1983).

181. H. Bildsoe, S. Donstrup, H. J. Jakobsen, and O. W. Sorensen, *J. Magn. Reson.*, **53**, 154 (1983).

182. O. W. Sorensen, S. Donstrup, H. Bildsoe, and H. J. Jakobsen, *J. Magn. Reson.*, **55**, 347 (1983).

183. J. M. Bulsing, W. M. Brooks, J. Field, and D. M. Doddrell, *J. Magn. Reson.*, **56**, 167 (1984).

184. O. W. Sorensen and R. R. Ernst, *J. Magn. Reson.*, **54**, 122 (1983).

185. U. Piantini, O. W. Sorensen, and R. R. Ernst, *J. Am. Chem. Soc.*, **104**, 6800 (1982).

186. A. J. Shaka and R. Freeman, *J. Magn. Reson.*, **51**, 169 (1983).

187. T. H. Mareci and R. Freeman, *J. Magn. Reson.*, **51**, 531 (1983).

188. J. Boyd, C. M. Dobson, and C. Redfield, *J. Magn. Reson.*, **55**, 170 (1983).

189. H. Kessler, H. Oschkinat, O. W. Sorensen, H. Kogler, and R. R. Ernst, *J. Magn. Reson.*, **55**, 329 (1983).

190. P. H. Bolton, *J. Magn. Reson.*, **57**, 427 (1984).

191. A. Bax, R. Griffey, and B. L. Hawkins, *J. Am. Chem. Soc.*, **105**, 7188 (1983).

192. A. Bax, R. Giffey, and B. L. Hawkins, *J. Magn. Reson.*, **55**, 301 (1983).

193. R. Griffey, C. D. Poulter, A. Bax, B. L. Hawkins, Z. Yamaizumi, and S. Nishimura, *Proc. Natl. Acad. Sci. USA*, **80**, 5895 (1983).

194. P. H. Bolton, *J. Magn. Reson.*, **54**, 333 (1983).

195. O. W. Sorensen and R. R. Ernst, *J. Magn. Reson.*, **55**, 338 (1983).

196. L. Braunschweiler, G. Bodenhausen, and R. R. Ernst, *Mol. Phys.*, **48**, 535 (1983).

197. R. M. Lynden-Bell, J. M. Bulsing, and D. M. Doddrell, *J. Magn. Reson.*, **55**, 128 (1983).

198. M. A. Thomas and A. Kumar, *J. Magn. Reson.*, **56**, 479 (1984).

199. S. L. Patt and B. D. Sykes, *J. Chem. Phys.*, **56**, 3182 (1972).

200. A. G. Redfield, S. D. Kunz, and E. K. Ralph, *J. Magn. Reson.*, **19**, 114 (1975).

201. P. Plateau and M. Gueron, *J. Am. Chem. Soc.*, **104**, 7310 (1982).

202. G. M. Clore, B. J. Kimber, and A. M. Gronenborn, *J. Magn. Reson.*, **54**, 170 (1983).

203. H. Bleich and J. Wilde, *J. Magn. Reson.*, **56**, 154 (1984).

204. P. Plateau, C. Dumas, and M. Gueron, *J. Magn. Reson.*, **54**, 46 (1983).

205. V. Sklenar and Z. Starcuk, *J. Magn. Reson.*, **50**, 495 (1982).

206. D. L. Turner, *J. Magn. Reson.*, **54**, 146 (1983).

207. P. J. Hore, *J. Magn. Reson.*, **55**, 283 (1983).

208. K. Roth, B. J. Kimber, and J. Feeney, *J. Magn. Reson.*, **41**, 302 (1980).

209. G. A. Morris and R. Freeman, *J. Magn. Reson.*, **29**, 433 (1978).

210. G. A. G. Haasnoot, *J. Magn. Reson.*, **52**, 153 (1983).

211. R. V. Hosur, R. R. Ernst, and K. Wuthrich, *J. Magn. Reson.*, **54**, 142 (1983).

212. G. Wider, R. V. Hosur, and K. Wuthrich, *J. Magn. Reson.*, **52**, 130 (1983).

213. C. A. G. Haasnoot, H. Heerschap, and C. W. Hilbers, *J. Am. Chem. Soc.*, **105**, 5483 (1983).

214. P. J. Hore, *J. Magn. Reson.*, **56**, 535 (1984).

Chapter 6

GAS CHROMATOGRAPHY

Roy A. Keller

1   INTRODUCTION

1.1   Chromatography as Differential Migration

All chromatographic methods are strictly separation techniques for both molecular and ionic species. The methods are utilized without knowing the number of species present, their identities, or their concentrations in the sample. The power of chromatography is awesome. We, infants in an infant field, watched the analysis of the essence of coffee progress from some 30 components to some 350 components in a single pass in a gas chromatographic system. If one contemplates the uncountable compounds in the universe, it is ridiculous to think that one chromatographic configuration will suffice. By combining the techniques of gas chromatography (GC), liquid chromatography (LC), size-exclusion chromatography (SEC), and field-flow fractionation (FFF) (although FFF is not regarded by some as a form of chromatography), it is possible to perform separations that range from the molecular dust motes (hydrogen and helium) to the molecular boulders (viruses and particulates).

Chromatographic techniques separate; they do not identify or quantitate. Identification is based on the assumption that a particular species will behave the same way on identical chromatographic systems. It is very difficult to identify those parameters, such as column length and diameter, packing particle size, amount of stationary phase present, flow rate, and temperature, that determine whether two systems are the same, and if an unknown component matches an authentic sample of the suspect. Chromatography will yield a best guess. One must ask how many "matched behaviors" on how many different chromatographic configurations must be employed to pass from conjecture to certainty. Even the use of mass spectrometry requires the spectrum of an authenticated sample. It is a reasonable desire to stretch the technique to its limit, but such stress must be applied with great caution.

Quantitation is a detector problem. It is rational to assume that a species, to be seen, must generate a signal. The signal intensity must be related to the amount of compound present. The smallest amount the detector will see, that is, its *lower limit of detection* (LLD), must be determined. Any upper limit, where the detector is "swamped," which means that the signal intensity is constant when the amount of the compound is increased, must be determined. The increase in signal intensity with a unit increase in the amount of the compound in question, that is, its *sensitivity*, must be determined. (Many authors use the term *sensitivity* when discussing LLD. This is confusing to the neophyte. An older generation of chemists raised on double-beam balances consistently avoids this ambiguity and uses sensitivity as it was originally defined.) The detector's *linear range* must be established, that is, the range of *solute* (a particular component of a mixture) concentration where the signal intensity varies linearly with the concentration so that doubling the solute concentration doubles the signal intensity. Does the detector see everything or only a select few solutes, such as, halogenated compounds (*selective* detector)? If two different solutes generate equal signal intensities, does this mean that they are present in equal amounts (*response factors*)?

The early literature is a jungle of terms, symbols, and recommendations. In an effort to standardize nomenclature and symbolism, the American Society for Testing and Materials (ASTM)† sponsored Committee E-19 on chromatography. This was reported as the *Recommended Practice for Gas Chromatography Terms and Relationships, Designation E355*. A reprint of this publication should be in the file of anyone reporting GC results whether for an in-house report or a journal publication.

A whimsical analogy of Henry Eyring (1953) combined with the ideas of H. H. Strain [1–3] offer a useful concept. Imagine a group of boats at a boat dock situated on a river, the *migration medium*. The boats are anchored. Now imagine that the river goes into flood and the boats begin to slip their anchors and migrate down the stream. If the boats have poor anchors they spend most of their time moving with the river. The river organizes their migration. Certainly they migrate swiftly along the river and movement from side to side (*lateral migration*) is unimportant to their progress down the river if the river (*column*) is sufficiently narrow. We will call the direction down the river the *migration coordinate*, x. For example, a 6-ft, $\frac{1}{8}$-in.-diameter (o.d.) *column*, the chromatographic counterpart of the river, has a length-to-diameter ratio of 576. The flowing river, necessary to produce the migration, is the *mobile phase* (it has also been called the *carrier*), which must be a fluid. The anchors catch on the river bottom, and when this happens, the boats are fixed. The river bottom is the *stationary phase*. Of course the boats are molecules and the anchors are functional groups within the molecules. The analogy is inexact because it fails to include properties of the boats—the size (molecular weight or volume) and shape (isomers). The great power of chromatography is the separation of isomers.

The quality of anchors may be compared two ways. Perhaps the next morning, after the flood, one strolls down the river and notes the positions of the boats. Those still near the dock must have fine anchors while those far down the

†ASTM, 1916 Race Street, Philadelphia, PA 19103.

river must have poor anchors. There has been a separation in space for a fixed time period, overnight in the analogy. This is *development chromatography*, such as thin-layer (TLC) or paper chromatography.

The alternate method is to take a position downstream with a clock in hand. A signal is sent when the boats begin to migrate (*instant of injection*). Boats with poor anchors are soon swept past and the time recorded (*retention time*). Boats with increasingly better anchors require greater time periods. This is a separation in time for a fixed migration distance, the distance between the dock and the observer (*detector*). This is *elution chromatography* (PC).

It is common in both development and elution chromatography to compare migration to an anchorless boat, a piece of driftwood, an unretained solute. In development chromatography, the ratio of the distance moved by a solute relative to an unretained species (generally the advancing front of the solvent) is the R_f value. In elution chromatography, it is a form of a *retention time* [GC and high-performance liquid chromatography (HPLC)]. Many feel that elution chromatography is superior because all of the solutes experience the full separating power of the system, whereas in development systems solutes of very similar behavior are exposed to the separation experience only through the distance migrated from the point of application. An objection to elution methods is the nagging doubt that every sample component has migrated through the column. The molecular species migrate at different rates (*differential migration*). To this concept, Strain added that the mixture must start from a narrow zone so that each species finishes in a narrow zone. *Chromatography is differential migration from a narrow zone.*

The *apparent velocity* u_i of a solute species i is the distance moved divided by the time taken. In reality there are only two velocities: zero velocity, when the molecules are attached to the stationary phase, and the velocity of the mobile phase u. The important parameter is the fraction of any time period an average molecule of the particular species spends in the mobile phase R_i, sometimes called the *general migration parameter*, which was emphasized to be most significant by Giddings [4]. Thus

$$u_i = R_i u \tag{1}$$

These are linear velocities along the migration coordinate, which may be the vector component of a very complex meandering.

If the stationary phase is a solid, for example, graphite, silica, or alumina such that the solutes attach directly to the surface (often referred to as *adsorption chromatography*) and the fluid mobile phase is a liquid, it is termed *liquid–solid chromatography* (LSC). If the fluid is a gas, it is *gas–solid chromatography* (GSC). If the stationary phase is a liquid (often called *partition chromatography*) and the mobile phase is also a liquid, it is *liquid–liquid chromatography* (LLC); and if the mobile phase is a gas, it is *gas–liquid chromatography* (GLC). Solution forces provided by the stationary phase are responsible for retention in GLC. Gas–liquid chromatography is emphasized here because it is the predominant GC system.

Some materials, such as, organic polymers, silica, and molecular sieves, can be prepared in such a fashion that the surface of the particle is pitted with pores of a diameter that can be controlled. Figure 6.1 is an idealized schematic of the surface of such a particle. This conical pore model is amenable to mathematical treatment [5]. The mobile phase, gas or liquid, enters these pores to form stagnent pools. The trapped mobile phase species is now the stationary phase. Small molecules A can diffuse to the very bottom of the pore and wander back out. While they are doing this, larger molecules B may enter the pores; but they cannot penetrate as deeply because of their size and because their diffusion paths are shorter. The small molecule has a small R value. This R value increases as the molecular size increases until a molecular size is reached such that the molecule C cannot enter the pore at all and the R value is unity; the molecular giants tumble by their smaller colleagues, which are exploring the pores. This is *size-exclusion chromatography* (SEC). In reality it is impossible to prepare a substance whose pores are of the same diameter, shape, and depth, but the description is apt. The solid should be perfectly inert so that size exclusion is the exclusive governing factor, which is impossible. Some adsorption (silica) and solution (organic polymers) occur. A problem with GLC, to be referred to later, is adsorption or chemical interaction of the solute with the stationary phase, which leads to tailing or chemical degradation. The first is generally caused by adsorption of the solute on the solid coated with stationary liquid phase. This is certainly the case with hydrogen-bonding solutes, such as water, ammonia, and amines. This effect is minimized with SEC. The slight reactivity of SEC materials allows separation of corrosive substances, for example, the oxides of sulfur and nitrogen.

1.2   Molecular Interactions

Selectivity is an ill-defined, qualitative, and vague apellation that refers to the ability of a chromatographic system to separate the members of a multicomponent sample. If no separation occurs, then the system is nonselective and useless.

Mobile Phase

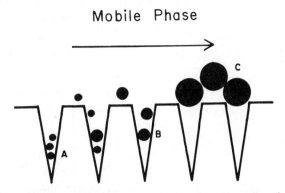

Figure 6.1   The mechanism of size-exclusion chromatography: A, small molecules, total pore penetration; B, intermediate molecules, limited pore penetration; C, large molecules, total pore exclusion.

If it separates all of the members of a very complex multicomponent mixture regardless of their chemical natures, then it is "infinitely" selective, which is impossible. Reality lies somewhere between. Selectivity is controlled by manipulation of the R values. The movement of the mobile phase along x is the *driving force* (strain). An increase in R means the average molecule spends more of its time in motion and has a greater apparent velocity and a smaller retention time. A decrease in R means the molecule spends more time in the stationary phase, which is the *retarding force*. The apparent velocity—the resultant of these two forces—decreases, and the retention time increases.

Figure 6.2 illustrates the conceivable interactions that occur. In GLC interaction of the mobile phase with the stationary phase is considered insignificant (very slight solubility of the carrier gas in the liquid and the liquid is saturated with the gas). The assumption that this interaction has no effect on the R values of solutes is satisfactory. This is not the case in GSC, where solute and carrier gas molecules compete for active adsorption sites and displace one another. The interaction is very important in LSC. In both GLC and LLC it is a tacit assumption that the solution in the stationary liquid is sufficiently dilute that there is no solute–solute interaction and Henry's law is obeyed (constant activity coefficient). This may not be the case in GSC and LSC (monolayer adsorption with or without neighboring solute molecules, multilayer adsorption, clusters, etc.). In LLC interaction of the solute with the mobile phase is a major factor. The R value is altered by using liquids that have different polarities or solute solubilities. The R value of a solute can be increased during the separation if solvents are mixed by a regulated composition change program. This is *gradient elution*. Thus, in LC there are a limited number of substances used for the stationary phase and a large number of liquids and their mixtures used for the mobile phase. Selectivity is engineered mainly by altering the mobile phase composition. In GLC and GSC *a mixture of solutes and mobile phase is assumed to be a mixture of ideal gases where there is no interaction among any of the species.* This is not exactly true, but very precisely controlled equipment is required to determine the effect on the R value of a particular solute as one changes from helium to hydrogen to nitrogen to carbon dioxide. *In GLC selectivity is governed by the choice of the stationary liquid and by the temperature.* This is the reason for so many, too many, liquid phases reported in

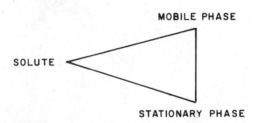

Figure 6.2 Interactions in chromatographic systems.

the literature. The ASTM gives data for about 350 liquids. In the counterpart for gradient elution the value of R is increased during the separation by continuously increasing the temperature. This is *programmed temperature gas chromatography* (PTGC).

1.3   The Sample

Chromatography can deal with a multicomponent mixture of unknown compounds at unknown concentrations. This is the most difficult problem for the analyst. It is eased in direct proportion to the amount of preinformation that is available about the sample.

1.3.1   Ionic or Molecular Species

If the sample contains ionic species, then the analyst must use LC, most likely with an ion exchanger as the stationary phase. The use of GC is inappropriate.

1.3.2   Volatility

The components must show some degree of volatility at the temperature of the column. This knowledge is impossible with unknown solutes. Even if the identity is known, vapor pressure data in the form of tables or curves is greatly limited. A property that parallels volatility is the normal boiling point. Although much more data exists, the identity, or at least the compound class, must still be known or suspected.

Both volatility and normal boiling point suffer the severe limitation that they are determined for the *pure* solute where vaporization involves breaking intermolecular bonds with identical neighboring molecules. It was assumed earlier that the solution in GLC is sufficiently dilute that the neighbors to the solute molecule are all strangers and the intermolecular forces are altogether different. With GSC the adsorptive forces holding the solute molecule on the surface may be vastly different from those involved on vaporizing the pure solute. Variation of the vapor pressure of the solute with temperature above the dilute solution in the stationary liquid or above the adsorbent is needed. This would require vapor pressure data for an enormous number of compounds dissolved in what at first appears to be over 300 liquids. Most likely such a compilation will never exist. The molecular weight of methane, 16, compares well with that of water, 18. The molecular sizes are also comparable. The normal boiling points are vastly different, $-164°$ and $100°C$, respectively. The explanation is the hydrogen bonding in pure water. In GLC one can select a stationary phase where the light hydrocarbons emerge before water and then, with ease, select a liquid phase where the water precedes the same hydrocarbons. This is not to imply that boiling points are completely useless. They are important in any homologous series of compounds where the functional group is the same for all. One must turn to chromatographic behavior. Most instruments are dual column, that is, two columns in the same oven, and one selects a column to use by injecting the sample into one of two injection ports. One column might be relatively polar while the other might be relatively nonpolar. This feature, coupled with PTGC can yield an appreciation of the sample. The sample is chromatographed with an initial temperature near ambient, for example, 60°C, and programs a temperature change up to about 200°C *if the temperature limits of the two stationary phases allow this*. This *temperature limit* is that at which the stationary phase becomes sufficiently volatile to escape from the column (*column bleed*). The final temperature is held constant for a reasonable period of time to insure that all components have passed through the column. This is repeated with the other column. With some thought, intuition, and experience, a feeling for the nature and complexity of the sample can be achieved and a stationary phase and column temperature or temperature program selected.

With low load columns, that is, those with a small amount of stationary liquid, one can elute a solute about 200°C below its normal boiling point. If the oven and columns can tolerate 200°C, then solutes of normal boiling points of about 400°C can be chromatographed by GLC.

1.3.3   Sample Stability

In the past the majority of columns were metal. If this hot metal or the support for the stationary phase acts as a catalyst for rearrangement or decomposition of the solute, the analyst has a problem. Thermal stability is another threat. If the solutes in the sample are known, the first step to avoid all this is to seek out the experience of others as reported in the literature. Hot metal perhaps may be solved by using glass or fused silica columns. A catalytic support requires a change of support material. Generally, decomposition of organic solutes proceeds with finite rates. The fragments will be lighter than the parent and move ahead. Because they are continuously produced they appear as peaks that severely tail into and join with the parent peak. To detect such decomposition it is also useful to carry out chromatography on the sample at a variety of flow rates to alter the residence time and thus the contact time with the catalyst or the high temperature. If peaks vanish, new peaks appear; some peaks are grossly distorted; relative peak heights change; then there is a strong indication of reaction on the column.

Decomposition in the injection port, if complete, is more difficult to detect. A glass or quartz line injection inlet is a must and is very common with modern gas chromatographs. Variation of the injection port temperature, if possible, may detect this.

1.3.4   The General Elution Problem

With samples where the solute boiling point range is 100°C or more or if the sample contains many different compound classes, such as ethers, alcohols, carbonyls, or polyfunctional compounds, then it is common for a low-temperature isothermal chromatogram to give a suitable separation and resolution of the low boilers. The problem arises with the high boilers, which appear much later: (1) They may not have appeared at all when the analyst terminates the chromatography. They appear in the next sample or during the *start-up* period the next day. (2) It is the universal experience, substantiated by theory, that the longer a solute remains on a column, the broader and flatter is its peak. It may not be seen at all, depending on the LLD of the detector, or may be attributed to long-term baseline drift by the analyst or, particularly, by an electronic integrator. If the temperature is increased, peaks undetected at the lower temperature may now appear, but the early peaks are crowded together and unresolved. The answer is PTGC and is discussed in Section 10.

A corollary to this is *wasted column* where suitable peaks are obtained for all solutes but groups of peaks are widely separated in retention time so that for long periods between groups nothing happens and the analytical cost increases.

1.3.5   Trace Quantities

Does the sample contain trace quantities of materials and must they be seen (environmental and toxicological samples)? This is a detector problem.

1.3.6   Water

If water is present, it will very likely yield a peak with extensive tailing with other solute peaks superimposed on it. Perhaps a stationary phase can be selected where the water appears as the last peak. The flame ionization detector (FID) does not respond to water; it is a *carbon counter*. Thus a mixture of alcohol and water need not be separated for the alcohol to be detected. However, the signal intensity generated by the alcohol mixed with water will differ from that of the same amount of alcohol alone. The water affects the ionization of the alcohol in the flame. The noise level will also increase and affect the LLD.

1.4   Information Desired

A goal must be established to select a chromatographic system that will provide the information required.

1.4.1   Purity

Some indication of the purity of a substance is often needed. It is now almost common for the supplier of spectrophotometric solvents to include a gas chromatogram of the product. Often it is not necessary to know the identities or the amounts of impurities. If this is the problem, then resolution and a superior LLD is needed.

1.4.2   Qualitative Analysis

It may be desirable to identify or at least make a reasonable guess about the species present from chromatographic behavior. The route to be taken, that is, the use of specific retention volumes, relative retentions, or the Kováts' retention index, must be planned, depending on the data available in the literature. Advanced planning will indicate the possible need for the recorder chart speed, the column inlet pressure, a controlled and measured flow rate, a carefully controlled and accurately measured oven temperature at a temperature corresponding to the available literature, the total amount of stationary liquid phase on the column, the liquid phase load, and so forth, to insure that a comparison with literature values can be made. Under certain circumstances the system and experiment may be designed so that only retention distances on the recorder chart are needed. Identification depends on an accurate determination of the position of the peak maximum. Snyder [6] has shown that peak overlap can shift this value. Thus good resolution is desirable.

1.4.3 Quantitative Analysis

In this case the identity of the solutes is known. It is meaningless to report the quantitative analysis of a mixture of unknown solutes. The information needed differs from that required for qualitative analysis. With some detectors it is essential that the carrier flow rate be constant but its numerical value is not needed. A flow controller is required. Calibration curves and response factors can be established with authentic samples. An analyst of any merit will establish in advance the number of significant figures desired in the result—the permissible standard deviation. A 5% relative error is easier and less costly than one of 0.5%.

Good resolution is a must. Overlapping peaks or a peak superimposed on the tail of a preceding peak or on serious baseline drift do not contribute to confidence in results obtained. Methods of peak deconvolution—mathematical and graphical methods of estimating peak areas for overlapping peaks—are the marks of poor chromatography.

1.4.4 Ancillary Techniques

Gas chromatography is used to separate and purify solutes to be examined subsequently by other methods, for example, IR and NMR spectroscopies. A gas chromatograph coupled to a mass spectrometer is a very powerful and popular combination. Knowledge of the minimum size sample required by the ancillary method and whether the gas chromatograph will accept this amount in a single pass is mandatory. Sample sizes may border on those of preparative GC. Ancillary techniques are discussed in [7, 8].

1.4.5 Preparative Scale

It was once envisioned that GC could be scaled upward by increasing column diameter to prepare large quantities of pure compounds (200 g, 500 mL). This proved elusive because of the rapid loss of resolution as column diameter exceeded about 1 in. Sample collection problems compounded the difficulties. If gram quantities are desired, it is preferable to program injection and collection to collect a particular peak on consecutive chromatograms. The problem is that of acceptable sample size. Cost of carrier gas, even if recovered, purified, and recompressed is no small factor. References [9–13] contain additional information.

1.4.6   Process Control

The sampling of a process stream, its analysis, and a decision about whether the process is operating at optimum are of great industrial importance and represent a very large portion of the GC instrumentation industry [14, 15]. The analyzer is located on-line in the plant where it may be subjected to severe changes in temperature; for example, seasonal temperatures may range from well below 0 to above 100°F. Vibration may be involved. The goal is ruggedness, reproducibility with continuous operation, simplicity, easy and infrequent maintenance, and speed of analysis. The problem is eased because it is common that a knowledge of the relative proportions of only a few solutes is sufficient for acceptable control.

1.5   The Scott Triangle

Scott and Kucera [16] summarized the situation simply and succinctly by the triangle (or tetrahedron) shown in Figure 6.3. Positions within the triangle may vary but they cannot include all three vertices simultaneously. A choice must be made.

It is essential to *define the problem and decide what information is needed.* It is advantageous to *know as much about the sample as possible before analysis.* These features determine the instrumentation needed and its operation.

Gas chromatography is powerful when employed within the range of its capabilities. With the development of HPLC, two very powerful separation techniques are available, each with its own realm of application.

2   EQUIPMENT

For a *conventional packed* (CP) *column*, particles of a porous solid, which is as chemically inert and as nonadsorptive as can be realized, are coated with a nonvolatile stationary liquid phase. Almost all of the liquid phase resides in the

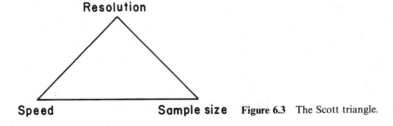

Figure 6.3   The Scott triangle.

pores. The coated particles are then packed in a metal or glass tube. The other popular column type is the *conventional open tubular* (COT) *column.* Here the inside wall of a long small diameter metal or glass column is coated with the liquid. In GSC and SEC the particulate stationary phase, without a liquid coating, is packed as with the CP column. Conventional packed and COT columns require different instrument modifications. It is a major error to use a COT column in an adaption of an instrument designed for a CP column. The COT column is always the loser.

Conventional packed columns with a thermal conductivity (TC) cell appeared first and this basic instrument has been modified to accept other detectors without radical design changes.

The TC cell is essentially a Wheatstone bridge (see Figure 6.4). The carrier gas always passes over a heated filament, which is controlled by the *bridge current* setting. This is the reference side. The sensing side, with the second heated filament, is downstream from the column. Both filaments develop a steady-state temperature and hence a steady electrical resistance as heat is conducted from the hot filament to the cooler wall of the TC cell block. The bridge is brought into balance with carrier flowing over both filaments by a variable resistance in the circuit—the *zero adjust.* This brings the recorder pen to the baseline, and the signal is "carrier gas only in both chambers." Now the sample is injected, and the individual separated solutes pass into the sensing side. The thermal conductivity of the solute, different from that of the carrier, causes a change in temperature of the filament; hence, its resistance is changed, and the bridge becomes unbalanced. A potentiometer, a null-seeking device, brings the bridge back into balance, essentially by moving a contact along a slide wire resistor; and the displacement of this contact is drawn on a strip-chart recorder as a peak, which registers the entrance and exit of the solute in the sensing side.

The sensing filament must be compared to the reference filament; the difference of resistances is the basis of the signal. This is the *single column configuration.* There are objections to it. To change columns, the column oven

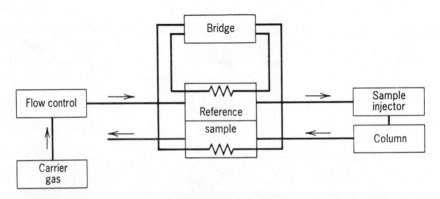

Figure 6.4   The single column configuration with a thermal conductivity cell.

must first be cooled, the column changed, and the desired temperature reestablished. Time is lost. Also, the pressure and flow rates in the reference and sensing sides differ because of the pressure drop across the column. If this is appreciable, it is sometimes difficult to balance the bridge.

The *dual column configuration* is shown in Figure 6.5. The carrier gas stream is split. Each branch has its own flow controller, sample port, and column. One branch is used as the reference side of the resistance bridge while the other branch is used for analysis by injecting the sample into its port. Column change involves changing from one sample port to the other. The advantages are (1) a change of columns is simple and fast; (2) the difference of pressure and flow rates between the reference and sensing chambers is much less than with the single column configuration. Disadvantages are (1) the additional cost of two flow controllers and two sample ports; the additional cost is quickly recovered by the technician time saved in changing columns; (2) the maximum column temperature is set by the temperature of the more volatile stationary phase.

Some detectors do not require a reference side. There is a single-filament TC detector that employs modulation of the mobile gas-phase flow and demodulation of the electrical signal followed by signal amplification. The FID requires only one burner. Commonly, dual column instruments are furnished with two flames even though only one is used at a time. With a single flame, column change is still tedious and dual flames are an advantage. Two electrometers are not required. We change from one set of electrical connections on one detector to the other when changing columns.

2.1 Mobile Phase

Figure 6.6 is a block diagram of a GC system. References [17, 18] contain further information about the mobile phase.

The identity of the carrier gas is set by the detector, not by the chromatographic process. Commercial gases used are furnished at about 2000 psi depending on the gas. A two-stage reduction valve is in order. Matching valves

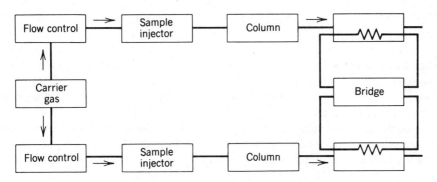

Figure 6.5 The dual column configuration with a thermal conductivity cell.

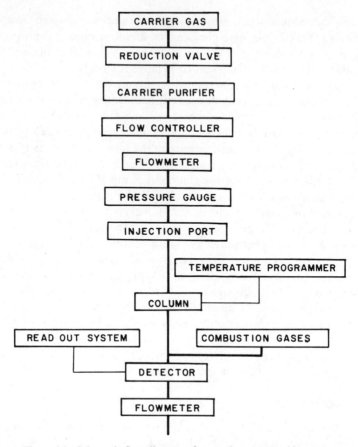

Figure 6.6   Schematic flow diagram of a gas chromatographic system.

to the tanks, which depends on the gas and all too often on the supplier, is a constant nuisance. The outlet pressure of the second stage should be at least 60 psi. This outlet pressure is set by the requirements of the flow controller. If the FID is used, then two combustion gases with appropriate reduction valves are required.

The purity of the gas depends on the detector and the problem. The TC cell has an LLD such that it does not qualify as a trace detector. The purity of the commercial gas, as received, is usually sufficient. The FID is a trace detector. If trace analysis is a part of the analytical program, then the rule of analytical chemistry that the reagents used must have a contaminant level much less than the component quantities in the sample applies to the carrier and combustion gases in GC; they are the reagents. Most practitioners of GC prefer to clean up their gases rather than purchase "ultrapure" gases. Water and hydrocarbons are removed by cartridges that commonly contain molecular sieves. These are readily available from GC suppliers. The removal of oxygen is a special problem and is not done routinely except when there are indications that components of the sample are easily oxidized. Stationary phases may also oxidize. Many of these have been discovered and dismissed from the ranks of readily available phases. If relative volumes begin to change during column use, something is happening to the column and stationary phase oxidation is a possibility. Generally it is completed within the first few inches of the column so that the several feet remaining are unaffected. If serious, change to a redundant phase, which may correct the situation (see Section 12.10 on stationary phases). The removal of oxygen is accomplished by readily available cartridges containing a catalyst. The efficiency of these devices depends on the extent of their contact with the active agent, that is, on the flow rate. Many suppliers furnish data on efficiency versus flow rate. Some instrument manuals suggest an oxygen trap to prolong the lifetime of the filaments of a TC cell. A 215-ft^3 cylinder at 1 ppm oxygen contains 6 mL of pure oxygen.

All tubing between the reduction valve and the chromatograph must be cleaned to remove the machine oil introduced during manufacture. Ether, acetone, methanol, water, and clean dry nitrogen, respectively, are drawn through the tubing. Very special care must be taken with the electron-capture detector to insure that the liquids used are free of chlorinated compounds. Plumbers tape used to seal the connectors will bleed enough to give some very fine chromatograms with some detectors.

2.2   Flow Controller

It is desirable to have a constant flow rate [19, 20] in all cases. It is essential in some. The TC cell is flow sensitive; that is, the area enclosed by a peak depends on the flow rate. The same amount of solute chromatographed at two different flow rates will give *different* peak areas. This presents no particular problem if relative peak areas are used or if an internal standard is used as long as the flow is constant during a particular chromatographic determination. If it is not, quantitative results are suspect. The hydrodynamics of flow shows that gas flow through a long capillary or across a small orfice is constant with minor vacillations of inlet pressure if there is an appreciable drop in pressure across this buffer, hence, the advice that the two-stage reduction valve be able to deliver the outlet gas at 60 psi. A precision-machined needle valve is the least one should demand, and it should be treated with the care warrented by fine tooling. However, this does not solve all problems. During a temperature program run, the temperature of the column is increased and the carrier gas viscosity increases. A needle valve will not maintain constant flow in this situation unless the pressure drop is very large. A PTGC from 50 to 300°C on a CP column at a constant inlet pressure of 14 psig showed a 50% decrease relative to the initial flow rate. When a capillary flow buffer with an inlet pressure 10 times that of the pressure drop across the column was used, there was a 2% variation in flow. A capillary inlet pressure of about 100 atm, a dangerous situation, would give a 0.2% variation in flow [19]. Also, measured flow rates are sometimes required if qualitative analysis is the problem. This cannot be done with a flame ionization detector because the column effluent is mixed with hydrogen before it reaches the flame.

With a differential flow controller the column may be detached from the detector; the flow rate established and measured at the outlet, all at room temperature; the column attached to the detector; and the oven temperature established. The flow rate remains the same. A Brooks Model 8744 flow controller for PTGC, again from 50 to 300°C on a CP column, gave an average variation of 0.06% relative to the initial flow rate with a column head pressure change from 16.2 to 26.5 psig [19].

2.3   Inlet Flowmeter

Many gas chromatographs are furnished with rotameters (falling ball flowmeters). The gas passes through a small diameter tube containing a ball. The position of the ball in the tube, as measured on a calibrated scale, is a measure of the inlet flow rate. There are a number of objections: (1) The position depends on the carrier gas identity. Nitrogen and helium at the same flow rate will give different readings. The flowmeter must be calibrated for each gas. (2) The same gas at the same flow rate but at different pressures within the tube will give different positions of the ball. (3) The precision of the reading is wholly inadequate for good qualitative work. Recent models are sold without rotameters, which is no great loss.

Column outlet flow rates are preferred.

2.4   Pressure Sensing

Qualitative identification, depending on the plan of attack, sometimes requires a correction for the pressure drop across the column. This requires a knowledge of the inlet pressure. The reduction valve gauge is neither accurate nor precise. A simple and ingeneous device is a pressure gauge sealed to a syringe needle. The needle is inserted into the injection port. This device is available for several pressure ranges.

2.5   Sample Injection and Injection Port

It is now assumed that the purified mobile phase is delivered to the injection port at a constant flow rate at a known pressure. The goal of all devices is to inject the sample as a pulse and get it onto the column as swiftly as possible to meet the requirement of migration from a narrow initial zone.

Gas samples are relatively simple. A bypass sampling loop, shown schematically in Figure 6.7, is one device. In the load position (solid line) the carrier gas passes through the valve and onto the column while the sample loop is being filled. A quick rotation of the valve passes the carrier gas through the loop and onto the column (dotted line). A return to the original position allows the loop to be recharged while the first sample is still on the column. The loop need not be

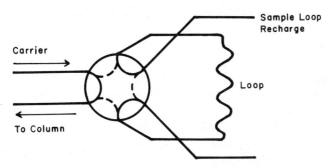

Figure 6.7 Gas-sampling loop: solid line, the carrier gas passes directly onto the column while the loop is filled from the recharge line; dotted line, the carrier gas passes through the loop and onto the column while the recharge line is flushed.

heated. Multiple loops may be used and the injection automated. There are some distinct advantages. The loop may be a coiled metal tube. This can be calibrated and changed in size if the volume per unit length of tubing is determined. It is the only way a precise amount of sample may be injected. Knowledge of the loop volume, gas pressure, and temperature yields a simple gas law problem. Such knowledge is difficult to justify in quantitative analysis where results are most often reported in relative amounts, but this method is very valuable for establishing detector characteristics. Gas-tight syringes are also available for syringe injection. Leaks are a hazard.

Liquids are injected most commonly by a form of a small volume hypodermic syringe; that is, by a calibrated glass barrel with a stainless steel plunger. Macro-size syringes (in GC terms) have the needles sealed into the barrel or have detachable needles. There are various needle sizes and lengths available. Bent needles or plungers are a plague. Needle guides that can be attached to the sample port and plunger guides that can be attached to the syringe are available. One may purchase a syringe cleaner, which is essentially a small high-temperature oven. The rinsed needle is inserted through a septum into the oven and clean dry nitrogen is drawn through the needle to bake it out.

The technique is to first draw clean air into the syringe and then some liquid sample. On injection, the air between the plunger and liquid purges the needle of the unmeasured volume of liquid retained in the needle. After a suitable liquid sample is drawn in, more air is pulled in. This gives a plug of liquid between an upper and lower meniscus that can be read against the calibration marks. The estimated error based on the combination of readings at both locations yields a very large relative error. This sampling method should never be used for quantitative analysis without an internal standard. Faced with the challenge of producing syringes to deliver smaller and smaller samples, a syringe was developed where the needle is the barrel and a very fine wire that fits the needle is the plunger. Inadvertent removal of the wire from the needle by pulling the plunger back too far is a disaster.

Liquid syringes are generally used to introduce solid samples as solutions. The volatile solvent appears first, and the detector is able to stabilize before the solid appears.

Sampling consists of every step from touching the needle to the surface of the liquid sample to passage of the vapor onto the column. Glass and metal are adsorptive surfaces. The syringe should be stored in a manner that prevents its exposure to contaminants in the laboratory atmosphere. It should be cleaned immediately after sample injection. The solvent used must be appropriate and pure. A syringe rinsed with carbon tetrachloride and baked out will show chlorine for months when used with an electron-capture detector. The syringe surfaces should be equilibrated with the current sample by rinsing them several times with the sample before injection. If one type of sample is routinely analyzed for certain components, it is wise to reserve a syringe for that sample type.

Resolution improves with small stationary liquid-phase loads. This requires a concomitant reduction in sample size (Scott triangle) to prevent *overload*. An equilibrium between the solute in the mobile gas phase and that in the stationary liquid phase is mandatory. If the sample is too large, the stationary phase becomes saturated with the solute. An amount in excess of the equilibrium concentration remains in the mobile phase. Very broad, malformed peaks result. This is a common experience with small-diameter COT columns. Perhaps no syringe delivers a sufficiently small sample to such a column and a heated *sample stream splitter* must be used to admit a fraction of the injected sample to the column C and then to discard the remainder to either the laboratory or a safety trap V. Crude examples are shown in Figure 6.8. The fraction, *split ratio*, admitted to the column depends on the diameter of the tubes used. Splitters with an adjustable split ratio are offered. A number of designs have been proposed to achieve a *linear split*; that is, the same fraction (for example, 10%) of every solute

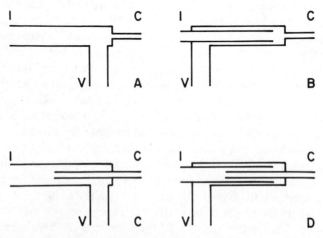

Figure 6.8   Sample stream splitters: I, inlet; C, column; V, vent to laboratory.

present is admitted to the column. This ideal is rarely realized, particularly when the components differ a great deal in boiling point, molecular weight, or molecular size. Careful calibration with a standard sample with a standardized procedure is the only safe recourse.

Van Deemter and colleagues [9, 21, 22] suggested that overload occurs when the vapor volume of the sample exceeds $0.5\ V_R/n^{1/2}$, where V_R is the retention volume and n is the plate number (see Section 6.1).

The sample injection port is heated to achieve flash vaporization of the sample. A liquid solution in the injection system, which slowly fractionally distills into the column, will never give suitable results. A suitable port has an independent heater to control the temperature. It is desirable to operate the block at a temperature 100°C above the boiling point of the highest boiler, *sample permitting*. It is common to find ports routinely operated at 200–250°C. Figure 6.9 is a schematic of a nicely designed port. The block is relatively large with a large heat capacity to insure temperature stability and to avoid cooling by sample vaporization. The carrier gas enters the port and immediately encounters a glass inner sleeve. The syringe needle passes through the septum into a glass tube. Even if the sample vaporizes to give a relatively large vapor volume and backs up in the port, it is highly unlikely that it will be exposed to hot metal. The glass liner and tube can be removed and cleaned. They should be inspected periodically for evidence of sample pyrolysis in the port. A sufficiently long syringe needle will pass through the inner tube down to the head of the column and the liquid sample injected directly onto the packing (*on-column injection*). If there is a battle for resolution, this is a good way to achieve a few more "theoretical plates." It is not done routinely because it is hard on columns.

Septums are generally self-sealing; that is, a viscous liquid is sandwiched between two layers and seals the puncture when the needle is withdrawn. A wide variety of septums are available that vary in sizes, price ranges, temperature ranges, and so on. Septum technology changes constantly, so specific advice is impossible. A reputable supplier is the best source of current septum information and specifications. Septum bleed is worrisome. Some analysts will "cure" their septums in place by heating them with carrier flow at the port

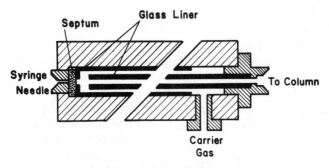

Figure 6.9 Sample injection port.

temperature for a period of time. It is wise to have a pristine syringe and needle available to puncture a septum to see whether a chromatogram results. The lifetime of a septum is short; it is good for about five injections. After multiple punctures, the septum begins to leak. This is particularly true when a needle guide is used. It always steers the needle to the same hole. When erratic behavior is encountered, *first check for septum leaks.*

Environmentalists dealing with air samples often encounter mandated species levels beyond the analytical capabilities of any instrument. Hence, they must preconcentrate their samples. One way to do this is to draw a monitored volume of air through a short section of tubing packed with an adsorptive substance. This tube is attached to the head of the chromatographic column, and injection consists of heating the tube rapidly so the trapped materials are flushed onto the column. The phantoms are all sampling problems involving the efficiency of the trap and complete expulsion of the trapped materials onto the column.

In large amounts, water can spoil chromatographic performance. If the solutes are volatile, one can withdraw a sample of the atmosphere above the water sample (*head space analysis*) and inject it.

A predetermination of the partition coefficients of the solute between the vapor and the liquid solution at the temperature of the sampling is required. Preconcentration steps are also involved in clinical and toxicological analysis.

2.6   Column Oven

The column is connected to the injection port and detector. Special connectors are required for glass columns. The column is housed in a separate oven with its own temperature control. The column is operated at a temperature very much lower than the sample port. The selectivity of the stationary phase operates only when the solutes are trapped in it and the sensitive intermolecular forces distinguish between slight differences in solute properties. Temperatures that encourage evaporation of the solute to place it in the rapidly moving mobile phase deter separation. Also, the number of employable liquid stationary phases decreases as the temperature is increased for reasons such as column bleed, thermal degradation, and oxidation. There is little reason for an oven temperature above 200°C. A "first" column temperature is midway between the lowest and highest boiler. If the mixture consists of many low boilers, they may still be poorly resolved. The temperature should be decreased or PTGC used.

All commercial ovens employ circulating air. The fan should provide turbulent flow. The oven compartment should be sufficiently spacious that the coils of the column are separated to contact the circulating air.

Good isothermal GC requires a high heat capacity oven; that is, the temperature does not change much with vacillation of current input to the heater. Such ovens are slow to reach the desired temperature and slow to cool down; but once the desired temperature is reached, they stay there. Programmed temperature gas chromatography requires a low heat capacity oven—the oven temperature must change fairly rapidly as energy is added. Instruments that provide both options are a compromise. A word of caution: *The thermometer measures the oven temperature and not the temperature within the column.* Thus fast program rates may be impressive, but they are meaningless.

Good temperature control is always desirable. Much effort has gone into this feature. One should demand $\pm 1.0°C$ even with a low budget. Claims of $\pm 0.5°C$ are reasonable, but those of $\pm 0.1°C$ and less stretch credibility.

Digital readout solid-state thermometers are a blessing. Dial readouts cannot be read to any desirable precision. However desirable a linear world may be, reality is nonlinear. If an accurate knowledge of temperature is essential, calibrate the thermometer.

2.7   Detector

The separated solutes now pass from the column into the detector, which has its own oven and temperature controls. The primary purpose of a heated detector is to prevent solute condensation in the detector. Thus, the injection port and detector are often housed in the same oven and operated at the same temperature. If the injection port block is heated directly with a cartridge heater, then the detector must have its own oven. For best results the TC cell should be operated at the lowest possible temperature that will still prevent condensation; the injector should be operated at the highest possible temperature. The two temperatures can differ.

2.8   Extra Column Contributions to Band Broadening

Ideally the injection port should be (1) a flash evaporator of zero volume, (2) flush with the column with zero volume (no tubing) between them, and (3) separated from the column oven by a perfect insulator of zero thickness. The same is true at the detector end. The injection port discussed closely approximates this, and on-column injection detours the problem. Any tubing volume leading from the column to the detector invites remixing of the separated solutes by diffusion. All connecting tubing should be of small diameter and as short as technically possible. The problem of extra column contributions to band broadening was treated in a definitive paper by Sternberg [23]. Instruments with multiple detectors fail in this respect because of the tubing involved leading to the various detectors. Such instruments serve a purpose, for example, an instructional instrument where experience, not results, is the goal. The problem and desired information should be defined and the appropriate instrument employed.

2.9   Flowmeter

Outlet flow rates are preferred. The simplest and almost perfect device is the *soap film flowmeter* shown in Figure 6.10. The carrier from the column enters the meter and passes through the calibrated tube. When a measurement is desired,

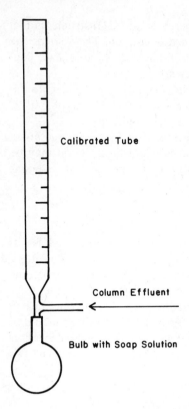

Calibrated Tube

Column Effluent

Bulb with Soap Solution

Figure 6.10   Soap film flowmeter.

the bulb containing a soap solution is squeezed to raise the liquid above the column entrance. Films rise up the tube. Once wet so that the film persists, the time is noted for a film to pass two calibration marks. The reciprocal is the flow rate. The device has never been improved. It is true that the volume measured includes the partial volume of water vapor. This depends on the partial pressure of the water vapor at the temperature of the flowmeter. Vapor pressure data for the soap solution, if known, are of doubtful use because equilibrium has not likely been established. If the retention index system is used, the question becomes largely an academic exercise.

Thermal mass flowmeters are available [19].

2.10   Pressure (Flow) Programming

To solve the general elution problem with thermally unstable solutes, investigations have been made into the programmed increase in inlet pressure and hence carrier gas flow rate (*pressure or flow programming* [24, 25]). Mechanical pressure control does not have the latitude that is achievable with temperature control. Low-load columns on deactivated supports have minimized the thermal stability problem.

2.11   Readout Systems

"The recorder will never be eliminated ... [19]." It is an honest visual display of the chromatogram. Regardless of the convenience of computer printouts, it is desirable to have a record of the chromatogram *before* the data has been manipulated. A recorder may be used alone or with a printer except when the printer furnishes recorderlike records.

The recorder's first requirement is that it be possible to match its input to the chromatograph's output; that is, the recorder must be able to accept the chromatograph's signal (voltage, impedance, etc.). For example, some recorders will have a *fixed range*, 1 mV, or a *variable range*, 0–1 V. *Response time* is very important. It is generally stated as either the time for the pen to travel from zero signal to the top of the recorder paper, which is the *full-scale response*, or for a specified instantaneous step to occur in the signal intensity. A full-scale response of 1 s is common. Slow recorders will distort the peak because the recorder pen cannot keep up with the signal. Peak positions, widths, and shapes are unreliable. If there is a rapid succession of many peaks, as with COT columns, a slow recorder is totally inadequate. Hence, response times from 0.25 to 0.50 s are mandatory. The *dead band* is the range of signal intensity that produces no response in the recorder. It should be about 0.1% of the full-scale signal. A wide dead band will miss small peaks that are important in trace analysis. As stated, the pen registers the position of a contact with a slide wire. The position of the pen may not be a linear function of the signal intensity. The *recorder linearity* should be 0.1% or less. A variable chart speed is a highly desirable option.

Recorders are limited by the width of the recorder paper. If sample components vary over a wide range, major components will be off-scale so that minor constituents can be seen. If the signal is *attenuated* (a fraction of the generated signal is passed to the recorder), a major component peak may be seen in its entirety, but small peaks are missed completely. Integration of the peak areas from the chart is a frustration.

Modern electronics has provided a remarkable solution in the form of the devoted computer—the reporting integrator—that can accept a very large signal and peaks are never off scale. A variety of options are available. Figure 6.11 is the printout from a reporting integrator. The signal is stored and recalled first as a record much like that of a recorder. The start of the chromatogram is indicated (event marker START) and each peak's retention time is printed above it and also in the first column of the table except where the area is ignored. Thus, the first peak is at RT (retention time) 0.09; however, this number and the area of this peak are not listed as per instructions to the computer—it is the air peak. The second column is the area in arbitrary area units. The peak type is a description of the peak as coded by the manufacturer, in this example, Hewlett-Packard Company. The notation BB indicates that the computer has seen the peak as beginning and ending on the baseline. The downward "ticks" on the record indicate when integration began and the upward "ticks," when it ended. The areas have been converted to area percentages in the last column. The

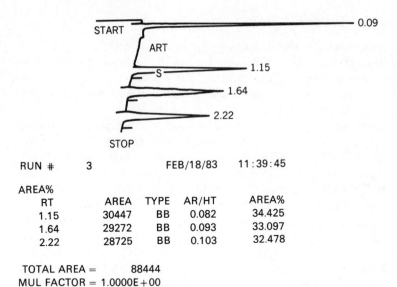

Figure 6.11 Printout from a reporting integrator, Hewlett-Packard Model 3390A.

column AR/HT approximates the width of the peak at half-height, presuming an isolated and symmetric peak on a flat baseline. It may be used to determine plate heights. Devices such as this will gain very wide use. Because of its ability to plot the signal, the recorder has not been "eliminated."

3 DETECTORS

Detectors are as important as the GC column. They are the eyes that see the separated components. General references are Perry [26], the books by David [27] and Ševčík [28]; and chapters by Gudzinowicz [29], Guichard and Buzon [30], and Sullivan and O'Brien [37]. Detectors may be organized in many ways. A *nondestructive* detector senses the solute without destroying it in the process; this is desirable when sample collection is planned. The thermal conductivity (TC) detector is an example. *Destructive detectors* consume the solute. *Ionization detectors* [32] form such a group, for example, flame ionization detector (FID) [33], electron capture (ECD), and mass spectrometer (GC/MS). Also included are the flame photometric detectors (FPD), which monitor the light emission of thermally excited species, and the coulometric detectors (CD) [34], which produce an ionic species with electrochemical properties. Nonionization detectors have been grouped [35]. *Nonselective detectors* sense all molecular species, whereas *selective detectors* [36] respond to compounds containing specific elements, such as, halogens, sulfur, phosphorous, and nitrogen. The signal intensity of *concentration detectors* depends on the solute concentration in the detector, whereas *mass flow detectors* depend on the mass delivered per unit time to the detector. Figure 6.12 demonstrates the difference between an *integral* and a *differential* detector. The integral detector sums the signal. It is exemplified by the titration cell used by A. T. James in the very early work with volatile organic acids. The effluent stream passed into the cell. The pH remained constant (first portion of the signal curve at small retention times). When a solute arrived in the cell, the pH changed, and the solution was returned to the original pH by adding base from a buret. When the solute had eluted, the pH again remained constant until another acid appeared (first plateau of the signal curve). The plot corresponds to the buret readings against time. The differential curve records the events: no solute in the detector; solute enters the detector and increases; solute reaches a maximum amount; solute decreases in the detector; and, finally, no solute in the detector. Mathematically, the differential record is the first derivative of the integral record where the maximum of the former corresponds to the point of inflection of the latter.

The detector should be viewed as the actual sensing device plus all associated electronics and the readout system. Each component contributes to the performance. Erratic random variation of the signal is called *noise*, and it is inherent in any electronic device. This is shown in Figure 6.13. A noise band of width n can usually be estimated from a strip-chart record. A signal is taken as signifying a solute peak if the signal-to-noise (S/N) ratio is 2:1 or greater. This is an important consideration when doing trace analysis with a COT column where one must distinguish solute spikes from noise spikes.

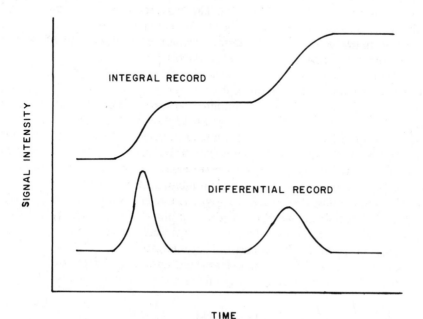

Figure 6.12 Integral and differential detector signals.

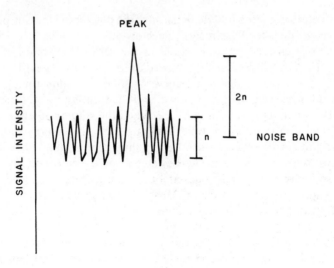

Figure 6.13   Signal-to-noise ratio.

Figure 6.14 demonstrates some important detector parameters. The *lower limit of detection* (LLD) is that amount of solute at which the signal can no longer be distinguished from the noise. If the noise can be reduced from level 1 to level 2, the LLD is improved. Much of modern detector technology is devoted to noise reduction with existing detectors. The increase in the signal intensity ΔI with an increase in the amount of solute Δm is the detector *sensitivity S*. As remarked, many authors use the term *sensitivity* when discussing LLD. The *linear range* is the range of variation of *m* where *S* is constant. The *response time* is the time required to reach the final deflection for an instantaneous unit increase of the solute in the mobile phase stream.

Here *response* will mean the signal intensity for a certain amount of solute, mass, or concentration that is being sensed by the detector. Ideally it is determined for a known amount of solute in the carrier stream at a particular flow; that is, it is a steady-state parameter. The detector parameters are then varied. Such detailed studies are rarely performed in the applications laboratory, because it is presumed that the instrument manufacturer has done this and has incorporated the information in the specifications. The wary will inquire about whether the specifications are for the detector alone (perhaps determined with "top-line" electronics and readout) or for the total detection system of the instrument proposed (with less costly electronics).

The peak area A is the integral under the response curve $R(t)$ from time t_1, the start of the deflection, to time t_2, baseline recovery, as shown in (2)

$$A = \int_{t_1}^{t_2} R(t)\, dt \qquad (2)$$

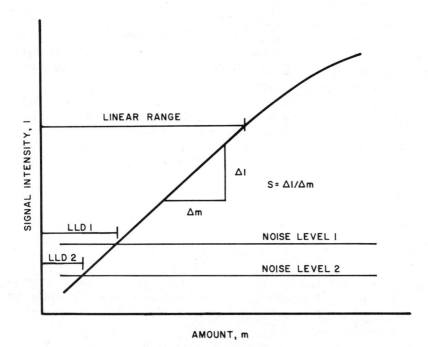

Figure 6.14 Detector parameters.

Ideally the response $R(t)$ as a function of t is a Gaussian curve. To simplify the argument, it will be assumed that $R(t)$ is a step function; that is, $R(t) = 0$, until t_1 is reached, when it instantly changes to $R(t) = R$ at t_1 and remains at this value until t_2, when it again changes to $R(t) = 0$. The response is constant from t_1 to t_2 and can be removed from the integral. Thus

$$A = R\,\Delta t$$

For a concentration detector the response is proportional to the concentration of the solute in the detector; that is,

$$R = K_c C$$

where K_c is a proportionality constant, the *response factor*, which depends on the detector system and often on the nature of the solute; this is an important feature to be discussed. Then

$$A = K_c C\,\Delta t$$

The concentration is $C = M/V$, the mass per unit volume of gas. The volume sensed is determined by the flow rate F; that is, $V = F\,\Delta t$, so that

$$A = K_c(M/F\,\Delta t)\,\Delta t = K_c(M/F) \tag{3}$$

The conclusion is that for a particular solute mass the area generated depends on the flow rate—the faster the flow rate, the smaller is the area. The detector is *flow sensitive*. Two conclusions arise: (1) Peak areas from two different separations cannot be compared quantitatively unless the two chromatograms are performed at identical flow rates. (2) Peak areas cannot be compared on the same chromatogram if the flow rate changes sporadically or continuously during the separation, for example, PTGC. This is the justification for previously emphasizing the differential flow controller.

A mass flow detector depends on the mass of the solute in the detector; that is,

$$R = K_m M$$

hence

$$A = K_m(M/\Delta t) \Delta t = K_m M \qquad (4)$$

where $M/\Delta t$ is the mass flow rate, M being the amount of mass entering the detector in time Δt. The area is now proportional to the mass. Flow rates may change and peaks differ in widths and degrees of sharpness, but the areas will be the same for the same masses.

The *attenuator* is a group of resistors so wired that a fraction of the signal may be directed to the readout device. It does this by taking 1/2, 1/4, 1/8, ..., $1/2^n$ of the signal generated by the detector. The primary intent was to allow the analyst to keep peaks on the recorder chart; that is, to insure that they did not go off scale, which occurs when the recorder pen reaches its upper limit and remains there until solute is swept from the detector. This is no longer a restriction with an electronic integrator that can accept and store a large range of signal intensity. However, the record shown in Figure 6.11 can still go off scale because it is restricted by the width of the tape; but the area units are still reliable. The disadvantage of attenuator use is that if the large peaks are brought on scale, small peaks, reduced by the same amount, may be lost. It is tempting to use the attenuator to reduce noise to provide a visually more appealing record, that is, a straight baseline. The danger is to attenuate so that a straight line is drawn even when a small amount of solute passes through the detector. Very uncluttered chromatograms can be obtained by using a high attentuation. The final error is then to increase the sample size to obtain useful peaks. The analyst must pretend that operation is still within the linear range of the detector and that the column has not been overloaded. A detector system should be "apprehensive" and not "complacent" or sluggish.

3.1 Thermal Conductivity Detector

The thermal conductivity (TC) detector is a nondestructive, flow-sensitive, nonselective detector. It is not considered to be a trace detector; that is, its LLD is 10^{-6}–10^{-7} g; it is linear over about four decades [27]. Such numbers are not absolute because they are very dependent on the system at hand. The basic principles were described earlier.

Heat may be conducted from the hot filament to the wall by radiation (trivial in this case); convection, which is a macroscopic movement of gas from the hot wire to the wall; and thermal conductivity, which is a kinetic molecular transport process. Molecules near the hot wire at a high temperature have a large kinetic energy that they transfer by energetic collision to neighboring molecules, which in turn transfer it to their cooler neighbors. Heat is transferred from the hot filament to the cold wall but molecules do not migrate across this space. Figure 6.15 shows two cell designs. In the flow-through cell, all of the carrier stream passes over the filament. The flow-through cell responds immediately to solute in the stream, but because of convection currents created by the flowing stream, the detector is noisy. In the diffusion cell the carrier stream does not flow through the filament chamber, because there is no outlet, and the solute must diffuse from the stream to the filament and then diffuse back out as the solute passes the chamber entrance. The diffusion cell is quiet in terms of noise but has a very slow response time. A compromise incorporates both flow patterns.

The response is proportional to the relative difference in thermal conductivities of the solute and mobile phase compared to that of the carrier gas; that is,

$$R \propto (\gamma_c - \gamma_s)/\gamma_c = R_\gamma \tag{5}$$

where γ_c and γ_s are the thermal conductivities of carrier and solute, respectively. Table 6.1 shows R_γ for ethane in selected carriers [27]. Hydrogen and helium are the preferred mobile phases because they have thermal conductivities larger than any other species (smallest molecular weight). The R_γ values are sufficiently

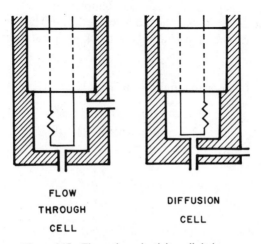

FLOW
THROUGH
CELL

DIFFUSION
CELL

Figure 6.15 Thermal conductivity cell design.

Table 6.1 Contribution to the Response R_γ for Ethane

Carrier	Thermal Conductivity [(cal/cm·s·deg)($\times 10^5$)]	R_γ
Ethane	7.3	
Hydrogen	53.4	0.863
Helium	41.6	0.824
Nitrogen	7.5	0.027
Carbon dioxide	5.3	−0.378

close that helium is preferred for reasons of safety. Nitrogen demonstrates that the response is very poor for solutes of thermal conductivities near that of the carrier. A similar situation exists for carbon dioxide. The negative value for ethane in this gas indicates a negative deviation from the baseline; this is a great inconvenience. Table 6.2 shows the R_γ values for selected species in helium [27]. This shows that equal responses may not exist for equal concentrations of different solutes; for example, methyl iodide differs from ethane by a factor of 1.16. This difference may not be important for qualitative identification, but the relative peak area for quantitative analysis will be about 0.85 instead of 1 for a 50:50 mixture.

The response depends directly on the resistance of the filament and the current that runs through it, the *bridge current*. The upper limit of this current is reached when the filament melts. The bridge current is set by the composition of the filament and the identity of the carrier gas. The instructions accompanying the instrument *must* be consulted and obeyed. If the bridge current is turned on when there is no carrier flow, the lifetime of the filament is only a fraction of a second.

The response also depends on the temperature difference between the filament and the detector wall, the block temperature. The filament temperature must be below the melting point and the block temperature must be above the

Table 6.2 Contribution to the Response R_γ for Solutes in Helium

Solutes	Thermal Conductivity [(cal/cm·s·deg)($\times 10^5$)]	R_γ
Nitrogen	7.5	0.820
Ethane	7.3	0.824
n-Nonane	4.5	0.892
Benzene	4.1	0.901
Chloroform	2.5	0.940
Methyl iodide	1.9	0.954

boiling point of the highest boiling component of the mixture. This point should be remembered when dealing with permanent gases. The block temperature can be equal to room temperature thus enhancing the response. A high heat capacity block is desirable to maintain a constant temperature.

3.2   Flame Ionization Detector

The flame ionization detector (FID) is a destructive, mass flow (flow-insensitive) detector. Some classify it as nonselective, others as selective. It will not respond to substances of high ionization potential, such as, water, hydrogen sulfide, oxides of carbon, nitrogen, and sulfur. It responds to organic compounds, which are of the greatest interest to the analyst. It has an LLD of about 10^{-10} g, which makes it a trace detector, and a linear range of six to seven decades [27], which is the best range of all detectors.

Figure 6.16 is a schematic of the FID. The effluent stream is mixed with hydrogen and burned in an envelope of air or oxygen. Solute combustion produces charged fragments. An electric field is established between the flame tip

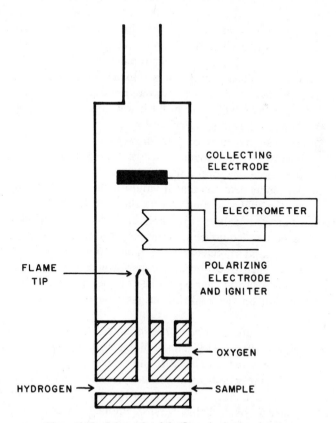

Figure 6.16   Schematic of the flame ionization detector.

and a collecting electrode. In this field ions produce a current, which is monitored. The direction of the field may be reversed by a polarity switch.

At one time much attention was given to the nature of the ions and the production mechanism. Combustion chemistry, still much a mystery, has had little impact on detector technology. Efforts to improve its performance by altering both the collector shape and spacing contributed little. When kept clean and used properly it is a superb and rugged detector. It is generally supplied with an igniting coil, which should be used. A cotton swab soaked in acetone, ignited, and thrust down the stack will produce a lovely yellow flame and noise for days. A colleague maintains that if you peer down the stack and see any flame, even light blue, the detector is dirty. Condensation of water on a watch crystal held above the stack is a suitable positive test for a flame.

The preferred carrier gas is nitrogen. Helium, with its high thermal conductivity, cools the flame and the ion yield is lower than with nitrogen. The combustion gases must be at the same purity level as the carrier. When used as a trace detector, a most appropriate application, carrier, and combustion gas purity is mandatory. The greatest diligence, short of using ultrapure grade gases, will still leave some hydrocarbon in the gases (subnanogram purity is required), which will produce a steady ion current. The "zero adjust" introduces a bucking voltage so a zero baseline can be achieved, but the background ions are still there. Imagine injecting a pulse of ultrapure nitrogen into a stream of "dirty" nitrogen carrier. When this pulse reaches the detector, the standing current is reduced and a negative deflection will be produced on the readout. If the sample contains a hydrocarbon level identical to that of the carrier, it will not be seen at all. All sample responses will be reduced below the potential response by an amount equivalent to the standing current.

The response also depends on the flow rate of the combustion gases. Generally the hydrogen and carrier gas flow rates should be equal, and the oxygen or air flow should be several times larger than this. The values depend so much on the particular detector that it would be a presumption to state such numbers; instead the analyst should consult the operations manual. Calibration would involve a stream of dilute hydrocarbon in nitrogen introduced into the detector and responses would be determined at a variety of combustion gas flows—a project not undertaken eagerly in a busy analytical laboratory but one that is worthwhile when trace analysis is performed. It is a puzzle why some instruments will have a rotameter positioned on the carrier gas stream, where it is ineffective, but not on the combustion gas lines, where precise metering is essential.

The FID is essentially a *carbon atom* counter. An awareness of response factors is critical. For example [37], if the molar response of octane is set at 800, then that of methane is 104; ethane, 201; and n-propane, 305. Equal peak areas of methane and ethane would mean the moles of methane are twice that of ethane. The ratio is not exactly the ratio of the number of carbon atoms, for example, 1.93 instead of 2. Table 6.3 lists molar responses of four-carbon compounds when heptane equals 700. Both the oxidation state and position

Table 6.3 FID Relative Molar Response
Factors of Four-Carbon Compounds[a]

Compound	Relative R
Diethyl ether	300
Butanol-1	356
Butanol-2	336
Butanone-2	327
Butanal	297

[a]Heptane = 700.

influence the response. A molecule can be thought to consist of structural groups; for example, CH_3, CH_2, 1°, 2° alcohols, and so on. A contribution to the response of the FID can be assigned to each of these groups (the group constants) based on an experimental study of a limited number of representative compounds. The response of a compound that is not a member of the group is calculated from the sum of the appropriate group constants. When the calculated values for compounds are compared with the experimental values of the same compounds, the error is greater than the experimental error of the determinations involved. The method has not been successful and if response factors are required, they should be determined directly for the compound of interest with the instrument at hand.

3.3 Alkali Flame Ionization Detector

The designation of this group is inconsistent; and the terms *thermionic detector* (TID) and *alkali flame ionization detector* (AFID) have been used interchangeably. Farwell, Gage, and Kagel [38] separate AFID from TID. The early literature refers to the *N/P detector*. All are based on introducing an alkali metal salt into the FID to enhance the detector's response to certain elements—the halogens, nitrogen, and phosphorous. It also responds to Si, S, B, As, Sb, and Bi. Previous detectors were not selective for nitrogen. Development of this group of detectors filled a need in pesticide and clinical, toxicological, and forensic drug analyses. It is not flow sensitive but this parameter is lost among the many others that must be controlled for optimum operation. This variability coupled with the range of response factors results in an almost meaningless LLD, but 10^{-12} g is about right. It is linear over about three decades.

The alkalis produce a greatly enhanced ion yield when particular elements are present. Rubidium seems to yield the best results. The following are representative of the several ways possible to introduce the alkali salt into the flame [27]: (1) directly into the flame by a probe that is holding the salt, (2) above the flame on a coated wire screen or bead, (3) at the base of the jet in a cup, or (4) by using a bushing as the flame jet. Two flames, *stacked flames*, have been employed; the lower flame is a common FID configuration. It is followed by an alkali-coated screen, above which is a second FID with separate flow controls.

As many as seven mechanisms have been proposed for ion production [27], and not all of the detector configurations appear to operate by the same mechanism.

The single-flame AFID is erratic and unstable because the single flame must produce both sample decomposition and alkali ionization; and the alkali salt is depleted rapidly.

The next development was the flameless nitrogen/phosphorous (N/P) version; there is an enhanced response to nitrogen when a heated rubidium- or cesium-containing bead is placed in the gas stream when the hydrogen–air ratio is too low to sustain a flame [38]. This is the TID version: For nitrogen compounds, it is thought that the CN-free radical is produced; ionizes the alkali, that is, Rb^+; and produces CN^-.

Response factors for both the AFID and TID are approximately proportional to the number of atoms of the species of interest. To obtain reliable quantitative results, response factors must be determined from authentic standards.

Farwell and his colleagues [38] summarize the situation: "The best practice, considering the current state of affairs, is to optimize carefully the particular detector in the user's laboratory by standards which contain the compounds of interest." David [27] catalogs many of the parameters to be considered.

3.4   Flame Photometric Detector

Certain species, particularly sulfur and phosphorous, are chemiluminescent in a hydrogen-rich flame. The emitted light is passed through suitable filters and the characteristic wavelength sensed with photomultiplier tubes, one for S (394 nm) and another for P (526 nm). The emission is delayed so the halo above the flame is largely free of background light in the flame proper (*flame noise*). It is a selective flow-dependent detector. It shows some response to the halogens N, B, Cr, Se, and Ge. The LLD is about 10^{-11} g [27] and it is linear (log/log scale) over two (S) to three (P) decades. It is most commonly used for sulfur. The response to S is 10^3–10^4 times that of P. The flame photometric detector (FPD) is limited by the flame noise, which cannot be eliminated completely. Nitrogen as the carrier is preferred. The FPD often has an additional collecting electrode to produce a simultaneous FID response.

The emitting species seems to be the excited S_2 [38]. The mechanism is bimolecular, which necessitates the log/log plot. The intensity of the emitted light I is given by

$$I = I_0[S]^n$$

where I_0 and n are constants and $[S]$ is the mass flow rate. In principle $n = 2$ for a bimolecular reaction. In practice n varies between 1.5 to 2 depending on the flame conditions and the specific compounds [38]. Careful calibration is required.

Although the FPD is a very fine detector, some [38] feel the TID is superior, particularly for phosphorous detection.

3.5 Electron-Capture Detector

The FID, AFID, TID, and FPD are grouped together because their ionization is thermal. The *electron-capture detector* (ECD) does not involve ionic fragments but is based on groups within the molecule capturing electrons. Of these groups, the halogens are outstanding, so the ECD is halogen selective. In view of some of the eccentricities of the detector, it would likely have remained obscure were it not for the importance of halogenated pesticides and the toxic and carcinogenic properties of chlorinated hydrocarbons. The LLD is so superior that halogenated derivatives are often formed of nonresponsive parent compounds to detect them at the trace level [28].

The LLD is about 10^{-12}–10^{-13} g [27]. This is a poor number because of the large variation of response factors. The linear range for the commonly used radioactive ECD is one to three decades [27]. A nonradioactive thermionic ECD has been reported with a six-decade linear range [39]. This is a significant advance. A small linear range coupled with the LLD presents severe calibration problems, that is, the preparation of reliable calibration standards at very dilute concentrations. Argon or nitrogen with 10% methane is the carrier.

A weak radioactive β emitter establishes a steady current in an electric field across the column effluent stream. Any molecule with a group of great electron affinity captures some of these electrons; the high-molecular-weight molecules are not deflected by the weak electric field, and the charge is carried out of the field. The monitored current is reduced, and a "negative" deflection results. The detector is easily "swamped," so the standing current is reduced rapidly, and an increase in sample does not further reduce the already weak current by any proportional amount. This continuous current method has been replaced by pulsing the electric field, and the performance is improved.

The early tritium foil source, a health hazard, was replaced by a ^{63}Ni source.

The reviews of David [27] and Farwell and colleagues [38], and Farwell [40] are required reading for a thorough understanding of the ECD and an appreciation of its eccentricities. Electron-capturing contaminants must be avoided. The carrier gas must be free of water and oxygen. Septum and stationary phase bleed must be avoided. The laboratory atmosphere and all solvents must be halogen free. Glassware in general and syringes in particular must not be contaminated. Flow regulators with plastic diaphragms must be avoided.

The electron-capture mechanism is not understood completely. Such studies are often suspect because of the inability to guarantee that the equipment used is not contaminated. Table 6.4 lists the relative response factors (1-chlorobutane = 1) selected from David [27] and demonstrates the variability and the lack of a simple correlation between the factors.

Table 6.4   Relative Response Factors of the ECD[a,b]

Compound	Relative Response
1,4-Dichlorobutane	15
1,1-Dichlorobutane	110
Carbon tetrachloride	4×10^5
Acetone	0.5
2,3-Butanedione	5×10^4
Benzene	0.06
Toluene	0.2

[a]See David [27].
[b]1-Chlorobutane $= 1$.

3.6   Electrolytic Conductivity Detector

The effluent organic substance is converted to an inorganic ion, and the ion is monitored. Coulson [34] reviewed the early work. Modifications and improvements by Hall and colleagues produced a version that has become known as the *Hall electrolytic conductivity detector* (HECD); it has been well reviewed by Farwell and colleagues [38]. This detector has been used for halogens, nitrogen, and sulfur compounds. The LLD is $10^{-9}–10^{-10}$ g and is linear over two to four decades, depending on the element sensed and the equipment used. When this method is employed, three essential steps, each with its accompanying technical minutiae, are followed.

1.   *Furnace chemistry.*   The parent compound must be reproducibly converted to the ionic species. The furnace may be operated in a *reductive mode* with hydrogen—from nitrogen to ammonia; an *oxidative mode* with oxygen or air—from sulfur to its oxides; or a *pyrolytic mode* with an inert gas—from a halogen to its hydrohalide.

2.   *Cell chemistry.*   The converted species must be dissolved in ionizable form. This requires effective contact between the gas and a suitable conductivity solvent (*scrubber*).

3.   *Electrochemistry.*   There must be a sensitive electrochemical method for detecting the ionic species, such as, halide ion or acidity.

Steps (1) and (2) require consideration of interfering substances.

As an example, chloride in the form of hydrogen chloride is passed into a cell containing silver ion and sensed with silver and reference electrodes. The change in cell potential is detected and a servomechanism operates a silver ion-generating electrode, which restores the silver ion concentration. The electrical charge required is measured (current–time curve).

An initial reaction to all this is that there must be an easier way. The same remark may be heard from those encountering the ECD. Farwell and colleagues

[38] state that lindane has been determined at the 10-pg level and that the HECD "has been applied to nearly all those analyses which were previously accomplished only by ECD." Although the sample for ECD must be water free, those for the HECD and toxic chlorinated organics can be determined in water, air, and biological samples without being dehydrated.

For nitrogen as ammonia, the scrubber is KOH, $BaCO_3$, or $Sr(OH)_2$ in a 1:1 n-propanol/water electrolyte. The pH is sensed. It has been applied to herbicides, pesticides, and drugs.

For sulfur as SO_2/SO_3, the scrubber is Ag^+, and the solvent is methanol. Chlorine interference can occur, and results are comparable to those obtained with FPD. The detector has been used for sulfur-containing pesticides, sulfur in fuels, drug analysis, and sulfur dioxide in air.

3.7   Mass Spectrometer

The mass spectrometer (GC/MS) [7, 41–44] is the only identifying detector in existence. The struggle to wed these techniques has produced a formidable literature. The quadrupole mass spectrometer is the unit of choice. The major problem encountered with this instrumentation occurs when trying to reduce the outlet pressure of 1 atm from the GC to the inlet pressure of about 10^{-6} torr without sample loss. Since the MS is not a trace detector, sample input to the MS must be matched with that to the GC. This does not dictate CP columns, however, as COT columns have been used [45]. Where the GC outlet flow rate defies pressure reduction by the best pumping technology, an interface must be used. The goal is to remove the carrier gas to reduce the pressure without depleting the sample. Two interfaces exploit the large diffusion coefficient of helium. Figure 6.17 is a schematic of the Becker–Ryhage or jet separator. The effluent stream enters an evacuated chamber through a small orifice; the stream is then directed at another orifice, which leads to a second evacuated chamber; and so on. Each chamber is at a successively lower pressure. In each chamber the very low-molecular-weight helium diffuses rapidly out of the jet stream to be pumped away to enrich the stream in solute in each chamber. Of course some solute is lost in proportion to its diffusion coefficient. Figure 6.18 is a schematic of the Watson–Bieman or pore-effusion separator. The helium passes through

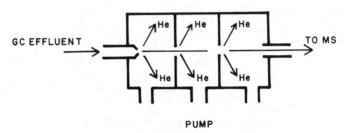

Figure 6.17   Becker–Ryhage or jet separator.

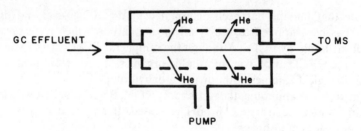

Figure 6.18 Watson–Bieman or pore-effusion separator.

the walls of the glass tube (0.1–1.0-μm pore diameter). Some other materials are Teflon and metal membranes. Figure 6.19 is the Llewellyn–Littlejohn or solution separator. A thin (0.002-in.) polymer membrane acts as a solvent for the organic solute but not for the helium. The carrier gas sweeps by while the solute diffuses across the membrane to enter the mass spectrometer. McFadden [45] reviewed interfacing. These interfaces demand large column resolution because the solute must clear the interface before the next solute enters. Memory effects occur. An important technological advance has made it possible to use PLOT and COT columns without an interface, which has reduced sample loss and is less demanding of solute resolution.

The solute enters the ionization chamber. If sufficiently energetic, the solute fragments to produce a characteristic pattern of ions. Mass spectra consist of "lines" at particular mass/charge ratios, or, if singly ionized, at particular fragment mass values. These numbers coupled with the intensity of each line, combine to give the unique pattern for solute identification. A panel of participants at a symposium was once asked for its top priority wish, and the GC/MS most closely satisfies that of Purnell: a recorder that is able to stamp out the identity of each peak. The "however" of all this is that the pattern depends on the ionization source, as discussed by Milberg and Cook [46].

An alternate to fragmentation is chemiionization where a low-energy electron source, such as a β emitter, produces ion molecules in the ion chamber, such as $(H_2O)_nH^+$, which then undergo charge transfer with solute molecules to

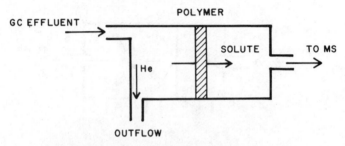

Figure 6.19 Llewellyn–Littlejohn or solution separator.

produce solute ion–molecules that enter the mass spectrometer. Essentially, the solute is tagged with a charge. Ion–molecule production is performed at atmospheric pressure and is termed API. Even if more than one solute ion–molecule species is produced, the spectrum is much simplified, of course, at the sacrifice of the information provided by a fragmentation pattern.

Identification requires a library of mass spectra and inclusion of a computer in the system is mandatory for full exploitation [47, 48].

The mass spectrometer has its own demands. Mass spectral resolution depends on the mass range scanned and the scan speed. These parameters are limited by the time the solute is in the ionization chamber. Compromises must be made. Rapidly emerging solutes require a rapid scan of a small mass range. The analyst cannot expect an optimum mass spectra. Septum and column bleed must be minimized.

The result of coupling retention data, mass spectra, and a computer is possibly the ultimate analytical tool. The computer can compare the stored signals with a library of spectra to propose solute identities that can be verified with an authenticated sample in the particular instrument in terms of both spectra and retention. Or, the computer can recall only certain mass/charge-relative intensity data for fragments characteristic of a particular solute, such as lindane, or a pattern of fragments characteristic of a family of compounds, such as chlorinated pesticides. The system becomes a selective detector. Or the computer can also combine ion currents and, with a time parameter, produce a typical chromatographic solute peak; that is, it is used as a general detector. Integration of the curve area and inclusion of response factors yields quantitative data. When data is stored, these manipulations can be applied to a single separation.

The answer to the inevitable question of what is wrong with it, is cost. A conservative estimate of the initial cost for a "top-line" system is $100,000; in addition 20% of this amount must be expanded per year for supplies and maintenance, and a highly qualified operator must be paid a yearly salary to operate the system.

4 CHROMATOGRAPHIC DATA

The terms and symbols used here are those recommended by ASTM E-19 and are based on the readout of a strip-chart recorder as shown in Figure 6.20.

4.1 Column and Operating Parameters

These parameters are measured without reference to the chromatogram and are listed with their adopted symbols.

Column length L.
Column inside diameter d_C.

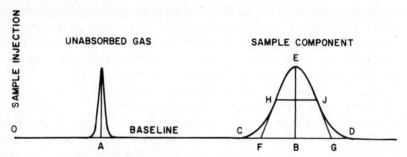

Figure 6.20   Strip-chart chromatogram.

Column inlet pressure p_i.

Column outlet pressure p_o (laboratory pressure).

Absolute temperature of carrier gas T (K). This temperature is determined at the position where the gas flow rate is measured, for example, outlet flowmeter temperature.

Absolute temperature of the column T_C (K).

Weight of liquid phase in column w_L. Unfortunately this determination is often ignored, because the procedure is considered too time-consuming to perform in laboratories doing routine analysis. Two weight parameters are important. The first is the percentage by weight of the liquid load in the packing material used. This can be vital when either attempting to reproduce separations reported in the literature or studying the properties of the stationary liquid phase. Liquid and support material appropriate for the desired liquid load are weighed out, and the support is then coated. A weighed amount of the final packing is extracted with a solvent, which dissolves the stationary phase, and the remaining support material is weighed. This gives the liquid load. When preparing the column, the packing material, taken in excess, is weighed before and after packing; the difference is the weight of the packing in the column. This yields the second important parameter w_L.

Density of liquid phase in column ρ_L at the column temperature. This number is not commonly available. If the number of liquids used is reduced (preferred stationary phases) there is some hope that these values will be determined.

Volume of liquid phase in column $V_L(w_L/\rho_L)$ at the column temperature.

Volume of supporting solid in column V_S. This is, as advised by ASTM, the volume expansion of methanol or its equivalent when the column weight of supporting solid is submerged in it at column temperature. This is not simple. Popular solid supports are very porous and to be accurate the support should be degassed to remove trapped air. Methanol at a temperature much above room temperature is unrealistic because of its volatility.

Volume of mobile phase (carrier gas) in column V_G. In principle

$$V_G = (\text{empty column volume}) - V_L - V_S \qquad (6)$$

In practice V_G is determined from chromatographic measurements and will be discussed. It is sometimes far easier to determine $(V_L + V_S)$ by applying the procedure for V_S to the loaded packing using a liquid in which the stationary phase is not soluble.

Gas:liquid volume ratio in column $\beta(V_G/V_L)$.

Cross-sectional area of mobile phase in column $A_G(V_G/L)$.

4.2 Sample Injection

The *instant of injection* or *origin, O*, is the position on the chromatogram that corresponds to the instant the sample is injected into the sample port. It may be marked by a co-worker on signal; or an *event marker*, which sends a small pulse to the recorder when a button is pressed, may be used. Attachments are available for the syringe so that the pulse is generated when the plunger is pressed. This is a useful reference point in all cases, but is essential with the FID, which does not respond to air included in the syringe. *Parameters measured from this origin bear no superscript.*

4.3 Baseline

The baseline is the signal generated when no solute is present in the detector. The zero adjust on the detector is used to bring the baseline to zero on the recorder paper. The electron-capture detector has a baseline at the top of the chart that signals the maximum ion current in the detector (no solute).

4.4 Unabsorbed Gas Peak or Gas Holdup

This peak is the driftwood of "boat chromatography." It is not retained by the stationary phase or the solid support. It is common practice with the TC cell that there be air included in the syringe, hence, the synonym, *air peak*. In terms of volumes, it represents the gas in the interstitial spaces of a CP column or the gas volume in the tube of a COT column, and *all extraneous volumes (extra-column volumes) contributed by injector, tubing, and detector*. This gas precedes the unretained solute; it is the *gas-holdup volume V_M*, and it is equal to V_G only if the extraneous volumes prove negligible.

4.5 Retention Distances: d_R, d_M, d'_R

All peak positions are determined by dropping a perpendicular from the peak maximum to the baseline and measuring the position of the intercept relative to the instant of injection or to the gas holdup. All measurements made from the gas holdup bear the adjective "adjusted," and the symbol bears a prime.

Peak distances are measured directly from the chart paper in appropriate units. A uniform chart speed is essential. If measured from the instant of injection, this is the *retention distance on chromatogram d_R*. If measured from the gas holdup, it is the *adjusted retention distance d'_R*. If d_M is the retention distance of the gas holdup (air peak) from the origin, then

$$d'_R = d_R - d_M \tag{7}$$

4.6   Retention Times: t_R, t_M, t'_R

To translate from distances to times, the chart speed must be known and some estimate of the error is useful.

Retention time t_R is the time from sample injection to the maximum concentration (peak height) of the eluted compound, extrapolated to zero amount injected. In GLC it is reassuring to determine that the retention time is independent of the sample size, which is expected. If it is not, then something is amiss, for example, the sample is overloaded. In GSC it is essential. A later section will discuss the partition coefficient, which is the ratio of the concentrations of the solute in the two phases. In *linear chromatography* this coefficient is independent of the solute concentration in either phase. When this is not the case, the chromatography is *nonlinear*; such is often the case in GSC where nonlinear adsorption isotherms are common and t_R does depend on sample size. The best number to report is that defined.

The *gas-holdup time* t_M is the observed time of unabsorbed gas.

The *adjusted retention time* t'_R is the retention time measured relative to the unretained solute; that is,

$$t'_R = t_R - t_M \tag{8}$$

4.7   Retention Volumes: V_R, V_M, V'_R

Later algebra will show that volumes are the most important parameters. To determine these one must have the *gas-flow rate from column* F_o at the column exit and its temperature and pressure.

The *retention volume* V_R is $t_R F_o$.
The *gas-holdup volume* V_M is $t_M F_o$.
The *adjusted retention volume* V'_R is measured from the air peak.

$$\begin{aligned} V'_R &= V_R - V_M \\ &= F_o t'_R = F_o(t_R - t_M) \\ &= F_o c d'_R = F_o c(d_R - d_M) \end{aligned} \tag{9}$$

where c is the chart speed. Because V_R and V_M include volume contributions of the injection port, detector, and connecting tubing, V'_R is independent of these features.

4.8   Martin's j Factor

The carrier gas enters the column at the inlet column pressure p_i and is driven to the lower pressure p_o. On undergoing the pressure drop, it expands so that its volume flow rate and linear velocity at a point within the column depend on the position along the migration coordinate. Retention volumes must be corrected for this pressure drop. This was done by James and Martin [49] for the region where d'Arcy's law applies. The correction, the *column pressure gradient correction factor j* involves only the inlet and outlet pressures for a uniform column; that is,

$$j = \frac{3}{2}\left[\left(\frac{p_i}{p_o}\right)^2 - 1\right]\bigg/\left[\left(\frac{p_i}{p_o}\right)^3 - 1\right] \tag{10}$$

Figure 6.21 shows the variation of j with inlet pressure when an outlet pressure of 1 atm is assumed. The pressure $p_i = 2.2$ is about 60 psi. Election to make this

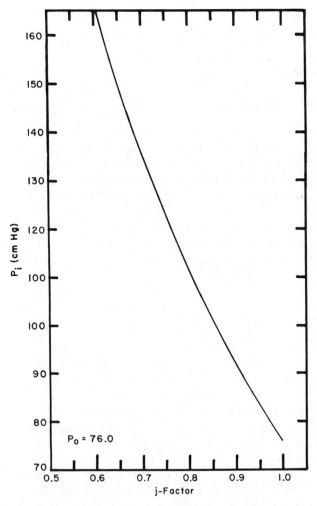

Figure 6.21 j Factor as a function of inlet pressure ($p_o = 1$ atm).

correction depends on the precision of the retention volumes. Essentially the correction reduces the pressure drop across the column to zero, which is physically unrealizable; that is, there is no driving force.

4.9   Corrected Retention Parameters: t_R^0, V_R^0, V_N, and Velocity

Multiplication of the retention parameters by j yields corrected values designated, for the most part, by a zero superscript.

Corrected retention time t_R^0 is jt_R.

Corrected retention volume V_R^0 is jV_R.

Net retention volume V_N is jV_R' (relative to the air peak and independent of extra column volumes). This parameter is of particular interest.

The *average linear gas velocity in column $\bar{u}$ is*

$$\bar{u} = jF_o/A_G \tag{11}$$

The units must be consistent.

4.10   Specific Retention Volume: V_g

The retention volume of a solute depends on the total volume of the stationary liquid in the column. It is desirable that a parameter exist that is independent of construction parameters, such as column length, diameter, packing particles size, amount of packing, liquid load, inlet pressure, and flow rate. Ideally, this parameter would depend only on the natures of the solute and the stationary phase and on the temperature. Theoretically the volume of the liquid phase V_L is desired. However, it is easier and more accurate to determine weights from the liquid load and the amount of packing in the column. In addition, it is convenient to reduce the retention volume to a reference temperature. Thus the *specific retention volume V_g is*

$$V_g = V_N(273\text{ K})/w_L T \tag{12}$$

which is the number of milliliters of carrier gas corrected to 0°C at the effective column pressure (outlet pressure). It is a puzzle why the volume was not reduced to STP. The analyst, given V_g, can calculate the amount of liquid required to give a desirable retention volume (peak position).

With the introduction of j, the volume of the mobile phase can be determined from the chromatogram if the sample inlet and detector dead volumes are known; that is,

$$V_G = jV_M - (p_o/p_i)\left(\begin{array}{c}\text{sample inlet}\\\text{dead volume}\end{array}\right) - \left(\begin{array}{c}\text{detector}\\\text{dead volume}\end{array}\right) \tag{13}$$

4.11   Peak Widths: Y_d, Y_t, Y_v

Resolution, the ability to distinguish between two solute peaks, depends on zone disengagement—the separation of peak maxima—and the broadness of the peaks—the width. Both experimentally and theoretically the peaks are Gaussian in shape; they are normal error distributions. Their widths should be reported properly in terms of the standard deviation. However, this is inconvenient. A satisfactory alternative is the *peak width*. Referring to Figure 6.20, tangents to the sides of the curve are drawn at the points of inflection and extended to intercept the baseline at F and G. The distance FG is the peak width. This is accomplished easily. The sides of the peaks are often very nearly straight lines so the inflection points are of little consequence. The signal trace is free from noise, which is not the case at points C and D where the Gaussian curve begins and ends, and minor tailing near the baseline (near D) is excluded. Such tailing is almost always encountered with the FID, not because of poor chromatography, but because it is such a superb detector that it detects the small number of ill-mannered molecules that are ignorant of the rules of "good behavior." Peak widths may be expressed in terms of length of chart paper Y_d, time Y_t, or gas volume Y_v. The units must be consistent with those used for peak position when calculating resolution. A complete discussion of resolution follows the section on band broadening.

Before the advent of electronic integrators, measurements were made directly from the recorder chart. Quantitative GC requires the area of the peak or suitable approximations to it. One of these involves the *peak width at half-height*. The perpendicular EB from the peak maximum to the baseline is bisected and a line is drawn through this point parallel to the baseline. The distance HJ between the intersection and the sides of the peak is this quantity.

When properly understood and applied, electronic integrators and subsequent modifications have proved to be major advances. They will print out the actual areas under the curve rather than approximations made from the chart paper. Prior to GC, analysis time was very much longer than that for data reduction. After GC, where multicomponent mixtures could be resolved in minutes, data reduction, especially quantitative, became the slow step in the process. Electronic devices have again reversed the situation. However, when these devices first focused on peak areas, peak widths were ignored and could only be determined if a recorder was used in conjunction with the integrator. Soon the recorder was abandoned, and a column of numbers on a paper strip was taken as the chromatogram. In this evolution very valuable peak-width data were lost. Not only information was lost, but also erroneous results were reported; for example, peak areas of very poorly resolved peaks were reported as a single peak. Modern reporting integrators correct this.

4.12   Relative Retention Parameters

The retention of solutes may be reported relative to that of another, which is particularly useful when one is a known solute. This is done in terms of adjusted parameters to obtain a simple ratio (exclude the gas-holdup volume). Relative retention may be calculated from distances, times, or volumes. If the solutes are designated i and s (standard), the *relative retention* r_{is} is

$$r_{is} = d'_{Ri}/d'_{Rs} = t'_{Ri}/t'_{Rs} = V'_{Ri}/V'_{Rs} = V_{Ni}/V_{Ns} = V_{gi}/V_{gs} = r_{im}/r_{sm} \qquad (14)$$

where m designates a third solute.

Relative retention parameters are important with COT columns with FID (no air peak) if it is reasonable to assume that $V_M \ll V_R$.

4.13 Kováts' Retention Index System

When reporting relative retention data, the standard s should have a peak near that of the solute i to partially compensate for vacillations during the separation (flow rate, temperature, etc.).

This is not possible with a single standard. Kováts [50] suggested that the n-alkanes be used. He further refined this by incorporating the observation that the logarithm of a retention parameter varies linearly with the number of methylene groups in any homologous series. This has been observed experimentally and will be justified later. The *retention index* I_x *(isothermal GC)* is

$$I_x = 100[z + (\log V_{Nx} - \log V_{Nz})/(\log V_{N(z+1)} - \log V_{Nz})] \qquad (15)$$

where z is the alkane carbon number and $(z + 1)$ is the alkane of a carbon number one greater. Best results are given when the solute x is between these standard alkanes. Under certain circumstances, any adjusted retention parameter may be used. The retention index is valid if comparisons are made at the same temperature, which must be stated. Whatever parameters are used, they must be independent of the liquid load, that is, I_x must be unaffected by a change of liquid load. However, the support should be described in terms of identity, deactivation, particle size, and liquid load.

Some individuals have objected to using the hydrocarbons as standards. Haken [51] has argued this point. Ettre [52–54] has reviewed the retention index system.

For a linear temperature program, the *retention index* *(linear PTGC)* I_{PT} is

$$I_{PT} \sim 100[z + (V_{Nx} - V_{Nz})/(V_{N(z+1)} - V_{Nz})] \qquad (16)$$

The retention index for methane is 100; for ethane, 200; for propane, 300; and so on, regardless of the amount of liquid present and the temperature. Propane will have an index of 300 whether its retention volume is 2 or 150 mL. A solute index of 385 indicates only that for the solvent and temperature employed, the solute will appear between propane and butane (closer to butane). Even with this limitation it is becoming common practice to report retention data in terms of this index system *providing that other properties of the column and its operation are also reported.*

5 PEAK POSITION

5.1 Equilibrium and Peak Position

Solute molecules exchange constantly between the mobile phase and the stationary phase because of their unceasing random thermal motion. As a cloud of molecules migrates along the migration coordinate, those at the front of the advancing zone encounter a stationary phase devoid of solute. Some deposit in the stationary phase. As these accumulate, some escape to reenter the moving phase. However, there is a net rate of deposition at the solute zone front as the rate of condensation from the highly populated vapor exceeds the rate of escape from the sparsely populated condensed phase. The reverse situation exists at the back of the zone. Carrier gas, devoid of solute, encounters a stationary phase containing molecules, and there is a net rate of escape. On passing from a net rate of deposition at the front to a net rate of escape at the tail there must be some position within the zone where the rate of deposition equals the rate of escape; that is, there is a *kinetic molecular equilibrium*. In linear chromatography this is the position of the peak maximum. This is often equated to a *thermodynamic equilibrium*, although thermodynamics does not recognize the existence of molecules nor does the theory require them. Statistical mechanics interprets equilibrium properties as the average behavior of a large collection of molecules or the behavior of an average molecule. Thus the peak position represents a kinetic equilibrium, the behavior of an average molecule, and a thermodynamic equilibrium. This last identification allows the application of solution thermodynamics. Similar considerations may be applied to GSC but now with the added complication of nonlinearity.

5.2 Zone Velocity

To treat (1) in more detail, it is rewritten as

$$u_i(x) = R_i u(x) \tag{17}$$

where $u_i(x)$ is the linear velocity of solute i in an increment of the migration coordinate dx and $u(x)$ is the mobile phase velocity in this same increment. The velocities are functions of x because of the compressibility of the mobile phase. Average velocities (not to be confused with average molecules) are employed by use of the j factor. An interpretation is that the actual velocity distribution, a function of x, may be replaced by an average velocity, which is not a function of x, to yield the same result; that is,

$$\bar{u}_i = R_i \bar{u} \tag{18}$$

The time for migration through the column length L is

$$t_{Ri}^0 = \left(\frac{1}{\bar{u}}\right) \int_0^L dx/R_i \tag{19}$$

If, in addition, R_i is independent of x, which may not be true because of stationary phase bleed or chemical alteration of the stationary liquid near the column inlet, then

$$t_{Ri}^0 = L/R_i \bar{u} \tag{20}$$

5.3 Partition Coefficient: K_i

If R is large, then the average molecules spend most of their time in the mobile phase. With such reasoning one may pass from a time average to a population average and claim that R also measures the fraction of *all* average molecules in the increment dx, not taken at the peak maximum, in the mobile phase. Thus

$$R_i = \frac{c_{Gi} A_G \, dx}{(c_{Gi} A_G + c_{Li} A_L + c_{1i} A_1 + c_{2i} A_2 + \cdots)} \, dx \tag{21}$$

where c_{Gi} is the concentration of solute i in the gas phase; and A_G is the amount of gas phase per unit length of column, that is, its cross-sectional area. The denominator must represent all of the molecules in all phases. The subscript L represents the liquid spread on the support where the liquid properties are those of the bulk liquid (the interior of the liquid as contained in a beaker) and the solute is retained by solution forces—their distribution is uniform and their orientation is relatively random. Solution thermodynamics would apply. However, as Martin and Synge [55] emphasized, the key to what was then called partition chromatography is that the stationary liquid be spread as a very thin film to insure rapid achievement of equilibrium of solute distributed between immiscible and stationary phases. Thus in the chromatographic situation the surface/volume ratio of the liquid is very much larger than for the bulk situation. There are two interfaces involved: the gas–liquid interface and the liquid–solid interface at the support surface. *If the concentration of solute at these interfaces differs from that in the immediately contiguous bulk phase, adsorption is said to occur.* This is demonstrated in Figure 6.22. Thus c_{Li} is the concentration of solute i in the liquid of bulk properties where the amount of bulk liquid per unit length of column is A_L (also a cross-sectional area); and c_{1i} is the concentration of solute i in the interfacial phase 1 expressed in suitable units; and A_1 is the amount of this interfacial phase per unit length of column. The subscripts 2, 3, ... refer to other retentive interfaces.

Adsorption at the gas–liquid interface was proposed by R. L. Martin, and much research was conducted by several groups [56]. This adsorption is in evidence when the solute is only slightly soluble in the stationary liquid and the column temperature is low. This situation would yield little retention, and such solutes would appear with very small retention times, that is, crowded into the air peak, which is undesirable from a separations viewpoint. When such retention is apparent the data are undesirable at the outset so little attention is given to this effect.

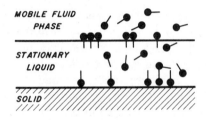

MOBILE FLUID PHASE

STATIONARY LIQUID

SOLID

Figure 6.22   Retention in the bulk and interfacial regions.

Having rationalized away this term and presuming that the liquid–solid interface, designated by S is active, (21) may be written

$$R_i = \frac{dx}{[1 + (c_{Li}/c_{Gi})(A_L/A_G) + (c_{Si}/c_{Gi})(A_S/A_G)]\, dx} \qquad (22)$$

The ratios of the concentrations are *thermodynamic distribution* or *partition coefficients* K and K_S, respectively. The coefficient for the solid phase is unusual in that it is not for immediately contiguous phases. Then

$$R_i = \frac{dx}{[1 + K(A_L/A_G) + K_S(A_S/A_G)]\, dx} \qquad (23)$$

The dx term cancels if it refers to the same increment in the column whereupon R_i applies only to this increment. To make this expression applicable to the entire column: (1) A_L and A_G must be known functions of x or uniform and independent of x, and (2) the partition coefficients must be independent of x. If these conditions are met

$$t_{Ri}^0 = [1 + K(A_L/A_G) + K_S(A_S/A_G)]\frac{L}{u} \qquad (24)$$

The retention time depends on the liquid and the support surface. This point is not widely appreciated. Peak asymmetry is attributed to adsorption on the support, and peak position is not suspect. Efforts are made to make the surface term negligible; that is, $K_S A_S \ll K A_L$. The term K_S is a measure of surface activity. The surface cannot be completely inert and K_S cannot vanish. The stationary phase would not wet or spread on the surface if this were the case. The contribution can be reduced by deactivation so that solution forces are more influential than surface forces. If $A_S \ll A_L$, the surface term is further diminished. Supports of small surface area are used, for example, nonporous glass beads (now mostly abandoned) or a porous support of suitable pore volume but with large pores to reduce the contribution of the surface within the pore. A large liquid load enhances the inequality. The disadvantage of this is that resulting peaks are broader and resolution is reduced. Tabulated literature values universally identify the stationary liquid most often with no explicit guarantee that the specific retention volumes or retention indices are independent of the liquid load. To be useful all retention data must include a statement of the support used, its particle size, any deactivation procedures, and the liquid load. Heavily liquid loaded columns (20%) are now employed rarely, yet loads of about this amount should be used in gathering data on the partitioning properties of the liquid. Low-load columns are much preferred in practice for reasons to be discussed. Reduced loads may lead to retention behavior different from that gathered with greater liquid loads.

$$t_{Ri}\bar{u} = jF_o t_{Ri} = V_R^0 = V_G + KV_L \tag{25}$$

$$V_N = KV_L \tag{26}$$

The equation divides column design into (1) a thermodynamic term that involves the solute and solvent identities and the temperature, and (2) a construction feature that involves the total amount of stationary phase on the column.

Column bleed will reduce V_R^0; and bleed, which favors support participation in retention by exposing it, will likely alter K. These are matters that affect the column lifetime as they alter the zone disengagement term in resolution.

5.4   Capacity Ratio: k

The *capacity* or *partition ratio* k is K/β, which is the amount of compound in column liquid divided by the amount of compound in column gas.

5.5   Conventional Open Tubular Columns

These considerations should apply to conventional open tubular (COT) columns, but the transfer of partitioning information from CP to COT columns has not yet been tested sufficiently. For example, in (25), if V_G is known, one may assume that a partition coefficient for a particular compound as determined on a CP column is still valid for a COT column and thus determine V_L: (1) V_L is likely to be very small and inaccurate; (2) the liquid spreads on the inside wall as a very thin film and surface effects of the tubing may be active; (3) the interfacial region has a finite thickness, and the solvent molecules may be oriented to yield solvent properties different from those shown by the bulk liquid. This situation was discussed by Keller and Stewart [57]. This subject recalls the early work with LC on the silica gel and paper chromatography of amino acids. The K value of an amino acid as distributed between bulk quantities of water and chloroform was taken to be valid, and V_L can be determined by the same equations. Partition coefficients of other amino acids determined chromatographically using this V_L did not all agree with bulk liquid values. This early exchange was reviewed by Keller and Giddings [58].

6 BAND BROADENING

6.1 Theoretical Plates and Plate Height

Zone disengagement has been discussed. Attention is now given to the second feature of resolution, zone broadening. Figure 6.23 illustrates the problem. Figure 6.23a is a chromatogram of a sample on a CP column. The peak maxima are well separated; but because of their width, there is peak overlap and incomplete area separation. Figure 6.23b is the same sample analyzed on a COT column. Now zone maxima are very close together, but the peaks are narrow spikes, and baseline return is almost complete despite the crowding. The

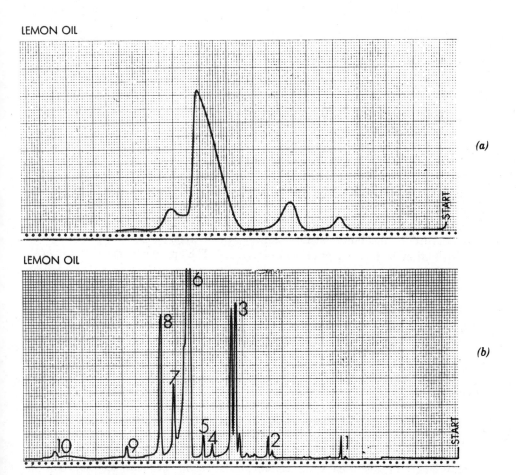

Figure 6.23   (a) Zone disengagement and (b) zone broadening. Reprinted, with permission from *Sci. Am.*, **205(4)**, 65 (October, 1961). Copyright © 1961 by Scientific American, Inc. All rights reserved.

temptation is to employ COT columns at the outset; however, there are difficulties:

1.   Coating a liquid on the wall of a long, small diameter column is a technique requiring much skill and experience. Few laboratories are interested in the occasional preparation of a COT column and prefer to purchase their material from a GC supplier, even at some expense. The vendor will want to know the column material desired, the length and diameter of the column, and the identity of the liquid. A vendor's advice is valuable. Not all liquids will wet the walls of a COT column, and selection of a particular phase may be ill advised.

2.   Because of the small amount of liquid on the column, a small sample must be used to avoid overload. This requires a detector with a superior LLD, for example, the FID, a destructive detector, and often a sample splitter that is likely nonlinear.

3.   Fast read-out systems are required to register the peaks. The CP column is not threatened by retirement.

The longer a solute remains on the column (large retention time), the broader is the peak. It is reasonable to ask whether the ratio of these quantities is constant. This is indeed the situation; that is, the ratio of the peak position, measured from the instant of injection, to the peak width is very nearly constant. There is some variation depending on the solute identity. The original publication of Martin and Synge dealing with the partition chromatography of amino acids on silica gel used a model much like that employed for countercurrent distribution and fractional distillation, one of Martin's original research interests, where equilibrium stages between the solute in counterflowing fluid streams were envisioned—the theoretical plates of a distillation column. This model led to the binomial distribution, which then passed to the Gaussian distribution as the number of plates increased. The approach led to the *number of theoretical plates n* given by

$$n = 16(d_R/Y_R)^2 = 16(t_R/Y_t)^2 = 16(V_R/Y_v)^2 \qquad (27)$$

This expression is still retained although the essential element is the constancy of the ratio. The interpretation is simple. A very poor column has broad early peaks (small d_R and large Y_d). Thus n is small and the *column efficiency* is said to be poor. The better columns have very narrow (small Y_d) late-appearing (large d_R) peaks so that n is large and the column is highly efficient. As with fractionation columns, n increases with the length of the column. A large n is desirable, but long columns often are not because of technical difficulties (packing long columns, high inlet pressures, poor thermal equilibrium with the mass of coiled metal, etc.). Attention was focused on the number of plates per unit length or, historically, its reciprocal, the *height equivalent to a theoretical plate* HETP or *plate height H*, where

$$H = L/n \qquad (28)$$

A concerted attack on H was launched by van Deemter and colleagues [21] using the partial differential equations of mass transfer. The approach here is that led mostly by Giddings [4, 59].

6.2   Random Walk

The Gaussian nature of a chromatographic peak is the logical consequence of molecular behavior being described by the laws of probability. The distribution curve is the result of a large number of random independent probability events (e.g., tossing a coin). Independent means that the result of a toss is unaffected by the result of the preceding toss and has no influence on the result of any subsequent toss. Random means that the results are equally probable. The width of the Gaussian distribution is most properly measured by the *standard deviation*, not by the peak width, but the difference is slight. The statistician is more interested in the square of the standard deviation, the *variance*, because it has been shown that if a number of different independent probability events contribute to a process, the variance for the total process is the sum of the variances of each contributor. Analysis of experimental error is the application of this principle to laboratory experience. The task becomes one of discovering the molecular dynamic behavior within the chromatographic column, described by probability, which contributes to band broadening. A rigorous application of the random walk model leads to the result that the plate height is the variance per unit length of the column [4]. As a consequence, the plate height is the linear sum of plate-height contributions from each process. The processes involved were indicated by van Deemter and colleagues [21]. The random walk model gives the detailed analysis of each contribution in terms of physical entities.

6.3   Molecular Diffusion

Molecules diffuse from a region of high to one of low concentration. Treatment of the process as a random walk probability event has existed for some time. The event is the outcome of a coin toss, perhaps a head being a step forward along the migration coordinate and a tail being a step in the opposite direction. The average molecule at the zone center (now *average* in terms of probability and not in terms of thermodynamics) will achieve as many heads as tails in a large number of trials and they dance about the zone center. Some molecules will elect more heads than tails and move ahead of the zone center; others will have the opposite experience and move in the opposite direction. These fluctuations about average behavior lead to the Gaussian distribution about the peak position, the most probable and mean value of the normal curve. The number of coin tosses, the number of random steps, can be related to the collision frequency of kinetic molecular theory (each collision involves a decision as to a change of direction, i.e., a new coin toss), and the length of a step is related to the mean free path. The result is

$$H_D = 2\gamma D_g/\bar{u} \qquad (29)$$

The terms are easily understood. The *diffusion coefficient in the gas phase* D_g summarizes the results of the random walk. Diffusion in a CP column is not physically along the migration coordinate. The molecules must dodge and dart around the grains of the coated support particles of the packing. A corrective term, the *obstruction factor* γ, is inserted. That the average linear velocity $\bar{u}$ appears in the denominator reflects the fact that the smaller this velocity, the longer the solute remains on the column and the more time there is for diffusion. Diffusion in the stationary liquid along the migration coordinate is justifiably ignored because liquid-phase diffusion coefficients are several orders of magnitude less than those for the gas phase. Equation (29) indicates that to reduce this contribution to plate height, $\bar{u}$ should be large and D_g and γ small. Little is done about D_g because the carrier gas identity is generally dictated by the detector; γ depends on how both the column is packed and the analyst accepts the results. The conclusion is valid whether the zone is stationary or moving.

6.4   Eddy Diffusion

This diffusion is also called the path difference term. The situation is depicted in Figure 6.24. Two identical molecules start out at the same position on the left. When they reach a junction point, a decision must be made about which fork in the road each is to take. One branch may be thought to be a fast branch and the other a slow branch because either the carrier gas velocity is swifter in one than in the other or the selection of one over the other advances the molecule a greater distance along the migration coordinate. Identical molecules are separated as shown by their final position on the right. A coin toss must be made at each junction point, the result of which is independent of the outcome of the toss at the preceding junction point and does not affect the result at any subsequent junction point. It is a random independent probability event and thus is described by a normal distribution curve involving molecules that consistently elect fast channels, molecules that consistently elect slow channels, and average molecules that select fast channels as often as slow channels. Channels vary in degrees of fastness and slowness, but this complexity is ignored at this elementary level. A junction point occurs about every particle diameter.

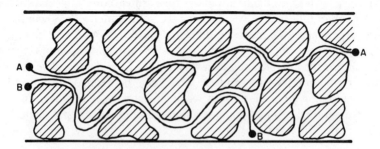

Figure 6.24   Eddy diffusion.

The number of junction points, related to the number of random steps, and the length of each step must intuitively depend on the diameter of the particle d_p. Thus the eddy diffusion contribution to the plate height H_E is

$$H_E = 2\lambda d_p \tag{30}$$

The organization of the particles as determined by the circumstances of packing will affect this contribution. Little can be done about it. That it exists is attested by the *tortuosity factor* λ.

6.5   Gas-Phase Mass Transfer

Hydrodynamics of fluid flow concludes that there is a parabolic velocity profile for any fluid through a pipe. This is shown in Figure 6.25. The fluid at the center of a channel flows faster than that at the wall. To be caught by the stationary phase at the wall, the molecules must diffuse or be transferred to it. This is a random walk, now normal or lateral, to the migration coordinate. It is complicated by the velocity profile. A probability outcome that favors molecules remaining in or moving toward the center of the channel will favor their migration through the column, while a probability outcome that favors residence near the wall or moving toward the wall will slow their progress. This concept is the heart of field-flow fractionation. Solute molecules must move or be transferred to the wall to be captured. The detailed mathematics yields

$$H_G = (\omega d_p^2 / D_g)\bar{u} \tag{31}$$

The gas-phase diffusion coefficient, now in the denominator, measures how fast lateral diffusion occurs. The distance the molecules must diffuse to reach the wall depends on the channel diameter and this depends on the particle size, now squared. To demonstrate, compare the channels available when marbles are packed in a jar to those when shotgun pellets are used. The carrier gas velocity,

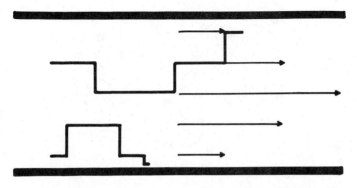

Figure 6.25   Resistance to gas-phase mass transfer.

now in the numerator, determines the sharpness of the velocity profile. Having exhausted descriptive vocabulary, ω, a constant, remains anonymous.

6.6   Coupled Eddy Diffusion

Eddy diffusion and gas-phase mass transfer are not independent. The physical basis is that a molecule entering a junction point at very fast velocity has sufficient momentum to carry it across the branching channel to reach a slow stream near the wall. This introduces a bias to the probability of being near the wall. This is called *coupled eddy diffusion*. The combined contribution H'_E is

$$H'_E = 1/[(1/2\lambda d_p) + (D_g/\omega \bar{u} d_p^2)] \tag{32}$$

6.7   Liquid-Phase Mass Transfer

Solute movement from the liquid to the moving gas is another independent random walk process. Figure 6.26 is a schematic cross section of a liquid-coated porous support particle. Adsorbed and dissolved molecules are located at appropriate positions. The random walk leads a dissolved molecule either to the surface to reenter the mobile phase or deeper into the liquid. The problem is transfer out of the liquid. The contribution, the *resistance to liquid-phase mass transfer* H_L, is

$$H_L = [q(1 - R)Rd^2/D_l]\bar{u} \tag{33}$$

The *diffusion coefficient in the liquid phase* D_l determines how fast a molecule will move about in the stationary liquid. Little attention is paid to it because the value of the diffusion coefficient is secondary to the selection of a liquid for its selectivity. How far it must move before reaching the moving gas will depend on

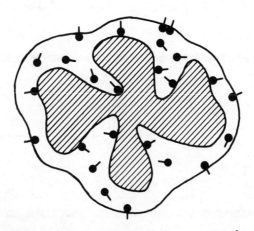

Figure 6.26   Resistance to liquid-phase mass transfer.

the *average film thickness d*. A glance at Figure 6.26 is sufficient to intimate that even with a detailed microscopic knowledge of the pore structure and pore distribution of support materials, a calculation of d will be enormously complicated and of doubtful utility. The small diameter pores fill first, and the governing principle is that the liquid surface with its variety of curvatures in the pores distributes itself so that the surface free energy is constant over the entire surface. The liquid cannot be considered as spread on the support as a uniform thin film. To this unassignable term is added a second unassignable, q, the *liquid configuration term*. One can confidently state only that the average film thickness must vary directly with the liquid load so that the smaller the liquid load, the smaller are the values of d and the plate height. Excape from the liquid is no longer a heads or tails situation because of the deliberately invoked affinity of the stationary phase for the solute. The probability is biased, and the situation is akin to drawing from a box of an unequal number of red and blue balls with replacement. The ratio of the differently colored balls reflects the R value. This is one reason why plate heights are solute identity dependent. That $\bar{u}$ appears in the numerator is understood because while the molecule is trapped in the liquid, identical molecules in the mobile phase are moving down the column. The greater this velocity, the greater is the separation of trapped and free molecules and the broader is the band.

6.8   Van Deemter Equation (Uncoupled Form)

Combination of these contributions yields

$$H = 2\lambda d_p + 2\gamma D_g/\bar{u} + [\omega d_p^2/D_g + q(1 - R)Rd^2/D_l]\bar{u} \tag{34}$$

Since the diffusion coefficients are secondary considerations they should be ignored, and particle sizes and liquid load should be kept small. Substitution of constants for the detailed coefficients yields

$$H = A + B/\bar{u} + (C_G + C_L)\bar{u} \tag{35}$$

$$= A + B/\bar{u} + C\bar{u} \tag{36}$$

Historically these terms are known as: A, eddy diffusion term; B, molecular diffusion term; C_G, gas-phase mass transfer term; C_L, liquid-phase mass transfer term; and C, mass transfer term. The important operating parameter is the mobile-phase velocity. Figure 6.27 is a stylized plot of H versus $\bar{u}$. There is a *minimum plate height* H_{min} at a particular value of $\bar{u}$, the *optimum velocity* u_{opt}. This represents the best achievable plate height for the column and instrument employed. With reference to the Scott triangle, the velocity (analysis time) is set when optimal resolution is desired (H_{min}). The low-liquid load mandates a small sample size.

The rising portion of the curve at values of $\bar{u}$ less than $\bar{u}_{opt}$ is the *diffusion branch*, while the portion where $\bar{u}$ greater than $\bar{u}_{opt}$ is the *mass-transfer branch*.

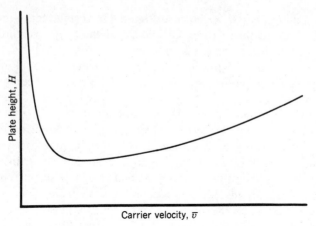

Figure 6.27   Van Deemter equation plot.

6.9   Excluded Considerations

The theory does not include other factors that affect plate height.

6.9.1   Support Identity

The early work of Kirkland [60] with various fluorocarbons (rarely used now) as supports for the same liquid at the same liquid load, illustrated in Figure 6.28, is instructive. The Chromosorb-W solid support is a commonly used diatomaceous earth. Fluoropack-80 packing behaves very badly. The plate height is so sensitive to flow rate that even a very small variation in flow rate will affect resolution. Plate heights are poor, and a minimum is not anticipated over the conditions employed. It will appear at a very small velocity. Kel-F packing is unusual. There is no ascending mass-transfer branch over the conditions studied, and resolution will not be injured by high flow rates. We do not pretend to understand the origins of these differences but can only allege the liquid configuration term.

6.9.2   Particle-Size Range

Experience has shown that the size range should be as narrow as is feasible [61]. Feasibility involves economics. The narrower the particle-size range specified, the more material must be size graded by the supplier and the greater will be the vendor cost. Popular size ranges are a compromise, that is, acceptable behavior at an acceptable cost.

6.9.3   Extra Column Contributions

Poorly designed, large volume injection ports; low injection port temperatures; and poor syringe techniques do not favor narrow initial solute zones. The zone starts out broad and does not repair.

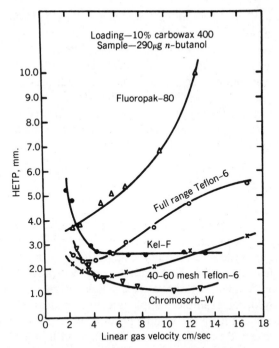

Figure 6.28 Van Deemter plot for fluorocarbon supports. Reprinted, with permission, from J. J. Kirkland, *Anal. Chem.*, **35**, 2003 (1963). Copyright © 1963 American Chemical Society.

All extraneous volumes act as chambers where diffusion mixes the separated zones, and the work of the column is undone. This is an instrument design feature. The definitive paper is that by Sternberg [23].

6.10   Rewards of Plate-Height Data

The determination of plate height is performed so simply that there should be no reluctance to make it routine. The inlet pressure is altered, the flow rate is measured, and chromatograms are performed. The remaining calculations are arithmetic.

A large plate height indicates either a poor column, a poor instrument, or both. Presuming that both a suitable instrument and a good technique are employed, the plate height of a new column can be compared with those of other columns, commercial or homemade, if such information has been logged. A comparison of the efficiency of the new column with the history of previous columns on that particular instrument will detect a faulty column. If the plate height is characteristic of the column and does not reflect extra column contributions, then the very best one may expect is about one particle diameter at optimum velocity for a perfect column (very narrow particle-size range, low liquid load, etc.) and the internal column diameter for a COT column.

Determination of the optimum velocity by constructing a van Deemter plot will allow us to answer questions about whether the resolution required for a separation or reliable quantitative analysis will ever be achieved with a particular column. With an understanding of the contributions to plate height, a new plan for a new column can be made on a rational basis.

Plate heights are solute dependent and are not strictly constant. However, if a peak has a plate height greater than the immediately preceding and following peaks, it is a very good indication of incomplete resolution of severely overlapping peaks.

7   RESOLUTION

Resolution is a measure of the ability to distinguish between signals. In elution chromatography the signal intensity is most often transformed to a concentration distribution of the solute along the migration coordinate (concentration profile), exemplified by a strip-chart record, as concentration is determined in the column outlet effluent by the detector. Resolution will depend on the technique employed, for example, sample injection, and on all of the components of the instrument from the sample injector port to the final record on a tape or recorder chart. Resolution is only important with immediately adjacent peaks. Widely separated peaks and a signal that returns to baseline present no problem except that when resolution is very large (widely separated narrow peaks), analysis time and equipment are being squandered. Figure 6.29 shows adjacent peaks with a degree of overlap. A situation has been deliberately selected where the peaks overlap such that the front edge of the second peak begins at the peak maximum plus two standard deviations for the first peak $(x_1 + 2\sigma_1)$. By doing this, the separation of the peak maxima $(x_2 - x_1)$ is

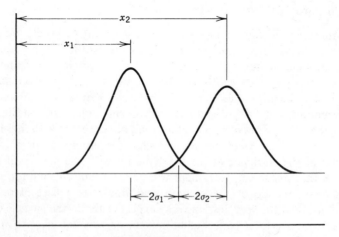

Figure 6.29   Resolution.

$(2\sigma_1 + 2\sigma_2)$, where σ_2 is the standard deviation of the second peak. The *resolution* R_{ij} is defined as

$$R_{ij} = (x_2 - x_1)/(2\sigma_1 + 2\sigma_2) \tag{37}$$

Hence in Figure 6.29, $R_{ij} = 1$. Statistics states that $(\bar{x} \pm 2\sigma)$ ($\bar{x}$ is the mean of the Gaussian distribution) includes 95% of the area of the normal curve (95% confidence limit), which means that 2.5% of the area of the tail of the first peak overlaps 2.5% of the area of the second peak. A reasonable approximation is that the peak width w, in suitable units, is

$$w = 4\sigma \tag{38}$$

so that

$$R_{ij} = 2(x_2 - x_1)/(w_1 + w_2) \tag{39}$$

In chromatographic parameters

$$\begin{aligned} R_{ij} &= 2(d_{Rj} - d_{Ri})/(Y_{dj} + Y_{di}) \tag{40} \\ &= 2(t_{Rj} - t_{Ri})/(Y_{tj} + Y_{ti}) \\ &= 2(V_{Rj} - V_{Ri})/(Y_{vj} + Y_{vi}) \end{aligned}$$

Because the peaks are adjacent, they experience very similar band broadening so that their widths are nearly the same. Then

$$R_{ij} = (x_2 - x_1)/w \tag{41}$$

where w is the peak width of either peak.

A resolution of unity is satisfactory; a resolution less than unity, which is still acceptable, depends on the problem. This is particularly true in quantitative analysis where peak areas are important.

If (40) is expressed in terms of V_R and if it is assumed that n is the same for both solutes, then some algebra yields the number of *theoretical plates required for resolution of peaks i and j, n_i*

$$n_i = 4R_{ij}^2 \left[\left(r_{ji} + 1 + \frac{2}{k_i} \right) \middle/ (r_{ji} - 1) \right]^2 \tag{42}$$

(the desired resolution is R_{ij}) and

$$n_i = 4R_{ij}^2 [(2 + k_j + k_i)/(k_j - k_i)]^2 \tag{43}$$

Recall that r_{ij} is the relative retention and that k is the capacity factor. If, in addition to assuming that the closely spaced peaks have nearly equal widths, it is also assumed that V_{Ri} nearly equals V_{Rj} (equally well expressed in terms of times or distances) so that $(V_{Ri} + V_{Rj} = 2V_{Ri})$ (or V_{Rj}), then (40) can be written

$$R_s = [1/4][(r_{ij} - 1)/r_{ij}][k_2/(1 + k_2)]n^{1/2} \tag{44}$$

In LC this appears as

$$R_s = [1/4][(\alpha - 1)/\alpha][k_2'/(1 + k_2')]n^{1/2} \tag{45}$$

It is unfortunate that different symbols ($\alpha = r_{ij}$ and $k = k'$) and terms (α is called the selectivity factor) have been introduced because the implication is that something new has been proposed. Each of the bracketed terms is treated separately.

A plot of R_s as a function of the capacity factor alone yields a curve that shows the best results lie in the range of 2 and 10 for the capacity factor. Small values of k reduce the resolution drastically, while large values of k enhance resolution only slightly at the price of analysis time. The following steps will improve the value of the capacity ratio (Section 5.4).

1. Increase V_L by increasing the column length but not the liquid load.
2. Decrease the temperature. This will also affect r_{ji} but perhaps not very dramatically. This is the most easily adjusted parameter.
3. Elect a different stationary liquid, and employ a different column.

The relative retention (selectivity factor) exerts a strong influence on resolution. There is no resolution if r_{ij} is unity. A change of r_{ij} from 1.05 to 1.50 (the numerator is the second peak) changes the relative retention term sevenfold from 0.05 to 0.33. If the solutes or solute functional groups are known, then a stationary phase may be rationally elected to yield a suitable relative retention. The relative retention is the ratio of partition coefficients. Although it anticipates later remarks, as a thermodynamic equilibrium constant

$$K = \exp(-\Delta\mu/\mathscr{R}T) \tag{46}$$

where $\Delta\mu$ is the change in the Gibbs chemical potential or Gibbs-free-energy change on transfer of solute from one phase to another and $\mathscr{R}$ is the universal gas constant. The Gibbs-free-energy change is

$$\Delta G = \Delta H - T\Delta S \tag{47}$$

Then

$$K = \exp(-\Delta H_v/\mathscr{R}T)\exp(\Delta S/\mathscr{R}) = g\exp(-\Delta H_v/\mathscr{R}T) \tag{48}$$

where g is understood to contain the entropy. Equation (48) is not rigorous (apologies to the thermodynamicists). Written for this application ΔH_v is the enthalpy of vaporization of the solute from the *solution* in the stationary phase. Then, assuming that $g_i = g_j$

$$K_i/K_j = \exp[-(\Delta H_{vi} - \Delta H_{vj})/\mathcal{R}T]$$

$$= \exp[-\Delta(\Delta H_v/\mathcal{R}T] \tag{49}$$

which states that the difference of the vaporization heats of the two solutes from solution should be large. If this is true at the outset, there is likely to be no resolution problem. If it is not true, one way to enhance this difference is to change stationary phases. If $\Delta(\Delta H)$ is small, the situation may be improved by reducing the temperature, which will increase the exponential quotient. This is particularly apt for members of a homologous series, where $\Delta(\Delta H) = \Delta n \Delta H(CH_2)$ and $\Delta H(CH_2)$ is the enthalpy of vaporization of the methylene group and Δn is the difference in the number of these groups for j and i.

Consideration of both features indicates that the column temperature should be as low as feasible. Low-load columns verify this contention. It is notable that column length is a consideration for both the capacity factor and the plate terms.

8 COLUMN TYPES

This section is based on papers by Halász and Heine [62] and Guiochon [63]; also, see Horváth [64]. Comparisons are based on a large number of published results. Three column types will be discussed in detail after a comparison has been presented.

8.1 Summary of Column Types

Columns are classified first with respect to the presence of a packing material and then with respect to column diameter.

8.1.1 Packed Columns

Adsorbent particles or liquid-coated support particles are packed in tubes.

CONVENTIONAL PACKED COLUMNS
These columns CP are of relatively large diameter tubing, $\frac{1}{8}$-in. o.d. and above; of modest length; and packed with particulates. As Guiochon remarks, there have been no dramatic changes beyond the columns used by James and Martin in their first paper.

GLASS BEAD B COLUMNS

The particles are glass beads. The early interest in glass beads was the desire for an inert support to reduce chemical alteration of solutes and reduce adsorptive tailing. However, the liquid load with nonporous glass beads is so small that sample injection and detector problems made them unpopular. The now routine deactivation procedures for porous diatomaceous earths have eliminated much of the motivation for their use. They were of great interest to theoreticians because the beads can be produced with an almost unbelievable uniform and measurable particle diameter. If, in addition, the beads are deactivated by silylating the surfaces, the liquid spreads as a film very nearly identical to the geometrical surface of the beads with little "puddling" at the particle contact point; that is, accumulation of liquid at these contact points. These "puddles" are equivalent to small diameter pores. Hishta and colleagues [65] reviewed much of the experience with lightly loaded B columns. They are not in common use in GLC.

Coated B Column Packing.   The liquid used as the stationary phase spreads on the beads by surface tension forces. Not all liquids will wet such beads.

Bonded Phase B Column Packing.   Deactivation of supports involves bonding an organic molecule to the siloxyl groups of silica-containing supports and thus covers those groups judged responsible for adsorption by hydrogen bonding. It seemed obvious to experiment with these bonded molecules as a stationary phase for GLC. This was vigorously investigated by Halász and Sebastian [66, 67], and they became known as "Halász brushes," because it was presumed that the organic end of the bonded molecule stood away from the surface like bristles on a brush. They have not been used to any great extent in GLC, but they solved a major problem in LLC where coated phases are stripped off the support by the mobile phase.

Porous Layer PLB Columns.   To increase their stationary phase capacity, glass beads were etched with reagents to produce shallow pores. Another procedure was to precipitate very finely divided substances on the bead surface to give a higher liquid load capacity. This very thin layer kept the liquid-phase mass-transfer term small. The same technique is employed in HPLC with great success. Phases may also be bonded to these materials. They are not used often in GC.

Small-Diameter Conventional Packed Columns.   There are two major members of this group. One with a column diameter of $\frac{1}{16}$ in. o.d. is prepared in a modified conventional manner. These columns were investigated, but because of technical difficulties that are encountered during their preparation, they are not common. Cramers and Rijks [68] reviewed *micropacked* columns. The other member of this group is the *packed capillary* PC *column.* A glass tube is packed loosely with a support material, and then it is drawn down to capillary diameters in a glass drawing machine. If a liquid is used, it must be deposited on the support materials after drawing. Understandably, these columns are not in common use, but they are most interesting to the theoretician. They were investigated by Halász and Heine [69].

8.1.2   Open Tubular Columns

CONVENTIONAL OPEN TUBULAR COLUMNS

Conventional open tubular COT columns are described in more detail in Section 8.4.

POROUS LAYER OPEN TUBULAR COLUMNS

Porous layer (PLOT) and *support-coated open tubular (SCOT) columns*, reviewed by Nikelley [70] and by Ettre and Purcell [71], are basically the same. As with glass beads, the sample capacity is increased by etching the wall of a small diameter tube or by depositing a porous solid on it and then coating the solid with the stationary phase. These columns have proved very useful when coupled with a mass spectrometer. A sample size compatible with the mass spectrometer is also compatible with the column. Also, the column gas volume flow is sufficiently small that the required pressure reduction can be achieved with high-speed pumps, and an interface between the instruments is no longer required.

8.2   Comparison of Column Types

Table 6.5 is a modification of that proposed by Guiochon. More than 90% of the data from the literature references surveyed by Guiochon fell within the ranges reported in the table. Column permeability is a measure of the difficulty of driving the mobile phase through the column (it is the proportionality constant in Darcys' law). The least permeable columns are the packed columns so they are, of necessity, short to keep inlet pressures manageable. Increasing the column length increases the number of theoretical plates, but it is a difficult way to obtain large plate numbers. Equation (40) shows that the number of plates, and hence the length of the column, varies as R_{ij}^2. To double the resolution, the length of the column must be quadrupled; for example, a 6-ft CP column with a resolution of 0.5 would be increased to 24 ft to obtain a resolution of unity. This may not be feasible. Open tubular columns present no such difficulty because of their permeability, and lengths from 300 to 600 ft are common. The data for plate heights demonstrates clearly that H_{min} values are very nearly equal for all

Table 6.5   Comparison of Column Types[a]

Type	Permeability ($\times 10^{-7}$)(cm^2)	H_{min} (cm)	u_{opt} (cm/s)	Sample Size (μg)
CP	1.0–10	0.05–0.2	5–20	10.0–1000
B	1.5–15	0.10–0.3	10–20	1.0–100
PLB	1.5–15	0.05–0.2	20–60	1.0–100
PC	5.0–40	0.05–0.2	10–40	1.0–50
COT	50.0–800	0.03–0.2	10–100	0.1–50
PLOT	200.0–1000	0.06–0.2	20–160	1.0–50

[a]Adapted from Guiochon [63], p. 186, by courtesy of Marcel Dekker, Inc.

Table 6.6   Comparison of CP and COT Columns[a]

Properties	CP	COT
Length (ft)	10	164
Internal diameter (mm)	2.2	0.25
Permeability ($\times 10^{-7}$)(cm^2)	1.96	195
Outlet flow (mL/min)	40.0	1.5
Outlet linear velocity (cm/s)	43.85	51.75
Inlet pressure (atm)	5.1	2.5
HETP (mm)	0.60	0.59
Number of plates	5000	85,034
Resolution	1.5	5

[a]Adapted from Ettre [72], with permission from Plenum Publishing Corporation.

column types (B columns are limited by the bead size). Packed columns have the greater sample capacity, and PLOT columns have the greater speed and permeability.

Table 6.6, adapted from Ettre [72], gives further emphasis. Conventional packed and COT columns constructed to give nearly the same plate height and linear velocity are compared. The COT column is over a decade longer and a decade smaller than the CP column. The plate heights are equal, but there are 17 times as many theoretical plates and a threefold increase in resolution is achieved for the COT column at one-half the inlet pressure. All of this is possible because permeability differs by a factor of 100.

8.3   Conventional Packed Columns

Conventional packed (CP) columns [73–75] will very likely remain the type used in the majority of separations. They are simple to construct, their sample sizes are easily handled, there are no restrictions on the detector, and they are inexpensive.

8.3.1   Column Material

Metal columns are the most common. Aluminum has been used, but because hot aluminum oxide, which cannot be avoided, is a catalyst, aluminum was abandoned. Copper tubing is widely used. It is inexpensive and readily available as $\frac{1}{4}$-in. o.d. refrigerator tubing and $\frac{1}{8}$-in. o.d. automotive fuel-line tubing. It is easily worked, that is, uncoiled, straightened, cut, and recoiled. Care must be taken to maintain the circular cross section (elliptical tubing rivals the Mikado's curse of elliptical billiard balls), and "kinks" must be avoided in coiling. Hot copper or its oxide threatens catalysis so stainless steel was introduced because it is more chemically inert. However, it is very difficult to manipulate because of its mechanical strength. Nickel columns are also very inert. All metal tubing must be cleaned to remove machine oil, starting with a nonpolar solvent and always ending with water (e.g., ether, acetone, alcohol, and water). If the electron-capture detector is to be used, a halogenated solvent or one containing a halogenated impurity cannot be used. The column should be dried with clean air or nitrogen, and it is best stored with dry nitrogen and with the column ends capped.

Glass and fused silica columns are popular. They have special requirements: (1) They are fragile and must be handled carefully. (2) Special glass–metal connectors are required. (3) They must be packed after coiling. (4) If glass has been selected because of its inertness, the glass surface should be silylated and a glass-lined injection port is required. Such columns are generally ordered from a supplier, who will require detailed specifications.

8.3.2 Column Length

Although resolution depends on column length, ease of handling and column permeability often obtrude as more important considerations. Six- to 8-ft columns are popular choices because inlet pressures are manageable.

8.3.3 Column Diameter

Column diameter and packing particle size are linked. Once the particle size is selected, perhaps because of a desired plate height, the diameter is approximately fixed (see Section 8.3.5).

8.3.4 Support Identity

Supports and their effects have been reviewed by Ottenstein [76, 77] and Urone and Parcher [78]. The most popular supports are those produced from fused diatomaceous earth. Among these, Chromosorb-W (white) solid support and Chromosorb-P (pink) solid support are the most widely used. The origin of the color difference is unimportant. The supports are produced by heating the diatomaceous earth with a suitable flux that fuses the diatoms into a highly porous structure. Microscopic examination indicates that the diatoms retain much of their delicate structure. Table 6.7 [76, 79, 80] summarizes important chromatographic properties.

Chromosorb-P solid support is mechanically the stronger and more robust. When aqueous washings are tested, the material is acidic. Occasionally, cases of acid-catalyzed reaction with solutes and stationary phases (depolymerization) have been noted. The pore volume of Chromosorb-P solid support is less than that of the white support. The great difference is the surface areas. When these data are coupled with the pore sizes, it is concluded that Chromosorb-P support is penetrated by many deep small diameter pores and the internal pore surface that provides the much greater area. Chromosorb-W support has large shallow pores. Adsorptive tailing of the Chromosorb-P support is consistent with the larger surface. Tailing effects are best explained by adsorptive sites of a wide

Table 6.7 Properties of the CHROMOSORB Supports

	Chromosorb	
Property	P	W
Mechanical	Sturdy	Delicate
Peak shape	Tail	Reduced tail
pH	6.5	8.5
Area (m^2/g)	2.80	0.91
Silanol groups (groups/m^2)	4×10^{19}	2.5×10^{19}
Pore size (μm)	ca. 0.8	ca. 5
Liquid capacity	High	Moderate
Pore volume (mL/g)	1.04	1.56

range of adsorption energies. Sites of nearly the same energy will yield a representative average energy affecting peak position to some extent but will not be responsible for dramatic peak asymmetry. This conclusion explains the success of tailing reducers in GLC and modified adsorbents in GSC.

Deactivated Chromosorb-W support seems to yield a more inert surface than deactivated Chromosorb-P support, and it is by far the more popular support for analytical columns. Chromosorb-P support retains some popularity when large samples are involved because of the feasibility of a high liquid load.

Chromosorb-G support is a compromise between the pink and the white.

Gas–solid chromatography employs active adsorbent particles (e.g., alumina, activated silica, and carbon). Graphite has received a great deal of attention [81]. It provides sensitive specific interactions with certain structural features not shared by solution forces (unsaturation, aromaticity). It is also very useful with permanent gases that are insoluble in liquids but do adsorb. Gas–solid chromatography adsorbents can be used at high temperatures. Graphitized carbon black is a treated amorphous carbon of enhanced mechanical strength and is also free of very active sites that are responsible for severe tailing. Silica is a siloxyl surface and acts by hydrogen bonding. Alumina has also been used. Its mechanism has been debated. Organic catalysis indicates a "positive" surface, which is about all that can be said about it. Gas–solid chromatography is a very powerful technique in certain specific separations that are not accomplished easily by GLC and deserves careful consideration in such situations. Its disadvantages are: (1) lack of generality in separations, (2) asymmetric peaks (much corrected), (3) nonlinear adsorption isotherms resulting in nonlinear chromatography, (4) catalytic effects, and (5) easily deactivated adsorbents (e.g., water vapor).

A modification called *gas–liquid–solid chromatography* has aroused interest. If a support is coated with a liquid, as is common in GLC, and then extracted off, a very small amount of liquid remains. The chromatographic behavior is a combination of GSC and GLC with many of the advantages of both and without some of their disadvantages [82].

Size-exclusion chromatography is powerful for low-molecular-weight gases or volatiles, particularly those with highly interactive or reactive groups that make them highly adsorptive or chemically reactive (corrosive), such as, water, oxides of carbon, nitrogen, and sulfur, ammonia and amines. The most popular are organic polymeric materials [83]. Pore size can be controlled by the degree of cross-linking that occurs. The retention order is not strictly by size so some adsorptive and solution effects are undoubtedly present and magnified by a generous internal pore surface. Silica is a fascinating material [84] because its adsorptive activity can be controlled by heating. Initially, condensed water is driven off. Then the adsorbed water that is probably clustered about the siloxyl groups (—Si—OH) is driven off. The exposed siloxyl groups are highly adsorptive sites. Higher temperatures deactivate the surface probably by the condensation of siloxyl to siloxane (—Si—O—Si—) groups. This yields a highly porous, relatively inert molecular sieve material. Peak shapes are remarkably sharp. Interested readers should consult the literature dealing with areas such as gas analysis, permanent gases, and atmospheric environmental analysis. Information about silica is given by Snyder [85] and Scott [86], some of which is controversial.

8.3.5   Particle Size

Size range is designated by mesh number (number of wires of standard diameter per linear inch of screen). Large particles will remain on a screen of small mesh number, and small particles will pass through the screen of large mesh number. Particle size is selected to obtain desirable plate heights in particular applications. Preparative GC, where resolution is sacrificed in favor of sample size, requires a large total volume of stationary liquid on the column, which means large diameter columns ($\frac{1}{4}$-in.) with large porous particles capable of holding a high load (Chromosorb-P support near 30% load). Table 6.8 summarizes popular size ranges.

Table 6.8   Common Support Particle Sizes

Scales	Mesh Range	Column Diameter (in.)
Preparative	40–60	1/2–3/4
Preparative and some analytical	60–80	1/4
Analytical	80–100	1/8
	100–120	1/8

8.3.6   Support Pretreatment

This information [87, 88] is presented should the reader intend to prepare a column using a Chromosorb support.

The material received should be resieved to remove small particles and possible dust that are produced in shipping, or it can be acid washed and then resieved.

Washing the support with acid (AW) is widely accepted as being beneficial. There is disagreement about the reasons for the improved performance. It is likely that heavy metals at the support surface are removed. The materials have a slight but measurable ion-exchange capacity, and acid washing protonates these sites. The acid concentration has not been a point of contention; for example, 6 M hydrochloric acid followed by rinsing the material with deionized water until the rinse is chloride free to silver nitrate seems to be the most popular. Hydrochloric acid should not be used if the electron-capture detector is to be employed. For this detector, nitric acid should be used, and the material should be rinsed to either neutrality or a constant pH. The support should be air dried, resieved, dried in an oven at 110–120°C, and stored over a drying agent. If moist support is used, the popular deactivation agents will react with the condensed or adsorbed water to produce a hydrocarbon "skin" that is loosely attached to the support.

Chemical reagents that react with siloxyl groups are used, and the reaction schemes are shown. Figure 6.30 shows the hexamethyldisilizane (HMDS) reaction. The reagent in an anhydrous solvent is available commercially. The support is covered with the reagent solution in a loosely stoppered, thoroughly dried flask for about an hour; rinsed several times with the anhydrous solvent of the HMDS reagent; air dried; oven dried; and stored in a dry atmosphere. The reaction scheme for dimethyldichlorosilane (DMCS) (2% solution in anhydrous toluene) is shown in Figure 6.31. Now, however, after the reagent is rinsed away, the packing must be treated with anhydrous methanol (dried over anhydrous sodium sulfate), washed several times, air and oven dried, and stored. Dimethyl-dichlorosilane cannot be used if the electron-capture detector is to be employed. The majority opinion is that DMCS is more effective.

The commercial designations for packings should be obvious, for example, 80 to 100 mesh Chromosorb-W-AW-DMCS support.

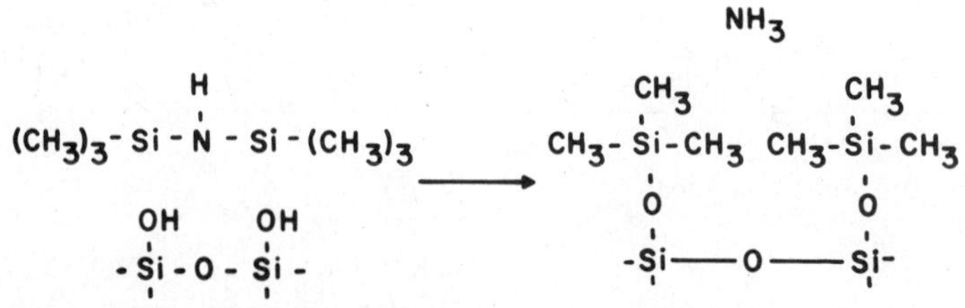

Figure 6.30   Hexamethyldisilizane (HMDS) reaction scheme.

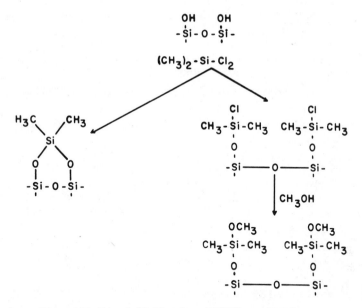

Figure 6.31   Dimethyldichlorosilane (DMCS) reaction scheme.

If glass columns are used, the walls should be silylated. Routine silylation is not always necessary. The treated support surface carries a hydrocarbon umbrella, and very polar liquids may not wet the surface. However, as with high temperatures, the use of very polar stationary phases is not advised.

8.3.7   Stationary Liquid-Phase Selection

Detailed discussion of this topic is deferred. Highly polar solutes will likely adsorb on the support to produce tailing. Deactivation alleviates this. A polar stationary liquid will also cover the active sites, but one may encounter a situation where a polar solute and the polar solvent compete for these active sites so support participation occurs and deactivation is ruled out, perhaps because of nonwetting. As an alternative to deactivation, a small amount of a highly adsorptive compound may be included in the stationary liquid, which tenaciously combines with these sites. This is a *tailing reducer*. The amount is insufficient to alter the partitioning properties of the liquid.

8.3.8   Liquid Load

Liquid loads were originally 20–30% by weight of the final packing. Theory and experience soon showed that resolution could be much improved if this amount were reduced, that is, a small average film thickness to reduce the liquid-phase mass-transfer contribution. A low load decreases both the retention volume of the solutes and the solute zone disengagement. The improved resolution showed that the effect on peak widths more than compensated for crowding the peaks together. Consider the following argument:

1. Let V_i and V_j be the retention volumes of solutes i and j at a column temperature T at a particular liquid load.
2. Now reduce the liquid load to produce retention volumes V_i' less than V_i and V_j' less than V_j at temperature T.
3. Reduce the temperature. The retention volumes increase. Continue this until $V_i' = V_i$ and $V_j' = V_j$ now at T' less than T.

The same separation is achieved at a lower temperature *with improved resolution* (i.e., narrower peaks). This is the basis of the successful GC of slightly volatile or termally unstable solutes, such as the steroids. It is also how one may separate solutes 200°C or more below their normal boiling points. The sample size must be reduced. Low-liquid load invites support participation and peak positions may not agree with data gathered on more heavily loaded columns.

Popular liquid loads for analytical columns are 3–5% on deactivated supports.

Karger and Cooke [89] reviewed the topic.

8.3.9   Coating the Support

A weighed portion of dry support is taken and an amount of the stationary phase weighed out to yield approximately the load desired.

Solution loading involves dissolving the stationary phase in a suitable volatile solvent to give enough solution to cover the support when it is spread on the bottom of a large evaporating dish. Supplier catalogs commonly list appropriate solvents. The solvent is slowly evaporated at room temperature or on a water bath. The stationary phase deposits on the support and fills first the smallest pores and then those that are successively larger. Occasional gentle stirring may be employed. Ideally, the support should be degassed to prevent the formation of trapped air bubbles. With slow evaporation the liquid surface free energy will be uniform over the support. Finally the packing is heated in an oven below the upper temperature limit of the stationary liquid. The packing should flow easily and behave as the support material before loading. Almost all the liquid is in the pores. The packing is often resieved. For the best results, the material should be stored in an opaque bottle under dry nitrogen in a cool cabinet. If necessary, the liquid load may be determined by extraction of a weighed portion of the packing.

If the stationary phase is not viscous, a weighed portion may be added directly to the support, stirred, and allowed to stand. This is *dry loading* and is preferred by some.

8.3.10   Column Packing

One end of the straight column is plunged with *silylated* glass wool (excluded by the ECD). A funnel is attached to the opposite end, and the packing is added very slowly, a small amount at a time, while the column is vibrated constantly and simultaneously tapped on the floor. Mechanical vibrators are available commercially. The vibrator should be continuously moved along the column while packing. If held in one place, a standing sine wave is established along the column and is reflected in the packing density. When no more packing will flow into the column, a bit is tapped out and the end is plugged. The true artist will reverse the column, remove the first plug, vibrate and tap, and add more packing if necessary. The weight of the packing used is determined by the difference between the weight of the initial and remaining amounts.

One end of a precoiled column (glass) is plugged and attached to a vacuum pump, and the packing is then added to the open end, a bit at a time, with tapping until the column is filled. A glass column has an advantage in that the packing is visible and gaps can be observed and corrected immediately. Some individuals prefer to pack precoiled columns (stainless steel) in this fashion.

8.3.11   Column Coiling

The packed straight column is carefully coiled by gentle but firm bending. Small coils should be avoided. The column should be as large as can be accommodated by the oven, and the coils should be separated so they will be in contact with the circulating air. Because the packing is crushed and the desirable narrow particle-size range desired is destroyed, kinks must be avoided.

8.3.12   Column Conditioning

If chromatograms are performed on newly made columns, vapors will likely appear in the effluent, and the plate heights will decrease, reach a minimum, and very slowly increase as the column is used. There are several changes that occur.

1. If solution loading has been used, some of the effluent is residual solvent.
2. Very nearly all stationary phases are polymeric, and low molecular weight monomers and polymers escape.
3. Stationary phase will migrate to the contact points between the particles, and this will continue until the surface free energy is uniform.

Columns are conditioned by passing carrier gas through the column at a temperature 50°C and greater above the highest temperature intended for the column, if the temperature limit of the stationary phase permits this increase. For volatile, low-viscosity liquids, such as squalane, a few hours is sufficient. For slightly volatile, highly viscous fluids, such as silicone greases, overnight is appropriate. The effluent is not passed through the detector.

8.3.13   Column Storage

Columns should be filled with the carrier gas and capped at both ends for storage. Their positions should be changed periodically, that is, turned over, rotated, and so on. In time, even the most viscous fluid will flow in the gravitational field.

8.3.14   Column Testing

It is wise to run a test mixture periodically to determine whether peak positions and plate heights are changing because of use or storage. A definite *column lifetime* cannot be stated. A column has exceeded its life span when it no longer yields the resolution for the separation problem of concern.

8.4   Conventional Open Tubular Columns

Conventional open tubular (COT) [26, 72, 73, 90–92] columns are traceable to Marcel Golay. He began by envisioning the tortuous channel of a CP column to be a "pipe" whose walls were coated with the stationary phase. To treat this difficult situation mathematically, Golay resorted to a model of a bundle of very long straight capillary diameter tubes. His mathematical results, which forecast a previously unobserved efficiency, were at first disbelieved. Fortunately the FID had just appeared as the only detector available with a suitable LLD, whereupon Golay and the group at Perkin-Elmer, Inc. verified his mathematical speculations experimentally. These results were first presented at a 1957 symposium in East Lansing sponsored by the Instrument Society of America [93]. There again skeptics doubted the results. Fortunately, A. J. P. Martin chaired the session where the paper was presented and verified Golay's remarks by saying: "First of all, I would like to congratulate you, Dr. Golay. I have been thinking on exactly the same lines as this and you've beaten me to it—very thoroughly." Golay [94] described his own experience. Desty [95, 96] has written of the early history of what have been called *Golay, capillary*, and *small diameter open tubular columns*. Conventional open tubular columns were the first and perhaps the only major advance in column design.

Some problems with COT columns have been mentioned previously. For example, there is the need for sample splitters that are complicated by the frequent nonlinearity of the split (splitless injection has been reported). There is also the need for a detector of superior LLD, and this led to the use of the FID, or some modification of it. These are ionization detectors and thus are destructive detectors. To these difficulties must be added the technical problem of coating very long small diameter columns. "Homemade" columns are rare so commercial columns are used commonly. However, they are expensive.

The column material is most often stainless steel or glass. The fragility of brittle glass has been solved by the introduction of fused silica, a material of surprising flexibility and adsorptive inertness. Special metal–glass fittings are required. A common internal diameter is 0.25 mm. Columns as small as 0.1 mm i.d. and as large as 1.55 mm i.d. have been reported. The 15-fold variation motivated the term "open tubular" so that the name no longer implied capillary dimensions. Long columns are expected because efficiency is achieved by length. Column lengths vary from about 150 to 600 ft. Because of this, precoiled columns are commonly used, which makes cleaning and loading difficult. In the *dynamic coating method* a solution of the stationary phase, about 10% by weight, is forced through the column until either it appears at the outlet or the amount of stationary phase in the effluent is the same as that of the solution introduced. There is not much control over the amount deposited. Determining factors are tubing radius and coating velocity. The problem has been reviewed and studied by Novotny and colleagues [97] and commented on by Guiochon [98]. The *static coating method* involves filling the tubing with the solution of partitioning liquid; plugging one end; and drawing the tube through an oven, open end first, to evaporate the solvent. Equipment is complicated, and the tubing cannot be precoiled. The liquid load can be determined. This is probably the only existing reliable method that can be used to evaluate the load. Any determination based on weighing the column before and after coating involves the error associated with a small difference between large numbers. Wall deactivation procedures are common. Parameters that are easily determined with CP columns and a TC detector become very difficult to evaluate with a COT column and a FID. Outlet flow rates are often insufficiently precise for a reliable determination of retention volumes. Retention data, such as the retention index, as determined with CP columns may prove inapplicable because of the low-liquid load and surface effects of COT columns. Quantitation is by peak heights, which presents a problem. Programmed temperature gas chromatography is common. Column lifetime may be short because the film breaks up to form small drops.

The same factors that affect band broadening and are encountered with CP columns also apply to COT columns but are now mathematically different because of this new configuration. This has been discussed by Ettre [72].

8.5   Porous Layer Open Tubular Columns

The highly efficient porous layer open tubular (PLOT) columns [70, 71] have greater liquid loads than COT columns, which make it possible to use larger samples, avoid sample injection splitters, employ detectors other than FID, and use shorter columns (100 ft seems common). In addition, the van Deemter curve is flatter or more shallow so that higher mobile phase velocities are possible without much variation of plate height, and analysis time is reduced. The Scott triangle is not equilateral.

Column diameters range from 0.01 to 0.02 in. i.d. Glass columns may be etched with hot ammonia solution to yield a porous silica layer about 5–20 μm thick depending on the glass and the procedure, hydrochloric acid forms a 100-μm-thick layer, and solium hydroxide forms a layer 10–50 μm thick. The silica may be used directly for GSC; it may have a phase bonded to it; or it can be coated with a liquid for GLC. Aluminum tubing has been treated to yield an alumina layer 5 μm thick. Copper treated with nitric acid, washed, and oxidized yields a spongelike layer. Metal columns are difficult to reproduce.

Solids may be deposited on the walls by filling the tube with a suitable suspension and then depositing the particles on the wall. Some of the particles used are clay, alumina, silica, molecular sieves, diatomaceous earths (SCOT columns), iron oxide, and carbon. For GLC the suspended particles may be coated with the desired liquid and deposited or a solution of the phase may be evaporated in contact with the support. Liquid loads range from 1 to 2.5 mg/ft.

Columns are discussed and compared in terms of the gas–liquid volume ratio in the column, β, and the partition ratio, k.

9   GAS–SOLID CHROMATOGRAPHY

Gas–solid chromatography is mentioned where it is appropriate within the general concepts shared by all chromatographic techniques and is not treated as a separate topic. Those interested in more information about GSC should consult the general references [81, 99, 100].

10   PROGRAMMED TEMPERATURE GAS CHROMATOGRAPHY

More extensive treatments of this topic exist elsewhere [101, 102].

The general elution problem is solved in GLC and GSC by increasing the oven temperature continuously during the chromatography. Figure 6.32 shows a gas chromatogram of either a mixture of solutes with a large range of boiling points or one that contains solutes of different chemical types, that is, different functional groups, at a relatively low column temperature T_C. The early peaks are well resolved but performance deteriorates for the later peaks, which are broad and flat, as expected, and may be mistaken for long-term baseline drift by the operator or by an ignorant integrator or missed completely because the concentration in the gas stream falls below the LLD of the detector. Figure 6.33 shows an isothermal chromatogram of the same sample on the same column at a higher temperature T_C' greater than T_C. The early peaks are crowded together and are poorly resolved. It is true that the previous late peaks now appear earlier

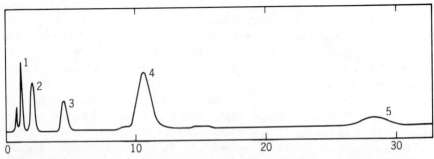

Figure 6.32   Isothermal gas chromatogram at temperature T_C [101]. Reprinted, with permission, from W. E. Harris and H. W. Habgood, *Programmed Gas Chromatography*, John Wiley & Sons, Inc., New York, 1966, p. 10. Copyright © 1966 by John Wiley & Sons, Inc.

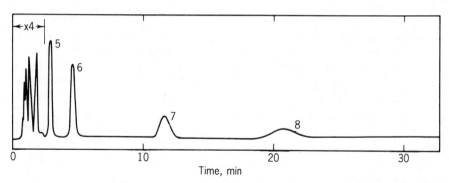

Figure 6.33 Isothermal gas chromatogram at temperature $T_c'' > T_c$ [101]. Reprinted, with permission, from W. E. Harris and H. W. Habgood, *Programmed Gas Chromatography*, John Wiley & Sons, Inc., New York, 1966, p. 10. Copyright © 1966 by John Wiley & Sons, Inc.

and are sharper with improved resolution. In fact some peaks, not evident at T_c, appear at the higher temperature. Neither isothermal result is wholly satisfactory. Figure 6.34 is a programmed temperature chromatogram of the same sample on the same column. The resolution is excellent at all peak positions, low and high boilers. It is not obvious immediately why the peak widths are almost the same when the expected experience is that the longer a solute remains on the column, the broader is the peak. The conceptually very simple explanation by Giddings [103] follows.

The sample represents a wide range of R values. The low boilers spend most of their time in the mobile gas phase and emerge quickly from the column at the

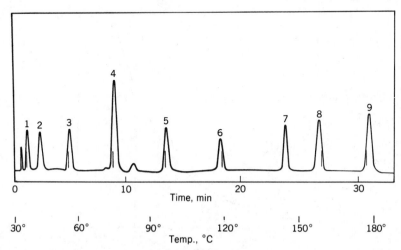

Figure 6.34 Programmed temperature gas chromatogram [101]. Reprinted, with permission, from H. W. Habgood and W. E. Harris, *Anal. Chem.*, **32**, 450 (1960). Copyright © 1960 American Chemical Society.

low temperature. The high boilers spend most of their time trapped in the liquid phase. Diffusional behavior in the liquid phase is inconsequential. The small number of molecules in the gas phase do undergo diffusional broadening; but because there is a very small concentration gradient, the effect on the zone is minimal. There is not much transfer between phases. That is not to say these diffusional processes are wholly absent. Given much time, a solute with a small R value will eventually appear at the column outlet. However, if the time period is the retention time of the low boilers, they are essentially "frozen" in the stationary phase for that period. To demonstrate this, Giddings took (48) and calculated the temperature increment that would double the value of the migration parameter R. His result is

$$\Delta T = 0.693 \mathscr{R} T^2 / \Delta H_v \qquad (50)$$

where T is the geometrical mean of T_1 and T_2. A further simplification is that Trouton's rule is obeyed (Giddings discussed the justification and limitation of this assumption) with the conclusion that the relative migration rate doubles for every 30°C increase in temperature. This is demonstrated in Figure 6.35 as modified [103]. The actual increase in rate is shown by the continuous curve for a *linear program rate*; that is,

$$T = T_o + \beta t \qquad (51)$$

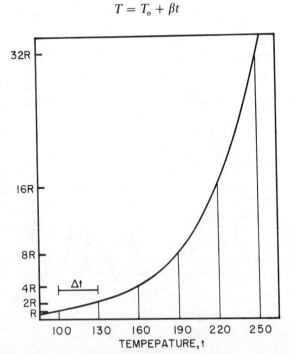

Figure 6.35 Zone migration as a function of temperature [103]. Adapted, with permission, from J. C. Giddings, *J. Chem. Educ.*, **39**, 569 (1962).

where T_o is the initial temperature, β is the program rate in degrees Celsius per minute, and t is the time that has elapsed from the initiation of the program ($t = 0$). The step function is an approximation where the temperature is instantly increased by 30°C. The *retention temperature* T_r is the temperature at which a solute appears at the outlet. It seems that little migration appears until $(T_r - 90)$°C. This is even more apparent in Figure 6.36. Half the column is traversed in the last 30°C temperature change. It is as if the different solutes left the start line of a race at different times and hence they *all* spend about the same time period in the mobile phase. This is largely the reason for uniform peak widths. Giddings defines the *significant temperature* as the isothermal temperature that will lead to the same amount of peak spreading and the same degree of separation as obtained with a programmed temperature chromatogram. For two closely spaced peaks where resolution may be of particular interest

$$T_C = T_r - 45°C \tag{52}$$

The parameter most easily controlled is the *program rate, β*. There are rigorous means of calculating this from a consideration of particular solutes; for example

$$R_r = \beta(L/\bar{u})/50 \tag{53}$$

where R_r is the value of R at the retention temperature, which is a function of the heat of vaporization from the solution, a property of the solute and its interaction with the stationary phase. Practical considerations are often more to the point. A continuous range of program rates is generally available, and the temptation is to use a fast rate to save time. Oven, not internal column, temperatures are controlled and monitored, and any relation between them at fast program rates is an illusion. There must be some approximation to thermal equilibrium between the oven and column interior. The rate should be "reasonably" slow but not to an extreme. Harris and Habgood [101] point out that the important parameter is the ratio of the heating rate to the volume flow

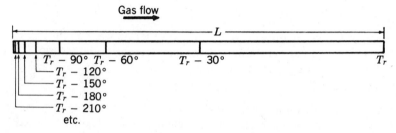

Figure 6.36 The distance migrated by a peak in successive 30°C-temperature increments [103]. Reprinted, with permission, from J. C. Giddings, *J. Chem. Educ.*, **39**, 569 (1962).

rate per gram of stationary phase corrected for the pressure drop $\beta/\bar{F}$, and this should be between 0.1 and 0.5 deg·g/mL. This assumes there is thermal equilibrium. Bearing in mind that broad generalizations are dangerous in direct proportion to their scope, a rate of 5°C/min is an estimate. Their survey states that about 55% of reported results employ rates between 1 and 5°C/min. A more thorough understanding of the principles allows a wiser selection of a rate.

Most ovens are operated at a low isothermal temperature for a selected period of time, temperature programmed to the final temperature, and held at this temperature for another selected period of time. The final temperature must not exceed the temperature limit of the stationary phase. This is a great advantage with unknown samples because it gives some insurance that all solutes have eluted from the column. The oven is cooled to its initial temperature. The complaint is often made that programmed temperature gas chromatography (PTGC) results are not reproducible. Usually the faults are that the "blow down" period is too short and the erroneous belief that the oven temperature is the same as column temperature. Programmed temperature gas chromatography is not fast if "blow down" time is included in analysis time. This is the reason that PTGC is not used often in process control.

Constant pressure flow controllers are wholly inadequate because of the viscosity increase of the carrier gas with an increase in temperature. Constant mass flow controllers are essential.

An experienced operator can conclude a great deal about an unknown sample with a dual column instrument by performing PTGC on both polar and nonpolar columns.

11   SOLUTE PROPERTIES

11.1   Thermodynamics

The intermolecular forces between the solute molecule and the stationary phase are the basis of separation. These forces are expressed by the partition coefficient and the value of R. In turn, R depends on the partial pressure on the solute vapor above the solution at T_C. This is described by Raoult's law

$$P_i = \gamma_i^* P_i^0 X_i \tag{54}$$

where P_i is the partial pressure of the solute vapor above the solution; P_i^0 is the vapor pressure of the pure solute at the temperature of the solution; X_i is the mole fraction of the solute in the solution; and γ_i^* is the symmetrical standard state activity coefficient of the solute in solution.

Textbooks are often unclear about the standard state, activity coefficients, and ideality. The nomenclature used here is that of Prigogine and Defay [104] and Prigogine [105]. The *symmetrical standard state* for both the solute and solvent is the pure liquid solute and solvent at the temperature of the solution. In this standard state, $\gamma_i^* = 1$ and $X_i = 1$ because P_i must reduce to P_i^0 for the pure substance. A *perfect solution* is one for which $\gamma_i^* = 1$ for *all* X_i. An *ideal solution* is one for which γ_i^* is a constant but not necessarily unity, which is most often the case for very dilute solutions. When true, this is the *Henry's law* region. In this dilute situation a solute molecule is surrounded by solvent molecules only; hence there are no solute–solute interactions, only those that are solute–solvent and solvent–solvent. In this standard state, the *symmetrical standard state activity coefficient* γ_i^* measures intermolecular interactions in terms of energy of interaction and entropy effects (solvation, order, molecular size differences, etc.). If $\gamma_i^* > 1$, *positive deviations from Raoult's law*, the solute escapes more easily from the solution than from an environment of neighbors of its own kind. Solute–solvent interactions are weak and/or solvent–solvent interactions are strong. If $\gamma_i^* < 1$, there is a *negative deviation from Raoult's law*, the escaping tendency is less than that from neighbors of the same kind, solute–solvent interactions are strong and/or solvent–solvent interactions are weak. The partition coefficient in terms of mole fraction K_{Xi} takes the form

$$K_{Xi} = \gamma_i^*(1)/\gamma_i^*(2) \tag{55}$$

where $\gamma_i^*(1)$ and $\gamma_i^*(2)$ are the activity coefficients in phases (1) and (2). The *unsymmetrical standard state* is the infinitely dilute solution. Its name arises from the asymmetry of the mole fraction of the solute approaching zero while that of the solvent approaches unity. The concept of a perfect solution and an ideal solution are the same except that some texts use ideal when the term *perfect* is more appropriate. The *unsymmetrical standard state activity coefficient* γ_i^0 is unity for an ideal solution in the Henry's law region; and when it is not unity, it measures deviation from ideal behavior, not from perfect behavior. Some tedious algebra concludes that

$$\gamma_i^*(1)/\gamma_i^*(2) = \exp(-\Delta\mu_i^0/\mathscr{R}T) \tag{56}$$

which is a restatement of (46). Now intermolecular forces in the Henry's law region are reflected in the change in the chemical potential when solute is transferred from one phase to another.

   Equation (47) defines the Gibbs free energy in terms of an energy or work term (bond breaking and forming), the enthalpy change, and an entropy change. Although $T\Delta S$ has the units of energy, the entropy change is best interpreted as an organizational term, a drive to the most probable state or the most random state. It is difficult to deal with entropy on any terms. Hildebrand and colleagues [106–108] defined *regular solutions* as those where the molecular distribution is that encountered in an ideal gas so that the entropy contribution to ΔG is that of mixing ideal gases and is independent of solute and solvent identity; that is, ΔS is the same for all solutions. This certainly simplified the whole problem except that in reality such solutions are of little interest in chromatography, and one discusses activity and partition coefficients for nonregular solutions as if they depend only on energy and work and generally pretends or hopes that the entropy is not there. Such rationalization is involved in (48)–(50).

11.2   Martin Equation

The following presentation is often practiced by physical chemists and is generally known as the *linear free energy relation*. Its application to chromatography was initiated by Martin [109] and has been a great success. The utility of energy to the physicist is that it is a scaler quantity obeying the laws of algebraic addition. If the chemical potential is treated as an energy term, which is reasonable for regular solutions, or the entropy is ignored, then $\Delta\mu^0$ can be treated as a sum of terms representing structural features of the molecule that interact energetically with the solvent. Suppose the solute molecule is represented by

$$H(CH_2)_n X$$

where X is a functional group, such as alcohol, halide, carboxyl, or aldehyde. Then

$$\Delta\mu_i^0 = \Delta\mu_H^0 + n\Delta\mu_{CH_2}^0 + \Delta\mu_X^0 \tag{57}$$

From (26)

$$\ln V_{Ni} = \ln V_L - (1/\mathscr{R}T)[\Delta\mu_H^0 + n\Delta\mu_{CH_2}^0 + \Delta\mu_X^0] \tag{58}$$

For a homologous series

$$\ln V_{Ni} = [\ln V_L - (1/\mathscr{R}T)(\Delta\mu_H^0 + \Delta\mu_X^0)] - (n/\mathscr{R}T)\Delta\mu_{CH_2}^0 \tag{59}$$

The number of methylene groups is the *carbon number n*. If $\ln V_{Ni}$ is plotted versus the carbon number, a straight line results with a slope of $(-\Delta\mu_{CH_2}^0/\mathscr{R}T)$. The intercept depends on the functional group with its chemical potential contribution $(-\Delta\mu_X^0)$ because for any one column V_L is fixed and all molecules contain a terminal hydrogen. Different homologous series, that is, different X groups, will have different intercepts, but the slopes of the plots will be equal or nearly so (i.e., nearly parallel lines) for a particular stationary phase. The Clausius–Clapeyron equation, written for the normal boiling point T_b ($P = 1$), is

$$\ln 1 = -\Delta H_v/\mathscr{R}T_b + C \tag{60}$$

or

$$\Delta H_v = -(\ln P - C)\mathscr{R}T_b$$

If $\Delta H_v = \Delta \mu_i^0$, then

$$\ln V_N = A + BT_b \tag{61}$$

and expresses the same conclusion as (59) where the argument of the function is the normal boiling point, and the intercept still depends on the functional group.

11.3   Martin Equation Limitations

Equation (59) yields identical results for all isomers of the compound, a feature fully appreciated by Martin, whether isomerism arises from other positions of the functional group or from chain branching. This can be corrected by assigning $\Delta \mu_X^0$ values for primary, secondary, and tertiary positions and including alkyl chain branching. Such a compilation is virtually endless if all such structural nuances are considered and the simple conceptual elegance of the approach is drowned in detail.

Certainly if attention is given to the identity and position of functional groups and the resultant effects on chromatographic behavior and if this is coupled with information supplied by the detector, excluding mass spectrometry, a relatively reliable qualitative analysis can be performed [26, 100–116]. Use of GC as a preseparator and sample injection port to a mass spectrometer is the most unequivocal identification application.

The thermodynamics presented here are the most basic and simple possible and lack detail and rigor. The reader will find greater satisfaction in the precise papers by Locke [117, 118].

12   SELECTIVITY

Selectivity is the ability to distinguish among chemical species. It cannot be assigned a number except in the trivial case of zero where there is no separation of anything. The concept exists but quantitation is elusive.

12.1   Solubility and Boiling Points

Solvent polarity and solubility have had an enduring association. Strictly, *solubility* is the concentration of a species in a saturated solution at a specified temperature; it is measured at equilibrium. It is a number but includes infinity; for example, gases are "infinitely soluble" in one another as is ethanol in water (complete miscibility). It is said that ethanol is "highly soluble" in water, but solubility as a number ceases to exist. Solubility is also often confused with the speed of dissolving. Ammonia and hydrogen chloride dissolve rapidly in water and are often said to be highly soluble even though a limit exists on their solubility. Sulfur trioxide dissolves to form sulfuric acid, which is miscible with water in all proportions and has virtually no vapor pressure; yet, because of a slow rate of solution it is not considered to be very soluble. Because solubility deals with a saturated solution, it refers to a concentration region to be avoided in chromatography because it is near the overload concentration and is also in a region that threatens nonlinear chromatography (nonideal solutions). The experienced and serious chromatographer tacitly understands these language errors, but the neophyte is likely to think that we know what we are talking about. The property desired is the affinity of the solute for the stationary phase where it forms a dilute solution (GLC) or is slightly sorbed, as in low surface coverage (GSC). This is done by the activity coefficient or the chemical potential change for very dilute solutions, the ideal region where γ_i^* is a constant or $\gamma_i^0 = 1$. When it is said that the solute should show some solubility in the stationary phase, it is meant that γ_i^* should not be very much greater than unity (no large deviation from Raoult's law) for small X_i. Locke is even more rigorous and in his notation uses $\gamma_i^{*\infty}$, which emphasizes an infinitely dilute solution; that is, $X_i \to 0$. Likewise γ_i^* should not be very much less than unity (negative deviation from Raoult's law) for small X_i because the solute would be retained for far too long. The popularist warns that "the solute should not be too soluble in the stationary phase." Equivalently, $\Delta\mu_i^0$ should not be large or small.

Hopefully, illusions about normal boiling points have already been shattered. The reason that the retention order of water relative to hydrocarbons can be shifted about so easily is the functional group term of the Martin equation. The interaction of this group with the stationary phase largely determines retention. If the stationary phase is a hydrocarbon, there are no interactive forces beyond the dispersion forces shown by all molecules. These are the weak interactions between electron clouds that are responsible for the liquification of helium, hydrogen, methane, and so on. The amount of interaction depends on the size of the molecule; that is, the interaction depends on such factors as the molecular size or weight or the carbon number. For hydrocarbons where only dispersive forces operate for the hydrocarbon solutes dissolved in a hydrocarbon solvent and in the pure hydrocarbon solute, the retention proceeds according to boiling point. Size would also dominate when the solute contained a suitable measure because stronger interactive forces are present in the pure solute; that is, hydrogen bonding in pure water and pure alcohols. Hydrocarbon stationary phases are often termed boiling point separators, because retention order is roughly determined by increasing boiling points, but the immediately preceding discussion shows the invalidity of this concept. They should be called dispersive separators to divorce their properties from those of the pure solute. All stationary phases are dispersive separators for members within a homologous series because, as shown by the Martin equation, retention depends on the number of methylene groups.

Langer [119] and Langer and Sheehan [120] summarized the situation very well. To separate solutes i and j, the ratio of adjusted retention volumes and hence partition coefficients must differ from unity; that is,

$$K_j/K_i = \gamma_j^* P_j^0 / \gamma_i^* P_i^0 \neq 1 \qquad (62)$$

Several cases arise.

1.   If i and j are members of a homologous series, then because of structural similarities γ_i^* nearly equals γ_j^* and a dispersive interaction (boiling point separator) is indicated; that is, one exploits the fact that $P_i^0 \neq P_j^0$. Since all phases act in this way, it is best to select one with little or only weak interaction with the functional group, that is, small $\Delta\mu_X^0$. This will yield a reasonable retention volume, and T_C can be kept low. The latter is desirable because the slope of the Martin equation depends on $\Delta\mu_{CH_2}^0/\mathcal{R}T_C$ and a large slope is desirable (large change in $\ln V_N$ per methylene group).

2.   If solutes i and j have nearly the same boiling points; that is, $P_i^0 = P_j^0$, then interactive stationary phases where $\gamma_i^* \neq \gamma_j^*$ are indicated. The popular statement is that the phase should be polar.

3.   If $\gamma_i^* = \gamma_j^*$ and $P_i^0 = P_j^0$, then there is a problem, which will most likely arise with isomers. Chromatography has its real power in this situation. It is most likely that the activity coefficients that differ slightly are not identical because of slight differences in steric effects and electron distributions so that the interacting solute function group responds slightly differently to a function group within the solvent. This is magnified by encouraging many encounters with the solvent functional group by employing a modest temperature and a stationary phase with a high concentration of this slightly selective functional group. One may also appeal to the entropy effect and seek departure from regular solutions. A stationary phase with a molar volume very much different from that of the solute is one approach, but the molar volumes of i and j must show some difference.

12.2   Polarity

An approach that resembles solubility is based on "polarity," which has its roots in solubility concepts established in college sophomore organic chemistry by the rule that "like dissolves like." Hildebrand and Scott [106] emphasized the difficulty of defining "likeness." It is again the difficulty of precisely defining a concept that everyone presumes to be obvious. Methyl chloride is polar because it has an asymmetric distribution of electronic charge as shown by the dipole moment and, because like dissolves like, anything that dissolves in methyl chloride must also be polar. Water is said to be polar; it too has a dipole moment. A comparison of solvent properties of alcohol with methyl chloride exposes some major differences. This is avoided by referring to water and alcohol as hydrogen-bonding solvents. Thus two kinds of polarity arise: a dipole–dipole polarity (classical polarity) and a hydrogen-bonding polarity. Long ago, Cassidy [121] pointed out an additional problem. When p-dichlorobenzene and o-dichlorobenzene are compared, the former is found to be nonpolar in terms of dipole moment but more polar in terms of chromato-graphic behavior (LSC). It must be that *chromatographic polarity is meaningful only if defined within chromatographic experience and the chromatographic system involved*, a theme strongly stated by Martire and Pollara [122].

12.3   Rohrschneider Polarity Scale

If a hydrocarbon phase is accepted as entirely nonpolar and capable of dispersive interactions, then any liquid other than a hydrocarbon must be more polar. Experience established an intuition for ranking stationary phase polarities from a knowledge of the functional group present in the phase. Solutes with functional groups have retentions that increase as this intuitive stationary phase polarity increases. Rohrschneider selected butadiene to measure this polarity possibly because of the nonspecificity of the delocalized π electrons as contrasted with a hydroxyl group and its similarity in size and boiling point to butane. Rohrschneider defined his polarity [123–125] as

$$P_p = a[\log(V_p^{\text{butadiene}}/V_p^{\text{butane}}) - \log(V_u^{\text{butadiene}}/V_u^{\text{butane}})] \tag{63}$$

where the retention volumes are V_N and the subscripts p and u indicate polar and nonpolar phases, respectively. The constant a is determined by setting a polarity of squalane (2,6,10,15,19,23-hexamethyltetracosane), a hydrocarbon, at $P_p = 0$ and $P_p = 100$ for β,β'-bispropionitrile ether, which is an extremely polar phase. When squalane is used to determine $V_u^{\text{butadiene}}/V_u^{\text{butane}}$, it measures the difference only in either molecular weight or dispersive interaction. This ratio for the polar phase contains all interactions that are loosely termed polar (dipole, hydrogen bonding, etc.) and dispersive interactions. The difference of these terms should isolate the polar contributions to retention.

This polarity scale was not developed to any great extent because it was quickly replaced by a more refined treatment, largely introduced by Rohrschneider. The basic concept remains definitive. Polarity is measured by chromatographic behavior and involves a particular solute to monitor polarity. This standard has become known as a *solute, molecule,* or *polarity probe.* A suspected shortcoming of a single probe is its inability to differentiate those interactions responsible for polarity, and modifications were introduced to correct this. This modification sacrificed the desire for an all-encompassing single characteristic number to gain additional information. Rohrschneider has described his work and ideas [126].

12.4   Test Mixtures

The order of peak position can be used as a qualitative gauge of polarity. Ettre [72] suggests the following (numbers are volume percentages): ethanol (EtOH), 40; methyl ethyl ketone (MEK), 20; cyclohexane (CyHx), 5; and benzene (Bnz), 10. The retention orders are (the least-retained solute is listed first)

Nonpolar	EtOH—MEK—Bnz—CyHx
Slightly polar	EtOH—MEK—CyHx—Bnz
Polar	CyHx—MEK—Bnz—EtOH

No effort is made to quantitate the change. The suggestion again involves the use of standard probes.

If used periodically, the mixture is also useful in detecting changes in the stationary phase during column use. A simultaneous determination of plate height adds to the utility of the data.

12.5   Selectivity Coefficient

The *selectivity coefficient* σ is described by Bayer [127]. The retention volumes of members of different homologous series: A, B, C,..., relative to a hydrocarbon (e.g., pentane) are plotted versus the normal boiling point on a particular stationary phase. A series of straight lines result (a phase acts as a boiling point separator within any homologous series). This is shown in Figure 6.37. At any one boiling point, a perpendicular to the boiling point axis is constructed. The selectivity coefficient is the ratio of the relative retentions at this arbitrary temperature; for example,

$$\sigma_{AB} = r_{AP}/r_{BP} = r_{AB} \tag{64}$$

The relative retention of A to B is now for hypothetical probes of the same boiling point, each representing a particular homologous series and measures the separation of intercepts as determined by the response of the stationary phase to a particular functional group and its position. For $\sigma = 1$, there is no separation of the probes. The greater the difference from unity, the better the promise of separation. Table 6.9 presents a number of selectivity coefficients taken from the data of Bayer. A few sample conclusions are

1.   Paraffin oil and DPM are not responsive to positional isomerism for the alcohols, but ODP does respond.

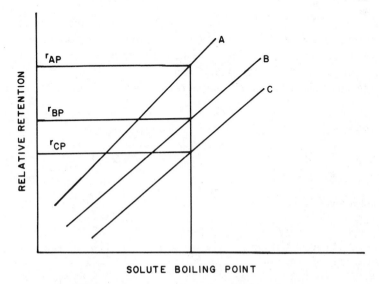

Figure 6.37   Determination of selectivity coefficients.

Table 6.9 Selectivity Coefficients on Various Stationary Phases[a]

Stationary Phases	Liquid Phase			
	Par	*ODP*	*PG4*	*DPM*
Diolefins/paraffins		3.3	1.3	2.0
Alkylbenzenes/methyl paraffins	1.2	19.0	2.6	6.0
1°/2°Alcohols	1.1	1.4	1.2	1.0
Ketones/ethers	0.6	8.0	1.7	2.4

[a]Par, paraffin oil; ODP, $\beta\beta'$-bispropionitrile ether; PG4, poly(ethylene glycol) 400; DPM, diphenylformamide.

2. β,β'-Bispropionitrile ether responds to aromaticity—in fact, too much so; this is the wasted column problem.

3. All liquids seem suitable for ketone–ether mixtures.

Separation is not guaranteed because a single relative retention will yield a hypothetical member from each series with a different boiling point. It would be very unusual if a real compound existed in each series with this exact boiling point, but there might be members in this proximity.

Unfortunately, the use of selectivity coefficients is not popular. If a particular laboratory deals with a relative limited number of compound types, tables of selectivity coefficients at different temperatures can be established in-house for use with particular separation problems.

12.6   Retention Index

The retention index [50] is a relative retention scale (alkanes), but we now consider specific compounds. If detailed composition (e.g., quantitative analysis) is known for the sample, a suitable phase may be selected. It is particularly useful when deciding on an internal standard where it is desirable that the peak falls in one of the vacancies between the sample peaks. Table 6.10 lists some retention

Table 6.10 Retention Indices *I* of Acetone and Diethyl Ether

Stationary Phase and T_C	$I\ (Et_2O)$ bp, 34.6°C	$I\ (Me_2CO)$ bp, 56.5°C
Silicone high grease DC (100°C)	505	510
Apiezon grease (130°C)	484	450
Diisodecyl phthalate (100°C)	526	590
Poly(ethylene glycol) 1500 (100°C)	642	865
Poly(ethylene glycol) succinate polyester (100°C)	725	1073

index values for acetone (Me$_2$CO) and diethyl ether (Et$_2$O). The first three liquids seem unsuitable because the ketone and ether fall between adjacent alkanes that differ by one methylene group. Note the reversal of appearance for the Apiezon (high boiler first). Poly(ethylene glycol) 1500 (PG 1500) seems satisfactory, the ether appears between hexane and heptane, and the ketone is between octane and nonane. The selectivity coefficient for PG 1500 (Table 6.9) indicates this same suitability for ethers and ketones in general. The polyester probably provides too much separation, which would yield the wasted column problem.

12.7 Retention Dispersion

The retention index meters retention on a scale of hydrocarbon behavior. If solute i is chromatographed on a nonpolar stationary phase, then the retention index measures the dispersive interaction of the solute, to a first approximation, independent of its specific functional group and relative to the dispersive interaction of the neighboring alkane solutes. If the stationary phase is changed to a polar liquid, the retention index will change, presumably because the solute functional group interacts with the solvent functionality in addition to the ever-present dispersive interactions, while the neighboring alkanes continue to interact through dispersive forces only. The change in the retention index, the *retention dispersion* [50] ΔI; that is,

$$\Delta = I_p - I_u \tag{65}$$

should measure the polar interactions regardless of the origin of these forces. With this concept the retention dispersion can now be used to describe stationary phase properties rather than solute properties.

12.8 Rohrschneider and McReynolds Polarity Probes

These ideas were combined by Rohrschneider [128, 129], and Supina and Rose [130] who began to characterize stationary phases by the retention dispersion of a select group of solutes. Table 6.11 lists these probes and suggests the interactions that might be involved. A more detailed discussion of potential intermolecular forces of these probes is needed. A *soft Lewis base* is an electron cloud donor while a *hard Lewis base* is an electron pair donor (Brønsted base). The conjugate interaction is termed a *soft* or *hard Lewis acid*, respectively. *Dipole interaction* indicates a permanent dipole moment. The *α-hydrogen effect* is the tautomerism associated with a methylene hydrogen alpha to a carbonyl; that is,

Table 6.11   Rohrschneider Polarity Probes

Solute Probes	Interaction
Benzene	Delocalized π electrons, aromaticity; soft Lewis base
Ethanol	Hydrogen bond, proton donor and acceptor, amphoteric; hard Lewis acid and base
Methyl ethyl ketone	Dipole interaction, hard Lewis base, α-hydrogen effect
Methyl nitrate	Dipole interaction
Pyridine	Hard Lewis base

McReynolds [13] suggested a more extensive set of probes that seemed to provide more information as indicated by a statistical analysis of the behavior of 68 solutes on 25 liquid phases. Those substituted for the Rohrschneider probes were less volatile, easier to measure and inject, and were usable over a wider temperature range. The McReynolds probes are benzene; *n*-butanol; methyl propyl ketone (2-pentanone); propyl nitrate (nitropropane); pyridine; 2-methyl-2-pentanol; 1-iodobutane (*n*-butyl iodide); 2-octyne; 1,4-dioxane; *cis*-hydrindane. These optimized probes were used to characterize over 200 liquids [131].

There is some agreement that most of the information desired can be furnished by the first five probes—a salute to Rohrschneider's original ingenuity. Alternate probes [132, 133] have been suggested; it has also been suggested that benzene alone is adequate, which reminds us of Rohrschneider's single probe scale where dispersion is related essentially to a single alkane, butane.

12.9   Phase Redundancies

The demonstration of redundancies alone is sufficient to warrent admiration of the probe system. Table 6.12 lists retention dispersions for the McReynolds probes for selected popular phases listed by their trade designations.

It is not necessary to maintain an extensive library of stationary phases. If a separation reported in the literature has been achieved with Apiezon M grease, it seems that Apiezon L grease will also serve. The phases of the last four columns seem interchangeable. This may not be the case for all separations. Chromatography is so successful because it is highly sensitive to minor differences of molecular architecture, and very minor differences can be exploited. Nonetheless, the first step at organization is a significant start.

12.10   Preferred Stationary Phases

The existence of obvious redundancies suggested that other stationary phases might fall into groups where one member of the group would represent the desired separation properties sought to be measured by the ever elusive polarity [134]. If one takes the McReynolds data for the 200 phases examined and averages the retention dispersion for the first five probes and orders them

Table 6.12   Probe Values of Redundant Phases

McReynolds Probe	Retention Dispersion on Selected Phases					
			Apiezon			
	M	L	SE-33	DC-200	OV-101	DC-410
Benzene	31	32	17	16	17	18
n-Pentanol	22	22	54	57	57	57
2-Pentanone	15	15	45	45	45	47
Nitropropane	30	32	67	66	67	68
Pyridine	40	42	42	43	43	44
2-Methyl-2-pentanol	12	13	33	33	33	34
n-Butyl iodide	32	35	4	3	4	5
2-Octyne	10	11	23	23	23	24
1,4-Dioxane	28	31	46	46	46	48
cis-Hydrindane	29	33	−1	−3	−2	0

according to increasing values of this polarity, there are no sharp breaks in the scale. The problem and the availability of copious data invited the application of pattern-recognition techniques [135–144]. A group of participants at an Eastern Analytical Symposium formed a committee and developed a list of 24 preferred stationary phases. Six of these were selected as components of a primary list. Table 6.13 gives this list with the primary members in italics [145]. Because the retention dispersion of benzene parallels the average polarity as calculated from the McReynolds probes, it is included in this table to supply some rough measure of polarity. Other suggested lists have been made [146–150].

Risby and colleagues [151, 152] employed thermodynamic properties to characterize phases.

The polarity range can be extended by mixing phases in one of three ways.

1. Separate columns that can be prepared with the proportion arranged to yield a desired weighted average are connected; they are termed *coupled columns*. These columns do not seem desirable because a reversal of the order alters peak position. This seems to be a pressure gradient effect.

2. The packings of the two columns may be mixed and packed in a single column and are termed *mixed bed columns*.

3. A mixture of the stationary phases, *mixed solvent columns*, may be dissolved in a single solvent and the solution loading used to prepare a packing coated with the liquid mixture.

All the physical chemistry of these configurations has not been settled. The situation has been reviewed by Pilgrim and Keller [153]. Laub and Purnell [154] suggest criteria for the use of mixed solvents. Most recently, Gröbler and

Table 6.13   Preferred Stationary Phases[a]

Compound	ΔI (Benzene)
Squalane	0
Dimethyl silicone (OV-101, SP-2100, SF-30, SF-96)	15
Apiezon L grease	33
Carborane-dimethyl silicone (Dexsil-300 GC)	37
20% Phenyl-methyl silicone (OV-7)	69
Dinonyl phthalate	83
50% Phenyl-methyl silicone (OV-17, SP-2250)	119
Poly(propylene glycol) (UCON LB-550X)	130
Trifluoropropyl-methyl silicone (OV-210, SP-2401)	145
Poly(phenyl ether)	176
25% 2-Cyanoethyl-methyl silicone (AN-600)	204
25% 3-Cyanopropyl—25% phenyl-methyl silicone (OV-225)	228
Nonylphenoxy(ethyleneoxy)ethanol (Igepal CO-880)	259
Cyclohexanedimethanol succinate	270
50% 3-Cyanopropyl—50% phenyl-methyl silicone (Silar-5CP, SP-2300)	319
Poly(ethylene glycol), MW exceeds 4000 (Carbowax)	323
Butanediol succinate	370
80% 3-Cyanopropyl—20% phenyl-methyl silicone (Silar-7CP, SP-2310)	440
90% 3-Cyanopropyl—10% phenyl-methyl silicone (Silar-9CP, SP-2330)	489
Diethylene glycol succinate	495
3-Cyanopropyl silicone (Silar 10C, Apolar 10C, SP-2340)	523
1,2,3-Tris(2-cyanoethoxy)propane	593
Cyanoethyl-cyanopropyl silicone (OV-275)	629
N-N-Bis(2-cyanoethyl)foramide	690

[a]Hawkes and colleagues [145]. Reproduced from the *Journal of Chromatographic Science* by permission of Preston Publications, A Division of Preston Industries, Inc.

Bálizs [155–157] studied mixed phases using the retention index system. Mixed solvents will likely become a close adjunct to preferred stationary phases.

Much of the work dealing with selectivity, the retention index, and probes has been based on very valuable data compilations [131, 158–160].

13   SUPERCRITICAL FLUID CHROMATOGRAPHY

Selectivity in GC has been limited to the stationary phase because of the near ideality of the gas-phase solution, that is, its lack of intermolecular forces. Such forces can be enhanced by using a dense gas or vapor at high pressure above the critical temperature [supercritical fluid chromatography (SCFS)]. Densities may be as high as 90% of the liquid density. Compression of gases (carbon dioxide, ammonia, etc.) to this density was studied by Giddings' group [161, 162]. Alternatively a vapor of a substance that is liquid at room temperature, may be used. Several reviews exist [163–168]. The technology is difficult, at least for dense gases. The detector problem is very difficult with compressed permanent gases because when the pressure is reduced at the column outlet, the solute precipitates as a smoke. The technique is applicable to nonvolatile species. Interest decreased with the development of HPLC and little more was done. It is a pleasant surprise to discover recent renewed interest [169–172]. Commercial SCFC equipment is available [169].

14   QUANTITATIVE GAS CHROMATOGRAPHY

Other discussions of quantitative GC are the books by Kaiser (in German) [173] and Novák [174] and chapters in Perry [26] and by Ettre [113], Kaiser and Debbrecht [115], Lebbe [175], Johnson [176], and Novák [177, 178]. The remarks made here apply to general LC and GC elution techniques.

14.1   Requirements

Ideally the peaks should be well separated, that is, with a resolution of at least unity, and Gaussian. Deviation from such situations requires great caution in dealing with the readout and possible sacrifice of accuracy. The basis of all quantitative calculations is that the *area under the peak and the baseline is proportional to the amount of the solute*. This area may be related to the absolute amount (i.e., weight) if sample injection is sufficiently precise, which is never the case with syringe injection. Sampling loops can be trusted but then only when used by skilled operators. Determination of relative amounts is far more common. Because calibration is common practice, either weight or mole percentage may be used. The former is more direct. Reproducibility is assumed too often. Ottenstein [76] described a situation where peak areas changed and reached constancy only after several successive injections of the same sample. This was thought to be caused by the removal of components by the slightly adsorptive support until it was saturated. The cautious analyst will replicate the determinations of both standards and the unknown. Adsorption often involves displacement; that is, an adsorbed solute will be displaced by a more strongly adsorbed solute following it. In anticipation that this might occur, the standard should qualitatively and quantitatively closely parallel the sample in composition. A variation of sample size will indicate whether the analyst is operating within the linear range of the detector. The analyst must be aware of the vagaries of the detector.

14.2   Graphical Approximations to the Normal Curve

This section applies to Gaussian concentration profiles with the peaks of interest on scale.

14.2.1   Peak Heights

Peak heights are seductive because they are measured easily and swiftly. However, their use is not advised. Peak width is determined by the kinetics of the chromatographic process. A solute with a particular retention time will show the same peak width if sample injection is nearly instantaneous; that is, if the sample is not slowly distilled onto the column from the injector. Then, for this solute, different amounts will show different areas with the same peak width, and the height will be proportional to the area. If retention times vary from chromatogram to chromatogram for the flow rate changes during a single chromatogram, then peak heights will vary as the width varies for a constant area. Relative peak heights are reliable at constant flow rates for peaks closely positioned because it is assumed that peak broadening is nearly the same for both if their retention times are nearly the same.

There are two situations where peak heights are used consistently. The first is COT chromatograms. The peaks are commonly so sharp that the width of the pen line is an appreciable part of the peak width and it is not possible to obtain accurate areas by graphical means. Even an electronic integrator can be in error. Such an integrator opens an electronic "window," that may be likened to a slit on a spectrophotometer, to sense the signal intensity. A nearly closed slit or a very narrow window does not admit sufficient signal. If the width of the window becomes sufficiently large compared to the peak width, the area is proportionally distorted; that is, the histogram no longer measures the area under the curve. Kipiniak [179] specifies at least ten sensings per peak. The second occasion is process control. Automatic process analyzers repeatedly measure the same sample, with only minor variations in solute quantities. The processor is designed to perform at constant conditions, and very precise analyses are generally not required. The relative peak heights of a limited number of sample components is often sufficient.

Peak heights may be used for overlapping peaks. Snyder's study [6] indicates that if a minor peak overlaps another and the height of the valley h_r between the peaks relative to that of the minor peak h_2; that is, h_r/h_2 is 0.25, then the area of the minor peak, as measured by the area from the perpendicular dropped at the valley, to the true area is 0.99. This is shown in Figure 6.38. Other values are 0.98 for $h_r/h_2 = 0.30$, 0.95 for 0.57, and 0.90 for 0.75. Very often this is the area measured by an electronic integrator. In such cases, relative peak heights are superior if the leading edges and tails of the peaks do not overlap the peak position. Snyder states that relative peak heights are satisfactory (95% accuracy) if the resolution is 0.6–0.8, depending on the relative size of the peaks, but must be 0.9 if peak areas are used for the same accuracy. The best quantitative results are obtained by first performing good chromatography. Kipiniak [179] discusses areas and heights in detail.

14.2.2   Triangulation

Gaussian curve areas may be approximated by drawing tangents to the curve at the points of inflection and extending them until they intersect the baseline and each other. Hence, area = $(\frac{1}{2})$(base)(height).

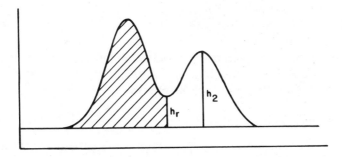

Figure 6.38 Peak-height–area relations for unresolved peaks.

14.2.3   Height × Width at Half-Height

This is a modification of the triangulation method. The width at half-height was defined in Section 4.11. It is faster than triangulation and more precise because less arbitrary judgment is involved when constructing the tangents.

14.3   Manual Graphical Integration

These methods are employed for non-Gaussian peaks.

14.3.1   Cut and Weigh

A measured area of the chart paper is cut out and weighed to establish a conversion factor, for example, square centimeter per milligram (cm^2/mg). The peak is cut out and weighed, and the area is calculated. The technique is laborious, destructive, and expensive in terms of technician time.

14.3.2   Planimeter

This "instrument of the devil" was invented by cartographers to measure irregular areas on maps. A jointed arm is anchored to a drafting table, and the free end, equipped with a stylus, is used to trace out the area. An attached wheel in contact with the board turns a counter and the reading is proportional to the area. Several tracings, a very steady hand, and superior eyesight are required. It is time-consuming and hence expensive.

14.4   Automatic Integration

These methods apply to peaks of any shape.

14.4.1   Ball-and-Disk Integrator

This is an automatic mechanical device that is attached to the recorder. As the recorder pen is displaced from the baseline, a ball is displaced along the radius of a spinning disk. The greater the displacement of the ball from the axis of the disk, the greater is its rotational velocity. The ball's rotation is transferred to a cam that translates the motion to a reciprocating arm to draw a jagged line on the chart paper. When the peak has passed, the pen returns to the baseline, which moves the ball back to the nonspinning position at the axis of rotation of the disk, and integration stops. The number of "blips" under the curve is proportional to the area. It is automatic in operation but counting the number of oscillations of the pen is not. It is not satisfactory for narrow peaks and integrates the entire areas of overlapping peaks.

14.4.2   Electronic Integration

As indicated, the electronic integrator samples the detector signal at rapid intervals and stores the intensity to form a histogram closely approximating the peak (as shown in Figure 6.39). When the peak has passed, the stored data is summed to yield a number, area counts, proportional to the area under the curve and the baseline. If sufficiently sophisticated, these areas are summed, a suitable detector response factor is included, and the percentage by weight is printed out. The situation sounds ideal, which it is, but reality is less than euphoric.

The operator must instruct this devoted computer of the signals to ignore as noise and those signals to recognize as peaks. This becomes difficult with COT chromatograms. The *noise-suppression control* must be understood. The integrator must be told when to start and when to stop integration. This is done by the computer, which compares signals by means of a logic circuit and decides, when there is a sufficient increase in the slope, that it is indeed dealing with a peak. This is shown in Figure 6.39. If the *slope sensor* is set too high, low flat peaks are ignored. If it is too shallow and long-term drift is present; that is, a baseline is not horizontal but slowly increases (as is often the case with PTGC

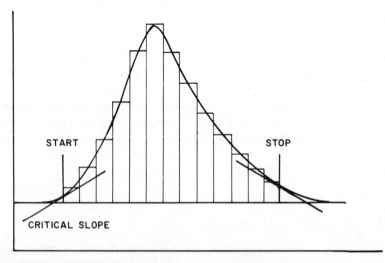

Figure 6.39   The histogram as a measure of peak area (electronic integration). The preset critical slope determines when the integration starts and stops.

where the column bleeds as the temperature is increased), the integrator may activate and not terminate the integration for the entire chromatogram. The analyst must be aware of the area that has been integrated. Figure 6.38 shows valley-to-baseline areas. Figure 6.40 shows a small peak superimposed on the tail of a large peak. The area desired is P. The determination of several areas is possible, depending on the program, for example, $(P + A)$ or $(P + A + B)$. Some integrators will draw a tangent to the major peak and report the area above the tangent. This is *tangent skimming*, also shown in Figure 6.40. The situation has best been summarized by Kipiniak [179] of Computer Inquiry Systems, and his conclusion deserves posting on every integrator.

> In situations as complex as these there is no single right answer. A computer program, no matter how powerful, sophisticated, and well designed, cannot take the place of the watchful eye of the chemist. The computer can carry out analysis and do so by several procedures within its repertoire, but it has to show the user, preferrably graphically, what it did, how it did, and what are the consequences of its assumptions. Be it in QC or research environment, the final responsibility for the accuracy and reliability of analysis rests with the human. Many of those who have intimate experience with computers think that perhaps that is just as well.

14.5   Total Relative Analysis

Presuming that reliable areas are available, the calculations are best demonstrated by an example. Suppose that four solutes, A, B, C, and D, are present and their identities are known so a standard of known composition can be prepared

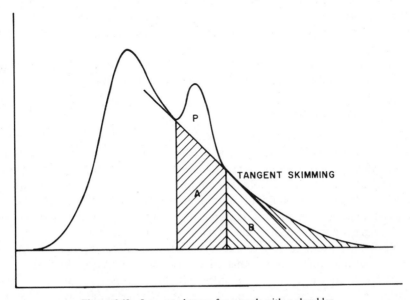

Figure 6.40   Integrated areas for a peak with a shoulder.

by weighing and mixing macroquantities using an ordinary analytical balance. Suppose the percentages by weight in the standard are $Q_A^s, Q_B^s, \ldots$. The chromatogram is performed, resulting in areas $A_A^s, A_B^s, \ldots$. The analyst, aware that the detector responds differently to each solute, for example, the FID as a carbon counter, inserts response factors $f_A, f_B, \ldots$ to "correct" the areas for the detector response differences. Then Q_A^s is proportional to $f_A A_A^s$. For relative amounts, the corrected "normalized" area percentage is formed, that is,

$$Q_A^s = f_A A_A^s / (f_A A_A^s + f_B A_B^s + f_C A_C^s + f_D A_D^s) = f_A A_A^s / \sum f_i A_i^s$$
$$Q_B^s = f_B A_B^s / \sum f_i A_i^s$$
$$\vdots$$

There are four unknowns, the f_i values, and four equations relating to them. To avoid tedious algebra, the following fractions are formed

$$Q_B^s / Q_A^s = f_B A_B^s / f_A A_A^s$$
$$Q_C^s / Q_A^s = f_C A_C^s / f_A A_A^s$$
$$\vdots$$

Setting $f_A = 1$

$$f_B = Q_B^s A_A^s / Q_A^s A_B^s$$
$$\vdots$$

The response factors $f_B, f_C,$ and f_D are relative to $f_A = 1$.

This calculation with these response factors is repeated for the unknown, where $A_A, A_B, \ldots$, are the areas of the peaks measured and $Q_A, Q_B, \ldots$ are calculated.

The method is reliable if the standard and sample are within the linear range of the detector and the critical detector parameters are constant.

14.6 Internal Standard

Often it is not necessary to have a total analysis, particularly for a complex multicomponent mixture. Only a limited number of components need be known. An *internal standard* is used; that is, a measured amount of a known pure substance is added to the sample. The sample should not originally contain the standard selected. The standard should be similar in response factor to the component species of interest. The standard should fall in a vacant region of the chromatogram but still be near the species of interest. Suppose such a standard S is selected to determine the amount of A in a sample. A binary mixture of S and A of known amounts is first prepared by weight in a macroquantity and f_A determined relative to $f_S (= 1)$ as described. In this example, weights will be used rather than percentages. Now a weighed amount of the standard w_S is added to a weighed amount of the unknown W. This weight should be such that A_S and A_A are nearly the same. The chromatography is performed, and A_A and A_S are determined. This yields w_A, the weight of A in the unknown; that is,

$$w_A = (f_A A_A / A_S) w_S$$

and

$$\% A = w_A / W$$

Acknowledgments

This material is the outgrowth of a short course on basic gas chromatography that was sponsored by the American Chemical Society and begun in 1969. The co-lecturer, collaborator, and a close friend is Dr. Michael F. Burke of the Department of Chemistry, University of Arizona at Tucson. I am grateful to the ACS (Moses Passer, Harold Walsh, and Nick Relacion in particular) for the opportunity to give the course and to the students who endured. I am indebted to Dr. Burke for his education of the author on more than one point. My continuing interest and activity in chromatography would not have been possible without the support of the Research Corporation, the National Science Foundation, the National Institutes of Health, Sigma Xi, and the Petroleum Research Fund.

References

1. H. H. Strain, "Differential Migration Methods of Analysis," in E. Heftman, Ed., *Chromatography*, Van Nostrand Reinhold, New York, 1961, p. 11.
2. H. H. Strain, "Differential Migration Methods of Analysis," in E. Heftman, Ed., *Chromatography*, 2nd ed., Van Nostrand Reinhold, New York, 1967, p. 11.
3. H. H. Strain and W. A. Svec, "Differential Migration Methods of Analysis," in E. Heftman, Ed., *Chromatography*, 3rd ed., Van Nostrand Reinhold, New York, 1975, p. 14.
4. J. C. Giddings, *Dynamics of Chromatography*, Dekker, New York, 1965.
5. K. H. Altgelt, "Theory and Mechanics of Gel Permeation Chromatography," in J. C. Giddings and R. A. Keller, Eds., *Advances in Chromatography*, Vol. 7, Dekker, New York, 1968, p. 3.
6. L. Snyder, *J. Chrom. Sci.*, **10**, 200 (1972).
7. L. S. Ettre and W. H. McFadden, Eds., *Ancillary Techniques of Gas Chromatography*, Wiley-Interscience, New York, 1969.
8. R. Teranishi, R. E. Lundin, W. H. McFadden, and J. R. Scherer, "Ancillary Systems," in L. S. Ettre and A. Zlatkis, Eds., *The Practice of Gas Chromatography*, Interscience, New York, 1967, p. 407.

9. G. W. A. Rijnders, "Preparative-Scale Gas Chromatography," in J. C. Giddings, and R. A. Keller, Eds., *Advances in Chromatography*, Vol. 3, Dekker, New York, 1966, p. 215.

10. A. Zlatkis and V. Pretorius, Eds., *Preparative Gas Chromatography*, Wiley-Interscience, New York, 1971.

11. M. Verzele, "Preparative Gas Chromatography," in E. S. Perry, Ed., *Progress in Separation and Purification*, Vol. 1, Interscience, New York, 1968, p. 83.

12. J. R. Conder, "Production-Scale Chromatography," in H. Purnell, Ed., *New Developments in Gas Chromatography*, Wiley, New York, 1973, p. 137.

13. D. T. Sawyer and G. L. Hargrove, "Preparative Gas Chromatography," in J. H. Purnell, Ed., *Progress in Gas Chromatography*, Interscience, New York, 1968, p. 325.

14. I. G. McWilliam, "Process Control by Gas Chromatography," in J. C. Giddings and R. A. Keller, Eds., *Advances in Chromatography*, Vol. 7, Dekker, New York, 1968, p. 163.

15. E. L. Szonntagh, "Automatic Process Gas Chromatography," in L. S. Ettre and A. Zlatkis, Eds., *The Practice of Gas Chromatography*, Interscience, New York, 1967, p. 511.

16. R. P. W. Scott and P. Kucera, *J. Chrom. Sci.*, **12**, 473 (1974).

17. *Scott Tech Newsline*, No. 4, Scott Specialty Gases, Plumsteadville, PA, February, 1982.

18. *Scott Tech Newsline*, No. 5, Scott Specialty Gases, Plumsteadville, PA, 1983.

19. R. Schill, "Instrumentation," in R. L. Grob, Ed., *Modern Practice of Gas Chromatography*, Wiley-Interscience, New York, 1977, p. 289.

20. J. Q. Walker, M. T. Jackson, Jr., and J. B. Maynard, *Chromatographic Systems-Maintenance and Trouble Shooting*, 2nd ed., Academic, New York, 1977.

21. J. J. van Deemter, F. J. Zuiderweg, and A. Klinkenberg, *Chem. Eng. Sci.*, **5**, 271 (1956).

22. A. Klinkenberg, "A Prepared Contribution," in R. P. W. Scott, Ed., *Gas Chromatography, 1960*, Butterworths, Washington, 1960, p. 182.

23. J. C. Sternberg, "Extracolumn Contributions to Chromatographic Band Broadening," in J. C. Giddings and R. A. Keller, Eds., *Advances in Chromatography*, Vol. 2, Dekker, New York, 1966, p. 205.

24. R. P. W. Scott, "Flow Programming," in J. H. Purnell, Ed., *Progress in Gas Chromatography*, Interscience, New York, 1968, p. 271.

25. L. S. Ettre, L. Mázor, and J. Takács, "Pressure (Flow) Programming in Gas Chromatography," in J. C. Giddings and R. A. Keller, Eds., *Advances in Chromatography*, Vol. 8, Dekker, New York, 1969, p. 271.

26. J. A. Perry, *Introduction to Analytical Gas Chromatography*, Dekker, New York, 1981.

27. D. J. David, *Gas Chromatographic Detectors*, Wiley, New York, 1974.

28. J. Ševčík, *Detectors in Gas Chromatography*, Elsevier, Amsterdam, 1976.

29. B. J. Gudzinowicz, "Detectors," in L. S. Ettre and A. Zlatkis, Eds., *The Practice of Gas Chromatography*, Interscience, New York, 1967, p. 239.

30. N. Guichard and J. Buzon, "Detectors," in J. Tranchant, Ed., *Practical Manual of Gas Chromatography*, Elsevier, Amsterdam, 1969, p. 166.

31. J. J. Sullivan and M. J. O'Brien, "Detectors," in R. L. Grob, Ed., *Modern Practice of Gas Chromatography*, Wiley-Interscience, New York, 1977, p. 213.

32. A. Karmen, "Ionization Detectors," in J. C. Giddings and R. A. Keller, Eds., *Advances in Chromatography*, Vol. 2, Dekker, New York, 1966, p. 293.

33. V. V. Brazhnikov, M. V. Gur'ev, and K. I. Sakodynski, *Chromatogr. Rev.*, **12**, 1 (1970).

34. D. M. Coulson, "Electrolytic Conductivity Detection in Gas Chromatography," in J. C. Giddings and R. A. Keller, Eds., *Advances in Chromatography*, Vol. 3, Dekker, New York, 1966, p. 197.

35. J. D. Winefordner and T. H. Glenn, "Nonionization Detectors and Their Use in Gas Chromatography," in J. C. Giddings and R. A. Keller, Eds., *Advances in Chromatography*, Vol. 5, Dekker, New York, 1968, p. 263.

36. M. Krejčí and M. Dressler, *Chromatogr. Rev.*, **13**, 1 (1970).

37. R. G. Ackman, *J. Gas Chromatogr.*, **2**, 173 (1964).

38. S. O. Farwell, D. R. Gage, and R. A. Kagel, *J. Chrom. Sci.*, **19**, 358 (1981).

39. J. J. Sullivan, *Hewlett-Packard Technical Paper No. 82*, Hewlett-Packard, Avondale, PA, 1972.

40. S. O. Farwell, *Standard Recommended Practice for Electron Capture Gas Chromatography*, ASTM-E-697-79, American Society for Testing and Materials, Philadelphia, PA, 1980.

41. W. H. McFadden, *Techniques of Combined Gas Chromatography/Mass Spectrometry: Applications in Organic Analysis*, Wiley-Interscience, New York, 1973.

42. C. Merritt, Jr., *Appl. Spectrosc. Rev.*, **3**, 263 (1970).

43. C. Merritt, Jr., "Techniques of Combined Gas Chromatography and Mass Spectrometry," in R. I. Reed, Ed., *Recent Topics in Mass Spectrometry*, Gordon & Breach, London, 1971, p. 195.

44. G. A. Junk, *Int. J. Mass Spectrom. Ion Phys.*, **8**, 1 (1972).

45. W. H. McFadden, *J. Chrom. Sci.*, **17**, 2 (1979).

46. R. M. Milberg and J. C. Cook, Jr., *J. Chrom. Sci.*, **17**, 17 (1979).

47. F. W. McLafferty and R. Venkatarghaven, *J. Chrom. Sci.*, **17**, 7 (1979).

48. J. Meili, F. C. Walls, R. McPherron, and A. L. Burlingame, *J. Chrom. Sci.*, **17**, 29 (1979).

49. A. T. James and A. J. P. Martin, *Biochem. J.*, **50**, 679 (1952).

50. E. sz. Kováts, "Gas Chromatographic Characterization of Organic Substances in the Retention Index System," in J. C. Giddings and R. A. Keller, Eds., *Advances in Chromatography*, Vol. 1, Dekker, New York, 1965, p. 229.

51. J. K. Haken, "Retention Indices in Gas Chromatography," in J. C. Giddings, E. Grushka, J. Cazes, and P. R. Brown, Eds., *Advances in Chromatography*, Vol. 14, Dekker, New York, 1976, p. 367.

52. L. Ettre, *Chromatographia*, **6**, 489 (1973).

53. L. Ettre, *Chromatographia*, **7**, 39 (1974).

54. L. Ettre, *Chromatographia*, **7**, 261 (1974).

55. A. J. P. Martin and R. L. M. Synge, *Biochem. J.*, **35**, 1358 (1941).

56. D. Martire, "Liquid Surface Effects in Gas–Liquid Chromatography (GLC)," in J. H. Purnell, Ed., *Progress in Gas Chromatography*, Interscience, New York, 1968, p. 93.

57. R. A. Keller and G. H. Stewart, *Anal. Chem.*, **34**, 1834 (1962).

58. R. A. Keller and J. C. Giddings, "Theoretical Basis of Partition Chromatography," in E. Heftman, Ed., *Chromatography*, 3rd ed., Van Nostrand Reinhold, New York, 1975, p. 110.

59. J. C. Giddings, "Theory of Chromatography," in E. Heftman, Ed., *Chromatography*, 3rd ed., Van Nostrand Reinhold, New York, 1975, p. 27.

60. J. J. Kirkland, *Anal. Chem.*, **35**, 2003 (1963).

61. J. Bohemen and J. H. Purnell, "Some Applications of Theory in the Attainment of High Column Efficiencies in Gas–Liquid Chromatography," in D. H. Desty, Ed., *Gas Chromatography, 1958*, Butterworths, London, 1958, p. 6.

62. I. Halász and E. Heine, "Optimum Conditions in Gas Chromatographic Analysis," in J. H. Purnell, Ed., *Progress in Gas Chromatography*, Interscience, New York, 1968, p. 153.

63. G. Guiochon, "Comparison of the Performance of the Various Column Types Used in Gas Chromatography," in J. C. Giddings and R. A. Keller, Eds., *Advances in Chromatography*, Vol. 8, Dekker, New York, 1969, p. 179.

64. C. Horváth, "Columns in Gas Chromatography," in L. S. Ettre and A. Zlatkis, Eds., *The Practice of Gas Chromatography*, Interscience, New York, 1967, p. 129.

65. C. Hishta, J. Bomstein, and W. D. Cooke, "Advances in the Technology of Lightly Loaded Glass Bead Columns," in J. C. Giddings and R. A. Keller, Eds., *Advances in Chromatography*, Vol. 9, Dekker, New York, 1970, p. 215.

66. I. Halász and I. Sebestian, *J. Chrom. Sci.*, **12**, 161 (1974).

67. I. Sebestian and I. Halász, "Chemically Bonded Phases in Chromatography," in J. C. Giddings, E. Grushka, J. Cazes, and P. R. Brown, Eds., *Advances in Chromatography*, Vol. 14, Dekker, New York, 1976, p. 75.

68. C. A. Cramers and J. A. Rijks, "Micropacked Columns in Gas Chromatography: An Evaluation," in J. C. Giddings, E. Grushka, J. Cazes, and P. R. Brown, Eds., *Advances in Chromatography*, Vol. 17, Dekker, New York, 1979, p. 101.

69. I. Halász and E. Heine, "Packed Capillary Columns in Gas Chromatography," in J. C. Giddings and R. A. Keller, Eds., *Advances in Chromatography*, Vol. 4, Dekker, New York, 1967, p. 207.

70. J. G. Nikelley, *Sep. Purif. Methods*, **3**, 423 (1974).

71. L. S. Ettre and J. E. Purcell, "Porous-Layer Open Tubular Columns—Theory, Practice, and Applications," in J. C. Giddings and R. A. Keller, Eds., *Advances in Chromatography*, Vol. 10, Dekker, New York, 1974, p. 1.

72. L. S. Ettre, *Open Tubular Columns in Gas Chromatography*, Plenum, New York, 1965.

73. A. Prévôt, "Columns," in J. Tranchant, Ed., *Practical Manual of Gas Chromatography*, Elsevier, Amsterdam, 1969, p. 86.

74. W. R. Supina, *The Packed Column in Gas Chromatography*, Supelco, Inc., Bellefonte, PA, 1974.

75. W. R. Supina, "Columns and Column Selection in Gas Chromatography," in R. L. Grob, Ed., *Modern Practice of Gas Chromatography*, Wiley-Interscience, New York, 1977, p. 113.

76. D. M. Ottenstein, "The Chromatographic Support," in J. C. Giddings and R. A. Keller, Eds., *Advances in Chromatography*, Vol. 3, Dekker, New York, 1966, p. 137.

77. D. M. Ottenstein, *J. Chrom. Sci.*, **11**, 136 (1973).

78. P. Urone and J. F. Parcher, "Support Effects on Retention Volumes in Gas–Liquid Chromatography," in J. C. Giddings and R. A. Keller, Eds., *Advances in Chromatography*, Vol. 6, Dekker, New York, 1968, p. 299.

79. G. Blandenet and J. P. Robin, *J. Gas Chromatogr.*, **2**, 225 (1964).

80. G. Blandenet and J. P. Robin, *J. Gas Chromatogr.*, **4**, 288 (1966).

81. A. V. Kiselev, "Adsorbents in Gas Chromatography," in J. C. Giddings and R. A. Keller, Eds., *Advances in Chromatography*, Vol. 4, Dekker, New York, 1967, p. 113.

82. A. DeCorcia and A. Liberti, "Gas–Liquid–Solid Chromatography," in J. C. Giddings, E. Grushka, E. Cazes, and P. R. Brown, Eds., *Advances in Chromatography*, Vol. 14, Dekker, New York, 1976, p. 305.

83. O. L. Hollis, *J. Chrom. Sci.*, **11**, 335 (1973).

84. C. L. Guillemin, M. LePage, and A. J. deVries, *J. Chrom. Sci.*, **9**, 470 (1971).

85. L. R. Snyder, *Principles of Adsorption Chromatography*, Dekker, New York, 1968.

86. R. P. W. Scott, "The Silica Gel Surface and Its Interactions with Solvent and Solute in Liquid Chromatography," in J. C. Giddings, E. Grushka, J. Cazes, and P. R. Brown, Eds., *Advances in Chromatography*, Vol. 20, Dekker, New York, 1982, p. 167.

87. R. J. Leibrand and L. L. Dunham, *Res. Dev.*, 32 (September, 1973).

88. D. M. Ottenstein, *J. Gas Chromatogr.*, **6**, 129 (1968).

89. B. L. Karger and W. D. Cooke, "Lightly Loaded Columns," in J. C. Giddings and R. A. Keller, Eds., *Advances in Chromatography*, Vol. 1, Dekker, New York, 1965, p. 309.

90. R. Kaiser, *Gas Phase Chromatography*, Vol. II, *Capillary Chromatography*, Butterworths, London, 1963.

91. W. Jennings, *Gas Chromatography with Glass Capillary Columns*, 2nd ed., Academic, New York, 1980.

92. W. Jennings, "New Developments in Capillary Columns for Gas Chromatography," in J. C. Giddings, E. Grushka, J. Cazes, and P. R. Brown, Eds., *Advances in Chromatography*, Vol. 20, Dekker, New York, 1982, p. 197.

93. M. J. E. Golay, "Theory and Practice of Gas–Liquid Partition Chromatography with Coated Capillaries," in V. J. Coates, H. J. Noebels, and I. S. Fagerson, Eds., *Gas Chromatography*, Academic, New York, 1958, p. 1.

94. M. J. E. Golay, "Marcel J. E. Golay," in L. S. Ettre and A. Zlatkis, Eds., *75 Years of Chromatography—A Historical Dialogue*, Elsevier, Amsterdam, 1979, p. 109.

95. D. H. Desty, "Capillary Columns: Trials, Tribulations, and Triumphs," in J. C. Giddings and R. A. Keller, Eds., *Advances in Chromatography*, Vol. 1, Dekker, New York, 1965, p. 199.

96. D. H. Desty, "Denis H. Desty," in L. S. Ettre and A. Zlatkis, Eds., *75 Years of Chromatography—A Historical Dialogue*, Elsevier, Amsterdam, 1979, p. 31.

97. M. Novotny, L. Blomberg, and K. D. Bartle, *J. Chrom. Sci.*, **8**, 390 (1970).

98. G. Guiochon, *J. Chrom. Sci.*, **9**, 512 (1971).

99. A. V. Kiselev and Y. I. Yashin, *Gas-Adsorption Chromatography*, Plenum, New York, 1969.

100. C. Vidal-Madjar and G. Guiochon, *Sep. Purif. Methods*, **2**, 1 (1973).

101. W. E. Harris and H. W. Habgood, *Programmed Temperature Gas Chromatography*, Wiley, New York, 1966.

102. L. Mikkelsen, "Advances in Programmed Temperature Gas Chromatography," in J. C. Giddings and R. A. Keller, Eds., *Advances in Chromatography*, Vol. 2, Dekker, New York, 1966, p. 337.

103. J. C. Giddings, *J. Chem. Educ.*, **39**, 569 (1962).

104. I. Prigogine and R. Defay, *Chemical Thermodynamics*, Longmans Green, London, 1954.

105. I. Prigogine, *The Molecular Theory of Solutions*, North-Holland, Amsterdam, 1957.

106. J. H. Hildebrand and R. L. Scott, *The Solubility of Nonelectrolytes*, 3rd ed., Reinhold, New York, 1950.

107. J. H. Hildebrand and R. L. Scott, *Regular Solutions*, Prentice-Hall, Englewood Cliffs, NJ, 1962.

108. J. H. Hildebrand, J. M. Prausnitz, and R. L. Scott, *Regular and Related Solutions*, Van Nostrand Reinhold, New York, 1970.

109. A. J. P. Martin, *Biochemical Society Symposia (Cambridge, England)*, **3**, 4 (1949).

110. D. A. Leathard and B. C. Shurlock, *Identification Techniques in Gas Chromatography*, Wiley-Interscience, New York, 1970.

111. D. A. Leathard and B. C. Shurlock, "Gas Chromatographic Identification," in J. H. Purnell, Ed., *Progress in Gas Chromatography*, Interscience, New York, 1968, p. 1.

112. D. A. Leathard, "Qualitative Analysis by Gas Chromatography," in J. C. Giddings, E. Grushka, R. A. Keller, and J. Cazes, Eds., *Advances in Chromatography*, Vol. 13, Dekker, New York, 1975, p. 265.

113. L. S. Ettre, "The Interpretation of Analytical Results: Qualitative and Quantitative Analysis," in L. S. Ettre and A. Zlatkis, Eds., *The Practice of Gas Chromatography*, Interscience, New York, 1967, p. 373.

114. J. Tranchant, "Qualitative Analysis: Separation and Identification," in J. Tranchant, Ed., *Practical Manual of Gas Chromatography*, Elsevier, Amsterdam, 1969, p. 214.

115. M. A. Kaiser and F. J. Debbrecht, "Qualitative and Quantitative Analysis by Gas Chromatography," in R. L. Grob, Ed., *Modern Practice of Gas Chromatography*, Wiley-Interscience, New York, 1977, p. 151.

116. G. Schomburg, "Identification by Retention and Response Values," in J. C. Giddings and R. A. Keller, Eds., *Advances in Chromatography*, Vol. 6, Dekker, New York, 1968, p. 211.

117. D. C. Locke, "Thermodynamics of Liquid–Liquid Partition Chromatography," in J. C. Giddings and R. A. Keller, Eds., *Advances in Chromatography*, Vol. 8, Dekker, New York, 1969, p. 47.

118. D. C. Locke, "Physicochemical Measurements Using Chromatography," in J. C. Giddings, E. Grushka, J. Cazes, and P. R. Brown, Eds., *Advances in Chromatography*, Vol. 14, Dekker, New York, 1976, p. 87.

119. S. H. Langer, *Anal. Chem.*, **39**, 524 (1967).

120. S. H. Langer and R. J. Sheehan, "Theory and Principles for Choosing and Designing Selective Stationary Phases," in J. H. Purnell, Ed., *Progress in Gas Chromatography*, Interscience, New York, 1968, p. 289.

121. H. G. Cassidy, *Fundamentals of Chromatography*, Interscience, New York, 1957.

122. D. E. Martire and L. Z. Pollara, "Interactions of the Solute with the Liquid Phase," in J. C. Giddings and R. A. Keller, Eds., *Advances in Chromatography*, Vol. 1, Dekker, New York, 1965, p. 335.

123. L. Rohrschneider, "The Polarity of Stationary Liquid Phases in Gas Chromatography," in J. C. Giddings and R. A. Keller, Eds., *Advances in Chromatography*, Vol. 4, Dekker, New York, 1967, p. 333.

124. G. E. Baiulescu and V. A. Ilie, *Stationary Phases in Gas Chromatography*, Pergamon, Oxford, 1975.

125. R. A. Keller, *J. Chrom. Sci.*, **11**, 49 (1973).

126. L. Rohrschneider, "Lutz Rohrschneider," in L. S. Ettre and A. Zlatkis, Eds., *75 Years of Chromatography—A Historical Dialogue*, Elsevier, Amsterdam, 1979, p. 351.

127. E. Bayer, *Gas Chromatography*, Elsevier, Amsterdam, 1961.

128. L. Rohrschneider, *J. Chromatogr.*, **22**, 6 (1966).

129. L. Rohrschneider, *J. Chrom. Sci.*, **11**, 160 (1973).

130. W. R. Supina and L. P. Rose, *J. Chrom. Sci.*, **8**, 214 (1970).

131. W. O. McReynolds, *J. Chrom. Sci.*, **8**, 685 (1970).

132. A. Hartkopf, *J. Chrom. Sci.*, **12**, 113 (1974).

133. A. Hartkopf, S. Grunfeld, and R. Delumyea, *J. Chrom. Sci.*, **12**, 119 (1974).

134. R. A. Keller, *J. Chrom. Sci.*, **11**, 188 (1973).

135. H. C. Andrews, *Introduction to Mathematical Techniques in Pattern Recognition*, Wiley-Interscience, New York, 1972.

136. P. C. Jurs and T. L. Isenhour, *Chemical Applications of Pattern Recognition*, Wiley-Interscience, New York, 1975.

137. D. L. Massart and L. Kaufman, *The Interpretation of Analytical Chemical Data by the Use of Cluster Analysis*, Wiley, New York, 1983.

138. E. R. Malinowski and D. G. Howery, *Factor Analysis in Chemistry*, Wiley-Interscience, New York, 1980.

139. T. L. Isenhour, B. R. Kowalski, and P. C. Jurs, *CRC Crit. Rev. Anal. Chem.*, **4**, 1 (1974).

140. B. R. Kowalski, Ed., *Chemometrics: Theory and Application* (ACS Symposium Series 52), American Chemical Society, Washington, DC, 1977.

141. D. L. Massart and H. L. O. DeClercq, "Numerical Taxonomy in Chromatography," in J. C. Giddings, E. Grushka, J. Cazes, and P. R. Brown, Eds., *Advances in Chromatography*, Vol. 16, Dekker, New York, 1978, p. 75.

142. S. Wold, *J. Chrom. Sci.*, **13**, 525 (1975).

143. J. K. Haken, M. S. Wainwright, and N. Do Phuong, *J. Chromatogr.*, **117**, 23 (1976).

144. D. L. Massart, P. Lenders, and M. Lauwereys, *J. Chrom. Sci.*, **12**, 617 (1974).

145. S. Hawkes, D. Grossman, A. Hartkopf, T. Isenhour, J. Leary, S. Wold, and J. Yancey, *J. Chrom. Sci.*, **13**, 115 (1975).

146. O. E. Schupp, III, *J. Chrom. Sci.*, **9** (6), 12A (June, 1971).

147. J. R. Mann and S. T. Preston, Jr., *J. Chrom. Sci.*, **11**, 216 (1973).

148. J. J. Leary, J. B. Justice, S. Tsuge, S. R. Lowry, and T. L. Isenhour, *J. Chrom. Sci.*, **11**, 201 (1973).

149. D. H. McClosky and S. J. Hawkes, *J. Chrom. Sci.*, **13**, 1 (1975).

150. S. Wold and K. Anderson, *J. Chromatogr.*, **80**, 43 (1973).

151. T. H. Risby, P. C. Jurs, and B. L. Reinbold, *J. Chromatogr.*, **99**, 173 (1974).

152. B. L. Reinbold and T. H. Risby, *J. Chrom. Sci.*, **13**, 372 (1975).

153. G. W. Pilgrim and R. A. Keller, *J. Chrom. Sci.*, **11**, 206 (1973).

154. R. J. Laub and J. H. Purnell, *J. Chromatogr.*, **112**, 71 (1975).

155. A. Gröbler and G. Bálizs, *J. Chrom. Sci.*, **17**, 631 (1979).

156. A. Gröbler and G. Bálizs, *J. Chrom. Sci.*, **17**, 671 (1979).

157. A. Gröbler and G. Bálizs, *J. Chrom. Sci.*, **19**, 46 (1981).

158. O. E. Schupp III and J. S. Lewis, Eds., *Compilation of Gas Chromatographic Data*, ASTM Data Series Publication No. DS`25A, American Society for Testing and Materials, Philadelphia, PA, 1967.

159. O. E. Schupp III and J. S. Lewis, Eds., *Gas Chromatographic Data Compilation Supplement 1*, ASTM AMD 25A S-1, American Society for Testing and Materials, Philadelphia, PA, 1971.

160. W. O. McReynolds, *Gas Chromatographic Retention Data*, Preston Technical Abstracts Co., Evanston, IL, 1966.

161. M. N. Myers and J. C. Giddings, *Sep. Sci.*, **1**, 761 (1966).

162. J. C. Giddings, M. N. Myers, L. McLaren, and R. A. Keller, *Science*, **162**, 67 (1968).

163. M. N. Myers and J. C. Giddings, "High-Pressure Gas Chromatography," in E. S. Perry and C. J. van Oss, Eds., *Progress in Separation and Purification*, Vol. 3, Wiley-Interscience, New York, 1970, p. 133.

164. T. H. Gouw and R. E. Jentoft, "Practical Aspects of Supercritical Fluid Chromatography," in J. C. Giddings, E. Grushka, R. A. Keller, and J. Cazes, Eds., *Advances in Chromatography*, Vol. 13, Dekker, New York, 1975, p. 1.

165. M. Novotny, S. R. Springston, P. A. Peaden, J. C. Fjeldsted, and M. L. Lee, *Anal. Chem.*, **53**, 407A (1981).

166. J. C. Fjeldsted and M. L. Lee, *Anal. Chem.*, **56**, 619A (1984).

167. U. van Wasen, I. Swaid, and G. M. Schneider, *Angew. Chem. Int. Ed., Engl.*, **19**, 575 (1980).

168. L. G. Randall, *Sep. Sci.*, **17**, 1 (1982).

169. D. R. Gere, R. Board, and D. McManigill, *Anal. Chem.*, **54**, 736 (1982).

170. P. A. Peaden, J. C. Fjeldsted, M. L. Lee, S. R. Springston, and M. Novotny, *Anal. Chem.*, **54**, 1090 (1982).

171. R. D. Smith, W. D. Felix, J. C. Fjeldsted, and M. L. Lee, *Anal. Chem.*, **54**, 1883 (1982).

172. H. H. Lauer, D. Manigill, and R. D. Board, *Anal. Chem.*, **55**, 1370 (1983).

173. R. Kaiser, *Chromatographie in der Gasphase. IV. Quantitative Auswertung*, Bibliographisches Institut, Mannheim, 1965.

174. J. Novák, *Quantitative Analysis by Gas Chromatography*, Dekker, New York, 1975.

175. J. Lebbe, "Apparatus," in J. Tranchant, Ed., *Practical Manual of Gas Chromatography*, Elsevier, Amsterdam, 1969, p. 257.

176. H. W. Johnson, Jr., "The Quantitative Interpretation of Gas Chromatographic Data," in J. C. Giddings and R. A. Keller, Eds., *Advances in Chromatography*, Vol. 5, Dekker, New York, 1968, p. 175.

177. J. Novák, "Quantitative Analysis by Gas Chromatography," in J. C. Giddings and R. A. Keller, Eds., *Advances in Chromatography*, Vol. 11, Dekker, New York, 1974, p. 1.

178. J. Novák, "Problems of Quantitation in Trace Analysis by Gas Chromatography," in J. C. Giddings, E. Grushka, J. Cazes, and P. R. Brown, Eds., *Advances in Chromatography*, Vol. 21, Dekker, New York, 1983, p. 303.

179. W. Kipiniak, *J. Chrom. Sci.*, **19**, 332 (1981).

Chapter **7**

SIZE-EXCLUSION
CHROMATOGRAPHY

Elizabeth V. Patton

1 INTRODUCTION

Size-exclusion chromatography (SEC) is a chromatographic technique in which the elution volume of polymer molecules is dependent on their molecular size in the mobile phase rather than on any form of interaction between the molecules and the packing material. The column-packing material consists of porous particles that are inert to the polymer dissolved in the eluent. When a polymer sample, a mixture of polymer molecules, is injected onto an SEC column, those whose size in solution is large are sterically prevented from entering most of the pore volume and are eluted sooner than those molecules with smaller volumes, which have access to more pore volume. The mixture is thus separated as it passes through the column, and the polymer molecules with the largest solution volume are eluted first, with progressively smaller molecules eluted progressively later. This separation process is illustrated in Figure 7.1, which is an electron micrograph of a gel particle, one type of porous stationary phase used in SEC. A molecular size distribution is recorded by the chromatographic detector and can be related by calibration to a molecular weight distribution (MWD) for the polymer sample, from which the moments of the distribution may be calculated. The separation range of SEC extends from small molecules with molecular weights of about 100 daltons to polymers with molecular weights of tens of

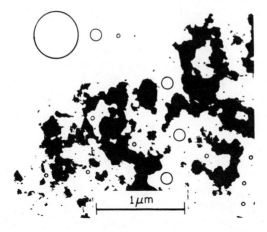

Figure 7.1 Electron micrograph of a gel particle, the stationary phase in SEC. The white areas are the pore structure; the black areas, the gel matrix. The circles are drawn to the average radii of gyration for solute molecules. The largest is totally excluded, and $K_{sec} = 0$. In order, the next two have $K_{sec} = 0.3$, and $K_{sec} = 0.5$. The smallest represents a solvent molecule, which has access to all of the pore volume, $K_{sec} = 1$. Reprinted with permission from K. H. Altgelt and J. C. Moore, in M. J. R. Cantow, Ed., *Polymer Fractionation*, Academic, NY, 1967, p. 123.

millions. The technique is simple, relatively inexpensive (especially if basic chromatography equipment is already available), and, when the currently available high-performance columns are used, rapid. Complete characterization of a polymer may be carried out in 10–30 min.

The technique was first described by Porath and Flodin [1] for the analysis of water-soluble polymers using cross-linked polydextran gels. In 1964, Moore [2] developed a semirigid polystyrene gel that was compatible with organic solvents, thus permitting analysis of the MWD of synthetic organic polymers. Current technology has focused on the development of packing materials consisting of smaller diameter particles, with which increased resolution and decreased band-broadening effects are obtained, and more rigid materials that withstand higher flow rates and can accommodate changes in solvent while maintaining their structures.

The volume of solvent present in the column may be divided into two parts; that which is between the particles, the *interstitial* (or *void*) *volume* V_o, and that contained within the pores of the packing material, called the *internal pore volume* V_i. The total accessible volume in the column V_t (exclusive of solids) is the sum of these

$$V_t = V_o + V_i \tag{1}$$

This total volume is accessible to solvent molecules and to solute molecules small enough to penetrate all of the pore volume. Such molecules, injected onto an SEC column, elute at volume V_t. Polymer molecules whose solution volumes exclude them from all of the pore volume diffuse through the interstitial volume alone and elute at V_o. If a polymer molecule has a size that allows it partial access to the pore volume, it elutes between V_o and V_i. This is the region of selective permeation. The larger the molecular size, the more pore volume is sterically inaccessible and the earlier the polymer elutes. The elution volume may be expressed as

$$V_e = V_o + K_{sec}V_i \qquad (2)$$

where K_{sec} is the distribution coefficient describing the ratio of polymer concentration in the pores to that in the interstitial volume. It ranges from 0 to 1. Larger polymer molecules have smaller values of K_{sec}. The separation range of the SEC column is within one column volume of eluent (Figure 7.2). In traditional liquid chromatography (LC) the separation is dictated by interaction between the solute and the stationary phase, and the separation range is beyond the total column volume of eluent; elution at V_t indicates no interaction. In LC solvent conditions can be modified to extend the separation range and increase the separation and resolution of solute molecules. In SEC that is not possible; the solvent is selected to solvate the polymer well and to insure that there is no interaction with the column packing. The separation range is entirely within the total column volume. To maximize this elution volume column-packing materials are constructed to have as large a pore volume as possible, while still maintaining enough mechanical strength to prevent compacting or crushing. The pore volume of SEC columns is typically about 50% of the total volume.

An example of elution profile of a polymer is depicted in Figure 7.3a. This is the recording of the response of a concentration-sensitive detector as a function of elution volume. Even without calibration, this profile contains information about the molecular size distribution of the polymer. For example, a peak near

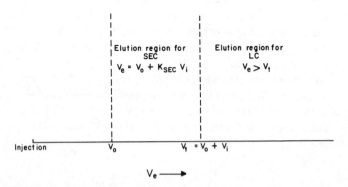

Figure 7.2 Chromatograph elution in SEC lies between V_o, the volume of the interparticle or interstitial areas of the column, and V_i, the total column volume, which is the sum of V_o and V_i, the volume contained in the pores of the stationary phase. Elution in other forms of LC that involve interaction with the stationary phase takes place at volumes greater than V_i.

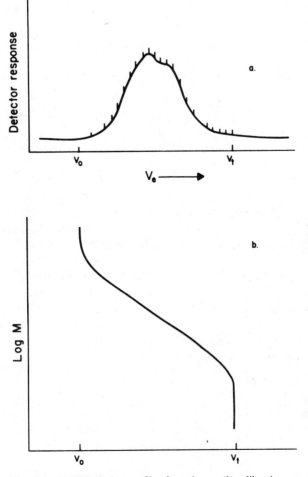

Figure 7.3   (a) SEC elution profile of a polymer; (b) calibration curve.

V_t signals the presence of low-molecular-weight material present in a synthetic polymer sample. The MWDs of a series of synthetic polymers may be compared by simply overlaying the chromatograms. The relationship between elution volume and molecular weight is established by calibration using standard polymer materials (Figure 7.3b). The form of this calibration curve and how it is obtained are described in Section 7.3. Once calibration is established molecular weights of incremental fractions of the elution curve can be assigned.

The molecular weight moments of the sample are typically found by dividing the elution curve into N slices of equal width, assuming the height from the baseline of each slice h_i is proportional to the concentration eluting at that volume, and assigning a molecular weight M_i to that slice from the calibration curve. Then $\bar{M}_n$, the number-average molecular weight, and $\bar{M}_w$, the weight-average molecular weight, are calculated using the equations

$$\bar{M}_n = \sum_{i=1}^{N} h_i \Big/ \sum_{i=1}^{N} \frac{h_i}{M_i} \tag{3a}$$

$$\bar{M}_w = \sum_{i=1}^{N} h_i M_i \Big/ \sum_{i=1}^{N} h_i \tag{3b}$$

These calculations can be made by manual measurement of h_i or by computer reduction of the data. The ratio $\bar{M}_w/\bar{M}_n$, called the *polydispersity* of the polymer, is a measure of the breadth of MWD. For a monodisperse polymer sample all molecules have the same molecular weight and $\bar{M}_w/\bar{M}_n = 1$. Theoretically, the SEC peak for such a polymer should be very sharp, since all of the polymer molecules should elute at a single volume. In fact, band-broadening effects, discussed in a later section, artificially broaden all SEC elution peaks so that polydispersities are artificially high. Proteins and other biological macromolecules are the only truly monodisperse polymers. Anionically polymerized synthetic polymers, which are used as narrow MWD standard materials, typically have polydispersities in the 1.01–1.10 range. The elution of the peak maximum for such samples is assumed to correspond to $\bar{M}_w$ and $\bar{M}_n$ (Figure 7.4). For polymer samples with broad MWD, $\bar{M}_w$ is much larger than $\bar{M}_n$ (Figure 7.5). Notice that for this broad sample, neither molecular weight average corresponds exactly to the molecular weight of the peak.

The z-average molecular weight can be calculated

$$\bar{M}_z = \sum_{i-1}^{N} h_i M_i^2 \Big/ \sum_{i=1}^{N} h_i M_i \tag{3c}$$

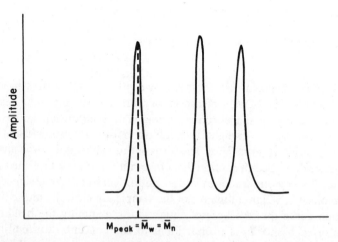

$$M_{peak} = \bar{M}_w = \bar{M}_n$$

Figure 7.4 Elution chromatograms of anionically polymerized, narrow MWD standards. The peak molecular weight is assumed to correspond to $\bar{M}_w$ and $\bar{M}_n$.

from the experimental chromatograms. Its position on the MWD distribution curve is listed in Figure 7.5. The viscosity-average molecular weight can be calculated using (3d)

$$\bar{M}_v = \sum_{i=1}^{N} h_i M_i^a \bigg/ \sum_{i-1}^{N} h_i^{1/a} \qquad (3d)$$

if the Mark–Houwink parameters $\mathscr{K}$ and a, which relate the intrinsic viscosity of a polymer to its molecular weight

$$[\eta] = \mathscr{K} M^a \qquad (4)$$

are known. The value of a depends on the amount of interaction between the polymer and the solvent in which it is dissolved. This value ranges from 0.5 for a polymer in a theta solvent to about 0.8 for one in a highly solvating solvent, and it may be as high as 1.0 in special cases; when this occurs, $\bar{M}_v$ is equal to $\bar{M}_w$.

Chromatograms of SEC separations are generally drawn in the order of increasing elution volume (Figure 7.6a). However, the convention for displaying the MWD of a polymer as a function of molecular weight is to plot the weight fraction in order of increasing molecular weight (Figure 7.6b). Since high-molecular-weight material is eluted first, these two curves are plotted in the opposite direction. This text follows these conventions.

Traditionally SEC has been divided into the areas of gel-permeation chromatography (GPC), for the fractionation of synthetic polymers in organic

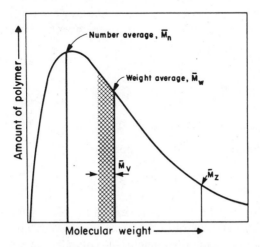

Figure 7.5 Molecular weight distribution of a broad MWD polymer and the molecular weights corresponding to the moments of the distribution. The breadth of a MWD is generally described by the polydispersity: the ratio $\bar{M}_w/\bar{M}_n$. Note that none of the averages corresponds exactly to the molecular weight at the peak of the curve.

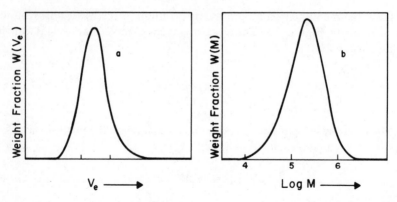

Figure 7.6 (*a*) By convention, the distribution of a polymer, when plotted as a function of elution volume V_e, is from low to high V_e, which is from high M to low M. (*b*) When plotted as a function of log M, the distribution is from low M to high M.

solvents using rigid gels, and gel-filtration chromatography (GFC), for the fractionation of biological macromolecules in aqueous solvents using soft gels. The theory of separation is common to both, but experimental techniques vary. The material covered is restricted to treatments of synthetic polymers. Biomedical applications of SEC have been reviewed recently [3]. The techniques described in this work are also restricted mainly to those commonly referred to as *high pressure* or *high performance*; that is, those employing column packings with particle diameters of about 10–30 μm, which allow rapid, high-resolution separations.

Reference [4] is a comprehensive publication on SEC, which has been the subject of several other books, chapters, and review articles. A complete compendium of work published in the field covering December, 1977, to November, 1981, is a useful resource [5].

2   THEORETICAL CONSIDERATIONS

2.1   Separation on the Basis of Size

In the equation that describes SEC fractionation

$$V_e = V_o + K_{sec}V_i \tag{2}$$

K_{sec} ranges from 0 to 1 and describes the permeation of the pores by the polymer solute. The mechanism of separation is a partitioning of the solute between the pore volume and the interstitial volume of the column. The term K_{sec} describes the ratio of solute concentrations in these two volumes. Experiments have demonstrated that the elution of the center of mass of polymer peaks is independent of flow rate, indicating that this partitioning is an equilibrium and not a kinetic phenomenon [6]. Therefore, K_{sec} is an equilibrium constant. Since the partitioning is a function of size rather than enthalpic interactions

$$K_{sec} \cong \exp(\Delta S^0/R) \tag{5}$$

in which ΔS^0 is the entropy change experienced by the polymer when it moves from the interstitial to the pore volume. Therefore the separation process is largely independent of temperature, although it may be affected by temperature through secondary interactions; for example, by affecting the polymer-solubility in the eluent, and thus its size. If there are any enthalpic interactions between the polymer and the stationary phase, these will be affected by temperature; in fact, change in elution volume with temperature is one method for detecting such interactions.

In the SEC process, polymer molecules are separated over a wide range of size, even if there is only one pore size in the stationary phase. An exact theoretical model is difficult to derive, but an approximate picture may be achieved by assuming a column packing with cylindrical pores of identical radius a and unlimited length. If the solute is further assumed to consist of spheres of radius r, then it is obvious that spheres with radius r greater than a will be excluded from the pores and have $K_{sec} = 0$. Permeating spheres of different sizes will be further separated in a column with uniform pore sizes, because there is a difference in the actual pore volume available to them. A sphere with radius r_1 has an effective pore radius that it can occupy of $(a - r_1)$ (Figure 7.7). A smaller sphere has a larger effective pore radius of $(a - r_2)$ available to it. The effective cross-sectional area available to a sphere of radius r is $\pi(a - r)^2$, and the fraction of the cross-sectional area of the pore that it can occupy is

$$\frac{\pi(a - r)^2}{\pi(a)^2} = \left(\frac{a - r}{a}\right)^2 \tag{6}$$

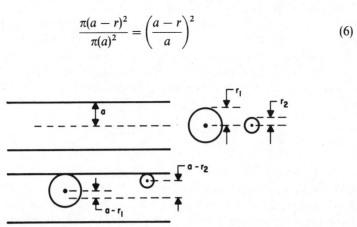

Figure 7.7 The pore radius available to permeating spheres of different size is equal to the difference between the actual radius a and the sphere radius r. Sphere 1 has an available radius $a - r_1$, smaller than that for sphere 2, $a - r_2$.

The probability of a sphere occupying a volume outside the pore is unlimited by sphere size because of the relatively large volumes available; this concentration outside the pore can only be achieved inside the pore within the $(a - r)$ radius. Therefore the ratio

$$\left(\frac{a - r}{a}\right)^2$$

represents the decrease in the probability of a sphere occupying a pore compared to an interstitial volume. This is the equilibrium constant K_{sec}.

$$K_{\text{sec}} = \left(\frac{a - r}{a}\right)^2 \tag{7}$$

The smaller sphere, with radius r_2, can achieve a higher concentration in the pore than the larger sphere and will have a larger K_{sec}.

$$K_{\text{sec},1} = \left(\frac{a - r_1}{a}\right)^2 < K_{\text{sec},2} = \left(\frac{a - r_2}{a}\right)^2 \tag{8}$$

In Figure 7.8 the dashed line is a plot of the log of the ratio r/a as a function of K_{sec}, illustrating the form of the separation of spheres on a column of uniform cylindrical pores. For $r/a \geqslant 1$, the sphere is totally excluded and $K_{\text{sec}} = 0$. For very small values of r/a, total permeation is approached, and K_{sec} approaches 1. Between these two limits, the single pore column separates spheres over a range of approximately one decade of radius.

The separation of random coil polymers in a column containing uniform cylindrical pores has been described by Cassasa [7–9]. His model takes into account the restrictions on polymer conformation imposed by the presence of the walls. This decreases the entropy of polymer molecules and therefore decreases their concentration in a pore. For example, in Figure 7.9 conformation a is allowed, conformation b is not. Cassasa predicted that the equilibrium constant K_{sec} with these restrictions may be expressed [7–9]

$$K_{\text{sec}} = 4 \sum_{m=1}^{\infty} \beta_m^{-2} \exp\left(-\beta_m^2 \frac{Nb^2}{6a^2}\right) \tag{9}$$

where β_m are the roots of the Bessell function of the first kind, order zero; and $Nb^2/6$ is the mean-square molecular radius of gyration of the polymer molecule R_G. The solid line in Figure 7.8 is $\log(R_G/a)$, the ratio of the polymer radius of gyration to the pore radius versus K_{sec} according to (9). The form of the size dependence is similar to that predicted by the hard sphere model, but with decreased values of K_{sec} because of the decreased probability of polymer occupation of the pore.

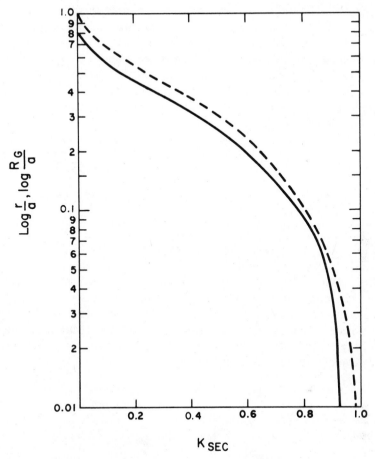

Figure 7.8 Theoretical size-separation curves for a stationary phase consisting of cylindrical pores of radius a. The dashed line represents the variation of $\log r/a$ with K_{sec} for a sphere of radius r; the solid line is the variation of $\log R_G/a$ with K_{sec} for a random coil with radius of gyration R_G, according to [7–9].

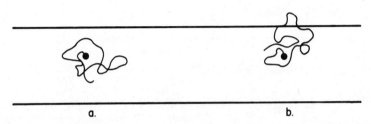

Figure 7.9 The presence of the pore walls restricts the conformations available to a polymer. Conformation a is allowed; conformation b is not.

Cassasa's model was tested by Yau and Malone [10], who compared the elution of polystyrene standards with the predicted calibration curve for a stationary phase containing cylindrical pores (see Figure 7.10). The average pore radius $\bar{a}$ was estimated by comparing measured values of pore volume and surface area. The experimental results lie close to the predicted curve.

The separation range in molecular weight afforded by a single pore size column depends on the relationship between molecular weight and R_G, which varies with the shape of the solute [11].

$$R_G(\text{sphere}) \propto M^{1/3} \tag{10a}$$

$$R_G(\text{rod}) \propto M \tag{10b}$$

$$R_G(\text{coil}) \propto M^a \tag{10c}$$

The value of a is approximately 0.5 for a random coil with unperturbed dimensions, and lies between 0.5 and 1.0 for a polymer in good solvent. If polymer molecules are separated over a range of one decade of R_G by a single pore size packing material, then this theory predicts that spheres are separated over three decades of molecular weight, rod-shaped molecules over one decade of molecular weight, and random coil molecules over two decades of molecular weight. Therefore the slope of the calibration curve should vary depending on the polymer concentration. The expected relationship between $\log M$ and K_{sec} for each conformation is illustrated in Figure 7.11.

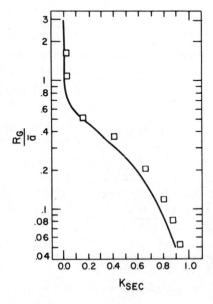

Figure 7.10 Theoretical curve for the separation of random coil molecules on stationary phase consisting of cylindrical pores of average radius $\bar{a}$; experimental data from the elution of polystyrene standards. The average pore radius was calculated from measured pore volumes and surface areas. Reprinted, with permission, from [10]. Copyright 1971 American Chemical Society.

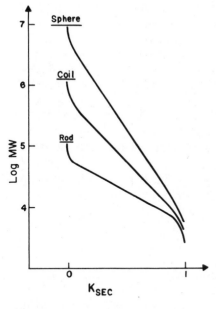

Figure 7.11 Theoretical SEC separations for polymers of different shapes. Reprinted, with permission, from [11]. Copyright 1980 American Chemical Society.

2.2   Column Dispersion

A polymer sample is introduced onto a column as a narrow band. As it passes through the column, dispersive processes occur that tend to spread this band. These take place simultaneously with size separation, and the net effect is that polymer molecules eluted from the column at certain V_e are not made up of a single molecular size, but are a distribution of molecular sizes caused by spreading from adjacent bands. The source of the dispersive effects in SEC will be considered as well as approaches to treating them to determine true molecular weight averages for eluted polymer samples.

Classically there are three types of dispersive forces operative in chromatography. These are random; therefore, each results in Gaussian band spreading causing a variance σ_i^2 in the eluted peak. If the processes are independent, the total variance in the eluted peak is the sum indicated by

$$\sigma^2 = \sigma_1^2 + \sigma_2^2 + \sigma_3^2 \tag{11}$$

The variance can be shown by classical plate theory to be proportional to H, the chromatographic height equivalent to a theoretical plate. In SEC this relationship is [4]

$$H = \sigma^2(L/V_e^2) \tag{12}$$

in which L is the column length and V_e is the elution volume. Therefore, the empirically determined quantity H can be expressed as a sum of independent contributions

$$H = H_1 + H_2 + H_3 \tag{13}$$

The first dispersive effect arises because individual molecules take different paths as they travel through a packed particle bed, so there will be a variation in the total path length traveled by each member of a group of molecules of the same molecular weight by the time it exits the column. This process is referred to as *eddy diffusion* and is related to the packing geometry of the column. The more irregular the particles and the more poorly packed the bed, the wider the variety of paths of different lengths through the bed that are available. A perfectly packed bed containing spherical particles will minimize the band broadening caused by eddy diffusion. The plate height contribution from this source is independent of flow rate and can be expressed [12]

$$H_1 = 2\lambda d_p \tag{14}$$

in which λ is a constant related to the structure of the packed bed, and d_p is the diameter of the packing particles.

The second dispersive force contributing to the plate height is caused by the diffusion of molecules from regions of high to low concentration. When this occurs in a band passing through the column, it results in *longitudinal diffusion*. This term is time dependent and is inversely proportional to the flow rate of the molecule through the column. At low flow rates, the residence time is long, and longitudinal diffusion may become significant depending on the size of the molecules. The plate height caused by longitudinal diffusion is expressed

$$H_2 = \frac{2\gamma' D_m}{v} \tag{15}$$

where v is the flow rate, D_m is the diffusion coefficient for the molecule in the mobile phase, and γ' is a tortuosity factor describing the restriction of diffusion caused by the column packing.

The third source of dispersion is caused by mass-transfer effects. In SEC two types of mass transfer are operative. The first is the interstitial, or mobile phase, mass transfer. The flow rate of a polymer in a given solvent stream depends on its nearness to a particle surface; those at the center of a solvent stream travel faster than those near the packed bed surfaces, and there is a difference in the amount of time required for the polymers to travel through the same solvent stream. The second is a stagnant mobile phase mass-transfer term that depends on the rate of diffusion of solute molecules in and around the pore. The plate height caused by diffusion terms is expressed

$$H_3 = \frac{C_m d_p^2}{D_m} v + \frac{C_s d_p^2}{D_s} v \tag{16}$$

in which C_m and C_s are the constants describing mass transfer in the mobile and stagnant phases, and D_m and D_s are the diffusion coefficients for the molecule in the two phases. These terms are particularly important for polymers, which have small diffusion coefficients, and therefore are not able to transfer rapidly between areas of solvent stream with different flow rates.

The plate height of the separation system is the sum of these contributions

$$H = 2\lambda d_p + \frac{2\gamma' D_m}{v} + \left(\frac{C_m}{D_m} + \frac{C_s}{D_s} \right) d_p^2 v \tag{17}$$

and this corresponds to the classic van Deemter equation [13]

$$H = A + \frac{B}{v} + Cv \tag{18}$$

In liquid chromatography, especially of polymers, the rates of molecular diffusion are so low that longitudinal diffusion is relatively unimportant; therefore, this term will be ignored. Giddings [14] has shown that the processes of eddy diffusion and mobile phase mass transfer are not independent, since eddy diffusion causes transfer of molecules between solvent streams of different rates. Hence the equation defining plate height should be written

$$H = \frac{C_s d_p^2 v}{D_s} + 1 \left/ \left(\frac{1}{2\lambda d_p} + \frac{D_m}{C_m d_p^2 v} \right) \right. \tag{19}$$

the sum of a stationary phase contribution H_s and an interstitial contribution H_m

$$H = H_s + H_m \tag{20}$$

The magnitude of H_s and H_m varies with the molecular weight of the solute, as well as with v and d_p. The diffusion coefficients D_s and D_m must be inversely dependent on solute molecular weight; these will cause the plate height to increase with increasing molecular weight. However, since molecules with larger molecular weights are more excluded from the pore volume, the contribution of the stationary phase term H_s to H must decrease with increasing molecular weight.

2.3 Measurement of Band Broadening

The plate height for a given solute on a specific column is calculated using (12), with σ determined from the eluted peak. If the broadening is assumed to be Gaussian, the elution band may be described by the equation

$$h = \frac{A}{\sigma \sqrt{2\pi}} \exp \left[-\frac{(V - V_e)^2}{2\sigma^2} \right] \tag{21}$$

in which h is the normalized chromatogram height at V, A is the peak area, V_e is the elution volume of the peak center, and σ is the variance of the chromatogram peak. The value of σ can be measured from the elution peak as

$$\sigma = \frac{W_b}{4} \qquad (22)$$

in which W_b is the baseline distance between tangents drawn to the inflection points, from the width at half-height W_h, or from the width at the inflection point W_i (Figure 7.12).

The variance measured for a monomeric material using this method is entirely caused by band broadening, but for polymers it contains a contribution from size separation as well. Synthetically prepared narrow MWD standard materials have polydispersities that range approximately from 1.02 to 1.10. These polydispersities cannot be known precisely since the nonchromatographic

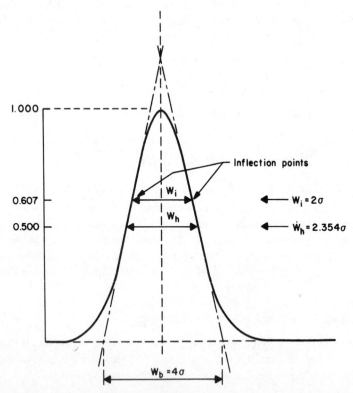

Figure 7.12 Gaussian band broadening. If the eluted solute is a monodisperse or narrow MWD polymer σ, the variance caused by band broadening can be determined by measuring the width either at the baseline between tangents drawn to the peak inflection points or at the inflection points.

methods used for measuring $\bar{M}_w$ and $\bar{M}_n$ are not precise enough. Therefore, the polymers used to determine the variance of a column as a function of molecular weight give rise to peaks that are broadened both by dispersion and size separation and differing amounts of size separation in each case. Methods have been devised to separate these two effects. In one, the solute is eluted halfway through the column, the flow stopped, and then reversed [15]. When the polymer is eluted, all effects of size separation have been eliminated and only dispersion broadening remains. This and other methods for measuring pure dispersion broadening for polymers are tedious and complicated. Frequently, the contribution of the polydispersity of narrow MWD standard polymers is ignored, and dispersion parameters are calculated using the peak widths of these materials.

Equation (19) divides band broadening into a pore process and an interstitial process. Several types of experiments have been carried out to determine the relative magnitudes of these terms for the SEC separation of polymers. Dependence of H on v has been compared for columns packed as identically as possible with porous and nonporous particles [16] and for excluded and permeating polymers on the same column containing porous particles [17]. In Figure 7.13 the reduced plate height $h = H/d_p$ is plotted as a function of the reduced velocity $v = vd_p/D_m$ for toluene and for a series of polystyrene solutes of different molecular weight on a column packed with porous and nonporous

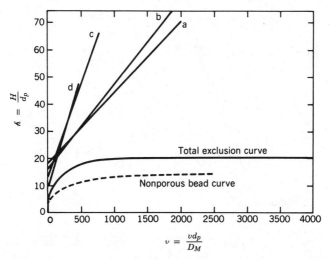

$$v = \frac{vd_p}{D_M}$$

Figure 7.13   Reduced plate height versus reduced velocity for toluene and for polystyrene polymers. The solid lines were obtained on a column packed with porous glass beads; the dashed line was obtained on a column packed with nonporous beads. The solvent is 1,1-dichloroethane except as indicated. The molecular weights of the polystyrene polymers are (a) 4000 (1,2-dichloroethane); (b) 4000; (c) 600; (d) toluene. The total exclusion curve is of polystyrene with molecular weight 20,000. Polystyrene with molecular weight 4000 was eluted on the nonporous bead curve. Reprinted, with permission, from [17]. Copyright 1977 American Chemical Society.

beads. When totally excluded solutes are eluted from the porous bead column, only interstitial effects, the last term in (19), should be operative. At low reduced velocities, h increases with v because of the effects of mobile-phase diffusion. At higher reduced velocities, the effects of eddy diffusion, which is independent of velocity, become predominant, and the curve flattens. The same curve should be observed when a solute is eluted from the same column repacked with nonporous beads. The discrepancy in position between the two in this case is caused by the inability to repack a column identically.

The curves for permeating polymers on porous beads are a sum of the interstitial and pore-broadening effects. Comparison with the total exclusion curve indicates that the magnitude of dispersion in the pore is much greater than interstitial dispersion. The reduced plate height caused by pore-broadening effects alone may be estimated by subtracting the interstitial curve from the total-exclusion curve (Figure 7.14).

The variation in plate height with molecular weight for the stagnant phase mass-transfer term was measured in an elegant experiment [18]. Groh and Halasz measured the variation in H with flow rate for benzene and polystyrene narrow MWD polymers on porous columns. The experiments were carried out with CH_2Cl_2 as the eluent, first under normal conditions to determine H as a function of v. The pores were then filled with water to make them inaccessible to the benzene and polystyrene solutes. In this way, H_m versus v was determined on

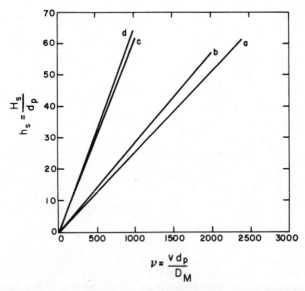

Figure 7.14 Reduced plate height caused by pore diffusion versus reduced velocity for the system illustrated in Figure 7.13; obtained by subtracting total exclusion curve h from total h. Curves identified as in Figure 7.13. Reprinted, with permission, from [17]. Copyright 1977 American Chemical Society.

the same column with all packing parameters the same. The plate height caused by stagnant phase mass transfer alone

$$H_s = H - H_m \tag{23}$$

could then be calculated. This is plotted (as $\Delta C'$, the stationary phase mass-transfer coefficient) as a function of M_w in Figure 7.15. The two effects of molecular weight on plate height are clearly seen by the presence of a maximum.

The actual value of D_m, the diffusion coefficient of a polymer within a pore, is difficult to measure. It is assumed to be proportional to the diffusion coefficient in the mobile phase.

$$D_s = \gamma D_m \tag{24}$$

Various models that take into account the tortuosity and constriction of a pore have been described to estimate the value of γ; it has been measured experimentally to be in the range of 0.1 to 0.3 [19], much lower than expected from model predictions [12].

2.4   Extra-Column Band Broadening

Equation (19) indicates that band-broadening processes are dependent on d_p, the diameter of the particles that make up the stationary phase. Development of high-performance column packings with d_p less than 10 μm has greatly reduced the amount of dispersion occurring during the separation. High-performance systems have much smaller columns; the total eluent volume over which the separation takes place has been reduced from hundreds of milliliters to tens of milliliters. Sample volumes are correspondingly smaller. Sources of band broadening occurring outside the column have become relatively more important since such small volumes are involved. Any point in the passage of the sample from injection to the outlet of the detectors that contains a large volume is a potential source of solute mixing and band broadening. The most likely places for this to occur are detector cells, column end fittings, tubing, and connectors. It is important to minimize the volume of cell fittings and tubing and to use detectors with low-volume cells designed for use with high-performance chromatography.

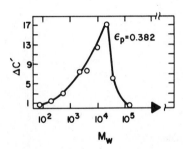

Figure 7.15 Mass transfer inside the pores, $\Delta C'$, versus $\bar{M}_w$ for columns containing porous beads; ε_p is the porosity of the beads. Reprinted, with permission, from [18]. Copyright 1981 American Chemical Society.

2.5 Resolution in Size-Exclusion Chromatography

The presence of dispersion causes the polymeric material eluted at a given volume to be a mixture of molecular weights. For molecular weight separations, the resolution of a column set, R, may be defined as the difference in the elution volume of polymers of narrow MWD of two different molecular weights compared to their peak widths (Figure 7.16).

$$R = \frac{V_{e,2} - V_{e,1}}{\frac{1}{2}(W_{b,1} + W_{b,2})} \approx \frac{\Delta V_e}{4\sigma} \tag{25}$$

The slope of the calibration curve that defines the change in M for a given change in V_e will affect the resolution of two molecular weight peaks. If the calibration curve is linear, the linear portion is generally expressed

$$M = D_1 \exp(-D_2 V_e) \tag{26}$$

in which $-D_2$ is the slope. The larger the magnitude of D_2, the more compressed are the molecular weights and the poorer is the resolution. The magnitude of D_2

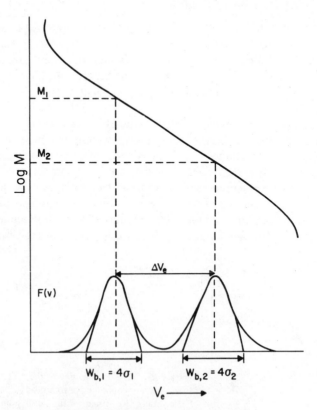

Figure 7.16 Resolution of two peaks in SEC.

is controlled by the total pore volume of the column; the larger this volume, the better is the separation and the smaller is the value of D_2. Yau and colleagues [20] suggested a method for defining column resolution using this parameter. The specific resolution R_{sp}, independent of the molecular weights of the peaks that are used to measure it, is

$$R_{sp} = \frac{\Delta V_e}{4\sigma} \left[\frac{1}{\log(M_1/M_2)} \right] \qquad (27)$$

and, defining M_1 and M_2 in terms of (26)

$$R_{sp} = \frac{\Delta V_e}{4\sigma} \left[\frac{1}{(D_2/2.303)(\Delta V_e)} \right] = \frac{0.576}{D_2\sigma} \qquad (28)$$

Resolution in SEC can thus be maximized by minimizing σ caused by band broadening and by minimizing D_2 by using large total pore volumes.

2.6   Effect of Column Dispersion on Molecular Weight Average

The general equation describing the effects of band broadening on the molecular weight distribution measured in an SEC experiment was given by Tung [21]

$$F(v) = \int_0^\infty W(y)G(v, y)\,dy \qquad (29)$$

in which $W(y)$ is the true distribution of molecular weights; that is, the form the elution curve would have if no band broadening took place; $G(v, y)$ is the instrumental spreading function; and $F(v)$ is the distribution recorded by the detector, the experimental chromatogram. Complete numerical solution of Tung's integral involves assuming the form of the spreading function $G(v, y)$, modeling $F(v)$ as a polynomial, and then deconvoluting the integral to determine $W(y)$. Tung's original solution assumes that $F(v)$ and $W(y)$ are fourth-degree polynomials and that $G(v, y)$ is Gaussian [22]

$$G(v, y) = \frac{1}{\sigma\sqrt{2\pi}} \exp\left[\frac{-(v - y)^2}{2\sigma^2} \right] \qquad (30)$$

The parameters describing the polynomial $F(v)$ are determined by computer fitting the experimental chromatogram. Using these and experimentally determined values of σ, $W(y)$ is generated point by point.

Numerical solutions to Tung's integral are time-consuming and difficult to carry out. Analytical solutions do not generate the corrected MWD curve, but provide corrections to the calculated molecular weight averages. An analytical solution that assumes $G(v, y)$ is Gaussian and the calibration curve for the separation system is linear, as described in (26), was devised by Friis and Hamielec [23]. The following expressions for the number- and weight-average molecular weights are generated by this approach

$$\bar{M}_n(t) = \bar{M}_n(u) \exp[\tfrac{1}{2}(D_2\sigma)^2] \tag{31a}$$

$$\bar{M}_w(t) = \bar{M}_w(u) \exp[-\tfrac{1}{2}(D_2\sigma)^2] \tag{31b}$$

in which $\bar{M}_n(u)$ and $\bar{M}_w(u)$ are the uncorrected molecular weight averages, as calculated in the normal way from the chromatographic output, and $\bar{M}_n(t)$ and $\bar{M}_w(t)$ are the "true" values corrected for Gaussian dispersion. The term D_2 is the slope of the linear calibration curve from (26). The parameter σ varies as a function of elution volume, as discussed in the preceding section. To determine its value at various values of V_e, standard polymeric materials, which have narrow molecular weight distributions, are used; and σ is calculated as described in Section 2.3 for standard polymers that elute at various V_e. From these results the variation of σ with V_e can be plotted.

This is the basis of the American Society of Testing and Materials (ASTM) suggested procedure for the determination of column dispersion [24]. Narrow MWD polystyrene standards are eluted from the column set to be characterized, and a calibration curve established by plotting $\bar{M}_n(t)$ or $\bar{M}_w(t)$ versus the peak-elution volume. If both are known, $[\bar{M}_n(t) \cdot \bar{M}_w(t)]^{1/2}$ is the preferred calibration parameter. The terms $\bar{M}_n(u)$ and $\bar{M}_w(u)$ are determined chromatographically using (3a) and (3b). A quantity X_1 is defined

$$X_1 = \frac{1}{2}\left[\frac{\bar{M}_n(t)}{\bar{M}_n(u)} + \frac{\bar{M}_w(u)}{\bar{M}_w(t)}\right] \tag{32}$$

If the dispersion is Gaussian as described in (30), then

$$X_1 = \exp[\tfrac{1}{2}(D_2\sigma)^2]$$

Because of the variation in dispersion with molecular weight, σ, and therefore X_1, varies with elution volume. The correction for an unknown sample

$$\bar{M}_n(t) = \bar{M}_n(u)X_1 \tag{33a}$$

$$\bar{M}_w(t) = \bar{M}_w(u)X_1 \tag{33b}$$

is calculated using a value of σ from the region in which the unknown elutes.

The ASTM procedure also suggests a way to identify and correct molecular weight averages for non-Gaussian or skewed dispersion. An empirical factor Φ is calculated from the polystyrene standard elution curves

$$\Phi = \frac{\bar{M}_n(t)}{\bar{M}_n(u)} \cdot \frac{\bar{M}_w(t)}{\bar{M}_w(u)} \tag{34}$$

This should equal 1 if only Gaussian dispersion is operative but be greater than 1 if the chromatograms are skewed. A correction factor X_2 is defined

$$X_2 = \frac{\Phi - 1}{\Phi + 1} \tag{35}$$

Corrections of molecular weight averages for both Gaussian dispersion and skewing are

$$\bar{M}_n(t) = (1 + X_2)X_1\bar{M}_n(u) \tag{36a}$$

and

$$\bar{M}_w(t) = \frac{\bar{M}_w(u)}{(1 - X_2)X_1} \tag{36b}$$

The terms X_1 and X_2 are determined in the elution range of the unknown polymer.

General guidelines for the use of these corrections are [24]

1. If $X_1 < 1.05$ and $X_2 < 0.5$, corrections to $\bar{M}_n$ and $\bar{M}_w$ are less than the precision of the independent methods used to measure these quantities, and no corrections should be made.
2. If $X_1 > 1.05$ and $0.15 > X_2 > 0.05$, correct for both X_1 and X_2.
3. If $X_1 > 1.05$ and $X_2 < 0.05$, correct for X_1 only.
4. If $X_2 > 0.15$, skewing is excessive, and this method cannot be used.

3 CALIBRATION OF SIZE-EXCLUSION-CHROMATOGRAPHY SYSTEMS

3.1 Calibration Using Standards With Narrow Molecular Weight Distribution

Calibration is required to establish the relationship between molecular weight and elution volume for a given SEC system. The most straightforward and commonly used method is to measure the elution volume of a set of anionically polymerized, narrow MWD polymer standards that have been characterized previously for molecular weight by light scattering and osmometry. If the distribution is narrow ($\bar{M}_w/\bar{M}_n$ less than 1.1), the molecular weight corresponding to the peak elution volume may be assumed to correspond to the measured molecular weight moments

$$M_{\text{peak}} \cong \bar{M}_w \cong \bar{M}_n$$

and the calibration curve is established by plotting $\log M$ against the peak-elution volume. If both $\bar{M}_w$ and $\bar{M}_n$ are known, M_{peak} is more accurately equated to $(\bar{M}_w \cdot \bar{M}_n)^{1/2}$.

A mathematical description of the calibration to be used in data reduction may be established in either of two ways. If the plot is definitely nonlinear, *S-shape*, it can be represented by a continuous analytical function [25–27].

$$\log M = \sum_{i=0}^{n} C_i V_e^i \tag{37}$$

If calibration of the column set gives a large central portion of the curve that is linear, this portion can be represented by (26)

$$M = D_1 \exp(-D_2 V_e) \tag{26}$$

or, using $\log M$, the quantity that is typically plotted

$$\log M = C_1 - C_2 V_e \tag{38}$$

in which

$$C_1 = \log D_1 \tag{39}$$

and

$$C_2 = \frac{D_2}{2.303} \tag{40}$$

Column sets can be designed for linearity; methods for doing this are discussed in Section 4.4. If the entire elution curve of the sample being analyzed falls within the linear range of the calibration curve, then the use of a linear approximation is justified [28]. If part of the elution curve falls outside this region, the slope of the calibration curve changes rapidly, and major errors can be made in the interpretation of the MW and MWD of the unknown [29]. In this case it is possible to approximate the center section as linear and curve fit the ends with polynomials such as (37).

Ideally, calibration is carried out with standards of the same chemical structure as the sample to be analyzed. The relationship between the volume of the polymer in solution and the polymer molecular weight is then the same for both, and real molecular weights are calculated. However, narrow MWD standards are not available for most synthetic polymers. It is a common practice to calibrate molecular weights of organic polymers of any structure with standards of polystyrene. These polymers are available as well-characterized, narrow MWD samples with molecular weights ranging from a few hundred to several million. The molecular weights calculated for the sample polymers are not their true values but *polystyrene equivalent* molecular weights. Such numbers are useful for comparing a series of samples of synthetic polymers of identical structure, but cannot be used to compare the molecular weights of two different types of polymers. A few narrow MWD standards of other organic polymers are available [4], but these are generally not available over a wide enough molecular weight range to be useful as calibrants. Polymer calibrants may be fabricated by fractionation of broad MWD synthetic polymers, with subsequent characterization for $\bar{M}_w$ and $\bar{M}_n$. These often do not have a narrow, or symmetric distribution as do anionically polymerized standards, and M_{peak} will lie between $\bar{M}_n$ and $\bar{M}_w$. It is also difficult and time-consuming to prepare and characterize such standard materials.

3.2 Universal Calibration

Benoit and co-workers [30] suggested a method of calibration based on hydrodynamic volumes rather than molecular weights. This method allows standardization with a polymer of one chemical composition such as polystyrene and calculation of real molecular weights for a polymer of a different chemical composition. According to the Flory–Fox equation, the intrinsic viscosity of a polymer in solution $[\eta]$ is equal to

$$[\eta] = \Phi \frac{(\bar{R}_G^2)^{3/2}}{M} \tag{41}$$

in which $\bar{R}_G^2$ is the root-mean-square radius of gyration of the polymer, and Φ is a universal constant. Therefore the product $[\eta]M$ for any polymer is proportional to $(\bar{R}_G^2)^{3/2}$, the volume displaced by an equivalent sphere in solution. It is sometimes referred to as the hydrodynamic volume J. Since it is a true size parameter, its use as a calibration quantity should be universal; two polymer samples with the same value of the product $[\eta]M$ should have the same volume in solution. These samples should elute at the same elution volume from a given column and have the same value of K_{sec} for that column. A plot of log $J = \log[\eta]M$ versus V_e should be the same for polymers of any composition eluting by size. Such a plot is called a *universal calibration curve*. Benoit has shown that a plot of $\log[\eta]M$ versus V_e for polystyrene, poly(methyl methacrylate), poly(vinyl chloride), and graft copolymers of polystyrene and poly(methyl methacrylate), as well as branched polystyrenes, is colinear [31]. This plot is contained in Figure 7.17.

For a sample S calibrated with a standard such as polystyrene PS, the relationship

$$[\eta]_S M_S = [\eta]_{PS} M_{PS} \tag{42}$$

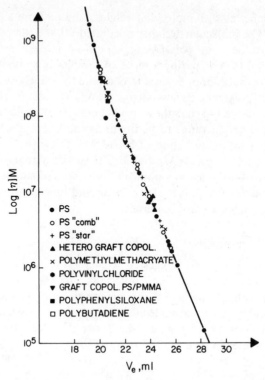

Figure 7.17 Universal calibration curve. Reprinted, with permission, from [31]. Copyright 1967 John Wiley & Sons, Inc.

is true at the same elution volume. The Mark–Houwink constants $\mathcal{K}$ and a describe the relationship between intrinsic viscosity and molecular weight.

$$[\eta] = \mathcal{K} M^a \qquad (43)$$

and

$$[\eta]M = \mathcal{K} M^{(1+a)} \qquad (44)$$

Therefore, at a common elution volume

$$\mathcal{K}_S M_S^{(1+a_S)} = \mathcal{K}_{PS} M_{PS}^{(1+a_{PS})} \qquad (45)$$

where $\mathcal{K}_S$ and a_S are the Mark–Houwink constants for the sample S, and $\mathcal{K}_{PS}$ and a_{PS} are those for polystyrene. The molecular weight of the sample is then equal to

$$\log M_S = \left(\frac{1}{1 + a_S}\right) \log\left(\frac{\mathcal{K}_{PS}}{\mathcal{K}_S}\right) + \left(\frac{1 + a_{PS}}{1 + a_S}\right) \log M_{PS} \qquad (46a)$$

This equation can be used to generate a calibration curve for the sample S. The term M_{PS} is measured from the polystyrene calibration curve at a given volume; M_S is then calculated for that volume from (46a). This is repeated at enough different elution volumes to establish the curve $\log M_S$ versus V_e.

The main drawback to this method is the lack of availability of reliable Mark–Houwink constants for polymers in the eluents used. Examination of the literature reveals a wide variety of numbers for any given polymer–solvent combination. The value found for M_S is strongly influenced by the values of $\mathscr{K}_S$ and a_S used, so these must be very accurate. A method has been described that can be used if the constants are not known; this will be discussed in Section 6.3.

3.3 Calibration Using Standards With Broad Molecular Weight Distribution

Calibration of a separation system for polymers of a specific chemical composition to obtain true molecular weights may be carried out with characterized standards of that composition having broad MWD. These are easier to synthesize and are therefore more available than narrow MWD standards.

3.3.1 Standards With Known Molecular Weight Distribution

For a polymer that contains a broad range of molecular weights, the MWD describes the weight fraction corresponding to each incremental molecular weight $W(M)$ (Figure 7.18a). This MWD can be determined empirically by column fractionation [32]; or, if the polymerization kinetics are known, it is often possible to describe the MWD of a polymer based on classical kinetic models [33–35]. The SEC elution curve (Figure 7.18b) is a plot of weight fraction corresponding to elution volume V_e. Since SEC elutes the polymers with highest M first, the highest molecular weight at the end of the $W(M)$ curve corresponds to the first fraction of polymer eluted by SEC, the beginning of the $W(V_e)$ curve. The relationship can be seen clearly by comparing the cumulative weight fractions $\Sigma W(M)$ and $\Sigma W(V_e)$ in Figures 7.18c and d; the molecular weight at $\Sigma W(M) = 0.05$; point A, corresponds to $\Sigma W(V_e) = 0.95$; $\Sigma W(M) = 0.10$ corresponds to $\Sigma W(V_e) = 0.90$, and so on. In this way the value of M as a function of V_e, a calibration curve, is generated. The correctness of the calibration is only limited by the accuracy with which $W(M)$ is known.

3.3.2 Linear Calibration Methods

In this approach, often referred to as the Hamielec method, broad MWD standards of the desired chemical composition are characterized for $\bar{M}_w$ and $\bar{M}_n$ using nonchromatographic methods and are then used to generate a calibration curve [36]. The calibration is assumed to be linear; the method must only be used with column sets that have been shown to be linear by the elution of a set of narrow MWD standards. The nonchromatographically measured values of $\bar{M}_n$

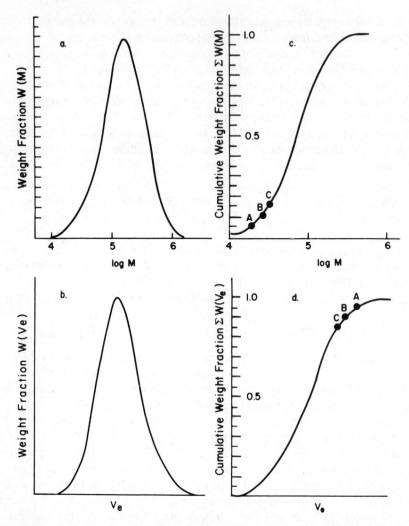

Figure 7.18   Generation of an SEC calibration curve using a characterized broad MWD polymer. Curve *a* is the weight fraction of polymer as a function of log *M*; this can be used to generate curve *c*, the cumulative weight fraction of polymer as a function of log *M*. Curve *b* is the SEC elution curve, the weight fraction of polymer as a function of V_e; this is used to generate curve *d*, the cumulative weight fraction as a function of V_e. Comparison of corresponding points on curves *c* and *d* define the relationship between log *M* and V_e, the calibration curve.

and $\bar{M}_w$, referred to as $\bar{M}_n(t)$ and $\bar{M}_w(t)$, are written

$$\bar{M}_n(t) = 1 \Big/ \int_0^\infty \left[\frac{F(V)}{M'(V)} \right] dV \qquad (46b)$$

$$\bar{M}_w(t) = \int_0^\infty F(V) M'(V)\, dV \qquad (46c)$$

where $F(V)$ describes the experimental chromatogram as a function of elution volume and $M'(V)$ is the sought-after calibration relation

$$M'(V) = D'_1 \exp(-D'_2 V) \tag{47}$$

Since $F(V)$ is an experimental chromatographic curve, which contains effects of band broadening as well as size separation, $M'(V)$ also contains effects of band broadening and size separation; therefore, it is an effective calibration curve relating real values of $\bar{M}_w$ and $\bar{M}_n$ to a separation-containing dispersion.

Expressed in terms of the effective constants D'_1 and D'_2, $\bar{M}_n(t)$ and $\bar{M}_w(t)$ may be written

$$\bar{M}_n(t) = D'_1 \Big/ \int_0^\infty [F(V)\exp(D'_2 V)]\, dV \tag{48a}$$

$$\bar{M}_w(t) = D'_1 \int_0^\infty F(V)\exp(-D'_2 V)\, dV \tag{48b}$$

The ratio $\bar{M}_w(t)/\bar{M}_n(t)$

$$\frac{M_w(t)}{M_n(t)} = \left[\int_0^\infty F(V)\exp(-D'_2 V)\, dV \right]\left[\int_0^\infty F(V)\exp(D'_2 V)\, dV \right] \tag{49}$$

eliminates D'_1 and allows calculation of D'_2 by a computer curve-fitting procedure; substitution of this value into (48a) or (48b) gives D'_1.

It is not necessary to make assumptions about the magnitude of σ or how it varies across the elution curve for this calculation. The effective linear calibration method assumes that the effects of dispersion are the same for both the calibrant and the sample. This is only valid if they have similar polydispersities and molecular weights, so use of the method is limited. Yau and associates [4] have shown that the effective calibration curve is rotated with respect to the true curve; if the calibrant has a lower molecular weight than the sample, the calculated molecular weight will be too low; if the calibrant has a broader MWD than the sample, the calculated polydispersity of the sample will be too narrow.

A first correction to this method [37–40], often referred to as GPCV2, assumes that the band spreading is Gaussian, with a constant σ, and calculates a true calibration curve. It was shown in Section 2.6 that the relationship between true and uncorrected molecular weights for Gaussian band spreading is

$$\bar{M}_n(t) = \bar{M}_n(u)\exp[\tfrac{1}{2}(D_2\sigma)^2] \tag{50a}$$

$$\bar{M}_w(t) = \bar{M}_w(u)\exp[-\tfrac{1}{2}(D_2\sigma)^2] \tag{50b}$$

The molecular weight calculated from the experimental elution curve $F(V)$ may be expressed

$$\bar{M}_n(u) = 1 \Big/ \int_0^\infty \left[\frac{F(V)}{M(V)} \right] dV \tag{51a}$$

$$\bar{M}_w(u) = \int_0^\infty F(V)M(V)\,dV \tag{51b}$$

It follows that

$$\bar{M}_n(t) = \exp[\tfrac{1}{2}(D_2\sigma)^2]\Big/ \int_0^\infty \left[\frac{F(V)}{M(V)}\right] dV \tag{52a}$$

$$\bar{M}_w(t) = \exp[-\tfrac{1}{2}(D_2\sigma)^2] \int_0^\infty F(V)M(V)\,dV \tag{52b}$$

in which $M(V)$ is the true calibration curve

$$M(V) = D_1 \exp(-D_2 V) \tag{26}$$

The unknowns are now σ, D_1, and D_2. An average value for σ can be found as described earlier using polystyrene standards. Hence, D_1 and D_2 can be determined using nonchromatographically measured values of $M_n(t)$ and $M_w(t)$ for the broad MWD calibrant; the procedure is similar to that described for the original Hamielec method. The calibration constants found, D_1 and D_2, are the true constants and should be valid for the calibrated column set for all polymers of the same composition as the standard. Further corrections, which consider the effect of skew in the column dispersion, have also been developed [37, 38, 40, 41].

If several characterized broad MWD standards are available, it is possible to determine the calibration constants and the actual σ for the separation system without having to determine it from the elution of polystyrene standards [42]. In this case, combining (26) with (52a) and (52b)

$$\bar{M}_{ni}(t) = \exp[\tfrac{1}{2}(D_2\sigma)^2]D_i\Big/ \int_0^\infty F_i(V)\exp(D_2 V)\,dV \tag{53a}$$

$$\bar{M}_{wi}(t) = \exp[\tfrac{1}{2}(-D_2\sigma)^2]D_1 \int_0^\infty F_i(V)\exp(-D_2 V)\,dV \tag{53b}$$

where i refers to the broad MWD standard. Calculation of the product $\bar{M}_{wi}(t)\bar{M}_{ni}(t)$ eliminates σ.

$$\bar{M}_{wi}(t)\bar{M}_{ni}(t) = D_1^2 \frac{\int_0^\infty F_i(V)\exp(-D_2 V)\,dV}{\int_0^\infty F_i(V)\exp(D_2 V)\,dV} \tag{54}$$

This product is calculated for two different standards, $i = 1$ and $i = 2$. The ratio of these products

$$\frac{M_{w1}(t)M_{n1}(t)}{M_{w2}(t)M_{n2}(t)} = \frac{\int_0^\infty F_1(V)\exp(-D_2 V)\,dV \int_0^\infty F_2(V)\exp(D_2 V)\,dV}{\int_0^\infty F_1(V)\exp(D_2 V)\,dV \int_0^\infty F_2(V)\exp(-D_2 V)\,dV} \tag{55}$$

eliminates D_1, and D_2 can be calculated; D_1 is found by substitution into (54), and σ can be determined using (53a) or (53b). This method is particularly useful for systems that are not amenable to polystyrene calibration, for example, SEC in aqueous solution.

If several broad MWD standards are available and universal calibration with polystyrene can be achieved, then a calibration curve and values of σ can be generated without assuming linearity. This method is discussed in detail in [42].

4   EXPERIMENTAL CONSIDERATIONS

4.1   Chromatographic Equipment

The instrumentation for high-performance size-exclusion chromatography (HPSEC) is similar to that used for isocratic high-performance liquid chromatography. The components are outlined schematically in Figure 7.19. Solvent is pumped through a set of one or more columns; the sample is injected into the solvent stream, generally passing through an in-line filter, then separated during passage through the column set. The effluent is monitored by one or more detectors. If the detectors are sensitive to concentration, data reduction can be carried out with (3a) and (3b) by using hand measurements of curve heights. Elution volumes are generally calculated by measuring elution times and assuming a constant flow rate. Computerized data reduction may be used; many standard programs are available to accomplish this.

4.1.1   Solvent

The solvent in a reservoir is placed at least several inches above the inlet manifold of the pump to maintain a positive hydrostatic pressure on the system.

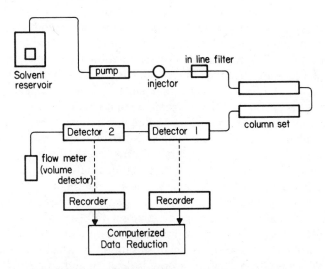

Figure 7.19   Schematic of SEC experimental apparatus.

The solvent should be extremely clean to avoid contamination of the pumps and columns and to avoid baseline drift. Organic solvents should have been distilled in glass. Aqueous solvents should be freshly prepared to avoid bacterial growth and then filtered through a 0.45-μm filter. Solvents should be degassed before use to prevent the formation of air bubbles in the pump. This can be accomplished by filtration through a vacuum filter, boiling slowly with stirring, aspiration with stirring, or sonication. Sparging of the solvent with helium during chromatography prevents redissolution of more soluble gases.

The solvent is transferred to the pump by Teflon tubing, which is fitted with a sintered metal filter to remove dust and other large particles. Solvent changeovers should be made only between comiscible solvents; if a change between nonmiscible solvents is desired, an intermediate solvent, comiscible with both, is used.

4.1.2   Solvent Delivery

The pump selected for solvent delivery in HPSEC must be capable of producing smooth reproducible flows at the high backpressures (often up to 6000–10,000 psi) created by columns packed with 5–10-μm particles. Since elution volumes are generally calculated from elution times, assuming constant flow rate, errors in flow rate have an extremely large effect on calculated molecular weights, which vary exponentially with elution volume. To maintain calculated values of $\bar{M}_n$ and $\bar{M}_w$ to less than 6% error, repeatability of flow rate between calibration and sample elution must be better than 0.3%; and drift in flow rate, from the injection to the completion of the chromatogram, must be held to less than 1% [43].

The high-pressure pumps currently available may be divided into two general types: constant pressure and constant displacement. Constant pressure pumps require a steady column backpressure to maintain a constant flow rate; therefore, they are not practical for SEC use. Constant displacement, or constant volume delivery systems, include reciprocating pumps and syringe pumps. Reciprocating pumps are most commonly used. These operate by drawing solvent into a low volume chamber by the retraction of a plunger, then expelling it toward the column when the plunger extends. The flow is driven in the proper direction by a set of check valves that block flow to the injector when the plunger is retracting and to the reservoir when it is extending (see Figure 7.20). This reciprocating motion creates a pressure fluctuation in the system that may affect baseline stability, especially with pressure-sensitive detectors such as the differential refractive index detector, and may crush column packings. The pressure fluctuations can be dampened by using a pulse damper and restrictor coil between the pump and the injector. Many pumps include high-pressure filters to control fluctuation. Currently the best pumps use two reciprocating pump heads set 180° out of phase, or three set 120° out of phase, to create a constant delivery of solvent and to minimize pressure fluctuations. The pumps may also provide compressibility compensation, by which the solvent compression, which occurs at high pressures, is detected and compensated for by a

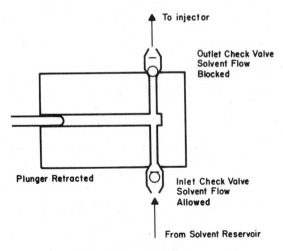

Figure 7.20   Reciprocating pump schematic.

change in the pump motor speed. The flow rate precision is about 0.1% at 1 mL/min.

Syringe pumps utilize a large-volume (ca. 250-mL) chamber, which is emptied by a motor-driven piston. The piston moves at a constant rate, controlled by the speed of the motor. The flow rate is initially variable as the solvent is compressed by the piston; once it has equilibrated, it is steady and pulseless. The pump is limited by the volume of the charge: when it is emptied, solvent delivery must be stopped to refill.

4.1.3   Flow-Rate Measurement

To assure precise and reproducible flow rates during individual experiments and between experiments performed on different days, it is necessary to monitor solvent flow. This can be done by periodically measuring the solvent leaving the last detector either volumetrically or by weight. Flowmeters have been developed recently that accurately measure the small volumes associated with high-performance chromatography. Two types are currently available. One flowmeter measures the time necessary to collect predetermined volumes of solvent. This operates at flow rates from 0.100 to 10.0 mL/min, with an accuracy of $\pm 1\%$. It is designed to be attached to the output of the last detector. The other flowmeter designed for use with high-performance equipment measures the time required for a thermal pulse to travel a given distance [44]. A pair of glass-coated thermistors is installed in the flow stream (Figure 7.21). A pulse of electrical current warms the liquid flowing past the upstream thermistor. The downstream, sensing thermistor detects the arrival of the heated liquid by measuring d^2T/dt^2. Since rate of temperature change is measured, slow drifts in ambient temperature will not affect the measurement. Evaluations have shown the precision to be 0.1% or better.

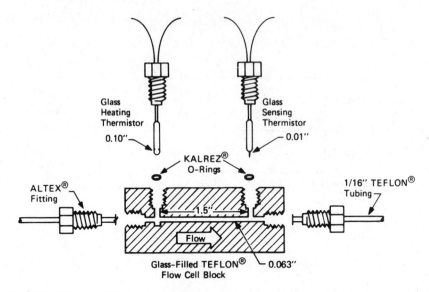

Figure 7.21 Diagram of flowmeter that measures flow rate by measurement of the time required for the passage of a thermal pulse across a known distance. Reprinted, with permission, from T. E. Miller, Jr. and H. Small, *Anal. Chem.*, **54**, 907 (1982). Copyright 1982 American Chemical Society.

The output of these flow-rate detectors can be interfaced directly to a computer accumulating the output of mass-sensitive detectors. In this way, chromatograms can be measured directly in a volume- rather than a time-based mode, and flow-rate variations will not affect the determination of $\bar{M}_w$ and $\bar{M}_n$.

Another approach to the measurement of flow rate is the addition of an internal standard to each solution of polymer sample injected [45]. The elution volume of the internal standard, a low-molecular-weight substance, is determined. Its elution volume is measured in all subsequent injections of polymer samples, and the values of elution volume measured for the polymer peak are corrected using that of the internal standard. This method corrects for changes in flow rate from one run to the next, but not for changes that occur during one run. It has the drawback that the peak corresponding to the internal standard may interfere with detection of low-molecular-weight species; it is, however, a valuable technique for monitoring and correcting for slow drifts in flow rate.

4.1.4 Injectors

Sample injectors must be capable of delivering narrow bands of sample solution into the eluent stream without disrupting the established flow rate or backpressure. This may be accomplished with a six-part sample injection value such as the one illustrated in Figure 7.22. The sample is injected into a loop with a syringe; the injector is then switched from the LOAD position to the INJECT position, and the sample is placed in the eluent stream. Loops are available in a

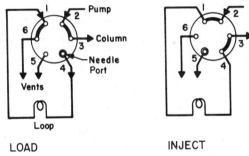

LOAD   INJECT

Figure 7.22 Rheodyne Model 7125 Injector valve. In LOAD position, injection at position 4 fills loop. Transfer to INJECT position causes eluent to flow through loop. Reprinted, with permission, from *Operating Instruction for Model 7125 Syringe Loading Sample Injector*, Rheodyne, Inc., P.O. Box 996, Cotati, CA, 94928.

variety of exact volumes, and may be easily changed. The entire loop volume may be injected; or if only small quantities of sample are available, the loop can be partially filled, with the amount injected determined by the syringe volume.

4.2   Detectors

4.2.1   Mass-Sensitive Detectors

A detector used to measure the concentration of polymer eluted as a function of elution volume should have the following characteristics:

1. Stable baseline; baseline drift makes interpolation of the baseline under the chromatographic peak difficult (Figure 7.23). Determination of MWD requires knowledge of concentration over the entire range of polymer elution. An error in baseline position at the high V_e, low M, end of the chromatogram (Figure 7.23b) leads to significant errors in the calculation of $\bar{M}_n$; an error at the low V_e, high M end leads to errors in the calculation of $\bar{M}_w$ and higher moments. The detector should be minimally sensitive to changes in temperature and backpressure, which could cause such drift.

2. Linear response over a wide range of concentration. Determination of $\bar{M}_n$ and $\bar{M}_w$ requires determination of concentrations at the wings of the elution curves that have low concentrations. The detector response should also be independent of the polymer molecular weight.

3. High sensitivity, rapid response.

4. Low volume tubing and sample cells that cause minimal mixing of eluted fractions.

5. Ability to be used with a wide variety of solvents.

DIFFERENTIAL REFRACTOMETERS

The differential refractometer (DRI) is the most commonly used detector in SEC. The detector response is proportional to the difference in refractive index (RI) between the mixture of solvent plus sample in the sample cell and the

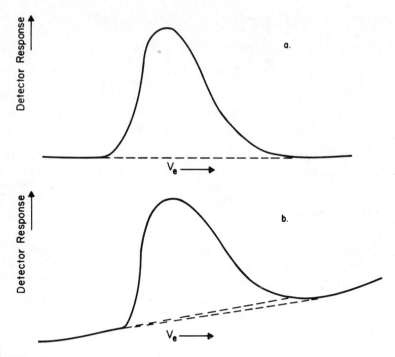

Figure 7.23   Drawing a baseline for a chromatogram with a (*a*) flat baseline, (*b*) sloping or irregular baseline.

solvent in the reference cell. Figure 7.24 is a diagram of a deflection type DRI. A beam of light is generated by the light source and collimated by the lens so it passes through the sample and reference cells to the mirror, which reflects it back through the cells and lens to the photocell. The position of the beam is deflected as it passes through the cell, and the location of the beam on the detector is determined by the amount of deflection it undergoes. An electrical signal, which is proportional to the beam position, is amplified and sent to the recorder. The optical zero allows the beam to be moved on the detector to adjust for a zero output signal.

This refractometer has a full-scale range of 3×10^{-3} RI units at the lowest sensitivity and 6×10^{-6} RI units at the highest sensitivity, with 10^{-7} RI units the minimum change that can be detected. Typical refractive index increments range from 0.05 to 0.15 mL/g; assuming a value of 0.1 mL/g, 10^{-7} RI units correspond to 10^{-6} g/mL concentration. Taking into account dilution caused by band spreading, this corresponds roughly to a sample size of 2×10^{-6} g injected for minimum detection. Under typical conditions, sample sizes injected in SEC are about 10^{-4} g, which puts them into a range that is easily detected by RI; however, the method cannot be used for trace concentrations.

Refractive-index measurements are very temperature sensitive. This effect is minimized by using differential measurements. A room temperature change of

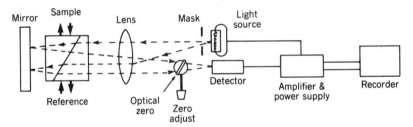

Figure 7.24 Schematic of DRI detector (Waters R-400 Series). Reprinted, from [46], courtesy of Waters Chromatography Division, Millipore Corporation.

$10°C$ causes a change of less than 7×10^{-6} DRI units [46]. Newer, high-resolution DRI detectors are thermostatted to minimize temperature sensitivity further.

The RI of a polymer solution has a small dependence on molecular weight for low-molecular-weight polymers. For example, the RI of polystyrene in toluene varies up to about 20,000 MW, at a rate of approximately 10% change per decade of MW, and more slowly after that to approximately 30,000 MW, above which it is essentially constant [4, 47]. This may have a significant effect on MWDs calculated for low-molecular-weight samples.

The DRI detector is generally employed as the last in a series of detectors because the cells are very sensitive to fluctuation in backpressure. Careful degassing of solvents is important when these detectors are used; the formation of a bubble in the DRI cell can occur because of the pressure drop. The presence of a bubble in this position can cause the cell to expand and contract with the pump cycle and greatly increase reciprocating pump noise in the baseline.

The baseline position for an experiment using this type of detector is established by filling the reference cell with solvent, closing it off with compression fittings, and zeroing the baseline when only solvent is present in the sample side of the detector. Since all materials have a refractive index, the DRI detector is sensitive to variation in the concentration of all components. For example, if the solvent is an aqueous buffer and is eluted with a high-molecular-weight polyelectrolyte, the concentration of salt in the pores will be different from that in the interstitial volume; the composition of eluent at the total volume V_t will reflect the composition in the pores. This gives rise to the *salt peaks* often observed in aqueous exclusion experiments (see Figure 7.25). Peaks in this position can also occur when organic solvents are employed, because of differences in concentrations of components such as of water, dissolved gases, or solvent-oxidation products. These may overlap the polymer elution peaks and interfere with chromatogram analysis.

Other types of DRI detectors are those based on Fresnel's law, such as the one manufactured by LDC, and the shearing interferometer, a design by Optilab, which is an order of magnitude more sensitive than current RI detectors [4].

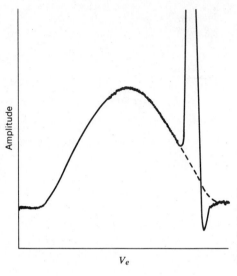

Amplitude

V_e

Figure 7.25 The DRI chromatograms may contain peaks that appear at the total column volume and are caused by small differences in composition between the injected eluent and the eluent from the solvent reservoir. Dashed line indicates correct polymer-elution curve.

ULTRAVIOLET PHOTOMETERS

Ultraviolet (UV) detectors are not as generally useful as DRI detectors because they require an eluent that is transparent to the detected wavelength and a polymer that absorbs at that wavelength. However, where they can be used they offer the advantage of very high sensitivity and selectivity for components of the sample. Both variable and fixed wavelength detectors are available. Variable wavelength models generally employ a deuterium lamp that extends from 190 to 600 nm. Fixed wavelength models typically use the 254-nm emission band of a mercury lamp. The sensitivity ranges from approximately 0.005 absorbance units full scale (AUFS) to approximately 2.0 AUFS, with excellent signal-to-noise (S/N) and linearity. This allows the detection of polymers containing vinyl groups that have only a small tailing absorbance at 254 nm. If the polymer is a strong chromophore, trace amounts can be detected down to a sample size of about 1×10^{-8} g.

These detectors are very stable to changes in pressure and temperature. Waters Associates offers a model (440 Series) with a tapered cell designed to eliminate artificial absorbance effects caused by RI phenomena that can occur in flow-through cells. Other options with some models include filters that allow a mercury lamp to be used at the other mercury bands (280, 313, 340, 365, 405, 436, 546, and 658 nm), a zinc or cadmium lamp to allow detection at 214 or 229 nm, and two-cell instruments that allow simultaneous detection of two wavelengths. Ultraviolet detectors are frequently used in series with DRI detectors. This combination can provide information about the compositional heterogeneity of a copolymer when only one component is UV absorbing.

MASS DETECTOR

The mass detector, or evaporative analyzer, employs a destructive technique for determining the amount of nonvolatile solute dissolved in a volatile eluent. The efflux from the column is sprayed into a heated zone in which the solvent is evaporated, and the nonvolatile material remains as a fine cloud of particles. These flow in a gas stream past a light source; the light deflected at a fixed angle is measured. This detector has been evaluated for SEC [48]. Molecular weights for polystyrene and epoxy resin samples in chloroform determined using a mass detector and RI detector in parallel were comparable. Baseline stability was good, but the authors found that results were very dependent on instrument parameters such as lamp age.

INFRARED PHOTOMETERS

Flow-through infrared IR detectors are available for use with HPSEC. Their sensitivity is approximately comparable to that of RI detectors. In two situations IR detection is the method of choice for SEC: (1) Because the detectors are quite insensitive to temperature, they can be operated with a heated cell up to approximately 150°C, for example, in the MWD measurement of polyolefins [49]. (2) Since the IR absorbance wavelengths of a compound are very specific for composition, this type of detection is very useful for copolymer analysis. For example, in Figure 7.26, elution curves at two different wavelengths (using two IR detectors) are monitored for an ethylene–vinyl acetate copolymer [50].

4.2.2 Detectors Sensitive to Molecular Size

The greatest technical advances in the use of SEC in recent years have been in the development of detectors that are sensitive to polymer size rather than concentrations, that is, the low-angle laser light-scattering photometer and the viscosity detector. These detectors are used in conjunction with a concentration

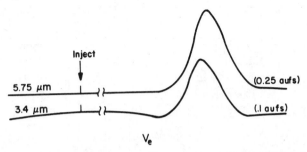

Figure 7.26 An analysis of an ethylene–vinyl acetate copolymer using two IR detectors; monitoring at two different wavelengths. Reprinted, with permission, from S. D. Abbot, *American Laboratory*, **9**, No. 8, 41 (1977). Copyright 1977 International Scientific Communications, Inc.

detector and eliminate the need for preparing calibrants through tedious fractionation procedures or for calibrating with standards of composition different from the polymer being studied. There are theoretical and experimental limits to the use of both techniques, but they will expand the usefulness of SEC in the analysis of polymer molecular weights.

LOW-ANGLE LASER LIGHT-SCATTERING PHOTOMETER

The low-angle laser light-scattering photometer LALLSP [51, 52] measures the intensity of light scattered by polymer solutions at low forward angles. This quantity plus the polymer concentration from a second detector allow calculation of the weight-average molecular weight of fractions of polymer eluted from an SEC column.

Figure 7.27 is an optical diagram of the Chromatix KMX-6 LALLSP. Light from a 2-mW helium–neon laser (633 nm) is focused on the sample, which is contained between two 5-cm fused silica windows. The scattering volume is very small (ca. 5 μL), so scattering caused by dust particles passing through the cell is minimized. A series of annuli defines the solid angle detected by the photomultiplier. The ratio of scattered light to transmitted light is measured. The difference between this ratio for the sample plus solvent and for the solvent alone is the excess Rayleigh scattering factor $\bar{R}_\theta$, which is related to $\bar{M}_w$ of the polymer at low angles by the equation

$$\frac{Kc}{\bar{R}_\theta} = \frac{1}{\bar{M}_w} + 2A_2c \tag{56}$$

in which c is the concentration of polymer in the cell in grams per milliliter; A_2 is the second virial coefficient of the polymer in the eluent. The optical constant

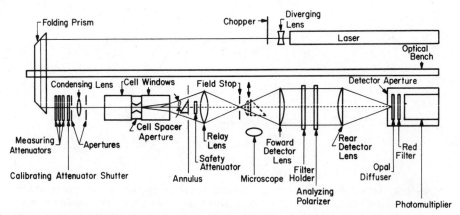

Figure 7.27 Schematic of Chromatix KMX-6 low-angle laser light-scattering photometer. Reprinted, with permission, from *KMX-6 Instruction Manual*, Chromatix, Sunnyvale, CA.

is K

$$K = \frac{2\pi^2 n^2}{\lambda^4 N} \left(\frac{dn}{dc}\right)^2 (1 + \cos^2 \theta) \tag{57}$$

in which n is the refractive index of the eluent; λ is the wavelength of the laser, 633 nm; N is Avogadro's number; and dn/dc is the refractive index increment of the polymer in the eluent. Since θ is close to 0, the quantity $(1 + \cos^2 \theta)$ can be approximated as 2, and K becomes

$$K = 4.08 \times 10^{-6} \, n^2 (dn/dc)^2 \tag{58}$$

The value of A_2 may be determined by measuring $\bar{M}_w$ at a series of sample concentrations, using the detector in a static, off-line mode.

When used as a detector, the LALLSP monitors $\bar{R}_{\theta,i}$ of the polymer fractions from the SEC column. The molecular weight of each fraction eluted is calculated using

$$\frac{Kc_i}{\bar{R}_{\theta,i}} = \frac{1}{M_i} + 2A_2 c_i \tag{59}$$

A DRI, or other concentration-sensitive detector, monitors c_i simultaneously. This is done easily by normalizing the detector response

$$c_i = mh_i / V_i \Sigma_i h_i \tag{60}$$

where h_i is the amplitude of the detector response, V_i is the volume of the fraction, and m is the total mass injected. The quantity c_i is used to solve for M_i. The DRI curve may then be reduced in the normal way

$$\bar{M}_n = \sum_{i=1}^{N} h_i \Big/ \sum_{i=1}^{N} \frac{h_i}{M_i} = \sum_{i=1}^{N} \frac{1}{c_i/M_i} \tag{61a}$$

$$\bar{M}_w = \sum_{i=1}^{N} h_i M_i \Big/ \sum_{i=1}^{N} h_i = \sum_{i=1}^{N} c_i M_i \tag{61b}$$

using the M_i determined by LALLSP, rather than that determined with a calibration curve.

Figure 7.28 illustrates the output from such an experiment. The DRI response is sensitive to concentration only; the KMX-6 response is sensitive to molecular weight and concentration. In the LALLSP curve there is a discontinuity in the relationship between molecular weight and elution volume [53], which may be caused by a small amount of high-molecular-weight material being excluded from the column-pore volume.

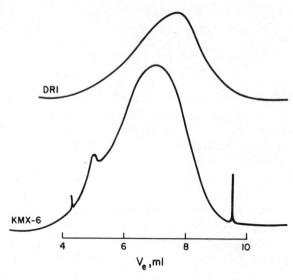

Figure 7.28 Simultaneous detection of molecular weight and concentration of SEC fractions. Reprinted, with permission, from *KMX-6 Application Note L25*, Chromatix, Sunnyvale, CA.

The minimum detectable concentration of polymer, c_{min}, in grams per milliliter is related to its molecular weight [53]

$$c_{min} = \frac{\bar{R}_{\theta min}}{(4.08 \times 10^{-6})(dn/dc)^2 n^2 \bar{M}_w} \text{ g/mL} \tag{62}$$

The minimum detectable excess Rayleigh factor $\bar{R}_{\theta min}$ is approximately 2–10×10^{-7}, depending on the clarity of the sample solutions. Chromatix has demonstrated that MWDs of polystyrene standards 3900, 2940, and 962 can be detected using sample sizes of 3–9 mg [54]. Higher molecular weights may be detected with smaller sample sizes.

VISCOSITY DETECTOR

This detector measures the pressure drop as the eluent stream passes through a capillary. The pressure drop is related to the viscosity η of the eluent stream for laminar flow systems by

$$\eta = \frac{\Delta P \pi R^2}{8Ql} \tag{63}$$

in which ΔP is the pressure drop, R the capillary radius, Q the flow rate, and l the capillary length. If the capillary geometry and flow rate are constant

$$\frac{\Delta P}{\Delta P_0} = \frac{\eta}{\eta_0} \tag{64}$$

where ΔP and η are the values for the polymer solution, and ΔP_0 and η_0 are those for the solvent. The intrinsic viscosity $[\eta]$ can be determined because the concentration is calculated by simultaneous use of a concentration-sensitive detector.

$$[\eta] = \left[\frac{\ln(\eta/\eta_0)}{c}\right]_{c \to 0} \tag{65}$$

Viscometry detectors have been described by Benoit and associates [55] and Ouano [56]. Figure 7.29 is a drawing of the viscometry detector designed by Ouano. The temperature variation is kept to $\pm 0.01°C$ with a recirculating bath. The sensitivity of the pressure transducers is one part per thousand. Letot and others [57] evaluated the performance characteristics of a high-speed viscometry detector, which they built and have had in use for several years. They found that the system is extremely sensitive to flow-rate variations; this parameter must be kept within $\pm 10^{-3}$ mL/min. To eliminate pump noise, it was necessary to employ two high-sensitivity filters and to keep flow rates greater than approximately 2 mL/min. They also found the detector sensitive to noise from flow-rate fluctuations originating in the column set; this was most serious at flow rates greater than 2 mL/min. Operation was therefore limited to a flow rate of approximately 2 mL/min.

The use of a continuous viscometry detector allows calculation of $\eta_i(V)$ for eluted fractions. Together with universal calibration, which gives $\eta_i M_i(V)$, the calibration curve for a polymer of a specific chemical composition $M_i(V)$ can be determined, and the elution curves can be integrated for molecular weight moments.

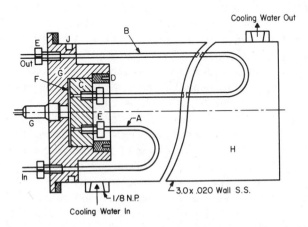

Figure 7.29 Diagram of the viscometer system: (A) 0.020 in ID stainless steel inlet connection to the viscometer; (B) capillary viscometer; (C) pressure cell upper platten; (D) upper platten retaining ring; (E) Swagelok fitting; (F) pressure cell chamber; (G) pressure transducer; (H) thermostat; (J) thermostat O-ring seal. Reprinted, with permission, from [56]. Copyright 1972 John Wiley & Sons, Inc.

4.3   Column-Packing Materials

4.3.1   Styrene–Divinylbenzene Gels

This column packing is widely used for the routine analysis of organic polymers. It is a polystyrene matrix containing cross-linked divinylbenzene groups. Polymerization is carried out in the presence of an organic diluent that causes swelling of the gel and maintains pores in the polymer matrix as the monomers polymerize around it [58]. Pore size is controlled by varying the amount or type of diluent. This technique allows a high degree of cross-linking to take place so that the gels will be rigid enough to withstand high-column pressures. The gels come prepacked in a wide variety of pore sizes; some of those available are listed in Table 7.1.

Figure 7.30 is an electron micrograph of a 10-μm styrene–divinylbenzene gel particle [59] (μStyragel, Waters Associates). The particle is spherical, but the surface is bumpy, and pieces of it can be sheared off at excessively high solvent flow rates or viscosities. This limits the flow rates that can be used for specific solvents. The most commonly used solvents for μStyragel supports and the recommended maximum flow rates for each are listed in Table 7.2. Pores in the small pore size materials (those nominally listed as 100–500 Å) are formed by solvent swelling; these are more limited in the solvents with which they are compatible than the large pore size materials. Pore sizes may vary slightly with change of solvent for all of the materials because of differences in the swelling of the gels, so parameters such as total pore volume must be determined for each solvent. Some manufacturers recommend never changing solvent for their materials; in any case, solvent changeovers should always be carried out at very low flow rates.

The surface polarity of styrene–divinylbenzene gels is low and makes them quite inert to adsorption of organic polymers [60]. An extensive review of uses of these gels with different classes of organic polymers has been included in a recent book [61]. The columns may be run at temperatures up to 145°C, which is useful for reducing the viscosity of some solvents such as DMF and rendering polymers such as polyolefins soluble. Styrene–divinylbenzene column packings with particle sizes less than 10 μm, such as Ultrastyragel column packings manufactured by Waters Associates, are now available. These have the same range of pore sizes and compatibility characteristics as μStyragel columns, with higher efficiencies. They are especially useful for analysis of low-molecular-weight polymeric materials. Figure 7.31 illustrates the linearity available in different molecular weight ranges using combinations of columns containing these packing materials. The small-pore-size columns can resolve some of the oligomeric components of polystyrene standard materials with molecular weights less than 1000.

Table 7.1 Polystyrene–Divinylbenzene Gels for High-Performance Size-Exclusion Chromatography

Name[a]	Nominal Pore Size (Å)	MW Range (PS Equivalent)
μStyragel[b]	100	0–700
	500	500–10,000
	10^3	1,000–20,000
	10^4	10,000–200,000
	10^5	100,000–2,000,000
	10^6	$10^6 - 2 \times 10^7$
μSpherogel[c]	50	$< 2,000$
	100	100–5,000
	500	$4,000 - 10^4$
	10^3	$10^3 - 5 \times 10^4$
	10^4	$10^4 - 5 \times 10^5$
	10^5	$10^5 - 5 \times 10^6$
	10^6	$> 10^6$
PL Gel[d]	50	
	100	
	500	
	10^3	
	10^4	
	10^5	
	10^6	
HSG[e]		4×10^2
		4×10^3
		4×10^4
		4×10^5
		2×10^6
		8×10^6 (est)
		10^7 (est)
MicroPak[f]	50	50–1,000
TSK Type H[f]	2.5×10^2	50–6,000
	1.5×10^3	1,000–60,000
	1×10^4	0.5–40,000
	10^5	200–4,000,000
	ca. 10^6	10^7
	10^7	10^8

[a]The names given are trademarks used by the companies for their polystyrene–divinylbenzene gels.
[b]Waters Associates.
[c]Altex Scientific, Inc.
[d]Polymer Laboratories, Ltd.
[e]Shimadzu.
[f]Toyo Soda Manufacturing Company.

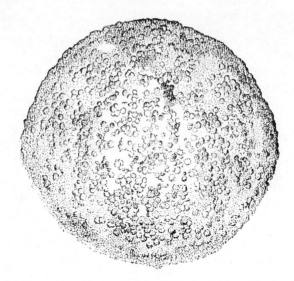

Figure 7.30 An electron photomicrograph of a μStyragel particle, enlarged to 20,000 times its original size. Reprinted from *Waters Associates Liquid Chromatography School Manual*, D-64, 1976, courtesy of Waters Chromatography Division, Millipore Corporation.

Table 7.2 μSTYRAGEL Styrene–Divinylbenzene Gel: Most Commonly Used Solvents

Solvent	UV Cutoff (nm)	Viscosity at 20°C (cP)	Maximum[a] Flow (mL/min)
N,N′-Dimethylformamide[b]	295	0.924	2.5
Ethylene chloride	225	0.84	2.8
p-Dioxane	220	1.439	1.6
Trichloroethane	225	1.20	2.0
Cyclohexane[b]	220	0.98	2.4
Carbon tetrachloride	265	0.969	2.4
Toluene	285	0.59	3.0
1,1,1-Trifluoroethanol[b]		1.996	1.1
Benzene	280	0.652	3.0
1,1,3,3,3-Hexafluoroisopropanol[b]		1.021	1.2
Chloroform	245	0.58	3.0
Hexane[b]	210	0.326	3.0
Xylene	290	0.81	2.9
Tetrahydrofuran	220	0.55	3.0
Methylene chloride	220	0.44	3.0

[a]Maximum flow should not exceed 3.0 mL/min when using 100-Å μStyragel columns.
[b]These solvents should not be used with 100- or 500-Å μStyragel columns.

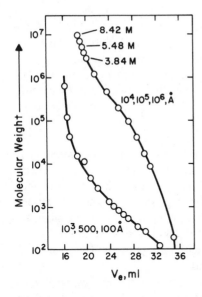

Figure 7.31 Calibration curves of column sets containing Ultrastyragel column packings. Calibrations were carried out using polystyrene standard and *n*-hydrocarbons in toluene. Reprinted from Waters Associates, Brochure No. 82337-B51, July, 1982, courtesy of Waters Chromatography Division, Millipore Corporation.

4.3.2   Silica

Porous silica microspheres are formed by the agglutination of colloidal silica particles into a three-dimensional lattice [62]. The pore size and surface area of the microspheres are determined by the size of the colloidal silica particles and are highly reproducible. These inorganic supports have high mechanical strength and are compatible with any solvent, including water (except at basic pH). Their use as exclusion materials has been described [62–65]. The major drawback of this material is its adsorptive properties, caused by the chemical structure of the surface, which contains the active silanol group:

$$\equiv\!Si\!-\!OH$$

These groups may be deactivated, or silanized, by treatment with short-chain chlorosilanes such as trimethylchlorosilane:

$$\equiv\!Si\!-\!OH + ClSi(CH_3)_3 \rightarrow \equiv\!Si\!-\!O\!-\!Si(CH_3)_3 + HCl$$

This derivatization causes the pore surface to become covered with a hydrophobic coating, which is less interactive with organic polymers dissolved in nonpolar solvents. Approximately half the silanols are sterically available to react; the rest become shielded by an umbrella of trimethylsilyl groups [4].

Silica is also available as irregularly shaped porous particles. Because of the irregularity, the interstitial spaces in these packings are of a wider volume range than in the microspheres, and the resolution is not as good. A list of commercially available underivatized high-performance silica supports is included in Table 7.4, Section 5.3.2.

4.4   Column Combinations

A single-pore-size column will provide separation over 1 to 1.5 decades of MW. Broad MWD polymers may contain a much wider range of molecular weights; in general a separation system should cover several decades to be useful. It is also valuable to maintain a linear calibration over this range, both for ease of data handling and for the use of broad MWD calibration methods, which may require assumption of linear calibration.

There are two approaches to achieving a broad range, linear separation. The first is to combine stationary phases containing a wide range of pore sizes. This can be done by connecting several columns, each of a single pore size. A linear set is generated if the pore volumes of each column are approximately the same; that is, the slope of the individual columns will be the same. Most manufacturers design their columns to match in this way. The same type of separation can be achieved using mixing bed columns—these each contain a range of pore sizes, giving a wide molecular weight separation with one column. For increased resolution, the number of columns is increased.

An alternative method of establishing a linear separation has been recommended by Yau and co-workers [66]. Using Cassassa's model of the separation of random coil molecules by cylindrical pores of uniform size, he showed that the combination of only two pore sizes, which differ in radius by a factor of about 10, provides a fairly linear fit over a wide range of MW. Figure 7.32 compares three such *bimodal sets*, showing that as the pore sizes become further apart (increasing $\Delta \log$ PS), the range of the separation increases while the linearity of the fit decreases. The loss of linearity is improved somewhat by broadening the pore-size distribution of each column. Figure 7.33 illustrates the linearity and molecular weight range of a bimodal set when the two pore sizes are not uniform but have a log normal distribution with a standard deviation of 0.15, approximately the normal pore-size distribution for porous silica [67].

Such bimodal sets may be readily constructed by combination of single pore-size columns with pore sizes that differ by a factor of approximately 10 and with similar total pore volumes. The pore volume determines the slope of the individual column calibration; these must be matched or the slopes of the two ends of the calibration will be different.

4.5 Experimental Design

4.5.1   Nonsteric Interactions

The SEC of organic polymers on both styrene–divinylbenzene gel and silica gel porous packings may give rise to elution patterns not completely explained by size separation [68]. These occur because of the existence of enthalpic interactions between the polymer and the pore surface. The elution of polymers from a weakly retentive SEC column was described by Dawkins and Hemming [69] using the relationship

$$V_e = V_o + K_p K_{sec} V_i \qquad (66)$$

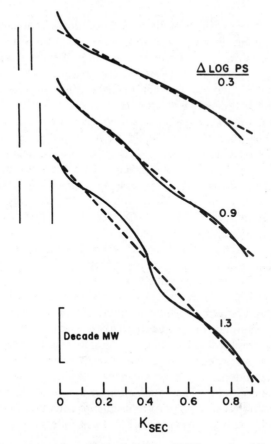

Figure 7.32 Effect of Δ log PS (pore size) on linearity and MW range for a bimodal column set with uniform pore sizes. Reprinted, with permission, from [66].

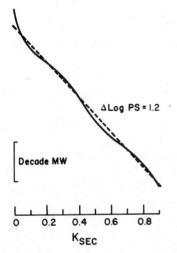

Figure 7.33 Linearity and MW range for a bimodal column set with Δ log PS = 1.2, and log normal pore size distribution, standard deviation = 0.15. Reprinted, with permission, from [66].

in which K_p is a coefficient describing the distribution of polymer between the stationary and mobile phases. When separation takes place by a single size-exclusion mechanism, $K_p = 1$; if retention because of enthalpic interactions is occurring, $K_p > 1$.

The coefficient K_p describes either an adsorption process in which the interaction occurs between the polymer and the pore surface or a partition process that occurs when the support matrix is a gel or derivatized phase swollen by the solvent. The pore "surface" in the partition process is covered with a phase comprising solvent plus swollen gel or organic molecule. The regions available to the polymer as it passes through the column are now interstitial, pore, and swollen gel phase; the polymer may be enthalpically retained in the gel phase, in the same manner as in liquid–liquid chromatography.

Dawkins and Hemming [69–71] and Dawkins [72, 73] verified this relationship for several systems and showed that universal calibration may be achieved between systems with different values of K_p when K_p is low and measurable. The magnitude of the interactions for these experiments gave a K_p between 1 and 2, which indicates that they are extremely weak; in conventional LC they range from 1 to 10^{15} [69].

The relationship between the elution volume of a polymer and its molecular weight when it is simultaneously undergoing size exclusion and adsorption or partition is quite complex when the interactions become stronger than described by (66). Size exclusion elutes high-molecular-weight species first; adsorptive- or reverse-phase interactions cause high-molecular-weight species to be eluted last [74, 75] (see Figure 7.34). The intervening region is difficult to interpret. Adsorptive interactions cause polymer molecules to enter pores from which they would be sterically excluded in normal SEC. Thus, unless interactions are very

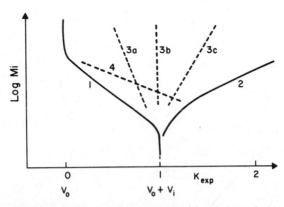

Figure 7.34 Schematic demonstration of typical molecular size-elution volume curves for different modes of chromatography: 1, SEC; 2, LC; 3a to c and 4, different amounts of LC–SEC interaction. Reprinted, with permission, from J. Klein and K. Treichel, *Chromatographia*, **10**, 604 (1977). Copyright 1977 Pergamon Press, Ltd.

weak and interpretable by (66), real molecular weight information cannot be obtained from experiments in which adsorption and size exclusion are occurring simultaneously.

4.5.2   Solvent Selection

Interactions between the polymer solute and the stationary phase can be minimized by selecting good solvents for the polymer as eluents. If the polymer is near theta conditions in the eluent, partitioning into the gel stationary phase or derivatized phase will be enthalpically more favorable than if it is well solvated [4]. The *goodness* of a solvent is indicated by the Mark–Houwink coefficient a for the polymer in that solvent; a value of a near 0.5 means that theta conditions are approached, a value of 0.7–0.8 means that the polymer is well solvated and fairly expanded in the solvent. Several workers have suggested the use of solubility parameters δ for the polymer, gel, and eluent to predict good solvent-stationary phase combinations for specific polymers [4, 76, 77]. The elimination of interactions between water-soluble polymers and stationary-phase material is a more complicated issue and is discussed in Section 5.1.

Several other factors affect the choice of solvent. It should be compatible with the stationary phase, which, for example, eliminates some solvents for styrene–divinylbenzene gels (see Table 7.2) and basic eluents for silica. It should not attack stainless steel (halide ions in aqueous eluents must be avoided, for example). It must be transparent to the detector and preferably should have a low viscosity, to minimize column backpressure and shearing effects on the packing material and on high-molecular-weight polymers.

4.5.3   Sample Size

Injection of high concentrations of polymer sample in SEC, especially those of high-molecular-weight polymer, can lead to significant differences between the viscosity of the injected solution and the eluent, giving rise to a severe band-broadening effect known as *viscous fingering*. Even below the range of viscous fingering, concentration effects are noted. In experiments with narrow MWD polymers, higher injection concentrations lead to greater elution volumes [78]. This effect is caused by the decrease in hydrodynamic volume experienced by polymers at high concentrations, because of steric crowding. It is more severe for high-molecular-weight polymers. The analysis of high-molecular-weight, broad MWD polymer samples can give substantial errors in apparent molecular weight distribution [79].

The choice of sample concentration used is a balance between two phenomena. The volume of solution injected should be as small as possible, as high injected volumes contribute to extra column-band broadening. Minimization of injected volume requires high-sample concentrations. The minimum sample mass that can be used is dictated by the sensitivity of the detector. Janca has determined a range in which both effects can be minimized [79]. If the injection volumes are 4.5% or less of the total width of the eluted peak, band broadening caused by the injection will be less than the detectable limits of the SEC experiment. For example, if the eluted peak has a baseline width of 3 mL, the injection volume should be limited to about 135 μL. Concentration effects were found to be negligible in this range.

5   AQUEOUS EXCLUSION CHROMATOGRAPHY

5.1   Column-Packing Interactions

The size-exclusion chromatography of organic polymers is a well-understood, fairly routine analytical procedure in which the elution of a polymer is controlled by its size, and nonsize separation effects are small. In contrast to this, the exclusion chromatography of water-soluble polymers (historically called gel-filtration chromatography) is still somewhat of an art in which nonsize separation effects, including interactions with stationary-phase materials and ionic-strength-dependent polyelectrolyte expansion, are not fully eliminated. This category will be considered to include synthetic polyelectrolytes, synthetic water-soluble neutral polymers, biopolymers, and copolymers containing very polar portions.

The stationary-phase materials available for the high-performance chromatography of such polymers can be divided into three categories: bare silica; derivatized silica; and semirigid, hydrophilic gels.

In this section, the techniques for SEC of aqueous polymers will be covered by first describing the interaction characteristics of these materials, including adsorption and electrostatic exclusion and intramolecular expansion, that can affect the elution volume of a polymer. The major progress in this field is in the development of new packing materials. The use of these for specific types of polymers is described, as well as the mobile phase parameters, which may be adjusted to minimize polymer–stationary-phase interactions.

5.1.1   Adsorption

Adsorption of aqueous polymers takes place through electrostatic interactions, hydrogen bonding, and hydrophobic interactions. Bare, underivatized silica in aqueous solution carries a net negative charge because of the dissociation of surface silanols [1].

$$\equiv\!\text{Si}\!-\!\text{OH} + \text{H}_2\text{O} \rightleftharpoons \,\equiv\!\text{Si}\!-\!\text{O}^- + \text{H}_3\text{O}^+$$

This occurs in eluents with pH greater than 2; below 2 the silanols become undissociated. Polycations and polybases will adsorb to these sites electrostatically; since there are multiple cationic sites on the polymer the adsorption is cooperative and may be irreversible. Polymers such as poly(ethylene oxide) and poly(vinyl alcohol) adsorb irreversibly to silica by the formation of hydrogen bonds [80]. If the silica surface is derivatized, even with a fairly hydrophilic organic molecule, the use of aqueous eluents containing salts may lead to hydrophobic partitioning of the polymers [81]. The amount of the interaction depends on the polarity of the polymer and that of the derivatized phase. When hydrophilic-derivatized phases are used, the retention may not be irreversible, and the elution volume of the polymer will depend on a combination of equilibration through size and hydrophobic partitioning. This effect is observed on hydrophilic gels. Such effects are extreme and can be recognized by elution of the polymer after the total volume of the column.

5.1.2 Electrostatic Exclusion

If the stationary phase carries a charge, ions of similar charge will tend to be excluded from the pores. The negative charges present on bare silica columns in aqueous solution, derivatized silica with residual underivatized silanol groups, and some hydrophilic gels will cause early elution of salts and anionic polyelectrolytes. This effect was first noted by Neddermeyer and Rogers [82, 83] on Sephadex gels, and it has been studied in detail by Rinaudo and others [83–87]. They found that when a constant volume of a low-molecular-weight electrolyte (salt) is injected onto a silica column in pure water or eluents of low-ionic strength, its elution volume depends on the concentration of electrolyte injected. This is illustrated in Figure 7.35a. Curve 1 is the elution volume of varying injected concentrations of salt C_S when the eluent salt concentration C_e is equal to $10^{-1} N$. No dependence of V_e on C_S is seen when C_e is this large; all elution is at V_t, the total volume, which is determined by the elution of D_2O in water. Curves 2 and 3 are the elution of varying injected concentrations of electrolyte in an eluent containing $10^{-4} N$ salt and in an eluent of pure water, respectively. The injected electrolyte is electrostatically excluded from the pores in the low-ionic strength media; as the injected concentration is increased, the ionic strength in the pores increases as the salt passes through the column, and it is partially screened from exclusion. The elution volume of the partially excluded electrolyte can be linearly related to $\mu^{-1/2}$ (μ = ionic strength), which is proportional to the Debye screening parameter for a given pore size [86].

The peak shapes at the intermediate elution volumes are indicative of electrostatic exclusion. The elution curves for a wide variety of injected salt concentrations in an eluent of pure water are shown in Figure 7.36 [84]. The curves tail strongly toward the column void volume (in this case at the right of the figure). The salt at the leading edge of the injected volume is totally excluded from the pores and elutes at the void volume. As the column fills with the injected salt, the ionic strength rises and some permeation occurs. The back sides of the peaks are much sharper because the charged sites in the pores cause the last salt through the column to diffuse rapidly from the pore volume [4].

If the ionic strength of the eluent is increased at a constant injected electrolyte concentration, electrostatic exclusion effects at any injected concentration may be overcome. Rinaudo and others [87] found that for porous silica, eluent ionic strengths greater than 5×10^{-2} are sufficient to achieve this (Figure 7.35b).

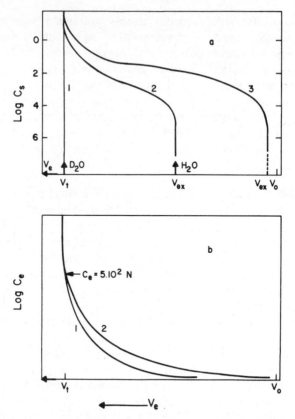

Figure 7.35 (a) Dependence of elution volume of a salt on the concentration injected, C_S. Curve 1: eluent salt concentration; C_e, 10^{-1} N. Curve 2: C_e, 10^{-4} N. Curve 3: eluent, water. (b) Dependence of the elution volume of a salt on the eluent salt concentration, C_e. Curve 1: C_S, 10^{-2} N. Curve 2: C_S, 10^{-4} N. Reprinted from [87] by courtesy of Marcel Dekker, Inc.

5.1.3   Donnan Equilibrium

If eluent-containing low-molecular-weight salt and polyelectrolyte are eluted through a column that contains pores too small for the polyelectrolyte to enter, the stationary phase acts as a semipermeable membrane, allowing passage of low-molecular-weight salt ions into the pore, but excluding the polymer. A Donnan equilibrium is established between the pore volume and the excluded, interstitial volume, in which the charge on the polyelectrolyte in the interstitial volume is partially balanced by an increase in concentration of salt ions of like charge in the pore volume and a decrease of such charges on the excluded volume (see Figure 7.37). This leads to a chromatogram with two peaks, one corresponding to the elution of the polymer and one corresponding to the elution of the excess salt in the pores at V_t, as illustrated in Figure 7.25. The excess amount of electrolyte in the pores Q_S can be related by (67) to the osmotic

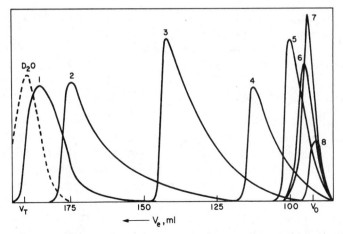

Figure 7.36   Elution peaks for $NaNO_3$ eluted in pure water (1) C_S, injected concentration, 10^1 M; (2) C_S, 1 M; (3) C_S, 10^{-1} M; (4) C_S, 10^{-2} M; (5) C_S, 10^{-3} M; (6) C_S, 10^{-4} M; (7) C_S, 10^{-5} M; (8) C_S, 10^{-6} M. A constant volume was injected in each case. The D_2O peak measured by refractometry, salt peaks by conductimetry with sensitivities adjusted to match size. Reprinted, with permission, from C. Rochas, A. Domard, and M. Rinaudo, *European Polymer Journal*, **16**, 135 (1980). Copyright 1980 Pergamon Press, Ltd.

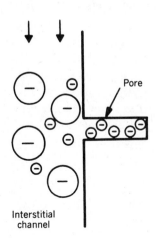

Figure 7.37   The gel matrix acts as a membrane that allows passage of low-molecular-weight electrolyte, but not high-molecular-weight polyelectrolyte. The concentration of low-molecular-weight electrolyte is higher in the pore volume than in the interstitial volume. Reprinted, with permission, from [88]. Copyright 1973 John Wiley & Sons, Inc.

coefficient of the polymer Φ and the charge density on the polymer Z

$$Q_S = \frac{\Phi Z Q_P}{4} \tag{67}$$

in which Q_P is the weight of polymer injected [84]. This relationship can be used to calculate the osmotic coefficient of the polyelectrolyte [85].

If a polyelectrolyte containing an excluded fraction and a permeable fraction is eluted in pure water, the retention volume of the permeating fraction will be increased; the excluded polyelectrolyte creates a Donnan potential that forces it further into the pores. This effect has been called ion inclusion [88, 89]. The presence of low-molecular-weight salt suppresses the effect; the electrolyte, rather than the permeable polyelectrolytes, is forced into the pores.

5.2 Intramolecular Electrostatic Effects

Because exclusion chromatography separates according to hydrodynamic volume, which is proportional to $[\eta]M$, the elution volume of a polymer is sensitive to changes in its intrinsic viscosity. The intrinsic viscosity of polyelectrolytes may change drastically with eluent composition. For example, in Table 7.3 are listed the intrinsic viscosities of a sample of sodium carboxymethylcellulose, with average degree of substitution 0.7, in solutions of varying ionic strength [90]. With no electrolyte present, the polymer is very extended and therefore has an extremely high viscosity. In the presence of salt, the polymer takes on a random coil conformation and the viscosity decreases. At higher ionic strengths the polymer coil becomes more contracted, indicated by a decrease in solution viscosity. This contraction causes the elution volume of the polymer to increase. Size-exclusion chromatography of polyelectrolytes must be carried out with some salt present in the eluent, and the concentration of salt must be maintained when comparing different polymers of the same structure.

5.3 Column-Packing Materials

Traditionally, aqueous exclusion chromatography (AEC) has been carried out on nonrigid gels such as agarose, polyacrylamide, and dextran. These have poor mechanical strengths and therefore must be operated at low flow rates. Recent developments in column technology have led to the development of rigid and semirigid supports that are compatible with high-pressure systems.

Table 7.3 Effect of Ionic Strength on Intrinsic Viscosity of Sodium Carboxymethylcellulose at pH 3.7[a]

Ionic Strength (M)	$[\eta]$
0	80
0.01	7.5
0.05	5.6
0.10	5.2
0.35	4.9
0.70	4.6

[a]Reprinted, with permission, from [90].

5.3.1   Silica

It has been suggested that AEC can be carried out on an underivatized silica stationary phase [91]. The elution behavior of neutral dextran standards on LiChrosphere silica HPSEC packing material is illustrated in Figure 7.38. Addition of salt to the eluent suppresses the electrostatic exclusion effects for anionically charged polyelectrolytes. Spatorico and Beyer [92] were able to achieve a universal calibration curve for narrow MWD poly(styrene sulfonate) sodium salts that was approximately colinear with that for polystyrene in several organic solvents (see Figure 7.39). The calibration was independent of ionic strength in the range studied (0.2–0.8 M sodium sulfate). The experiment was carried out on controlled pore glass, which has similar, but less active, surface characteristics to silica. Butenhuys and van der Maeden [91] chromatographed the polyanions poly(acrylic acid), poly(methacrylic acid), and sodium carboxymethylcellulose on silica in an eluent of 0.5 M sodium acetate. Use of

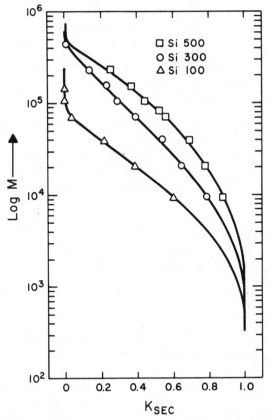

Figure 7.38 Molecular weight calibration curves for LiChrosphere columns with dextran standards in 0.5 M sodium acetate solution (pH 5). Conditions: flow rate, 0.5 mL; sample 20 μL of 1% solution. Reprinted, with permission, from [91].

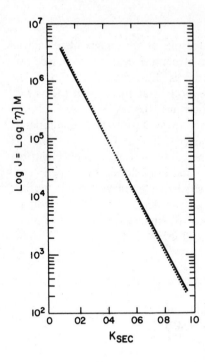

Figure 7.39 Universal calibration plot for: ——, polystyrene in THF, CHCl₃, and benzene; ---, poly(styrene sulfonate) sodium salts in 0.2 M sodium sulfate and 0.8 M sodium sulfate. Reprinted, with permission, from [92]. Copyright 1973 John Wiley & Sons, Inc.

very high ionic strength may induce adsorption of some polymers to the silica surface.

The elution of cationic polymers from silica columns is complicated by the presence of ionized silanols, which irreversibly adsorb such polymers by electrostatic interactions. Even operation at very low pH (ca. 2.0) does not totally suppress these effects [93]. It has been suggested that the addition of cationic mobile-phase modifiers, such as tetraalkylammonium salts, to the eluent may mask the silanols sufficiently to allow elution of cationic polymers [91], but this has not been demonstrated.

Characteristics of commercially available underivatized silica gels are listed in Table 7.4.

5.3.2 Derivatized Silica

The silanol groups on the surface of silica may be reacted with organic molecules to form a coated phase; for example,

$$\equiv Si\!-\!OH + (C_2H_5O)_3Si\!-\!R \rightarrow \equiv Si\!-\!O\!-\!\overset{\displaystyle OC_2H_5}{\underset{\displaystyle OC_2H_5}{\overset{|}{\underset{|}{Si}}}}\!-\!R$$

Table 7.4 Underivatized High-Performance Silica Gels for High-Performance Size-Exclusion Chromatography

Name[a]	Particle Size (μm)	Nominal Pore Size (Å)	MW Range (PS Equivalent)
LiChrosphere[b]			
Si100	10	100	$<5.0\text{--}8 \times 10^4$
Si300	10	300	$<1.5\text{--}3 \times 10^5$
Si500	10	500	$<3.0\text{--}6 \times 10^5$
Si1000	10	1000	$<0.6\text{--}1.4 \times 10^6$
Si4000	10	4000	$<2.5\text{--}8 \times 10^6$
Zorbax SE[c]			
SE60	9–11	60	$2 \times 10^2\text{--}1 \times 10^4$
SE100	9–11	100	$5 \times 10^3\text{--}7 \times 10^4$
SE500	9–11	500	$3 \times 10^3\text{--}5 \times 10^5$
SE1000	9–11	1000	$2.5 \times 10^4\text{--}3 \times 10^6$
SE4000	9–11	4000	$10^5\text{--}10^7$
Zorbax PSM[c]			
PSM60	5–7	60	$2 \times 10^2\text{--}4 \times 10^6$
PSM1000	5–7	750	$2 \times 10^4\text{--}6 \times 10^6$
Spherosil[d]			
XOA400	7	100	$<4 \times 10^4$
XOA200	7	150	$<2 \times 10^5$
XOB075	7	300	$<4 \times 10^5$
XOB030	7	600	$<1 \times 10^6$
XOB015	7	1000	$<1.5 \times 10^6$
XOB005	7	1500	NA[e]

[a]The names given are trademarks used by the companies for their silica HPSEC packing materials.
[b]E. Merck.
[c]E. I. du Pont de Nemours & Company.
[d]Rhone-Poulenc.
[e]Not available.

The derivatization decreases the activity of the surface, while maintaining the mechanical strength and pore characteristics of silica. As in the derivatization with chlorotrimethylsilane for use with organic solvents, only a fraction of the silanols are sterically available to react; the rest are shielded by an umbrella effect from interaction with eluting polymers.

The characteristics of the derivatized phase depend on the nature of the R group. For aqueous systems, R should be hydrophilic; that is, it should consist of short carbon chains and polar groups so that the derivatized phase can be solvated by water. The longer the chain and the less polar the substituted group, the more hydrophobic partitioning will occur [94, 95]. There are several commercially derivatized column types available. Most have been designed for use with biopolymers that may contain cationic, anionic, and hydrophilic groups and may often be used with water-soluble synthetic polymers and polyelectrolytes.

DIOL-TYPE DERIVATIZED PHASES

This derivatized phase was originally developed by Regnier and Noel [96, 97] and provides a short alkyl chain and diol group to maximize hydrophilicity. Glycidoxypropyltrimethoxysilane is reacted with bare silica and subsequently acid hydrolyzed to open the epoxide to the diol.

$$\equiv\!Si\!-\!OH + (MeO)_3Si(CH_2)_3OCH_2CH\overset{O}{\underset{}{\diagup\diagdown}}CH_2 \xrightarrow{H^+}$$

$$\equiv\!Si\!-\!O\!-\!\underset{\underset{OMe}{|}}{\overset{\overset{OMe}{|}}{Si}}\!-\!(CH_2)_3OCH_2\underset{}{\overset{\overset{OH}{|}}{CH}}\!-\!\overset{\overset{OH}{|}}{CH_2}$$

Several commercially available diol phase columns are listed in Table 7.5.

Schmidt and colleagues [98] measured the surface coverage of Merck's Lichrosorb Diol chromatographic support, which is commercially available, and found 2.5 μmol/m^2 by measurement of the carbon content and 2.1 μmol/m^2 by hydroxyl content. The concentration of available silanol groups on the silica surface is 9 ± 1 μmol/m^2 [99]. Therefore, there is a large fraction of underivatized sites that could leave a net negative charge in the pores. This may vary from one commercial preparation to another. Schmidt and colleagues [98] found that at ionic strengths greater than 0.2, electrostatic effects are neutralized. Pfannkoch and others [100] confirmed this for Lichrosorb Diol, TSK-SW, and SynChropak GPC 100 chromatographic supports, but found that hydrophobic partitioning was significant at these ionic strengths. Herman and others [101] noted that addition of methanol to the derivatized phase will reduce this hydrophobic partitioning. With derivatized and underivatized columns it is necessary to optimize conditions for specific polymer–stationary phase–eluent combinations.

ETHER-TYPE DERIVATIZED PHASES

A monolayer of organic ethers has been bonded to μ-Bondagel silica-based column support; the exact structure of the derivatized phase is proprietary. It is compatible with both organic and aqueous solvents in a pH range from 2 to 8. The surface coverage is good because these columns elute polyvinyl alcohol in water [102]. Tests with monomer probes indicate that electrostatic interactions are slightly greater than those found with diol-type columns, and hydrophobic

Table 7.5 Hydrophilic Derivatized High-Performance Silica Gels for High-Performance Size-Exclusion Chromatography

Name[a]	Particle Size (μm)	Nominal Pore Size (Å)	MW Range (PS Equivalent)
Diol[b]			
100	10	100	Similar to LiChrosphere[a,b]
300	10	300	
500	10	500	
1000	10	1000	
4000	10	4000	
SynChropak[c] (a diol type)			
100	10	100	Similar to LiChrosphere[a,b]
300	10	300	
500	10	500	
1000	10	1000	
4000	10	4000	
TSK-SW[d,e]			
2000SW	10	130	$5 \times 10^2 - 6 \times 10^4$
3000SW	10	240	$1 \times 10^3 - 3 \times 10^5$
4000SW	13	450	$5 \times 10^4 - 1 \times 10^6$
Protein Column[f] (irregular shape)			
I-6000	10	60	$1 \times 10^3 - 2 \times 10^{4\,g}$
I-125	10	125	$2 \times 10^3 - 8 \times 10^4$
I-250	10	250	$1 \times 10^4 - 5 \times 10^5$
μ-Bondagel[f,h] (irregular shape)			
E125	10	125	$2 \times 10^3 - 5 \times 10^4$
E300	10	300	
E500	10	500	$5 \times 10^3 - 5 \times 10^5$
E1000	10	1000	$5 \times 10^4 - 2 \times 10^6$
ELinear	10	Blend	$1 \times 10^4 - 2 \times 10^6$
EHigh	20		
CATSEC[c]			
CATSEC100	10	100	Similar to LiChrosphere[a,b]
CATSEC300	10	300	
CATSEC1000	10	1000	
CATSEC4000	10	4000	

[a]The names given are trademarks used by the companies for their HPSEC packing materials.
[b]E. Merck.
[c]SynChrome, Inc.
[d]Actual structure highly hydroxylated.
[e]Toyo Soda Manufacturing Company.
[f]Waters Associates.
[g]Based on proteins in aqueous mobile phase.
[h]Actual structure contains ether groups.

interactions are much greater [100]. Hydrophobic partitioning may be suppressed by the addition of ethylene glycol or sodium dodecyl sulfate to this mobile phase [102].

Polystyrene calibration curves in methylene chloride eluent (Figure 7.40) illustrate the separation range of the available pore sizes. The pore sizes given are nominal, based on those for the silica before derivatization. This derivatization process reduces the available pore volume by about 45% [102], indicating a very thick phase and probable closure of some pores.

TOYO SODA TSK-SW SUPPORT

The TSK-SW series of columns consists of silica derivatized with a hydrophilic phase that is thought to be a highly hydroxylated glycol ether, similar to the diol-type derivatized phase. It was designed to be most applicable for the chromatography of proteins and other biopolymers. The columns can be operated over a pH range from 2.5 to 7.5. Ionic strengths greater than 0.1 M are preferred for reducing electrostatic effects; ionic strengths greater than 0.3 M cause hydrophobic partitioning [103]. The chromatographic performance and resolution of these columns have been reviewed [103, 104]. Calibration curves using poly(ethylene glycol) standards are included in Figure 7.41.

SYNCHROME CATSEC SUPPORT

Cationic polyelectrolytes are especially difficult to chromatograph on silica-based packings that have been modified with neutral phases. The electrostatic interactions between the polymer and underivatized silanols lead to irreversible adsorption or peak tailing. An approach to the analysis of this type polymer was

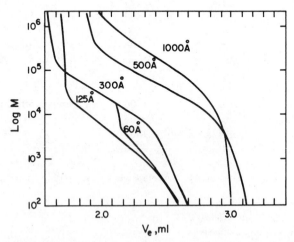

Figure 7.40 Polystyrene calibration curves of μ-Bondagel columns. Mobile phase; methylene chloride. Reprinted from [102] by permission of Preston Publications, Inc.

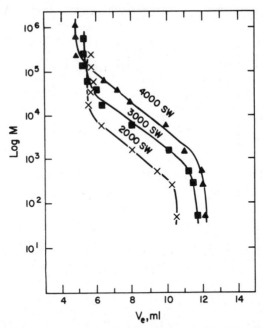

Figure 7.41 Poly(ethylene glycol) standard calibration curves for Toya Soda TSK-SW type columns. Mobile phase: H_2O; flow rate, 1 mL/min. Injection volume: 100 μL of 0.1% PEG. Reprinted with permission from [103] by courtesy of Marcel Dekker, Inc.

suggested by Talley and Bowman [105], who derivatized the surface of controlled pore glass with 3-aminopropyltriethoxysilane, which they subsequently quarternized. Cationic polyelectrolytes were electrostatically excluded from close contact with the glass surface and thus were not adsorbed. Size separation of poly(2-vinylpyridine) standards in aqueous solution at low pH was achieved. SynChrome, Inc. has developed a high-performance silica-based support derivatized with a polymeric amine [106] that uses the same approach. Aqueous elutions are carried out at low pH, which gives a positive charge to the derivatized phase polymer and allows elution of cationic polyelectrolytes without adsorption by the silica.

The pores of this column are positively charged, and ion exclusion of cationic polyelectrolytes occurs in the same way as it does for anionic polyelectrolytes on silica. Figure 7.42 illustrates this effect. Poly(2-vinylpyridine) standards are eluted later in an eluent containing 0.1% trifluoroacetic acid (TFA) and 0.2 M NaCl than in 0.1% TFA alone. The salt shortens the range of the electrostatic effects and allows the polycations more access to pore volume.

The exclusion characteristics of commercially derivatized high-performance silica columns are listed in Table 7.5.

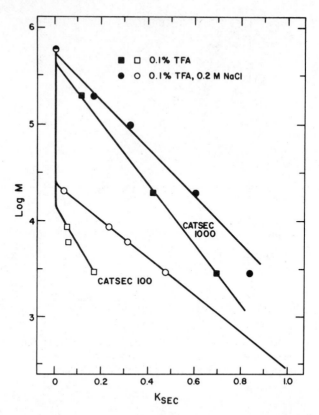

Figure 7.42 Elution of poly(2-vinylpyridine) standards from CATSEC columns. These polymers are cationic in 0.1% trifluoroacetic acid, TFA; with no electrolyte present in the eluent, they are partially excluded from the pore volume and elute early. The addition of 0.2 *M* NaCl shortens the range of the electrostatic effects and allows the polymers more access to the pore volume. Reprinted, with permission, from [106] by courtesy of Marcel Dekker, Inc.

5.3.3 High-Performance Aqueous Gels

The surface of derivatized supports is a mixture of organic phase and underivatized silica, and interaction characteristics may change if the derivatized phase hydrolyzes with time. The recent development of rigid, hydrophilic gels offers an option to derivatized silica stationary phases.

TOYO SODA TSK-SW SUPPORT

This support material is available in a wide range of pore sizes and has a proprietary structure believed to be a cross-linked, hydroxylated polyether gel [104]. It is designed for use primarily with water-soluble synthetic polyelectrolytes and may be used over a pH range from 2 to 12, much wider than silica based columns, which hydrolyze above about pH 8.0. The particle sizes range from 10 μm for the small pore size material to 25 μm for the largest. Because of the large particle size and dispersity present in a set of columns, the efficiency is lower than for porous silica microsphere based columns, and analysis times are longer. There are some electrostatic charges present, apparently because of residual groups on the polymeric matrix. It is suggested that mobile phase ionic strengths be greater than 0.3 M for analysis of very polar polymers [103]. This support may be used with cationic polyelectrolytes [107].

SHODEX ION PAK SUPPORT

This material is a sulfonated poly(styrene divinylbenzene) support that may be used for the analysis of neutral, synthetic polymers [103].

SHODEX OH PAK SUPPORT

This is a proprietary material, believed to be a glycerol methacrylate copolymer that may be employed in the analysis of polysaccharides and biopolymers.

POLYMER LABORATORIES AQUAGEL SUPPORT

This is a macroporous polyacrylamide gel that is currently available in two pore sizes. It is stable over a pH range from 2 to 10 and has small (10-μm) particles with a narrow size distribution for high efficiency [108]. Most of the materials designed to be used with aqueous media can also be used with polar organic solvents.

The exclusion characteristics of these gels are listed in Table 7.6.

5.4   Calibration in Aqueous Exclusion Chromatography

Several water-soluble standards are available commercially for calibration in aqueous systems and are listed in Table 7.7. Dextran is the traditional aqueous calibrant. The polydispersity of dextran standards is high ($\bar{M}_w/\bar{M}_n$ greater than 1.5) [109], and the peak molecular weight will not correspond exactly to $\bar{M}_w$ or $\bar{M}_n$. These standards are often used for calibration by peak position, but should more correctly be employed as broad MWD standards, using one of the methods described in Section 3.3. They are neutral polymers with a small number of negatively charged groups, which causes them to elute early from glass or silica columns in an aqueous eluent with no salt added.

Poly(ethylene oxide) narrow MWD standards have recently become available in a wide range of molecular weights. These are completely neutral. They may not be used on underivatized silica columns because they will bind to the surface through hydrogen bonding.

Poly(styrene sulfonate) salts are made by the sulfonation of narrow MWD polystyrene standards. Poly(2-vinylpyridine) standards are synthesized by anionic polymerization. The availability of these two polymers provides a set of anionic and cationic standards for AEC that may be used for probing the electrostatic behavior of column-packing materials or for calibration.

Other water-soluble polymers currently available as standards are biopolymers, including synthetic polypeptides and proteins. These are complicated molecules with multiple-charged sites and hydrophobic characteristics. In addition, many have secondary structures that differ from the random coil, and

Table 7.6 Polymeric Gel Supports

Name	Particle Size (μm)	Nominal Pore Size (Å)	MW Range (PS Equivalent)
TSK-PW[a]			
G1000PW	10		ca. 100–1,000[b]
G2000PW	10	50	200–5,000
G3000PW	13	200	ca. 1,000–50,000
G4000PW	13	500	4,000–800,000
G5000PW	17	1000	40,000–8,000,000
Shodex ION PAK[c]			
S-801	10	55	> 1,000[d]
S-802	10	100	> 5,000
S-803	10	160	> 50,000
S-804	ca. 15	220	> 500,000
S-805	ca. 15	350	> 5,000,000
Shodex OH PAK[e]	10	NA	> 400,000
Aquagel[f]			
P-2	10		
P-3	10		

[a]Hydroxylated polyether copolymer; contains residual CO_2H groups and NH_2 groups; pH range 2–12 (Toyo Soda Manufacturing Company).
[b]Poly(ethylene glycol)/H_2O.
[c]Sulfonated poly(styrene divinylbenzene) copolymer; pH range 2–11 (Showa Denko).
[d]Polysaccharides/H_2O.
[e]Methacrylate glycerol copolymer; pH range 4–12 (Showa Denko).
[f]Polyacrylamide gel (Polymer Laboratories, Ltd.).

thus the hydrodynamic volume–molecular weight relationship will differ from that of a synthetic polymer. They are generally useful as calibrants only for other biopolymer systems.

5.5   Universal Calibration in Aqueous Exclusion Chromatography

The possibility of universal calibration in aqueous systems has been the subject of some controversy. The difficulty of removing all interactions, both adsorptive and exclusion, between the polymers and the stationary phase is significant. The existence of even a mild charge on the pore walls changes the effective pore radius for polymers of like charge and makes the surface an adsorbent for polymers of opposite charge.

Some evidence of universal size-exclusion behavior in aqueous systems has been reported. Spatorico and Beyer [92] plotted the hydrodynamic volume of several water-soluble polymers, both anionic and neutral, as a function of elution volume from a column set consisting of controlled pore glass (see Figure 7.43). The plots were very close to colinear. Hester and Mitchell [110] replotted

Table 7.7 Molecular Weight Standards for Aqueous Exclusion Chromatography

	$\bar{M}_w$	$\bar{M}_n$	$\bar{M}_w/\bar{M}_n$
Neutral			
1. Dextran[a,b]			
500	510,000	185,000	2.75
250	236,000	109,000	2.17
150	150,000		
110	100,500	62,000	1.69
80	85,800	43,700	1.95
40	39,800	25,600	1.55
20	21,800	14,500	1.50
10	11,200	5,700	1.94
2. Poly(ethylene oxide)[c]			
SE-2	25,000		1.14[d]
SE-5	40,000		1.03
SE-15	148,000		1.04
SE-30	278,000		1.05
SE-70	661,000		1.10
SE-150	1,200,000		1.12
3. Poly(styrene sulfonate, sodium salt)[e]			
Lot 19	1,600		1.25[f]
18	4,000		1.10
13	6,500		1.10
20	16,000		1.10
11	31,000		1.10
21	65,000		1.10
14	88,000		1.10
15	195,000		1.10
12	354,000		1.10
16	690,000		1.10
17	1,060,000		1.10
Cationic			
4. Poly(2-vinylpyridine)[e]			
PVP-2	31,000		
PVP-3	92,000		
PVP-5	230,000		
PVP-4	460,000		
PVP-7	600,000		

[a]Source: Pharmacia Fine Chemicals A.B.
[b]Characterization from [33].
[c]Source: Toyo Soda Manufacturing Company.
[d]SEC values.
[e]Source: Pressure Chemical Company.
[f]From comparison of SEC curves.

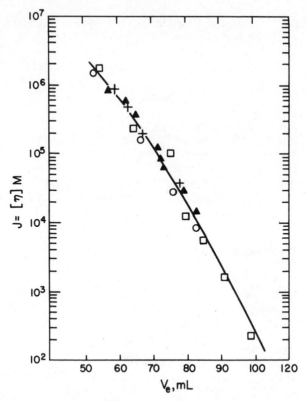

Figure 7.43 Universal calibration plots for hydrophilic polymers in 0.2 M sodium sulfate: □, poly(styrene sulfonate, sodium salt); ▲, dextran; ○, 1.1 copolymer of acrylic acid and ethyl acrylate; +, poly(acrylic acid, sodium salt). Reprinted, with permission, from [92]. Copyright 1973 John Wiley & Sons, Inc.

the data of Spatarico and Beyer using $([\eta]M_w)^{1/3}$ versus $\ln K_{sec}$, and found a slightly better fit than using the Benoit-type universal plot. Kuga [111], using a relationship similar to that of Hester and Mitchell, compared the dependence of K_{sec} on R_G for poly(ethylene oxide) and dextran standards eluted from Sephadex gel in pure water and found that the plots were not colinear. Desbrieres and others [112] compared the elution of sodium poly(styrene sulfonate) and dextran standards in 0.1 M NaNO$_3$ from silica columns. Dextran was partially retained; addition of 1% ethylene glycol eliminated this retention and rendered both plots of $\log[\eta]M$ versus V_e colinear (see Figure 7.44). It should be noted that Desbrieres found a strong concentration effect in the elution and measured the elution volumes for the polymer by extrapolation of the results of several experiments at different concentrations to infinite dilution.

It is difficult to reach any conclusions regarding the validity of universal calibration in aqueous solution. Enthalpic effects are so strong that their existence can never be assumed to be entirely absent.

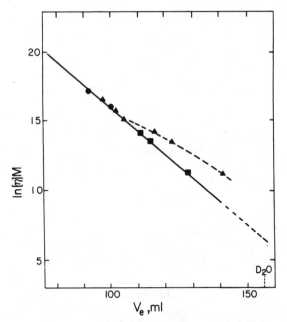

Figure 7.44   Calibration curve for ▲ dextran in 0.1 M NaNO$_3$ and H$_2$O; ■ dextran in H$_2$O + 10% ethylene glycol; ● poly(styrene sulfonate, sodium salt) in 0.1 M NaNO$_3$ and H$_2$O. The adsorption of dextran in H$_2$O because of hydrophobic partitioning is eliminated by the addition of ethylene glycol. Reprinted, with permission, from J. Desbrieres, J. Mazet, and M. Rinaudo, *European Polymer Journal*, **18**, 269 (1982). Copyright 1982 Pergamon Press, Ltd.

6   SPECIAL APPLICATIONS OF SIZE-EXCLUSION CHROMATOGRAPHY

6.1   Determination of Random Long-Chain Branching

In addition to MW and MWD, the amount of branching present in a polymer that contains multifunctional monomer residues such as divinylbenzene or butadiene in radical polymerizations, or glycerol in condensation polymerizations, is an important factor in determining its physical properties. Size-exclusion chromatography, combined with universal calibration and intrinsic viscosity data, can provide branching parameters as well as molecular weight information on branched polymers [113–116].

The greater the degree of branching in a polymer, the smaller is its radius of gyration. This reduction is expressed by a ratio

$$g = \bar{R}_b^2 / \bar{R}_l^2 \tag{68}$$

in which $\bar{R}_b^2$ and $\bar{R}_l^2$ are the mean square radii of gyration of branched and linear polymers of the same molecular weight. The parameter g has been related to the number of branch points for polymers with long-chain branching by Zimm and Stockmayer [117].

$$\langle g_3(m)\rangle = [(1 + m/7)^{1/2} + (4m/9\pi)]^{-1/2} \qquad (69)$$

$$\langle g_4(m)\rangle = [(1 + m/6)^{1/2} + 4m/3\pi]^{-1/2} \qquad (70)$$

where $\langle g_3(m)\rangle$ is the value for trifunctional branching and $\langle g_4(m)\rangle$ for tetrafunctional branching in monodisperse systems. Other equations describe the relationships for polydisperse systems [117]. The parameter m is the number of branch points in the polymer. A branching density λ is defined

$$\lambda = m/M \qquad (71)$$

in which M is the polymer molecular weight; λ is generally assumed to be constant over the entire molecular weight distribution of a polymer, which implies that there will be some $M = M^*$ such that $m = 1$ and

$$M^* = 1/\lambda \qquad (72)$$

Therefore M^* is the molecular weight at which just one branch occurs. Below M^*, $m = 0$, there is no branching, and the chains are linear.

The reduction in radius of gyration is calculated by measuring the reduction in intrinsic viscosity

$$\left(\frac{[\eta]_b}{[\eta]_l}\right)_M = g_M \qquad (73)$$

for branched and linear polymers of the same molecular weight. The ratio of these is g_M. The relationship between g_M and g is generally empirically determined. For lightly branched, or star-branched polymers, they are assumed to be related by

$$g_M = g^{1/2} \qquad (74)$$

while for highly branched polymers the relationship

$$g_M = g^{3/2} \qquad (75)$$

is preferred [118]. If a plot of the intrinsic viscosity of fractions of a branched polymer versus M is made, those fractions with M less than M^* will lie on the linear curve typical of linear polymers. For values of M greater than M^*, the value of $\log [\eta]_b$ will drop below this linear plot; as the branching becomes more significant at high M, the difference between the measured value of $[\eta]_b$ and the linear plot will increase. This is illustrated in the top half of Figure 7.45.

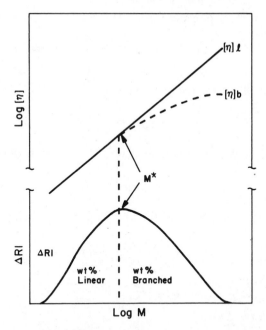

Figure 7.45 Relationship between MWD and Mark–Houwink plot for branched polymers. Reprinted, with permission, from [115]. Copyright 1974 John Wiley & Sons, Inc.

6.1.1   Size-Exclusion Chromatography—Intrinsic Viscosity Method

The decrease in intrinsic viscosity with branching will cause the hydrodynamic volume J of the polymer to decrease. The hydrodynamic volume of a fraction J_i is related to the molecular weight of that fraction by

$$J_i = [\eta]_i M_i = \mathscr{K} M_i^{a+1} \tag{76}$$

The $\bar{M}_w$ and $\bar{M}_n$ for a linear polymer may be found using (77a) and (77b) [115]

$$\bar{M}_w = \mathscr{K}^{-1/(1+a)} \sum_{i=1}^{N} w_i J_i^{1/(1+a)} \tag{77a}$$

$$\bar{M}_n = \mathscr{K}^{-1/(1+a)} \left[1 \bigg/ \sum_{i=1}^{N} w_i J_j^{-1/(1+a)} \right] \tag{77b}$$

in which w_i is the weight fraction of chromatographic fraction i and may be determined from the heights of chromatographic slices of equal widths as

$$w_i = h_i \bigg/ \sum_{i=1}^{N} h_i \tag{78}$$

The viscosity of a linear polymer is

$$[\eta]_l = K^{1/(1+a)} \sum_{i=1}^{N} w_i J_i^{a/(1+a)} \tag{79}$$

If the polymer is branched, then

$$[\eta]_b = g_M [\eta]_l = g_M \mathcal{K} M^a \tag{80}$$

The value of g_M will vary with molecular weight; the viscosity of each fraction is

$$[\eta]_{b,i} = g_{M,i} \mathcal{K} M_i^a \tag{81}$$

and the equations analogous to (77) and (79) for a branched polymer are

$$\bar{M}_{w,b} = \mathcal{K}^{-1/(1+a)} \sum_{i=1}^{N} w_i J_i^{1/(1+a)} g_{Mi}^{-1/(1+a)} \tag{82}$$

$$\bar{M}_{n,b} = \mathcal{K}^{-1/(1+a)} \sum_{i=1}^{N} \frac{1}{w_i J_1^{-1/(1+a)} g_{Mi}^{1/(1+a)}} \tag{83}$$

$$[\eta]_b = \mathcal{K}^{-1/(1+a)} \sum_{i=1}^{N} w_i J_i^{a/(a+1)} g_{Mi}^{1/(1+a)} \tag{84}$$

The value of g_{Mi} can be found from (84) as follows: Using values of $\mathcal{K}$ and a for the linear polymer, and assuming universal calibration for the separation system to determine J_i, various values of the branching density λ are used to generate g_{Mi} until the value calculated for $[\eta]_b$ is equal to the experimental value. Once g_{Mi} is known, (82) and (83) provide the molecular weight information of the sample.

Since λ is determined through this iterative procedure, M^* may be found using (72). When this information is combined with the SEC elution curve, the weight percentage of polymer below and above M^* can be calculated. This is equal to the weight percentage of linear and branched polymer present in the sample (Figure 7.45). Also, $\bar{M}_{bp}$, the molecular weight between branch points, can be calculated using the relationship

$$\bar{M}_{bp} = \frac{1}{(f-1)\lambda} \tag{85}$$

where f is equal to the functionality of the branch points [118].

6.1.2   Size-Exclusion Chromatography—Low-Angle Laser Light Scattering Method

If on-line low-angle laser light scattering (LALLS) detection of the effluent from SEC is available, the molecular weight of the polymer fractions can be measured directly. Again assuming universal calibration for the linear and branched polymer, at a given elution volume V [119]

$$([\eta]_b M_b)_V = ([\eta]_l M_l)_V \tag{86}$$

An intrinsic viscosity ratio g_V is defined as

$$g_V = \left(\frac{[\eta]_b}{[\eta]_l}\right)_V = \left(\frac{M_l}{M_b}\right)_V \tag{87}$$

the ratio of the intrinsic viscosity of branched and linear polymers at the same elution volume. The value of g_V can be determined if branched and linear samples are available by comparing the molecular weights of the two species at a given elution volume. The quantity of interest g_M

$$g_M = \left(\frac{[\eta]_b}{[\eta]_l}\right)_M \tag{88}$$

can be related to g_V by using the Mark–Houwink relationship [119] and is found to be

$$g_M = g_V^{(a+1)} \tag{89}$$

This is calculated from the LALLS results, and the various branching parameters determined as described previously.

6.2   Determination of Polymer Viscosities and Mark–Houwink Coefficients

6.2.1   Size-Exclusion Chromatography Measurement of Viscosity

Hellman [120] has shown that polymer viscosities may be determined accurately using SEC. If the Mark–Houwink constants for a polymer are known accurately, the viscosities of the eluting fractions may be calculated

$$[\eta]_i = \mathcal{K} M_i^a \tag{90}$$

and the bulk viscosity of the polymer is then

$$[\eta] = \sum_{i=1}^{N} w_i [\eta]_i = \mathcal{K} \sum_{i=1}^{N} w_i M_i^a \tag{91}$$

where w_i is the weight fraction of chromatographic fraction i. Yau and others [121] have suggested that if several narrow fractions of the polymer are available, they may be used to calibrate the SEC system directly for viscosity, without the necessity of measuring molecular weights or calculating Mark–Houwink parameters. If the column set is linear in log M versus V_e, there will also be a linear relationship between log $[\eta]$ and V_e. This may be expressed as

$$[\eta] = E_1 \exp(-E_2 V_e) \tag{92}$$

Such a calibration was generated using polystyrene standards with viscosities measured in THF and chromatographed on a column set designed for linearity. The calibration curve is illustrated in Figure 7.46. Intrinsic viscosities of unknown polystyrene materials may then be calculated directly from their SEC curves.

$$[\eta] = \sum_{i=1}^{N} w_i [\eta]_i(v) \tag{93}$$

with the values of $[\eta]_i$ taken from the calibration curve.

The values of E_1 and E_2 can be related to D_1 and D_2. Using the linear

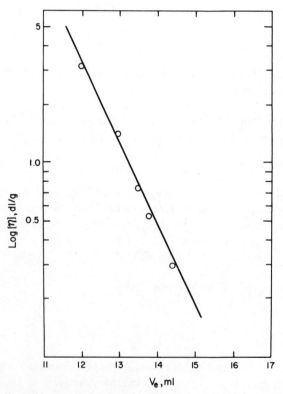

Figure 7.46 Intrinsic viscosity calibration of a set of narrow MWD polystyrene standards in THF. Reprinted, with permission, from [121].

calibration relationship

$$M = D_1 \exp(-D_2 V_e) \tag{94}$$

and substituting

$$[\eta] = \mathcal{K} M^a \tag{95}$$

results in a definition for $[\eta]$ in terms of D_1 and D_2

$$[\eta] = \mathcal{K} D_1^a \exp(-aD_2 V_e) \tag{96}$$

thus

$$E_1 = \mathcal{K} D_1^a \tag{97}$$

$$E_2 = aD_2 \tag{98}$$

If narrow MWD viscosity standards are not available, broad MWD standards with known $[\eta]$ may be used for calibration in a manner completely analogous to broad MWD molecular weight calibration [121]. Assuming that a linear calibration according to (92) is valid, then

$$[\eta] = E_1 \sum_{i=1}^{N} w_i \exp(-E_2 V_{ei}) \tag{99}$$

and the ratio $[\eta]_1/[\eta]_2$ for two polymers generates E_2 by a computer-fitting procedure; E_1 is found by substitution. Methods of generating E_1 and E_2 using assumed values of column dispersion σ are described in [121].

6.2.2 Size-Exclusion Chromatography Measurement of Mark–Houwink Coefficients

The calculation of Mark–Houwink parameters for a polymer using broad MWD standards follows easily from this method [121]. The computer-fitting programs for determining D_1 and D_2 using one of the linear methods for molecular weight calibration with broad MWD standards are identical to those used to determine E_1 and E_2. If two or more broad standards, characterized for two values of molecular weight moment and $[\eta]$, are available, then D_1, D_2, E_1, and E_2 may be determined. From (97) and (98),

$$a = E_2/D_2 \tag{100}$$

and

$$\mathcal{K} = E_1/D_1^a \tag{101}$$

and the Mark–Houwink parameters are generated.

Mark–Houwink parameters may also be determined for broad MWD polymers that have known values of $[\eta]$ but not $\bar{M}_w$ or $\bar{M}_n$ in a method utilizing universal calibration, described by Weiss and Cohn–Ginsberg [122]. Polymer molecular weights are also calculated. Two or more samples are required. The polymer samples are chromatographed in a column set that has been calibrated with polystyrene, or another narrow MWD standard that may be used as a universal calibrant with the unknown. The molecular weights of the eluting fractions are expressed

$$M_i = \left(\frac{1}{\mathscr{K}}\right)^{1/a} [\eta]_i^{1/a} \tag{102}$$

from the Mark–Houwink relationship. The hydrodynamic volume of each fraction is

$$J_i = [\eta]_i M_i = \mathscr{K} M_i^{(a+1)} \tag{103}$$

$$= \mathscr{K}^{-1/a} [\eta]_i^{1+(1/a)} \tag{104}$$

and the viscosity of each fraction is

$$[\eta]_i = J_i^{a/(a+1)} \mathscr{K}^{1/(a+1)} \tag{105}$$

The viscosity of a polymer sample is the weighted sum of the viscosities of each fraction

$$[\eta] = \sum_{i=1}^{N} w_i[\eta]_i = \mathscr{K}^{1/(a+1)} \sum_{i=1}^{N} w_i J_i^{a/(a+1)} \tag{106}$$

The ratio of the viscosities of two samples of polymer is

$$\frac{[\eta]_A}{[\eta]_B} = \sum_{i=1}^{N} w_{A_i} J_{A_i}^{a/(a+1)} \bigg/ \sum_{i=1}^{N} w_{B_i} J_{B_i}^{a/(a+1)} \tag{107}$$

If the intrinsic viscosities of the two samples are sufficiently different, (107) can be used to determine a, with J values assigned from the universal calibration curve, and w_i values from the SEC elution curve of the sample polymer. Values of a ranging between 0.5 and 1.0 are substituted into the right half of (107) [123]. The resulting value is compared with the predicted value for $[\eta]_A/[\eta]_B$. The best value of a is then used to calculate $\mathscr{K}$ from (105) rearranged

$$\mathscr{K} = \left([\eta] \bigg/ \sum_{i=1}^{N} w_i J_i^{a/(a+1)}\right)^{(a+1)} \tag{108}$$

These methods for determining polymer viscosities and Mark–Houwink coefficients are summarized in Table 7.8.

Table 7.8 Methods for the Determination of Polymer Viscosities and Mark–Houwink Coefficients Using Size-Exclusion Chromatography

Quantity Determined	Information Needed	Method	References
Polymer viscosity	Mark–Houwink coefficients, calibrated elution curve	(91)	[120]
	Narrow MWD standards of known [η]	Linear calibration parameters E_1 and E_2 describing [η](V_e) generated; viscosity calculated from (93)	[121]
	Two or more broad MWD standards of known [η]	Linear calibration parameters E_1 and E_2 describing [η](V_e) generated; viscosity calculated from (93)	[121]
Mark–Houwink coefficients	Two or more broad MSD standards of known [η] and $\bar{M}_w$ or $\bar{M}_n$	Linear calibration parameters D_1 and D_2 describing $M(V_e)$ and E_1 and E_2 describing [η](V_e) generated; $\mathscr{K}$ and a calculated from (100), (101)	[121]
	Two or more broad MWD standards of known [η]; universal calibration	Values of a and $\mathscr{K}$ that give the best calculated values of [η] for the standards are computer generated using (107), (108)	[122], [123]

6.3   Analysis of Copolymers

Copolymers are polymers synthesized from more than one type of monomer unit. In addition to molecular weight polydispersity, copolymers may have a compositional polydispersity. Three types of compositional makeup for a copolymer may be defined. An alternating copolymer is defined by the sequencing

$$A—B—A—B—A—B—A—B—$$

Such a sequencing is only observed occasionally unless special synthetic conditions are used. Random copolymers have a sequencing

$$A—A—B—A—B—B—B—A—A—B—A—A—$$

in which the sequence length of similar units has a statistical distribution. Block copolymers have long sequences of a single component, such that regions of the polymers in solution have solvation and structural characteristics of the homopolymer of that component. A complex copolymer may contain components with more than one of these copolymers and may also contain some fraction of homopolymer of one or more of the species. The molecular weight distribution of each component may be quite different. Complete characterization of a copolymer would require analysis of weight fraction as a function of composition and molecular weight simultaneously and would generate a three-dimensional plot such as the one illustrated in Figure 7.47. To obtain such a plot, it is necessary to fractionate the polymer both by composition and by molecular weight.

Two approaches have been considered for experimentally generating such a plot. The first is to separate the copolymer into fractions of uniform composition and subsequently analyze each fraction for MWD using SEC. The initial separation must be independent of molecular weight. Obtaining such separations is a complex experimental problem but methods such as TLC (thin layer chromatography) and other forms of adsorption chromatography show promise [124–127].

Once the polymer has been fractionated by composition, the MWD of each must be determined. If a universal calibration approach is used, the equation

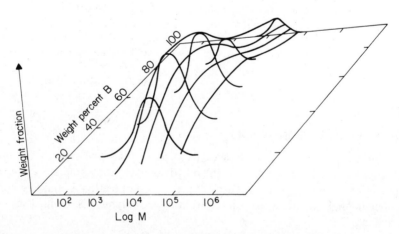

Figure 7.47 Three-dimensional plot of the MWD and compositional distribution of an A—B copolymer.

establishing the calibration curve for a copolymer of a specific composition X is [128]

$$\log M_{COP}(X) = \left[\frac{1}{1 + a_{COP}(X)}\right] \log \frac{\mathscr{K}_{PS}}{\mathscr{K}_{COP}(X)} + \left[\frac{1 + a_{PS}}{1 + a_{COP}(X)}\right] \log M_{PS} \quad (109)$$

This is analogous to (46) for homopolymers. The terms M_{PS}, $\mathscr{K}_{PS}$, and a_{PS} are the molecular weights and Mark–Houwink constants for polystyrene or another calibrant. The molecular weights comprising the copolymer calibration curve M_{COP} are generated as described in Section 3.2, assuming that the Mark–Houwink coefficients of the copolymer $\mathscr{K}_{COP}(X)$ and $a_{COP}(X)$ are known. These coefficients vary greatly with composition and generally are not predictable from the Mark–Houwink coefficients of the homopolymers (except for simple diblock or triblock systems [129, 130]). The variation of $\mathscr{K}_{COP}$ and a_{COP} with composition will depend on how different these parameters are for the homopolymers in a given solvent system, on interactions that exist between components along the polymer chain, and on the sequencing of the copolymer. For example, Garcia–Rubio [131] has determined the value of a_{COP} as a function of composition for random copolymers of styrene–acrylonitrile in several solvents (see Figure 7.48), showing that this parameter varies significantly with small changes in X in some solvents.

It is apparent that after compositional fractionation each copolymer composition must be treated as a separate polymer and be characterized for universal calibration as described previously or characterized using one of the broad MWD methods described in the calibration section (3.3). In the absence of adsorption effects, it is then theoretically possible to generate a complete MWD compositional distribution plot (Figure 7.47).

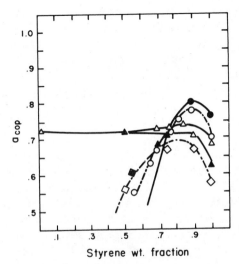

Figure 7.48 Dependence of the Mark–Houwink exponent, a_{COP}, on composition of styrene–acrylonitrile random copolymers: (●) CHCl$_3$ (20°C); (○) THF (25°C); (◇) MEK (20°C); (◆) MEK (30°C); (△) DMF (20°C); (▲) DMF (30°C). Reprinted with permission, from [131]. Copyright 1983 American Chemical Society.

The second approach suggested is to fractionate the copolymer using SEC and analyze for composition. It is not possible to fractionate a complex copolymer exactly according to molecular weight because the hydrodynamic volume varies with chemical composition as well as molecular weight. Therefore the material eluted at a specific V_e will comprise copolymers with the same hydrodynamic volume but different molecular weights, compositions, and sequence length distributions [132].

Despite this, the fractionation of a copolymer using SEC, with simultaneous detection of composition using multiple detectors, is a widely used technique, first introduced by Alliet and Pacco [133]. It is quite straightforward and can generally be carried out with equipment and techniques available in any SEC laboratory. It generates a size-exclusion chromatogram that qualitatively describes the molecular size distribution of the copolymer and also gives information about compositional drift across this size distribution. Both results are very important for the analysis of copolymer synthetic techniques and comparisons of repeat syntheses of copolymer materials.

In general, one detector is necessary for each component present in the copolymer. A common combination of detectors has been UV and RI, connected in series. The difference in dead volume between the two is corrected for, the response of each detector to each of the components determined using homopolymers, and then the combination of the two signals as a function of elution volume is translated into composition as a function of elution volume. The technique has also been used with IR detection, which is very specific for composition.

Two assumptions are made in translating detector response to copolymer composition [134]. The first is that Beer's law is valid and there is a linear relationship between detector signal and concentration; the second is that the signal can be considered to be a sum of the signals of each component. These assumptions are not always valid. Refractive index response may be a function of molecular weight, as discussed in Section 4.2, or of copolymer composition [135]. Deviations from Beer's law of up to 60% have been reported in the UV analysis of styrene copolymers [134–136]. Experimental evidence indicates that the UV extinction coefficient in fact varies linearly with styrene sequence length [134, 136]. It must be concluded that for compositional analysis of copolymers, careful measurements of detector response factors for specific components must be made before quantitative conclusions can be made.

6.4   Particle-Size Analysis

Two chromatographic methods have been developed for the separation of colloidal particles by size. The first of these, called *hydrodynamic chromatography* (HDC), utilizes a bed packed with a nonporous material. The velocity at which a particle travels through the interstitial volume of the bed depends on the flow streams to which it has access. Larger particles are restricted by their size to the fast streams in the center of interstitial channels and are thus eluted early. Smaller particles can also travel in the slower moving streams close to the "walls" formed by the packing material and are eluted relatively later. The second technique is based on conventional SEC. Particles are separated on the basis of their access to the pore volume of a porous stationary phase.

6.4.1   Hydrodynamic Chromatography

Hydrodynamic chromatography (HDC), originally developed by Small and others [137, 138], is carried out by injection of a dilute suspension of latex particles onto a bed packed with nonporous beads. Separation is characterized by a factor R_F

$$R_F = \bar{V}_p/\bar{V}_E \tag{110}$$

in which the average axial velocity of the particle is compared to the average axial velocity of the eluent, or an ionic marker. Since particles are restricted from the very slow-moving streams, their average velocity will always be larger than that of an eluent molecule, and R_F will always be greater than 1. According to the capillary model [137, 139–141], which considers the interstitial space to be a system of capillaries of equal radius, the average particle velocity may be expressed

$$\bar{V}_p = \frac{\int_0^{(R_0 - R_p)} V_p(r)\exp[-\Phi(r)/kT]\,r\,dr}{\int_0^{(R_0 - R_p)} \exp[-\Phi(r)/kT]\,r\,dr} \tag{111}$$

in which R_0 is the equivalent capillary radius, R_p is the particle radius, and $\Phi(r)$ is the energy of interaction between the particle and capillary wall. This interaction term is a combination of a repulsive force caused by the double layer on the surface of the particle and the capillary wall (stationary phase) and an attractive force arising from van der Waals attraction between them [142].

Figure 7.49 illustrates the effect of ionic strength on R_F [137]. At low values of ionic strength, the repulsive force of the double layer is important, the particles are electrostatically prevented from approaching the wall and consequently occupy the central, faster flowstreams and have higher R_F. As the ionic strength increases, repulsion is overcome, the particles occupy a wider cross section of the capillary, and R_F decreases. At very high ionic strengths, the particles are able to approach the walls so closely that van der Waals forces become operative and the larger particles become selectively retained.

In the regions of low ionic strength, the surface characteristics of the particle are less important, and universal calibration of particle size; that is, separation of all particles on the basis of size rather than chemical nature, should become operative. This is illustrated in Figure 7.50 for three types of latex particles [143], in which the particle diameter is plotted as a function of the difference in elution volume between the particle and the marker.

The total volume difference over which separation is observed in Figure 7.50 is 4 mL, which is only about 6% of the total void volume of the column [143].

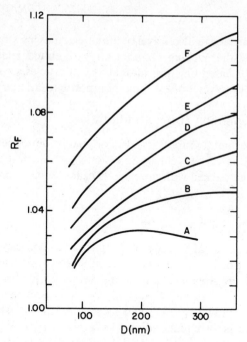

Figure 7.49 The effect of eluent strength on the R_F of polystyrene latex particles. Eluent concentrations in moles per liter of NaCl. A, 1.76×10^{-1}; B, 9×10^{-2}; C, 2.96×10^{-2}; D, 4.6×10^{-3}; E, 1.7×10^{-3}; F, 4.25×10^{-4}. Reprinted, with permission, from [137].

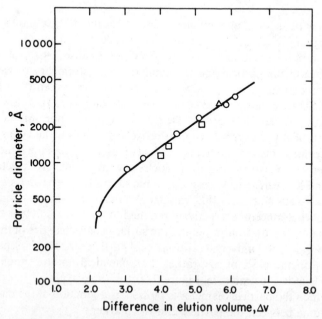

Figure 7.50 HDC universal calibration curve (eluent ionic strength 1.29 mM); monodisperse latices; (○) polystyrene; (□) poly(vinyl chloride); (△) poly(styrene)cobutadiene. Reprinted, with permission, from [143] by courtesy of Marcel Dekker, Inc.

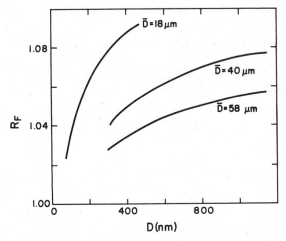

Figure 7.51 The dependence of R_F on latex particle diameters, D(nm), and packing diameter, $\bar{D}$. Reprinted, with permission, from [137].

Although the separation capacity of HDC is low, the resolution is excellent. The percentage of void volume involved in the separation, and therefore the values of R_F, may be increased by decreasing the size of the packing particles (Figure 7.51). Because this causes the interstitial channels, the capillaries, to become smaller, it also limits the upper end of particle sizes that can be chromatographed.

6.4.2 Size-Exclusion Chromatography

In SEC, particles are separated on the basis of their steric access to the porous volume of the packing particles. The original experiment of this type was carried out by Krebs and Wunderlich [144], who chromatographed polystrene and poly(methyl methacrylate) latex particles on silica gels of pore size 50–500 nm. Hamielec and Singh [145] demonstrated that universal calibration of polystyrene, silica, polystyrene–methyl methacrylate, and poly(butadiene)–acrylonitrile particles could be achieved. They used columns packed with porous silica or glass, and eluent containing both surfactant to prevent adsorption of the particles to the stationary phase and electrolyte to prevent exclusion of the particles from the pores. This mobile phase had been suggested by Coll and Fague [146]. This universal calibration plot is shown in Figure 7.52.

Kirkland [147] investigated the use of porous silica microspheres (less than 10 μm) for the separation of small inorganic silica particles (5–200 nm). He found the linear separation range of a single pore size to be approximately 0.7 decades in size. Figures 7.53 and 7.54 show calibration curves for three narrow pore size distribution columns. Figure 7.55 illustrates the resolution available using high-performance packings; distinction can be made between particles of aluminosilicate sols with silica gel equivalent sizes of 8 and 10 nm.

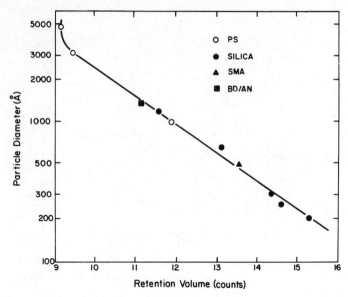

Figure 7.52 Universal particle diameter-retention volume calibration curve. Column set: CPG; particles: PS, polystyrene; silica; SMA; styrene–methacrylate copolymer; BD/AN, butadiene–acrylonitrile copolymer. Reprinted, with permission, from [145].

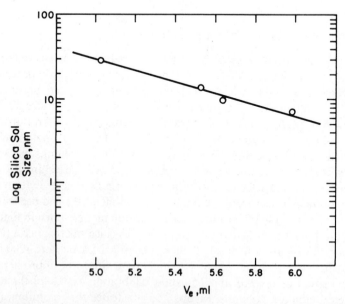

Figure 7.53 Silica sol particle-size calibration plot for porous silica microsphere column with 75-nm probes; 8.9 μm packing particles; mobile phase: 0.1 M Na$_2$HPO$_4$—NaH$_2$PO$_4$; pH 8.0; detector: UV, 254 nm; sample: 25 μL 2% colloid in mobile phase. Reprinted, with permission, from [147].

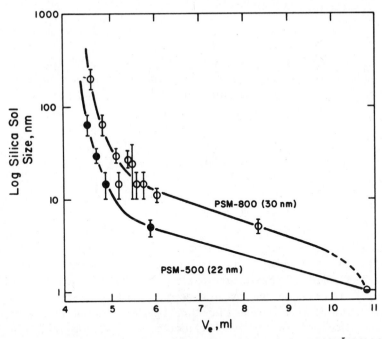

Figure 7.54 Silica sol particle-size calibration for porous silica microspheres with (⬮) PSM-800: 30-nm pores, 6.0-μm packing particles; (⬤) PSM-500: 22-nm pores, 7.7-μm packing particles; mobile phase 0.001 M NH$_4$OH; samples, 25 μL 2% colloid in mobile phase, acetone. Reprinted, with permission, from [147].

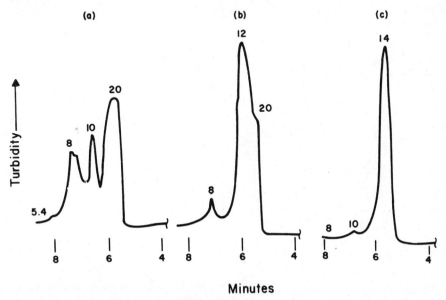

Minutes

Figure 7.55 Fractionation of three experimental alumino-silicate sols: *a*, *b*, and *c*. Mobile phase: 0.001 M NH$_4$OH; column: PSM-800 (6.0-μm, 30-nm pores); detector: UV, 254 nm. The numbers above the peaks indicate calibration values in nanometers from silica sol calibration curve. Reprinted, with permission, from [147].

6.4.3   Detection

Detection of eluted particles can be accomplished using a UV photometer to measure the turbidity of the solution in the detector cell. The turbidity at a given volume of noninteracting suspensions is given by [143]

$$\tau = NR_{\text{ext}}X \tag{112}$$

in which N is the number density of the particles, and X is the cell path length. The extinction cross section, R_{ext}, is the sum of two cross sections

$$R_{\text{ext}} = R_{\text{scat}} + R_{\text{abs}} \tag{113}$$

that caused by scattering R_{scat} and that caused by absorption R_{abs}; R_{scat} depends on particle diameter and the wavelength of detection. For example, measurement of the turbidity of a mixture of two samples of different size is illustrated in Figure 7.56. To calculate true particle concentrations and particle-size distributions, R_{ext} must be determined as a function of particle size and this relationship taken into account in the analysis of elution curves [143, 148]. Hamielec and colleagues [149–151] developed a theory for converting the detector response $F(v)$ from a turbidity detector into particle-size distributions, including number, surface volume, specific surface, weight, and turbidity averages. They include methods of correction for Gaussian band broadening for all of these averages. Band broadening is significant for particle SEC because of the low diffusion constants for particles.

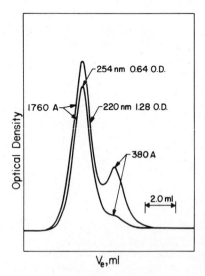

Figure 7.56 The HDC separation of a synthetic bimodal mixture of 380 and 1760-Å polystyrene standards (weight ratio 1.00/1.20) at 220 and 254 nm. Reprinted, with permission, from [143].

References

1. J. Porath and P. Flodin, *Nature (London)*, **183**, 1657 (1959).
2. J. C. Moore, *J. Polym. Sci. Part A*, **2**, 835 (1964).
3. L. Fischer, *Gel Filtration Chromatography*, 2nd ed., Elsevier/North-Holland Biomedical, Amsterdam, 1980.
4. W. W. Yau, J. J. Kirkland, and D. D. Bly, *Modern Size-Exclusion Liquid Chromatography*, Wiley, New York, 1979.
5. G. L. Hagnauer, *Anal. Chem.*, **54**, 265R (1982).
6. W. W. Yau, C. P. Malone, and H. L. Suchan, *Sep. Sci.*, **5**, 259 (1970).
7. E. F. Casassa, *J. Polym. Sci. Part B*, **5**, 773 (1967).
8. E. F. Casassa and Y. Tagami, *Macromolecules*, **2**, 14 (1969).
9. E. F. Casassa, *J. Phys. Chem.*, **75**, 3929 (1971).
10. W. W. Yau and C. P. Malone, *Polym. Prepr.*, **12**(2), 797 (1971).
11. W. W. Yau and D. D. Bly, "Effect of Solute Shape or Conformation in Size Exclusion Chromatography," in T. Provder, Ed., *Size Exclusion Chromatography* (ACS Symposium Series, Vol. 138), American Chemical Society, Washington, DC, 1980, p. 197.
12. J. C. Giddings and K. K. Mallik, *Anal. Chem.*, **38**, 997 (1966).
13. J. J. Van Deemter, F. J. Zuiderweg, and A. Klinkenberg, *Chem. Eng. Sci.*, **5**, 271 (1956).
14. J. C. Giddings, *Dynamics of Chromatography, Part* 1, Wiley, New York, 1965.
15. L. H. Tung and J. R. Runyon, *J. Appl. Polym. Sci.*, **13**, 2397 (1969).
16. R. N. Kelley and F. W. Billmeyer, Jr., *Anal. Chem.*, **42**, 399 (1970).
17. J. C. Giddings, L. M. Bowman, Jr., and M. N. Myers, *Macromolecules*, **10**, 443 (1977).
18. R. Groh and I. Halasz, *Anal. Chem.*, **54**, 1325 (1981).
19. J. Klein and M. Grüneberg, *Macromolecules*, **14**, 1411 (1981).
20. W. W. Yau, J. J. Kirkland, D. D. Bly, and H. J. Stoklosa, *J. Chromatogr.*, **125**, 219 (1976).
21. L. H. Tung, *J. Appl. Polym. Sci.*, **10**, 375 (1966).
22. L. H. Tung, *J. Appl. Polym. Sci.*, **13**, 775 (1969).
23. N. Friis and A. Hamielec, "Gel Permeation Chromatography: A Review of Axial Dispersion Phenomena, Their Detection and Correction," in J. C. Giddings, Ed., *Advances in Chromatography*, Vol. 13, Dekker, New York, 1975, p. 41.
24. ASTM Method D3593-77, *Standard Method of Test for the Determination of Molecular Weight Averages and Molecular Weight Distribution of Certain Polymers by Liquid Chromatography Using Universal Calibration*, ASTM, Philadelphia, PA.
25. D. J. Pollock and R. F. Kratz, *Proceedings of the Fourth International Seminar on Gel Permeation Chromatography*, Miami Beach, FL, 1967.
26. S. T. Balke, A. E. Hamielec, B. P. LeClair, and S. L. Pearce, *Ind. Eng. Chem., Prod. Res. Dev.*, **8**, 54 (1969).
27. L. H. Tung, *Sep. Sci.*, **5**, 339 (1970).

28. J. Janca, "Calibration of Separation Systems in Gel Permeation Chromatography for Polymer Characterization," in J. C. Giddings, Ed., *Advances in Chromatography*, Vol. 19, Dekker, New York, 1980, p. 37.

29. W. W. Yau and S. W. Fleming, *J. Appl. Polym. Sci.*, **12**, 2111 (1968).

30. H. Benoit, Z. Grubisic, P. Rempp, D. Decker, and J. C. Zilliox, *J. Chem. Phys.*, **63**, 1507 (1966).

31. Z. Grubisic, P. Rempp, and H. Benoit, *J. Polym. Sci. Part B*, **5**, 753 (1967).

32. M. J. R. Cantow, R. S. Porter, and J. F. Johnson, *J. Polym. Sci. Part A-1*, **5**, 1391 (1967).

33. T. D. Swartz, D. D. Bly, and A. S. Edwards, *J. Appl. Polym. Sci.*, **16**, 3353 (1972).

34. A. H. Abdel-Alim and A. E. Hamielec, *J. Appl. Polym. Sci.*, **18**, 297 (1974).

35. A. R. Weiss and E. Cohn-Ginsberg, *J. Polym. Sci. Part A-2*, **8**, 148 (1970).

36. S. T. Balke, A. E. Hamielec, B. P. LeClair, and S. L. Pearce, *Ind. Eng. Chem., Prod. Res. Dev.*, **8**, 54 (1969).

37. W. W. Yau, H. J. Stoklosa, and D. D. Bly, *J. Appl. Polym. Sci.*, **21**, 1911 (1977).

38. S. T. Balke and A. E. Hamielec, *J. Appl. Polym. Sci.*, **13**, 1381 (1969).

39. A. E. Hamielec, *J. Appl. Polym. Sci.*, **14**, 1519 (1970).

40. T. Provder and E. M. Rosen, *Sep. Sci.*, **5**, 437 (1970).

41. W. W. Yau, H. J. Stoklosa, C. R. Ginnard, and D. D. Bly, *Paper P*013, 12th Middle Atlantic Regional Meeting, American Chemical Society, Washington, DC, April 5–7, 1978.

42. A. E. Hamielec and S. N. E. Omorodion, "Molecular Weight and Peak Broadening Calibration in Size Exclusion Chromatography," in T. Provder, Ed., *Size Exclusion Chromatography* (ACS Symposium Series, Vol. 138) American Chemical Society, Washington, DC, 1980, p. 183.

43. D. D. Bly, H. J. Stoklosa, J. J. Kirkland, and W. W. Yau, *Anal. Chem.*, **47**, 1810 (1975).

44. T. E. Miller, Jr., T. A. Chamberlin, and H. E. Tuinstra, *Am. Lab.*, **15**, No. 1, 74 (1983).

45. E. Kohn and R. W. Ashcraft, "Calibration and Data Processing in High Speed Gel Permeation Chromatography," in J. Cazes, Ed., *Liquid Chromatography of Polymers and Related Materials*, Vol. 8, Dekker, New York, 1977, p. 105.

46. *Series R-400 Instruction Manual*, Waters Associates, Inc., Framingham, MA, 1976.

47. E. M. Barrall, II, M. J. R. Cantow, and J. F. Johnson, *J. Appl. Polym. Sci.*, **12**, 1373 (1968).

48. C. E. M. Morris and I. Grabovac, *J. Chromatogr.*, **189**, 259 (1980).

49. J. H. Ross and M. E. Castro, *J. Polym. Sci. Part C*, **21**, 143 (1968).

50. S. D. Abbot, *Am. Lab.*, **9**, No. 8, 41 (1977).

51. A. C. Ouano and W. Kaye, *J. Polym. Sci. Part A-1*, **12**, 1151 (1974).

52. M. L. McConnell, *Am. Lab.*, **10**, No. 5, 63 (1978).

53. *Chromatix KMX-6 Application Note LS-5*, Chromatix, Sunnyvale, CA, 1978.

54. *Chromatix KMX-6 Application Note LS-10*, Chromatix, Sunnyvale, CA, 1978.

55. Z. Grubisic-Gallot, M. Picot, Ph. Gramain, and H. Benoit, *J. Appl. Polym. Sci.*, **16**, 2931 (1972).

56. A. C. Ouano, *J. Polym. Sci. Part A*-1, **10**, 2169 (1972).

57. L. Letot, J. Lesec, and C. Quivoron, *J. Liq. Chromatogr.*, **3**, 427 (1980).

58. J. C. Watters and T. G. Smith, *Ind. Eng. Chem. Process Des. Dev.*, **18**, 591 (1979).

59. *Liquid Chromatography School Manual*, Waters Associates, Inc., Milford, MA, 1976, p. KS4.

60. R. V. Vivilecchia, B. G. Lightbody, N. Z. Thimot, and H. M. Quinn, *J. Chromatogr. Sci.*, **15**, 424 (1977).

61. A. R. Cooper, "Analysis of Polymers by Gel Permeation Chromatography," in L. S. Bark and N. S. Allen, Eds., *Analysis of Polymer Systems*, Applied Science Publishers, London, 1982, p. 243.

62. J. J. Kirkland, *J. Chromatogr.*, **125**, 231 (1976).

63. A. R. Cooper and E. M. Barrall, II, *J. Appl. Polym. Sci.*, **17**, 1253 (1973).

64. R. P. W. Scott and P. Kucera, *J. Chromatogr.*, **125**, 251 (1976).

65. F. A. Butenhuys and F. P. B. van der Maeden, *J. Chromatogr.*, **149**, 489 (1978).

66. W. W. Yau, C. R. Ginnard, and J. J. Kirkland, *J. Chromatogr.*, **149**, 465 (1978).

67. K. K. Unger, *Porous Silica*, Elsevier, Amsterdam, 1979.

68. R. Audebert, *Polymer*, **20**, 1561 (1979).

69. J. V. Dawkins and M. Hemming, *Makromol. Chem.*, **176**, 1795 (1975).

70. J. V. Dawkins and M. Hemming, *Makromol. Chem.*, **176**, 1777 (1975).

71. J. V. Dawkins and M. Hemming, *Makromol. Chem.*, **176**, 1815 (1975).

72. J. V. Dawkins, *J. Liq. Chromatogr.*, **1**, 279 (1978).

73. J. V. Dawkins, *J. Polym. Sci. Part A*-2, **14**, 569 (1976).

74. J. Klein and K. Treichel, *Chromatographia*, **10**, 604 (1977).

75. J. G. Bergman, L. J. Duffy, and R. B. Stevenson, *Anal. Chem.*, **43**, 131 (1971); J. Lecourtier, R. Audebert, and C. Quivoron, *J. Chromatogr.*, **121**, 173 (1976).

76. D. H. Freeman and D. Killion, *J. Polym. Sci. Part A*-2, **15**, 2047 (1977).

77. J. V. Dawkins, *Pure Appl. Chem.*, **51**, 1473 (1979).

78. J. Janca, *Anal. Chem.*, **51**, 637 (1979).

79. J. Janca, *J. Liq. Chromatogr.*, **4**, 181 (1981).

80. K. J. Bombaugh, W. A. Dark, and J. N. Little, *Anal. Chem.*, **41**, 1337 (1969).

81. C. Horvath, W. Melander, and I. Molnar, *J. Chromatogr.*, **125**, 129 (1976).

82. P. A. Neddermeyer and L. B. Rogers, *Anal. Chem.*, **40**, 755 (1968).

83. P. A. Neddermeyer and L. B. Rogers, *Anal. Chem.*, **41**, 94 (1968).

84. A. Domard, M. Rinaudo, and C. Rochas, *J. Polym. Sci. Part A*-2, **17**, 673 (1979).

85. C. Rochas, A. Domard, and M. Rinaudo, *Eur. Polym. J.*, **16**, 135 (1980).

86. M. Rinaudo and J. Desbrieres, *Eur. Polym. J.*, **16**, 849 (1980).

87. M. Rinaudo, J. Desbrieres, and C. Rochas, *J. Liq. Chromatogr.*, **4**, 1297 (1981).

88. K. G. Forss and B. G. Stenlund, *J. Polym. Sci. Part C*, **42**, 951 (1973).

89. B. Stenlung, *Adv. Chromatogr.*, **14**, 37 (1976).

90. H. G. Barth and F. E. Regnier, *J. Chromatogr.*, **192**, 275 (1980).

91. F. A. Butenhuys and F. P. B. van der Maeden, *J. Chromatogr.*, **149**, 490 (1978).

92. A. L. Spatorico and G. L. Beyer, *J. Appl. Polym. Sci.*, **19**, 2933 (1975).

93. E. Patton, unpublished results.

94. J. Engelhardt and D. Mathes, *J. Chromatogr.*, **142**, 311 (1977).

95. J. Englehardt and D. Mathes, *J. Chromatogr.*, **185**, 305 (1979).

96. F. E. Regnier and R. Noel, *J. Chromatogr. Sci.*, **14**, 316 (1976).

97. F. E. Regnier, U.S. Patent 3,983,299, September 28, 1976.

98. D. E. Schmidt, Jr., R. W. Giese, D. Conron, and B. L. Karger, *Anal. Chem.*, **52**, 177 (1980).

99. K. K. Unger, *Porous Silica*, Elsevier, Amsterdam, 1979, p. 67.

100. E. Pfannkoch, K. C. Lu, F. E. Regnier, and H. G. Barth, *J. Chromatogr. Sci.*, **18**, 430 (1980).

101. D. P. Herman, L. R. Field, and S. Abbott, *J. Chromatogr. Sci.*, **19**, 470 (1981).

102. R. V. Vivilecchia, B. G. Lightbody, N. Z. Thimot, and H. M. Quinn, *J. Chromatogr. Sci.*, **15**, 424 (1977).

103. T. V. Alfredson, C. T. Wehr, L. Tallman, and F. Klink, *J. Liq. Chromatogr.*, **5**, 489 (1982).

104. Y. Kato, K. Komiya, H. Sasaki, and T. Hashimoto, *J. Chromatogr.*, **193**, 311 (1980).

105. C. P. Talley and L. M. Bowman, *Anal. Chem.*, **51**, 2239 (1979).

106. D. L. Gooding, M. N. Schmuck, and K. M. Gooding, *J. Liq. Chromatogr.*, **5**, 2259 (1982).

107. P. L. Dubin and I. J. Levy, *J. Chromatogr.*, **235**, 377 (1982).

108. J. V. Dawkins and N. P. Gabbot, Polymer, **22**, 291 (1981).

109. K. J. Bombaugh, W. A. Dark, and J. N. Little, *Anal. Chem.*, **41**, 1137 (1969).

110. R. D. Hester and P. H. Mitchell, *J. Polym. Sci. Part A-1*, **18**, 1727 (1980).

111. S. Kuga, *J. Chromatogr.*, **206**, 449 (1981).

112. J. Desbrieres, J. Mazet, and M. Rinaudo, *Eur. Polym. J.*, **18**, 269 (1982).

113. E. E. Drott and R. A. Mendelson, *J. Polym. Sci. Part A-2*, **8**, 1361 (1970).

114. E. E. Drott and R. A. Mendelson, *J. Polym. Sci. Part A-2*, **8**, 1373 (1970).

115. M. R. Ambler, R. D. Mate, and J. R. Purdon, Jr., *J. Polym. Sci. Part A-1*, **12**, 1759 (1974).

116. M. R. Ambler, R. D. Mate, and J. R. Purdon, Jr., *J. Polym. Sci. Part A-1*, **12**, 1771 (1974).

117. B. H. Zimm and W. H. Stockmayer, *J. Chem. Phys.*, **17**, 1301 (1949).

118. M. R. Ambler, *J. Appl. Polym. Sci.*, **21**, 1655 (1977).

119. R. C. Johnson and M. L. McConnell, "Characterization of Branched Polymers by Size Exclusion Chromatography With Light Scattering Detection," in T. Provder, Ed., *Size Exclusion Chromatography* (ACS Symposium Series, Vol. 138) American Chemical Society, Washington, DC, 1980, p. 107.

120. M. Y. Hellman, "Intrinsic Viscosity by Gel Permeation Chromatography," in J. Cazes, Ed., *Liquid Chromatography of Polymers and Related Materials*, Dekker, New York, 1977, p. 29.

121. W. W. Yau, M. E. Jones, C. R. Ginnard, and D. D. Bly, "Polymer Viscosity Characterization by Size Exclusion Chromatography," in T. Provder, Ed., *Size Exclusion Chromatography* (ACS Symposium Series, Vol. 138), American Chemical Society, Washington, DC, 1980, p. 91.

122. A. R. Weiss and E. Cohn-Ginsberg, *J. Polym. Sci. Part B*, **7**, 379 (1969).

123. C. J. B. Eobbin, A. Rudin, and M. F. Tchir, *J. Appl. Polym. Sci.*, **25**, 2985 (1980).

124. H. Inagaki and T. Tanaka, *Pure Appl. Chem.*, **54**, 309 (1982).

125. H. Inagaki, *Adv. Polym. Sci.*, **24**, 190 (1977).

126. B. G. Belenkii and E. S. Gankina, *J. Chromatogr.*, **53**, 3 (1970).

127. S. Teramachi, A. Hasegawa, Y. Shima, M. Akatsuka, and M. Nakajima, *Macromolecules*, **12**, 992 (1979).

128. L. H. Garcia-Rubio, J. F. MacGregor, and A. E. Mamielec, *Polym. Prepr.*, **22**, No. 1, 292 (1981).

129. F. S. C. Chang, *Adv. Chem. Ser.*, **125**, 154 (1973).

130. F. S. C. Chang, *J. Chromatogr.*, **55**, 67 (1971).

131. L. H. Garcia-Rubio, J. F. MacGregor, and A. E. Hamielec, "Size Exclusion Chromatography of Copolymers," in C. D. Craver, Ed., *Polymer Characterization* (Advances in Chemistry Series, Vol. 203), American Chemical Society, Washington, DC, 1983, p. 311.

132. A. E. Hamielec, *Pure Appl. Chem.*, **54**, 293 (1982).

133. D. F. Alliet and J. M. Pacco, *Sep. Sci.*, **6**, 153 (1971).

134. L. H. Garcia-Rubio, A. E. Hamielec, and J. F. MacGregor, "UV Spectrophotometers as Detectors for Size Exclusion Chromatography of Styrene–Acrylonitrile (SAN) Copolymers," in T. Provder, Ed., *Computer Applications in Applied Polymer Science* (ACS Symposium Series, Vol. 197), American Chemical Society, Washington, DC, 1982, p. 151.

135. K. F. O'Driscoll, W. Wertz, and A. Husar, *J. Polym. Sci. Part A*-1, **5**, 2159 (1967).

136. R. J. Brussau and D. J. Stein, *Angew. Makromol. Chem.*, **12**, 59 (1970).

137. H. Small, *J. Colloid Interface Sci.*, **48**, 147 (1974).

138. H. Small, F. L. Saunders, and J. Solc, *Adv. Colloid Interface Sci.*, **6**, 237 (1976).

139. A. J. McHugh, C. Silebi, G. W. Poehlein, and J. W. Vanderhoff, *Colloid and Interface Science*, Vol. 4, Academic, New York, 1976, p. 549.

140. C. A. Silebi and A. J. McHugh, *AIChE J.*, **24**, 204 (1978).

141. F. R. Stoisits, G. W. Poehlein, and J. W. Vanderhoff, *J. Colloid Interface Sci.*, **57**, 337 (1976).

142. A. Husain, A. E. Hamielec, and J. Vlachopoulos, *J. Liq. Chromatogr.*, **4**, S295 (1981).

143. A. J. McHugh, D. J. Nagy, and C. A. Silebi, "Particle Size Analysis," in T. Provder, Ed., *Size Exclusion Chromatography* (ACS Symposium Series, Vol. 138), American Chemical Society, Washington, DC, 1980, p. 1.

144. K. F. Krebs and W. Wunderlich, *Angew. Makromol. Chem.*, **20**, 203 (1971).

145. A. E. Hamielec and S. Singh, *J. Liq. Chromatogr.*, **1**, 187 (1978).

146. H. Coll and G. R. Fague, *J. Colloid Interface Sci.*, **76**, 116 (1980).

147. J. J. Kirkland, *J. Chromatogr.*, **185**, 273 (1979).

148. D. J. Nagy, *Column Chromatography of Polymer Latexes*, unpublished doctoral dissertation, Lehigh University, PA, 1979.

149. A. E. Hamielec and S. Singh, *J. Liq. Chromatogr.*, **1**, 187 (1978).

150. A. Husain, J. Vlachopoulos, A. E. Hamielec, *J. Liq. Chromatogr.*, **2**, 193 (1979).

151. A. Husain, A. E. Hamielec, and J. Vlachopoulos, "Particle Size Analysis Using Size Exclusion Chromatography," in T. Provder, Ed., *Size Exclusion Chromatography* (ACS Symposium Series, Vol. 138), American Chemical Society, Washington, DC, 1980, p. 47.

Chapter 8

FIELD-FLOW FRACTIONATION

J. Calvin Giddings and Karin D. Caldwell

1   INTRODUCTION

1.1   Historical View

A new approach to separation, with similarities to chromatography, was introduced in 1966 [1]. Termed *field-flow fractionation* (FFF), the approach achieves separation by differential migration in a narrow channel through which a fluid flows unidirectionally, as in chromatography. However, there is no packing in the channel and no stationary phase. Field-flow fractionation has been described as one-phase chromatography because it employs only one phase.

The partitioning effect of the stationary phase is replaced by an external field that causes retention by forcing solute species into the slow-flow region near the wall of the generally ribbonlike channel. The use of an external, solute-enriching field aligned in a direction perpendicular to the flow stream places FFF in the $F(+)cd$ category of separation methods (defined as a flow-induced separation taking place under the influence of a potential gradient that is perpendicular to the direction of flow [2]).

When first introduced, the FFF concept was predicted to have considerable potential for analytical scale work. It would be applicable to chromatography-type separations, but more importantly, it would exhibit a distinct advantage in separating macromolecules and colloids because of its essentially one-phase nature. Subsequent work has borne out these predictions, and recent experiments have extended the concept to even larger particles, up to about 100 μm. External fields, such as thermal gradients, electrical, magnetic, and centrifugal forces, were anticipated from the beginning. Also suggested were several programming techniques and the possibility of a parametric pump mode.

Since the concept was presented, an active program at the University of Utah has sought experimental realization of the basic concepts. Other laboratories have followed suit, and to date thermal, electrical, centrifugal, magnetic, and cross-flow gradients have been applied with success, and programming methods have been developed and found advantageous. The FFF technique has now been refined adequately to give it a place in the stable of practical analytical separation tools, particularly for characterizing macromolecules, colloids, and particles.

While the great bulk of the published literature (ca. 100 papers through mid-1985) has come from the University of Utah project, several contributions have been made by other groups. In 1967 Berg, Purcell, and Stewart [3–5] began publication of a group of three papers on a form of FFF now termed *sedimentation FFF*. A University of Wisconsin group headed by Lightfoot has developed techniques equivalent to *electrical* and *flow FFF* [6,7], while Subramanian and colleagues at Clarkson College of Technology have undertaken a detailed theoretical study of FFF [8]. Kirkland, Yau, and co-workers [9–13] extended and improved sedimentation FFF technology, and Oppenheimer and Mourey [14] facilitated the interpretation of fractograms in terms of particle size data by the use of novel detection methods. *Thermal FFF* has been investigated by Martin and colleagues [15–17] in France, and by Janča and co-workers in Czechoslovakia [18], and lately this group has extended its work to include some novel sedimentation FFF techniques [19, 20]. At the University of Arizona, Burke and co-workers [21] recently reported retention in an FFF system influenced by a magnetic field. Despite these important advances and others being developed elsewhere, the University of Utah has the most extensive project. Consequently, the majority of results described in this chapter derive from that work.

1.2   Description of the Field-Flow Fractionation Channel

Figure 8.1 shows a side view of a generalized FFF channel. The channel cross section is rectangular and open so that the only absolute constraints on a migrating species are the well-defined channel walls. The channel is generally very thin, 0.05–0.5 mm. The perpendicular field may be of any kind that will induce the migration of molecules or particles. We will elaborate on particular field types later. The field, by driving different kinds of particles to different degrees against the wall, where the slow flow leads to slow downstream displacement, sets up the separation, which evolves along the flow axis.

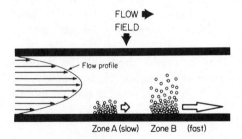

FLOW ▶
FIELD
▼

Flow profile

Zone A (slow)   Zone B   (fast)

Figure 8.1 Illustration of the FFF principle. Side view of channel.

1.3   Scope of Field-Flow Fractionation

The advantages of FFF for separating macromolecules and colloids, anticipated about two decades ago, have proven valid in most respects. Examples of polymer separations and of the separation of colloidal particles abound; some of these are shown in later figures. These results make it increasingly apparent that the versatility of FFF is one of its outstanding attributes. We have successfully worked with both aqueous and nonaqueous systems, charged and uncharged species, random coil as well as globular molecules, and particles of spherical and irregular shape. Applicability has been extended over an extremely broad mass range, from 10^3 to almost 10^{18} in effective molecular weight.

In addition to its basic use as a separation method, we have recently used FFF techniques to calculate molecular weights of discrete species, such as virus particles [22–25], and molecular weight distributions of polydisperse materials, such as polyethylene polymers [26]. Particle size distributions have also been determined for colloidal suspensions and emulsions [10, 27–31]. Diffusion coefficients of proteins and spherical particles [32, 33] have been measured, and particle densities have been determined [30, 34].

1.4   Advantages of Field-Flow Fractionation

Although the procedures and the resulting fractograms of FFF are superficially similar to those of chromatography, the unique one-phase characteristic of FFF and the dissimilarity of the forces responsible for fractionation create fundamental differences in the way solutes migrate and separate. The selectivity in FFF is based on differential and reversible levels of coupling between the external field and the components of the mixture and in some cases on differential values of the diffusion coefficients. The selective solvent–solute interactions of chromatography are, of course, absent. The elution sequence in FFF is that often observed for chromatographic techniques: Small species are eluted first and large ones last. However, it is important to note that this sequence is opposite to that of size-exclusion chromatography (SEC), which is presently the principal separation tool for macromolecules in solution.

The forces exerted by the field—responsible for retention in FFF—are continuous and generally uniform throughout the channel. There are no abrupt discontinuities (like those at active interfaces) that may prove damaging to fragile species. The level of retention may be varied at will during a run, or from one run to the next, because field strength is immediately and widely variable. This means that optimum resolution for different samples can be conveniently provided by adjusting the field strength.

Such flexibility in retention control is generally impossible in chromatography because it is difficult to alter the basic retention forces, namely, the phase-distribution forces. In chromatography, some control of elution may be gained by varying (programming) temperature or mobile phase composition, but there are important limits to the practical extent of the variations as well as substantial time lags to be considered. In certain cases, sample or phase instabilities do not permit large variations. Often, then, a different chromatographic column must be substituted to provide proper retention conditions. In FFF, by contrast, retention within a single column extends unbrokenly from zero to very high levels for a wide range of solutes.

Programming techniques exploiting this versatile retention control effectively extend the molecular weight range of FFF in the course of single runs [35]. When field strength is initially applied at a high level and is continuously (or periodically) reduced during a run, low molecular weight species elute first and successively higher molecular weight components follow in rapid sequence as the field is reduced. If very high molecular weight species are present, this strategy will shorten their elution time without sacrificing the resolution of the smaller species eluted earlier. In fact, by reducing the field strength to zero and allowing a few extra void volumes to pass through the channel, one can, in principle, sweep out any residually retained material and avoid the need for techniques such as backflushing to clean the column of tenacious species.

The uniform, open structure of the FFF channel lends itself to direct theoretical calculations. It is possible, for example, to obtain a precise mathematical prediction of retention in terms of the basic parameters controlling retention. The usefulness of this tractability is twofold: one can predict optimal operating conditions very simply, and—perhaps more important—species in complex systems can be characterized to a first approximation by their measured retention without resorting to an empirical calibration curve. Thus a species appearing at a given point in the elution spectrum can be directly associated by theory with some value of a physicochemical parameter or a simple group of such parameters. These parameters can be related to molecular weight, size, charge, and other important properties of the sample. For continuous distributions of macromolecules or particles, the elution spectrum provides a distribution curve for the mass, size, or charge of the species present. Size and molecular weight distribution spectra are increasingly important for predicting and correlating the properties of polymer and particulate systems.

Also as a result of retention predictability, the qualitative analysis of mixtures can be reinforced and confirmed by the agreement of retention data with theoretically predicted values. Quantitative analysis, of course, is available from peak area measurements providing the detector response is known. In both regards FFF is much like conventional chromatography and is subject to the same advantages and limitations. However, in size-exclusion chromatography (SEC)—the only chromatographic method widely applicable to macromolecules—a precise relationship between molecular size or mass and retention cannot be established on theoretical grounds and must therefore depend on empirical calculations.

2 THEORY

2.1 Physical Basis

Four subtechniques of FFF have been rather extensively developed in the University of Utah laboratories. They have a similar physical and theoretical basis but utilize different external fields. A separate technique, called *steric field-flow fractionation*, constitutes the upper field limit case of each subtechnique. This upper limit is used for separating larger particles, greater than about 1 μm, and arises from processes that are negligible in normal FFF but become dominant when larger particles are involved.

In the four normal types (subtechniques) of FFF, particles of different size, or of other properties appropriate to the specific technique, are forced into steady-state layers against one channel wall termed the *accumulation wall*. The mean thickness of each individual layer is governed by the interaction between the compressive force of the external field on the one hand and the dispersive effect of Brownian motion (diffusion) on the other. Usually the largest particles form the thinnest layers, which are confined to the slow-flow regions close to the channel wall; smaller particles form thicker layers, which extend further into the channel where flow is faster. Larger particles are thus retarded more than smaller particles in their migration downstream (see Figure 8.1).

Large particles in the range of 1 to 100 μm behave differently. The steric FFF method [36–41] for separating particles in this size range exhibits an inversion of the normal elution order. These particles are sufficiently large to project into faster flow regions of the channel even though they are held against the channel wall by the external field. Thus, within certain limits, larger particles protruding into faster flow regions are caused to migrate downstream more rapidly by the higher velocity of the fluid phase (Figure 8.2). Although any external field can be used for steric FFF, gravitational or centrifugal forces are usually the most practical unless particles are suspended in a carrier medium where they are neutrally buoyant.

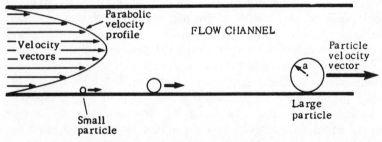

Figure 8.2 Diagram of the side view of a steric FFF channel, where large particles are transported downstream at higher velocities than small ones because of their extension into faster moving flow lines.

The four basic subtechniques, each employing a different external field, have been used to separate many kinds of particles less than a micrometer in size. Sedimentation FFF is based on a gravitational force or on a centrifugal force field created in a spinning channel. In thermal FFF, a temperature differential existing between the upper and lower surfaces of the channel works through the effects of thermal diffusion. Electrical FFF establishes an electrical potential gradient across the column. Flow FFF utilizes a cross flow of the carrier liquid, where channel walls are constructed of a semipermeable membrane to allow passage of the carrier while retaining the solute.

The FFF theory describes the distribution of particles within the FFF channel under the influence of the external field and the internal flow. Theory enables us to predict (or at least approximate) the retention of particles within the channel, and to determine the ability of the system to separate the components of a sample and to establish optimal operating conditions for a particular subtechnique or sample type. The FFF techniques have the capacity to separate many peaks in a single run, with elution *space* ranging from 1 channel void volume up to 10, 20, or even 50 void volumes. The theory rests on the basic concept of laminar flow between two parallel plates whose breadth, practically speaking, is infinite compared to their distance apart. The flow profile is generally parabolic or bullet shaped, as was shown in Figure 8.1. The applied external field is perpendicular to the plates.

The ensuing discussion describes first the nature of the steady-state distribution of particles against the channel wall and introduces the important parameter λ, the ratio of mean layer thickness to channel width. This is followed by the theory of retention, peak dispersion, resolution, and peak capacity (interested readers may wish to refer to [2] for a more generalized treatment of these concepts). Procedures for optimizing experimental conditions (including programming) are considered at the end of Section 2.

2.1.1 Steady-State Sample Layer

We look first at the sample layer formed by application of the field, because its character determines the differential migration of species in the field-flow channel. In the absence of flow, the field-induced forces normal to the channel impose a velocity on the sample molecules or particles, herding them toward the accumulation wall of the channel. The rate of movement toward the wall is governed by the magnitude of the field, by its level of interaction with the sample, and by the frictional resistance to motion of sample through carrier. The field-induced solute migration toward the accumulation wall is opposed by a dispersive flux caused by diffusion or Brownian motion. At steady state, where these two processes balance, the sample layer has been shown to assume an exponential distribution [42]

$$c/c_0 = \exp(-x/l) \tag{1}$$

where c is sample concentration at distance x from the accumulation wall, c_0 is concentration at that wall, and l is essentially the mean thickness of the layer. Figure 8.3 shows this relationship. Should the magnitude of l approach channel width w (defined as the distance between upper and lower plates), the exponential distribution would be truncated by the upper wall. We generally confine ourselves to situations where $l/w \ll 1$ [43]. Practical values of l/w are <0.2 and preferably <0.1. A value of 0.2 gives a retention ratio (discussed later) of around 0.7, the point at which practical resolution begins.

A given field will cause molecules with different diffusion characteristics and different field susceptibilities to distribute unequally in the field direction x, thereby establishing differential values of l. The basis of this follows.

The steady-state layer thickness l is determined by an equilibrium between the outward movement of particles away from the wall, measured by the diffusion coefficient D for the particular sample-carrier system, and the inward movement of particles toward the wall, characterized by the average drift velocity U induced by the field force

$$l = D/U \tag{2}$$

A most important ratio in FFF is that of l to w, the channel width. We express this as the basic dimensionless *retention parameter* λ

$$\lambda = l/w \tag{3}$$

Equations (2) and (3) can be combined to give

$$\lambda = D/Uw \tag{4}$$

The value of D may be most fundamentally obtained from the relationship [2]

$$D = \mathcal{R}T/f \tag{5a}$$

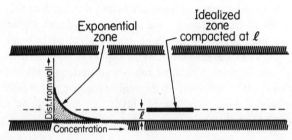

Figure 8.3 Concentration profile of an exponential zone compared to an idealized zone concentrated at altitude l.

where $\mathscr{R}$ is the gas constant, T the absolute temperature, and f the molar friction coefficient. For particles that are large compared with carrier molecules, molar friction coefficient f may be obtained from the Stokes equation, which relates f to particle diameter d and viscosity η,

$$f = 3\mathscr{N}\pi\eta d \tag{5b}$$

where $\mathscr{N}$ is Avogadro's number. It is frequently useful, in practice, to deal with the properties on a molecular basis, which requires the use of the Boltzmann constant k in place of the gas constant $\mathscr{R}$ in (5a) and the omission of $\mathscr{N}$ in (5b).

Drift velocity U is linearly dependent on effective molar force F but inversely dependent on frictional resistance to motion [42]

$$U = F/f \tag{6}$$

In terms of (5a) and (6), λ becomes

$$\lambda = \mathscr{R}T/Fw \tag{7}$$

Force F can be resolved into two components: one is field strength S and the other is a molar *sample–field interaction constant* ϕ. Parameter ϕ will depend in some predictable way on a particular property of the sample, for example, on mass and density in sedimentation FFF and charge in electrical FFF. The value of ϕ is defined by the following expression for force per mole

$$F = \phi S \tag{8}$$

Equation (7) can be modified by (8) to yield

$$\lambda = \mathscr{R}T/\phi Sw \tag{9}$$

Since the values of $\mathscr{R}$, T, S, and w are constant within the system, only ϕ will vary between components. In a mixture of samples, each component possesses a unique ϕ value and thus a characteristic λ value. This leads to the observation that for an arbitrary component, say the ith sample, λ can be expressed as $\lambda_i = \mathscr{R}T/\phi_i Sw$.

The inverse proportionality of λ and S are well worth noting. A plot of λ versus $1/S$ for a given sample should give a straight line through the origin for any subtechnique.

For sedimentation FFF, S is simply G, the centrifugal acceleration, which in the laboratory is $\omega^2 r_0$ with r_0 being the rotor radius and ω the angular velocity determined by the speed of centrifugation. Using subscript S to denote applicability to sedimentation FFF, we have

$$S_S = G = \omega^2 r_0 \tag{10}$$

The ϕ term may be written

$$\phi_S = M\left(\frac{\rho_s - \rho}{\rho_s}\right) = M\frac{\Delta\rho}{\rho_s} \tag{11a}$$

where ρ_s is the density of the sample particle, ρ is the carrier density, and M is the molecular weight of the species under consideration [22, 44]. Combining (9) and (11a), we find the general expression for λ under sedimentation FFF

$$\lambda_S = \frac{\mathscr{R}T}{M(\Delta\rho/\rho_s)Gw} \tag{11b}$$

A spherical sample with particle diameter d has a mass $m = \frac{1}{6}\pi d^3\rho_s$, which corresponds to a molecular weight M of $m\mathcal{N}$. For such particles, parameter ϕ_S can be expressed as

$$\phi_S = \frac{1}{6}\pi d^3\Delta\rho\mathcal{N} \tag{12a}$$

yielding

$$\lambda_S = \frac{6kT}{d^3\pi\Delta\rho Gw} \tag{12b}$$

The analogous situation in an electrical FFF system (denoted by subscript E) is

$$S_E = E \tag{13}$$

where E is the electric field strength. The ϕ_E term depends on the effective charge per molecule $\dot{z}$ and the charge per mole of electrons, the Faraday $\mathscr{F}$

$$\phi_E = \dot{z}\mathscr{F} \tag{14a}$$

Thus

$$\lambda_E = \frac{\mathscr{R}T}{\dot{z}\mathscr{F}Ew} = \frac{D}{\mu Ew} \tag{14b}$$

where μ is the electrophoretic mobility ($\mu = \dot{z}\mathscr{F}/f$).

Because the presence of a temperature gradient leads to the transport of macromolecules toward regions of low temperature, we may define a phenomenological transport parameter, the *thermal diffusion coefficient*, D_T [45]. Thermal diffusion is often expressed in terms of the *thermal diffusion factor* α, which characterizes the distribution of solutes in a temperature gradient. This factor is defined as

$$\alpha = D_T T / D \tag{15}$$

where D is the ordinary diffusion coefficient and T is the temperature [45].

Cast in terms of the thermal diffusion factor α, the reduced layer thickness λ takes the following form for the thermal FFF subtechnique (subscript T)

$$\lambda_T = \frac{T}{\alpha w (dT/dx)} = \frac{D}{D_T w (dT/dx)} \tag{16}$$

where dT/dx is the temperature gradient across the channel, which is taken as the field strength S_T

$$S_T = dT/dx \tag{17}$$

The product of thermal diffusion coefficient D_T and the sample's coefficient of friction f is recognized as the field interaction constant ϕ_T

$$\phi_T = D_T f = \alpha \mathscr{R} \tag{18}$$

In practice, D_T stays more or less invariant with molecular weight for a given polymer type, whereas f is shown to vary as molecular weight to some fractional power whose magnitude depends on the ideality of the solute–solvent pair. To an approximation, f varies as the square root of molecular weight [46]

$$f = A(M)^{0.5} \tag{19}$$

where A is a constant.

When the *field* of FFF is a flow of the carrier, as is the case in flow FFF (subscript F), the force on a particle is caused entirely by the drift velocity U, which is rigorously constant for all species regardless of size, density, or charge

$$S_F = U \tag{20}$$

This nonspecific *field strength* is experimentally determined as the volumetric cross-flow rate $\dot{V}_c$ divided by column area bL [32]

$$S_F = \dot{V}_c / bL \tag{21}$$

The flow field is unique in that the unequal values of l, which are required for differential displacement, must depend entirely on diffusivity D [see (2)]. We can see from (5a) that, within a given system, the only variable affecting D will be the sample's friction coefficient f. Therefore

$$\phi_F = f \tag{22}$$

Following (9), we have

$$\lambda_F = \mathscr{R}T/fUw \tag{23a}$$

The unique dependence of retention (discussed later) on differential values of friction coefficient f leads to special expressions for λ, which always contain f or some term, like D, which depends on f. The substitution of (5a) and (21) into (23a) and the replacement of product bL by column volume V^0 divided by w gives various forms of these expressions

$$\lambda_F = \frac{\mathscr{R}TbL}{f\dot{V}_c w} = \frac{\mathscr{R}TV^0}{f\dot{V}_c w^2} = \frac{DV^0}{\dot{V}_c w^2} \tag{23b}$$

The Stokes equation relates the friction coefficient f to particle diameter d and solvent viscosity η, as shown in (5b). This assumes a spherical molecule or particle. However, even for nonspherical species an *effective* diameter may replace d in the equations. This diameter is usually represented in the literature as the *Stokes diameter* of a particle. It is frequently an important piece of information in characterizing macromolecules, and the ease with which it can be obtained in flow FFF from experimental values of λ contributes greatly to the analytical power of this technique.

To reach parameter d, the Stokes equation is used to replace f by d in (23b), giving

$$\lambda_F = \frac{\mathscr{R}TV^0}{3\mathscr{N}\pi\eta w^2 \dot{V}_c d} \tag{24}$$

which indicates an inverse relationship between λ_F and particle diameter d.

The expressions for ϕ and S for the different subtechniques are summarized in Table 8.1. They will have an important bearing on our discussion of resolution in FFF. The table includes expressions for reduced layer thickness λ derived from functions S and ϕ through use of (9). Note that λ depends on $M \cdot \Delta\rho/\rho_s$, α, μ, and f. Thus, these values (or a distribution in their values for polydisperse materials) may be obtained by measurements of FFF retention.

2.1.2   Retention

Once the steady-state sample layer has formed, carrier flow along the channel moves individual layers at different rates along the flow path; the specific rates depend on whether the layers have equilibrated in generally slow-flow regions close to the wall or in faster flow regions located closer to the center of the channel.

Retention in FFF is measured in terms of the retention ratio R, which, as in chromatography, may be specified in terms of relative velocity or of relative retention volume. The retention ratio defined in velocity terms is the ratio of the

Table 8.1 Parameters S, ϕ, and λ, for the Principal Subtechniques of Field-Flow Fractionation

FFF Subtechnique	Field Strength (S)	Solute–Field Interaction Constant (ϕ)	Retention Parameter (λ)
Sedimentation FFF	$G = \omega^2 r$	$M\dfrac{\Delta\rho}{\rho_s}$	$\dfrac{\mathscr{R}T}{M(1-\Delta\rho/\rho_s)Gw}\,;\ \dfrac{6kT}{\pi Gwd^3\Delta\rho}$
Electrical FFF	E	$\mathscr{F}\dot{z}$	$\dfrac{\mathscr{R}T}{F\dot{z}Ew}$
Thermal FFF	dT/dx	$\alpha\mathscr{R} = D_T f$	$\dfrac{T}{\alpha w(dT/dx)} \simeq \dfrac{T}{\alpha\Delta T} = \dfrac{D}{D_T\Delta T}$
Flow FFF	$U = \dot{V_c}/bL$	f	$\dfrac{\mathscr{R}TbL}{\dot{V_c}fw}\,,\ \dfrac{DV^0}{\dot{V_c}w^2}\,;\ \dfrac{kTV^0}{3\pi\eta w^2\dot{V_c}d}$

aSymbols are defined in the text and in the Glossary of Symbols, Section 5.

velocity of the solute zone v to the mean velocity of the carrier stream $\langle v \rangle$

$$R = \frac{v}{\langle v \rangle}, \qquad 0 \leqslant R \leqslant 1 \tag{25}$$

Of course, parameter R is related to the velocity profile for laminar flow between parallel plates, normally in the parabolic form [47]

$$v = 6\langle v \rangle \left(\frac{x}{w} - \frac{x^2}{w^2} \right) \tag{26}$$

The temperature gradients imposed across the channel in thermal FFF result in nonuniformities in the carrier viscosity, which in turn cause deviations from parabolic flow [48]. In the interest of clarity we choose to neglect this distortion in the present discussion. The velocity of a zone imagined to be concentrated at $x = l$, as suggested by Figure 8.3 [43], is therefore

$$v \approx 6\langle v \rangle \left(\frac{l}{w} - \frac{l^2}{w^2} \right) \tag{27}$$

or, since $\lambda = l/w$

$$v \approx 6\langle v \rangle (\lambda - \lambda^2) \tag{28}$$

Hence, for this model $R \approx 6(\lambda - \lambda^2)$, and in the limiting case in which $\lambda \to 0$, R becomes

$$R = 6\lambda \tag{29}$$

This simplified expression for R overestimates R by 8% when R is as large as 0.25, but becomes increasingly accurate for lower R values (higher retention).

A more rigorous expression for R is obtained by integrating the sample distribution multiplied by the velocity profile over the channel cross-sectional area. For an exponential sample distribution (1) and a parabolic velocity profile between infinite parallel plates (26) the value of R is

$$R = 6\lambda[\coth(1/2\lambda) - 2\lambda] \tag{30}$$

Figure 8.4 shows the curve for (30) and compares it with various approximations for R.

It is often convenient to express retention in terms of retention volume. Retention volume V_r has the same relationship to void volume V^0 as does carrier velocity to zone velocity in (25). At constant flow rate the same relationship also holds for the time t_r required for elution of the sample peak and the time t^0 for

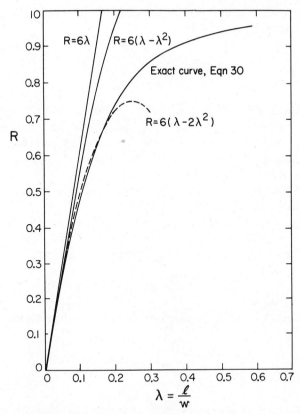

Figure 8.4 The exact retention curve for FFF in an infinite parallel plate channel, as obtained from (30), and various approximations valid at high retention (low R values). The exact curve is based on the assumption of parabolic flow with an exponential concentration profile.

the passage of one void volume

$$R = V^0/V_r = t^0/t_r \tag{31}$$

In terms of (30), the retention volume ratio is

$$V^0/V_r = 6\lambda[\coth(1/2\lambda) - 2\lambda] \tag{32}$$

or, using the simplifying approximation of (29)

$$V^0/V_r = 6\lambda \tag{33}$$

A better approximation is [49]

$$\frac{V_r}{V^0} = \frac{1}{6\lambda} + \frac{1}{3} \tag{34}$$

(Note that the volume ratio is now inverted.) Even greater accuracy is obtained by adding an empirical nonlinear term [49]

$$\frac{V_r}{V^0} = \frac{1}{6\lambda} + \frac{1}{3} + \frac{2\exp(-1/4\lambda)}{3} \tag{35}$$

Since R is a function of λ [(30)] and λ is a function of field strength S [(9)], R must always depend on field strength. The relationship is always such that R decreases with increasing S. For example, the retention ratio R is shown as a function of field strength $S = G$ in Figure 8.5 for polystyrene beads of different diameters in a sedimentation FFF channel.

For steric FFF, which operates at the upper limit of particle sizes normally resolvable with FFF techniques, the particles are large enough that their size overwhelms the magnitude of the Brownian displacement [50]. Hence, particle radius $d/2 = a$ replaces mean layer thickness l in dominating the retention process, and, to an approximation, λ in (29) and (33) is replaced by a/w

$$R = 6(a/w) \tag{36}$$

Retention volume $V_r = V^0/R$ is therefore given by the expression

$$\frac{V_r}{V^0} = \frac{w}{6a} \tag{37}$$

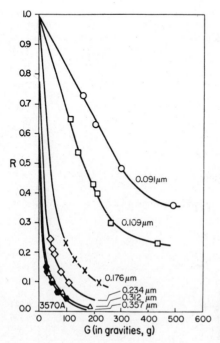

Figure 8.5 Retention parameter R versus field strength (measured in gravities) for various polystyrene beads. Bead diameters are indicated in micrometers (μm).

which shows that retention volume is inversely proportional to particle radius.

For spheres touching the wall the zone velocity will ideally be that of the carrier stream at the x coordinate corresponding to the center of gravity of the particle. However, viscous drag on the particles will reduce this velocity and affect observed retention. The shear on the particle also produces a torque, which results in a velocity component perpendicular to, and directed away from, the wall [51]. This lift force can be overcome by strong field-induced settling forces.

Equation (36) may be modified by a factor γ, which accounts for nonideal effects on particle velocity, and the expression becomes [51]

$$R = 6\gamma\frac{a}{w} = 3\gamma\frac{d}{w} \tag{38}$$

Experimental evidence indicates that a value of $\gamma = 0.7$ applies for silica beads under typical circumstances [51], but the variations in γ can be very wide, depending especially on flow conditions.

From the foregoing it is clear that solute molecules or suspended particles of small size compared to l are sorted according to size or net charge. As a result the smallest particles, which form less compact equilibrium layers, elute ahead of more massive ones, which move as compact zones. From (30)–(33), together with λ expressions from Table 8.1, it is evident that for such particles retention ratio R is normally inversely related to some power of the particle diameter. By contrast, particles of significant size, and thus dominated by steric effects, display a retention ratio that varies linearly with particle diameter, as shown by (38).

The combined effect of these two retention mechanisms is illustrated in Figure 8.6, where retention ratio R is plotted as a function of particle diameter d for various centrifugal fields in sedimentation FFF [52].

2.1.3   Plate Height

Plate height H is a measure of the relative width of eluted peaks in a fractogram [2]. Plate height in FFF is defined in the usual way for zonal migration systems, including chromatography, as the rate of generation of variance σ^2 per unit distance of migration in the downstream or z direction. If a column of length L is uniform throughout, the differential expression can be replaced by a constant

$$H = \frac{d\sigma^2}{dz} = \frac{\sigma^2}{L} \tag{39}$$

In common with other methods of separation, plate height in FFF is of great practical importance, ultimately imposing limits on component resolution and peak capacity.

In FFF systems, the factors comprising H can be divided into ideal and nonideal categories. In an ideal system there are no contributions from instrumental imperfections, dead volumes, solute inhomogeneities, injection, or

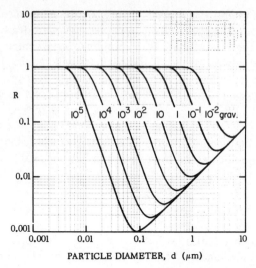

Figure 8.6 Theoretical plots of retention ratio R versus particle diameter d for eight different field strengths (expressed in gravities). The following parameters were used for the calculations: $w = 0.0254$ cm, $T = 298$ K, $\gamma = 0.7$, and $\Delta\rho = 1.0$ g/cm^3.

irregularities in sample behavior at commencement of flow; the only contributions to plate height are those caused by dispersion effects found in perfect parallel plate systems. There are two such contributions.

The first is caused by longitudinal molecular diffusion, in which σ^2 increases in proportion to the length of time the zone spends in the channel, implying an inverse dependence on zone velocity. This effect, expressed by $H_D = 2D/R\langle v\rangle$, is negligible at practical flow rates because of sluggish diffusion in systems of macromolecules [44].

The second, a nonequilibrium effect, is usually the major contributor to peak broadening in well-designed systems. If we exclude sample inhomogeneities (polydispersity), nonequilibrium constitutes the major contribution to H in contemporary systems. The nonequilibrium dispersion is caused by the random movement of particles back and forth across their mean equilibrium position l in the column. Such movement places particles erratically in faster flow regions when they are further than the distance l from the wall, and in slower flow regions when they are closer to the wall. The fluctuating variations in migration velocity lead to zone broadening. This nonequilibrium effect is described by the plate-height term H_n.

The equation for plate height can be rewritten more explicitly to account for these ideal terms and the nonideal terms yet to be described [53]

$$H = \frac{2D}{R\langle v\rangle} + \chi\frac{w^2\langle v\rangle}{D} + \sum H_j + H_p \qquad (40)$$

where the first and second terms are the ideal contributions H_D and H_n, and the nonideal terms are implicit in the summation term ΣH_j. Term H_p is caused by the polydispersity of the sample. As previously, D, R, $\langle v \rangle$, and w designate diffusion coefficient, retention ratio, average flow velocity, and channel thickness (width), respectively. Parameter χ is a complicated function of λ; the relationship has been derived elsewhere [54]. Because of this complexity, approximations for the relationship have been sought. In the case where λ approaches zero

$$\lim_{\lambda \to 0} \chi = 24\lambda^3 \qquad (41)$$

For values of $\lambda \ll 1$, a valid approximation is a three-term power series

$$\chi_{approx} = 24\lambda^3 [1 - 8\lambda + 12\lambda^2] \qquad (42)$$

Alternatively, the first two terms of the power series can be approximated by $\exp(-8\lambda)$, resulting in [54]

$$\chi_{approx} = 24\lambda^3 \exp(-8\lambda) \qquad (43)$$

The various approximations are compared in Figure 8.7.

Nonideal contributions to plate height have several sources. Among these are the relaxation process, the dead volume of the apparatus, and the finite injection volume and time. Small departures from perfect smoothness and flatness in channel geometry and drag at the outer edges of the channel where walls are joined will, of course, create inequalities from point to point in carrier flow velocity and therefore in sample velocity, thus leading to additional nonideal zone broadening. Minimizing these nonideal effects is an aim of FFF technology; normally such effects are now small. Next we show how various terms originate and thus how they can be minimized.

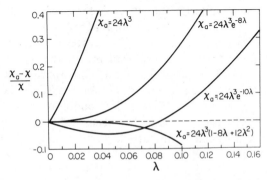

Figure 8.7 The fractional difference between various approximate χ values (χ_a) and the exact value (χ) for different values of retention parameter λ.

The first nonideal contribution to peak spreading derives from the process termed *relaxation*, in which the sample initially forms its steady-state layer upon being subjected to the field. We define one relaxation time τ as the average time required for a particle to travel, under the influence of the field, across the channel from top to bottom. Expressed in terms of λ this is

$$\tau = w^2\lambda/D \tag{44}$$

Relaxation times are typically about a few seconds to several minutes.

To eliminate zone broadening that results from the relaxation process, the longitudinal flow can be interrupted during the relaxation process. In practice, this is advisable in all instances where τ is estimated to be longer than a few seconds. In the absence of stop-flow procedures, the plate-height contribution caused by relaxation may be estimated by [53]

$$H_r = \frac{17}{140}\left[\frac{1}{L}\left(\frac{w^2\langle v\rangle\lambda}{D}\right)^2\right] \tag{45}$$

Real polymer and particle fractions invariably consist of a distribution of molecular weights. Because the different molecular weights migrate at different rates as part of the separation process, the zone is continually widened as migration in the channel proceeds. The polydispersity contribution to peak broadening is represented by [53]

$$H_p = L\left(\frac{d\ln V_r}{d\ln M}\right)^2\frac{\mu-1}{\mu} \tag{46}$$

where M is molecular weight and $\mu = \bar{M}_w/\bar{M}_n$; $\bar{M}_w$ and $\bar{M}_n$ are, respectively, the weight average and number average of the molecular weight. Quantity L is the distance migrated (column length), and V_r is the retention volume.

The factor $(d\ln V_r/d\ln M)$ is a measure of the system's *selectivity*, and it will vary between the value of zero at the void volume and some maximum value, characteristic for each subtechnique, which is reached at high retention [55]. For thermal FFF the selectivity maximum ranges between 0.5 and 0.7 depending on the nature of sample and carrier. The corresponding expression for the case of sedimentation FFF takes the value of unity in the limit of high retention.

In studies of colloidal particles, we find it more appropriate to express the polydispersity in terms of the standard deviation in particle diameter σ_d rather than μ. It has been shown [44] that under conditions of high retention the polydispersity contribution to plate height in sedimentation FFF can be expressed in terms of σ_d and mean particle diameter d as

$$H_p = 9L(\sigma_d/d)^2 \tag{47}$$

The multipath effect is related to the nonequilibrium effect H_n discussed earlier. In the limit of zero diffusivity, every flow streamline will carry its load of sample through the channel in a time characteristic of that streamline. However, because diffusivity is finite, particles originating at a given point will diffuse a lateral distance approximately $(Dt)^{1/2}$ during travel time t [53]. Hence streamlines within the approximate range $(Dt)^{1/2}$ will exchange sample back and forth. This includes streamlines above and below one another in the flow velocity profile and also streamlines a similar lateral distance apart (usually ca. 1 mm) if any lateral velocity differences exist. Inequalities in sample velocity are reflected in the variance in flow velocity σ_v^2. In the process of passing through the channel, the sample's variations in velocity translate into variations in migration distance z as measured by σ_z^2. In the mean retention time t_r these variances are related by

$$\sigma_z^2 = t_r^2 \sigma_v^2 \tag{48}$$

The contribution to plate height is

$$H_m = \frac{\sigma_z^2}{L} = \frac{t_r^2 \sigma_v^2}{L} \tag{49}$$

and is independent of mean carrier velocity $\langle v \rangle$ because t_r and σ_v are, respectively, proportional to $1/\langle v \rangle$ and $\langle v \rangle$ [53].

Ideally, a sample is introduced into the FFF channel as an infinitely narrow concentration pulse, in which case any zone broadening from the injection can be ignored. In practice, one is often forced to work with dilute samples where injection volumes V_f are substantial and add to the broadness of the solute zone. The contribution to plate height from this effect H_f can be expressed as [56–58]

$$H_f = \frac{L}{12} \left(\frac{V_f}{V^0} \right)^2 \tag{50}$$

The expressions for H_D, H_n, H_m, H_r, H_f, and H_p can be cast in the general velocity dependent form [53]

$$H = \text{const} \langle v \rangle^{\dot{r}} \tag{51}$$

where $\dot{r}$ is a value (usually an integer) that depends on the specific plate-height term under consideration. The $\dot{r}$ values associated with each mechanism are summarized in Table 8.2.

Equation (51) can be written in logarithmic form

$$\log H = c + \dot{r} \log \langle v \rangle \tag{52}$$

where c is a constant. A plot of $\log H$ versus $\log \langle v \rangle$ should yield a straight line whose slope equals $\dot{r}$ over any range where a single term is dominant. In this way, a log–log plot of H versus $\langle v \rangle$ becomes a diagnostic tool for reducing the number of candidate mechanisms for the major plate-height effect in FFF. This

Table 8.2 Contributions to Zone Broadening and Their Dependence on Carrier Velocity[a]

Plate-Height Mechanism	Equation	r Value
Longitudinal diffusion	$H_D = \dfrac{2D}{R\langle v \rangle}$	-1
Nonequilibrium	$H_n = \chi \dfrac{w^2 \langle v \rangle}{D}$	1
Multipath	$H_m = \dfrac{\sigma_z^2}{L}$	0
Relaxation	$H_r = \dfrac{17}{140L} \left(\dfrac{w^2 \langle v \rangle \lambda}{D} \right)^2$	2
Polydispersity (alternate forms)	$H_D = L \left(\dfrac{d \ln V_r}{d \ln M} \right) \dfrac{\mu - 1}{\mu}$	0
	$H_p = 9L \left(\dfrac{\sigma_d}{d} \right)^2$	
Injection volume	$H_f = \dfrac{L}{12} \left(\dfrac{V_f}{V^0} \right)^2$	0

[a]The sum of these contributions, evaluated for a given sample, constitutes the sample's expected plate height H.

method can be applied to different parts of the velocity range, with the possibility that different principal mechanisms will emerge upon passing from one region to another. The diagnosis is further reinforced because all the factors in Table 8.2 except H_m are subject to explicit evaluation in terms of known parameters.

Our observations [59, 60] of the velocity dependence of zone spreading in sedimentation FFF indicate that the only first-order term listed in Table 8.2, that is, the term describing nonequilibrium effects, is dominant in the velocity region where the bulk of our data has been collected (1–120 mL/h).

Figure 8.8 shows plate-height measurements on a polystyrene latex sample with a bead diameter 0.357 μm as specified by the manufacturer. The sample was studied under conditions of high retention where the nonequilibrium parameter χ in (40) can be assumed to equal $24\lambda^3$ [(41)]. The Stokes–Einstein relationship [(5a) and (5b)] gives a value for sample diffusivity D, and it is thus possible to calculate the theoretical slope of the line if the nonequilibrium term is assumed to be overwhelmingly responsible for zone broadening in this velocity range. The theoretical and experimental slopes agree to within 5% in this case, which is a good indication that the assumption concerning the dominance of the nonequilibrium term is valid.

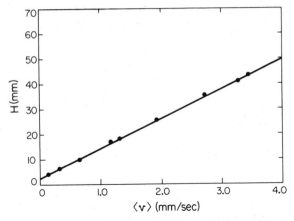

Figure 8.8 Experimental plot of plate height versus flow velocity for a sedimentation FFF system spinning under a field of 67 gravities. The sample consists of polystyrene latex spheres with a nominal diameter of 0.357 μm, retained with an $R = 0.124$ in the 83.3-cm-long channel whose thickness $w = 0.0254$ cm.

An extrapolation of the plate-height curve to zero velocity will yield information on the total effect of all terms (including nonideal terms arising, for example, from flaws in injection technique and channel construction) whose $\dot{r}$ values equal zero (see Table 8.2). The manufacturer's data for polydispersity of the sample give a value for $\sigma_d/d = 0.016$, to be compared with the value 0.018 obtained experimentally when it is assumed that the intercept in the plate-height curve (Figure 8.8) arises from polydispersity alone. The very good agreement between these numbers leads us to assume that the polydispersity contribution to plate height is the dominant velocity-independent term listed in Table 8.2, and that nonideal system effects can often be ignored.

Although the FFF techniques are analytical in nature with sample requirements in the submilligram range, they lend themselves to characterization of large volumes of highly diluted sample [57]. Such samples may be fed into the FFF channel under the condition of an extremely high field. Immediately upon entering the channel, the sample relaxes into a highly compressed layer at the accumulation wall. During this feed cycle the front of the sample will migrate downstream at a zone velocity v, which is inversely dependent on the chosen field strength. At the end of the feed cycle the field is reduced to an appropriate working level, and the zone becomes exposed to all the broadening effects discussed previously. The broadening during the feed cycle can be treated as an injection contribution H_f, which is independent of the flow velocity chosen for the actual separation ($\dot{r} = 0$). This injection contribution has two components, one caused by the downstream migration of sample prior to complete relaxation (H_r) and one caused by migration after the sample specific equilibrium position has been assumed (H_z). The H_r value may be evaluated through use of (45) where the appropriate λ parameter represents the reduced layer thickness under feed-field conditions.

We derive H_z from the finite length of the sample slug and evaluate it from knowledge of sample volume V_f and retention under feed conditions [57]

$$H_z = \left(\frac{R_f V_f L^2}{V^0}\right)\Big/ 12L \tag{53}$$

The total injection contribution to plate height during this type of feed injection becomes

$$H_f = H_r + H_z \tag{54}$$

2.1.4   Resolution and Peak Capacity

The resolution R_s of two peaks is defined as the distance between their centers (Δz) divided by the sum of their half-widths at baseline. If the peaks are Gaussian, they will each have a baseline half-width that is approximately twice the standard deviation of the respective peaks (σ_1 and σ_2), and the expression for R_s becomes [56]

$$R_s = \frac{\Delta z}{2\sigma_1 + 2\sigma_2} = \frac{\Delta z}{4\bar{\sigma}} \tag{55}$$

where $\bar{\sigma}$ represents the average standard deviation for the pair of components to be resolved.

The difference in migration distance at any time can be determined from the differences in migration velocity for the two components. Through (25) one may cast differences in migration velocity in terms of the average carrier velocity $\langle v \rangle$ and differences in the dimensionless retention ratio R, which is characteristic for each component

$$\Delta v = \langle v \rangle \Delta R \tag{56}$$

The average time the solute pair spends in the column is

$$\bar{t}_r = L/\langle v \rangle \bar{R} \tag{57a}$$

Since Δz equals the product of time and velocity difference

$$\Delta z = \bar{t}_r \Delta v \tag{57b}$$

the resolution achieved during this time amounts to

$$R_s = \frac{\bar{t}_r \Delta v}{4\bar{\sigma}} = \frac{L(\Delta R/\bar{R})}{4\bar{\sigma}} \tag{58}$$

Equation (39) defined plate height H in terms of peak variance σ^2 at the end of a column of length L. The efficiency of a column is, however, often measured not by the plate height but by the total number N of theoretical plates in the column, where

$$N = \frac{L}{H} = \left(\frac{L}{\sigma}\right)^2 \tag{59}$$

In terms of the total plate count the resolution expression [(58)] becomes

$$R_s = \frac{N^{1/2}}{4}\left(\frac{\Delta R}{\bar{R}}\right) \tag{60a}$$

This equation is an explicit statement of the dependence of resolution on column efficiency (through $N^{1/2}$) as well as on sample selectivity, reflected by $\Delta R/\bar{R}$. This ratio of dimensionless velocities is identical to the ratio of real migration velocities [(25)]

$$\Delta R/\bar{R} \equiv \Delta v/\bar{v} \tag{60b}$$

Cast in terms of $\Delta v/\bar{v}$, (60a) is an exact analogue to the resolution expression used in general chromatographic theory; see (135) of [2].

The minimum number of plates N_{min} required for *baseline* resolution ($R_s = 1$) of a component pair with specified selectivity is found from (60a)

$$N_{min} = \frac{16}{(\Delta R/\bar{R})^2} \tag{61}$$

In practice, the resolution of any pair of components is the result of a careful balance of the two experimental parameters (60a): field strength, which governs column efficiency N and to some extent $\Delta R/\bar{R}$, and flow velocity $\langle v \rangle$, which also influences column efficiency N.

The peak capacity n is the upper limit of resolvable components for a given technique under prescribed conditions. This upper limit can be increased, under certain conditions, by varying appropriate parameters during a run. Such programming methods are discussed later under optimization in Section 2.3.2. It is important to emphasize that a rigid limit on the capability of chromatographiclike columns (including those of FFF) is imposed by two parameters: The number N of theoretical plates and retention volume range V_n/V_1, the latter being the ratio of the elution volume of the final peak to that of the first peak. It was established early on that the peak capacity determined by these parameters is [61]

$$n = 1 + \frac{N^{1/2}}{4R_s}\ln\frac{V_n}{V_1} \tag{62}$$

Unlike SEC where V_n/V_1 has a well-defined upper limit of about 2.3 [62], there is no theoretical upper limit to this ratio in FFF. By assuming 50 column volumes as a practical upper limit to the elution range, n can be estimated as

$$n \simeq 1 + N^{1/2} \tag{63}$$

The corresponding peak capacity in an SEC column with identical plate count is about one-fifth of this value. Equation (62) is based on a constant N; if N varies from component to component, as in FFF, an effective average will be needed for N. More detailed treatments of n have been given [32].

2.2   Measurement of Physicochemical Properties of Solutes by Field-Flow Fractionation

The retention of particles (or molecules) in FFF systems is governed by different particle parameters, depending on the FFF subtechnique employed. These parameters can be characterized for individual particle types by the use of appropriate retention measurements. To date FFF systems at the University of Utah have been used to characterize molecular weights, densities, Stokes' diameters, and diffusion coefficients of certain molecules and particles [23, 30, 34, 59, 63–65]. Theory shows that electrophoretic mobility μ, thermal diffusion coefficient D_T, thermal diffusion factor α, sedimentation coefficient, and partial specific molar volume may also be derived from measurements of FFF retention [66].

The theory for the calculation of the various parameters rests on their relationships to the basic retention parameter λ. Output from FFF systems is measured in terms of elution volume V_r, and in the simplest case V_r is related to λ by

$$R = V^0/V_r = 6\lambda \tag{29}$$

as has already been shown in (31) and (33). More precise relationships are listed as (30) through (35). Since such relationships exist, we may express λ in terms of different parameters depending on the subtechnique under consideration. Expressions for λ have been discussed earlier and presented in Table 8.1. Each physicochemical parameter will now be explained briefly.

2.2.1   Molecular Weight and Particle Diameter

Parameter λ in sedimentation FFF is a function of the known system parameters T, G, and w, which signify the absolute temperature, centrifugal acceleration, and channel width, respectively. However, λ is also a function of the particle parameters M and $\Delta\rho$

$$\lambda_S = \frac{\mathcal{R}T}{M(\Delta\rho/\rho_s)Gw} \tag{11b}$$

If the density ρ_s of the particle and the particle–carrier density difference $\Delta\rho$ are known, then the molecular weight of the particle can be readily obtained from its measured elution volume.

If the sample is a suspension of spherical particles, its reduced layer thickness λ may be expressed in terms of particle diameter. Recognizing M as the product of Avogadro's number $\mathcal{N}$ and particle mass, where the mass is calculated from particle volume and density, (11b) becomes

$$\lambda_S = \frac{6kT}{d^3\pi\Delta\rho G w} \tag{12b}$$

Particle diameter d is thus determined directly from retention measurements.

2.2.2   Density Determination

The density of sample particles is not always available from external sources. In such cases one can employ the remedy of studying retention in carriers of different density. Since carrier density is determined easily with great accuracy, it is, in principle, sufficient to study sample retention in carriers of two different densities to evaluate the two unknown particle characteristics: molecular weight (or particle diameter) and density.

In practice it is preferable to perform the study in a series of carriers with different densities to ensure that no nonidealities in particle mass, such as aggregation or dissolution, result from the differences in carrier composition. Taking the inverse form of (11b) and replacing $\Delta\rho$ with $|(\rho_s - \rho)|$, ρ being the carrier density, we obtain [34]

$$\frac{1}{\lambda} = \left| \frac{MGw}{\mathcal{R}T} \frac{\rho}{\rho_s} - \frac{MGw}{\mathcal{R}T} \right| \tag{64}$$

or

$$\frac{\mathcal{R}T}{\lambda Gw} = \left| M\frac{\rho}{\rho_s} - M \right| \tag{65}$$

Substituting the symbol M' for the left side of the equality, we write

$$M' = \left| \frac{M}{\rho_s}\rho - M \right| \tag{66}$$

Quantity M', the *effective molecular weight*, is measured experimentally through λ. In the absence of nonidealities, parameter M' varies linearly with solvent density ρ. From a plot of M' versus ρ, one finds the molecular weight M from the intercept at $\rho = 0$ and the particle density from the slope of the line.

A treatment similar to (64), which is valid for spherical beads, yields:

$$\rho = \rho_s \pm \left(\frac{1}{0.31534d^3}\right) M' \tag{67}$$

In this special case, a plot of carrier density versus the experimental (retention related) parameter M' yields particle density as the intercept and particle diameter from the slope of the line.

Figure 8.9 illustrates how this procedure was applied [34] to a sample of spherical polystyrene latex beads from Dow Diagnostics. The manufacturer's data for particle diameter and density were 0.357 μm and 1.05 g/cm^3, respectively. The density of the carrier solution was varied through additions of sucrose. From the intercept in Figure 8.9, we obtain a value for the particle mass that together with the slope of the line yields a particle density of 1.051 $\pm$ 0.001 g/cm^3. These values for mass and density correspond to a particle diameter of 0.362 μm, in good agreement with the manufacturer's specifications.

2.2.3 Diffusion Coefficient

Diffusion coefficient D is a component of the λ function in all FFF systems because of the general basic relationship $\lambda = D/Uw$, where U is field-induced velocity and w is channel width. Diffusion coefficients may be most conveniently

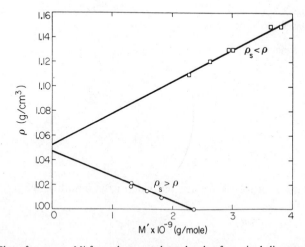

Figure 8.9 Plot of ρ versus M' for polystyrene latex beads of nominal diameter 0.481 μm. The upper line represents a linear least-square fit of data collected in carriers denser than the beads (beads floating), whereas the lower line relates observations in media of density less than the bead density. Both lines intersect the ordinate at a value close to 1.05 g/cm^3, which is the known density of the beads.

measured by flow FFF because of the relationship

$$\lambda_F = \frac{DV^0}{\dot{V}_c w^2} \tag{23b}$$

where V^0 is the void volume and $\dot{V}_c$ is the volumetric cross-flow rate. From this and the relationship of λ to retention ratio R, it is relatively simple to obtain values for D.

The discussion of factors that influence zone broadening led to the establishment of Table 8.2, where each factor is evaluated with respect to its dependence on flow velocity $\langle v \rangle$. Experience shows that the only significant velocity-dependent term is the nonequilibrium term H_n [59, 60, 63], which varies linearly with $\langle v \rangle$ and inversely with solute diffusivity D. Since the plate height is linear in flow velocity, one may evaluate D from the slope of a plot of H versus $\langle v \rangle$ (Figure 8.8) obtained at constant retention

$$H = \chi(\lambda)\frac{w^2}{D}\langle v \rangle + \sum H_j \tag{68}$$

The nonequilibrium coefficient $\chi(\lambda)$ is obtained from (39), (40), or (41) using the λ value, which is determined from the observed retention.

2.2.4   Thermal Diffusion Factor α and Coefficient D_T

Equation (16) casts λ in terms of the thermal diffusion factor α, as well as in terms of the thermal diffusivity D_T and the ordinary diffusion coefficient D. Retention measurements in thermal FFF can thus be converted directly into values for α, which are characteristic for a given polymer–solvent system at a given temperature [65]. When the ordinary diffusion coefficient D is available or can be estimated [65, 67] for a given solvent–solute pair at the experimental temperature, (16) can be used to convert an experimental α value into the corresponding thermal diffusion coefficient D_T.

2.2.5   Electrophoretic Mobility

For samples retained in an electrical FFF system [68], parameter λ is a linear function of the ratio of sample diffusivity to mobility μ, as specified by (14b). When diffusion coefficients are known from outside sources, retention ratio R becomes a direct measure of sample mobility.

2.2.6   Other Parameters

Given an appropriate FFF subsystem, it is possible to design experiments for obtaining other physiochemical parameters of sample particles from the relationships given for λ earlier in the theory section. Although many such experiments have been performed in our laboratories, they have not given sufficiently conclusive results to be treated here.

2.3   Optimization

Maximizing separation effectiveness in FFF systems requires optimizing peak resolution and separation speed. These optimizations are arrived at either by choosing steady operating conditions throughout the run or by continuously varying some parameter of the system such as flow rate or field strength, a method termed *programming*.

2.3.1   Peak Resolution

Equation (40) can be rewritten in terms of zone velocity v using the relationships in (3), (25), (29), and (41)

$$H = \frac{2D}{v} + \frac{4l^2 v}{D} + \sum H_j \tag{69}$$

The third term of the expression, $\sum H_j$, can be ignored for ideal systems and samples, and the plate-height expression can be approximately represented as

$$H = \frac{B}{v} + Cv \tag{70}$$

where $B = 2D$ and, in the limit of high retention, $C = 4l^2/D$. The minimum plate height (with respect to velocity changes) can be obtained by setting the derivative dH/dv equal to zero, which leads to

$$H_{min} = 2\sqrt{BC} = \sqrt{32}\,l = 5.66l \tag{71}$$

The corresponding optimum zone velocity is [69]

$$v_{opt} = \sqrt{\frac{B}{C}} = \frac{D}{\sqrt{2}\,l} = 0.707\frac{D}{l} \tag{72}$$

Since l cannot be observed directly, the substitution of $l = Rw/6$ [(3) and (29)] gives

$$H_{min} = \frac{\sqrt{32}}{6} Rw = 0.94Rw \simeq Rw \tag{73}$$

so that the minimum plate height is represented as a function of two observable parameters. It can be seen that the most highly retained peaks (lowest R values) provide the best resolution, as indicated by the minimal H_{min} values. In normal

FFF systems (i.e., excluding steric FFF) these are ordinarily the higher molecular weight components of a mixture. In practical terms, with $R = 0.1$ and $w = 0.025$ cm, the minimum plate height would be 0.0025 cm (25 μm), providing 12,000 plates per foot, a decidedly high efficiency [69]. However, the requirements for high speed and the onset of nonideal effects both mitigate against the practical achievement of such high plate counts.

2.3.2   Carrier Velocity

Various manipulations of the foregoing expressions [69] show that the optimum carrier velocity is greater for narrower channels and also greater for more retained components because of an inverse dependence on R. The implication is that increasing the flow rate *during* a run will keep the system closer to optimum resolution throughout the run. This procedure is called *flow programming* [70].

Optimum carrier velocity, for runs made at constant velocity, may be determined by using the expression in which l is the ratio of field-induced velocity to diffusion coefficient, $l = D/U$; see (2). By substituting this into (72), we obtain

$$v_{\text{opt}} = U/\sqrt{2} \tag{74}$$

and

$$\langle v \rangle_{\text{opt}} = U/\sqrt{2}\,R \tag{75}$$

Hence, we see that the optimum rate of migration of the peak is just slightly less than the field-induced drift velocity. Calculations using typical values show the optimum zone velocity to be about 1 cm/h, a very sluggish and rather impractical rate of transport.

2.3.3   Plate-Generation Rate

Accelerating the separation process is crucial to practical FFF utilization. A minimum number of theoretical plates is required for successful separation, given a specified set of retention conditions. The rate of generating theoretical plates, $\dot{N}$, increases with compaction of the sample zone (reduction of l) and is thus a function of retention. This can be seen by expressing $\dot{N}$ in the way customarily used in chromatography [56], that is, as the reciprocal of the time needed for the zone to pass through one plate

$$\dot{N} = \frac{v}{H} = \frac{R\langle v \rangle}{H} \tag{76}$$

The maximum plate-generation rate is shown to be a function of diffusivity and layer thickness l [69]

$$\dot{N}_{\text{max}} = \frac{D}{4l^2} \tag{77}$$

Thus separation speed increases with increasing diffusivity and with increasing compression of the sample layer.

The expression for $\dot{N}_{max}$ shows that a minimum friction coefficient (maximum D and U) is desirable, which in practical terms means using a low molecular weight, nonviscous carrier at the highest practical temperature.

Indirectly, the expression for $\dot{N}$ demonstrates that separation speed increases dramatically with decreases in both retention ratio R and column width w through their effect on the nonequilibrium plate-height term of (40). These are the same conditions that favor optimum resolution. Calculations of typical values of $\dot{N}$ give plate-generation rates of about 500–5000 plates per hour.

2.3.4   Limits

Before discussing specific limits eventually encountered with FFF subtechniques, we note some general practical limits that may crop up in any system. These are fivefold: (1) problems of channel wall nonuniformities, (2) overloading by virtue of overly concentrated sample, (3) limits to the strength of a field that can be generated [71], (4) practical separation times, and (5) sample properties.

Hurdles to effective separation may be encountered if the sample layer is compressed enough to multiply sample concentrations to the point where particle–particle interactions occur and column overloading results [28, 33, 46, 72]. Both retention and plate-height disturbances may occur. Also, this compression may trap sample particles in surface irregularities that may be present even in a carefully constructed system.

On the other hand, for some systems and some samples one simply cannot generate a field strength great enough to compress the sample layer sufficiently for effective separation. For example, sedimentation FFF is estimated to effectively resolve molecular weights only above some limiting value M_{lim} where

$$M_{lim} \simeq 10^{10}/G \tag{78}$$

where G is the acceleration expressed in gravities. When one reaches the limit of practical field strengths, it may be profitable to choose a different subtechnique.

A compact sample layer, characterized by a small layer thickness l (of about 10 μm or less) is the basic requirement for either effective fractionation or accurate determination of physical constants such as particle diameter or molecular weight. Two conditions signal the achievement of a satisfactory value of l. First, the system must have an l value small enough to exhibit considerable retention, of about $R < 0.5$, preferably smaller. Second, there is a crucial time element involved in parameter l, because the generation of sample peaks narrow enough to achieve good fractionation and physical constant accuracy requires several diffusional excursions over distance l [43]. Because the time required for each such excursion increases with the square of l, in accord with the Einstein equation, it is important to make l as small as possible to reduce the time required for an experiment [22].

Quite apart from limitations of the FFF system, properties of the sample species that will limit their effective separability must also be considered. Several are size related. Macromolecules and particles are inherently slow to separate because of sluggish diffusion and transport. This may increase relaxation and/or elution time and contribute to peak spreading. A special problem occurs when the radius of a particle occupies a significant fraction of channel width w or of layer thickness l. As discussed earlier in this section and later in the section on applications, the term steric FFF is applied when particle radius $a \gg l$. Steric effects are known to influence retention for much smaller particles than those migrating exclusively as a result of steric FFF. Figure 8.6 illustrates the variations in retention with the particle diameter under conditions of constant field [52].

2.3.5   Programming

The concept of field programming (gradually reducing the field strength) in FFF systems was enunciated in the first paper on FFF [1]. Later it was pointed out that this (and other) programming approaches would solve the same problems in FFF as programming solves in chromatography; with wide-ranging mixtures, early peaks are incompletely resolved and late peaks take too long to elute [73]. Finally, a theoretical analysis of the problem indicated that flow programming (gradually increasing flow during a run) might be more effective than field programming [70]. This idea has not yet been fully explored.

The advantages of programming are dramatically illustrated in Figures 8.10 and 8.11, the former demonstrating the programming of flow rate [70] and the latter the programming of field strength [74]. Figure 8.10 shows the result of separation in a sedimentation FFF system in which flow rate is held constant (A) and varied continuously (B). Separation time can be reduced by an order of magnitude by such methods.

Field strength programming in a thermal FFF system was used to separate a nine-component mixture of polystyrenes with molecular weights ranging from 4000 to 7,100,000. Without programming, only up to half of these components were resolvable in a single run: high molecular weights needed low thermal gradients; low molecular weights needed high gradients. Figure 8.11 demonstrates the effect of a parabolic reduction in temperature gradient during the course of the run. The improvement in separation efficiency compared to nonprogrammed operations is remarkable [74]. Field strength programming in a sedimentation FFF system is effected by decreasing the spin rate gradually during a run [73].

The general theory of field programming has been developed [73]; this was later extended to include flow programming. Retention time t_r is related to channel length L by the integral equation

$$L = \int_0^{t_r} R(t)\langle v \rangle(t)\, dt \tag{79}$$

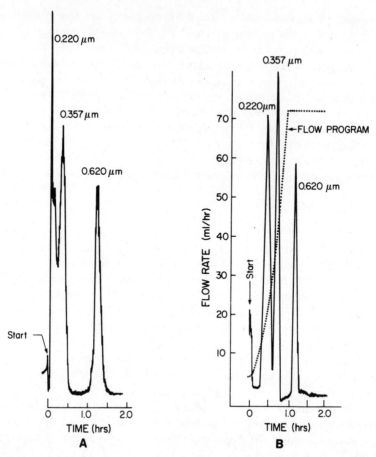

Figure 8.10 Sedimentation FFF separation of a three-component mixture of polystyrene latex particles under a field of 86 g. (*A*) The flow rate was held constant at 32 mL/h during the entire run. The most massive component is fully eluted in less than 1.5 h, but the smaller particles are not well resolved. (*B*) The flow was programmed to start at 4 mL/h and increase in a parabolic fashion to 72 mL/h. The slow flow rate early in the run helps resolve the less massive components, and the elution is completed in about 1.25 h.

With field programming, the field is varied in such a way that R changes with time in accordance with the function $R(t)$. The relationship between R and field strength presented in earlier sections allows one to express $R(t)$ as a function of the chosen time dependence of field strength S, namely, $S(t)$. For flow programming, flow velocity varies as $\langle v \rangle(t)$. Nothing precludes using both programming methods simultaneously; (79) still applies.

For field programming, field strength $S(t)$ can be varied during a run in essentially any desired manner since field strength is subject to almost instantaneous, controllable changes. One should choose $S(t)$ in such a manner that the

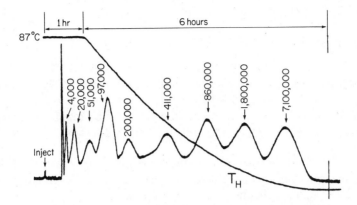

Figure 8.11 Thermal FFF fractogram of a polystyrene mixture obtained using a programmed (parabolic) decay in the applied temperature gradient. The carrier was flowing at 0.023 cm/s, and the initial ΔT was 70°C. The first spike after the injection represents the void peak.

desired peak spacing is obtained [35]. Accordingly, various parabolic, linear, and abrupt-change profiles have been used in our laboratory for sedimentation, flow, and thermal FFF. Yau, Kirkland, and Rementer [10] developed a special time-delayed exponential field program for sedimentation FFF that produces logarithmic spacing. Under such conditions retention time t_r becomes a linear function of the logarithm of molecular weight M

$$\ln M = \ln\left[\frac{6\mathcal{R}T\tau}{e(V^0/\langle v\rangle)G_0 w(\Delta\rho/\rho_s)}\right] + \frac{t_r}{\tau} \tag{80}$$

or of particle diameter d

$$\ln d = \frac{1}{3}\ln\left\{\frac{36kT\tau}{e[\pi(V^0/\langle v\rangle)G_0 w\Delta\rho]}\right\} + \frac{t_r}{3\tau} \tag{81}$$

where e is the base for the natural logarithm system, V^0 is the channel void volume, and G_0 is the field strength at the start of the run. The validity of (80) and (81) rests on the instantaneous adjustment of layer thickness l to the changing field strength. In practice, the field decay is often rapid, and sample constituents diffuse only slowly out of their initially compact distributions. Thus, it has proven necessary to introduce a correction [13] for the finite *secondary relaxation* times required to reestablish an equilibrium distribution at the new level of field strength.

3   INSTRUMENTATION

3.1   Instrument Design

The specific design and instrumentation of the different FFF systems has been described in detail elsewhere [75] and is not repeated here. Instead, we present a brief summary of the system design, and we describe recent advances, techniques, and solutions to specific problems. The reader should refer to [75] for additional details.

The FFF system comprises a separation column and auxiliary devices for carrier delivery, sample introduction, detection, and collection. The column includes the channel, channel walls, and apparatus for supplying the external field. The FFF systems are much like chromatographic systems; most FFF systems work well with ancillary units developed for chromatography, although pumping requirements are not as stringent since high pressures are not required. A diagram of a generalized FFF system is shown in Figure 8.12. Aside from the column, which will be described later, the system requires an injection port for introduction of the sample into the column and a carrier reservoir and pump to provide a steady, controllable flow of liquid at the chosen rate. The effluent from the column, containing zones of sample material with different values of the

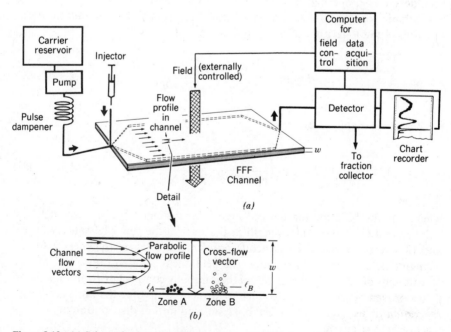

Figure 8.12   (*a*) Schematic view of a FFF apparatus. (*b*) FFF channel showing the parabolic flow profile in detail.

designated property (e.g., size, mass), passes through the detector whose signal is fed to a chart recorder or directly to a computer. Detectors are usually refractive index, UV, visible, or IR monitors. The eluted volume may then be collected in a flow-measuring device or in a fraction collector if the separated portions of the sample are to be examined by other means.

The channels used in flow, thermal, and electrical systems are flat and planar, and are usually operated in a horizontal position. The design of such channels can be seen in Figure 8.13, which shows an exploded view of a flow FFF channel. Similarly detailed designs for the other systems are presented in [75]. In general, channels are cut in a thin Mylar or Teflon spacer of known thickness that is clamped tightly between plates forming the channel walls. For sedimentation and steric FFF, the plates are of highly polished steel or glass, sometimes coated with a polymeric film such as polyimide to reduce sample adhesion to the surface. For flow FFF, they are porous to allow a flow-through of the field-producing solvent. Electrical FFF uses membranes permeable to small ions, and thermal FFF utilizes highly polished copper heating and cooling plates. These plates are sometimes chrome-plated for high corrosion resistance and reduced sample adsorption.

Our sedimentation FFF system uses a centrifugal force field and is curved to fit in a circular centrifuge basket (Figure 8.14). We employ both vertical and horizontal configurations for the channel, but a vertical channel (horizontal spin axis) is preferred so that particles do not have any tendency to settle out against the lower edge of the channel. E. I. du Pont de Nemours and Company is marketing a sedimentation FFF instrument [76] that differs somewhat in design

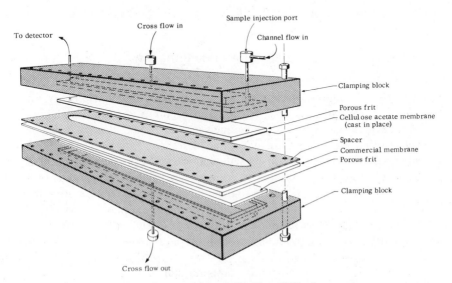

Figure 8.13   Flow FFF column system.

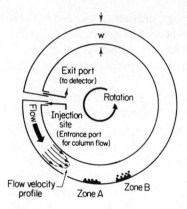

Figure 8.14 Principles of a sedimentation FFF column operating in a centrifuge.

from our unit described in [75]. Most significantly, the Dupont instrument is capable of operation at considerably higher fields, as in the following discussion.

Steric FFF channels are either flat, in which case they are operated in the horizontal plane with gravity as the external force, or curved to fit into a rotor basket like the customary sedimentation FFF channel.

The remainder of this section updates the description in [75].

3.1.1   General Instrument Assembly

In addition to the pumping systems used previously, we find the Gilson Minipuls 2 peristaltic pump to be quite acceptable for most applications.

When the injection port uses a septum-type injection, tiny chips of septum are sometimes carried onto the column, thereby disturbing flow lines near the column entrance. In some recent studies, Carle and Valco sampling valves have been used.

3.1.2   Sedimentation Field-Flow Fractionation

In newer channels, the wall toward the inside of the centrifuge basket is made of $\frac{1}{4}$-in. (0.64-cm) stainless steel with the sample port mounted directly on the column instead of on the seal, as shown in Figure 5 of [75]. This controls a problem experienced with earlier channels in which the centrifugal force on the solvent caused the channel to bulge and increase in volume nearly 5% upon starting rotation. Improvements in sedimentation FFF include the use of larger centrifugal baskets that will accept columns up to 16.5 cm in radius (around 90 cm in column length). Columns that can be taken apart to allow cleaning and coating of surfaces with various materials (such as Teflon and polyimide) have also been developed. Systems built of titanium or Hastelloy C are also in use for fractionations requiring physiological buffers that may be corrosive to stainless steel.

The centrifuge speed is currently computer controlled using a feedback loop to increase the accuracy of speed regulation.

Early O-ring seals were of such a chemical nature that only aqueous solvents could be used without damaging the seals. A construction utilizing spring-loaded Teflon seals (supplied by Bal-Seal) running against a chromium oxide-coated shaft has allowed the use of solvents other than water. However, the life of the seal is rather short. Recently, seals used for aqueous systems have become quite dependable and utilize self-lubricating O-rings and drive shafts that have been coated with ceramic by flame-plasma spraying. Our ceramic coatings have been highly polished aluminum oxide and chromium oxide. A great step forward in seal design by Kirkland and his co-workers [12] has yielded a face-seal operated at speeds in excess of 20,000 rpm, which translates into fields in excess of 50,000 gravities. The higher field strengths, used in conjunction with field programming, have extended the range of sedimentation FFF to much smaller particles and also reduced the necessary separation times.

3.1.3   Flow Field-Flow Fractionation

The pumping system for control of channel flow has been changed from the design presented in [75]. Currently, both the channel and cross flows are delivered by means of highly variable Kontron pumps. The matching of inlet and outlet flows in each of the two directions is done either by use of needle valves or by flow restrictors in the form of stainless steel tubing of specified length.

3.1.4   Electrical Field-Flow Fractionation

Essentially, instrumentation for this technique is as described in [75].

3.1.5   Thermal Field-Flow Fractionation

Larger temperature drops across the channel provide greater retention, but the temperature of the hot wall is limited by the boiling point of the solvent. This temperature limit can be raised by increasing the pressure in the channel, forcing the liquid to boil at a higher temperature. Our earlier systems used a metering valve at the channel exit for this purpose, but the internal volume of this device introduced some zone spreading. Recently, we used detectors that will take pressures up to 300 psi; the regulator valve can then be located behind the detector. This has proven very useful for studies of polyethylene samples that are only sparingly soluble below 100°C [26].

Corrosion of the copper plates has been a problem. Most of the plates are currently polished to a fine finish, then chrome-plated and repolished. This has increased the life of the plates by decreasing the corrosion problem and lessening the danger of accidentally scratching the column while handling the plates during assembly or disassembly. More recently, a copper–zirconium alloy (ca. 0.3% Zr) has been used; this alloy has an excellent combination of mechanical and thermal properties.

3.1.6   Steric Field-Flow Fractionation

We have changed our inlet and outlet ports in the flat (one gravity) steric FFF system from the design in [75]. The inlet and outlet to the column are made by drilling 0.055-in. (0.140-cm) holes in the glass plates and 0.25-in. (0.64-cm) holes in the Plexiglas blocks, and gluing 18-gauge stainless steel needle stock or $\frac{1}{16}$-in. (0.159-cm) Teflon tubing in place with epoxy cement. The Teflon tubing is stretched on one end to reduce its size sufficiently to allow it to pass through the hole; it is then pulled into the hole to make a very tight fit, and excess tubing is removed with a razor blade. We found that even when the glass plates have not been cleaned with strong cleaning agents the surface appears to leach to some extent after long usage, and replacement of the plates is desirable. For enhanced detectability of zones, a steric FFF channel was built with a split-flow outlet, utilizing exit ports in both the top and bottom walls at the end of the column. This arrangement permitted discarding the upper sample-free portion of the effluent, while channeling lower sample-rich flow lines through the bottom exit port by way of a detector to a fraction collector. Similar arrangements are currently applied to other FFF systems.

3.2   Calibration

Calibration of the various FFF systems is performed by running standard particles or macromolecules through the systems and measuring positions of elution peaks. Most materials have been obtained commercially. Materials we have used are polystyrene beads [40, 44, 60], linear polystyrenes [46, 77], polyethylenes [26], colloidal silica sols [27], glass beads [50], proteins [49, 63, 68], viruses [23, 24, 33], and whole cells [37].

3.3   Sample Preparation

On selection of a good solvent or suspension medium for the sample to be studied, solutions or suspensions are made up to concentrations normally of about 1% by weight. Sample volumes vary from 1 to 100 μL per injection. Where the detectability permits, it is desirable to reduce the sample load to reduce nonideal effects.

3.4   Data Analysis

Historically, most data reduction has been performed by hand, but data are now being computer analyzed. The necessary hardware and software are available to monitor detector output, find baselines, smooth data, and perform point-by-point calculations from peak data.

4   APPLICATIONS

4.1   Determination of Appropriate Technique

Although not all subtechniques are practicable for all materials, the FFF method is so broad in scope that the inadequacy of one approach does not preclude the success of another. While definite practical limits exist for the exploitation of each system, a carefully considered selection of the appropriate FFF technique plus proper programming can provide a remarkable range and versatility with respect to sample requirements.

Care must be taken when choosing a subtechnique to consider the particular property of the sample material that is expected to cause differential migration. For instance, species with different molecular weights but equal friction coefficients can be expected to separate in a sedimentation FFF system where λ depends on molecular weight, but not in a flow FFF system where separation is caused entirely by frictional differences. Table 8.1 provides a convenient reference for these specific dependencies. Figure 8.15 shows the nature and molecular weight range of species separated in our different systems.

One of the major advantages of FFF is its amenability to ultrahigh molecular weight separations, an area of particular difficulty for standard techniques of gas and liquid chromatography and SEC. While FFF technically does not belong in the class of liquid chromatography because partitioning occurs in one phase rather than in two, in many respects the theory of chromatography parallels that of FFF. Resolution, retention, and plate height may be directly compared; much FFF theory is built on the theoretical framework developed for chromatography.

One of the principal challenges in developing FFF methodology is to obtain reasonable separation speed, especially in the case of macromolecular and particle separations, without sacrificing resolution. Unlike chromatography, FFF is theoretically capable [according to (77)] of generating plates faster with highly retained peaks than with moderately retained peaks. One may expect that slower moving peaks will generate fewer plates per second, but this is not so in FFF because highly retained species are compressed by the field into very thin layers (small λ values). The reduced diffusion distance, and therefore the reduced time required to traverse this layer, explains the dramatic increase in the rate of acquiring plates [69].

Field-flow fractionation is effectively a multistage process in the sense that the same potential drop between walls is used over and over again as components migrate along the channel. This cascading effect makes it possible to achieve separation with much smaller potential drops than are needed in, for example, electrophoresis [1], especially since the FFF channel can be lengthened to achieve any desired number of theoretical plates [69]. This feature is particularly useful when utilizing thermal diffusion to achieve separation. While direct thermal diffusive separations are virtually impossible because the largest

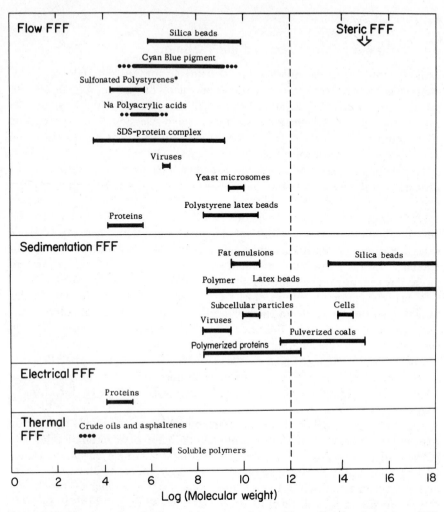

Figure 8.15 Nature and molecular weight range of materials studied by the various FFF subtechniques at the University of Utah project.

practical temperature drops cannot provide adequate fractionation, thermal FFF is capable of producing multicomponent polymer separations in this same gradient range [78].

4.2 Specific Applications

4.2.1 Sedimentation Field-Flow Fractionation

From the earlier discussion of sedimentation FFF, it is evident that this subtechnique is well suited for the separation and characterization of particles ranging in diameter from a few hundreds of a micrometer to about 1 μm, at which size steric effects become substantial.

MONODISPERSE LATEX SAMPLES

Much of the work performed with this technique has used samples consisting of well-characterized polystyrene latex beads of narrow polydispersity. By the systematic observation of retention and zone broadening of five such beads with particle diameters of 0.220, 0.357, 0.481, 0.620, and 0.945 μm, we have been able to demonstrate a convincing agreement between experiment and theory [59, 60]. Similar agreement is reported by Yau and Kirkland in work with polystyrene latex standards [13].

Particle diameters calculated from measured retention show little variation from instrument to instrument and are mostly within a few percent of the values given by the manufacturer. The accuracy in the nominal sizes should be questioned, however, since sizing is normally carried out using electron microscopy where the high-voltage electron beam has been shown to cause substantial shrinkage of some latex preparations [71, 79].

Through observation of zone broadening as a function of carrier velocity at constant field (Figure 8.8) one has an independent means of determining particle diameter. Equation (69) shows how the plate-height expression's mass-transfer term, which is linear in velocity, can be used to determine sample diffusion coefficient. From this value the diameter of a spherical particle is calculated using the Stokes–Einstein relationship [(5a) and (5b)].

The intercept of the plate height versus the velocity plot must, of course, contain all velocity-independent contributions to zone broadening, one of which is sample polydispersity [(46) and (47)]. From the manufacturer's reported polydispersity data for the five sample latices, one can estimate plate-height contributions that are comparable to the experimentally observed intercepts [59]. These results suggest that nonidealities, for example, in channel flow and injection technique, contribute negligibly to zone broadening.

In all retention and zone-broadening experiments discussed previously, the monodisperse sample beads were of known and constant density, and the only sample-specific parameter to influence retention was the particle diameter [(9), (12a), and (30)]. For cases where the density is not known, one can calculate both density and diameter simultaneously through a systematic study of particle retention in media of different density, as described in the theory section. For this purpose, the normally used 0.1% FL-70 detergent carrier was enriched with sucrose to varying degrees and the resultant solution densities measured pycnometrically. To demonstrate the technique, carriers were made to densities both above and below the density of the sample beads, which in this case was known as 1.05 g/cm^3 [34]. Bead retention was thus observed as a result of both sedimentation and flotation, and the determined density value was in good agreement with the manufacturer's value (see intercepts of Figure 8.9).

POLYDISPERSE LATEX SAMPLES

Broad size distributions are not adequately characterized by values for mean and standard deviation in particle diameter. For such samples, the sedimentation FFF fractogram represents a mass spectrum that can be converted into a conventional size-distribution curve. The conversion accounts for two principal distortions of the size distribution that are inherent in the detection and recording of particles in the column effluent.

The first correction is necessitated by the method of detection. Ideally, the detector response is a linear function of sample mass in the effluent, in which case no detector correction is needed. In reality, detectors are often light-transmission monitors whose response is dependent not only on sample mass, but also on particle size and composition. Thus, with polydisperse particulate samples, where particles of different sizes are eluting at different times, the detector response is strongly particle-size dependent.

Our particle-size determinations are ordinarily monitored by a UV detector with a 254-nm light source. When the particles consist of matter [e.g., poly(vinyl chloride)] with low UV absorbance, light scattering is the major source of the detector response, which, as a result, becomes a function not only of sample mass, but also of particle diameter [28, 80]. Such size dependence must be removed from the recorder trace to generate a true distribution function; this can be done either empirically [29] or by a deconvolution procedure based on the Mie theory [10, 28].

A second correction allows for the fact that retention varies in a nonlinear way with particle diameter [28]. Once corrected for scattering, the FFF fractogram constitutes a distribution of sample concentration c in terms of elution volume V_r. If the true particle-size distribution is represented by $m(d)$, where $m(d)dd$ is the mass of particles of diameters from d to $d + dd$, then $m(d)dd$ can be determined as the (mass) concentration of sample in the corresponding elution volume interval dV_r multiplied by the size of this volume element

$$m(d)dd = c(V_r)\, dV_r \qquad (82)$$

Thus, the size distribution $m(d)$ is obtained as the product of the (scattering-corrected) detector response $c(V_r)$ and the scale correction function dV_r/dd, evaluated at each V_r

$$m(d) = c(V_r)\frac{dV_r}{dd} \qquad (83)$$

Equations (30) and (31) and the limiting form, (29), indicate how R and thus V_r varies with λ and consequently with particle diameter (Table 8.1) at constant field.

These relationships are used to evaluate the scale correction function. To simplify this procedure, the chain rule of differentiation is used to cast dV_r/dd in terms of derivatives that are evaluated easily [77]

$$\frac{dV_r}{dd} = \frac{dV_r}{dR}\frac{dR}{d\lambda}\frac{d\lambda}{dd} \qquad (84)$$

where, according to (31),

$$\frac{dV_r}{dR} = -\frac{V_r}{R} \tag{85}$$

Differentiation of (30) with respect to λ gives

$$\frac{dR}{d\lambda} = \frac{3}{\lambda}\left[\frac{R^2}{36\lambda^2} + R - 1\right] \tag{86}$$

and the last derivative in (84) is obtained by differentiation of (12b)

$$\frac{d\lambda}{dd} = -3\frac{\lambda}{d} \tag{87}$$

The scale correction function is thus fully described as

$$\frac{dV_r}{dd} = 9\frac{V_r}{d}\left(\frac{R}{36\lambda^2} + 1 - \frac{1}{R}\right) = \frac{9}{d}\left(\frac{V^0}{36\lambda^2} + V_r - \frac{V_r^2}{V^0}\right) \tag{88}$$

where λ is derived from the particular V_r as a numerical solution to (32), and d derives from the value for λ by way of (12b).

The light scattering and scale corrections often work in opposite directions; thus, at times the true size distribution may closely resemble the uncorrected distribution, which is evaluated directly from the fractogram using (30) and (12b). The effects of these corrections are illustrated in Figure 8.16.

The encouraging results seen in work with synthetic spherical particles has prompted studies of several particles with less well-defined physical properties. A discussion of some of these studies serves to illustrate how FFF can be used to characterize complex colloidal materials.

VIRUS PARTICLES

To date, four different virus samples have been characterized [22–25]. Among the more massive of these particles is the gypsy moth nuclear polyhedrosis virus (NPV), a rodlike lipoprotein-enveloped structure with molecular weight around 950×10^6 dalton. In nature this virus appears embedded in large protein clusters called *inclusion bodies*, where each cluster contains a multitude of virus particles.

The particles are set free by alkaline dissolution of inclusion bodies, which yields a suspension of viral aggregates containing organized bundles of rods, each bundle surrounded by lipoprotein. Electron microscopy of the intact inclusion body shows that aggregates consisting of anywhere from 1 to about 10 rods are already present in this structure and should not be thought of as solubilization artifacts. The customary technique for fractionation of this type of sample into monomers, dimers, and so on, involves centrifugation in a sucrose

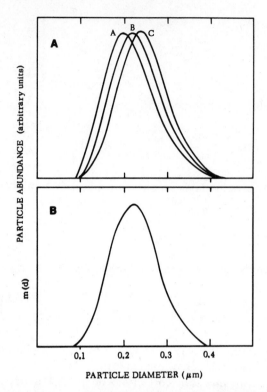

Figure 8.16 Size-distribution curves for PVC latex obtained by sedimentation FFF. The field was 194 g and the channel flow was 28.5 mL/h; the column had a thickness $w = 0.0127$ cm and a volume of 2.0 mL. Detection was performed by a UV monitor with a 254-μm light source. (*A*) Curve B represents the original chart recorder tracing of the fractionation; curve A represents the fractogram corrected for the size dependence in the detector response, whereas curve C represents the recorder trace corrected for scale compression. (*B*) Application of both scattering and scale corrections yields the true size distribution of the sample, which, in this case, is remarkably close to the original fractogram (curve B).

density gradient. For the purpose of comparing this *standard* procedure with the sedimentation FFF technique, a gypsy moth NPV preparation was split into two samples to be characterized by the two techniques. Figure 8.17 shows a fractogram obtained from sedimentation FFF at 173 gravities. Using a flow rate of 24 mL/h, the monomer appears after 53 min quite well resolved from dimers (elution time 82 min) and higher order aggregates. The morphology of the different fractions was verified by electron microscopy. The conventional purification method demanded about 6 h to obtain the various fractions, which then had to be subjected to removal of the sucrose density gradient former, a rather harsh procedure if done quickly. The monomer and dimer fractions from the density gradient sedimentation were injected separately into the sedimentation FFF channel and found to elute at the same positions as components from the original material. It was concluded that sedimentation FFF, which is a

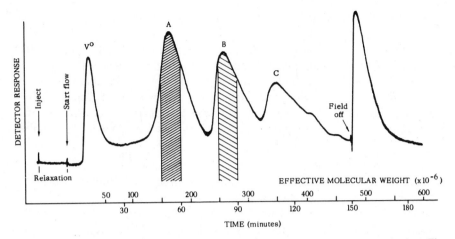

Figure 8.17 Sedimentation FFF of freshly prepared gypsy moth nuclear polyhedrosis virus. The field was 173 g and the flow rate was 23.7 mL/h. Shaded areas in peaks A and B represent material identified morphologically in the electron microscope as monomers and dimers, respectively. The abscissa is graduated in time as well as in effective molecular weight of the eluate. Following injection, a 15-min stop flow period for relaxation was observed.

continuous elution technique, is faster and more convenient for the operator as well as gentler for the sample than the conventional fractionation method [23].

Bacteriophage T2 was studied by sedimentation FFF at several different field strengths [22]. In this study, parameter λ was seen to vary linearly with the inverse of the applied gravitational acceleration, as predicted by (9) and (10). By relying on a literature value for the density of the virus (1.52 g/cm^3), it was possible to determine a molecular weight for the particle that was virtually invariant with field strength. This condition indicates absence of serious nonidealities in the separation procedure.

Bacteriophage T4D was likewise studied by the sedimentation FFF method [24]. Conversion of effective mass into actual molecular weight was hampered by the lack of reliable density values. (The literature reports values anywhere from 1.52 to 1.67 g/cm^3.) Through use of the graphic technique outlined in Section 2.2.2 for evaluating particle densities from retention in carriers of different density, a value for the T4D density of 1.47 g/cm^3 was obtained. From this value, the true particle molecular weight was calculated as $315 \pm 2 \times 10^6$ dalton. The sedimentation FFF study of T4D bacteriophage also demonstrated the nondestructiveness of the separation technique by monitoring a functional criterion of the sample. The numbers of infectious units contained in the sample before and after fractionation showed no significant difference, thus alleviating any fear of particle destruction, for example, as the zone passes through the channel or the rotating seal.

Recently, sedimentation FFF was used in conjunction with other conventional techniques to characterize a massive virus named PBCV-1, which infects chlorella-like algae [25]. Based on an outside value of 1.30 g/cm³ for the viral density, a molecular weight of 1.1×10^9 was determined both from measurements of retention [(11b) and (30)] and plate height [(68)]. In the latter case the diffusion coefficient determined from an H versus $\langle v \rangle$ plot was converted by means of (5a) and (5b) into a value for the particle diameter. From electron microscopic studies the particle was known to be approximately spherical, and thus the product of its diameter-based volume and assumed density gave the weight per particle, which, multiplied by Avogadro's number $\mathcal{N}$, gave the viral molecular weight.

SUBCELLULAR PARTICLES

Work currently in progress concerns the fractionation of subcellular particles by sedimentation FFF. Of particular interest in this project is the separation of lysosomes and mitochondria from livers of radiation-exposed animals for the purpose of determining the binding site for plutonium. Initial work with frozen samples of partially purified subcellular particles from rats showed complete recovery of the marker enzymes associated with lysosomes (acid phosphatase) and mitochondria (cytochrome c oxidase) following baseline separation of the two components in less than 10 min (Figure 8.18). The freezing technique yielded heavily damaged lysosomes, as shown by electron microscopy. Nonfrozen samples were fractionated less completely, but recoveries were still substantial, and the particle morphologies were unaltered by the separation procedure.

PROTEINS

Currently, in the University of Utah laboratories, the highest practical field strength that can be achieved is 1600 gravities. As mentioned in Section 3.1.2, Kirkland's group at the Dupont Experiment Station has reached considerably higher fields with their design [12]. Based on a density of around 1.4 g/cm³,

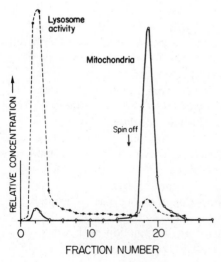

Figure 8.18 Sedimentation FFF of subcellular particles from rat liver. The first peak contained acid phosphatase activity associated with lysosomes, whereas the second peak, which emerged as the field was turned off, contained cytochrome c oxidase activity characteristic of mitochondria. The initial field strength was 258 g and the flow rate was 60 mL/h.

using our equipment, one finds that significant retention ($R = 0.7$ or lower) can be expected only for species with molecular weights above 12×10^6 dalton. This high mass value for the smallest retainable particles excludes most protein molecules from analysis in our laboratory. (Other FFF techniques such as flow or electrical are applicable to components whose molecular weights are four orders of magnitude lower than the sedimentation FFF limiting value.) However, protein aggregates can be studied. To date two protein samples have been studied by sedimentation FFF in our laboratory.

Serum albumin can be polymerized in a two-phase system to form spherical particles suitable for pharmaceutical use. Such particles are immunologically compatible with the species from which the albumin originated and can be used intravenously as carriers of a variety of chemical probes. A sample of bovine serum albumin (BSA) beads, prepared in the Department of Biology at the University of Utah, was retained in a Teflon-sealed sedimentation FFF system with ethanol as carrier [63], as shown in Figure 8.19A. Using known values for the densities of BSA and ethanol, an average particle diameter can be calculated

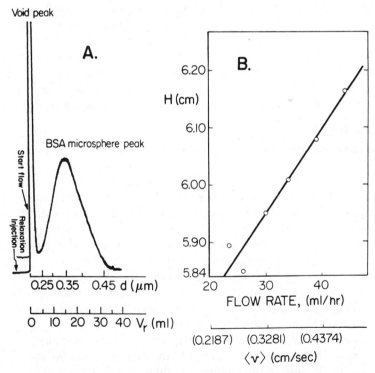

Figure 8.19 (A) Sedimentation FFF fractogram of BSA microspheres recorded at a field of 13.8 g and a flow of 30 mL/h. Retention volume and particle-size scale are both shown on the longitudinal axis. (B) Plot of plate height versus flow rate for albumin microspheres at 13.8 g. The sample was retained with an average $R = 0.6$.

from the observed retention through use of (12b) and (30). The polydispersity of the preparation was determined from plate-height studies, as described previously for the polystyrene latex beads. Both the average particle diameter and its standard deviation coincided with values determined by electron microscopy. Because of the operational reliability of the sedimentation FFF system, these two sample characteristics can be determined in one single run (rather than by the more elaborate establishment of a complete plate-height versus carrier-velocity plot as in Figure 8.19B) simply by measuring retention and zone broadening at the chosen field and flow. Equations (12b), (30), and (40) yield the size and polydispersity within a total analysis time of 1 h.

The formation of cataracts is intimately linked to the polymerization of lens protein into large aggregates [81]. In a collaboration with the Department of Ophthalmology at the University of Utah, we have recently studied the size distribution of particles present in human lenses. Because of the small sample requirements of the sedimentation FFF technique (0.1 mg or less per injection), one lens yields sufficient material for a large number of runs.

In view of the broad spectrum of sizes present in most samples, it was practical to perform the analysis with programmed field reductions. As discussed in the theory section, programming techniques are used to give rapid resolution of a wide range of particle sizes. A particularly attractive program is the time-delayed exponential program developed by Kirkland, Rementer, and Yau [10]. By using this program it was possible to scan a lens sample for its particle content in the 0.1–2-μm diameter range in 45 min. Figure 8.20 shows fractograms of normal and cataractous lenses with logarithmic particle-diameter scales superimposed on the linear effluent volume scales.

EMULSIONS

Because of the fragile nature of liquid–liquid suspensions, there are relatively few methods available for emulsion characterization. Nondestructive methods based on light microscopy can handle large, micron-sized droplets reasonably, although a sufficient number of representative particles must be photographed and measured to yield a statistically significant result. Droplets in the submicron range may be characterized by light scattering and other laser optical techniques, but such methods generally give only an estimate of sample polydispersity without the details of the droplet size-distribution curve.

Electron microscopy has been applied to the sizing of stainable droplets. The technique, however, suffers from the general drawback of not measuring the droplets directly in their suspension, but only after drying and staining procedures have been used that may seriously affect their size. Sedimentation FFF has now been applied to a wide variety of emulsions, including liposomes [11, 30], milk [80], artificial blood [31], and fat emulsions for intravenous nutrition [29]. Because of the relative instability of emulsions compared with many suspensions of solid particles, the droplet size distribution is a sample characteristic that is of particular interest if monitored as a function of time. Any droplet coalescence or other instability will manifest itself as a time-dependent

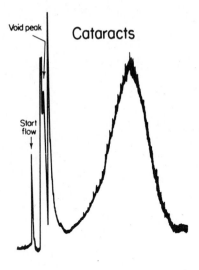

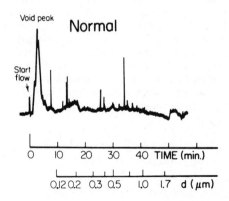

Figure 8.20 Raw particle size distribution in a normal and in a seriously cataractous human lens. (These fractograms are not corrected for either scattering or scale nonuniformities.) The samples were characterized by sedimentation FFF using a time-delayed exponential field decay. The flow rate was 60 mL/h; the initial field of 86 g was decayed with a time constant $\tau = 5$ min.

shift in the droplet size distribution, which makes the distribution curve a valuable stability diagnostic. In view of its potential instability, an emulsion sample must be characterized with particular care to avoid the introduction of artifacts. Since the sedimentation FFF technique functions by concentrating a sample into a narrow zone in the wall region of the flow channel, particular attention must be focused on the effects of sample size and field strength on the recorded distribution curve. Our work with intravenous fat emulsions proceeded along these lines. From the original fractogram (Figure 8.21), one obtains an *uncorrected* size distribution simply by converting the elution volume axis into a droplet diameter scale by means of (12b) and (30). To convert this uncorrected distribution into a true distribution curve, one needs to perform both a scale correction and a detector-response correction as described previously. A detector-response correction can be determined experimentally by

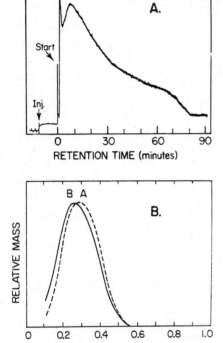

Figure 8.21 (*A*) Sedimentation FFF fractogram of Intralipid fat emulsion. (*B*) Droplet size distribution in Intralipid fat emulsion obtained from the fractogram in (*A*). The solid line shows the distribution without scattering correction, whereas the dotted line represents the scattering corrected distribution. The experimental conditions were: field = 194 g and flow = 60 mL/h. The channel void volume was 2 mL; $w = 0.0127$ cm.

measuring the absorbance of fractions collected after FFF sizing. The relationship between elution volume (fraction number) and particle diameter indicated the size of droplets present in a given fraction. By breaking the emulsion through addition of propanol and remeasuring the absorbance, the fat concentration in the sample could be determined and thus a scattering correction function analogous to the theoretical Mie correction described earlier would be established. Fully corrected size distributions, or those obtained after scale correction, can then be compared from run to run to evaluate any adverse effect of the separation procedure. Ideally, of course, alterations in experimental conditions should not lead to changes in the droplet size-distribution characteristic for a given sample. Figure 8.22 shows distribution curves obtained for a slightly different fat emulsion (Liposyn fat emulsion) under a variety of experimental conditions. The invariance in the distribution gives confidence in the accuracy of the FFF characterization. For simple diagnosis of emulsion stability, one does not need to use corrected distribution curves, provided experimental conditions are held constant during the observation period. A series of daily runs on a commercial milk sample clearly demonstrated the creaming that milk undergoes on storage; a coalescence of fat droplets was indicated by a shift toward larger droplets over a period of a few days [80].

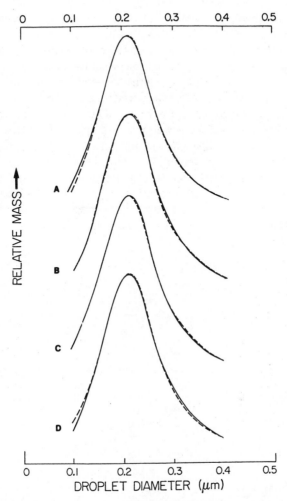

Figure 8.22 Raw droplet size distribution (without scale or scattering corrections) for Liposyn fat emulsion, recorded under a variety of conditions. (*A*) Flow = 30 mL/h, V^0 = 2.0 mL, injection size = 1 μL; the solid line represents a field of 194 g, and the dotted line a field of 345 g. (*B*) Field = 194 g, flow = 30 mL/h, V^0 = 2.0 mL; the solid line represents an injection size of 1 μL, and the dotted line an injection size of 10 μL. (*C*) Field = 194 g, V^0 = 2.0 mL, injection size = 1 μL; the solid line represents a flow of 30 mL/h, and the dotted line a flow of 15 mL/h. (*D*) Field = 194 g, flow = 30 mL/h, injection size = 1 μL; the solid line represents a channel with V^0 = 2.0 mL, w = 0.0127 cm; and the dotted line a channel with V^0 = 4.5 mL, w = 0.0254 cm.

METHODOLOGY FOR DILUTE SAMPLES

The FFF method has been adapted to the characterization of dilute samples [57]. In such cases, a large volume of the sample is fed to the FFF separation chamber under conditions of high field strength. A flow rate is chosen low enough to allow time for the migration of all particles to the wall region before they are displaced significantly by flow. In this way, they are concentrated within the first tenth or so of the channel length. Once the entire sample volume has been pumped into the channel, followed by a small portion of pure carrier fluid sufficient to sweep out the volume from the pump to the channel inlet, the liquid flow is stopped and the field strength adjusted to the value desired for subsequent characterization work. Such strategies have been particularly useful in studying the nature of colloidal particles in stream waters [82].

A clear demonstration of the power of this sample-concentration technique is given in Figure 8.23. Here, identical amounts of solids (100 μg polystyrene latex beads of 0.357-μm diameter) were injected into the sedimentation FFF channel, first in the conventional way by means of a syringe directly into the channel, and second by way of the solvent pumping system. In the first case (fractogram A in Figure 8.23), the injection volume was 1 μL, whereas in the second case (B), it was 50 mL. The working field used to create retention was 67 g in both cases,

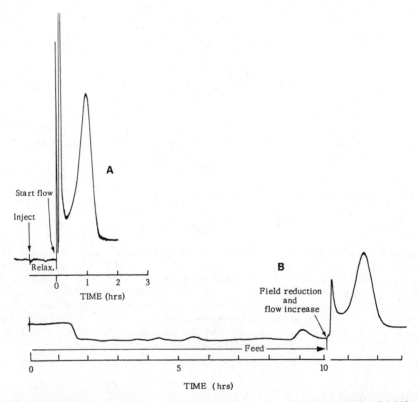

Figure 8.23 Sedimentation FFF retention of 0.1 mg of polystyrene latex beads of 0.357-μm diameter following (A) direct injection in 1-μL sample, and (B) feed injection diluted in 50-mL sample. The working field was 67 g. In case B the concentration step involved pumping in the sample at 5.8 mL/h with the channel spinning at 279 g. In the separation step the flow rates were 35.2 mL/h (A) and 30.8 mL/h (B).

and the feed (in case *B*) was made at 279 g. The working flow rate in each case was about 35 mL/h, whereas the feed in case *B* was made at 5.8 mL/h. Despite the 50,000-fold difference in injection volume, the level of retention was identical and the zone broadening remarkably similar in the two cases, with $H = 2.3$ cm for the small-volume injection compared with $H = 3.9$ cm for the large-volume injection.

4.2.2   Flow Field-Flow Fractionation

Flow FFF offers an extremely large size range for fractionation: Its lower limit is set by the cutoff in permeability of the semipermeable channel walls, and its upper limit is determined by the onset of the steric effect and gravitational disturbances. The permeability of the membrane and the applied pressure determines the range of practical cross flows in particular cases. Millipore's PTGC membrane is sufficiently permeable to allow cross flows of aqueous buffers in the 0–300 mL/h range for a membrane area of 50 cm^2 at very moderate pressures (< 30 psi). These membranes are also durable and perform reproducibly over extended periods of time (6 months or more). As a result of these characteristics, most of our flow FFF work today uses these membranes. Even at modest cross flows, the equipment is capable of creating an acceptable resolution of larger particles.

Earlier, a mixture of Qβ and P22 viruses was separated at room temperature in a 0.02 *M* phosphate buffer at pH 7.12 [33] using a cellulose acetate membrane cast in our laboratory onto a porous polypropylene frit. Bovine serum albumin was added as a low molecular weight marker, to indicate the position of the void peak. Figure 8.24 demonstrates the separation of this mixture using a cross flow of 8 mL/h.

Retention in flow FFF is controlled by sample diffusion coefficients or, equivalently, Stokes radii; these physicochemical quantities can therefore be obtained from measured retention values. Diffusion coefficients calculated from retention data conform quite satisfactorily with values determined by other methods over a wide range, as Figure 8.25 indicates.

By studying zone broadening (under a given cross flow) as a function of channel flow rate one has, of course, the means for determining the non-equilibrium contribution to plate height [(40)]. Since the slope of the plate height versus velocity plot is inversely dependent on diffusion coefficient *D*, this parameter can be determined in a way that is independent of retention measurements. From a zone-broadening study involving seven proteins, we obtained diffusion coefficients through these two approaches that were in good agreement with each other as well as with values from outside sources.

Figure 8.26 illustrates how a complex sample such as rat plasma can be separated by flow FFF and the resulting elution profile converted directly into a spectrum of diffusivities [49]. The carrier employed was 0.02 *M* phosphate buffer at pH 7.2.

In the analysis of many spherical colloidal particles, the parameter of primary

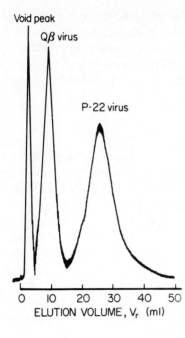

Void peak

$Q\beta$ virus

P-22 virus

ELUTION VOLUME, V_r (ml)

0   10   20   30   40   50

Figure 8.24 Flow FFF separation of bovine serum albumin (void peak), $Q\beta$, and P22 viruses. The channel flow was 8 mL/h and the cross flow was 22.1 mL/h. The channel had a void volume of 1.84 mL; $w = 0.038$ cm.

interest is not the diffusion coefficient but the particle radius or diameter. Since, according to the Stokes–Einstein equation [(5a) and (5b)], there exists an inverse relationship between these two properties, the elution diagram yields information on the size of an eluting particle. In fact, for moderately high to high retention levels, the elution volume is a linear function of particle (Stokes) diameter. This is reflected in the particle diameter d scale shown in parallel with the retention volume V_r scale in Figure 8.27.

A mixture of colloidal silica beads was separated and characterized according to size, as shown in Figure 8.27 [27]. The retention behavior of the early eluting particles showed good agreement with previously determined sizes, which are noted in the figure and positioned along the diameter d scale. The most highly retained silica beads were assumed to have undergone aggregation in that the corresponding peak was unduly broad, and the average particle radius calculated from the FFF observations was substantially larger than the reported values.

Recently we have worked with water-soluble, linear sulfonated polystyrenes and demonstrated good separation, although at higher cross flows the retention becomes affected by the flow rate used to transport the sample along the length of the channel. This effect is as yet poorly understood; for globular solute species it appears at significantly higher fields than for solutes in a random coil configuration.

Although most flow FFF applications to date have involved aqueous buffers, we have recently demonstrated that the technique is well suited also for work in nonaqueous systems [83]. A variety of polystyrene samples were examined

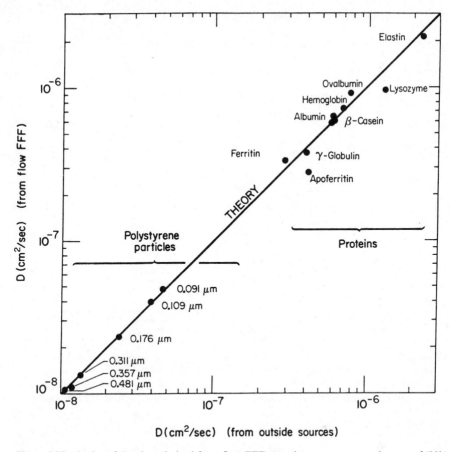

Figure 8.25 A plot of D values derived from flow FFF retention measurements by way of (23b) versus D values acquired from independent sources.

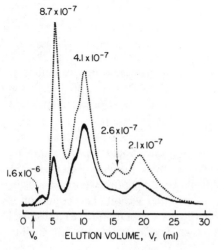

Figure 8.26 Flow FFF fractograms of plasma from two different rats. The values for the diffusion coefficient associated with each peak are calculated from (23b). The channel volume was 1.68 mL, w was 0.038 cm, and the cross flow was 58.7 mL/h.

923

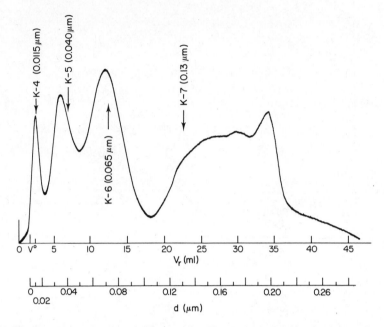

Figure 8.27 Flow FFF separation of silica particles. The nominal values for the average diameters of the various components in the mixture are given in parenthesis above each peak in the fractogram. From the elution volume scale, a Stokes diameter d scale is established through use of (24) and (30). The longitudinal flow was 3.16 mL/h and the cross flow 11.1 mL/h. The column volume was 1.85 mL, and w was 0.041 cm.

using cellulose nitrate membranes from Schleicher and Schuell as the semipermeable accumulation wall and ethylbenzene as the carrier fluid. Retention-derived diffusion coefficients were shown to depend on molecular weight in the manner predicted by (5a) and (19).

4.2.3   Electrical Field-Flow Fractionation

Charged species in polar solvents, notably proteins and nucleic acids, are amenable to electrical FFF. We have successfully demonstrated retention of albumin, lysozyme, cytochrome c, and γ-globulin [68, 84]. The level of retention was found in all cases to increase with field strength as expected. Retention values closely approximate theoretical expectations [85].

Electrophoresis has long been used to characterize and separate charged species on the basis of differentials in electrophoretic mobility μ. By exposing proteins to a charge-carrying surfactant, such as sodium dodecyl sulfate (SDS), which acts as a denaturant and binds uniformly to the peptide chain, one forms adducts of constant mobility irrespective of molecular weight. This condition is naturally present in nucleic acids, whose charge density is uniform as a result of the repetitious nature of the molecular structure. Species with uniform mobility

μ would exhibit electrical FFF retention based solely on differences in diffusion coefficient D. For the random coils of denatured protein, or for randomly coiled single-stranded nucleic acids, D becomes a simple function of molecular weight, thus theoretically permitting the acquisition of molecular weight values from retention data [85].

The size range that can be studied by the electrical FFF technique is limited on the lower end by the character of the semipermeable membrane, whose pore size must be small enough to retain the solute yet large enough to permit the passage of current in the form of small ions. Furthermore, the membrane must neither mechanically trap the material in surface irregularities nor make irreversible adsorptive interactions with solute molecules. If the membrane is selectively permeable to ions of one charge, one tends to polarize the membrane walls and thus reduce the field in the channel. An additional problem concerns the sample itself, in that it may alter conductivity and hence field strength. It is for these reasons and to account for potential loss outside the channel that we use an internal standard method to measure field strength [85].

A flexible membrane channel was the first to yield experimental results [68]. A mixture of commercially available purified bovine hemoglobin and serum albumin, together with human γ-globulin, was separated at pH 4.5 in 0.02 M acetate buffer (Figure 8.28). The order of elution was confirmed through runs of the separate components one at a time and was in qualitative agreement with known electrophoretic properties of these sample proteins. An increase in pH, effected through a change to 0.02 M tris-HCl buffer at pH 8.0, led predictably to a switch in the elution order between albumin and γ-globulin, as shown. The field strength used in these separations was of the order of 3–4 V/cm, which, in view of the small distance between the electrodes, involved minimal applied voltages and low heat evolution. The ease with which the *parallel wall* geometry

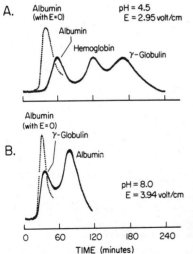

Figure 8.28 Protein separation by electrical FFF. At the ionic strength of 0.02 the isoelectric points of human serum albumin, hemoglobin, and γ-globulin are approximately 4.7, 7.0, and 8.2. The observed elution orders in two buffers of (*A*) pH 4.5 and (*B*) pH 8.0 are consistent with these isoelectric data.

can be distorted by small pressure disturbances in the flexible membrane channel led to a divergence from theory for both retention and zone broadening. By constructing a rigid membrane channel, these problems were reduced.

According to Table 8.1, the λ parameter should vary linearly with $1/E$ and should yield zero intercept, where E is the electric field strength expressed in volts per centimeter. Figure 8.29 reveals that a considerable improvement in conformity with this theoretical prediction was made by introduction of the rigid channel. However, the zone spreading is still large in comparison with theoretical calculations for many samples. Since data for lysozyme are in reasonable agreement with theory, the observed deviations for other substances cannot be related simply to faulty channel construction.

Simple protein separations have been achieved in the rigid channel system. However, the development of effective electrical FFF instrumentation has proved to be a more difficult task than the development of the other FFF subtechniques. Consequently, there are few demonstrated applications of this method. Nonetheless, the future is considered to hold a high potential for widespread applications among charged species.

4.2.4   Thermal Field-Flow Fractionation

The bulk of current applications of the thermal FFF technique involve separations of synthetic polymers in nonaqueous solvents. Commercial polystyrene fractions of narrow molecular weight distribution are available in a wide range of molecular weights. These fractions have served as samples throughout most of the developmental work, not only because of their low degree of polydispersity, but also because of their solubility in a large number of solvents [65]. The influence the carrier exerts on retention of a given polymer was

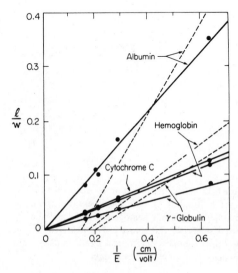

Figure 8.29 Plots of retention parameter $\lambda_E (= l/w)$ versus the reciprocal of field strength E in two electrical FFF systems. Solid lines represent data from a channel with rigid membrane walls, whereas dashed lines symbolize data from a channel with flexible walls.

explored using four polystyrene samples of different molecular weights in eight solvents. The samples were adequately soluble in all eight solvents selected for the study. Retention in all solvents increased with increased sample molecular weight, but each solvent gave different retentions for one and the same sample. Attempts at correlating the observed thermal diffusion factor α for a given sample with solvent dipole moment, Hildebrand solubility parameter, and molecular weight and size showed no clear pattern.

Besides polystyrene, retention in thermal FFF has been observed for many polymeric samples including polyethylene and polypropylene [26], polyisoprene, polytetrahydrofuran, and poly(methyl methacrylate) [15, 17, 86]. The selectivity [(60b)] of the thermal FFF technique, that is, the variation of retention with molecular weight, is strongly affected by the solvent in which the separation is carried out [65]. The nature of the polymer is likewise of importance, as shown in [38]. Recently, Dr. M. Schimpf in our laboratory has been able to demonstrate significant differences in retention between linear and star-branched polystyrenes of the same molecular weights.

To date, we have been unsuccessful in achieving separations in aqueous media, although there are indications in the literature that carbohydrates and lipids in aqueous solution possess reasonable levels of thermal diffusion [87–89]. In a thermal FFF column, operating with $\Delta T = 60°C$ and using toluene as the carrier liquid, a polystyrene fraction of 5000 molecular weight showed measurable retention ($R = 0.85$), whereas the same column, operated at the same ΔT but now with water as the carrier, failed to retain dextran of molecular weight 2×10^6. In an independent study using a static thermal diffusion cell, Bonner [90] observed substantial thermal diffusion for polystyrene in toluene, but none for dextran in aqueous salt solution.

The geometry of the thermal FFF channel is quite well defined because its solid copper block walls are polished to a high degree of flatness and smoothness. This has made it possible to work with very thin spacers and thus increase the speed of separation by reducing diffusion distances [36]. Typically, using an *ultrathin* (0.0051-cm) spacer, we are able to resolve completely a mixture of three polystyrene standards (ranging from 5000 to 160,000 in molecular weight) within a few minutes, as seen in Figure 8.30.

To enhance resolution of complex samples, one may either increase the field (i.e., the temperature differential across the channel) or increase the channel length, as indicated by (58). Since it is impractical to build extended straight units, we have lengthened the thermal FFF columns by coupling several channels in series and sandwiching them all between two flat copper blocks [91]. Resolution in this *hairpin* channel (Figure 8.31) is dependent on the design of the flow connections between the straight column segments, as these bends must be narrow to avoid zone broadening caused by the *race-track* effect.

For multicomponent samples with a large spread in molecular weights, programming strategies have been used successfully to decrease separation time [74]. In this case, the sample is injected under a high-temperature differential, which leads to the resolution of small components. The temperature gradient is

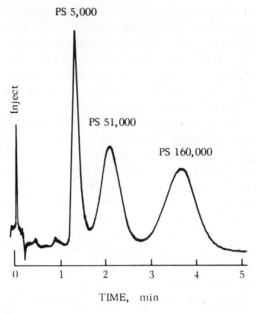

PS 5,000

Inject

PS 51,000

PS 160,000

TIME, min

Figure 8.30 Separation of a three-component mixture of polystyrenes by thermal FFF in an *ultrathin* channel, where $w = 0.0051$ cm; ΔT was 60°C and the longitudinal flow was 21.1 mL/h (0.56 cm/s); V^0 was 0.44 mL.

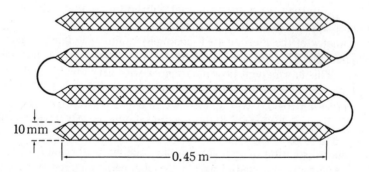

10 mm

0.45 m

Figure 8.31 Configuration of a four-pass hairpin channel.

then gradually decreased, which results in accelerated migration rates, still at adequate resolution, for all components left in the channel. Because of the system's high selectivity, heavily retained polydisperse samples move as very broad and often barely detectable zones. The temperature-decay program speeds up these zones, which results in less zone broadening and thereby in higher detectability. Figure 8.11 shows a programmed thermal FFF run in which a nine-component mixture of polystyrenes ranging in molecular weight from 4×10^3 to 7.1×10^6 was resolved in 7 h.

In theory, one can also use carrier composition programming, as outlined for sedimentation FFF. However, in sedimentation FFF the relevant carrier parameter, density, has a well-understood role, whereas in thermal FFF no single parameter can be identified predictably with changes in retention. Hence, rational solvent programs must await more fundamental developments in our understanding of thermal diffusion [74].

Recently, we have been able to extend the molecular weight range upward to 25×10^6 without indication of sample degradation during the thermal FFF process [77]. Extension of the range downward to molecular weights of around 500 is possible by use of a pressurized system to increase the solvent boiling point, and hence the potential ΔT, by almost $100°C$. This has the expected effect in that a doubling of ΔT results in a nearly fourfold decrease in the molecular weight of species that can be retained [26, 92]. This substantial increase in tractable molecular weight range prompted a brief investigation to see what characterizations might be performed on some naturally occurring compounds using the pressurized system. Samples of crude oil, asphalt, asphaltenes, and pine tar gave characteristic elution patterns. One can, it appears, readily identify specific crudes or, more generally, identify those samples with the largest percentage of the low-molecular-weight fraction [92].

Although the samples discussed thus far have been mostly mixtures of nearly monodisperse polymers, the thermal FFF technique has also been used to establish molecular weight distribution curves for highly polydisperse samples of both polystyrene [77] and polyethylene [26]. Well-characterized standards of low polydispersity polyethylene were unavailable, and calibration curves had to be established using samples with a large discrepancy between weight- and number-average molecular weights. Based on a smoothed calibration curve relating retention R to weight-average molecular weight M_w, it was possible to convert a fractogram for a sample from the National Institute of Standards and Technology (NIST, formerly the National Bureau of Standards) into a molecular weight distribution curve. The weight and number averages for the sample's molecular weight were within 5% of values given by NIST.

4.2.5 Steric Field-Flow Fractionation

The application of steric FFF to achieve the separation of particles is relatively recent. According to (36), the elution order observed for normal FFF, where smaller particles or molecules usually elute ahead of large particles, is reversed when particle migration is governed by steric effects.

The first qualitative demonstrations of steric FFF were obtained using a flat horizontal channel of the usual ribbonlike geometry. The walls were made of glass to permit visual inspection of the separation process. The earth's gravity was the field used to force sample constituents into the wall region. By this arrangement, we were able to size glass beads [50] and a variety of materials used as solid supports for high-performance liquid chromatography [64]. Figure 8.32 demonstrates a typical fractionation of a three-component mixture

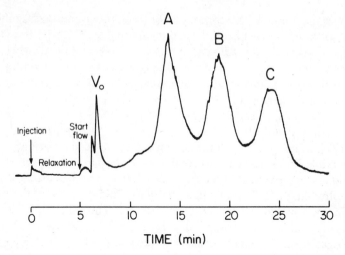

Figure 8.32 Steric FFF of silica beads. Peak A represents particles in the 10–14 μm size range; B, those of 7.5-μm average diameter; and C, particles 5.6 μm in diameter. The carrier, 0.017 M aqueous ammonia, was flowing at 39.1 mL/h. The channel volume was 1.25 mL, and w was 0.0127 cm.

of silica beads [51]. The content of each peak was verified by the light microscope.

Although (36) shows particle radius to be the only sample-specific parameter with influence on retention, it was demonstrated [51] that particle density plays an important role in the retention mechanism. Two bead types of different densities but essentially identical diameters were baseline separated, with the less dense polystyrene beads eluting ahead of the denser silica sample (Figure 8.33).

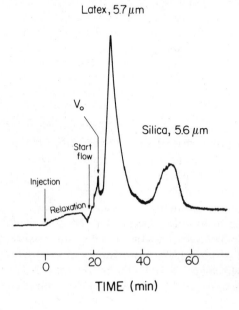

Figure 8.33 Separation of a mixture of silica beads of diameter 5.6 μm and polystyrene latex beads of diameter 5.7 μm at one gravity. The carrier liquid is water, containing 0.1% FL-70 detergent flowing at a rate of 23.5 mL/h; $V^0 = 1.25$ mL; $w = 0.0127$ cm.

This resolution showed a strong velocity-dependence with faster flows resulting in better separations. Obviously, the proportionality factor γ in (38) must be a function of both flow rate and particle density. The early elution of low-density particles indicates that migration occurs following flow lines somewhat closer to the center of the channel than the assumed distance of one particle radius from the wall. The displacement or *lift* away from the wall increases with velocity but can be overcome by an increase in the field, which forces sample particles to settle back into near contact with the wall. Such field increases are conveniently performed in an ordinary sedimentation FFF channel. Figure 8.34 shows the retention ratio R of human red blood cells under a variety of field strengths and flow conditions [37]. In working with live cells and other delicate samples, the possibility of reducing separation times to about a few minutes is particularly valuable.

Steric FFF has recently been applied to the analysis of fine-ground coal samples [38, 39] containing particles of widely differing shapes. Even samples of this complexity are sorted according to size, as predicted by (38) and demon-

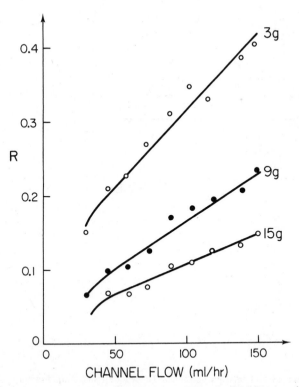

Figure 8.34 Retention ratio R for steric FFF of human red blood cells; R is seen to be influenced by both settling field (assigned in gravities for each plot) and channel flow; $V^0 = 2.85$ mL, $w = 0.0254$ cm and channel length 98.3 cm. The channel walls were Teflon-coated.

strated by Figures 8.35 and 8.36, but the relationship between retention and average particle diameter is not straightforward. The integrity of the fractionation has been demonstrated by reinjection of collected fractions from several samples. The elution positions of the rerun fractions agree well with their respective position in the original fractionation, as demonstrated in Figure 8.35. The steric FFF fractogram can thus be used directly as a *fingerprint* of a particular material in process control and other diagnostic work. A transformation of the fingerprint into a true size distribution curve for the sample involves calibration with known standards of comparable shape and density.

Additional work is clearly needed to formulate a theoretical model for

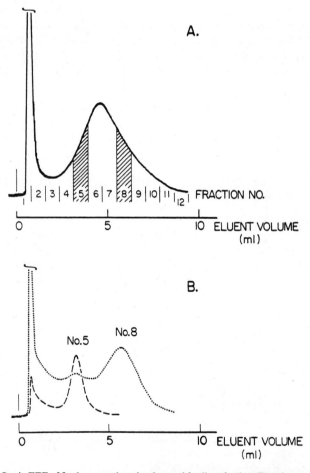

Figure 8.35 Steric FFF of finely ground coal to be used for liquefaction. Fractions were collected as indicated in (*A*). Numbers 5 and 8 were reinjected under the same experimental conditions as used in the parent run; (*B*) shows that the elution positions of the two fractions coincide with the position of original sampling; V^0 was 0.9 mL; w was 0.0127 cm; and the channel flow was 26 mL/h.

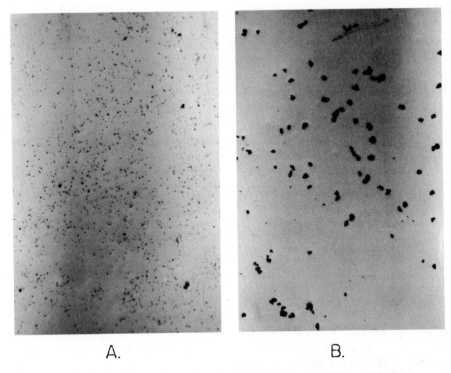

A.   B.

Figure 8.36 Photomicrographs of particles eluting in fractions 5 and 8 of Figure 8.35A. The coarser particles (*B*) elute in the earlier fraction, in qualitative agreement with the theory for fractionation in the steric regime.

retention and zone broadening in steric FFF. In view of its intrinsic speed and operational simplicity, this technique should lead to useful applications in many different fields, including medical, biological, geological, and environmental sciences.

5 GLOSSARY OF SYMBOLS

a	Particle radius
b	Channel breadth
c	Sample concentration (position dependent)
c_0	Sample concentration at channel wall
D	Coefficient of ordinary diffusion
D_T	Coefficient of thermal diffusion
d	Particle diameter
E	Electrical field strength
F	Effective molar force
f	Molar friction coefficient

$\mathscr{F}$	Faraday's constant
G	Gravitational acceleration
G_0	Field strength at the start of a sedimentation run
g	Field strength expressed in gravities ($980 \, \text{cm/s}^2$)
H	Plate height
$H_{\min}$	Plate-height minimum (at optimal velocity)
H_D	Plate-height contribution from longitudinal diffusion
H_f	Plate-height contribution from feed injection
H_m	Plate-height contribution from multipath effects
H_n	Plate-height contribution from nonequilibrium
H_p	Plate-height contribution from sample polydispersity
H_r	Plate-height contribution from relaxation
H_z	Plate-height contribution from finite length of sample slug
k	Boltzmann constant
l	Distance between accumulation wall and center of gravity of a zone
L	Channel length
m	Particle mass
M	Molecular weight
$\bar{M}_w$	Weight-average molecular weight
$\bar{M}_n$	Number-average molecular weight
M'	Effective molecular weight
$\mathscr{N}$	Avogadro's number
N	Number of theoretical plates
$N_{\min}$	Minimum number of theoretical plates needed for resolution
n	Peak capacity
$\dot{N}$	Rate of plate generation
$\dot{N}_{\max}$	Maximum rate of plate generation
R	Retention ratio
R_f	Retention during feed conditions
$\mathscr{R}$	Gas constant
R_s	Resolution
r_0	Rotor radius
$\dot{r}$	Power of the plate height's velocity dependence
S	General field strength parameter
T	Temperature
t_r	Retention time
t^0	Void time
U	Field-induced drift velocity
V^0	Void volume
V_r	Retention volume
v	Carrier velocity (position dependent)
$\langle v \rangle$	Mean cross-sectional carrier velocity
$\langle v \rangle_{\text{opt}}$	Optimum carrier velocity
υ	Zone velocity
υ_{opt}	Optimum zone velocity

V_f	Sample volume at feed injection
$\dot{V_c}$	Volumetric cross flow
w	Channel thickness
x	Position coordinate in field direction
z	Position coordinate in flow direction
$\dot{z}$	Effective charge per molecule or particle
α	Thermal diffusion factor
γ	Correction coefficient in steric FFF
$\Delta\rho$	Density difference between carrier and particle
ρ_s	Particle (solute) density
ρ	Density of carrier
λ	Dimensionless retention parameter
μ	Electrophoretic mobility
μ	Polydispersity index, $\bar{M}_w/\bar{M}_n$
η	Viscosity
σ^2	Peak variance
σ_v^2	Variance in flow velocity
σ_z^2	Variance in migration distance
σ_d	Standard deviation in diameter
τ	Relaxation time
τ	Time constant for exponential field decay
ω	Angular rotor velocity
ϕ	Sample–field interaction parameter
χ	Nonequilibrium parameter

Acknowledgment

This material is based on work supported by the National Science Foundation under Grant CHE82-18503.

References

1. J. C. Giddings, *Sep. Sci.*, **1**, 123 (1966).
2. J. C. Giddings, "Principles of Chemical Separations," in I. M. Kolthoff and P. J. Elving, Eds., *Treatise on Analytical Chemistry*, Part I, Wiley, New York, 1981, Chap. 3, p. 63.
3. H. C. Berg and E. M. Purcell, *Proc. Natl. Acad. Sci. USA*, **58**, 862 (1967).
4. H. C. Berg and E. M. Purcell, *Proc. Natl. Acad. Sci. USA*, **58**, 1821 (1967).
5. H. C. Berg, E. M. Purcell, and W. W. Stewart, *Proc. Natl. Acad. Sci. USA*, **58**, 1286 (1967).
6. E. N. Lightfoot, A. S. Chiang, and P. T. Noble, *Annu. Rev. Fluid Mech.*, **13**, 351 (1981).

7. E. N. Lightfoot, P. T. Noble, A. S. Chiang, and T. A. Ugulini, *Sep. Sci. Technol.*, **16**, 619 (1981).

8. S. Krishnamurthy and R. S. Subramanian, *Sep. Sci.*, **12**, 347 (1977).

9. J. J. Kirkland, W. W. Yau, W. A. Doerner, and J. W. Grant, *Anal. Chem.*, **52**, 1944 (1980).

10. J. J. Kirkland, S. W. Rementer, and W. W. Yau, *Anal. Chem.*, **53**, 1730 (1981).

11. J. J. Kirkland, W. W. Yau, and F. C. Szoka, *Science*, **215**, 296 (1982).

12. J. J. Kirkland, C. H. Dilks, Jr., and W. W. Yau, *J. Chromatogr.*, **255**, 255 (1983).

13. W. W. Yau and J. J. Kirkland, *Anal. Chem.*, **56**, 1461 (1984).

14. L. E. Oppenheimer and T. H. Mourey, *J. Chromatogr.*, **298**, 217 (1984).

15. M. Martin and P. Reynaud, *Anal. Chem.*, **52**, 2293 (1980).

16. M. Martin, *Chromatographia*, **15**, 426 (1982).

17. M. Martin and J. Hes, *Sep. Sci. Technol.*, **19**, 685 (1984).

18. J. Janča and K. Kleparnik, *Sep. Sci. Technol.*, **16**, 657 (1981).

19. J. Janča and J. Chmelik, *Anal. Chem.*, **56**, 2481 (1984).

20. J. Janča, F. Stehlik, and V. Slavic, *Sep. Sci. Technol.*, **19**, 709 (1984).

21. T. C. Schunk, J. Gorse, and M. F. Burke, *Sep. Sci. Technol.*, **19**, 653 (1984).

22. J. C. Giddings, F. J. F. Yang, and M. N. Myers, *Sep. Sci.*, **10**, 126 (1975).

23. K. D. Caldwell, T. T. Nguyen, J. C. Giddings, and H. M. Mazzone, *J. Virol. Methods*, **1**, 241 (1980).

24. K. D. Caldwell, G. Karaiskakis, and J. C. Giddings, *J. Chromatogr.*, **215**, 323 (1981).

25. C. R. Yonker, K. D. Caldwell, J. C. Giddings, and J. L. Van Etten, *J. Virol. Methods*, **11**, 145 (1985).

26. S. L. Brimhall, M. N. Myers, K. D. Caldwell, and J. C. Giddings, *Sep. Sci. Technol.*, **16**, 671 (1981).

27. J. C. Giddings, G. C. Lin, and M. N. Myers, *J. Colloid Interface Sci.*, **65**, 67 (1978).

28. F.-S. Yang, K. D. Caldwell, and J. C. Giddings, *J. Colloid Interface Sci.*, **92**, 81 (1983).

29. F.-S. Yang, K. D. Caldwell, M. N. Myers, and J. C. Giddings, *J. Colloid Interface Sci.*, **93**, 115 (1983).

30. K. D. Caldwell, G. Karaiskakis, and J. C. Giddings, *Colloids Surf.*, **3**, 233 (1981).

31. F.-S. Yang, K. D. Caldwell, J. C. Giddings, and L. Astle, *Anal. Biochem.*, **138**, 488 (1984).

32. J. C. Giddings, F. J. Yang, and M. N. Myers, *Anal. Chem.*, **48**, 1126 (1976).

33. J. C. Giddings, F. J. Yang, and M. N. Myers, *J. Virol.*, **21**, 131 (1977).

34. J. C. Giddings, G. Karaiskakis, and K. D. Caldwell, *Sep. Sci. Technol.*, **16**, 607 (1981).

35. J. C. Giddings and K. D. Caldwell, *Anal. Chem.*, **56**, 2093 (1984).

36. J. C. Giddings, M. Martin, and M. N. Myers, *J. Chromatogr.*, **149**, 501 (1978).

37. K. D. Caldwell, Z.-Q. Cheng, P. Hradecky, and J. C. Giddings, *Cell Biophys.*, **6**, 233 (1984).

38. K. A. Graff, K. D. Caldwell, M. N. Myers, and J. C. Giddings, *Fuel*, **63**, 621 (1984).

39. H. Meng, K. D. Caldwell, and J. C. Giddings, *Fuel Process. Technol.*, **8**, 313 (1984).

40. R. E. Peterson II, M. N. Myers, and J. C. Giddings, *Sep. Sci. Technol.*, **19**, 307 (1984).

41. M. R. Schure, M. N. Myers, K. D. Caldwell, C. Byron, K. P. Chan, and J. C. Giddings, *Environ. Sci. Technol.*, **19**, 686 (1985).

42. E. Grushka, K. D. Caldwell, M. N. Myers, and J. C. Giddings, *Sep. Purif. Methods*, **2**, 129 (1973).

43. J. C. Giddings, *J. Chem. Educ.*, **50**, 667 (1973).

44. J. C. Giddings, F. J. F. Yang, and M. N. Myers, *Anal. Chem.*, **46**, 1917 (1974).

45. H. J. V. Tyrrell, *Diffusion and Heat Flow in Liquids*, Butterworths, London, 1961, p. 45.

46. M. N. Myers, K. D. Caldwell, and J. C. Giddings, *Sep. Sci.*, **9**, 47 (1974).

47. J. Happel and H. Brenner, *Low Reynolds Number Hydrodynamics*, Prentice-Hall, Englewood Cliffs, NJ, 1965, p. 34.

48. J. J. Gunderson, K. D. Caldwell, and J. C. Giddings, *Sep. Sci. Technol.*, **19**, 667 (1984).

49. J. C. Giddings, F. J. Yang, and M. N. Myers, *Anal. Biochem.*, **81**, 395 (1977).

50. J. C. Giddings, and M. N. Myers, *Sep. Sci. Technol.*, **13**, 637 (1978).

51. K. D. Caldwell, T. T. Nguyen, M. N. Myers, and J. C. Giddings, *Sep. Sci. Technol.*, **14**, 935 (1979).

52. M. N. Myers and J. C. Giddings, *Anal. Chem.*, **54**, 2284 (1982).

53. L. K. Smith, M. N. Myers, and J. C. Giddings, *Anal. Chem.*, **49**, 1750 (1977).

54. J. C. Giddings, Y. H. Yoon, K. D. Caldwell, M. N. Myers, and M. E. Hovingh, *Sep. Sci.*, **10**, 447 (1975).

55. J. C. Giddings, *Pure Appl. Chem.*, **51**, 1459 (1979).

56. J. C. Giddings, *Dynamics of Chromatography*, Part I, Dekker, New York, 1965.

57. J. C. Giddings, G. Karaiskakis, and K. D. Caldwell, *Sep. Sci. Technol.*, **16**, 725 (1981).

58. J. C. Giddings and F.-S. Yang, *J. Colloid Interface Sci.*, **105**, 55 (1985).

59. G. Karaiskakis, M. N. Myers, K. D. Caldwell, and J. C. Giddings, *Anal. Chem.*, **53**, 1314 (1981).

60. J. C. Giddings, G. Karaiskakis, K. D. Caldwell, and M. N. Myers, *J. Colloid Interface Sci.*, **92**, 66 (1983).

61. J. C. Giddings, *J. Chromatogr.*, **125**, 3 (1976).

62. J. C. Giddings, Y. H. Yoon, and M. N. Myers, *Anal. Chem.*, **47**, 126 (1975).

63. K. D. Caldwell, G. Karaiskakis, M. N. Myers, and J. C. Giddings, *J. Pharm. Sci.*, **70**, 1350 (1981).

64. J. C. Giddings, M. N. Myers, K. D. Caldwell, and J. W. Pav, *J. Chromatogr.*, **185**, 261 (1979).

65. J. C. Giddings, K. D. Caldwell, and M. N. Myers, *Macromolecules*, **9**, 106 (1976).

66. J. C. Giddings, S. R. Fisher, and M. N. Myers, *Am. Lab.*, 15 (May, 1978).

67. S. L. Brimhall, M. N. Myers, K. D. Caldwell, and J. C. Giddings, *J. Polymer Sci. Polym. Phys. Ed.*, **23**, 2445 (1985).

68. K. D. Caldwell, L. F. Kesner, M. N. Myers, and J. C. Giddings, *Science*, **176**, 296 (1972).

69. J. C. Giddings, *Sep. Sci.*, **8**, 567 (1973).

70. J. C. Giddings, K. D. Caldwell, J. F., Moellmer, T. H. Dickinson, M. N. Myers, and M. Martin, *Anal. Chem.*, **51**, 30 (1979).

71. C. A. Daniels, S. A. McDonald, and J. A. Davidson, "Comparative Particle Size Techniques for Poly(vinyl chloride) and Other Latices," in P. Becher and M. N. Yudenfreund, Eds., *Emulsion, Latices, and Dispersions*, Dekker, New York, 1978.

72. K. D. Caldwell, S. L. Brimhall, Y. S. Gao, and J. C. Giddings, *J. Appl. Polymer Sci.*, **36**, 703 (1988).

73. F. J. F. Yang, M. N. Myers, and J. C. Giddings, *Anal. Chem.*, **46**, 1924 (1974).

74. J. C. Giddings, L. K. Smith, and M. N. Myers, *Anal. Chem.*, **48**, 1587 (1976).

75. J. C. Giddings, M. N. Myers, K. D. Caldwell, and S. R. Fisher, in D. Glick, Ed., *Methods of Biochemical Analysis*, Vol. 26, Wiley, New York, 1980, p. 79.

76. L. E. Schallinger and L. A. Kaminsky, *Biotechniques*, **3**, 124 (1985).

77. Y. S. Gao, K. D. Caldwell, M. N. Myers, and J. C. Giddings, *Macromolecules*, **18**, 1272 (1985).

78. G. H. Thompson, M. N. Myers, and J. C. Giddings, *Anal. Chem.*, **41**, 1219 (1969).

79. J. Pörstendorfer and J. Heyder, *Atmos. Environ.*, **4**, 149 (1970).

80. F.-S. Yang, "Sedimentation Field-Flow Fractionation for Characterization of Polydisperse Colloidal Samples," doctoral dissertation, University of Utah, Salt Lake City, UT, 1981.

81. K. D. Caldwell, B. J. Compton, J. C. Giddings, and R. J. Olson, *Invest. Ophthalmol. Vis. Sci.*, **25**, 153 (1984).

82. G. Karaiskakis, K. A. Graff, K. D. Caldwell, and J. C. Giddings, *Intern. J. Environ. Anal. Chem.*, **12**, 1 (1982).

83. S. L. Brimhall, M. N. Myers, K. D. Caldwell, and J. C. Giddings, *J. Polymer Sci. Polym. Lett. Ed.*, **22**, 339 (1984).

84. L. F. Kesner, K. D. Caldwell, M. N. Myers, and J. C. Giddings, *Anal. Chem.*, **48**, 1834 (1976).

85. J. C. Giddings, G. C. Lin, and M. N. Myers, *Sep. Sci.*, **11**, 553 (1976).

86. J. C. Giddings, M. N. Myers, and J. Janča, *J. Chromatogr.*, **186**, 37 (1979).

87. R. S. Johnson and W. Niedermeier, *Thermochim. Acta.*, **2**, 497 (1971).

88. A. T. Pawlowski, "Study of Thermal Diffusion of Some Aqueous Carbohydrate Solutions," doctoral dissertation, Rutgers—The State University, New Brunswick, NJ, Univ. Microfilm 66-1194, 1965.

89. C. W. Seelbach, "Thermal Diffusion of Lipids and Other Biochemical Substances," doctoral dissertation, Purdue University, Lafayette, IN, Univ. Microfilm 11-659, 1965.

90. F. J. Bonner, *Chem. Scr.*, **3**, 149 (1973).

91. J. C. Giddings, M. Martin, and M. N. Myers, *J. Polym. Sci.*, **19**, 815 (1981).

92. J. C. Giddings, L. K. Smith, and M. N. Myers, *Anal. Chem.*, **47**, 2389 (1975).

INDEX